WVU - PARKERSBURG LIBRARY

DATE DUE

GAYLORD — PRINTED IN U.S.A.

West Virginia University
PARKERSBURG

Presented by

Purchased with a
grant from the

Society of Manufacturing
Engineers
2003

TOOL AND MANUFACTURING ENGINEERS HANDBOOK

VOLUME VII
CONTINUOUS IMPROVEMENT

SOCIETY OF MANUFACTURING ENGINEERS
OFFICERS AND DIRECTORS, 1993

President
Frank H. McCarty, CMfgE, PE
Eliot & Carr, Inc.

President-Elect
Charles A. Templeton, CMfgE, PE
Charles A. Templeton Machine

Second Vice President
Gustav J. Olling, PhD, CMfgE, PE
Chrysler Corporation

Third Vice President
Jack L. Ferrell
TRW, Inc. (Retired)

Secretary/Treasurer
Alan T. Male, Ph.D., PE
Concurrent Technologies Corp.

Douglas E. Booth, CMfgE, PE
Bond Robotics, Inc.

Albert A. Brandenburg, CMfgE
Mohawk Valley Applied Tech. Commission

Joseph E. Brown
(Retired)

Gary L. Cline, CMfgE
Brunswick Corp.

Sidney G. Connor, CMfgE, PE
Wichita State University

Charles D. Cox, CMfgE, PE
Southern Mfg. Consultants

Leon W. Fortin, PE
L.W. Fortin Consultants

Donald G. Foster, CMfgE, PE
Lawrence Berkley Laboratory

Keith M. Gardiner, Ph.D., CMfgE, PE
Lehigh University

James M. Hardy
Hardy Associates

Ronald P. Harrelson, CMfgE
Caterpillar Institute

Barbara P. Knape, CMfgE, PE
Knape Engineering

James F. Lardner
Deere & Co. (Retired)

James K. Long, CMfgE, PE
AMP, Incorporated

Donald I. Manor, PE
Deere & Co.

Paul A. Misegades, CMfgE
General Electric Co.

Louis M. Papp, CMfgE, PE
Windsor-Essex County Dev. Comm.

John T. Parsons
John T. Parsons Co.

Vern L. Stumpenhorst
Ford New Holland, Inc.

Clifford Terry, PE
Consultant

Clarence H. Tombarge
Boeing Co., Inc. (Retired)

Carl R. Williams, CMfgE, PE
Productive Systems, Inc.

TOOL AND MANUFACTURING ENGINEERS HANDBOOK

FOURTH EDITION

VOLUME VII
CONTINUOUS IMPROVEMENT

A reference book for manufacturing engineers, managers, and technicians

Ramon Bakerjian, CMfgE
Handbook Editor

Philip Mitchell, CMfgT
Staff Editor

Produced under the supervision of
the SME Reference Publications
Committee in cooperation with the
SME Technical Divisions

Society of Manufacturing Engineers
One SME Drive
Dearborn, Michigan

TMEH®

ISBN No. 0-87263-420-5

Library of Congress Catalog No. 92-081156

Society of Manufacturing Engineers (SME)

Copyright ©1993, 1976, 1959, 1949 by Society of Manufacturing Engineers,
One SME Drive, P.O. Box 930, Dearborn, Michigan 48121

All rights reserved, including those of translation. This book, or parts thereof, may not be reproduced in any form without written permission of the copyright owner. The Society does not, by publication of data in this book, ensure to anyone the use of such data against liability of any kind, including infringement of any patent. Publication of any data in this book does not constitute a recommendation of any patent or proprietary right that may be involved. The Society of Manufacturing Engineers, as well as all contributors and reviewers of information in this volume, disclaim any and all responsibility for use of the information contained herein by readers and users of this Handbook.

First edition published 1949 by McGraw-Hill Book Co. in cooperation with SME under earlier Society name, American Society of Tool Engineers (ASTE), and under title *Tool Engineers Handbook*. Second edition published 1959 by McGraw-Hill Book Co. in cooperation with SME under earlier society name, American Society of Tool and Manufacturing Engineers (ASTME), and under title *Tool Engineers Handbook*. Third edition published 1976 by McGraw-Hill Book Co. in cooperation with SME under current Society name and under title *Tool and Manufacturing Engineers Handbook*.

Printed in the United States of America.

PREFACE

The first edition, published as the *Tool Engineers Handbook* in 1949, established a useful and authoritative editorial format that was successfully expanded and improved on in the publication of highly acclaimed subsequent editions, published in 1959 and 1976, respectively. Now, with continuing dramatic advances in manufacturing technology, increasing competitive pressure both in the United States and abroad, and a significant diversification of information needs of the modern manufacturing engineer, comes the need for further expansion of the Handbook. As succinctly stated by Editor Frank W. Wilson in the preface to the second edition: "...no 'bible' of the industry can indefinitely survive the impact of new and changed technology."

Although greatly expanded and updated to reflect the latest in manufacturing technology, the nature of coverage in this edition is deeply rooted in the heritage of previous editions, constituting a unique compilation of practical data detailing the specification and use of modern manufacturing equipment and processes. Other volumes in this Handbook series include: Volume I, *Machining*, published in March 1983, Volume II, *Forming*, in April 1984, Volume III, *Materials, Finishing and Coating*, in July 1985, Volume IV, *Quality Control and Assembly*, in January 1987, in April, 1989, *Manufacturing Management*, and in January, 1992, *Design for Manufacturability*.

The scope of this edition is multifaceted, offering a ready reference source of authoritative manufacturing information for daily use by engineers, managers, and technicians, yet providing significant coverage of the fundamentals of manufacturing processes, equipment, and tooling for study by the novice engineer or student. Uniquely, this blend of coverage has characterized the proven usefulness and reputation of SME Handbooks in previous editions and continues in this edition to provide the basis for acceptance across all segments of manufacturing.

In this, and other TMEH volumes, in-dept coverage of all subjects is presented in an easy-to-read format. A comprehensive index cross-references all subjects, facilitating quick access to information. The liberal use of drawings, graphs, and tables also speeds information gathering and problem solving.

The reference material contained in this volume is the product of incalculable hours of unselfish contribution by hundreds of individuals and organizations, as listed at the beginning of each chapter. No written words of appreciation can sufficiently express the special thanks due these many forward-thinking professionals. Their work is deeply appreciated by the Society; but more important, their contributions will undoubtedly serve to advance the understanding of manufacturing management throughout industry and will certainly help spur major productivity gains in the years ahead. Industry as a whole will be the beneficiary of their dedication.

Further recognition is due the members of the SME Reference Publication Committee for their expert guidance and support as well as the many members of the SME Technical Activities Board.

The Editors

SME staff who participated in the editorial development and production of this volume include:

EDITORIAL	TYPESETTING	GRAPHICS
Thomas J. Drozda Director of Publications	**Kathy Allison** Typesetter	**Donna Hicks** Adcomp Services, Inc. Canton, MI
Karen Wilhelm Publications Manager	**Marcia Theisen** Typesetter	
Ramon Bakerjian Handbook Editor		
Philip Mitchell Staff Editor		
Frances Kania Editorial Secretary		
Dorothy Wylo Editorial Secretary		

SME

The Society of Manufacturing Engineers is a professional society dedicated to advancing manufacturing through the continuing education of manufacturing managers, engineers, technicians, and other manufacturing professionals. The specific goal of the Society is to advance scientific knowledge in the field of manufacturing and to apply its resources to research, writing, publishing, and disseminating information. "The purpose of SME is to serve the professional needs of the many types of practitioners that make up the manufacturing community...The collective goals of the membership is the sharing and advancement of knowledge in the field of manufacturing for the good of humanity."

The Society was founded in 1932 as the American Society of Tool Engineers (ASTE). From 1960 to 1969 it was known as the American Society of Tool and Manufacturing Engineers (ASTME), and in January 1970 it became the Society of Manufacturing Engineers. The changes in name reflect the evolution of the manufacturing engineering profession and the growth and increasing sophistication of a technical society that has gained an international reputation for being the most knowledgeable and progressive voice in the field.

Associations of SME—The Society provides complete technical services and membership benefits through a number of associations. Each serves a special interest area. Members may join these associations in addition to SME. The associations are:

Association for Finishing Processes of SME (AFP/SME)
Computer and Automated Systems Association of SME (CASA/SME)
Machine Vision Association of SME (MVA/SME)
North American Manufacturing Research Institute of SME
(NAMRI/SME)
Robotics International of SME (RI/SME)
Association for Electronics Manufacturing of SME (EM/SME)
Composites Manufacturing Association of SME (CMA/SME)
Machining Technology Association of SME (MTA/SME)

Members and Chapters—The Society and its associations have 80,000 member in 73 countries, most of whom are affiliated with SME's 300-plus senior chapters. The Society also has some 8,000 student members and more than 150 student chapters at colleges and universities.

Publications—The Society is involved in various publication activities encompassing handbooks, textbooks, videotapes, and magazines. Current periodicals include:

Manufacturing Engineering
Manufacturing Insights (a video magazine)
SME Technical Digest
SME News
Journal of Manufacturing Systems

Certification—This SME program formally recognizes manufacturing managers, engineers, and technologists based on experience and knowledge. The key certification requirement is successful completion of a two-part written examination covering (1) engineering fundamentals and (2) an area of manufacturing specialization.

Educational Programs—The Society sponsors a wide range of educational activities, including conferences, clinics, in-plant courses, expositions, publications and other educational/training media, professional certification, and the SME Manufacturing Engineering Education Foundation.

CONTENTS

VOLUME VII—CONTINUOUS IMPROVEMENT

Symbols and Abbreviations	xi
Continuous Improvement	1-1
Total Quality Management	2-1
Continuous Improvement Teams	3-1
Continuous Improvement and Training	4-1
Implementing Continuous Improvement	5-1
Supplier Involvement in Continuous Process Improvement Efforts	6-1
Benchmarking	7-1
Activity-Based Costing	8-1
Continuous Improvement and Just-in-Time	9-1
Deming, Juran and Taguchi	10-1
Process Appraisal	11-1
The Role of ISO 9000 in Continuous Improvement	12-1
The Baldrige Criteria as a Self-Assessment Tool	13-1
General Productivity Improvement	14-1
Total Productive Maintenance	15-1
Machining	16-1
Forming	17-1
Finishing	18-1
Assembly	19-1
Accident Prevention and Continuous Improvement	20-1
Index	I-1

SYMBOLS AND ABBREVIATIONS

The following is a list of symbols and abbreviations in general use throughout this volume. Supplementary and/or derived units, symbols, and abbreviations that are peculiar to specific subject matter are listed within chapters.

A-B

ABC	Activity-based costing
ABM	Activity-based management
ABS	Anti-braking system
ABX	Activity-based information
AC	Alternating current
ADA	Americans with Disabilities Act
AFAQ	Association Francaise pour L'Assurance de la Qualite
AFNOR	Association francaise de normalization
AIAG	Automotive Industry Action Group
ANOVA	Analysis of variance
ANSI	American National Standards Institute
ANSI/ASQC Q90 series	US equivalent of the ISO 9000 series
APQC	American Productivity and Quality Center
AQAP-1	Allied Quality Assurance Publication 1
ARP	Activity-requirement planning
ASME	American Society of Mechanical Engineers
ASQC	American Society for Quality Control
ASTM	American Society for Testing and Materials
AWS	American Welding Society
BEST	Burr, Edge, and Surface conditioning Technology division of SME
BOM	Bill of materials
BPR	Business process reengineering
BSI	British Standards Institute
BTU	British thermal unit

C-D-E

C&E	Cause and effect (diagram)
CAM-I	Computer-Aided Manufacturing-International
CARC	Chemical agent resistant coating
CASCO	Committee for Conformity Assessment
CASE	Conformity Assessment System Evaluation Program
CBN	Cubic boron nitride
CBT	Computer-based training
CCT	Competitive cycle time
CD	Committee draft
CDCF	Continuous dress creep-feed
CDT	Cumulative trauma disorder
CE	Mark European Community Mark
CED	Cathodic electrodeposition
CEN	European Committee for Standardization
CENENLEC	European Committee for Electrotechnical Standardization
CI	Continuous improvement
CIE	Computer-integrated enterprise
CIM	Computer-integrated manufacturing
CIP	Continuous improvement process, continuous improvement program
CIUG	Continuous improvement users group
cm	centimeter
CM	Cell manufacturing
CMM	Coordinate measuring machine
CMMS	Computerized maintenance management systems
CNC	Computer numerical control
CPI	Corrugated plate interceptors
CPSC	Consumer Product Safety Commission Regulations, U.S. Government–Consumer Product Safety
CTD	Cumulative trauma disorders
CVD	Chemical vaporized deposition
DC	Direct current
DESC	Defense Electronics Supply Center, DOD
DFARS DOD	Federal Acqusition Regulation Supplement
DFM	Design for manufacturability
DHHS	Department of Health and Human Services
DIN	Deutsches Institute fur Normung
DIS	Draft International Standard
DITI	Department of Trade and Industry (UK)
DOC	Department of Commerce
DOD	Department of Defense
DOE	Department of Energy
DOT	Department of Transportation
DRF	Data reference frame
EAC	European Accreditation of Certification
EC	European Community
ECC	Engineering change control
ECN	Engineering change notice
EDA	Economic Development Administration
EDI	Electronic Data Interchange
EEA	European Economic Area
EFTA	European Free Trade Association
EN 29000 series	European equivalent of ISO 9000
EN	European Norm of Standard
ENV	European Pre-standards
EOQ	European Organization for Quality
EOQ	Economic order quantity
EOTA	European Organization for Technical Approvals
EOTC	European Organization for Testing and Certification
EPA	Environmental Protection Agency
EQNET	European Network for Quality System Assessment and Certification

EQS	European Committed for Quality System Assessment and Certification	MTC	Manufacturing technology centers
ETA	European Technical Approval	MTM	Motion and time methods
ETSI	European Telecommunications Standards Institute	NAC-QS	Comite National pour L'accreditation des Organismes de Certification
EVOP	Evolutionary optimization	NACCB	National Accreditation Council for Certification Bodies (Belgium)
		NATO	North Atlantic Treaty Organization
		NCSCI	National Center for Standards and Certification Information

F-G-H-I-J-K

FAA	Federal Aviation Administration	NEMA	National Electrical Manufacturer's Association
FAR	Federal Acquisition Adminstration	NFPA	National Fire Protection Association
FDA	Food and Drug Administration	NIST	National Institute of Standards and Technology
FMCA	Failure mode and critical analysis	NRC	Nuclear Regulatory Commission
FMEA	Failure mode and effects analysis	NVCASE	National Voluntary Conformity Assessment System Evaluation
FOD	Foreign object damage	NVLAP	National Voluntary Laboratory Accreditation Program
FTA	Fault tree analysis	OD	Outer diameter
FY	Fiscal year	OEE	Overall equipment effectiveness
GMP	Good Manufacturing Practice Guidelines (FDA)	OEM	Original equipment manufacturer
gm	gram	OJT	On-the-job (training)
GNP	Gross national product	ONA	Orbital nozzle assembly
GPa	Gigapascal	OSHA	Occupational Safety and Health Administration
GRR	Gage repeatability and reproducibility		
GSA	General Services Administration		
GT	Group technology		
HAZ	Heat affected zone		
HD	Harmonized document		

P-Q-R-S-T

HLVP	High-volume, low pressure	PAS	Performance appraisal system
HP	Hewlett-Packard	PC	Personal computer
HSM	High speed machining	PCD	Polycristalline diamonds
HSS	High speed steel	PCS	Plain carbon steels
ID	Inner diameter	PDCA	Plan-do-check-act
IEC	International Electrotechnical Commission	PDF	Probability density function
in.	inch	PDSA	Plan-do-study-act
IQA	Institute for Quality Assurance	PLC	Programmable logic controller
ISO	International Standards Organization	PM	Preventive maintenance
ISO/TC176	International Standards Organization Technical Committee 176	PPE	Personal protective equipment
ISR	Initial sample runs	PPH	Parts per hour
JAZ	Japanese Certification Organization	ppm	Parts per million
JIT	Just-in-time	PPM	Parts per minute
JSA	Job safety analysis	PTS	Performance to schedule
kg	kilogram	PVD	Physical vaporized deposition
kN	kilonewton	QCO	Quick change over
kPa	kilopascal	QFD	Quality function deployment
		QIS	Quality information system
		QIT	Quality improvement team
		QML	Qualified manufacturers list

L-M-N-O

LAN	Local area network	QPL	Qualified products lists
lb(s)	pound(s)	QSR	Quality System Registrar
LPS	Lean production systems	RAB	Registrar Accreditation Board
MAU	Medium access units	RCRA	Resource Conservation and Recovery Act
MBNQA	Malcomb Baldrige National Quality Award	RF	Radio frequency
MBO	Management by objective	RFDC	Radio Frequency Data Communications
MIL SPECS	U.S. Government Military Specifications	rms	Root mean square
mm	millimeter	ROI	Return on investment
MOU	Memorandum of Understanding	rpm	Revolution per minute
MP	Master performer	RPN	Risk priority number
MPa	Megapascal	RVC	Raad Voor de Certificatie (Holland)
MRA	Mutual Recognition Agreement	SAE	Society of Automotive Engineers
MRP	Material requirements planning	SC	Subcommittee
MSDS	Material Safety Data Sheets	SCC	Standards Council of Canada

SEM	Scanning electron microscopy	**VOC**	Volatile organic compounds
SIMS	Scanning ion mass spectroscopy	**WBS**	Work breakdown structure
SM	Synchronous manufacturing	**WC**	Workman's compensation
SME	Subject matter expert	**WCM**	World class manufacturing
SMED	Single minute exchange dies	**WD**	Working draft
SMT	Self managed team	**WG**	Working group
SOP	Standard operating procedure		
SPC	Statistical process control		
SQC	Statistical process control		
TAAC	Trade Adustment Assistance Centers		
TC	Technical Committee		
TCM	Total change management		
TEI	Total employee involvement		
TMU	Time measurement unit		
TOC	Theory of constraints		
TPM	Total preventive maintenance		
TQM	Total quality management		
TTT	Total team trainers		

U-V-W-X-Y-Z

UL	Underwriters Laboratories
UPC	Universal product code

SYMBOLS

α	Alpha
$\approx$	Approximately equal to
β	Beta
$°$	Degree
$>$	Greater than
$\geq$	Greater than or equal to
$<$	Less than
$\leq$	Less than or equal to
μ	Mu
Ω	Omega
$\%$	Percent
$\pm$	Plus or minus
Σ	Sigma (summation)

CHAPTER 1

CONTINUOUS IMPROVEMENT

INTRODUCTION

Continuous improvement (CI) is a term that defies a simple definition. Superficially, it means to continually seek out improvement opportunities in daily life, especially those aspects involved with work and productivity. However, when the term is examined in depth, it is found to be both a philosophy and a tool of productivity. It involves the active participation of many people in teams, but it is dependent upon the dedication of the individual.

Continuous improvement is a discipline. It is an individual's or group's ongoing dedication to making life, work, and personal outlook a little better each day. It is not a technology. Continuous improvement means a constant effort toward seeking opportunities for increased efficiency, effectiveness, and quality. At the same time, it implies that one is seeking, identifying, and eliminating barriers to these same improvements and efficient operations. This is true regardless if it is a product or a service.

The Japanese have developed a working definition of continuous improvement using the phrase *kaizen*. Loosely translated, *kaizen* means making "small incremental improvements or refinements involving everyone in the organization without spending much money." This distinguishes the practice of continuous improvement from innovation. Innovations represent large improvements in an organization through new and revolutionary ideas.

The Western approach to continuous improvement is to allow the scope and application of continuous improvement to vary by user. Thus, one company may adopt a formal position on continuous improvement while a competitor sees it as merely a tool used in a larger plan such as total quality management.

In an ideal situation, a company adopts an operational philosophy based on continuous improvement. A good example of this is the operational philosophy of the Ford Motor Company:

Ford Motor Company is a worldwide leader in automotive and automotive-related products and services as well as in newer industries such as aerospace, communications, and financial services. Our mission is to improve continually our products and services to meet our customers' needs, allowing us to prosper as a business and to provide a reasonable return for our stockholders, the owners of our business.

As can be seen, Ford Motor Company has expanded the definition of continuous improvement to include its purpose for Ford. This includes the need to listen to the needs of customers and the desire to be profitable through competitiveness.

The basic idea behind CI is that many people within an organization work each day to discover or test new improvements. The new improvements are then tested to see if they do indeed improve quality and productivity, or if they reduce costs. If so, the information gained is then passed on to management. Management compiles a database of information on successful improvements. It is the job of management to look for patterns of possible improvements within the data being gathered. Management is also responsible for the coordination and support of the people pursuing continuous improvement.

Inevitably, continuous improvement is linked to modern approaches to quality assurance. Total quality management (TQM), ISO 9000, automotive supplier requirements, and other standards specifically call out the use of continuous improvement to reduce costs, speed delivery, and increase quality.

In the final examination of the definition, continuous improvement is the idea of making progress in little steps. To magnify the effect, organizations involve all personnel in the effort. Thus, many small improvements eventually lead to large gains in productivity and quality.

CHAPTER CONTENTS:

INTRODUCTION 1-1

THE IMPORTANCE OF CI 1-1

THE MANY FACES OF CONTINUOUS IMPROVEMENT 1-4

THE MANY PRACTICES OF CI 1-5

CUSTOMER REQUIREMENTS FOR CI 1-9

TOOLS OF CONTINUOUS IMPROVEMENT 1-11

IMPLEMENTATION OF CI 1-13

THE IMPORTANCE OF CI

Continuous improvement grew from being a general philosophy into a modern competitive and strategic tool of industrial operations. Today, a company may compete on a global scale no matter how small or local it is. Regardless of the size or nature of a business organization, there exists a competitor

---

*The Contributor of this chapter is: **Richard Clements**, President, Solution Specialists, Alto, MI.*
*The Reviewers of this chapter are: **William S. Baxter**, Director of Quality Planning, Seating Division, Steelcase, Inc.; **Thomas Bradley, Ph.D.**, Senior Manager TQM/Excellence, Northern Telecom, Inc.; **William H. Cogwell III**, Senior Quality Engineer, Quality Technology Center, Northern Telecom, Inc.; **James Gooch**, Vice President of Operations, George Group, Inc.; **Rodrick B. Ma, PE, ASQC CQE**, Engineer, Quality Technology Center, Northern Telecom, Inc.; **Brian Margetson, PE, ASQC CQE**, Advisor, Strategic Quality Management, Northern Telecom, Inc.; **David Shipp**, Sales Manager, Weldon Machine Tool Co.; **Herb Shrieves, Ph.D.**, Advisor, Strategic Quality Management, Northern Telecom, Inc.; **Eugene Wolf**, Advisor, Strategic Quality Management, Northern Telecom, Inc.*

CHAPTER 1
THE IMPORTANCE OF CI

somewhere else in the world. Therefore, all businesses face real competition for a limited market.

To succeed at capturing and holding market share in a global economy, a company must:

1. Deploy its resources to maximize efficiency.
2. Insure that the quality of the resulting product and service are better than those of all competitors.

Thus, businesses today must continually improve quality and economic efficiency while maximizing the use of the existing work force. Therefore, an operational philosophy of continuous improvement is not only logical, but inevitable.

At the same time, continuous improvement must serve as a strategic tool, one of many such tools within an organization's arsenal. Long-term planning, financial restructuring, and other innovations can produce dramatic changes in a company's profit picture, but to sustain profitability and to maintain such strategic changes, continuous improvement is required. Continuous improvement takes care of the small opportunities for changing and upgrading the infrastructure of an organization. It helps to identify small problems which may become obstacles to the strategic plan.

THE BENEFITS OF CONTINUOUS IMPROVEMENT

As a result, several benefits are derived from the pursuit of continuous improvement. Any or all of the benefits can be realized by any organization. These include:

1. A higher quality product or service.

   During the 1980s and 1990s companies have come to realize that quality is an important competitive tool. With the price of resources so high (and thus products), customers rely on higher quality goods to return value to the marketplace. A competitive company must seek out the opinion of the customer in defining the quality characteristics that are critical to the customer's needs.

   Continuous improvement implies that the quality of the goods and services a company supplies are fair game for examination and modification. Teams of employees that are seeking opportunities for a company focus a large degree of their attention on product and service quality.

2. A lower cost for products or services—hence higher profit potential.

   Without profits a company will soon find itself out of business. Therefore, using continuous improvement to obtain more cost-efficient operations and competitively priced products and services will result in a company which is more robust. In addition, a company that is already competitive because of a recent innovation can sustain that profitability by employing continuous improvement as an operational strategy.

   Very few companies are aware of the large costs associated with scrap, rework, returned goods, and other quality related expenses. In fact, many have no idea of the proportion of manufacturing that results in nonconforming units of production. Internal studies have discovered that scrap rates of 10% to 15% can almost double the cost of production. Worse still, services provided by a company are rarely examined for their unnecessary costs due to delays, errors, and other performance deficiencies. This results in operations that are spending far too much money in the production of a product or service.

   Continuous improvement emphasizes that all aspects of a process be examined for effectiveness and efficiency. For example, scrap rates that are considered "normal for this business" should be re-examined to determine if they can be eliminated. In most cases they can, and the company benefits in several ways. Continuing with the scrap illustration, reducing a scrap rate from, say 10%, to zero means that:

   - More raw material is ending up as a product.
   - Less time is spent reworking products.
   - Production machinery only needs to work on a product once.
   - Extra manufacturing time does not have to be scheduled to make up for what had been thrown away.
   - The cost of disposing of wasted material is eliminated.

3. Less irritation in the execution of jobs.

   Continuous improvement does not have to be confined to the manufacturing or quantifiable aspects of improving a process. The flow and feel of a process can also be addressed. By communicating as a team about the execution of a process and the resulting quality and cost effects, a manufacturing cell team commonly discovers ways to better communicate their own needs to their internal customers and suppliers. The internal customer is the next stage in the process, while the internal supplier is the stage of the process this cell draws its work from. By treating other related areas of production as internal customers and suppliers, the continuous improvement team can use the same tools described below with similar positive results (see Fig. 1-1).

   The higher level of within-team and team-to-team communications usually results in greater opportunities for conflict resolution and additional problem solving. This is particularly true if the management encourages such resolution and trains its workers in negotiation, problem solving, and statistical skills. This, in turn, reduces the level of irritation associated with particular jobs and thus increases productivity.

4. Increased prestige in the marketplace.

   Increasing quality, decreasing costs, and improving the flow of products and information within the manufacturing environment typically result in visible improvements in the products and services from a company. Companies are strongly advised to exploit the gains realized by continuous improvement in the marketplace. If an improvement makes a product easier to use, then the advertising for the product should point this out. When the customer becomes aware of real improvements in the value of a product, that increases the prestige and perceived quality of the product.

5. Increased productivity through greater efficiency.

   As mentioned above, continuous improvement can increase productivity through the secondary effect of better communications and information flow. However, it can have a primary effect on improving productivity by having the continuous improvement teams directly examine questions of productivity. It is at this point that continuous improvement starts to be integrated with other techniques and technologies used in a modern manufacturing plant.

   For example, the issue of the quick change of dies and tools is a popular approach to increasing productivity. When a die change is accomplished in ten minutes instead

CHAPTER 1

THE IMPORTANCE OF CI

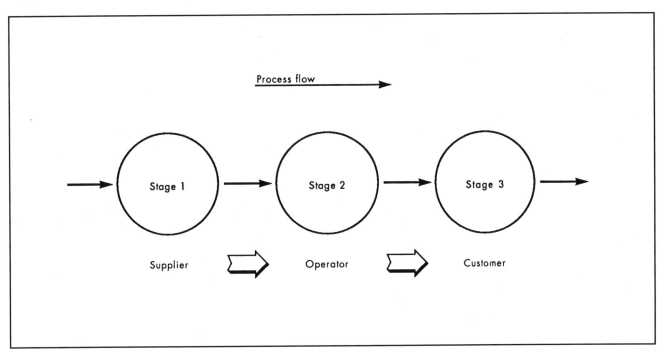

Fig. 1-1 The internal customer and supplier.

of eight hours, the productivity improvement is readily observable. (Most setup reduction yields time savings which translate to direct labor savings, but seldom are the batch sizes reduced. Batch size reduction will yield the greatest savings through inventory reduction, cycle time reduction, better quality, faster delivery of possible higher quality product to the customer. The direct labor savings is important but typically has less influence on product cost.) However, to get such a quick change requires that a quick-die or tool change team organize itself and execute its plan using many of the techniques described for continuous improvement. In reality, the process characteristic this continuous improvement team is addressing is time. Time is seen as a waste and must be used to maximum efficiency. Therefore, the time required to exchange die or tool is first measured and then monitored as a performance characteristic as the team brainstorms ideas, tests them, and implements successful changes.

6. Opportunities for reduced inventories and support of just-in-time (JIT) systems.

    Another aspect of using time as a performance measure is the implementation of a JIT system of production. Although many models of JIT exist, all are tied to the idea that timing between customers and suppliers, whether internal or external, is critical. Implementation of a JIT system is similar to most continuous improvement based projects. This difference lies in the fact that the teams are not looking for opportunities for improvement as much as they are seeking the elimination of barriers to the free flow of materials and information through the production area.

    Another possible focus for a continuous improvement team would be the reduction of inventories. These teams usually monitor inventory characteristics, such as turnovers and average available stock, as performance measures. In most cases, predictability during production is the main goal. By gaining a predictable count of inventory needs, stocks and reorders can be set to minimize the amount of inventory present at any time. This reduces the need for floor space, insurance, and the like, which thus reduces costs. Again, the continuous improvement team needs to actively increase routes of communication to make this type of effort successful.

HISTORY AND BACKGROUND

Although several ancient sources promoted the idea of improvement in man and society, the idea of continuous improvement as an integral part of management and production is a modern concept. Until the end of the eighteenth century, most societies had held fast to the idea of a static existence, that every day would be very similar to the day before.

However, at the beginning of the nineteenth century several historic forces came together to create the idea that active progress was possible. That is, individuals could take actions that would result in improvements in both life and work. This concept was born in Western Europe through the work of "naturalists" who discovered that nature was anything but static. Combined with Darwin's view of survival through adaptation and the American view of self-determination, an environment was created where people perceived the ability to improve their situation.

At the same time the industrial revolution was also maturing. Today, it seems only natural that industrialization would include the idea of progress. However, it took most of the nineteenth century for social moralities to change to support the idea of companies having the prerogative to expand and improve.

The labor movements of the early 20th century combined with the quantification of efficiency methods promoted by Henry Ford to set the stage for today's approach to continuous improvement. The technology of statistics was brought into the workplace with the "scientific management" studies of Fredrick W. Taylor. For

CHAPTER 1

THE MANY FACES OF CONTINUOUS IMPROVEMENT

the first time, the concept of improvement was given an analytical nature. Thus, it changed from being a management ideal to becoming an applicable tool. Job titles could be formed around specific functions, with the delegation of responsibilities based on maximum efficiency.

With the 1931 publication of W.A. Shewhart's *Economic Control of Quality of Manufactured Product*, the issue of the quality of products was joined with the idea of improvement. Shewhart was working with Bell Laboratories to invent ways to increase the quality of goods, produced by sister organizations such as Western Electric. However, besides discussing the ways to monitor quality on a daily basis—and thus develop control charts—he also introduced several ideas on how to actively improve quality by involving various key people in the process.

World War II saw the firm establishment of statistical methods to monitor production and quality of manufactured goods, combined with the active participation of labor groups in the management of companies. The final piece needed to create the modern era of continuous improvement would come from a defeated combatant whose country and economy were in ruins, Japan.

General Douglas MacArthur envisioned a new Japan based on the Western ideal of capitalism and modern manufacturing systems of management. To that end, he brought to Japan many American experts who taught Japanese managers new management techniques.

One of the most influential was Joseph Juran who had just published in 1951 his monumental work on quality control which stated that quality improvements represented increased profits and markets. At the same time in Japan, Joseph Dodge expounded the view that economic conservatism was the most efficient use of capital. Thus, the Japanese received the message that active improvements were needed for economic survival, but they could not "throw money at the problem."

To solve this dilemma, the Japanese employed the one resource it had in abundance—people. During the 1950s and 1960s, they developed what Westerners prefer to call "quality circles." Groups of workers from a similar functional area of the plant would gather to discuss how small improvements could be made. These groups used the simple tools of continuous improvement, such as charts, diagrams, brainstorming, and the like. Over time, the slow increase in productivity performed on a daily basis with great discipline began to be seen as an increase in the quality and competitiveness of Japanese goods.

It is during this time that the philosophy being employed by the Japanese was recognized by the name *kaizen*. Because of the lack of capital and resources in Japan during this time period, this ideal of continuous improvement also included the concept of improving without spending much money.

During the 1950s and 1960s, Americans were developing the idea of quality assurance instead of quality control. The 1956 publication of Armand Feigenbaum's book on Total Quality Control stated that "quality is everybody's job." Thus, the Americans had also taken the step of bringing all employees into the task of continuous improvement. However, it was the Japanese that would receive the majority of the credit, since they developed a set of simple tools and an established procedure for forming employee teams to seek opportunities for improvements. Most notable of the Japanese publications was Kaoru Ishikawa's *Guide to Quality Control*, that specifically assigned CI tasks to teams of workers with management lending support.

Thus, by the 1970s the modern era of continuous improvement had begun. The 1973 oil crisis would propel continuous improvement to the head of the priority list for most managers and company executives. The oil crisis had the effect of making the resources of production considerably more expensive. This, in turn, along with losing industries and market share to foreign competition, motivated Western companies to seek more efficient ways to produce goods. This was the modern pursuit of never-ending improvements in quality, and eventually the empowerment of employee work teams in searching for avenues of improvement was initiated.

This change in emphasis was most visible in the automotive industry. In 1982, for example, Ford Motor Company republished its supplier quality assurance procedure with the expressed mission of seeking "never-ending improvements." General Motors was even more specific a few years later with the introduction of its "Targets for Excellence" program. The first chapter of this book was devoted solely to the concept of continuous improvement. General Motors went as far as to specifically link continuous improvement to the sustained profitability of the company.

Today, the concept of continuous improvement is firmly implanted into the mentality of modern companies through its overt presence in benchmark references such as ISO 9000, the Malcolm Baldrige National Quality Award, and the U.S. Government's TQM program.

So, on a historic scale, continuous improvement is the product of several forces coming together to evolve the idea of progress into a toolbox for industrial application. There really was no single moment when the name continuous improvement was developed or specifically designed for a particular purpose. Instead, history shows that continuous improvement is a fluid philosophy because it builds on general principles that find different applications in history driven by the concerns of the times. This is why continuous improvement is usually linked with quality assurance in modern times. The need for increased quality is the most recent concern of industry.

THE MANY FACES OF CONTINUOUS IMPROVEMENT

Continuous improvement must be defined in terms of its application and scope. The wider the scope of application, the more people who are involved and the more effective the results. In addition, the applications can vary widely. Below are three common ways to group applications by the type of entity deploying the technique.

CONTINUOUS IMPROVEMENT IN SOCIETY

Dr. Geniichi Taguchi is famous for stating that quality is measured as the loss that society suffers when a product is shipped. This implies that industries are partially responsible for the economic well-being of the society at large. When a defective product is shipped, the customer must pay extra money to have it

CHAPTER 1

THE MANY PRACTICES OF CI

fixed or replaced. This represents a loss to the customer and society. Thus, it is poor quality.

To prevent this loss to society, the quality of the product must be improved and the design of the product must be changed to prevent unnecessary loss. To accomplish this, one of the responsibilities of a modern manufacturing plant is to continuously improve its productivity and quality while reducing costs. Large OEM customers are requiring their supplier companies to adopt methods of continuous improvement. Thus, industries such as automotive now enforce the use of continuous improvement.

CONTINUOUS IMPROVEMENT WITHIN A COMPANY

An individual company uses continuous improvement for more selfish reasons. Individual companies need to cut costs and improve quality to remain competitive. Such a company tends to use continuous improvement as an operational philosophy. This philosophy of continuous improvement is then implemented by requiring the use of employee involvement teams armed with simple problem solving tools. Such teams are given a free rein to attack common problems in their portion of the manufacturing process. By communicating with other teams, these groups are then able to improve work flow, prevent defects, and reduce costs. This, in turn, tends to make a company more competitive.

CONTINUOUS IMPROVEMENT ON A PERSONAL SCALE

On a personal scale, the philosophy and techniques of continuous improvement can be used by individuals to improve their own environment. Participants in employee work teams report that they "feel more involved" with their work and they generally have a higher state of morale.

In addition, the individual can also employ continuous improvement to help improve the quality of work life. Improvements in methods, environment, and management lead toward a better feeling about the job. Several studies report that this results in a more productive worker with fewer absentee or sick days.

THE MANY PRACTICES OF CI

Like any technology that is not based on a property of physical science, continuous improvement is open to interpretation. This, in turn, leads to a diversity of attitudes on how it should be approached. Although there are many interpretations, the following represent the alternatives that are most frequently used by organizations.

PLAN-DO-CHECK-ACT

The first of these common approaches is the Plan-Do-Check-Act (PDCA) cycle. This is a highly localized approach to continuous improvement where problem solving work teams use simple statistical and data gathering tools to attack tactical level problems of production. Recently, this industrial approach has been adapted for processes outside the production area, such as paperwork flow, computer networking of data, and delivery service.

While Dr. J. Edwards Deming is the most recognized name associated with the PDCA cycle (with others attempting to take credit for its development), it is the result of several groups discovering that most improvements are best realized through a common approach that can be summarized under the PDCA model.

The "Plan" stage of the process involves the identification and evaluation of potential areas of improvement. Typically, management provides guidelines as to the types of problems to be addressed, usually restricting a team to working on those processes for which they have direct responsibility. The result of the "Plan" phase is a list of potential problems and possible corrective actions.

During the "Do" phase, the team tries each corrective action, one at a time. As each corrective action is implemented it is up to the team to maintain a constant monitoring of process performance and other baseline measures.

The "Check" phase involves an evaluation of the effect of each corrective action on the performance measure being tracked. The team looks for significant improvements in quality, time, costs, or other selected responses.

Then "Action" is taken to put effective countermeasures into the process as a permanent change of process procedures or design. The potential benefits of the change are then estimated and reported to management for further action, if necessary. The team then return to the list of potential areas of improvement and selects the next problem to address. Thus, the PDCA cycle repeats. In this way, each potential or real problem eventually gets addressed. At the same time, improvement is seen as a slow spiral of increased performance with each cycling of the PDCA model. Figure 1-2 shows the PDCA cycle as a spiral function.

THE JURAN APPROACH

Dr. Joseph M. Juran produced a landmark work in the field of management and quality assurance based on the idea that real improvements come through a breakthrough in management thinking. He saw improvements as both increases in the fitness for use of products and decreases in the number of defects, for both internal and external customers.

Juran's approach is more strategic in nature and involves management in the central role of continuous improvement efforts. There are other similar methods developed by various groups and individuals, but Juran's is the best documented and organized.

Juran uses a simple chart (see Fig. 1-3) to point out the difference between continuous improvement and "fire fighting" within a company. The level of performance, in this case quality performance, is fairly steady around a 10% scrap rate. This is called the "chronic" condition of the company. Occasionally, a spike of poor performance occurs, also called the "sporadic" performance of an organization. Juran points out that many managers devote most of their time to correcting the sporadic behavior instead of working every day to reduce the chronic condition. In other words, modern managers tend to "fight fires" instead of pursuing continuous improvement.

Juran's technique has several distinct, but flexible stages. It begins with "proof of the need," that is convincing management

CHAPTER 1

THE MANY PRACTICES OF CI

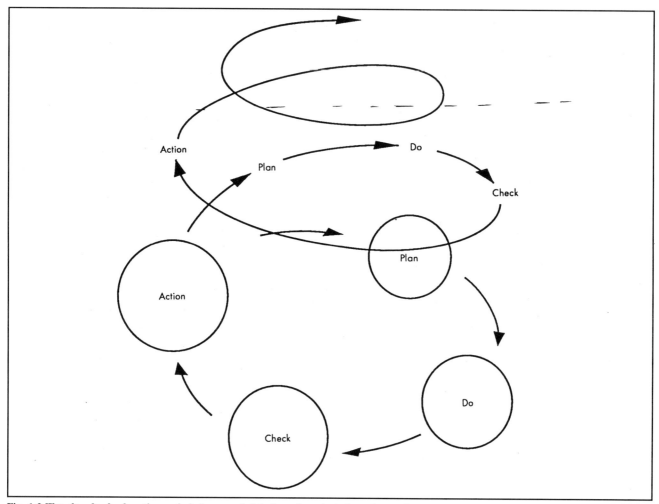

Fig. 1-2 The plan-do-check-action cycle being repeated to form the spiral of continuous improvement.

that a problem is severe enough to warrant action. Juran emphasizes that the "language of money" is very effective at communicating need to upper management. Next comes the identification of the projects needing attention. This list of projects is then broken down into areas that can be dealt with most effectively. This is accomplished by establishing teams that search from the symptom to the root cause. After root causes are discovered, various remedies and tests are suggested. The teams are free to use a wide range of statistical and problem solving approaches to quantify the problem and measure their progress toward a remedy.

Solutions are then tested during actual operations to verify their effectiveness and to estimate the potential improvements in quality and reductions in costs. At the same time, management is kept informed of progress through a series of presentations by the improvement teams to the managers. The result is usually several groups recommending a wide variety of changes to the process. Therefore, the final decision rests with management on how to best compromise and implement the potential solutions. The management breakthrough occurs when management, participating in this process, realizes that the teams are finding opportunities for improvement that can be exploited to reduce the chronic problems within the organization. Management then assumes a key role in this process.

KAIZEN

The Japanese system of *kaizen* is an all-inclusive approach to continuous improvement where every employee, including all members of management, assimilate the philosophy of gradual improvements over time. In fact, Masaaki Imai's book on this topic points out that *kaizen* is almost an unconscious instinct in Japanese business practices. Therefore, *kaizen* can be seen as an extension of Juran's method, just as the PDCA cycle can be expanded to form Juran's approach.

In Japan, one of the most common *kaizen* programs is called the five S's, which stands for *seiri*, *seiton*, *seiso*, *seiketsu*, and *shitsuke*. *Seiri* refers to the practice of straightening up the work area. Tools, work in progress, scrap, waste, documents, and the like must to be in their proper places. Anything that is unnecessary is discarded, or stored in a separate location. *Seiton* means to put things in order, more specifically within easy reach of the operator. It is very common under this system for commonly used tools to be located closest to the process, as in quick die change tool carts. *Seiso* means to keep the workplace clean. American companies have recently rediscovered that cleanliness in the workplace leads to greater productivity. This is seen in the housekeeping requirements now issued to supply companies by major customers, such as Ford Motor Company. *Seiketsu* stands

CHAPTER 1
THE MANY PRACTICES OF CI

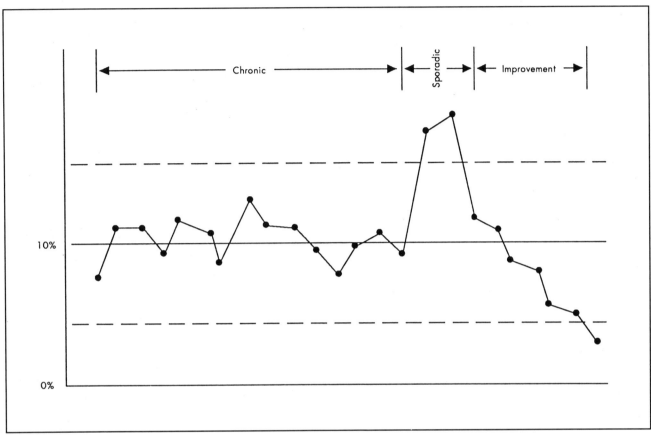

Fig. 1-3 Chronic and sporadic performance of a process.

for standardization, is in a sense a repetition of cleanliness, but considers colors, shapes, clothing, etc. to give a sense of cleanliness. Finally, there is *shitsuke*, discipline. It is seen as imperative that each employee follow workshop procedures.

To distinguish *kaizen* from the Western approaches to continuous improvement, one must realize that the West focuses on results, while *kaizen* is more concerned with the process. While a Western company's management will focus on sales and dividends, a Japanese manager focuses attention on the attitudes of the workers and the current operation of the process. The *kaizen* approach looks at the success of the process and then seeks ways to improve. The theory is that ongoing improvement of processes will eventually lead to increased performance of results-oriented measurements.

This can be seen in the widely used suggestion systems in Japan and the time managers spend on human issues such as discipline, training, worker participation, morale, and communications. It can also be seen influencing the use of cross-functional teams to solve problems and cross-functional managers to properly deploy company policies.

Thus, *kaizen* is more a discipline than a particular method. Companies that have adopted the *kaizen* philosophy have used various models of employee involvement to pursue continuous improvement. Some companies form permanent teams to solve area-specific problems, while others form temporary teams to address particular problems. These problem-specific teams are then disbanded after a solution is reached. Some companies place managers into the teams, others do not. In whole, *kaizen* is seen as an umbrella over a wide selection of continuous improvement methods and philosophies (see Fig. 1-4).

THE TAGUCHI APPROACH

Dr. Taguchi introduced a quantifiable method of continuous improvement with the introduction of loss-functions to the engineering and quality activities of a company. Specifically, Dr. Taguchi looked at the question of meeting customer requirements and how to constantly improve the design of goods and processes. His conclusion was that quality and improvement should be measured economically. This can be seen in his famous definition of quality as, "the loss to society after a product is shipped." In other words, when a process is not running perfectly or a product does not meet the exact needs of a customer, society suffers an economic loss.

For example, if the brakes on a car are not made to specifications, the customer will need to make frequent adjustments and will have to replace the brake pads and rotors prematurely. This represents an economic loss to the customer and society, and is thus seen as poor quality.

Dr. Taguchi illustrates this as a loss-function curve approximating a quadratic function, with its minimal point at the target value for customer satisfaction—usually the optimal of a specification or tolerance. If a product or process performance characteristic misses this target value, the loss to society increases exponentially (see Fig. 1-5).

To minimize this loss, Dr. Taguchi strongly encourages several methods. One is the use of the loss-function formula to set the in-house tolerance on a product to minimize the potential loss to a

CHAPTER 1

THE MANY PRACTICES OF CI

Fig. 1-4 The *kaizen* umbrella.

customer. Another is to center processes on the desired target before attempting to reduce process variation. This is the inverse of the procedure commonly practiced in Western companies.

However, his most controversial approach to continuous improvement is to change the designs of products and processes to make them robust against uncontrollable sources of variation—what he calls "noise factors." Three techniques are suggested for creating a robust design. The first is system design, where the management of a company uses engineering and other quantitative methods to determine the proper plant layout and machinery selection to assure the success of the manufacturing process. The second is parameter design, where uncontrollable factors in the environment (noise) are used in the replication of a designed experiment to find controllable factor settings that resist the noise factors. The third technique is tolerance design, where the tolerances for a product or process are studied scientifically to determine which need to be tightened to assure production success. At the same time, tolerances that have little effect on success are loosened to allow for the use of less expensive components and multiple suppliers. Combined, these three techniques produce lower cost goods at higher quality.

Therefore, Dr. Taguchi's approach focuses on the efforts of engineers and quality professionals using quantifiable measurements and scientific methods to make better products and more efficient processes. The use of teams is again encouraged in that Dr. Taguchi suggests that no experiments take place until at least ten people involved in the process have been interviewed. Such groups brainstorm possible factors causing variation and possible interactions between the factors. Thus, continuous improvement is seen as a logical path for technicians and engineers, while it is achieved primarily through design change.

HEWLETT-PACKARD'S BEST PRACTICE

The influences in America of Japanese techniques of continuous improvement were very evident in the 1970s and 1980s. However, some American companies stuck to their tendency to depend on innovations for improvement while also adopting continuous improvement philosophies. A notable case was Hewlett-Packard, which was continually seeking methods for improving itself as a company. It readily modified JIT techniques into its own unique, internal form. The same is true of continuous improvement.

Hewlett-Packard (HP) has a long history of participatory management with worker involvement. Their approach to continuous improvement has been to focus on the procedures used by each employee. Under the theory that the best expert on a particular job is the person performing that job, HP instituted a program called "best practices."

Each job at HP has standard operating procedures (SOPs) associated with it. These SOPs are called practices. If an employee can find a better way to accomplish a job, he or she is encouraged to document the new method. Management then reviews the new method and its estimated advantages and, if approved, it becomes a new "best practice." Each procedure guide is updated and the employee receives credit for the new method. Appropriately, the reward system is geared to benefit those employees who actively seek a better way to do their own jobs.

Thus, the HP method of continuous improvement focuses on the individual with some participation by management. Naturally, some of the other methods of continuous improvement described above are also used at HP, but the best practices program is an excellent illustration of how continuous improvement's philosophy can be adopted to most situations by changing the key players and the emphasis of the program.

CHAPTER 1

CUSTOMER REQUIREMENTS FOR CI

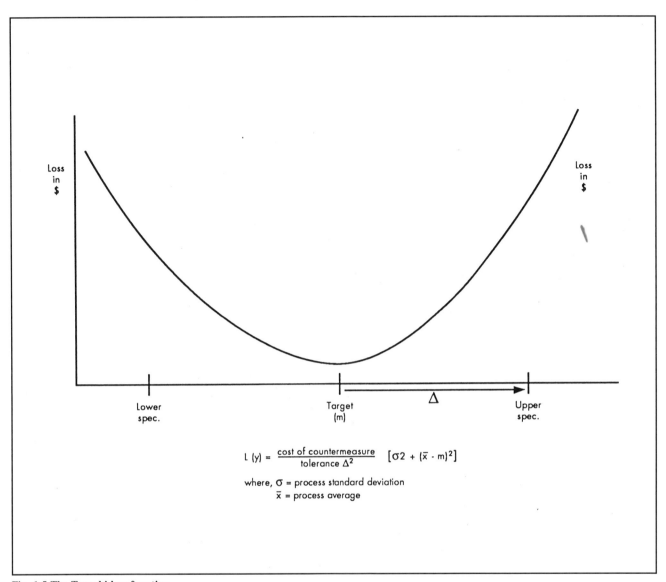

Fig. 1-5 The Taguchi loss function.

CUSTOMER REQUIREMENTS FOR CI

As noted above, many customer requirements now call out continuous improvement. Specifically, the largest purchasing customers want their suppliers to practice the techniques of continuous improvement, or even to adopt a philosophy of continuous improvement as an operational strategy. Most of these requirements then call for "management commitment" to continuous improvement. In fact, when audited, many supply companies discover that a key part of the audit is a close examination of policies and procedures to see if they fully support CI.

ISO 9000

The international standard of quality assurance within an organization, ISO 9000, states:

Efforts to develop new methods for improving production quality and process capability should be encouraged.

When a company undergoes ISO 9000 certification, each area of responsibility for overall quality of the product is examined for dedication to continuous improvement. A history of improvement based on factual information gathered from the process is required as part of ISO 9000 certification.

During assessment for certification for ISO 9000, companies have discovered that the requirements of the standard are evaluated using the PDCA cycle described above. For example, the requirement for production control is audited by first examining the plan for production control. Then the auditor checks to see if the plan was carried out as stated. Next, there is a search for documented

CHAPTER 1

CUSTOMER REQUIREMENTS FOR CI

evidence that the success of the production control efforts was checked by management. Finally, any unexpected deviations that were found by checking are noted to require corrective action. Therefore, continuous improvement is built into all aspects of the implementation of ISO 9000.

Currently, over 20,000 companies worldwide have adopted ISO 9000. The standard is used in over 90 countries and in the European Economic Community. In the United States, several industrial groups are actively requiring compliance with ISO 9000 or plan to do so in the future. Thus, continuous improvement should remain a critical requirement of company management for many years to come.

THE MALCOLM BALDRIGE NATIONAL QUALITY AWARD

The 1992 Malcolm Baldrige National Quality Award states in its criteria:

> Achieving the highest level of quality and competitiveness requires a well-defined and well-executed approach to continuous improvement.

Specifically, the award committee looks for companies that have a well defined philosophy of continuous improvement and evidence that information gathered by employee involvement teams is used by management to increase the competitive position of the company.

In 1992, the Malcolm Baldrige National Quality Award criteria were given out to over 250,000 companies and individuals. The award is presented each year by the President of the United States to no more than six companies, two each from service, large, and small companies.

THE UNITED STATES GOVERNMENT

An active program of TQM is now being implemented by the United States Government at the Federal level. A few state and many local governments are also pursuing the use of continuous improvement to increase the level of service they provide while reducing costs. In addition, the Administration and the Department of Defense have identified increased product and service improvement as national goals.

Prime contractors and their suppliers are required to implement some form of TQM within their organizations. ISO 9000 is also being integrated into this plan. By the end of the century, the view that continuous improvement is a requirement to do business will be firmly rooted within the defense industry and the agencies of the Federal government.

THE AUTOMOTIVE INDUSTRY

Perhaps the most visible industry involved with the use and development of CI is the automotive industry. All major U.S. automotive makers require continuous improvement within their own and their supply organizations. For example:

> Ford expects that producers will not only meet engineering specifications, but will continuously improve products and services by reducing process variation and optimizing performance to target values. (Ford Worldwide Quality System Standard Q-101.)

> Implementing a continuous improvement process will result in small improvements made in day-to-day operations as a result of group and individual analysis and efforts to improve processes. (General Motors' Targets for Excellence.)

It is the automotive industry that leads the way for making continuous improvement an integral part of manufacturing. The

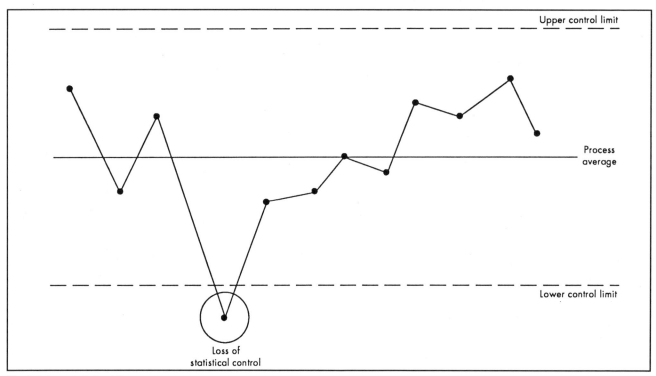

Fig. 1-6 A control chart for averages.

CHAPTER 1

THE TOOLS OF CONTINUOUS IMPROVEMENT

competitive challenge of the 1970s and 1980s presented by the Japanese forced the U.S. automakers to find new methods of management. The widespread adoption of problem solving tools, such as Pareto charts, control charts, scattergrams, and brainstorming, also made it necessary to integrate these techniques under the umbrella of continuous improvement. Eventually, management discovered that by expanding and exploiting continuous improvement, greater competitiveness could be achieved.

THE TOOLS OF CONTINUOUS IMPROVEMENT

Ostensibly, any problem-solving method, tool, or algorithm can be considered a tool of continuous improvement. The most common tools are discussed in detail below. Ultimately, having people within the organization well trained in the use of these tools and possessing a discipline for continuous improvement makes any improvement effort successful.

Ironically, the best tools for continuous improvement have proved to be the simplest. A simple count of defects, a map of defect locations, a flow diagram of the process, and the brainstorming of ideas from a group are effective because they are simple. Simplicity breeds familiarity with those applying these methods, and simply put, a tool that is used frequently with comfort is usually the most effective.

CONTROL CHARTS

Usually based on the central limit theorem, control charts measure the inherent variation of a process and portray it as a set of control limits. Operators of the process can then sample the process and evaluate whether it is maintaining statistical control. Points out of statistical control indicate an assignable cause of variation and thus an opportunity for adjustment. The knowledge gained from detecting these sources of added variation can lead to process improvement (see Fig. 1-6).

FISHBONE DIAGRAMS

Also called the Ishikawa or cause-and-effect diagram, the fishbone chart is based on the theory that there is a direct cause for

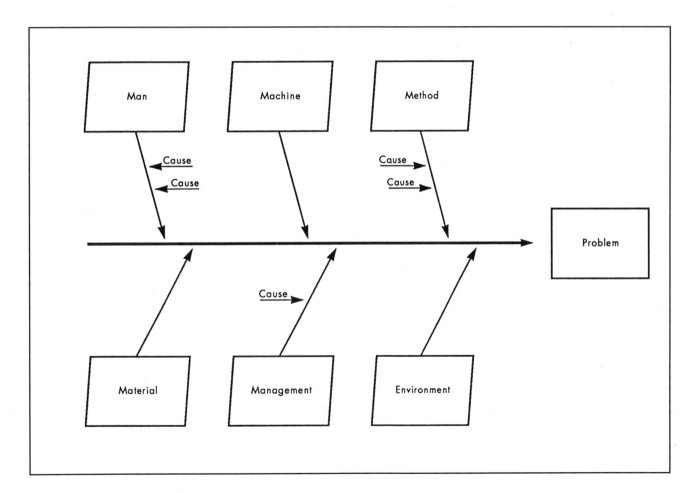

Fig. 1-7 A fishbone diagram.

CHAPTER 1

THE TOOLS OF CONTINUOUS IMPROVEMENT

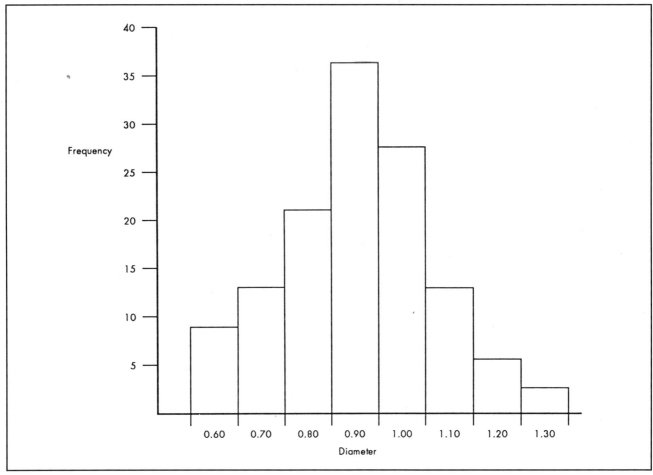

Fig. 1-8 A histogram.

a problem (see Fig. 1-7). The process is driven to develop this problem by the influences of man, machine, method, material, management, or environment. By identifying the sources of a problem, corrective action can be taken to create an improvement.

HISTOGRAMS

Data taken from a situation through sampling can be quickly summarized using the histogram (see Fig. 1-8). Essentially, a histogram is a bar chart of data categories portrayed by frequency. The distribution, variation, and central tendency of the histogram provide several important statistical summaries used to track an improvement effort.

CAPABILITY STUDIES

Formal statistical studies of processes can reveal the amount of variation to expect from production systems. This, in turn, can lead to discoveries of the source of common causes of variation and thus further improvements.

QUALITY FUNCTION DEPLOYMENT

Quality function deployment (QFD) is a matrix that captures the needs of a customer and then compares them to the proposed engineering specification. Each, in turn, is benchmarked against competing products. This leads to the discovery of opportunities for improvements in competitive features and services.

FAILURE MODE AND EFFECTS ANALYSIS

Failure mode and effects analysis (FMEA) is the documentation of the thought process of the designer: specifically, the detailed listing of points of potential failure in a product and the necessary countermeasures. This allows improvement by preventing potentially serious failures.

DESIGNED EXPERIMENTS

Perhaps the most powerful tool of continuous improvement is the designed experiment. Such experiments test combinations of factor settings in a product or process for optimal setup sheets and product designs. The use of designed experiments in industry is unique in that it identifies factors that can best change variation within a product or process, while also identifying external sources of data error.

CHECKSHEETS

A checksheet is a simple tool involving the listing of the characteristics that should be present in a process or product. By documenting the actual conditions and operational results of a process, a team or an individual can derive actions for improvement.

SCATTERGRAMS

A scattergram is the plotting of two variables on an X and Y chart to look for possible associations. This is also a tool to find

CHAPTER 1

IMPLEMENTATION OF CI

the relationships that exist within a process. When one factor correlates to another, there exists a possible source of greater control and predictability.

PROJECT MANAGEMENT

Project management techniques, including flow diagrams, PERT charts, and Gantt charts, help continuous improvement teams to organize their efforts for seeking incremental improvements. PERT charts are particularly useful because they show relationships between different stages of a project and the critical path of the project.

BENCHMARKING

Benchmarking is the process of quantifying the competitive features of similar products available in the marketplace. The technique of benchmarking gives the continuous improvement team valuable information on the course they need to steer to make continuous improvement a competitively effective tool.

IMPLEMENTATION OF CI

Continuous improvement is a discipline. A great deal of training and culture change within an organization is usually necessary to make continuous improvement a daily and permanent reality. Thus, a concerted, never-ending effort by management to promote continuous improvement is necessary.

WHERE TO IMPLEMENT

Continuous improvement can be implemented in virtually any department and cover any function within an organization. The best implementation for manufacturing is a plant-wide approach. Areas such as production, research and development, purchasing, and sales are particularly well suited for pilot projects.

However, a careful examination of the functions within each department should be made before implementing continuous improvement. This will provide a baseline for comparison to measure the actual improvements received. The following areas should be part of any improvement measurement system:

1. Profits.
2. Quality characteristics.
3. Delivery.
4. Inventory control.
5. Paperwork flow.
6. Safety.
7. Facilities management (that is, energy efficiency).
8. Labor relations.
9. Throughput (cycle time).
10. Technical and human skills development.
11. Gage accuracy, repeatability, and reproducibility.
12. Vendor relations.
13. Work in progress.
14. Customer complaints and compliments.
15. Downtime and tool change time.

HOW TO ASSURE SUCCESS OF A CI PROGRAM

Like any new technique or technology, change is part of the process of continuous improvement. Change is one of the greatest sources of resistance by the people involved with a new method. In addition, other obstacles lie in the path of a company seeking to implement continuous improvement. However, through education, careful planning, and measured implementation, a company can prevent many of these obstacles from interfering with implementation.

Many studies and the advice of experts identify management commitment as the most important element of a continuous improvement program. Without management's commitment to supporting problem solving teams, evaluations, and the search for opportunities, the daily requirements of everyone's job will quickly wear down their enthusiasm for pursuing continuous improvement. Although most managers will state that continuous improvement is an important element in any company, experience has demonstrated that many delegate or avoid participation. Critical to the success of continuous improvement is the full involvement of management. This means that every management decision made must include a moment to think of possible opportunities that could be exploited. It also means that managers must meet face to face with the teams of employees working on continuous improvement projects.

After management commitment is secured, an organized approach to continuous improvement is highly recommended. Such an approach should be complete and documented as part of the normal operations of the company. A typical approach is to make participation in continuous improvement teams part of a job description. To reinforce the need to participate, a reward system based on team performance is also recommended. It is not unusual today to find job positions with promotions and pay raises based on team participation and team success.

At the same time, continuous improvement cannot be projected as yet another task for an already overwhelmed workforce. Instead, it has to be pointed out, quite accurately, that continuous improvement can represent a way to discover easier and faster ways to perform tasks and complete jobs, while also eliminating problems. Thus, continuous improvement can actually reduce the work load and increase the flow of processes.

The most difficult obstacle to remove involves the attitude of the organization on how to get a job done. In other words, the most difficult task for management might be inducing the cultural change within the organization toward the philosophy of continuous improvement. This involves directly confronting the sources of resistance to continuous improvement while also working with the recognized leaders of the organization's culture, such as the union leader, the company president, and the head of human relations. After that, cultural change is accomplished by providing opportunities for participation in pilot groups that are, in turn, highly publicized within the organization for their successful efforts. This promotes greater participation by those first resistant to the idea and helps to spread continuous improvement from a pilot group based approach to a company wide method of operation.

Finally, time is another great enemy of continuous improvement. An over-scheduled workforce will not be able to participate in continuous improvement. Therefore, efforts must be made to free up time every day for participation in the system.

CHAPTER 1

IMPLEMENTATION OF CI

Bibliography

Brassard, Michael. *The Memory Jogger Plus: Featuring the Seven Management and Planning Tools*. Methuen, MA: GOAL/QPC, 1989.

Clements, Richard B. *Handbook of Statistical Methods in Manufacturing*. Englewood Cliffs, NJ: Prentice-Hall, 1990.

Continuous Improvement—The MGI Management Institute Newsletter of Quality Techniques. Harrison, NY: MGI Management Institute, various issues from 1988 to 1992.

Continuous Improvement. Bethlehem, PA: Bethlehem Steel Corporation, 1991.

Davidow, William H. and Uttal, Bro. *Total Customer Service: The Ultimate Weapon*. New York: Harper and Row, 1989.

Eureka, William E. and Ryan, Nancy E. *The Customer Driven Company: Managerial Perspectives on QFD*. Dearborn, MI: The American Supplier Institute, 1988.

Garvin, David A. *Managing Quality: The Strategic and Competitive Edge*. New York: The Free Press, A Division of MacMillan Inc., 1988.

Imai, Masaaki. *Kaizen, The Key to Japan's Competitive Success*. New York: Random House, 1986.

Ishikawa, Koaru. *Introduction to Quality Control*. Tokyo: 3A Corporation, 1990.

ISO 9000/Q-90 Series. American Society for Quality Control/American National Standards Institute: 1987.

Juran, J.M. *Juran on Leadership for Quality: An Executive Handbook*. New York: The Free Press, A Division of MacMillan Inc., 1989.

Juran, J.M. *Juran on Quality by Design: The New Steps for Planning Quality into Goods and Services*. New York: The Free Press, A Division of MacMillan Inc., 1988.

O'Connor, P.D.T. *Practical Reliability Engineering*. 3rd edition. Toronto: John Wiley, Q-101, 1991. *Worldwide Quality System Standard*. Dearborn, MI: Ford Motor Company, 1990.

Ryan, Thomas P. *Statistical Methods for Quality Improvement*. Toronto: John Wiley & Sons, 1989.

Scholtes, Peter R. *The Team Handbook: How to Use Teams to Improve Quality*. Madison, WI, 1988.

Targets for Excellence. Detroit, MI: General Motors, 1990.

Wadsworth, Jr. Harrison M., Stephens, Kenneth S. and Godfrey, A. Blanton. *Modern Methods for Quality Control and Improvement*. Toronto: John Wiley & Sons, 1986.

Zeithamel, Valerie A., Parasuraman, A., and Berry, Leonard. *Delivering Quality Service: Balancing Customer Perceptions and Expectations*. New York: The Free Press, A Division of MacMillan Inc., 1990

1992 Malcolm Baldrige National Quality Award Criteria. Gaithersburg, MD: NIST, 1992.

CHAPTER 2

TOTAL QUALITY MANAGEMENT

HISTORY OF MODERN ORGANIZATIONS

CHAPTER CONTENTS:

HISTORY OF MODERN ORGANIZATIONS 2-1

RESPONSES TO THE NEW REALITY 2-4

TQM AND TOMORROW'S ORGANIZATION 2-8

This chapter provides an overview and a brief history of total quality management (TQM). Implementation of continuous improvement (CI) and TQM is described elsewhere in this handbook. According to the American Society for Quality Control's (ASQC) TQM committee, TQM is a customer-driven process improvement approach to management. TQM is sometimes called total quality control (TQC), company wide quality control, total quality, total quality leadership, etc.

To understand the need for TQM, the reader must understand how modern organizations evolved—and why, in some ways, they are no longer performing satisfactorily. This section examines the history of American organizations from the dawn of the industrial revolution to the present. The next section examines TQM as a response to the new environment businesses find themselves in today. The final section describes the effect of TQM on future organizations.

The chapter uses a graphical method of portraying the structure of organizations that was developed by Henry Mintzberg. Figure 2-1 illustrates the basic parts of an organization using Mintzberg's approach.[1]

ORGANIZATIONS PRIOR TO 1840

Early American business organizations were faced with constraints primarily dictated by nature. Year-round transportation was impossible. Rivers and canals froze, roads were blocked with snow, and no all-weather modes of transportation existed. The state of production technology was quite primitive as well. Steam engines had not yet been invented; energy needs were served primarily by burning wood or using water wheels (on streams that froze and flooded periodically).

The economy of pre-1840 America was dominated by subsistence agriculture; 72% of all workers labored on farms. Commerce consisted mainly of the general merchant, and the family was the basic business unit. Large cities were almost nonexistent, with only 5% of the population living in towns with populations of more than 2500.[2] Businesses existed primarily to serve local markets with specialized products for the end user, such as cotton for weaving cloth. The small amount of manufacturing carried on was primarily done by artisans working alone.

Other than artisans working alone, the primary organizational form of the era was the putting-out system. Under the putting-out system there was a degree of division of labor. For example, a shoemaker would have the upper part of the shoe produced in one place, soles cut in another, and the shoe assembled somewhere else. This system provided more output than individuals working alone, but work was still done by hand at the individuals own time and pace. Other activities, such as tending the crops, often took precedence over the putting-out work. The simple organizational form of this early era is shown in Fig. 2-2.[3] Key management positions were generally held by family members. Management control systems, such as accounting systems, were simple variations of systems that had been in use since medieval times.

1840-1900

The second stage in the development of modern business organizations was precipitated by the development of new technologies, beginning with the steam engine. Steam produced a reliable, all-weather, year-round energy source that made it possible to operate despite the seasons. Unlike energy produced by water power, steam was a portable energy source. Although steam engines were large and ungainly, they could be integrated into large mobile structures, such as ships and locomotives. After 1840, steamships plied the waters of the Atlantic, vastly increasing the amount of trade and commerce between America and other nations. With the rise of railroads it suddenly became possible to transport merchandise during the entire year. A new era had begun for business, with changes at least as significant as those now taking place, including:

---

*The Contributor of this chapter is: **Thomas Pyzdek**, Ph.D. Candidate, Department of Management and Policy, University of Arizona.*

*The Reviewers of this chapter are: **Thomas Bradley, Ph.D.**, Senior Manager TQM/Excellence, Northern Telecom, Inc.; **William H. Cogwell III**, Senior Quality Engineer, Quality Technology Center, Northern Telecom, Inc.; **Rodrick B. Ma, PE, ASQC CQE**, Engineer, Quality Technology Center, Northern Telecom, Inc.; **Brian Margetson, PE, ASQC CQE**, Advisor, Strategic Quality Management, Northern Telecom, Inc.; **Brian McDermott**, Editor, Total Quality Newsletter, Lakewood Publications; **Herb Shrieves, Ph.D.**, Advisor, Strategic Quality Management, Northern Telecom, Inc.; **Eugene Wolf**, Advisor, Strategic Quality Management, Northern Telecom, Inc.*

CHAPTER 2

HISTORY OF MODERN ORGANIZATIONS

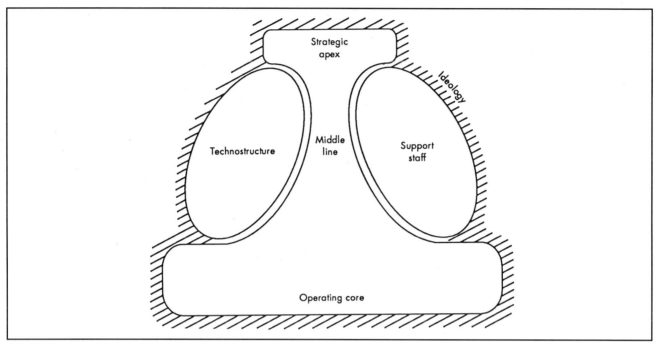

Fig. 2-1 The six basic parts of the organization.

- Other new technologies. Although recent inventions such as the digital computer are truly impressive and revolutionary, the period of 1840 to 1900 introduced to mankind such earth-shaking inventions as electricity, cast iron and steel, metal plating, artificial fertilizer, vulcanized rubber, insulation, the telephone and telegraph, and anesthesia, just to name a few.
- Full time employees. Prior to the all-season, all-weather business environment there was little need to have year-around employees. Most businesses operated with workers who were occupied elsewhere (primarily on their farms) most of the time. After 1840, full-time employment grew in popularity.
- Professional managers. As the size of businesses grew, it became necessary to have people other than family members take on critical management responsibilities as their lifelong profession.
- Corporate organization. Capital requirements outgrew the financial ability of the owners, and ownership became separated from daily operations management.

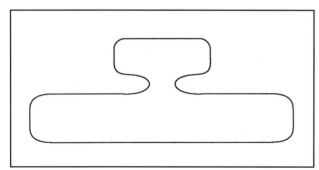

Fig. 2-2 Pre-1840 organization.

- Vertical integration of operations. Manufacturing changed its emphasis, from one-of-a-kind products created by craftsmen and sold to local markets, to mass-produced products marketed nationally and sometimes internationally. Producers extended their ownership back into raw materials and forward into distribution and merchandising; this reduced uncertainty and increased control.

In the industrialized countries, the world of organizations changed completely. One of the effects was the rise of the bureaucratic form of organization. This organizational form is characterized by the division of activities and responsibilities into departments managed by full-time management professionals who had no source of livelihood outside the organization. These people devoted their lives to building a career with the organization. The form of the new bureaucratic organizations is depicted in Fig. 2-3.

1900-1950

As explosive growth continued, one serious limitation on businesses was the availability of skilled labor. Although a large population of unskilled immigrant labor was available, the nature of work in the factory at the turn of the century still depended to a considerable extent on the skills of the individual craftsman. The craft tradition was dominant, and individual craftsmen performed complex tasks in the course of normal production. The professional manager and engineer of the day had little or no knowledge of the details of the activities performed by the work force.

Managers, and before them owners, have contested with workers for control of the workplace since the beginning of the industrial revolution. One of the workers' most effective weapons in this struggle was their intimate knowledge of how the work was actually done. Scientific management, the term applied to the methods advocated by Frederick W. Taylor, is an approach to management that attempts to obtain and standardize this knowledge to better control the work. The workers generally sought to prevent the spread of job-specific knowledge to management.

CHAPTER 2
HISTORY OF MODERN ORGANIZATIONS

F.W. Taylor was a man with a mission: spread scientific management throughout industry. Taylor believed that management could never effectively control the workplace unless it controlled the work itself, that is, the specific tasks performed by the workers to get the job done. Furthermore, Taylor believed, management could also improve the efficiency of work by applying the methods of science to the evaluation of tasks. The improvement would come from two primary sources:

1. Selecting the individuals best suited to a particular job.
2. Identifying the optimal way to perform the jobs.

Taylor believed that scientific management would benefit both management and workers. Management would benefit from increased worker productivity, and workers would receive higher pay (which management could afford due to the increased productivity) and would be assigned to jobs that suited them.

A casual look at the organization of work today reveals, at the very least, a strong scientific management influence. There seems to be agreement that worker opposition did have a dampening effect on the spread of scientific management in its raw form. There were efforts by Taylor to get his approach endorsed by the American Society of Mechanical Engineers (ASME). However, the engineers in ASME, as well as corporate liberal reformers, had difficulty with many of the properties of scientific management which they saw as dehumanizing. While not completely rejecting the scientific management approach, which had great appeal to many engineers, these groups worked to add consideration of human factors to it. Also, where Taylor steadfastly refused to deal with workers in groups, the engineers and reformers worked closely with labor groups, including unions, to develop new methods. The revisionists were able to spread their modified version of scientific management into areas ignored by Taylor, including management and government. The new approach formed the basis for a number of organizational innovations, such as the creation of the fields of personnel management and industrial relations.

Henry Ford introduced production innovations which had an impact comparable to that of F.W. Taylor. Where Taylor looked at analyzing the work being done by people, Ford mechanized the work. Under Fordism the worker was reduced to the role of machine-tender, while the machines did most of the actual work.

In its effort to gain control of the workplace by better understanding the work, management reduced the skill levels required for each job. This was no accident; Taylor describes this effect and heartily endorses it. The major resistance came from craftsmen, such as machinists, who understood the value of their knowledge and skill in terms of both monetary rewards and job security. The reduction of work to a series of simple tasks that can be done with a relatively small investment in training is one of the major results of scientific management. This has a number of effects. First, it makes it easier for management to monitor the work, one of Taylor's primary goals. Because the company's investment in each worker's training was smaller, it also made it easier to replace those who did unsatisfactory work, thus providing a negative incentive to perform well. However, the redesigned job is usually far more boring, leading to a variety of problems such as high levels of stress and employee turnover.

The legacy of "deskilling" is that the work force becomes less able to change as new conditions arise. Whereas a machinist could work for any number of companies in many industries, a machine tender at Ford had very limited prospects outside Ford. Thus, scientific management tended to increase the demands for job security. In the modern era, lack of generalized employee skills is a major impediment to the ability of firms to react quickly to rapidly changing market conditions. Employees in the modern organization possess skills that are task specific, rather than general. They can operate a machine and are correctly called machine operators, but they are not machinists in the sense that they can identify and perform multiple tasks with machine tools. When rapid change creates new tasks, the workers previous experience does not help them adapt to the new circumstance; they must constantly be retrained. What is needed in the modern workplace is not more training in the classical sense; it is what Deming[4] calls profound knowledge. Profound knowledge is *a priori* knowledge that transcends mere fact.[5] Profound knowledge provides a theoretical context in which the facts are interpreted and their meaning understood. Without it, the worker and firm alike become the victims of asset specificity, where this time the asset is the worker himself.

Organizationally, the introduction of scientific management resulted in a modified and enlarged bureaucratic form. On the technical side, organizational units were formed to codify, in detail, the knowledge of how work is to be performed. These organizational units included manufacturing engineering, industrial engineering, quality control, and other functions. On the human relations side, these organizations handled such activities as employee relations and employee health insurance. Even traditional functions such as accounting were affected as they became increasingly specialized. The structure of these new organizations resemble that shown in Fig. 2-4. This is still the

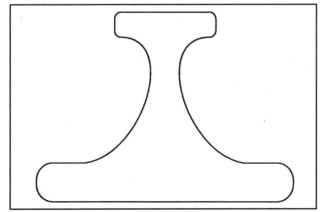

Fig. 2-3 Early bureaucratic organizational form.

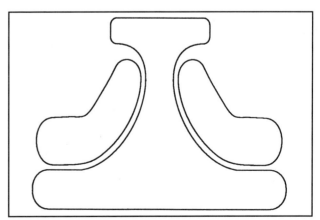

Fig. 2-4 The traditional machine organization.

CHAPTER 2

RESPONSES TO THE NEW REALITY

structure of many of today's manufacturing organizations, but the number of such organizations is steadily decreasing.

The essence of the development of manufacturing organizations between 1900 and 1950 was increasing specialization. Not only did labor become more specialized, so did management activities. Even capital equipment was increasingly specialized. Wherever possible, the focus was on controlling uncertainty by the imposition of rigid structure on the environment. The period was characterized by the ongoing development of mass-marketed goods. The scope of most firms was national, and most of the competition for American firms came from other American firms.

POST-WORLD WAR II TO PRESENT

This period is primarily one of transition. Table 2-1 summarizes the changing pattern of the marketplace.

As can be seen, in some ways the changing business environment involves a return to states similar to those seen in the past. For example, the return to more complex jobs and the resulting need for workers with a broader repertoire of skills. Other tendencies are continuations of past trends; international markets are the next logical step after moving from local markets to national markets. In other ways the new world of business is different; modern flexible systems differ in fundamental ways from previous systems.

TABLE 2-1
The Changing Business Environment

WAS	IS
National markets	International markets
National competition	International competition
Control the business environment	Adapt to the environment rapidly
Homogeneous product	Customized product
Deskilled jobs	Complex jobs
Product-specific capital	Flexible systems
Maintain status quo	Continuous improvement
Management by control	Management by planning

RESPONSES TO THE NEW REALITY

Table 2-1 makes it clear that yesterday's environment was fundamentally different from today's. It follows that yesterday's organizations, which evolved in response to the realities of the past, might not be suited to the changing reality. In fact, there is strong evidence to suggest that organizations that do not adapt will simply disappear. Over 40% of the 1979 list of the Fortune 500 had disappeared by 1990.[6] The organizations that have managed to progress have not stood still.

JAPANESE ADAPTATION OF TQC

Japan is well known for replacing its old reputation for poor quality with a new reputation for excellence. The system they employed to accomplish this impressive feat is a uniquely Japanese version of a system that originated in America known as total quality control. TQC is the predecessor of today's total quality management. TQC is a system of specialized quality control activities initially developed by Feigenbaum.[7] The Japanese took Feigenbaum's American version of quality control (which was very much a continuation of the scientific management approach) and Japanized it. The Japanese rendition of TQC is described by Ishikawa.[8] The Japanese system also draws heavily on contributions of other American quality experts, especially Walter A. Shewhart, W. Edwards Deming, and Joseph M. Juran. However, there are many elements of the Japanese system that are purely Japanese in character.

Cross-Functional Management

One of the uniquely Japanese features involves a fundamental modification of the bureaucratic model of organization. Under the bureaucratic form all activities of the organization are divided into functions such as manufacturing, engineering, marketing, and finance. It has long been known that this form of organization, sometimes called the chimney stack model, results in isolation of the various functions from one another. This in turn results in parochialism and in behavior that, while optimal for a given function, is detrimental to the enterprise as a whole. Most business texts address this problem superficially at best. The Japanese developed a formal approach for dealing with it, which is called cross-functional management.[9]

Cross-functional management is a variation of the matrix-management organizational form that was the rage in the business literature in America in the 1970s. However, where the matrix-management approach was advocated for use in the management of large projects that required long-term close cooperation between different functional organizations, cross-functional management is applied to a different situation and for a different reason. The Japanese identified certain specific performance criteria that suffered as a result of the chimney stack form of organization, such as quality and cost of delivery, and they created a high-level organization to deal with these issues on an organization-wide scale. The approach is illustrated in Fig. 2-5.

Cross-functional management is organized at the highest level in the organization. Generally the quality department will head the quality cross-departmental function, the accounting department the cost cross-departmental function, and production or sales the delivery cross-departmental function. These three functions, quality, cost and delivery, are not the only cross-departmental functions; however, they are usually present at a minimum.

Once organized, the cross-functional committee will operate within the departmental/functional matrix. When conflicts arise, departmental control is superseded by cross-functional policy. The purpose of cross-functional management is to replace the short-term focus with a long-term focus, to replace departmental goals with organizational goals, and to deal with issues that span several departments.

Cross-functional management takes a system-wide perspective. It looks beyond the individual perspective of given departments to examine the relationships between departments. It looks outside the organization to the critical role played by customers, suppliers, competitors, and regulators. It is an attempt to deal with nonlocal issues on a routine, ongoing basis.

CHAPTER 2

RESPONSES TO THE NEW REALITY

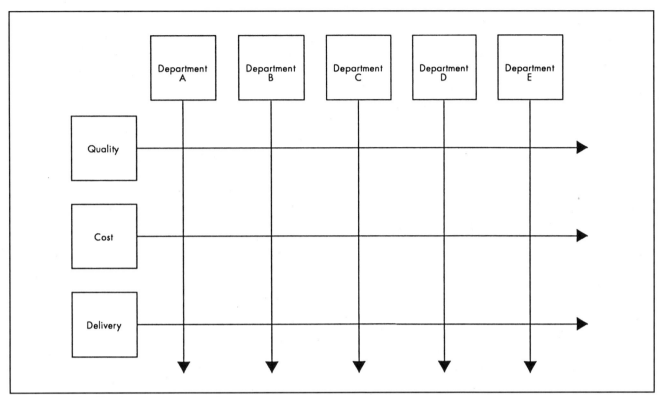

Fig. 2-5 Cross-functional management.

Focus On The Core

Another feature of the Japanese approach to management is concentration on the core business. The core business is the essence of what the company does; for example, Toyota might identify its core as the production of personal transportation vehicles. For full-time employees in the core, the company strives to provide a sense of family and belonging. Lifelong service to the employer is expected, and the employer demonstrates similar loyalty to the employee, for example by providing lifelong job security. While this is not done in all Japanese firms, most top tier firms do attempt to do this.

How can this be done? After all, the normal business cycle rises and falls periodically and a company that is adequately staffed for peak production is overstaffed when production downturns occur. The Japanese manage to provide security in good times and bad in a number of ways, one by massive outsourcing of noncore business activities to suppliers. It is not unusual to find as much as 80% of the value of the finished product in purchased materials. Another is by making use of a large buffer of part-time employees. In some Japanese companies as much as 50% of their employees during peak production periods are part-time workers.

Prevailing wisdom in America would suggest that this strategy would result in a great increase in uncertainty for the prime manufacturer. Also, most American managers believe that farming out production to suppliers and using part-time workers results in an overall decline in quality. Again, the Japanese have developed strategies for dealing with this. Suppliers are very tightly controlled using a variety of means. Suppliers to major firms are part of a system known as the *keiretsu* system. Members of the *keiretsu* often have board members from the parent company, and from other members of the *keiretsu*. Unlike American businesses, which tend to deal with their suppliers at arms length, parent companies play a very active role in the affairs of their suppliers. For example, in the U.S. a firm might simply judge the quality of the product delivered by a supplier. In Japan, the parent would carefully evaluate how the product was made. If appropriate, the parent would suggest better, more economical ways to produce the product. Parent companies provide larger, longer-term contracts than their American counterparts, and they often demand steady price decreases. The supplier might be expected to become dedicated to its parent, providing product only to the parent and not to competitors.

Continuous improvement in quality, cost, and delivery is expected of all Japanese suppliers, just as it is expected from the employees.

AMERICAN RESPONSE TO CHANGE

There is considerable doubt that much of what the Japanese have done so successfully can, or should, be done in America. Although a great many lessons can be learned from studying foreign businesses, the differences in culture and population make a direct transfer of systems and approaches a dubious undertaking at best. Accordingly, the most successful American firms have achieved success in ways that are distinctly their own.

For most of the post-WWII period, American business did not recognize the changing nature of the marketplace. This is not due to any character flaw on the part of American managers; it is simply the result of their being buffered from the forces of change. Part of the buffer was the structure of the organization itself. As described earlier, the history of organizations involves the extension of control backward to the sources of supply and forward to the distribution and merchandising. To the extent that organiza-

CHAPTER 2

RESPONSES TO THE NEW REALITY

tions succeeded in this endeavor, they reduced uncertainty in their environments and gained control over critical elements of their business. However, to a degree, they also insulated themselves from the larger environment. Thus the signals of change were somewhat muted. Also, unlike the rest of the industrialized world, the American infrastructure was undamaged by the war. While the rest of the world rebuilt, America supplied them. The result of all of this was that American managers received feedback that indicated that what they were doing worked. It is not easy to change when all signs indicate that there is no urgent need to change.

Today, however, the need for change is widely accepted, at least in the manufacturing sector of the economy. There is little doubt that improvement is possible, and competitive pressure is pushing more and more firms to consider even drastic changes. This began in the late 1970s and it has accelerated ever since. Thus, there are now numerous American models of successful change.

When studying TQM success stories in America, the winners of the Malcolm Baldrige National Quality Award are a good place to start.[10] Here is a brief review of the winning strategies of some of the early winners of the Baldrige award. Detailed reports can be obtained for all winners by contacting the American Society for Quality Control (1-800-952-6587).

Strategies of Baldrige Award Winners

One of the fundamental purposes of the Malcolm Baldrige National Quality Award is to promote quality improvement in the United States by publicizing the strategies used by the winners to achieve their success. The National Institute of Standards and Technology (NIST) sponsors forums conducted by the winners each year to describe their strategies in greater depth.

Globe Metallurgical, Inc. (1988)

Globe Metallurgical, Inc. is a major producer of silicon metal and ferrosilicon products. In 1985, Globe did a self-assessment of their quality system as it related to the requirements of Ford's Q-101 quality standard. They discovered deficiencies in that their system emphasized detection instead of prevention of quality problems, failed to use statistical process control (SPC), lacked quality planning, and did not involve their employees in the improvement process. They addressed these issues by:

1. Training the entire work force in SPC.
2. Establishing a Quality Manual and the mechanics of their quality system.
3. Educating and training Globe's suppliers in the necessity of SPC and a quality system. Outside consultants were used for SPC training and supervisors were trained in Deming's approach with a videotape program.

Management formed an executive steering committee and plant committees. The executive committee, chaired by the President and CEO, meets monthly to discuss broad issues of development and maintenance of the quality system. Plant committees, chaired by the plant managers, meet daily to discuss specific projects and implementation issues.

Milliken & Company (1989)

Headquartered in Spartanburg, SC, the privately owned Milliken & Company has 14,300 "associates" employed primarily at 47 manufacturing facilities in the United States. Annual sales for textile and chemical products exceed $1 billion.

In 1981, senior management implemented Milliken's Pursuit of Excellence (POE), a commitment to customer satisfaction that pervades all company levels. Since then, productivity has increased 42%.

Teams are a hallmark of the Milliken quality improvement process. In 1988, 1600 Corrective Action Teams were formed to address specific manufacturing or other internal business problems; 200 Supplier Action Teams worked to improve Milliken's relationships with its suppliers; and nearly 500 teams responded to the needs and aims of customers. Quality improvement measures are solidly based on factual information, contained in an array of standardized databases accessible from all Milliken facilities. Most manufacturing processes are under the scrutiny of real-time monitoring systems that detect errors and help pinpoint their causes. Milliken's successful push for quality improvement has allowed it to increase U.S. sales and enter foreign markets.

Motorola (1988)

At Motorola, the quest for quality involves every aspect of work: from SPC to product development cycles to the time it takes for an order to be entered. In 1987, Motorola invested $44 million in training, much of that strictly on quality improvement.

A participative management process which emphasizes employee involvement at all levels was a key factor in Motorola's achievements.

In 1981, Motorola launched an ambitious drive for a tenfold improvement in the quality of products and services by 1986. Having largely accomplished that goal in 1986, the company then set its sights even higher, striving for a hundred-fold improvement by 1991. The company's next goal is Six Sigma quality by 1992. Six Sigma means no more than 3.4 defects per million parts produced.

While the company expresses this goal in the language of statistics, the ultimate goal is zero defects in everything, and Motorola's overriding objective is Total Customer Satisfaction. The strategy is to refocus all elements of Motorola's business on serving the customer.

Westinghouse Commercial Nuclear Fuel Division (1988)

Westinghouse Commercial Nuclear Fuels Division is a division within the Energy Systems Business Unit of Westinghouse. It is responsible for the engineering, manufacturing, and supply of Pressurized Water Reactor (PWR) fuel assemblies for commercial nuclear power reactors. The fuel contained in these assemblies generates heat (through nuclear fission) which is converted to electricity. Westinghouse Commercial Nuclear Fuels Division supplies 40% of the U.S. light water reactor fuel market and 20% of the free-world market.

Quality standards are demanding. The reliability of Westinghouse nuclear fuel approaches 99.9995%. The quality challenge is compounded by the complexity of the fuel production process, which involves both chemical and mechanical operations and literally thousands of variables. By 1983, Westinghouse Commercial Nuclear Fuels Division recognized that quality could provide the most significant competitive edge. Consequently, the division made continuous, long-term quality improvement its number one management objective. The division was quick to identify and implement new and innovative management techniques and quality improvement processes, including human resource management concepts like participative management and quality teams, new technologies like robotics, artificial intelligence, and statistical in-process quality control, and advanced management concepts like cycle-time reduction.

CHAPTER 2

RESPONSES TO THE NEW REALITY

Each year, Westinghouse Commercial Nuclear Fuels Division increased its understanding of the quality management process and sharpened the focus and direction of its program. The plan focused on a different theme each year: Increased general employee awareness of the importance of quality to its business (1983); Began tracking and reporting quality failure costs (1984); Established division-wide quality goals and a measurement system to monitor progress (1985); Created a division-wide mission statement and added second- and third-tier quality measures (1986); Shifted primary focus to customer (1987); Adapted "Total Quality" theme (1988).

From its experience in quality improvement, Westinghouse Commercial Nuclear Fuels Division learned a number of lessons, which it believes can be applied with reasonable success to any business. First, an organization must have a common vision, and all people in that organization must embrace a common mission. Next, a framework for Total Quality is absolutely critical. This is the model or blueprint that keeps everyone focused on continuous improvement in all aspects of the business. The third lesson: measure, measure, measure. Westinghouse Commercial Nuclear Fuels Division discovered that no quality improvement program can be successful unless both employees and customers are intimately involved in the process. Finally, Total Quality requires a long-term commitment to continuous quality improvement. A Total Quality culture can not be built overnight. Total Quality is not a short-term proposition, but neither are the rewards; the benefits of Total Quality are long-term for a company willing to make the commitment.

Xerox Business Products and Systems (1989)

In 1983, Xerox Business Products and Systems launched an ambitious quality improvement program to arrest its decline in a world market it had once dominated. Today, the company can once again claim the title of the industry's best in nearly all copier-product markets. The company, headquartered in Stamford, CT, attributes the turnaround to its strategy of "leadership through quality." Through extensive data-collection efforts, Xerox Business Products and Systems knows what customers want in products and services. Planning of new products and services is based on detailed analyses of data organized in some 375 information management systems, of which 175 are specific to planning, managing, and evaluating quality improvement.

Benchmarking is highly developed at Xerox Business Products and Systems. In all key areas of product, service, and business performance, the company measures its achievement for each attribute and compares itself with the level of performance achieved by the world leader, regardless of industry.

Quality improvement and, ultimately, customer satisfaction are the job of every employee. Working with the Union, the company ensures that workers are vested with considerable authority over day-to-day work decisions. Employees are expected to take the initiative in identifying and correcting problems that affect the quality of products or services.

Xerox Business Products and Systems employs 50,200 people at 83 U.S. locations. The company makes more than 250 types of document processing equipment. U.S. sales exceeded $6 billion in 1988.

Other American Success Stories

In addition to the Baldrige winners there are many other American firms that have demonstrated remarkable sustained excellence. Sheridan[11] describes several manufacturing firms that fit this description. Most of these firms have not followed any TQM blueprint in accomplishing their impressive results. It is important to remember that there is no such thing as a TQM blueprint! Implementing TQM by the book is a sure path to failure.

Features of Tomorrow's Organizations

There are a number of trends that have been taking place since the end of the Second World War and seem likely to continue into the future. Tomorrow's business will have a smaller number of employees than today's. It will have fewer people in blue collar jobs, and far fewer people in middle management jobs. The organization will have fewer layers of structure, perhaps averaging five to seven layers of management instead of the 10 to 15 or more layers common in today's large organizations.

The focus of tomorrow's business will be external rather than internal. Rather than controlling the environment by bringing activities into the organization's domain, tomorrow's organization will aggressively use information to anticipate external changes so they can influence events with proactive management activity. For example, tomorrow's organization will not merely try to match the offering of a competitor; it will attempt to create new markets by providing products and services never before offered.

To facilitate fast response, tomorrow's organization will decentralize decision making. Small, local business units will act autonomously to satisfy the demands of their customers. To maintain organizational integrity under this system, tomorrow's organization will need a powerful ideology and a compatible long-term mission that bonds the autonomous units together.

Continuous improvement will be the byword. The status quo will be viewed with suspicion. Standard operating procedure will be perceived as merely providing a starting point. Dynamic change will be a prerequisite for survival.

Employees in tomorrow's organization will need to have a broad repertoire of skills. The skills needed will be constantly changing, so tomorrow's employee will be a continuous learner. The employer will provide ongoing education and training. In general, tomorrow's organization will be much more involved in and concerned with the welfare of each employee. Employees will be considered the key asset of the organization.

Tomorrow's organization will thrive on the rapid acquisition, analysis, and dissemination of information. A large part of the organizations resources will be devoted to these tasks.

Although tomorrow's organization will purchase a far larger share of their products value from outside sources, the number of suppliers used will be far smaller. Single source suppliers will be the norm for critical items. The relationships with suppliers will be much closer than they are today, and the suppliers will need to meet requirements that are far more stringent than those of today. The suppliers internal systems and processes will be a matter of concern—not just the product delivered.

The organization of tomorrow will involve constantly changing networks of partners. The networks will facilitate information exchange and they will enable the organization to utilize the best in any given field.

The capital used by tomorrow's organization will be flexible. The dynamic external environment will require dynamic internal systems that can readily adapt to new demands. The process of continuous improvement means continuous change, and hardware or systems that can not cope will not be tolerated.

CHAPTER 2

TQM AND TOMORROW'S ORGANIZATION

TQM AND TOMORROW'S ORGANIZATION

THE TECHNOLOGY OF TQM

Much of what has been written about TQM is written in a way that indicates that TQM is merely a collection of techniques. This is not the case. Treating TQM as a collection of techniques misses the point entirely: TQM is a response to a dynamic external environment. The definition of TQM given at the beginning of this chapter should be kept in mind when studying the techniques described in this section. In all likelihood, all of the techniques will eventually outlive their usefulness and need to be replaced or modified. Collecting up-to-date information on new management techniques will be one of many information oriented tasks of tomorrow's organization.

Daily Management Tools

The techniques provided in Table 2-2 are given in alphabetical order. They are from the ASQC-TQM subcommittees' Daily Management subcommittee.[12] The descriptions of the techniques are by the author and are not part of the ASQC report.

The TQM daily tool kit also includes the traditional quality control tools shown in Fig. 2-6.[13]

Long-Term TQM Tools

By definition, TQM's focus is the long term. It is no surprise, therefore, that a number of tools have been developed that facilitate long-term improvement. These are summarized in Table 2-3. Figures 2-7 and 2-8 illustrate the new 7M tools and the house of quality, respectively.

Deming[4] provides a framework for fusing all of the tools of TQM into an effective, long-term organizational strategy. Deming's 14 points for management (see Table 2-4) represent the essence of his prescription for change. The 14 points are not merely a collection of things that can be selectively used or ignored; they represent a unified whole that must be applied in total. People frequently have difficulty understanding the meaning of each point, and group discussion is needed to assure that everyone has reached the same conclusion.

TQM PHILOSOPHY

While one important contribution of TQM is in providing a framework for continuous improvement and an abundance of tools, it is not the most important contribution. The most important contribution is the *philosophy* of TQM, namely the philosophy of providing value to society. This is what separates TQM from its predecessors, such as matrix management, quality circles, and other attempts to improve short-term profitability. Arguably, these approaches failed precisely because they lacked an underlying unifying philosophy. The systems of the organization exist as a concrete implementation of the organization's philosophy; unless the philosophy changes the systems cannot, at least not permanently.

Vision

The successful organization will outlive the people who are currently its members. Thus, the mission of the successful organization must provide vision for the long term. Unsuccessful organizations (most organizations) don't have this problem. Typically, large organizations have a life expectancy of less than 40 years.[14] In contrast, at least one Japanese company is in the 75th year of its 250 year-plan.

What is meant by having a mission? The answer is simple: the organization's mission describes the reason that the organization exists. By the way, no organization exists merely to make a profit. Profits are not why any organization exists, they simply accrue to organizations that produce value in excess of their costs; that is, profits are an effect of productive existence, not a cause. Perhaps the best way to define the term *mission* is to illustrate it with some examples:

Matsushita Electric Industrial Company is one of the world's largest firms. Its stated mission is to eliminate poverty in the world by making their products available to the people of the world at the lowest possible cost.[3]

Henry Ford's mission was to provide low-cost transportation to the common man.

One might go a step further and ask why the organization was created to fulfill its mission. The answer, at least in the beginning, might lie in the values of the organization's founder. Henry Ford, for whatever reason, felt that it was important to provide the farmer with affordable and reliable motorized transportation. Just how he came to possess such values is beyond the scope of this chapter, but it seems crucial that these values exist. Furthermore, to elicit the cooperation of the members of the organization, the values of the organization (which are often difficult to identify or define) must be compatible with the values of its members.

Integrity

The TQM philosophy requires integrity. It is impossible to obtain the needed commitment without integrity. The external environment is much more important to TQM organizations than it is to non-TQM organizations. This is because the TQM organization has externalized a number of the operations that are done internally by non-TQM organizations. For example, the percentage of value added in purchased materials may increase from less than 50% in non-TQM organizations to over 80% in TQM organizations. Also, each supplier is more important because TQM organizations make suppliers part of their design team, and they act as full partners with their suppliers. In the past an arm's-length legal relationship with suppliers sufficed; in the future it will not.

Integrity is also necessary when dealing with customers. In the past, competition was less fierce and a firm that lost customers might survive simply because the customer could find no alternative—or when markets were expanding rapidly. No longer. If customers are alienated in today's business environment, a competitor will snap them up immediately. Just as with suppliers, the arm's-length legal arrangement will no longer work with customers. It is not enough to show the customer the limitations of the warranty, the TQM organization will do what is right for the customer.

Competitors are not enemies. In fact, the existence of competitors should be viewed as a good thing. Competitors may provide customers with options that a company may not offer, thus stimulating it to greater accomplishments. The stimulation provides an incentive for employees and it makes work a great deal more enjoyable than it would be if no competition existed. Competitors deserve help, provided that helping the competitor

CHAPTER 2
TQM AND TOMORROW'S ORGANIZATION

TABLE 2-2
Daily Management TQM Techniques

Technique	Description
Brainstorming	A creative thinking activity that involves suspending judgment and analysis and soliciting the maximum number of ideas and suggestions from a team or group without regard for the quality of the idea or suggestion.
Customer-supplier maps	A process analysis approach applied to operations both within the organization and outside the organization. Designed to identify the sources of process inputs and the users of process outputs and their interrelationships.
Cause and effect analysis	A rigorous analysis of the causes of a particular event. Often involves the use of a tool known as the cause and effect diagram or Ishikawa diagram (see Figure 2-6).
Checklist, check sheets	See Figure 2-6.
Control charts	See Figure 2-6.
Defect maps	The location of defects is shown on drawings, illustrations, or photographs, thus providing information on the causes of the defects.
Design of experiments	See Figure 2-6.
Employee involvement/participation/ empowerment	In the TQM organization authority is shared between management and the employees.
Flow diagrams	See Figure 2-6.
Histograms	See Figure 2-6.
I/O analysis	A method of systems analysis designed to show the relationship between system inputs and outputs. Rigorous interviews of "process owners" are a key feature of the approach.
Kaizen	A philosophy at the heart of TQM, the English translation is continuous never-ending improvement.
Pareto analysis	See Figure 2-6.
Plan-do-check-act cycle	Also known as the "Deming cycle" or "Shewhart cycle," this is the method of implementing the Kaizen philosophy at the process or system level.
Run charts	A line chart of data in a time-ordered sequence, the patterns on the run chart can be analyzed statistically.
Scatter diagrams	See Figure 2-6.
Suggestion programs	TQM suggestion programs are very active, often eliciting huge numbers of suggestions (Toyota's system generated 20 million ideas in 40 years). Unlike traditional suggestion programs, with TQM all suggestions are carefully studied and responded to in a timely manner.
Team building	The act of building trust and team spirit among a group and among all of the employees in the organization.
Team problem solving	A group effort to identify the root causes. Usually involves teams composed of individuals from different departments.
Tree diagram	See Figure 2-7.
Work flow analysis	An industrial engineering technique used to identify the process used to transform raw materials into finished products and services. Useful in process improvement planning.

does not cause an organization harm. If a prospective customer's needs cannot be met by an organization, but can be met by a competitor, the customer should be referred to the competitor. If a serious problem with the competitor's product is discovered, the competitor should be notified.

Employees are the essence of any organization, but TQM organizations realize just how valuable their employees are. Employees are treated with integrity, as responsible adults, and given the respect they deserve.

The isolated organizations of the past could sometimes survive in spite of a lack of integrity, but the TQM organization is much more closely linked to the outside world and it must not alienate its external contacts. Also, because the TQM organization vests so much more authority in the individual employee it must provide

CHAPTER 2

TQM AND TOMORROW'S ORGANIZATION

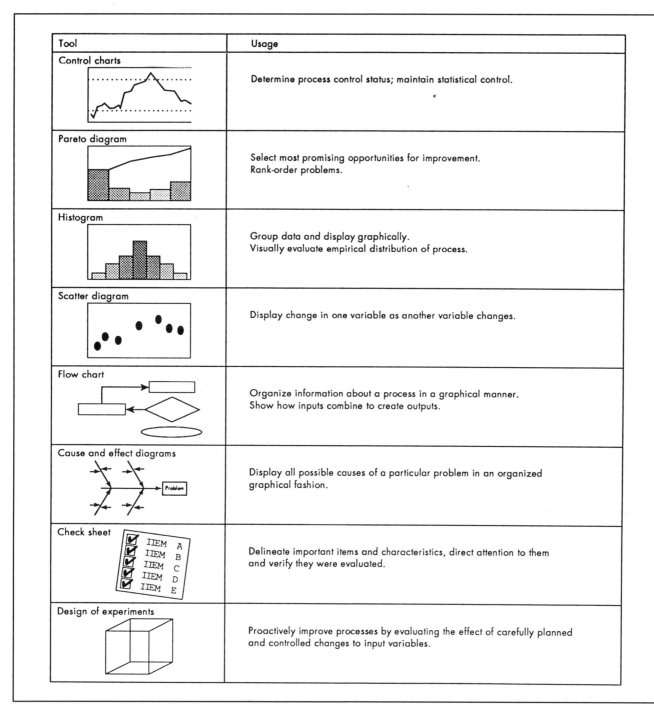

Fig. 2-6 Traditional quality control tools (copyright 1992 by Thomas Pyzdek, reprinted by permission).

more support for each employee. Where in the past integrity was sometimes viewed as an indulgence for the naive, in the future it will be an essential ingredient for survival.

TQM Success and Failure Factors

Many organizations that jumped on the TQM bandwagon have been disappointed with the results.[15] A survey of 500 American manufacturing and service companies found that only a third felt their total-quality programs were having a significant impact on their competitiveness. A similar study in Britain revealed that only one-fifth of the 100 firms surveyed believed that their quality programs had any tangible results.

In contrast, a General Accounting Office survey of 20 of the highest scoring applicants for the 1988 and 1989 Malcolm Baldrige National Quality Award found:

CHAPTER 2

TQM AND TOMORROW'S ORGANIZATION

Tool	Usage
Affinity diagram 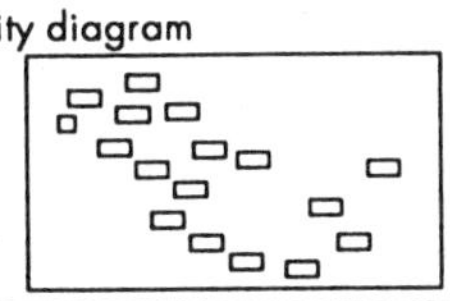	Organize ideas into categories, use both sides of the brain by working in silence, work in groups.
Interrelationship digraph 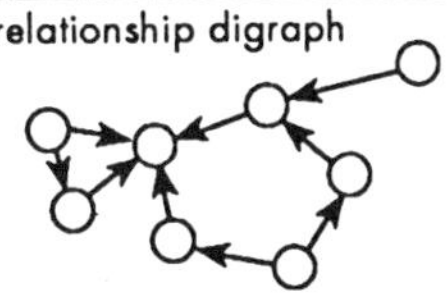	Define how factors in a situation relate to one another.
Tree diagram 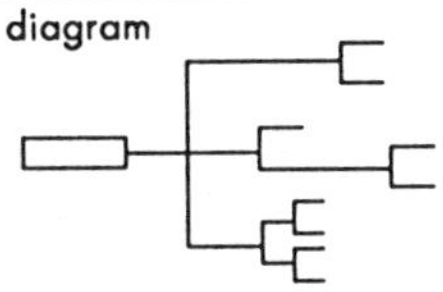	Analyze a situation by stratifying ideas by increasing level of detail.
Matrix chart 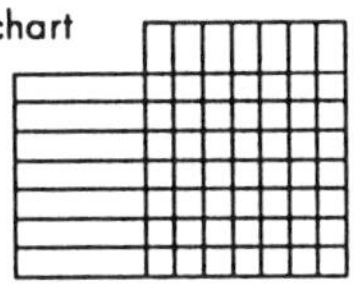	Analyze correlation between two related groups of ideas. (e.g., whats and hows)
Matrix data analysis 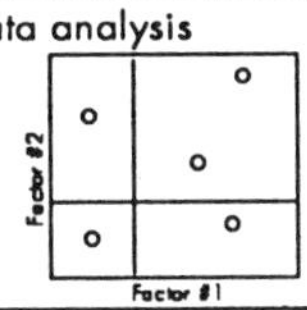	Portray correlations of multiple variables to selected important "factors" e.g., sales of dresses of different designs versus "fit" and "look".
Process decision program chart 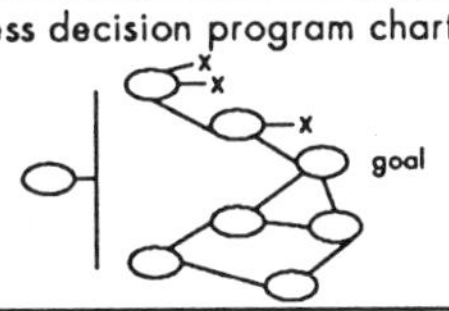	Show alternate paths to a given goal. Help in preparing contingency plans, keeping on schedule, etc.
Arrow diagram 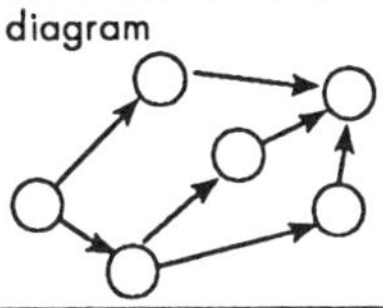	Determine the order in which different activities must be performed and estimate the time required to complete a project. Identify potential problems and opportunities for improvement.

Fig. 2-7 New 7M tools (copyright 1992 by Thomas Pyzdek, reprinted by permission).

Companies that adopted quality management practices experienced an overall improvement in corporate performance. In nearly all cases, companies that used total quality management practices achieved better employee relations, higher productivity, greater customer satisfaction, increased market share, and improved profitability.[16]

What accounts for the differences in the results? Obviously, the investment in TQM described in this chapter is huge. It is important that it not be wasted. While another chapter in this handbook describes the implementation of continuous improvement and TQM in detail, there are a number of factors that seem to be related to success and failure.

- Failure is likely if the techniques of TQM are implemented without a commitment to the underlying philosophy. TQM is a customer-focused, process-driven activity. Many of the firms that experienced failures forgot about the customer and devoted all of their efforts to studying their internal processes.
- TQM is a company-wide activity. Those firms that approached TQM by beefing up their quality departments sent the wrong message. TQM disperses the responsibility for quality outside of the quality department.
- TQM takes time. The GAO reports many different kinds of companies benefitted from putting specific total quality

CHAPTER 2

TQM AND TOMORROW'S ORGANIZATION

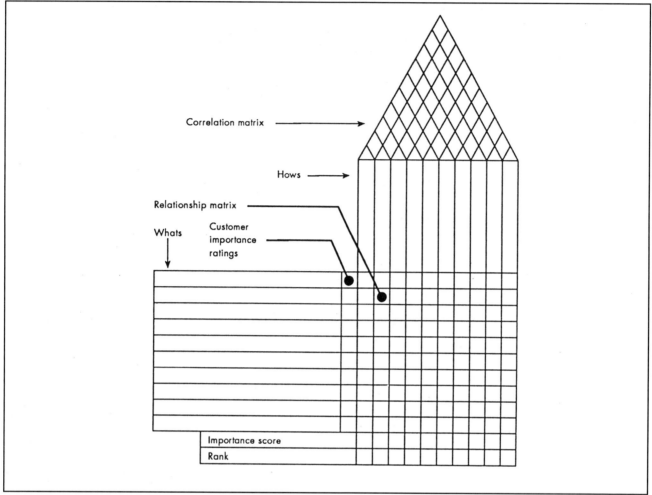

Fig. 2-8 House of quality, QFD (copyright 1992 by Thomas Pyzdek, reprinted by permission).

management practices in place. However, none of these companies reaped those benefits immediately. Allowing sufficient time for results to be achieved was as important as initiating a quality management program.
- Successful implementation of TQM usually involves the presence of a total quality leader (TQL).[17] These leaders, usually not the CEO, encourage the CEO to be an instrument for change and to function as an extension of the CEO as a change agent. The CEO and the TQL must be closely allied. The TQL is often the only person within the organization who embodies customer attitudes.

SUMMARY

Modern organizations exist to fulfill a mission that is compatible with the values of its members and society. In a free and competitive environment, organizations compete with each other for customers and resources. In the long run, the most effective organizations tend to prevail. This results in uncertainty and risk; most organizations ultimately fail and cease to exist.

Historically, organizations have tried to reduce uncertainty by expanding their control both forward and backward in the production sequence. For a time, this strategy was successful and it resulted in the domination of markets by a small number of very large organizations. However, the strategy of management by control seems to be less effective in today's environment. Firms which continue to adhere to the previous strategy appear to be less competitive than those adopting different management strategies.

A new management strategy, known by the general term TQM, has arisen as a response to changes in the competitive environment. TQM is a customer-driven process improvement approach to management that differs in fundamental ways from its predecessors. Although the implementation of TQM involves the use of a large number of tools and techniques, the essence of TQM is a refocus of the organization on its long-term mission, which is in turn based on the organizations values. The performance of organizations that have successfully adopted TQM, relative to their previous performance and that of competing non-TQM organizations, indicates that the TQM approach results in better employee relations, higher productivity, greater customer satisfaction, increased market share, and improved profitability. However, the successful implementation of TQM is difficult, and a majority of the organizations that attempt to implement TQM fail to accomplish it.

CHAPTER 2
TQM AND TOMORROW'S ORGANIZATION

TABLE 2-3
Long-term Management TQM Techniques

Technique	Description
7M tools	See Figure 2-7.
Baldrige criteria	The criteria used to evaluate those companies competing for the Malcolm Baldrige National Quality Award. The application guidelines provide detailed criteria for evaluation of a company's total quality system.
Benchmarking	A technique developed by Xerox Corporation for determining the best-in-class for any given process.[18]
Concurrent engineering	Sometimes called simultaneous engineering, this approach involves the creation of product designs by interdepartmental teams working together on the design. The traditional method, which involves each department performing its activity and then "handling off" the project to the next department, is known as sequential engineering.[19]
Cost of quality analysis	Accounting method of analyzing the total cost of quality, including failure costs, appraisal costs and prevention costs.[20]
Cross-functional management	See the section on this topic earlier in this chapter.
Customer focus	A rigorous and deliberate effort to direct the attentions and energies of the organization to its external customers.
Deming's 14 points for management	A set of guidelines to be used to transform the management of any organization; see Table 2-4. Reference 4.
Deming prize	Japanese industry's most coveted prize established in 1951 and named for Dr. W. Edwards Deming. Prize is awarded based on the evaluation of an organization's quality improvement processes.
Employee development	Long-term program designed to impart profound knowledge and technical knowledge to all employees. Profound knowledge is knowledge of how to acquire new knowledge.
Hoshin planning	A process focused system of long-term planning designed to convey the voice of the customer to everyone in the organization through the development and deployment of appropriate management policy. Also called management by policy.
Mission	A statement of how the organization plans to implement its ideology.
President's audit	A quality systems audit conducted by the head of the organization.
Process focus	An emphasis on managing the means of producing an outcome rather than on managing the outcome directly.
Quality function deployment (QFD)	A rigorous system designed to obtain customer input and convey ("deploy") the voice of the customer to all individuals and organizational units. Sometimes called policy deployment. One method of implementing QFD involves the creation of one or more "houses of quality." See Figure 2-8.
Value engineering	A problem-solving system implemented by the use of a specific set of techniques, a body of knowledge, and a group of learned skills. It is an organized creative approach that has for its purpose the efficient identification of unnecessary cost, i.e., cost that provides neither quality nor use nor life nor appearance nor customer features, i.e., waste.[21]

CHAPTER 2

TQM AND TOMORROW'S ORGANIZATION

TABLE 2-4
Deming's 14 Points for Management

1	Create constancy of purpose toward improvement of product and service, with the aim to become competitive, to stay in business, and to provide jobs.
2	Adopt the new philosophy. We are in a new economic age. Western management must awaken to the challenge, learn their responsibilities, and take on leadership for change.
3	Cease dependence on inspection to achieve quality. Eliminate the need for inspection on a mass basis by building quality into the product in the first place.
4	End the practice of awarding business on the basis of price tag. Instead, minimize total cost. Move toward a single supplier for any one item, on a long-term relationship of loyalty and trust.
5	Improve constantly and forever the system of production and service, to improve quality and productivity, and thus constantly decrease costs.
6	Institute training on the job.
7	Institute leadership. The aim of supervision should be to help people and machines and gadgets to do a better job. Supervision of management is in need of overhaul, as well as supervision of production workers.
8	Drive out fear, so that everyone may work effectively for the company.
9	Break down barriers between departments.
10	Eliminate slogans, exhortations, and targets for the work force asking for zero defects and new levels of productivity.
11	Eliminate work standards (quotas) on the factory floor. Eliminate management by objective, management by numbers, numerical goals. Substitute leadership.
12	Remove barriers that rob the hourly worker, and people in management and engineering of their right to pride of workmanship.
13	Institute a vigorous program of education and self-improvement.
14	Put everybody in the company to work to accomplish the transformation. It is everybody's job.

References

1. H. Mintzberg, "The Structuring of Organizations," in H. Mintzberg and J.B. Quinn, *The Strategy Process—Concepts, Contexts, Cases*, 2nd edition (Englewood Cliffs, NJ: Prentice-Hall, Inc., 1991).
2. A.D. Chandler, *The Visible Hand: The Managerial Revolution in American Business* (Cambridge, MA: The Belknap Press of Harvard University Press, 1977).
3. H. Mintzberg and J.B. Quinn, *The Strategy Process—Concepts, Contexts, Cases*, 2nd edition (Englewood Cliffs, NJ: Prentice-Hall, Inc., 1991).
4. W.E. Deming, *Out of the Crisis* (Cambridge, MA: MIT Press, 1986).
5. C.I. Lewis, *Mind and the World Order* (New York: Scribners, 1929).
6. Tom Peters, "The New Building Blocks," *Industry Week* (January 8, 1990), p. 101 (2).
7. A.J. Feigenbaum, *Total Quality Control*, 3rd edition (New York: McGraw-Hill, 1951, 1983).
8. K. Ishikawa, *What Is Total Quality Control the Japanese Way?* (Englewood Cliffs, NJ: Prentice-Hall, 1985).
9. *Cross-Functional Management: Research Report #90-12-01* (Methuen, MA: GOAL/QPC, 1990).
10. Thomas Pyzdek, *What Every Manager Should Know About Quality* (New York: Marcel Dekker, Inc., 1991).
11. John W. Sheridan, "America's Best Plants," *Industry Week*, v239, p. 27 (19) (1990).
12. J.W. Moran, Report to the Total Quality Management Committee on Daily Management, minutes of February 4, 1992 meeting (Milwaukee, WI: ASQC, 1992).
13. Thomas Pyzdek, *What Every Engineer Should Know about Quality Control*, (New York: Marcel Dekker, Inc., 1989).
14. P.M. Senge, *The Fifth Discipline—The Art and Practice of the Learning Organization* (New York: Doubleday, 1990).
15. "The Cracks in Quality," *The Economist* (April 18, 1992), p 67.
16. *Management Practices: U.S. Companies Improve Performance Through Quality Effort*, GAO/NSIAD-91-190 (Washington, D.C.: Superintendent of Documents, 1991).
17. J.J. Kendrick, "How to Spot Total Quality Leaders and Predict Success of Quality Programs," *Quality* (April 1992), p. 13.
18. Camp, Robert C. *Benchmarking*. Copublished by ASQC Quality Press, Milwaukee, WI and Quality Resources, White Plains, NY, 1989.
19. Campanella, J., ed. *Principles of Quality Costs*. 2nd edition (Milwaukee, WI: ASQC Quality Press, 1990).
20. Hartley, J.R. *Concurrent Engineering* (Cambridge, MA: Productivity Press, 1991).
21. Miles, L.D. *Techniques of Value Analysis and Engineering* (New York: McGraw-Hill, 1972).

CHAPTER 3

CONTINUOUS IMPROVEMENT TEAMS

TEAMWORK IN THE TOTAL QUALITY ORGANIZATION

Total Quality places significant emphasis on the role of teams. Teamwork is the basic underpinning of total quality management (TQM), and plays an essential role in a successful Total Quality business philosophy. Just as the family is the foundation structure of a healthy society, teams fulfill that role in the business organization. Every employee belongs to some team or group, either formal or informal. The organization will benefit if it can organize this energy within the business around common goals and objectives that are aligned with the goals of the total organization. There is no better way to effectively involve an organization than to create natural teams of people who work together to achieve a common goal developed by them for themselves.

There are several important characteristics that differentiate a team from the traditional work group.

- Team members share mutual goals or a reason to work together.
- Team members must share a need for an interdependent working relationship.
- Individuals on the team must be committed to the group effort.
- The team must be accountable to a higher level within the organization.

REQUIREMENTS FOR TEAMWORK SUCCESS

On highly productive teams, each team member must:

- Understand and be committed to group goals despite any individual objectives or measurements.
- Acknowledge and confront conflict openly.
- Listen to other members with understanding.
- Include others in the decision making process.
- Recognize and respect individual differences.
- Contribute ideas and solutions.
- Value the ideas and contributions of others.
- Recognize and reward team efforts.
- Encourage and appreciate comments about team performance.

Teamwork allows people to draw on their mutual strengths to gain a successful conclusion. Just as teamwork is the lifeblood of a world champion basketball or baseball team, or a world class symphonic orchestra, it is a natural part of the business process in world class businesses. Although it is a natural part of the business process, teamwork is a learned process. Teamwork requires the development of trust throughout an organization and within the team, and individual and team trustworthiness within the business organization. Teamwork requires team members to become interdependent, with a willingness to cooperate and with common goals. Such a transformation requires structure, training, and coaching of the organization and the teams.

Teamwork, the formation and use of teams, is not just a process. It must become a part of the business culture. The use of teamwork in business is a new paradigm, and may require a change in the business culture.

A NEW PARADIGM

Total Quality is continuous improvement of the total business system, including technical processes and social systems. The Total Quality paradigm includes team contributions in each of its three basic focus areas (see Fig. 3-1):

1. Long-term planning.
2. Daily control.
3. Business management processes.

Within these areas, there are functional, cross-functional, and multifunctional teams, each established as needed to meet the goals of the business.

A change in the business structure to support teams is a critical part of the social system or cultural change in the business. The new structure must be based on a clearly understood business mission and

CHAPTER CONTENTS:

Teamwork in the Total Quality Organization   3-1

Developing and Measuring High Performance in Teams   3-6

---

*The Contributors of sections of this chapter are: **Richard L. Kelbaugh**, Senior Engineer, Supplier Development, GE Appliances, General Electric Company; **John N. Younker**, Managing Director, Associates in Continuous Improvement, Houston, TX.*
*The Reviewers of sections of this chapter are: **Ron H. Cassell**, Vice President/Director Quality Assurance, ITT Higbie Baylock, ITT Automotive; **Mike Jury**, Consultant, Organizational Dynamics, Inc.; **Cathy E. Kramer, Ph.D.**, Executive Vice President, Association for Quality and Participation, Cincinnati, OH; **Michael S. Oakes**, Engineering Specialist, CAE, Electronic Data Systems.*

CHAPTER 3

TEAMWORK IN THE TOTAL QUALITY ORGANIZATION

Fig. 3-1 Elements of a customer-focused business strategy.

- Structured planning and priority setting
- Boundaryless natural teams
- Continuous improvement

Fig. 3-2 Long-term planning.

plan, and an understanding of that plan by everyone in the business. In addition to the workforce understanding the plan, business leaders must become versed in the process of teamwork. This understanding should include firsthand experience as a team member. Appropriately then, teamwork needs to start at the top of the organization and cascade down, rather than at the bottom and swell up. Where a labor union exists within the business organization, the labor leadership must be involved in this process as if it were a function or agency of the business. It is. Failing to meet these prerequisites has been a principal cause of the failure of many quality circle efforts initiated in the US in the 1970s and 1980s. Without the top-down culture change and support structure, and without labor's buy-in, teams cannot work.

Changing the business culture is the most difficult challenge in creating a Total Quality organization. Culture resists change because culture represents the sum of all of everyone's habits. It is difficult to change just one habit, much less change all. Language is a type of habit. People learn it at a young age and feel very comfortable with it as adults. There is usually not much thinking involved in the actual use of language. But what if they moved to Mexico or Japan, they would have to learn an entirely new pattern, a new habit.

Changes must be made in business habits that are similar in nature to learning a new language. People will need to change communication patterns, their need for control of outcomes, their willingness to allow others to do work and make decisions. There will have to be a sharing of information that had always been considered confidential. Expectations that were previously felt to be unachievable will be loosened, and people will be given the freedom to achieve using their own methods. The changes in old habits will ultimately determine the success or failure of any total quality initiative. The results in learning a new language generally come from having a good teacher, commitment, and application. To have a successful Total Quality initiative, the change in culture requires a strong process coach, a commitment to achieve, and personal application of the process to meeting the business needs.

Business leadership plays a major role in developing a long-term plan for the business as well as in carrying it out. In a business practicing Total Quality, this is the natural means for defining business direction and focusing operational objectives on achieving the business mission. Through the process of developing a long-range improvement plan, each participating member of the team provides focused inputs on organizational needs and understands how they fit within the larger organization. Figure 3-2 lists the elements of long-term planning.

In problem solving it is important that teams of individuals subrogate their own needs and objectives to those of the organization. Intra-staff communications during long-term planning sessions provide a means for assuring total business focus on the customer. If a strong team is formed at this level, it will help to lower the internal barriers that often hamper organizational effectiveness.

Daily control of business processes depends on the ability of the entire organization to operate as a single business team. Operational processes can be controlled for the long term only to the extent that systems are in place that allow the person who touches the process to control it. Process operators must be tasked to work together, to understand the process, and to interact with the process to control and improve it. Tools such as statistical process control, process standardization and documentation, and work team formation are typically used. The first step in achieving operator control is to confirm the concept that the operator is a part of the business team. Business leadership must then train and empower operators to control processes, and must release control to them. Figure 3-3 lists the elements of daily control.

Cultural changes to accommodate this structural imperative are significant. If the organization has not developed a culture in which there is trust for operational decisions to be made "on the floor," that trust must be developed throughout the organization. If the process operators have never been empowered to manage processes before, their trustworthiness must be assured. The formation of teams and initiation of education in business values will contribute to raising the level of trustworthiness of process operators. During the team development, operators must learn the meaning of *empowerment* and *ownership*. They must learn their role and their expected contribution to the business. Once the "one business team" concept is developed, and the roles within that team are defined, the subordinate organization work teams or process improvement groups can form and become effective.

- Standardization of processes and parts
- Use of statistical methods
- Work groups
- Process improvement teams

Fig. 3-3 Daily control.

CHAPTER 3
TEAMWORK IN THE TOTAL QUALITY ORGANIZATION

Cross-functional alignment of business management processes is essential to establishing a Total Quality concept. During long-term planning, the relationships focus on how the organization needs to structure its activities to achieve long-term plans. On a day-to-day basis, the business must be sure that production, engineering, purchasing, sales, and service all have consistent, integrated quality efforts which are aligned with the overall business mission and plan. Total quality focuses on meeting the customers' needs, so the understanding of customers and suppliers and their needs is essential. For this to be effective, teamwork must develop interdepartmentally and between businesses. This area of team development must not be overlooked during the restructuring for TQM, since without integration of *all* the business organizations into the Total Quality System, the process will fail. Figure 3-4 lists important business management process elements.

BECOMING AN EFFECTIVE TEAM

The process of becoming an effective team is developmental. This aspect of teamwork is a generally accepted norm in sports. It is an accepted norm for baseball and football teams to undergo an extensive practice period prior to playing games that count toward the season record. Yet often, businesses will bring a group of individuals together as a team and allow little or no opportunity for the team to develop before being expected to perform efficiently. When brought together for the first time, the team must accomplish several tasks before undertaking its objective:

1. Reach a mutual understanding of what the mission of the team is.
2. Decide the roles of the various members of the team.
3. Decide the methods by which the team will operate.

The Mission

The basic management model developed by Tom Peters and Bob Waterman in their best seller, *In Search of Excellence*, contains many of the same basic values that are needed by effective business teams. In the chapter, "Hands-On, Value Driven," the authors said, "Decide what your company stands for. What does your enterprise do that gives everyone the most pride?"[1] Just as successful businesses have a well defined purpose or reason for being, the team mission identifies the team's purpose in terms that both the team and manager understand and agree upon. The mission should be the constitution for the team. It must be adopted by every member of the team. When a team defines its mission, it defines the framework within which it will identify problems to solve.

A team mission provides natural carryover between projects and helps the team build identity and problem-solving skills. The mission of a Self-Managed Team (SMT) in a manufacturing plant may be to meet production, yield, and quality objectives in that area. The definition of the traditional Quality Circle, ". . . a small group of people working together to solve problems in the work area," is a mission. It simply says that the team is going to spend their time solving problems in the work area. In the case of the Quality Circle mission, members can strengthen the mission by changing it to read, "This team will work together to solve problems to reduce cost associated with scrap and rework in the area." This narrowing of the problem-solving focus helps to reduce the size of the project brainstorming list. This can help clarify the alignment of the team with business goals.

It is not sufficient for the leader to provide a mission to the team; rather the team must internally commit to achieving the mission individually and collectively. Accomplishing this requires

- Information systems
- Audit processes
- Customer / supplier relationships
- Quality function deployment

Fig. 3-4 Business management processes.

that the team have sufficient understanding of the business needs and values. The team should understand the key measures of business performance and the impact of their team's work on those measures. If the business has developed a plan, it can provide the basis for developing the team mission. The business leader having oversight responsibility for the team's efforts needs to provide feedback to the team concerning the mission. When the team mission establishes a priority for working on significant business issues, the business leader should express acceptance of the mission, and should extend additional assistance to the team in terms of information needed for successful problem solving. The business leader should not accept a mission that does not meet the needs of the business. Negative communication that is consistent and correct is as important to effective development of business trust and effective business teams as positive support of team decisions.

Roles of Team Members

On a business team, roles are defined by assignment and team consensus. There are four basic roles of the various members on a business team:

1. Team leader.
2. Team facilitator (Coach).
3. Team recorder.
4. Team members.

Teams may have additional tasks defined, but each team should define these roles as a minimum and select or assign people to do them.

Team leader. The team leader focuses on teamwork content. He or she will help maintain focus on the team mission and has the primary responsibility to act as liaison between the team and the business. As liaison, the team leader generally keeps the business informed about how the team is performing based on the team mission. Likewise, the team leader will usually be the means of communicating business news to the team. He or she arranges for the team's needs from the business and informs business leadership of the team's progress. In addition, the team leader is an active member of the team and should be elected by the team, not appointed.

Team facilitator. The facilitator focuses on the processes of the team. In this role, the team facilitator (coach) is an "outsider" to the team and should remain neutral. The primary task of the facilitator is to help the team operate smoothly. The facilitator is the key person to prepare for meetings, typically planning logistics for the meeting, creating a tentative agenda and plan for completing the meeting work, and informing team members of the meeting time.

CHAPTER 3

TEAMWORK IN THE TOTAL QUALITY ORGANIZATION

During the meeting, the facilitator helps the team keep on track by initiating discussion and making sure it stays focused and on schedule. Everyone should get a chance to participate. The facilitator must remain focused on the process of the meeting rather than the content of the discussion. When the team is ready for introduction of a new tool or skill of problem solving or process improvement, training in these tools or skills is a prime responsibility of the facilitator.

The roles of Team Leader and Facilitator may vary from business to business, and from process to process within businesses. The role of facilitator should be filled by a team dynamics expert during the developmental stages of team growth. During the early stages of development, the facilitator or team coach carries the major burden of team organization and development. This role shifts as the team matures. Once the team is mature, the role of the facilitator may be filled by team members on a rotating basis. Once the team has matured, who does the job is not so important as making sure the responsibilities of the job are fulfilled.

Team recorder. The team recorder is not only the team historian, but also the key to keeping track of the team's performance. Sometimes it works well to divide these duties among several people. The recorder maintains records of the meeting as it is happening, including action plans and commitments. Besides the meeting minutes and Problem and Action lists, the recorder should maintain team records, meeting agendas, research data, and other appropriate information such as mission measurement data. These records and information help in tracking the team's progress.

Team member. Team members, other than the leader and recorder, also have significant roles in accomplishing the team mission. It is essential that all team members "buy-in" to the team mission, and share in the ownership of that mission. If any team members have goals that are not aligned with those of the group, or if they have overriding accountability to others outside the team, they may have priority conflicts that can affect their contribution to the team goals and reduce effective team operation. As owners, all team members need to be involved and take responsibility for achieving the goal. Being involved will include volunteering for assignments and contributing knowledge to the team. Often, team members will have assignments requiring work to be completed between meetings. Completion of those tasks is a quintessential part of being a team member.

Method

In his book, *A Whack on the Side of The Head*, Roger von Oech talks about the creative problem solver and the need for separating a person from his or her mental blocks to be creative. He goes on to say that "Whacks come in all shapes, sizes and colors. They have one thing in common, however. They force you . . . to think something different."[2] The process of brainstorming was developed as a way to get creativity working, to come up with different ideas. The team will need to develop a process or method of operating, to include methods for problem investigation, decision making, and follow up, as well as a code of conduct for the team itself. These methods will provide a structure to help the team break away from current routines and find new ways of doing things. Teams can use any procedure to do this, and there are almost as many processes as there are employee involvement consultants. The common thread of these processes is that they provide a structure that maintains focus on the path to the goal and does not allow the team to lapse into unproductive discussion.

Teams need to be taught effective methods of problem investigation and project leadership. Then they need to become as well drilled in their method as that world class orchestra is in one of Mozart's symphonies. Training in the use of a method begins the day the team is formed and continues as long as the team exists. Early method training includes team-building exercises and discussions needed to define team operating guidelines. More basic techniques such as brainstorming, process mapping, and cause and effect diagramming are introduced as the team begins to work on the first team problems. As they become more mature, teams should continue to learn more complex and powerful techniques to apply within the framework of their problem solving process. The team facilitator and leader need to understand the process thoroughly. They will coach the team in process methods. How well they train in the tools will ultimately decide the success of the team.

TEAM FORMATION AND GROWTH

Teams go through a distinct developmental process. The most popular model for this process is "Form; Storm; Norm; Perform."[3] This is illustrated in Fig. 3-5.

Forming

The initial phase of team development, "Forming," focuses on gaining commitment from the team. During this phase, the team leader or coach will probably dominate the meeting agenda. The work of the team will focus on discovering what the mission of the team is, and who the team members are. Members in the forming stage may tread carefully so as not to offend another team member. Team accomplishments are usually superficial. During this phase of development, the new team members need to understand who is aboard and who is not. They need to understand individual differences and roles within the team. They need to gain understanding of the task expectations for the team and the team operating procedures.

When a new team is formed, the team coach must help the team through the process. During the Forming phase, teamwork may include debate about the mission, or physical, social, and psychological exercises. These activities help team members become more aware of individual differences and how these differences can make them an effective team. Physical team building can include group physical challenges and exercises that allow for working together to solve mutual problems. Social team building includes climate-setting exercises that enable team members to discuss individual characteristics and align various team members with a common goal. These exercises can be fun and help the members gain an attachment to their new team. Psychological exercises include values testing such as Myers-Briggs or Kiersey-Bates. Each team participant completes a self-evaluation. He or she then has the opportunity to work with the entire team to understand the relationships between individual values brought out through the evaluation. The team can discuss these differences and use the knowledge gained in this discussion as the basis for forming a bond.

Storming

Teams "Storm" to clarify individual and collective fit. During Storming, team members begin to experiment with more risky behavior. Members will challenge the leader and test the boundaries. As team members begin to gain comfort in their team roles, underlying personal values will influence member behavior. There will likely be some subgrouping as different viewpoints and values solidify. Conflict may develop between these groups. This phase represents a major test for teams.

CHAPTER 3
TEAMWORK IN THE TOTAL QUALITY ORGANIZATION

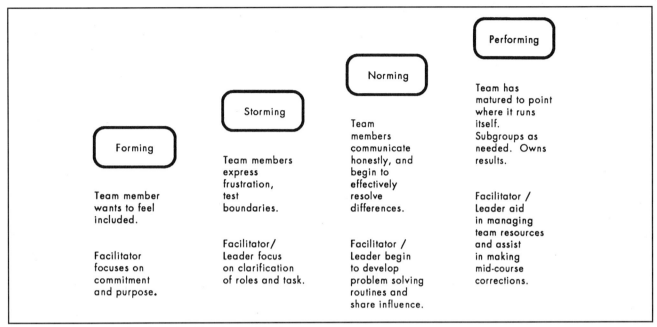

Fig. 3-5 Elements of Forming, Storming, Norming, and Performing.

Teams must work through the conflicts and positioning using techniques that reinforce the team concept. Conflict that occurs during this phase must not be sidestepped. Unresolved team conflict will be a barrier to further development of the team. Exercises that can help the team through rough stretches in this phase typically include problem solving of non-emotional issues as a team. The use of physical challenges or mental challenge exercises works well. The key in selecting exercises for teams in this situation is that the exercise should focus on total team inclusion rather than individual work.

These first two stages of team development are essential precursors to effective team functioning. If the team does not first work through these steps, it is likely there will be limited success as a team later. What is essential is that the business leader who plans to use teams as a keystone of Total Quality must allow teams to go through these phases. They are not usually periods of high team productivity, but are periods of internal growth. There is no set time span to allow for these stages to run their course. Work with the team coach to be sure the team is openly working through the conflict and is growing. If that is happening, allow time for the team to complete this important work.

Norming

Once the team has worked through its internal development, it will begin to function more effectively. The next phase of team development is Norming. While there will continue to be periods of conflict, the more experienced team members will be much more proficient in dealing with these situations and gaining energy from them. Once the internal struggles have been resolved and team members become more aware of their collective strengths, they will be more able to focus on achieving their team mission.

During the Norming phase of development, there is a more open climate. Team members are willing to express their true feelings, and deal with differences that develop. While differences do occur, there is more listening and greater acceptance of differing viewpoints. Team members can be expected to become more involved in team decision making. While a coach is still needed, the team is ready and able to be consulted and share responsibility for decisions.

When the team reaches this level of development, it is important to provide them with problem solving process methodology that provides structure for in-depth investigation and broad team involvement. Teams may be effective in accomplishing early team development goals with unstructured processes such as brainstorming. More advanced skills, such as structure trees, Pareto diagramming, and data collection are appropriate tools for introduction at this time.

Performing

As the team reaches maturity in its development, it can be expected to readily assume ownership of team results. Team members will subgroup as appropriate to achieve their goals and can be expected to distribute leadership roles among the members as expertise requirements dictate. The strong team will establish milestones and make necessary mid-course corrections, doing whatever it takes collectively or individually to achieve the desired results. Success at this level will depend on the use of skills learned earlier and the ability to maintain momentum as they transition from milestone to milestone.

Support of a team at this level of development should focus on clearing external barriers to performance. Business leaders and team coaches of teams at this level of development need only maintain an awareness of these barriers, such as funds availability or system barriers. Work to resolve these issues and enable the team to maintain momentum. In a Total Quality business environment, the team is formed with a business related mission. Expect the team to perform and measure its performance against these business goals. Provide regular feedback to team members on their performance against these goals.

These stages of team development do not represent a single pass process. Teams will revert from more advanced stages to developmental stages from time to time. For example, when a new

CHAPTER 3

TEAMWORK IN THE TOTAL QUALITY ORGANIZATION

team member joins an established group, there will be some period of time spent re-forming the group. When a new project is started, with new people having the key experience to lead, the team will likely do some re-forming. When a serious team conflict is uncovered on a team in the Norming or Performing phases, the team may revert to Storming. Usually these reversions do not require total rebuilding of the team, but they are a part of team dynamics. Failure to recognize a shift in team readiness could lead to team stagnation or failure. It is important that the team leader or facilitator be aware of the symptoms of the various stages of development and provide the appropriate process to accommodate the team's developmental needs.

ALLOW TIME FOR THE PROCESS TO WORK

The initiation of teams and the development of the needed support structure is a major challenge for the business organization that is just beginning the change to a total quality management structure. Change of this sort does not take place in six months, and in most circumstances, not even in a year. The cultural changes will take varying times, but in a traditional organization normally one to two years is a reasonable estimate for developing the communications networks and elevating trust levels to a point where the culture is receptive to the responsibilities of empowerment. It will take three to five years for the natural variations in organizational readiness to level out and the new process to be considered a business culture.

Until the groundwork on the business culture is complete, a mix of traditional and empowered styles is appropriate. Process Improvement Teams can be organized within a traditional business setting. Within the team, the operation is identical to the fully implemented process, but structurally, the team is responsive to a traditional management system rather than focused totally on its own mission.

Start the process by developing a strategic plan with the business leadership, as discussed in the Long-term Planning section. As part of the long-term plan, identify the processes where the business can most benefit from initiating teams and a Total Quality Approach. Long-term process commitment is an initial requirement. Hire a Total Quality Advisor for the business and task him or her to initiate teams in one or two of those process areas. Provide the teams with information about the business issues they should focus on. Communicate this plan to the managers of these areas and the operational people in the business. If there is a labor organization, involve them in all the planning for team initiation. Allow six months for these teams to establish themselves. Track their growth both socially and technically. By the time the team has been meeting for six months, they should be in the Norming phase of development or beyond. They should have an understanding of the process issues and have a tracking measure in place. They should be using a structured methodology for problem solving and using process data as part of their investigation information. They should understand the business goal they have adopted and have a plan for addressing improvement of the process issues affecting that goal.

After a year, the team should be able to show clear improvement in the issues undertaken, or have an understanding of process limitations which are preventing improvement. If process limitations are preventing the team from making the desired process improvements, the limiting factors must be addressed as soon as they are identified.

Typically, the formation of one team does not provide sufficient mass to bring about the social change needed to make teamwork a part of the business culture. But the formation of one or two teams enables the business to learn about the process and identify the training materials that work. Additional teams should be formed on an ongoing basis as support for the process grows, and as the business strategic plan dictates that a need exists. Once a critical mass of the business organization begins to trust the process and the new culture, the shift to a total quality organization with a teaming culture is a logical next step.

DEVELOPING AND MEASURING HIGH PERFORMANCE IN TEAMS

With the recognition that participative management is not simply a passing fad but rather a long-term strategy, senior management is working to effectively link the organization's people and technology in ways that optimize both resources. When this linkage is effective there exists an organizational condition known as "high performance."

In evaluating an organization's state of high performance, five key management practices are measured:

1. Delegation. The placing of the responsibility and authority for decisions and their implementation with those members who have the most relevant information and the skills to most effectively use it.
2. Empowerment. The provision to inviduals of critical business information, the training to effectively use it, and the opportunity to make valuable contributions. An expectation that everyone will accept and exercise the responsibility necessary to accomplish the organization's goals and objectives and to fulfill its mission.
3. Integration. The integrating of people with technology that allows for initiative and creativity in all parts of the organization. A clear sense that the people are in charge of the technology, not the reverse; people provide the intelligence and wisdom to maximize the machines' potential. It is the optimum alignment of the organization's support structures and systems.
4. Shared purpose. The sharing of a common vision of the organization's future that includes its mission, strategic priorities and business objectives, values, and the means for realizing the vision. This vision (shared purpose) provides energy and direction (focus). It empowers the individual employee as well as the various work units, and provides for an organizational cultural foundation.
5. Teamwork. The collaborative involvement of the right people in the organization at the right point in time. The focusing of the organization's human resources in an integrated effort to deliver high quality goods and services to its customers. Having a "customer orientation" that places the primary emphasis on supporting the customer (internal and external) rather than focusing on the function

CHAPTER 3
DEVELOPING AND MEASURING HIGH PERFORMANCE IN TEAMS

or professional discipline. Simply stated, there is a strong sense of "common fate," full commitment to a common mission, and mutual support present within the organization.

A strong and compelling body of knowledge from numerous field studies and cases leads to an inescapable conclusion: high performance cannot be copied from another organization; rather, it must be developed from within the organization by its members.

An organization desiring to become a high performance entity must develop its internal capabilities for designing and implementing improved processes and systems for conducting its day-to-day business operations. Essential to the development of these improved capabilities is the capacity for "self-(re)design." Self-(re)design is an organization's ability to continuously monitor its performance, identify opportunities for improvement, and implement innovative work process improvements.

Another essential capability found in a truly high performing organization is its ability to empower and directly involve all of its employees in the management and direction of the day-to-day business operations. High performing organizations have moved beyond traditional management practices (managers solving the problems and making the decisions). They have developed and empowered their employees to collaborate with managers in the day-to-day decision making and problem solving. They have extended their employee involvement practices beyond the parallel structures approach. The "Employee Involvement Continuum" is an evolutionary model of contemporary employee involvement practices (see Fig. 3-6). As this model indicates, the high performance team is a form of employee involvement that only exists in those organizations where operating management practices demonstrate a very high level of employee involvement and empowerment.

Two questions often come up when organizations first embark on their quest for high performance: "Where should an organization focus its efforts?" and "How best to manage the transformation?" The complex structures, systems, and processes found in most current organizations do not readily lend themselves to rapid, wholesale change. There is an old proverb that says, "When one is considering having elephant as the main course, one should not plan on eating the whole elephant at one sitting, but rather in small portions spread over several meals." Organizations would do well to consider a "pilot test approach" where key units are selected to serve as trailblazers and innovators for the whole organization. These pilot units could be existing "natural teams," work groups or units who are responsible for critical elements of the business, share common objectives, and have a sense of interdependency. To guide them in their journey toward becoming high performance teams, the pilot units will need a proven, structured transformation process. This process should be straightforward, customer (output) focused, and directly linked with the overall business strategy. Although this transformation process is based upon proven approaches to team development and performance improvement, it should be flexible to allow for unique cultural requirements. The "Team Performance Improvement Process" is just such an approach.

TEAM PERFORMANCE IMPROVEMENT PROCESS

The Team Performance Improvement Process is designed as a vehicle for improving organization performance (productivity, quality, and customer service) at the work unit or operations level. This structured process serves as a guide to teams in their efforts to improve upon their outputs (products and services), customer service, internal capabilities, and overall performance. Drawing upon performance-related information provided by its members, customers, and key interfaces—and the direction set by senior management—the teams undertake the identification and achievement of improvements of real value. During the implementation of the Team Performance Improvement Process, the teams will develop the internal capabilities (knowledge and skills in the use of selected tools and techniques) to support continuous performance improvement.

The Team Performance Improvement Process is based on extensive field research and application in a broad range of organizational settings. This field research, which began in 1981 at the American Productivity and Quality Center (APQC), continues to be carried out today by the APQC and other researchers. The process is the result of the knowledge and field experience gained from working with over 100 different teams from several different industrial and organizational settings.

The three primary phases of the Team Performance Improvement Process (Diagnosis, Direction, and (Re)Design) are depicted in Fig. 3-7. The two secondary pre- and post-process steps (Planning and Preparation and Review and Recycle) are also shown. In each of the phases, there is active involvement of all of the teams' key stakeholders; its members, customers, senior managers, and key interfaces (for example, vendors and suppliers). This broad-based involvement is in recognition of the organizational change stratagem, "Those who participate, tend to support." Collaboration and cooperation (teamwork) are essential to meaningful performance improvement; team members who are most responsible for the products and/or services need to be directly involved in the improvement process. Along with the customers and key interfaces, team members should be charged with determining the most effective way to address the technical and human resources aspects of the team's work systems and processes. In its entirety, the Team Performance Improvement Process supports the teams in their efforts to achieve the best fit between customers, resources, procedures, and structures. It does all of this while simultaneously providing for the development of the internal capabilities to sustain a culture of continuous performance improvement.

PLANNING AND PREPARATION

The initial effort is to assist the teams in planning and preparing for the implementation of the Team Performance Improvement Process. The teams are encouraged to develop a plan for implementing the process paying specific attention to their individual team's business cycle. This effort is an attempt to identify downstream obstacles and barriers that can adversely impact the process (for example, budget planning, performance reviews, or peak vacation periods). Additionally, the teams are asked to identify specific expected outcomes and/or performance improvement goals. Planning and preparation identifies key process roles and responsibilities, team support requirements, communication plans, key customers, and outputs (products and/or services). Finally, during this preliminary phase in the implementation of the Team Performance Improvement Process, the teams are encouraged to hold orientation sessions for all key stakeholders.

DIAGNOSIS

The first of the three primary phases in the Team Performance Improvement Process is Diagnosis (see Fig. 3-8). In this phase the

CHAPTER 3
DEVELOPING AND MEASURING HIGH PERFORMANCE IN TEAMS

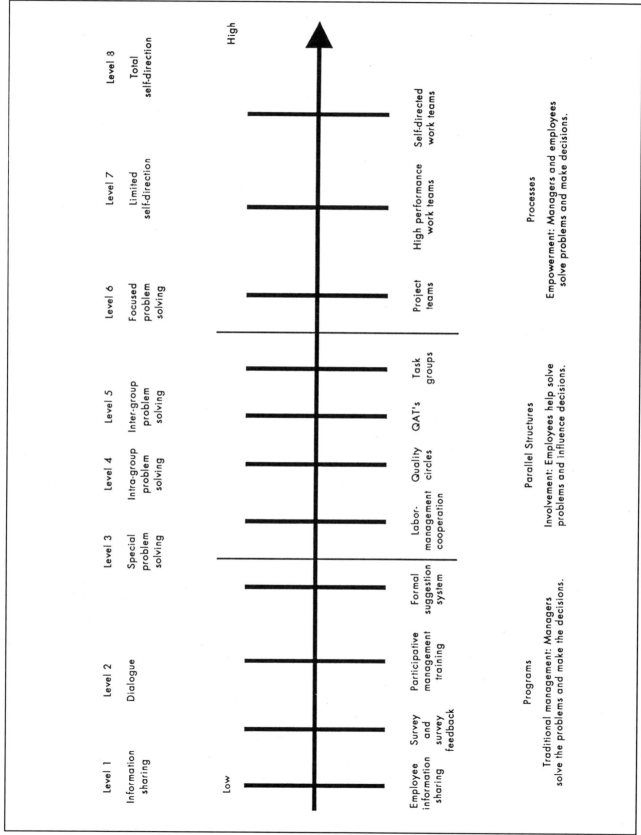

Fig. 3-6 The employee involvement continuum.

CHAPTER 3
DEVELOPING AND MEASURING HIGH PERFORMANCE IN TEAMS

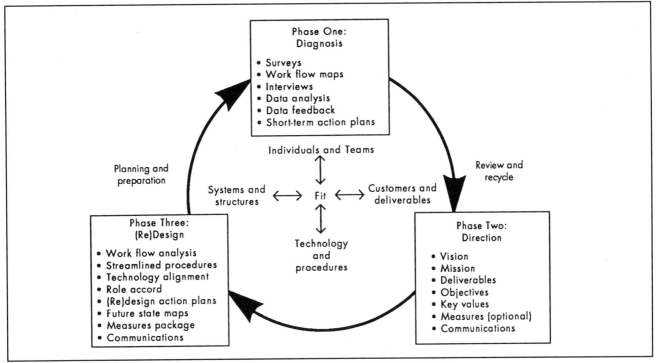

Fig. 3-7 Overall view of Team Performance Improvement Process.

team undertakes a data gathering and analysis effort that leads to the identification of specific performance improvement opportunities. Further, there will be a confirmation and clarification of customer needs, to focus the direction setting and work (re)design efforts in the subsequent phases. Key activities in this phase include the use of survey and interview questionnaires, work flow mapping, data analysis and feedback, and short-term (immediate opportunity) action planning.

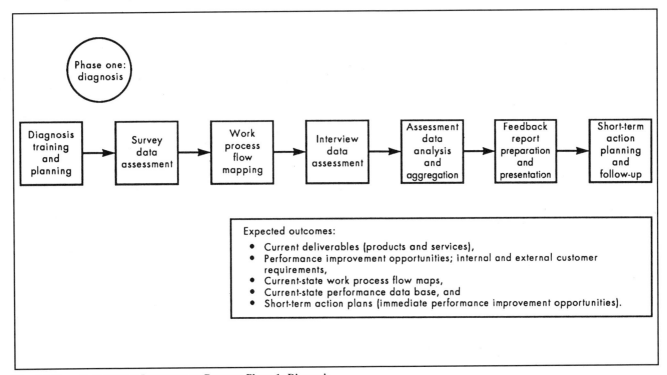

Fig. 3-8 Team Performance Improvement Process; Phase 1, Diagnosis.

3-9

CHAPTER 3

DEVELOPING AND MEASURING HIGH PERFORMANCE IN TEAMS

DIRECTION

The second primary phase, Direction, results in the team's defining its future direction; its vision, mission/purpose, products and/or services, performance objectives, and key supporting values or principles (see Fig. 3-9). It is a formal declaration of what the team intends to do in response the needs of its customers, the overall organization business plan, and the specific performance improvement opportunities identified in the Diagnosis Phase. Some specific action steps in this phase include the selection and training of a Direction Setting Task Team. The use of a representative task team may be more practical in larger size teams. However, all members of the team contribute to the data to be used in the direction setting process. Key parts of the direction statement are:

1. Vision. "What we want to become" . . . "our future state."
2. Mission/purpose. "This is why we exist." . . . "This is the business that we are in."
3. Products and services. "These are the products and services we provide to our customers."
4. Performance objectives. "This is what you can count upon us to accomplish."
5. Key supporting values and/or principles. "This is what we stand for."

The team's direction statement is then reviewed for completeness (in addressing the results of the Diagnosis Phase) and to determine the measurability of the performance objectives. A key measurement consideration is the question of *what* should be measured. It has been written that . . . "in American business today, are cultural paradigm shifts to world class quality . . . one of these changes is a shift from a task (only) focus to a process and customer focus." If a company is attempting to measure high performance, it must measure both the task and the process accomplishments of its teams. The company must focus on both the "tangible" (for example, outputs, quality, timeliness) and the "intangible" (for example, responsiveness, empathy, conflict management) accomplishments. Further, there must be a "family of measures" that provide these measures of task and process from both internal and external (customer's) points of view.

A decision can be made at this point in the process regarding what is the best time to select and train a Measures Development Task Team. There are conflicting results from various field research studies that tried to determine at what point in the Team Performance Improvement Process the team should develop its performance measures. There is data supporting the conclusion that performance measures are best developed by the team immediately after the development of the team's performance objectives. Other research findings support the belief that the performance measures are best developed after the team has completed its work (re)design efforts. The recommended guidelines for developing high performing team measures are:

1. Identify "Key stakeholders" (for example, team members, customers, key interfaces) and solicit their support and active involvement, and their participation on the measures development task team; "Those who are involved tend to support."
2. Train the task team members in measures development concepts and skills. Training topics might include:
 • Why measure? (The purpose of measures).
 • How will the measurement data be used?
 • Who will have access to the data?

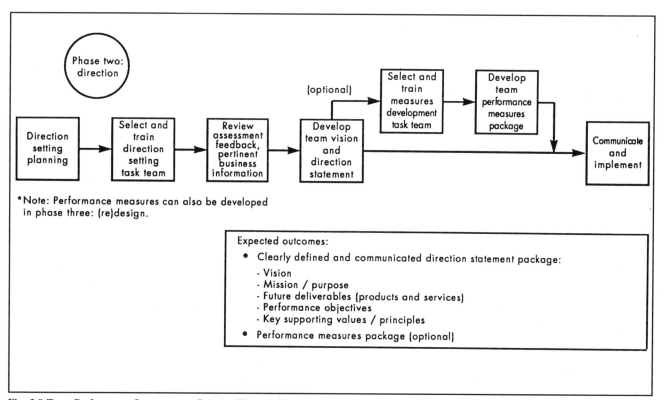

Fig. 3-9 Team Performance Improvement Process; Phase 2, Direction.

CHAPTER 3

DEVELOPING AND MEASURING HIGH PERFORMANCE IN TEAMS

- What is the "family of measures" concept?
- What is the step-by-step process for developing App<A measures in a task team (The modified NGT process)?

3. Clarify the measures development task statement. What performance indicators (measures) are needed to monitor this team's performance plan? (The measures of this team's task and process accomplishments.)
4. Initiate the modified NGT process:
   Step 1: Silent generation of performance indicators.
   Step 2: Round-robin listing of candidate performance indicators.
   Step 3: Group discussion, clarification, and combination of candidate indicators.
   Step 4: Vote ranking of candidate performance indicators.
   Step 5: Group discussion of rankings and consensus of final selection.
5. Conduct a feasibility review of the whole package (family of measures) to ensure their mutual compatibility and to determine if the selected measures are sufficiently comprehensive and well integrated.
6. Assign weights to individual measures (optional).
7. Communicate the selected performance measures to all key stakeholders.

In a effort to ensure that full value is realized from the team's performance measures, it is strongly recommended that a "Measures Utilization Plan" be developed. This plan would answer the following key questions:

1. What data will be collected?
2. What is the source(s) of the data?
3. How will the data be collected?
4. Who is responsible for collecting the data?
5. Who will see the data?
6. How will the data be disseminated?
7. What will occur as a result of the data?
   - Team and/or individual analysis.
   - Charting.
   - Data recalculation (as required).
   - Recognition and rewards.
   - Action planning.
   - Communications and/or information updates.
8. How frequently will the performance measures be reviewed and revised?

(RE)DESIGN

The third primary phase in the Team Performance Improvement Process, (Re)Design, is often perceived by team members as having the greatest impact upon the team's overall performance improvement efforts (see Fig. 3-10). The result of the (re)design efforts are well integrated performance improvement action plans. These action plans address the procedural, technological, and

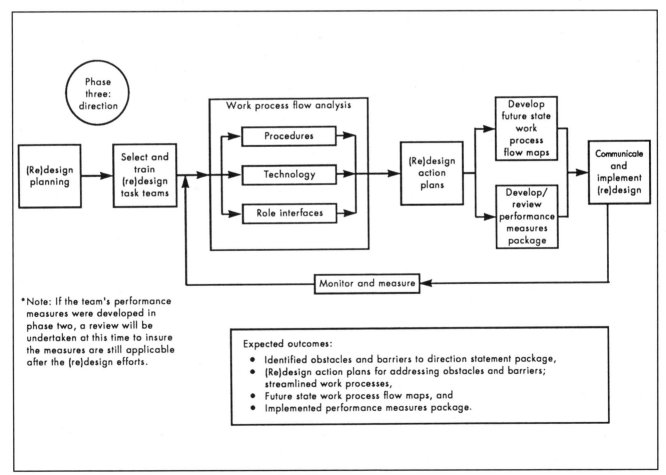

Fig. 3-10 Team Performance Improvement Process; Phase 3, (Re)design.

CHAPTER 3

DEVELOPING AND MEASURING HIGH PERFORMANCE IN TEAMS

interpersonal obstacles and barriers to the team's vision: its future state. Key activities in this phase include the selection and training of work process (re)design task teams, work flow analysis, identification of obstacles and barriers to the teams performance plan, action planning, the elimination of these barriers through work (re)design, performance measures development (or review), and the communication and review of the (re)design action plans with all key stakeholders.

REVIEW AND RECYCLE

The post-process step, Review and Recycle, addresses the need for the team to review the process and the performance improvement results they have achieved, and to make any adjustments to the process that will improve its effectiveness. It provides the team with an opportunity to formally review its progress and determine the immediate priorities for continuing the team performance improvement efforts. It is also a time for formal recognition of individual and team accomplishments.

During the team's progress review a set of "Performance Indicators of High Performance Teams" can serve as a benchmarking tool:

- Team members can readily and consistently articulate the team's mission and performance objectives.
- Team meetings have an atmosphere of high energy and a sense of comfort; the team's meeting norms are well attended to.
- Team member participation and idea exchange is very visible and the level of accomplishment is high.
- Active listening is strongly evidenced in the team's interactions.
- Disagreements are willingly surfaced and effectively and openly dealt with.
- Decisions are made in a timely manner with attention being paid to buy-in and quality.
- Trust levels between team members are high, as evidenced by the high degree of openness.
- Team member roles and responsibilities are clearly understood, and there is a strong sense of common fate and mutual support.
- There is a zest and urgency for serving the team's downstream customers and assisting them in fulfilling their mission and achieving their business objectives.
- The team is continuously measuring and improving its performance through its data-focused self-redesign efforts.
- The team enjoys a strong level of support for its mission, objectives, and overall success throughout the entire organization.

SUPPORT STRUCTURES AND TRAINING

The Team Performance Improvement Process is a straightforward process based upon proven performance improvement tools and techniques. However, without the provision of accompanying support structures and training, there will likely be a significantly adverse impact on the teams' results.

The required support structures and training include a senior management steering committee to provide strategic direction, a senior management champion to ensure the availability of the required resources and boundary management capability, and the provision of coaching, facilitation, and training. Training is needed in both the Team Performance Improvement Process and the related team/group dynamics skills (for example, problem solving, conflict management, team leadership). The process related training will focus on the knowledge and skills related to the three primary phases: Diagnosis, Direction, and (Re)Design.

CULTURAL INTEGRATION

If the teams are to improve their performance, the organization support structures and systems need to be aligned with the new "teams culture." To support the new culture, an analysis of current systems and procedures must be undertaken. Those systems and procedures that fail to support and reinforce the teams' new way of doing business must be replaced. Organizational systems such as performance appraisals, and rewards that are based on individual effort and achievement, must be redesigned to support team performance. An organization desiring to fully implement and utilize high performing teams will need to consider the use of nontraditional methods such as peer performance reviews, pay for performance, and gain sharing.

SUMMARY

Numerous organizations, as diverse as Bell Canada, Hoechst Celanese Corporation, and the Zurich-American Insurance Group, have all utilized customized versions of the Team Performance Improvement Process in their quest for high performance. While these organizations are diverse in their business focus, they share common concerns and priorities. All of them are faced with strong marketplace competition and a desire to be industry leaders. Their common response to this challenge has been their decision to embark on long-term efforts to develop well integrated, high performance work units.

These organizations have been rewarded in their efforts by the development of teams that are demonstrating their support and commitment to the parent organizations' business objectives. The high performing teams in these companies have developed their internal capabilities to optimize their technological and human resources; they are finding the "best fit." More important, the development of integrated systems and structures within these organizations ensures the presence of internal capabilities and capacities for sustaining continuous performance improvement.

References

1. Thomas J. Peters, and Robert H. Waterman, Jr., *In Search of Excellence* (Warner Books, 1982), p. 279.
2. Roger von Oech, PhD., *A Whack on the Side of the Head* (Warner Books, 1983), p. 12.
3. Bruce W. Tuckman, "Development Sequence in Small Groups," *Psychological Bulletin*, 1955.

CHAPTER 4

CONTINUOUS IMPROVEMENT AND TRAINING

INTRODUCTION

Training is critical to the success of most continuous improvement (CI) efforts. When processes are improved, maximizing the benefits of the improvement requires that those interacting with the process understand and can work within the context of the improved process. This may require a simple briefing or more extensive training, but training concerns are often overlooked and the costs of training ignored as improvement efforts are initially funded and undertaken. Are those costs then avoided? Not likely. It may be necessary to find funding later, when the need for training becomes an obvious requirement for implementing the changes.

CI is basically an attempt to reduce the inherent variability of a process, as well as to incorporate breakthroughs and innovations. Training can be an important factor in reducing the variation of the human performance component of a process. Continuous improvement has two distinct implications for training:

1. Personnel who are affected by the improved products and processes will require training on how to do their jobs/tasks differently given the improvements/changes made to the process.
2. Those who are to be involved in the efforts to make continuous improvements to products and processes will require training on how to do so. To create an organizational culture that embraces continuous improvement, training must give people the appropriate skills.

This chapter focuses on the issues and strategies for addressing these implications. First, it is important to establish a common understanding of what a training product is. Later, the process for creating a good training product will be addressed.

TRAINING PRODUCTS

Training is a product or service that is created to improve human performance by reducing its variation within work processes. As a human capital improvement effort, training investments should hold the same promise of sufficient return as any other capital improvement effort. Failure to ensure that return is an irresponsible use of an organization's limited resources. Determining that a training need exists does not by itself justify meeting that need. An analysis is required to determine if the investment in training will yield adequate return to justify the expenditure. The return can include costs avoided, increased revenues, and enhanced employee motivation and satisfaction.

Like any other product or service, training is best defined by the customer's requirements. Training formats extend beyond the typical notion of classroom education. The appropriateness of the training delivery method can only be judged based on the stakeholder requirements. Common delivery methods include:

- Unstructured OJT (on-the-job training).
- Structured OJT.
- Job performance aids.
- Self-paced reading or exercises.
- Group-paced, instructor-led reading or practice.
- Computer-based training (CBT).
- Interactive video.

The list is not all-inclusive. The best method for packaging and delivering training products depends on the needs and constraints of the target audience, as well as the requirements of those groups that have a stake or interest in the training. Effectiveness of the method must be assessed against its costs. If total

CHAPTER CONTENTS:

INTRODUCTION 4-1

TRAINING PRODUCTS 4-1

THE LEVELS OF CONTINUOUS IMPROVEMENT 4-2

TRAINING AS A RESULT OF CI EFFORTS 4-2

TRAINING TO SUPPORT CI EFFORTS 4-5

THE TRAINING DEVELOPMENT PROCESS 4-6

PHASE 1: PROJECT PLANNING AND KICK-OFF 4-7

PHASE 2: ANALYSIS 4-8

PHASE 3: DESIGN 4-9

PHASE 4: DEVELOPMENT 4-9

PHASE 5: PILOT-TEST 4-10

PHASE 6: REVISION AND RELEASE 4-10

THE STORY OF GPT MANUFACTURING 4-10

SUMMARY 4-11

---

*The Contributors of this chapter are: **Guy Wallace**, Partner, Svenson & Wallace, Naperville, IL; **Peter R. Hybert**, Associate, Svenson & Wallace; **Terri Knicker**, Associate, Svenson & Wallace.*
*The Reviewers of this chapter are: **Jeffrey S. Brody**, Manager, Human Resources, PRC, Inc., Arlington, VA; **Ellen Carnavale**, Editor, Technical & Skills Training, American Society of Training and Development, Alexandria, VA; **John D. Hromi**, Professor Emeritus, Center for Quality and Applied Statistics, Rochester Institute of Technology; **E. LaVerne Johnson**, President & CEO, International Institute for Learning, New York, NY; **Dr. John J. Tice, IV**, Manager, Science and Technology Issues, Manufacturers' Alliance for Productivity and Innovation, Washington, DC.*

CHAPTER 4

TRAINING PRODUCTS

costs of both development and delivery are considered, the most expensive to develop may be the least expensive to deliver and therefore, given a large enough audience and stable enough content, may very well be worth the extra initial cost.

Good, high-quality training will present concepts and develop behaviors that lead to improved job or task performance as measured by the appropriate metrics. The activities or behaviors of people in a process must be in control, just as any other process variable must be in control. Training can be thought of as the vehicle to optimize human performance or minimize the variations in overall performance. Tighter process performance will lead to improved process output.

Good training has no gaps in content and no excess content ("nice to know" versus "need to know"). Good training focuses on performance (what people need to do) rather than subject matter, and maintains a balance between providing information, performance demonstration, and skills development practice with feedback.

Good training is the result of a series of systematic processes analogous to any product development effort.

Good training does not just happen, it is engineered.

Who should be involved in engineering good training? It cannot be the sole responsibility of those in the training organization, although they may assume the lead role. Others need to be involved to influence the design and content of the training. The training organization must work in conjunction with representatives from the intended target audience(s) and their management, to understand the critical training requirements. Master performers, subject matter experts, and managers/supervisors of the target audiences are the best partners, teamed with professional trainers to define and build the training product. Without this teamwork, the training created may just be somebody's best guess at what is really needed and its support of continuous improvement may be negligible.

THE LEVELS OF CONTINUOUS IMPROVEMENT

There are three different levels at which improvements may occur. These can be classified as

- Job.
- Process.
- System.

Observe a chemical process in a plant which creates a cooling agent. In this process, an operator at a reactor combines specific chemicals to create the cooling agent. From the operator's perspective, anything that makes the job easier is an improvement. Most people performing any type of job will look for ways to improve the job by eliminating wasted efforts. This is the most basic level of improvement. Training to support these improvements could be accomplished through informal discussions on-the-job or captured in formal procedure manuals. The training requirements are typically minimal.

The Chemical Engineer who has responsibility for the process may be searching for ways to improve the efficiency, quality, reliability, etc., of the process. The engineer may decide to institute statistical process control (SPC) measures on the process and take intermittent samples to be tested to determine if there are any early warning signs of a bad batch. Or, perhaps another chemical could be used in the process which would not be subject to the same OSHA and EPA restrictions as a chemical currently used. These improvements change the process, which means that new tasks and skills may be needed for the people working in the process. Formal training may be the best way to adequately transfer this knowledge and skills.

If a company's culture supports continuous improvement efforts through empowered teams, improvements can also be the result of specific team efforts at the system level. For example, a team consisting of Operators, Chemical Engineers, and Industrial Engineers could have been formed to investigate the entire chemical process to find and implement improvements. Perhaps they discovered that it was possible to begin different phases of the reaction procedure in parallel so that there was less wasted operator time and an overall reduction in cycle time. This would be an example of a specific team effort directed at improving a process. These teams need training to enable them to work through the CI process effectively as a group. They need to understand how to work together, and they need to understand all of the elements of the CI process. Many different types of training content may be required to support this level of continuous improvement.

It is important to understand the different levels of improvements. Any of these efforts can lead to incremental or breakthrough improvement. Formal training may not be required with every slight modification to the process. Training is, however, likely to be required to support major modifications to the process and the efforts of an organization trying to turn itself into an organization which embraces continuous improvement.

TRAINING AS A RESULT OF CI EFFORTS

Continuous improvement efforts are typically not undertaken without a clear reason or goal for expending the time and effort. CI is usually driven by at least one of the following:

- Product changes that require changes in the process.
- A need or desire for incremental process changes to achieve tighter process control.
- A need or desire for breakthrough changes to increase competitiveness.
- A desire to alter an organization's values and approaches to work—that is, culture.

Regardless of the original stimulus leading to the CI effort, CI will inherently change how the work gets done within a process. It

CHAPTER 4

TRAINING AS A RESULT OF CI EFFORTS

may seem obvious that, as improvements are implemented within a process, the changes themselves will require that the people in the process do something different or new, and that some level of new awareness, knowledge, or skill will be required to support the new behavior requirements.

Too often, the training requirements and resources are not considered in the decision-making and planning processes for an improvement. This can lead to a situation where the change does not achieve optimal results. To counteract this, training issues need to be planned for and integrated into the overall CI process.

There are aspects surrounding any process improvement which need to be understood before training can be instituted. Important points are:

- What has changed and how much?
- Who is affected and how much?
- What is the best, most robust training approach to meet the situation?

WHAT HAS CHANGED AND HOW MUCH?

Assessing the extent of the change is done within the context of the CI effort. To improve any process, one needs to understand how it works currently. A detailed map of the process should be created and used as the basis for analysis. Any suggested improvements can be mapped, and a new process flow created which will show the extent of the changes. These maps can be used to identify training requirements.

How does one determine whether training is necessary when a process is changed? Look at the extent of the change. When an evaluation of the process is conducted, determine whether the tasks are simply rearranged or whether new methods or equipment are employed. For example, if a new step is added to the process, but it is a common procedure (such as taking a measurement using familiar tools or equipment), the person performing the step need only be made aware of the change. On the other hand, if the measurement is taken using a new type of tool or instrument, the very least that must happen is that the performer should be shown how it is done and should have the opportunity to practice if that is required to perform reliably.

If the learning required for awareness, knowledge, or skill levels is classified, then one can better estimate the difficulty of the change from the performer's perspective. This will help decide how critical the training piece is and later, the best training approach for the situation. Figure 4-1 describes the levels of learning that training can be developed to handle. Because each level has a different goal for learning, the training for each level would most likely be approached differently.

WHO IS AFFECTED AND HOW MUCH?

Determining who is affected by the process improvement, and who may require training at some level, is extremely important. The most obvious group are the employees working in the process—the process participants. But are there other audiences which should be targeted for training? Will other processes be affected by the changes made in the process? To answer this question requires a closer look at several groups—suppliers, customers, and management. Any changes or modifications which impact these groups need to be examined carefully for training implications.

For example, if a design team decides to use a concurrent engineering approach to reduce project cycle time and downstream rework, that will move a larger percentage of the costs into the initial phases of the effort as organizations, such as Production, participate earlier in the process than they normally would. If management personnel are not "trained" on how to manage budgets differently in a concurrent engineering effort, they may resist supporting the required up-front spending. Inadequate resourcing will cause the concurrent engineering effort to be less than successful.

Within each audience group, people also need to be aware that training needs to address two types of people in the target audience, incumbent employees and new employees. Their specific needs will differ. The incumbents need training only on the specific changes resulting from the CI effort. There is only a short-term need for this type of training. Any new employees for this targeted job will need training on the entire set of tasks inherent in the job they are assuming. The need for this training will be dependent on the future plans of the organization. Moreover, it is important to remember that new employees will need an orientation to the organization's approach to CI.

The method used to bring everyone up to speed will vary, depending on the process changes and the level at which they impact employee behavior. The training will potentially have to provide some level of awareness, knowledge, and skill to multiple target audiences. There might be a core element of training that is common to all audiences, and beyond that some unique training. Specifics need to be derived from an understanding of the changes in the required behaviors to support the process change. The old and new process maps can be compared to gain insight.

WHAT IS THE BEST, MOST ROBUST TRAINING APPROACH TO MEET THE NEEDS OF THE SITUATION?

When defining the training delivery strategy (for example, formal classroom training, self-paced learning materials), robustness should be a primary consideration. Continuous improvements to processes require continuous changes to training, as well as changes to documented practices, policies, etc. Ensuring the "robustness" of the training concept and designing the training system to facilitate these anticipated future changes is a challenge. The costs of training to meet identified needs must be viewed well beyond their initial costs; they must be considered in terms of their "life cycle costs." Considerations should be made for issues such as the ongoing maintenance costs of one training design concept compared to another.

As with any product designed and manufactured for robustness, the designers and developers of training need to understand the total picture, and be duly influenced by all the stakeholders before committing to a design, development, or delivery concept. Training must be designed to contend with its own "-ilities." Table 4-1 relates training "-ilities" to manufacturing "-ilities." Effectiveness is related to the level of interaction required in the training on the actual task. A video and a workbook will not be as

TABLE 4-1

Training "-ility"	As Related to the "-ilities of Manufacturing"
Effectiveness	Fitness for use
Deliverability	Useability/Flexibility
Update-ability	Serviceability
Affordability	Initial cost/Life cost/Return on the investment
Development ability	Manufacturability

CHAPTER 4

TRAINING AS A RESULT OF CI EFFORTS

Category	Definition	Example
Awareness	Convey the idea of the change or impacted process to the performer	The order of the steps performed is changed. A step in the procedure is deleted (e.g., eliminating a test or measurement). A step with standard procedures is added (e.g., waiting for QA approval before proceeding).
Knowledge	More detailed level of information given, often accompanied by examples or demonstrations	Decision-making criteria are changed (e.g., the operators are now allowed to inspect their own work).
Skill	Allow the performer to build skill through practicing the desired behavior	New methods are employed (e.g., a new machine is purchased or a manual procedure is automated).

Fig. 4-1 The levels of learning that training can be developed to handle.

effective in teaching interpersonal communications skills as a classroom environment with the opportunity for skill practice with coaching and feedback.

Deliverability focuses on the issues surrounding the delivery of training. For example, if training is being developed for a small audience which is widely dispersed throughout the United States and needs the training at various points in time, classroom training may not be the recommended alternative.

Designing for update-ability means that training covering topics that change frequently must be designed in a modular fashion so that the changes can be more easily incorporated prior to each delivery. Training must change as processes, procedures, and policies change.

The delivery strategy selected for training will always be governed by resource constraints—primarily budget constraints, but also including limited facilities (such as machinery, equipment) or resources (such as instructors). Training costs must be viewed in terms of total costs—not just the costs incurred to develop or acquire the training. That means delivery, support, and maintenance costs should be included (see Fig. 4-2). All aspects of investment costs need to be considered.

Other factors to be considered include the target audience size, geographic location and stability, and the level of risk associated with the failure for the training to "take." The training needs of a small target population with low turnover at a single location might be best met with structured OJT (a supervisor or leader is assigned to walk a trainee through the changes based on a checklist or other OJT guide). But if a mistake could cost lives, something more formal and substantial, with testing and certification, may be required. Table 4-2 will help assess the situation.

After a thorough assessment of the training needs, the instruction can be designed to help workers understand the changes in the process and learn the new skills required. A training solution, just like any solution to any identified problem, should be tested before being fully implemented. A pilot test of the training materials allows those who developed the training to get input and feedback from a selected group of participants on how well the training worked and whether it met the stated learning objectives. Prior to being broadly deployed, the training should be updated to reflect the changes and improvements suggested by the pilot group evaluation.

Training products should be looked at as dynamic products. Business and work environments are changing and the training that people receive should reflect whatever they are currently facing. Outdated training, manuals, or other job aids add little value. The same philosophy of continuous improvement that is applied to other products and processes in an organization should be applied to training products. If CI is *not* applied to the training product, the chances of decreasing the return on the training investment are greater.

CHAPTER 4

TRAINING TO SUPPORT CI EFFORTS

Method	Cost to develop	Cost to deliver*	Cost to update	Best for ... A	Best for ... K	Best for ... S
Classroom	M-L	H-M	L			X
Self-paced, text	H-M	L	M-L	X	X	
Self-paced, video	H-M	L	M-L	X	X	
CBT, interactive	H	L	H			X
CBT, non-interactive	M	L	L	X	X	

Key
A = Awareness   H = High
K = Knowledge   M = Medium
S = Skills   L = Low

*Assuming facilities/infrastructure (such as VCRs or computer hardware) is available

Fig. 4-2 Costs associated with delivery strategies.

TRAINING TO SUPPORT CI EFFORTS

The other major implication of CI to training is that people will require training to enable them to support the efforts to make CI happen in an organization. To create a culture which both embraces and is capable of continuous improvement, efforts to provide training and support must be proactive.

Whether the players are a homogeneous work team or pulled together in a cross-functional improvement team, they will need to be trained in the methods that will help them uncover opportunities for improvement and find solutions to current work problems. To understand what training will be required in an organization, one needs to begin by asking the following questions:

- What is the current plan and schedule for CI implementation?
- Is the company on schedule?
- What processes are targeted for improvement?
- Who is involved (what groups of employees) and to what extent?
- What methods should the people employ?
- Do they possess the skills to employ those methods, or is there a need to provide a way to give them those skills?

An analysis of the job tasks and requirements for CI implementation should be conducted to uncover training requirements. It is best to look at a company's training requirements as a collection of training products—a training Curriculum Architecture. A Curriculum Architecture provides a method for organizing the training requirements into logical segments which clarify the "big picture" of what pieces of training are required by various target audiences. A curriculum might be thought of as a work breakdown structure (WBS) for training—it contains all of the pieces needed to put a sound training system in place.

When thinking about the training required to implement a CI effort, most people think of common skill elements such as:

- How to work in a team environment.
- How to gather and analyze process or product data.
- How to conduct team-based, systematic problem solving.

These elements are crucial to the success of continuous improvement efforts. But a culture which plans to support continuous improvement for all levels of employees must look beyond, and conduct the necessary analyses of the target audiences to be involved in the processes. Depending on a company's CI process, the specific knowledge and skills required can be quite complex. For example, the training modules of the Curriculum Architecture to support CI could include the following courses or modules:

- The Quality/CI Story and Plans of the Company.
- The Quality/CI Story and Plans of a Function or Department.
- Team Leadership Training.
- Team Membership Training.
- Business and Quality Metrics.
- Financial and Economic Analysis.
- Quality Function Deployment (QFD).
- Value Engineering.
- Basic Statistics.

CHAPTER 4

TRAINING TO SUPPORT CI EFFORTS

TABLE 4-2
Training Delivery Strategies

Delivery Method	Can Be Used When...	Considerations
Unstructured OJT	• Performance is not critical • Target audience is very small • Target audience is widely dispersed	• No control over learning
Structured OJT	• Target audience is small • Target audience is widely dispersed • Learning must be guided • Need for guided skill practice	• Training is required for those responsible for "coaching" the learners • Need for process documentation to manage consistently
Job Aid	• Static content • Higher risks associated with the tasks • Tasks are nonrepetitive/nonfrequent so reference is required	• Job aids are usually used in conjunction with another form of instruction
Self-paced, Text or CBT, Non-interactive	• Static content • Medium to large target audience • Widely dispersed target audience • Only limited skill requirement	• Some audiences dislike reading • No feedback given during instruction; no place for learners to go with questions • Not the best method to use when skill practice is necessary
CBT, Interactive or Interactive Video	• Large target audience • Fairly stable content • Requirement that skill proficiency be demonstrated • Widely dispersed target audience • When simulated performance is needed but using actual equipment is prohibitive	• Cost prohibitive for smaller target populations • Equipment requirements to administer training
Classroom, Instructor-led	• Medium to large target audience • Need for opportunity to practice skills • Group interaction aids learning • Complex content • Target audience is concentrated in a few geographic areas • Higher risks associated with not comprehending content • Content changes frequently	• Expensive method for training—must include student time plus travel and living expense where necessary

- Concurrent Engineering/Integrated Product Development.
- Design for Manufacturability/Reliability/Serviceability/etc.
- Design of Experiments (DOE).
- Process Modeling/Mapping.
- Variability Reduction Methods.
- Cycle Time Reduction Methods.
- Process Capability Studies.
- SPC.
- Benchmarking.
- Diversity.
- Brainstorming.
- Problem Solving.
- Risk Analysis and Decision Making.
- Communication and Interpersonal Skills.

The list could go on, but one thing is certain: nothing should be included in the list if it has not been identified as necessary by a thorough analysis of the intended target audiences.

Ideally, the training that is provided to staff will be delivered just in time (JIT). To do that, a company needs to determine what the right training is, for the right people, at the right time. If that is done, then the modules of the total training architecture could be configured into unique learning paths appropriate to the specific target audience(s).

The ultimate goal is to link the training to the requirements of the performance situation. That involves specifying what will be done, by whom, where, and when. Any plans for training should be linked to the organization's plan for implementing CI, total quality management (TQM), etc.

THE TRAINING DEVELOPMENT PROCESS

It is necessary to use a systematic approach to develop the training required to support a CI or TQM effort. This is not something that can be approached in a random fashion with bits and pieces of training assembled when the need arises. Training

CHAPTER 4

THE TRAINING DEVELOPMENT PROCESS

should be in place to support the start-up efforts of deploying a new methodology. If the organization is serious about wanting to implement changes, it has to have the support systems in place to allow people to do what is asked of them. In other words, teach them the things they need to know so that they can succeed.

Good training does not just happen; it is engineered. The training development process offers a logical series of steps designed to create good training. Many models exist that portray the training development process. The model in Fig. 4-3 is one example.

It is useful for the individual or individuals responsible for providing training, or the customer of the training providers, to understand the process of training development. That is the only way to assess the present status of the process, what comes next, and what roles are assigned to whom.

A model is only valuable if it is understood by all involved parties and used consistently, but with a degree of flexibility. If the situation dictates, this model can be used as a platform to customize a process view. An overview of each of the six phases of the model follows.

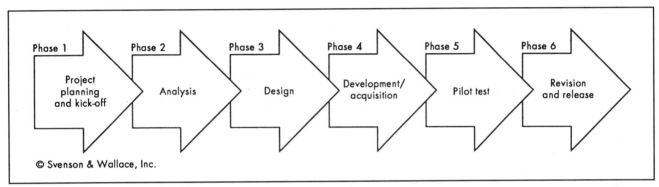

Fig. 4-3 An example of a training development process.

PHASE 1: PROJECT PLANNING AND KICK-OFF

The purpose of Phase 1 is to get the project defined, agreed to by all those who need to support the project, and off to the right start. If the up-front planning is done poorly, everything that occurs downstream will be subject to constant change, schedule slippage, and cost overruns.

Before planning any training efforts, determine the overall roll-out and implementation plans/schedule for CI (or TQM/TQC/etc.), segmenting the effort into phases if that has not been done already. A company's plan for training must reflect its plan for CI.

In every effort, crawling comes before walking, and walking before running. It is assumed that there will be a phased implementation approach to CI, which results in a phased implementation of the training to support CI. For example, the CI effort may initially focus on getting people to work as teams and getting the processes in control. Once they have mastered those skills, they can move to incremental or breakthrough improvement, and then on to integrated product design and development to influence the product design to better accommodate the manufacturing process constraints. If CI is approached in this fashion, Fig. 4-4 shows what a company's CI program roll-out may look like.

Using the phases and roll-out schedule as a guideline, determine which processes, functions, departments, or work centers will be targeted for change and the order in which that will happen. This gives a broad picture of the training needs and the time frames in which training must happen.

KEY ROLES

The Project Manager (or Management Team) and the Project Steering Team are the key players in Phase 1. The Project

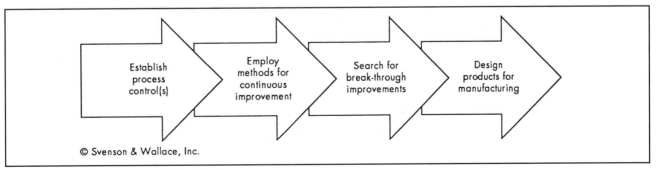

Fig. 4-4 An example of a company's CI roll-out.

4-7

CHAPTER 4

PHASE 1: PROJECT PLANNING AND KICK-OFF

Manager is responsible for laying out a detailed draft Project Plan which outlines all the tasks, roles, and responsibilities, and the schedule for completing the proposed project. The Project Plan must identify all of the major outputs and review points.

The Project Steering Team should be comprised of representatives from all key groups that require training. They should represent their areas and be capable of making decisions on behalf of their groups. Depending on the project and its importance, the representatives may have to be fairly high level managers from the various stakeholder organizations.

THINGS TO KNOW

Although this phase is critical, it is often ignored or short-changed in the desire to get things moving quickly. This shows very poor thinking, such as, "We don't have time to plan to do it right the first time, however, we are willing to find the time later to do things over and over again." This runs counter to the concepts of the quality movement.

Plans lacking detail often do not provide the sufficient level of visibility required for an honest assessment of the plan's logic and feasibility. Reviewing the plans with all of the parties who have a stake in the project is a good idea. It is certainly better than having them disagree later with what has been done. Remember, any training plans must be integrated into the overall CI implementation plans.

Make certain the right people serve on the Steering Team. These people must be able to make decisions for the organizations they represent and have enough influence so that the decisions are not continually challenged. The Steering Team decisions must represent the interests of all critical target audience areas.

PHASE 2: ANALYSIS

PURPOSE

The purpose of Phase 2 is to gather and analyze data, to understand:

- The target audience(s) to be trained.
- The performance to be affected by the training.
- The knowledge, skills, and personal attributes required of the target audience(s) for successful performance.
- The acceptability of any existing training as a part of, or as the whole training solution.

It is necessary to understand the populations that will be part of the CI effort: who they are in terms of their function in the CI process, their role in the CI effort, and their role in helping the organization satisfy customer requirements. Once it is understood who will need training, it must be determined exactly what training they will need. An analysis, to derive the required knowledge and skills for the target audiences based upon performance requirements, will lead to the content to be addressed in the training. However, training of a different nature may be necessary to achieve breakthrough improvement.

Figure 4-5 provides a way to capture the data required and prioritize the training needs appropriately. This will indicate which skills need to be developed or acquired most immediately. Priorities can be assigned by looking at those tools or techniques that are highly critical to success in each phase of the CI implementation.

KEY ROLES

An Analysis Team for this effort could include the following people:

- Master Performers to provide job/task knowledge and application of the content.
- Subject Matter Experts to provide content knowledge.
- Supervisors/Managers of the target audience(s).
- Novice Performers.
- Training Specialists.

Building a common understanding and framework for the diverse members will help them share in the decisions and ownership of the outputs of this phase, and more importantly, the outputs of the phases which follow.

CI Tools/ Techniques	Deptartment/Jobs							Tool/ Tech Criticality	CI Phase			Priority
	A			B		C	D		1	2	3	
	1	2	3	1	2	1	1					
• SPC	√	√	√	√	√	√	√	High	√	√	√	
• Design of experiments	√	√			√			Medium		√		
• Pugh	√	√	√	√	√	√	√	Low		√	√	
• Cause/effect diagrams	√	√		√	√	√	√	Medium	√	√	√	

Fig. 4-5 Capturing required data and prioritizing training needs appropriately.

CHAPTER 4

PHASE 2: ANALYSIS

THINGS TO KNOW

Facilitating an Analysis Team requires a strong facilitator who can deal with conflicting views and language that different groups may have. Group-based analysis is a tough approach, but well worth the effort in the long run. The dialogue between Analysis Team members will establish a more common understanding of the performance situation and the associated training requirements, and should ease downstream participation and project decision making.

PHASE 3: DESIGN

PURPOSE

The purpose of this phase is to create a design for the training that meets the desires, demands, and constraints that exist as a company implements its CI program. Of course, training has to be effective in preparing people to meet the performance requirements of their jobs, but not at any cost.

KEY ROLES

A Design Team should be assembled from the membership of the Analysis Team. The actual design work should be accomplished by skilled training designers with influence from the Design Team. The design will explain how the training will look and feel conceptually, the criteria to be used for evaluation, and the methods and approaches to consider.

The training design should be based on design concepts, criteria, and considerations embraced by the Design Team. Training must be designed with the company's culture and target audience groups in mind. For example, if the target population typically has poor reading and comprehension levels, self-paced reading would not be a good training strategy for teaching employees about SPC.

The design should also be flexible. If there are a number of different audiences who need to receive similar training on the same subject matter, chances are that pieces of one training event could be used in another.

Ideally, the design would also include an evaluation system for the training, including all five levels of evaluation:

1. Reaction to the training experience.
2. Mastery of the stated learning objectives.
3. Application on the job after the training is over.
4. Return on Investment calculating the returns compared to all the expenses incurred.
5. Customer Satisfaction with the total training experience (registration, scheduling, availability, length, etc.).

The training design must provide a level of information sufficient to enable the developers to produce the training effectively and efficiently. The evaluation system design must produce feedback which can be used to continuously improve the training program.

THINGS TO KNOW

The design for training must focus on performance first and subject matter second. "Nice-to-know" content must be eliminated, and time allocated for opportunity to practice the "need-to-use" content.

Training must be designed to contend with its own "-ilities:" effectiveness, deliverability, affordability, etc. Prerequisite knowledge and skills must be determined and addressed by the overall design.

PHASE 4: DEVELOPMENT

PURPOSE

The purpose of this phase is to create drafts of the training materials.

KEY ROLES

The key roles in Phase 4 include the developers, the Master Performers (MPs), and the Subject Matter Experts (SMEs). Developers provide the expertise in constructing good learning materials while the MPs and SMEs provide in-depth knowledge of the subject matter and its on-the-job application. For example, workers from the manufacturing line might share process knowledge with the developer for a Process Capabilities course.

The key outputs are the training material drafts. The developers create the units or lessons of the training and the MPs and SMEs review and provide feedback. The process is very similar to a shop creating a prototype unit. The unit is created and tested, and then adapted based on the test results. Training materials go through the same process.

THINGS TO KNOW

It should be expected, as with any product development effort, that multiple drafts may be required. The number of drafts will typically be dependent on the skills of the developers, the complexity and familiarity of the content, the amount of undefined content, etc.

It should also be expected that some amount of design change may be required as new information is learned during development. Changes during the development of a training product must be managed much like changes elsewhere. The more complex the training product, the more a formal change management system may be required.

CHAPTER 4
PHASE 5: PILOT-TEST

PHASE 5: PILOT-TEST

PURPOSE
The purpose of the Pilot-Test phase is to field test the training as it is intended to be delivered in the future.

KEY ROLES
The key roles in this phase include the learners and the instructors or facilitators. There are two types of learners who should be present in the pilot session: those who represent the target audience and those who represent management. Target audience members help assess whether learning takes place, and management representatives can help determine whether the material presented is accurate and comprehensive.

The pilot session is the time to test all of the materials developed, to see whether they are effective and accomplish the learning objectives. Extensive evaluations should be conducted during the pilot test and used as input and feedback for finalizing the training materials. A report of the pilot results and recommended revisions is usually created to be reviewed by the Steering Team for final revision and approval. Revision recommendations become the revision specifications used to guide activities in the next phase.

COMMENTS AND LESSONS LEARNED
Forewarn all Pilot-Test participants of the nature of the session, including the length of the day and the need to conduct extensive evaluations. Management may have viable concerns about this training, so allow management representatives to participate in the pilot session and evaluate the course. Their feedback is also valuable.

Try to eliminate any passive observers in the session. Everyone in the room should be an active participant in all exercises and discussions, and therefore share any risks that may be associated with the experience.

PHASE 6: REVISION AND RELEASE

PURPOSE
The purpose of Phase 6 is to revise the training according to the agreed Revision Specifications and release it into the training system. Depending on the formality of the training system, it may be necessary to have certain course data entered into various systems, such as the materials system, registration system, scheduling system, evaluation system, personnel system, etc.

KEY ROLES
The developers play the key role in making the updates, but the designers may be required to redesign aspects of the training. If the overall training development process was run correctly, changes in this phase should be kept to a total of 10% to 15% of the entire course.

THINGS TO KNOW
Anyone who has experience with training development may have experienced the typical pain associated with this phase. If not done properly, it can be a nightmare of extensive rework requirements. The best way to minimize the rework (not eliminate it) is to conduct the activities in the first two phases with diligence. Consider the rework required to product and/or process designs if the up-front work has not been done properly; training is subject to the same rework requirements.

All of the phases are important to do a quality job of defining and acquiring the right training for the right people at the right time. The story of GPT Manufacturing shows how one man, Lee Moore, used the phased approach to training development to successfully determine the training required to support continuous improvement teams in his organization.

THE STORY OF GPT MANUFACTURING

Lee Moore, a lead Project Engineer at GPT Manufacturing Company, was assigned to put the training in place to support continuous improvement teams. The directive came from his plant manager. Lee was told that the program he implemented must work not only in his plant, but in the nine other plants owned by GPT Manufacturing Company. Lee had worked on continuous improvement teams at his former company and could draw on some of his experience from that. He understood the need for training; after reading material on continuous improvement and training, he found a logical process which would lead to the training that the people would need to operate effectively on teams.

Lee decided that in order for this project to be a success, he needed to put together a Steering Team to get agreement and support for the project. He put together a Project Plan for the project and determined who had a stake in the training and needed to be a part of his Steering Team. He definitely wanted representation from each of the ten production centers, so he included a mix of senior hourly workers and supervisors. He needed a person from Safety Engineering, Quality Engineering, and Environmental Engineering. The Employee Relations person in his plant would eventually be responsible for the training, so she was invited to participate. Lee knew that the Plant Foreman should be behind this project 100% in order for it to succeed, so he made

CHAPTER 4

THE STORY OF GPT MANUFACTURING

certain that he could be a part of the team. At the kick-off meeting, he explained the project and the team discussed the implications of continuous improvement for the employees of GPT Manufacturing, giving their opinions on where the most support would be needed to get the CI Program off the ground. The Steering Team discussed the need for eventually having their own trainers at each of the ten production centers and the need to minimize the time taken away from people's jobs for training. The Plant Foreman was adamant about this for Lee's plant.

Lee received the green light from the Steering Team to move forward with his project. He had asked the Steering Team to appoint representatives from each of the ten production centers along with Subject Matter Experts (SMEs) who had worked on CI teams in the past. Together, the people in this group looked at the tasks of the continuous improvement teams and identified the knowledge and skills they would need to effectively do their jobs. The analysis uncovered high priority needs for training in team building, problem solving, interpersonal skills, and SPC. The analysis team decided that Lee ought to concentrate his initial efforts on finding a vendor with a team building course that would fit the needs of the target audiences at GPT Manufacturing.

Lee understood that even though GPT Manufacturing would not be developing the course, he would have to create a general "design" for the course so that he could evaluate the materials available on the market against what they needed internally. He and a selected group from those participating in the analysis formed a Design Team. They specified the learning objectives for the course along with the criteria for evaluation and selection. The Design Team focused on three key issues—affordability, the delivery method of the training, and the ability to customize the training to fit the GPT environment. The training acquired cannot exceed the allocated budget, it must be transferable to internal trainers so that they may deliver it at each of the ten production centers, and it must be relevant to GPT Manufacturing.

Acquisition of course materials required as much work as development would have. Lee had to contact training vendors to see what they had to offer. He found that he had many to choose from, but had to keep going back to the selection criteria to see whether the courses actually fit the need. He narrowed the choices down to three final vendors who sent complete sets of their training materials for evaluation, along with plans for transferring the course to GPT. Lee evaluated the materials and prepared a summary of his findings to his Design Team. The vendor, Total Team Trainers (TTT), was chosen to supply the course to GPT Manufacturing because they were able to demonstrate a better understanding of the customization needs that GPT Manufacturing had.

Lee assembled a group of target audience members, both experienced and inexperienced, to participate in the pilot session. The session was conducted by experienced trainers from TTT, and two trainer candidates from GPT Manufacturing participated in the pilot session. The materials and training experience were evaluated on two levels—reaction to the experience and mastery of the stated learning objectives. The results from the extensive participant comments revealed that the terminology in some of the lessons needed to be changed to reflect the language more common to GPT Manufacturing, and that the participants needed an introduction to the plant's policies and goals for improvement teams. These findings, along with other comments and suggestions for improvement, were reviewed with the Steering Team to arrive at the revision specifications for TTT.

TTT took responsibility for adapting the materials to accommodate the required changes. The release to the field was to be handled by the trainer candidates who participated in the pilot session, but GPT Manufacturing realized that the initial Train-the-Trainer session should be conducted by TTT instructors. Lee assumed responsibility for coordinating the training release efforts. This meant ensuring that all of the administrative tasks, such as getting the course into the GPT Registration and Scheduling System, were handled.

Lee found that his project was successful because of a number of key things. He planned the effort properly, got the right people involved and in agreement with the effort at the start, and he used the right mix of people from the target populations to make certain that the training would address their real needs, not the needs the population was assumed to have. Lee found out how important it was to communicate with all the people who have a stake in the training so that he understood their concerns and considered them when developing and acquiring the course. Lee drew this analogy: a training investment must be handled in the same manner as any capital equipment investment.

SUMMARY

Good training is engineered following a systematic development process analogous to any good product development process. Identifying critical issues and resolving them as early in the cycle as possible is key to achieving high quality training.

Regardless of whether CI is causing a need for training or training is being done to promote CI, one must understand the CI process and the performance requirements of the target audiences in order to provide the right training, to the right people, at the right time. Training needs to be based on these performance requirements to eliminate wasteful training.

When the goal is to improve process performance, there is always a human component that contributes to the potential variation. Training may offer a way to reduce variation of the human performance in a company's processes, and can give people the tools they need to participate effectively in the continuous improvement effort.

CHAPTER 5

IMPLEMENTING CONTINUOUS IMPROVEMENT

This chapter covers the topics important to planning and implementing a successful continuous improvement (CI) program. It also discusses how to reduce the number of operating procedures, or rules, in a company; the new concept of user groups (an American variation of *keiretsu*); and mistakes to avoid in implementing CI and total quality management (TQM) programs.

CHAPTER CONTENTS:

PLANNING 5-1

FEWER, BETTER RULES 5-8

IMPLEMENTATION OF CI 5-11

AVOIDING MISTAKES IN IMPLEMENTATION 5-15

CONTINUOUS IMPROVEMENT USER GROUPS 5-18

PLANNING

Being a vital element of TQM, CI requires adherence to TQM's fundamental principles: the use of facts and data and the plan-do-check-act (PDCA) cycle. Then, before implementing CI in an organization, there must be a planning step. Planning is another TQM tool. It allows the organization to understand the current situation, and to explore alternatives for the future. It is that exploration process that serves to educate the organization, and to build up a sense of ownership and a commitment to the anticipated changes.

PLANNING VERSUS PLANS

Planning is a process. Plans are the documentation developed during this process. It is important to remember that the plans are a result of the planning process, not the reason for the planning effort. Often the plan becomes the end objective and planning becomes subjugated to producing volumes of paper. One reason that this occurs is a lack of understanding about the planning process and its benefits. Another reason is that people do not understand the scope of planning and try to make a comprehensive document that covers all aspects of planning. This leads to mixed time horizons and multiple scopes.

Strategic Planning

Strategic planning relates to those activities that have a time horizon of three to five years and a global perspective. For a firm trying to implement a continuous improvement program, strategic planning should be the first step. Implementing CI is not a short-term effort. Because there must be cultural as well as technical changes within the organization, plenty of time must be allocated. Three years from concept to fully ingrained philosophy of business is not unusual and is considered fairly quick. Thus the time horizon test has been met.

The other aspect associated with strategic planning is the global nature of the planning effort. If the firm is a multinational or national firm, the earliest planning may have a global nature. More often, the strategic planning is being done at a single facility that is embarking on a CI effort. In this context, the global test is met by the planning effort examining all of the aspects of CI or the total impact CI will have on the organization. In this sense the strategic effort focuses on the use of CI to reshape the business and to gain significant leverage through change. Strategic planning requires the participation of all elements of an organization—Financial, Human Resources, Production, Engineering, Technical, Maintenance, and so forth. When the time horizon is longer term, the participation is across the organization, and the view is toward how CI will change the business. That is strategic planning.

Tactical Planning

Tactical planning has shorter time horizons than strategic planning. Typically, tactical planning relates to those activities that are expected to occur in the six-month to one-year time horizon. While this may seem to leave a gap in the one- to three-year window, it does not. A series of tactical planning efforts occur which address the issues raised in strategic planning. Also, the strategic planning effort should be updated annually so it provides a sliding window on the future.

Another characteristic of tactical planning is that it is more focused in scope. It does not take the

---

The Contributors of sections of this chapter are: **Bruce Brocka**, *Technical Director, Executive Sciences Institute, Davenport, IA;* **Richard Fleming**, *President, Continuous Improvement Associates, Northville, MI;* **Daniel Miklovic, CMfgE**, *Consultant, Issaquah, WA;* **Judd Prozeller**, *Quality Consultant, Training and Professional Development, Rochester Institute of Technology;* **Mike Rother**, *Research Engineer, Industrial Technology Institute, Ann Arbor, MI;* **Jerre Stead**, *President, AT&T Global Business Communications Systems.*

The Reviewers of sections of this chapter are: **Rick Gammache**, *Associate, Booz-Allen;* **Peter Hamm**, *President and CEO, Council for Continuous Improvement, San Jose, CA;* **Howard Sanderow**, *President, Management and Engineering Technology, Dayton, OH;* **Judy Stanger**, *Secretary to CEO, Square D Company.*

CHAPTER 5
PLANNING

global perspective. It concentrates on the "theater of operations." Tactical planning focuses on a specific element of the strategic CI plan. The strategic CI plan may state that the maintenance department will implement CI to reduce the unavailability of machines by reducing downtime and shortening repair cycles. Thinking how that strategic goal will be met, assigning expected benefits of achieving the targets—that is the role of tactical planning. The shorter time horizon results from the more focused effort. The piece of the pie is smaller so it takes less time to deal with it. The main results of tactical planning are an understanding of what tools and techniques are available to achieve the strategic vision, and the prioritization of the implementation of these CI processes.

The Elements of a Successful Plan

One of the most important aspects of the plan is that it is written in plain English. There must be a minimum of jargon about continuous improvement. If the document is littered with CI and quality-related terms and phrases it will be intimidating to those who did not participate in its creation. It is considered acceptable for the document to use common business jargon. If a company uses common industry terms to describe finishing or manufacturing processes, and those terms are understood by virtually everyone in the company, there is no reason not to use that terminology in the plan.

Another aspect of a successful plan is that it has plenty of pictures and a lot of white space. The plan should use figures and illustrations to make a point. Which of the following, Fig. 5-1 or the next paragraph, conveys the message more clearly?

Continuous improvement is like a rectangular box filled with a set of metal and wood devices that can be used to effect adjustments or repairs. It is not a single device; rather, there are a number of these devices within the concept of CI which are used to adjust or repair the organization.

Figure 5-1 illustrates one view of continuous improvement that can be used to adjust an organization's operation.

Obviously the meaning is easily conveyed by the simple sentence: CI is like an ambulance that can transport an ailing company to a better future. The key here is that the term *ambulance* is well understood. With continuous improvement, many of the concepts and terminology are not well understood. In those cases often a picture is the proper tool to convey the meaning.

The plan should be easy to maintain. The strategic plan needs updating every year, and tactical plans must accommodate shifting business conditions or the introduction of new technologies. The plan should not be bound tightly in a covered volume. It should be kept in a binder that facilitates easy page replacement. The plan may even be best kept on-line in an information system if access is available throughout the facility. Flexibility is the key. The plan must accommodate changes if it is to be useful over the long term.

Finally, the plan must focus on goals and objectives. It should be oriented to explaining how the facility can meet its goals and objectives by following the plan's recommendations. If a plan has all of these elements the only other critical element is that the plan must represent reality. Several sections in this chapter deal with the issue of capturing the actual situation and achievable steps toward improvement.

WHY PLAN?

Planning serves a number of vital roles. Today, quality methods are generally accepted as an essential part of modern manufacturing. With ever-increasing attention to quality, the need for continuous improvement is inherent in the manufacturing process. The trends are toward providing consistently higher quality. Not all facilities have the capability of meeting these demands. To adequately serve the customer they must implement new quality systems such as continuous improvement.

A total quality management approach dictates that before rushing in to solve "the problem" of the lack of these systems, the issues must be understood. Once they are understood, a plan of

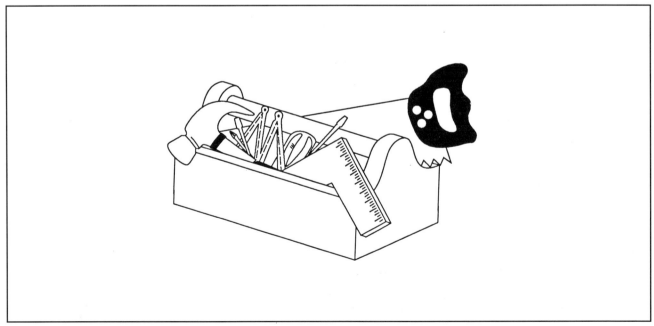

Fig. 5-1 Simple analogy to continuous improvement.

CHAPTER 5
PLANNING

action can be devised, and measurements can be taken to gage the effectiveness of the change. *Planning* as used in this sense is the exploration of the issues associated with introducing new systems technology. It also is the identification of opportunities for systems technology to change the business and the definition of the tasks required to achieve the new "vision." It must also include establishing criteria for measuring success.

Benefits in General

There are a number of benefits associated with planning in general. These include Communication, Participation, Education, Documentation, Validation, Financial, and Cultural.

Communication. One of the most important roles that planning plays in any organization is that of communicating the strategies and objectives of the business throughout the organization. While most businesses have strong statements of purpose and goals or objectives written on paper, the essence of the organization is its vision, or philosophy. This way of doing business originates with the senior management at the facility and reflects their own attitudes and philosophies. Where there is a strong team management approach, the philosophy often is shaped by other leaders at the facility, but it still bears a strong imprint from the person who is ultimately responsible for the unit's performance. The planning process, through the use of cross-functional teams and active management participation, helps to communicate the subtleties of the vision and operating philosophy to levels beyond the management team. Also, the planning effort can put key management people in contact with the staff that actually have the most impact on the business, the manufacturing workers. The planning effort often provides the first instance of bidirectional communication about strategic issues.

Participation. Through the use of cross-functional teams, and through data gathering techniques that maximize participation by all unit staff, planning increases the sense of ownership of technology in the organization. Again, a TQM approach dictates that those closest to the problem be on the team that fixes it. An old axiom is that "People do not resist change—they resist the way change changes their job." A well thought out, highly participatory planning process will serve to build new work teams or build upon existing teams. Planning, by itself, does not ensure that there is ownership throughout an organization. The chronicles of business are littered with stories of failed businesses where the senior staff did exceptional planning but execution never matched the planning effort. Numerous postmortems have shown that the "PLAN" that management created was never understood or believed in by the staff expected to execute it. By using a participatory approach, this pitfall of planning can be avoided, and planning can actually aid in the development of an organizationally "deep" commitment to the business goals and objectives. This does not release management from the critical roles of guiding and supporting the planning effort; management must *participate* to have a successful planning effort.

Education. Planning serves another vital role beyond communication and team building, and that is education. While a key planning premise should be "Business needs drive applications, applications drive technology," often the very people charged with implementing new systems select particular technologies first, then try to gain leverage from the tools at hand. This generally is because the demands of keeping a manufacturing facility operating preclude the opportunity to remain current with emerging technologies. The trade journals, vendor demonstrations, even many trade expositions, all generally focus on currently available, commercially viable technologies. The planning process provides a mechanism to expose the operating staff to these future technologies, to study how they can be used to leverage the business into a world class competitor.

In addition, the planning process is an educational tool itself. Planning can be used to train the staff in areas such as strategic thinking, improving group problem-solving skills, enhancing teamwork skills, and in the process itself. One of the key measures of the success of a planning effort is whether the team that prepared the original plan can modify it if required due to changing technology or business issues.

Documentation. As noted earlier, plans capture the knowledge gained in the planning process so it may be communicated to others. They also serve to capture and—often for the first time—document the current state of affairs. Very few facilities have the opportunity to build their systems "in a green field." Reality dictates that what works today should be saved as the foundation for the new systems. The documentation prepared during the planning process serves as the "blueprints" of the foundation for the new systems.

The plans also provide a tool to document the benefits expected from implementing new systems. While some businesses have implemented these computer-integrated manufacturing (CIM), TQM, or manufacturing systems on faith, it generally is not considered good practice to do so. In the *Harvard Business Review* article "Must CIM be Justified by Faith Alone?," Robert Kaplan argues that if all of the benefits of these systems are appropriately accounted for, enabling technologies, like CIM, can easily pass most investment analysis.[1] One key point is that the savings come from an extremely large number of opportunities for saving, each fairly small in its own right. It is the accumulation of all of these small, incremental benefits, which justifies the overall expenditure.

Validation. As mentioned, a very important part of the planning process is the identification of opportunities for improvement through the use of the new technologies, and the benefits that will be achieved. The planning process, if complete, will not only identify where these benefits will come from, but how they can be measured. The planning process helps to establish performance criteria and new measurement tools as an intrinsic part of the installed systems.

Also, planning helps to validate visions and strategies. A complete planning effort may often entail the use of simulation to test several alternatives. It is vital to identify alternatives and understand the particular strengths and weaknesses each offers. As the business environment changes during the course of implementing the plan, changes in available technology, or changes in economic variables may change a previously rejected alternative into the most attractive option. Without previous analysis, selection of the alternative could be time consuming, or the alternative may be overlooked.

Financial. While not a given, often planning efforts identify some straightforward changes that could be made without the expenditure of any significant capital. Sometimes these simple procedural changes have benefits that exceed the cost of the entire planning effort. Because the planning process helps people accommodate change and helps develop new ways of thinking, often process improvements and increased operating efficiencies occur, seemingly unexplained, during or shortly after the planning process. These financial benefits should not be overlooked, but should be, at least partially, credited to the cultural change that planning helps to foster.

Another financial benefit that results from good planning is generally lower overall total expenditures for systems technology.

CHAPTER 5
PLANNING

Without a detailed plan it is common to implement many redundant features. For example: all DCS and Process Management packages or PLC and Supervisory Control packages, as well as most statistics and analysis tools, have optional quality control charting features. Without an adequate understanding of where the SQC charting function will reside, individual purchasers of all of these systems often pay additional money to secure this functionality—"just in case it's needed later." Finally, someone will purchase a quality management system for even more money. At the end of the cycle, there are three or four quality tools, all costing money to purchase and maintain. A basic tenet of quality, the use of common, reliable methods, has been violated.

Cultural. Finally, given the education, communication, and participation which planning fosters, a cultural change will occur. Planning, as presented here, indicates that management supports changes, and this will enable the organization to change more rapidly. While it will affect those on the planning team more than nonparticipants, the cultural changes usually find their way into the organization as well. Planning, in the form described here, helps to foster a new view of the facility.

Benefits Specific to CI Planning

Continuous improvement requires a cultural change. The change is often best led by believers from within the organization. A properly constituted and properly trained CI planning team can become the seed that sparks change. The CI planning process can be the first example of a continuous improvement effort. Both the CI planning process and the creation of the plan lend themselves to the constant refinement inherent in CI. In no other aspect of manufacturing is the need for participation at the shop floor level as important as it is with quality in general and continuous improvement specifically. Planning builds that commitment and ownership more than any simple training program can. This is not to say that training is not needed, or that it does not play a vital role. Rather, planning sets the mood of the organization, making it receptive to the training efforts and actually helping to identify what training is required.

PREPARING TO PLAN

There are a number of steps in preparing to plan, but the two most important ones are making sure that there is a clear statement of the business goals and objectives, and selecting the proper planning team. If either of these conditions are not met, the planning effort will fail.

Business Assessment

To effectively gage how continuous improvement can be used to improve the business from a strategic perspective, there must be a very clear and well stated business plan. If there is not a clear understanding of the goals of the business and an understanding of what factors are critical to success, the resulting CI plan may prove to be more than useless—it may be dangerous. If meeting shipment dates with full orders is a critical factor of business success, a CI plan focused on perfection in meeting specifications at the cost of meeting delivery dates could result in a loss of business. It is quite possible that the customer can accept a less than perfect part and use it as a place holder until the correct part is delivered, but that if that customer does not have *any* part it cannot start a run. If a supplier cannot meet the delivery deadline, the customer will place the order elsewhere. Focusing on the wrong objectives to improve can put a firm out of business.

Equally important is the understanding of the philosophy of how the business is operated. The general manager has the greatest influence on how the organization works. Within the business plan there is usually an indication of the philosophy of the plant. It may be safety, maintenance, quality, customer service, or any of many other aspects of the workplace. A continuous improvement plan that focuses on an area of little interest to the general manager and secondary to critical success factors will not receive management's support. Without that support it is doomed.

There must be a clearly articulated business plan to identify the critical success factors for the organization and management's vision about how those factors will be achieved before there can be successful planning about the tools and systems that will be used to achieve the goals of the plan. If that plan does not exist it must be created.

Another important factor related to business assessment, but generally done as part of the information gathering phase, is *benchmarking*. Benchmarking is critical to identifying the opportunities for improvement and the benefits associated with making those improvements. The selection of facilities to benchmark against is very important. Because each firm has a specific way of processing orders and distributing them to the plants, it does not make sense to benchmark any given plant's performance in this area against a competitor's plant if order processing at that other company is done totally differently. It makes much more sense to benchmark this aspect of the plant's performance against either (1) a dissimilar, noncompetitive company or plant, or (2) at least other plants within the company. It may make sense at a corporate level to benchmark the way a firm processes orders against the way competitors do order processing. On the other hand, the chemistry and physics of a particular process tend to be very similar across companies, and it makes sense to benchmark control strategies, levels of control, and quality performance against one's competitors. Where a firm does not have adequate knowledge of competitor's performance, it is imperative that outside information be gathered.

Who Should be Involved?

There are two major groups associated with the planning effort. The first is the Planning Team. Depending on the approach, this may be a few key contact people at a facility supported by an outside team, or more likely a local team supported by a few key corporate staff or others with special skills. The second approach is generally the best way to achieve all of the human resources-related benefits associated with planning. If the first method is adopted, there will be minimal ownership at the facility. Generally, the team will need training, in the tools and methods that will be used in the planning process as well as the fundamental concepts associated with continuous improvement. For strategic planning the team should consist of a cross-section of the organization. It should contain members from every major functional group within the facility and from every level within the firm. The team should range in size from five to 15. It should not be more than 5% of the staff and not less than 1/2 of 1%. Obviously in very large or very small organizations both criteria could be hard to meet. It is probably better to err on the side of having a team that is too large rather than too small.

The second group that must be identified is the "Approving Authority." The Approving Authority is generally the General Manager and the Manager's staff who participate in the decisions on how capital funds are to be allocated. In some cases this group may be very small, even just the GM. In other cases the team might be quite large. The role of the approving authority is to

CHAPTER 5
PLANNING

review the results of the planning team at intermediate steps, and either accept the results and grant approval to proceed or ask the planning team to reexamine its work and provide additional details or clarification. Points to interact with the approving authority vary according to the planning methodology used. At a minimum there should be at least two intermediate review points and more is better than less. The process of review gives management the opportunity to provide a critique and helps in developing management's ownership of the plan.

PLANNING METHODOLOGIES

A number of firms have developed and marketed planning methodologies. Some refer to them as Planning, Advanced Planning, Systems Planning, Project Programming, etc. Generally they all have several points in common, revolving around a three-step process:

1. Document and understand the current situation.
2. Identify alternatives to reach the future state.
3. Select from the alternatives and set priorities.

These three activities basically require the gathering of information about the current situation and the alternatives, the building of consensus about the future state, and which alternatives offer the highest probability of reaching that future state.

Figure 5-2 is a flow chart which shows these activities in the context of the decision to implement CI in an organization. Figure 5-3 is a typical checksheet used to organize a planning effort.

Information Gathering

Information gathering is important from two perspectives. Accurate information must be gathered about the current situation. Without a full set of facts and data, a basic tenet of the quality paradigm has been violated. Also, information needs to be gathered about alternatives that will help move the organization forward toward a new state. Both of these information gathering steps can utilize a number of techniques. Two are fairly common.

Group techniques. Group interviewing techniques have a very high involvement factor, making them a good tool for developing group consensus and ownership. The down side is that they cost more both in staff involvement time and planning team resources. Yet these additional costs are offset by the benefits in some circumstances. Group interviewing is particularly effective at identifying the needs and wants that shape the future state. In a group environment there is synergy. One person may begin with a thought about a particular need and then others build upon it, ending with an overwhelming consensus about a very strong need or want. The key is to properly identify the issues as to whether they are truly *needs* and critical in reaching the new end state, or rather *wants* that may slow the progress of reaching the new end state, but not inhibit it. Finally, there is the issue of whether to use small group techniques or large group techniques.

Small group techniques. Small group interviewing techniques focus on the use of a facilitator to capture and document the recommendations from the group. Small group interviews are generally held with groups of five to 25 people and have a facilitator and scribe. The facilitator leads the interviewing process asking questions like: "Finish the following sentence—If I could only change one thing about how we do things around here it would be XXXXXXXXXX." The facilitator records the comments on a chartboard and posts the completed charts on the wall for all to see. Besides capturing the information on the chartboards, the facilitator serves a vital controlling role. The facilitator is charged with defusing gripe sessions. If in response to the line of questioning the replies start turning critical instead of constructive, the facilitator brings the comments back into focus. For example: A participant says that "Things never change, they just keep doing the same stuff over and over," the facilitator may then ask, "Well, if you were in charge what would you change?" It is quite possible that the reply might be "fire the bosses," to which the facilitator inquires "and how will that cause changes?" The facilitator must refocus the energy into answering the original question "If I could only change one thing about how we do things around here it would be XXXXXXXXXX." The scribe records details not captured on the chartboard notes and helps the facilitator capture multiple streams of information if they occur.

Large group techniques. Where the group size is more on the order of 10 to 50 people, or where the group is extremely committed and actively participating, the techniques used in small group interviewing become unwieldy. In those cases a modified interviewing technique for larger groups usually works well. In the large group techniques each member of the group serves as a scribe. Individuals write their contributions on sheets that are posted on the wall. The use of 5 x 8 cards is common, as is the use of large sticky notes. The 5 x 8 cards offer a larger writing surface and are easier to read from a distance, while sticky notes are easily rearranged into groupings as trends emerge. Whether using the large or small group technique, the keys to success are ensuring full participation, giving everyone a chance to be heard or seen, and recording information accurately. Use colors or paraphrasing, draw pictures or encourage the participants to do so, to ensure that what is written is what was meant.

One-on-one techniques. In some cases it is better to gather information in a more individualized manner. The techniques to do this are referred to as one-on-one techniques although in some cases it may two-on-one or three-on-two meetings. A very basic one-on-one technique is the questionnaire. This tool is particularly useful for gathering historical information about the current situation. It can also be used to gather information about potential future states from those who are unable to attend group interviews, although it does not offer others the opportunity to build upon ideas. Questionnaires should always be validated with some personal follow-up and inspection. This quality check is necessary, so people take the time to complete the questionnaire and do it accurately. It is human nature to do a job better if it is known that someone is likely to look at the results.

Another technique is the personal interview. This may be either a one-on-one arrangement or multiple interviewers interviewing one or more interviewees. A checklist or questionnaire should serve as the basis for the interview, but the interviewer should be free to explore additional topics as they arise. In all cases it is important that the interviewees understand the importance of their input to the planning process, and that they develop some sense of connection if not ownership with the findings of the questionnaire or the interviewing process.

Building Consensus

One of the most difficult parts of the planning process is developing a consensus regarding what continuous improvement strategy to adopt and which tools will help the organization best meet its goals. The tools mentioned above begin that process. One way to heighten their impact is to apply them to the largest possible cross-section of the facility. The more people interviewed, the greater the probability that the true needs of the organization will surface. If the true needs are surfaced, consensus is almost a given.

CHAPTER 5
PLANNING

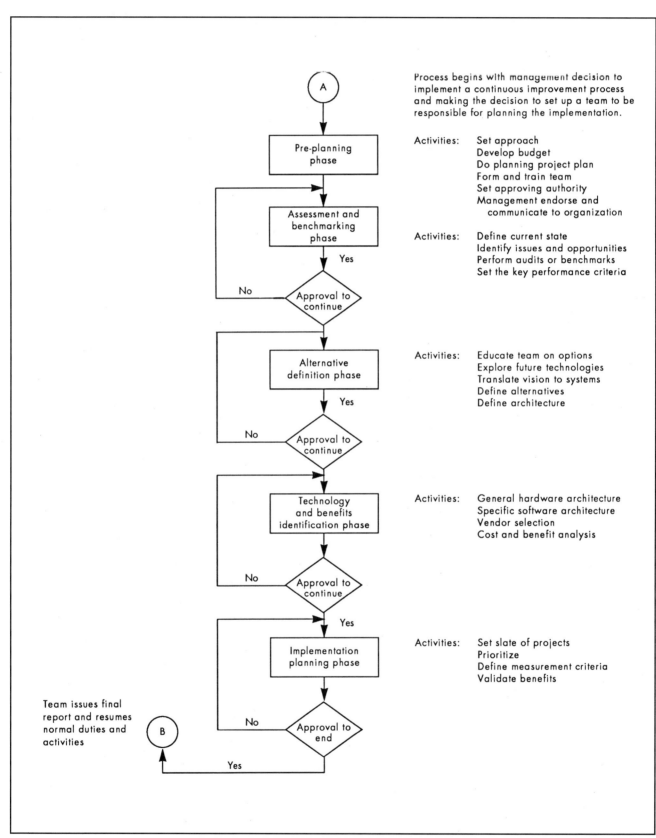

Fig. 5-2 Process flowchart.

CHAPTER 5
PLANNING

Phase Activity	Available Tools	Status
Pre-planning		
Set approach	Select fast track or methodical	
Budget	Use corporate accounting tools	
Project plan	CPM and/or Gantt charts	
Form and train team	Train the trainer, direct training	
Set approving authority		
Endorse/communicate	Meeting, letter	
Assessment/Benchmarking		
Define current state	Inventory checksheets, checklists, Questionnaires, diagrams	
Identify issues/ opportunities	Questionnaires, 1-on-1, large or small group, customer interviews	
Audits/benchmarking	Literature search, visit, consultant	
Key performance criteria	Define, allocate votes, consensus, brainstorming, diagramming	
AA approval	Report, presentation, meeting	
Alternative Definition		
Education	Experts, classes, self study	
Explore futures	Experts, classes, self study	
Vision>systems	Models, group consensus making	
Define alternatives	Models, descriptive text	
Application architecture	Models, diagrams	
AA approval	Report, presentation, meeting	
Technology/Benefits ID		
General hardware architecture	Models	
Specific software architecture	Lists, models	
Vendor selection	Presentation, RFI, RFP, standards	
Cost/benefit analysis	ROI, DCF, NPV, ROE, IRR, PAM	
AA approval	Report, presentation, meeting	
Implementation Planning		
Set slate of projects		
Priorities	Vote, consensus, C/B rank, Kepner-Tregoe	
Define measurement criteria	Checksheets, financial measures	
Validate benefits	Vote, discuss, report	
AA approval	Report, presentation, meeting	

Fig. 5-3 Planning checklist.

PLAN MAINTENANCE

As continuous improvement becomes an ingrained way of operating, the needs of the organization will change. Keeping with the PDCA model, the original CI plan will need to be improved. This plan maintenance is just another aspect of continuous improvement.

Update When?

Strategic plans should not be allowed to age beyond their original time horizon. A good approach is to update the plan every two years and keep the horizon at least three years in the future. In that way the plan stays current and reflects the foreseeable future. This does not mean that the strategic plan should only be updated every two years, however. Whenever technology or business conditions change significantly, the plan needs updating. Many firms find that after 12 to 18 months of implementation of the continuous improvement process, the business has changed so much that they need to update their strategic plans. Because of their CI efforts, the economics of the manufacturing process changes or the market they are selling into changes. New product lines also necessitate changes in the strategic plan.

Tactical plans should change as often as needed. The tactical planning effort is of such short duration or small scope that it truly reflects the essence of continuous improvement. Changes should be incorporated as they occur during implementation.

Measuring Success

One of the most important aspects of the quality PDCA cycle is the "C;" checking on the results achieved in planning should begin immediately. Even before the planning is complete, the benefits of planning should be verified. Benefits such as improved communication, participation, education, documentation, and cultural changes were presented earlier in this section. When the planning process was organized, specific benefits should have been identified for management to gain their approval of the planning effort. As the planning process proceeds, it becomes necessary to track which benefits are accruing and which are not yet apparent. If particular benefits predicted are not materializing, the continuous improvement model suggests that a modification to the planning process may be in order.

Another aspect of benefit measurement is related to achieving the results predicted in the planning process, for each aspect of continuous improvement that is implemented in a particular plant or area. This requires that the planning process itself identify what measurement techniques will be used to assess performance against targets. Planning is a tool to improve accountability. The most common methods of gaging performance are the classic financial techniques such as Return On Investment, Return On Equity, Discounted Cash Flow, and Net Present Value. Some plants are beginning to rely on newer models such as Productivity Accounting. The value of Productivity Accounting is that it uses dollars to weigh and normalize the calculations, but the actual results are reported in ratios. This comparison to a base year as a ratio removes market prices for finished products from the equation, and simplifies the actual impact of a given set of changes. This facilitates judging performance related factors instead of financial performance. This is not to say that financial performance is not important, but that the impact of process improvements should be judged separately.

ORGANIZATIONAL ISSUES

The planning process as well as the implementation of a continuous improvement program will fundamentally change the organization. If organizational issues are not addressed early on, some of the key benefits of planning will be lost and the implementation of the continuous improvement process will be hindered.

Information as Power

One of the key purposes behind planning is the dissemination of information and the building of consensus and ownership. The planning team members become the new owners of a great deal of information about the future of the organization. In the past the role of the foreman has been as a gatekeeper. The foreman retained power in an organization by passing on to employees only the information they wanted and providing management only information specifically asked for. Although the role of the foreman has changed to that of coach and information technology has made information more accessible throughout the organization, the first-line supervisor has acted as a conduit for feedback to management in many cases. The creation of a cross-functional planning team will open new channels of communication and threaten some people's positions of power. This is not to say that the planning should not be done. It merely indicates that there

CHAPTER 5

FEWER, BETTER RULES

should be a sensitivity to the fact that power bases will be changing and that the change process must be managed for a proper outcome.

Organizational Structures

Organizational design, specifically as it applies to continuous improvement is beyond the scope of this section. However, the planning process offers an opportunity to include organizational redesign as part of the implementation planning. Continuous improvement programs require a flatter organization to facilitate the problem solving process and the formation of teams. Hierarchical approval chains will slow the implementation of CI and can serve as a barrier to a successful CI program. The opportunity that planning offers in reshaping the organization should not be ignored. Because the focus of the strategic planning effort is across the entire organization and because it deals with time frames of three years and beyond, the recommendations that the plan makes can be revised by management before actual implementation. The CI process offers the opportunity to consider the feedback from the planning team, and the organizational input gathered during the information-gathering steps in forming the new structure.

Plan Acceptance

If all of the steps above are followed, there is a high probability that the plan represents the consensus of not only the planning team but the entire organization. By reviewing intermediate results with the ''Approving Authority'' regularly, management will also have developed ownership of the plan.

If there is not a high degree of acceptance and commitment to the plan there has been a breakdown in the process and corrective action is needed. Another potentially disruptive situation arises when the plan has acceptance by most of the facility but top management proceeds to implement the continuous improvement program in a different manner. This is a very dangerous situation. The likely outcome will be that the CI program will fail to achieve the intended results and that the organization's morale will deteriorate. It will deteriorate quickly, because the personnel will feel that their efforts in contributing to the plan are not appreciated (otherwise why would management reject it?), and the overall poor performance from the CI program will contribute to the problem.

Planning is not cheap, but it does not have to be expensive. A properly constituted team supported by selected outside resources can generally provide all of the expertise an organization needs to move toward an environment of continuous improvement.

FEWER, BETTER RULES

Many U.S. businesses literally have created enough manuals to cover walls. They are nothing but policies and procedures for high-level executives to hide behind. Following the rules has become an end in itself—tantamount to surrogate leadership. A company's primary goals—to make the highest quality products and satisfy its customers—become subordinate to the internal rules of the company.

Why are many American companies having difficulty competing in the new global marketplace? Maybe the decision-makers do not know what the marketplace is all about because they've stayed locked up in the executive suites far too long, letting the rules run the company.

Some companies—most, in fact—have numerous rules and regulations that govern everything they do. In most highly traditional companies, executives have dug in behind this fortress of rules and have forgotten—or possibly never learned—what it is to lead.

The sad thing is they've never had to. It was not part of the culture. Everybody who made it to the executive level ended up following the example of his or her predecessor, making sure to follow the rules, and possibly adding some new ones. They were rewarded for this behavior, so they weren't about to change.

FROM 760 TO 14

Acme, a composite example of many large corporations in America today, at one time had 360 corporate policies and procedures. The marketing and sales organizations, for good measure, had another 400 guidelines to live by. A total of 760 rules and regulations led Acme from its glory days back in the 1950s and 60s, through the roller-coaster times of the 70s and 80s. Nothing really changed, except that the number of rules grew and the executives became less and less visible. And revenues steadily declined as global competition became stronger.

Once the 90s arrived, the company knew it had to find new markets around the world. The problem was that nobody in management really knew how to do it. The 760 rules provided them no magic formula to surviving and thriving in the 90s.

Obviously, a change was needed. It began when a combination of old and new leaders decided that the wall of manuals should be toppled and critically examined for any valuable fossils. After none were found, the old ''760'' was exposed for what it was: dinosaur management.

Consultants were called in to facilitate the rule reduction. Members of the corporate staff each received between 20 and 40 rules to evaluate. Each was to report at the company's off-premises strategy meeting which policies and procedures were really needed. It was decided in the meeting that, despite a glorious corporate history, the rules and regulations that governed every action and movement of Acme Co. were redundant, overstated, confusing, out-of-date—and unnecessary.

New rules were created from the rubble. Some companies that went through similar transformations actually had fewer, some had more. In Acme's case, the new rules numbered 14.

CREATING A VISION, MISSION AND VALUES

The first eight new rules expressed the corporation's Vision, Mission, and Values. Its vision was to be a company dedicated to quality, and committed to the success of its customers and itself. Its mission was to be the global leader in its line of business. Its five values, a strictly arbitrary number, were:

1. Keep all customers satisfied.
2. Work together.
3. Be honest, fair, and ethical.
4. Be accountable.
5. Always strive for excellence.

CHAPTER 5

FEWER, BETTER RULES

Managers Must Become Coaches

The second part of the transformation was to encourage and teach the leaders of the company, the top 20 or 30 people, to change their entire mode of operation. They were required to "get out there" and start living the values. An executive's salary and bonus would depend solely on his or her newfound ability to live the values and, by example, to encourage everybody else to live them. They were to become visible role models.

Creating a New Leadership Model

The only rule of the new leadership model was for executives to get out of their offices and start moving around, visiting other corporate locations and customers, and to make sure that all their customers—all their associates as well as external customers—were having their needs satisfied. The leaders were to be out there making sure that they were being truthful and up-front in all their dealings, and making sure that they, as coaches, were encouraging the proper teamwork by trusting people in their organizations to do their jobs—and not wasting everybody's time by having a network of "checkers" keeping tabs on the next level down to make sure no mistakes were made and that all bases were covered. And by giving everybody in the organization the accountability they needed to do the best job possible, they would begin to see the excellent results that were always possible, but never allowed by all the old rules.

By living the values, the coaches would be able to help their teams break out of the ruts in which they were stuck.

People Power

What Acme found was that its only sustainable competitive advantage was its people. And if they were good at what they did (and they were), the only thing holding them back from being great performers and world-class competitors was management.

It was tough for some managers to learn to say, "No, we're not going to do that now because it's not our No. 1 priority." They would rather say, "We'll get back to you on that." That was wrong. It undersold the people who were looking to them for leadership—and they hurt their company.

In the new Acme, managers learned to say: "You know, your idea is great. You've done a good job of putting this plan together and asking for these resources, but we have to get these other three things done and then we'll talk about implementing your idea. That's a promise."

The employee felt a new sense of accomplishment: "All right, I can understand that. And I appreciate the vote of confidence and the fact that you were up-front—and didn't snow me."

Too many managers have learned to take the easy way out and make no commitment, and just to sweep it under the rug. Managers have to become coaches and cheerleaders to unlock the potential of their teams. The power and intelligence that the people of a company possess is awesome, and largely untapped.

Unfortunately, many managers have been taught by their corporate cultures to be condescending and noncommittal, and the attitude gets passed along. Others were "beat up" once by their managers, and they will not do this to their subordinates now that they have the chance. The "what-was-good-enough-for-me" attitude has hurt American industry, and it must change.

Creating New Management Systems

The final part of creating a new leadership model at Acme was for the new, enlightened management to develop new and simpler management systems. In this case, there were five elements:

Improve customer focus. Listen carefully to both existing and potential customers. Place strong emphasis on managing and improving relationships with them. Constantly strive to integrate their needs into products and services, and continuously improve commitment to them.

Improve processes. Place strong emphasis on continuous improvement of all of a company's processes, including prevention and elimination of defects and reduction of cycle times. Figure 5-4 contrasts quality control and quality improvement.

Improve employee development. Create an environment in which all associates excel in pursuit of business and personal goals. Develop and utilize the full potential of all associates through involvement in cross-functional teams, skill development, and reward and recognition linked directly to accountability and performance. Create "people power."

Improve planning. Ensure that both short- and long-term planning processes focus on achieving business success and competitive advantage. Make sure that the processes are based on facts, driven by customer requirements, and aimed at achieving best-in-class quality through the use of continuous improvement of processes, products, and services.

Improve facts-and-analysis procedure. Ensure that sound, actionable facts and data are the basis for all decision-making and continuous improvement activities.

There is a total of 14 rules dictated by the vision, mission, and values, the single rule of the new leadership model, and the rules of the five new management systems.

The bottom line is that everything that drives the corporation can literally be written on two sheets of paper. No manuals are needed. And if people are doing something that does not have anything to do with, or is out of sync with, what is on those sheets of paper...they should stop doing it. It is probably not moving the organization forward. Worse yet, it is probably impeding the corporation's progress.

SUMMARY

Changing the entire culture at a corporation, any corporation, is no mean feat. The motivation today in corporate America is that change really is not an option. Companies must change if they want to survive; it is really that simple. Change can only occur if someone cuts through the morass of rules and regulations and comes to an agreement on what is really important—such as a corporate vision, mission, and values. It can only occur if management fosters excellence and accountability by giving people what they need to do their jobs better, and by instituting new management systems like a quality improvement process that go from top to bottom.

It is not management's job to create more rules; in world-class companies, managers exist to eliminate rules and stumbling blocks so that the doers can do what they do best. Until a manager can say to all "I am proud to be part of this team. It is a great team that will grow together each and every day. Our potential is huge. Let's keep it rolling and have fun making it happen. Together we will make our vision of being the very best a reality. You are all making a tremendous difference." Until that happens positive change can not take place.

CHAPTER 5
FEWER, BETTER RULES

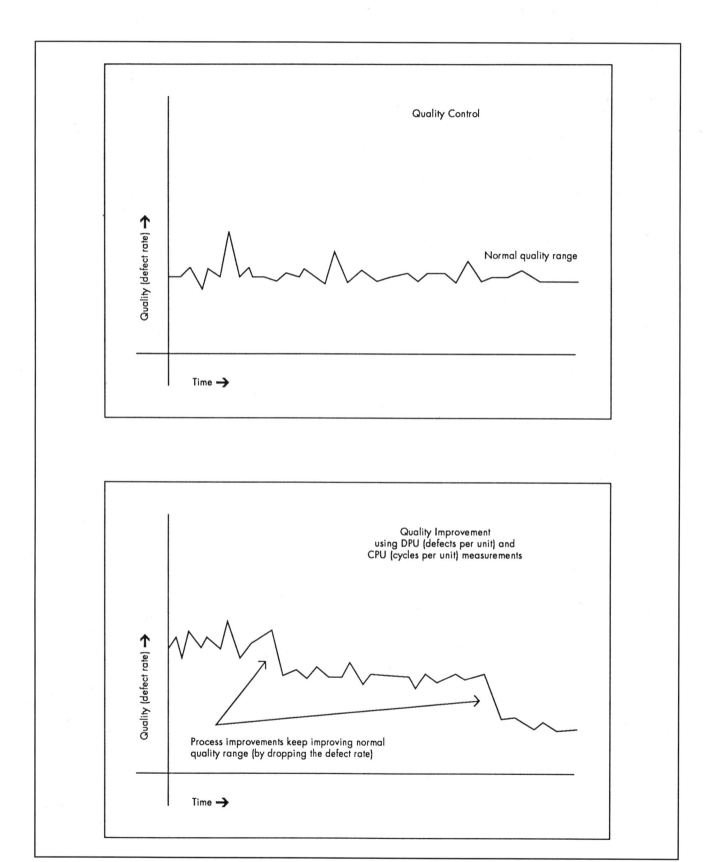

Fig. 5-4 Contrast between quality control and quality improvement.

CHAPTER 5

IMPLEMENTATION OF CI

IMPLEMENTATION OF CI

The philosophies, practices, tools, methods, and techniques of continuous improvement must be integrated into a coherent implementation plan. There are perhaps as many implementation schemes as there are CI practitioners. The reason for this lies in the individual needs of the organization. Some require tremendous changes in corporate culture to bring about change, and some do not. It is difficult to implement continuous improvement because it can be complex, and because it is very difficult to change a corporate culture that may be used to fatalism or disillusionment. Unlike total quality management (TQM), which requires a systemic approach and vast changes in corporate culture, CI can be implemented in a small, piecewise method and still bring about satisfactory results.

Continuous improvement practitioners often espouse a top-down commitment. Top-level managers become enlightened, and then pass the lamp of wisdom onward. Fortunately, it is not true that CI must come from the top—this is an ideal, but not a requirement. All employees can institute CI on whatever sphere of influence they have. As success increases, so will interest in exporting CI to other areas of the organization.

This bottom-up approach may not be as awkward as it seems. After all, how can the work force change unless it is either already willing, or encounters a significant or enabling event brought about by top management? If the employees are willing, but were unaware of how to proceed, continuous improvement can quickly revolutionize the way work is done. If they are not willing, coaching the work force has a slim chance of working due to barriers imposed by status quo attitudes. A significant event, such as a layoff, may be necessary.

However continuous improvement evolves, it is instructive to have some sort of implementation model. There are simply too many tools and concepts to proceed with all of it at once. There are many implementation models. No magic is involved in picking a model; the key is to adopt one, modify it for organizational needs, and persevere with the plan.

PROCESS

Continuous improvement is often at odds with current organizational practice. The status quo is cherished because it is known, and changes are often dismissed due to "liability;" suggestions are shunted aside because "that is not our area." Creativity and exploration are relegated to mysterious research and development departments. What is needed is a management approach that will unleash an employee's latent, possibly atrophied talent, providing benefit to both the worker and the firm. The goal may be continuous improvement, but the threshold of employee motivation and empowerment must be crossed first. The following process is a guide to achieving this.

Embrace Continuous Improvement Principles

Continuous improvement is a journey, not a destination. It must become a way of organizational life. It is a revolutionary way to invert the organizational hierarchy, put customers first, eliminate managerial deadwood, and overcome whatever stands in the way of fulfilling customer needs. There are four primary areas that must be addressed: Vision, Empowerment, Continuous Evaluation, and Customer Orientation. Each of these areas is briefly discussed below, and a summary of strategies presented in Fig. 5-5.

Vision. Organizational vision provides the framework that guides a firm's beliefs and values. This can be as simple as "making the best widgets at the lowest cost to the consumer," or as structured as the organizational culture of IBM. The gist of the corporate vision should be a simple, one sentence guide or motto that every employee knows, and more important, believes. If well crafted, the vision statement (which is typically one paragraph to no more than two pages) can serve through a torrent of change in product and service technology. If CI is being implemented on a small scale, the vision statement may be much more mundane, and indeed may be more of a goals statement than a sweeping corporate vision declaration.

Simply stating a vision is not enough. Become obsessed with implementing the vision. Sweat out the details. Observe how the customer perceives the vision. Is it obvious, or do customers have to be told that "Quality is Job One" (ignoring the manipulative aspects that advertising can have on perceptions). This vision must be a constant; dissonant goals will result in frustration.

Empowerment. In CI, a process orientation, rather than a (solely) result-centered orientation is required. The effect of this is to challenge the traditional roles of management toward workers. Workers are now part of the solution rather than the problem. This

---

- Vision
  Process must begin with topmost management
  Problems often must be solved by changing or renewing the process or system
  Rampant involvement, including suppliers and customers, is essential

- Empowerment
  Yield authority to the lowest possible level to resolve problems
  Communication and information dissemination are vital
  It is more important to be clear than correct
  Drive out management by fear
  Encourage and reward creative thinking-even if the ideas are not implemented
  Share the credit for success
  Revise and renew performance measurement systems
  Establish ownership of tasks and projects
  Enhance skills to measure quality and identify problems
  Involve every individual in improving his/her own work processes
  Recognize people as the most important resource

- Continuous Evaluation and Measurement
  Constantly use feedback
  Self-assess and reflect continuously
  Quantify and measure
  Measure the cost of quality
  Continuously monitor vital measurements of a product
  Reduce variation

- Customer Orientation
  Identifying customer's needs, wants, and desires is continuous
  Vendors are part of the solution, not the problem
  Customer requirements, desires, hopes, and fears must be continuously monitored
  Customers may be internal

Fig. 5-5 A summary of strategies for vision, empowerment, continuous evaluation, and customer orientation.

CHAPTER 5

IMPLEMENTATION OF CI

process orientation will lead to employee empowerment. Empowering means enabling a worker to achieve his or her highest potential. For most American companies this is new and may be the most powerful and useful concept in continuous improvement. Allowing and helping workers to achieve their highest potential may seem obvious or impossible, but it is in fact neither. Empowerment requires turning the organization chart upside down, recognizing that management is in place to aid workers in overcoming problems they encounter, not to place new roadblocks in the way. Key strategies in implementing an empowerment scheme are:

1. Effective, unfettered vertical and horizontal communication. Using this type of communication is essential to continuous improvement efforts. Continuous improvement practices aim at removing communication blocks, facilitating bidirectional communication between leaders and subordinates, and ensuring that the firm's goals and objectives are clearly delineated and disseminated throughout. A broad array of tools and techniques are available for enhancing vertical and horizontal communication. Virtually every technique in Fig. 5-6 is communication oriented or enhancing.
2. Two heads are better than one. Without teamwork, continuous improvement is finished before it can start. The modern team works together as a single entity, and not as a "committee" where one or a few members do or direct the work. This may require dramatic rethinking by management and employees as well. Teamwork is essential for continuous improvement. Team activities build communication and cooperation, stimulate creative thought, and provide an infrastructure supporting continuous improvement practices. Teams are a waste of time if management vetoes or substantially changes their recommendations. Teams must be allowed autonomy. If management is unable to trust the recommendations that come from the team, then management by fear rules, and productivity will spiral lower and lower.
3. Ownership. A key element in empowering employees is to allow them ownership of a tasking, project, or division. Ownership implies trust and it requires a delegation of authority commensurate with the responsibility of the task. Ownership can also be granted to a team. Ownership also demands that the final resolution of the tasking be in the hands of the owner. Nitpicking, rearranging, and otherwise finding fault with the tasking upon completion will undermine any attempt at empowerment via ownership. A simple concept, but hard to do—just as it is hard for a parent not to correct a child's first attempt at making a bed, putting things away, or cleaning the table. Any correction may ruin weeks of encouragement. Besides, to the child, the bed looks just fine. It was made to the best of the child's abilities. If the employees' abilities are none too impressive, it is time to train them.
4. Delegate authority. Always attempt to delegate authority to the lowest possible organizational level. Constantly ask: Why should it be done? If competent people are employed, let them do their job. No one knows more about the job than the person directly involved with it. Giving 20-year old advice on what it takes to get a job or assignment will fall on deaf ears. They will learn how to act within the new environment.
5. Constantly remove barriers to CI. It is inevitable that change will be resisted, and that organizational practices stand in the way of CI. In fact, a great deal of effort in CI is expended in overcoming such resistance, usually by allowing change to come from individuals directly involved, rather than as a directive from management. The whole idea of continuous improvement leads to continuous change. The following process is recommended for barrier removal:

- Identify barriers. Anything that stands in the way of implementing and realizing continuous improvement should be considered a barrier. This means examining internal procedures, customer relations and concerns, and personnel issues. Anything that is perceived to be a barrier deserves further consideration. At this initial stage, no judgment as to priority or validity should be made.
- Place into categories. Related barriers and their systemic causes may now be analyzed. Validity judgment should still be held in abeyance at this stage. Categorization may be facilitated by using cause-effect diagrams or other organizing tools. Be alert for barriers that mask or cause one another. It is not unusual for a myriad of problems to be caused by a few difficulties.
- Establish priority. At this stage barriers are judged on their validity in accordance with the severity of the problem. It can be difficult to compare relative barriers at this stage, unless a common denominator, such as dollars or number of defective units, is used.

```
Vision                              Continuous Evaluation
Leader task:                        Leader focus:
    Vision and culture definition       Strategic planning
Articulator task:                   Articulator task:
    Employee involvement                Performance management
Implementor toolkit:                Implementor toolkit:
    Benchmarking                        Auditing
    Force field analysis                Benchmarking
    Goal setting                        Brainstorming
    Systematic diagram                  Cause effect diagrams
                                        Control charts
Empowerment                             Data collection
Leader focus:                           Delphi technique
    Leadership                          Design of experiments
Articulator task:                       Evolutionary operation
    Team building                       EMEA analysis
Implementor toolkit:                    Flowcharts
    Auditing                            Nominal group technique
    Brainstorming                       Quality costs
    Cause effect diagrams               Sampling
    Creativity                          Statistical measures
    Data Collection                     Five W's, one Y
    Nominal group technique
    Pareto analysis                 Customer Orientation
    Process decision program        Leader focus:
        chart                           Change management
    Quality circles                 Articulator task:
    Service quality                     Motivation
    Time management                 Implementor toolkit:
    Work flow analysis                  Benchmarking
                                        Data collection
                                        Delphi technique
                                        Foolproofing
                                        Quality function deployment
                                        Nominal group techniques
                                        Sampling Service quality
```

Fig. 5-6 Techniques used in each of the strategies listed in Fig. 5-5.

CHAPTER 5
IMPLEMENTATION OF CI

- Problem solving. This means more than symptom removal! It is vital to address the root cause of the problem. By using cause-effect diagrams and quality function deployment it may become apparent that the elimination of one barrier may solve many problems. Do not be surprised if this "master cause" looks intractable, such as "poor communication among management and workers." These "soft" problems are the ones that plague employees for years, and may take years to solve.
- Goals and strategies for resolution. Resolution of problems may entail goals over a period of months or years. Goals should be realistic and attainable with the given resources. Strategies ensure that goals can be accomplished.

Evaluation. Continuous improvement requires continuous evaluation. How else does anyone know if the process is effective? Evaluation requires a benchmarking or baselining of the initial process. Process improvements can then be measured. While there are a myriad of data collection, presentation, and analysis tools, the very foundations of what and how to measure must constantly be called into question. This is due to the "Heisenberg Uncertainty Principle" of CI: Measuring the process influences it. For example, if the length of each phone call of order takers is measured, and espoused as a goal, then the call length will be kept to the average or established target value. Even if this means not answering any questions the customer may have, even if this means not fulfilling an order, the phone calls will be answered within that target.

The lesson here is to measure, but observe the effect the measurement has against other principles or measurements.

Measure everyone. This means administration as well as the production line. Develop a suite of measurement metrics appropriate to the vision and goals of the organization. Be sure to develop process measures as well as results-oriented measures. Automate the measurement as much as possible, and reduce its expense as an overhead item. Above all, be sure to treat defects and problems as opportunities, not as failures.

Customer orientation. The driving force in any organization is its customers. All CI efforts must be directed toward the customer. Some strategies for improving customer and vendor relations are:

- Link organizational vision to customer satisfaction.
- Move to single sourcing.
- Minimize the overall number of vendors.
- Identify internal and external customers.
- Identify end users and distributors.
- Establish routine dialogue with customers.
- Involve the customer in planning and development.
- View vendors as strategic partners.
- Give quality awards to customers who have improved their business.
- Reward vendors by giving them more business.

Start Small

Implementing a company-wide continuous improvement plan all at once is suicidal. A single division, department, or branch must first serve as a test site. In this stage, vision leadership is articulated and implemented. Try to transform the test site completely before transporting the plan company-wide. This means at least 12 to 18 months, and encountering and surmounting at least one crisis. Start change with small wins. Implement changes in a "lead-the-fleet" department or division. Once the bugs have been worked out, begin exporting this model elsewhere.

Rely on interplant or division rivalry: "If the guys in Yuba City can do it, we can do it too."

Do not pick a certain set of members within the test site. Pick everyone in an easily identifiable group, or pick another group. Continuous improvement will have to work on all employees, not just the best or the worst. Constantly communicate and provide feedback. Pursue the right way, not the quick way. Visit a Baldrige Award winner. Take courses on the Malcolm Baldrige National Quality Award and benchmarking. Read every relevant case study available.

Implement the Change Process

Change is debilitating when done *to* us, but exhilarating when done *by* us (Rosabeth Moss Kanter). The pace at which change can take place can be categorized in four ways:

All-at-once. Both social and technological changes can be brought about immediately. This method induces an enormous amount of stress. Liker describes an example of this type of chaotic change at Basic Industries, where even some of the key players were so disturbed that they left.

Technical systems first. In this scheme technical systems are changed and then the human resources aspect is changed.

Social systems first. Here the human resources aspects are changed first. Changing social systems first may obviate some changes in technical systems. Since changing human resources is often less expensive than technological change, the advantages to this scheme are obvious.

Staged approach. Change is implemented gradually. In computer-integrated manufacturing, for example, "islands of automation" (separately automated work cells with no links to other work cells) were common during the mid to late 1980s. These islands are now becoming integrated as sophisticated software and standards emerge. Due to the extraordinary complexity of integration, it is best to focus on the integration aspects alone, without worrying unduly about each station operating properly. This has already been done during the island stage.

Three stakeholder roles are involved in managing change:

- Sponsors, who are the individuals or group who authorize or are empowered with the change.
- Implementors, who set the changes in motion.
- Users, who are directly impacted by the change and/or are expected to achieve the desired change.

Three stages of change are identified in Table 5-1.

Celebrate Success

Show how well the test site did by making a video. Conduct tours of the test site, and allow test site employees to host discussion groups. Rewards are nice, but take it easy. Rewards are often monetary in nature, may prove to be poor motivators (especially among knowledgeable workers), and can actually demotivate others. A job well done is its own reward—work can be a fulfilling and rewarding experience when employees are involved and empowered.

Cascade Implementation

Continuous improvement should first be implemented by top leadership and flow through the management structure similar to a waterfall. This cascading deployment ensures that leaders understand, demonstrate, and can teach continuous improvement principles and practices before expecting them from, and evaluating them in, their staff. The cascade effect flows to the suppliers as well. Taking the continuous improvement process organization-

CHAPTER 5

IMPLEMENTATION OF CI

TABLE 5-1.
Change Process

Unfreeze Phase	1. Identify the need for a change.	Sponsor
	2. Define the what-how-why-when and who.	Sponsor
Move Phase	3. Consider how to announce the changes.	Sponsor/Implementor
	4. Promote correct interpretation of changes.	Sponsor/Implementor
	5. Determine if the users are able to implement the changes.	Implementor
	6. Decide if the users are prepared psychologically for the changes.	Implementor
	7. Prepare for and overcome pockets of resistance.	Implementor
Refreeze Phase	8. System changes accepted.	User

wide is a big step, and will require several years to implement—and that is just the beginning. Employees must understand that continuous improvement never stops, and that this is not a program or push. It is a way of life to be applied with almost religious zeal.

A SPIRAL MODEL

The spiral model depicted in Fig. 5-7 relates the concepts and principles of continuous improvement. While most models appear in linear form, the spiral model serves as a reminder that continuous improvement implementation needs to be cascaded through the company, and done iteratively. From the center of the spiral emanates the vision of the organization. The first layer consists of the foundational principles, the second the management dynamics required by midlevel managers and supervisors, and the third (or implementation) layer, contains some suggested tool kits. There are three shell layers:

- Vision leaders—top management.
- Vision articulators—middle and supervisory management.
- Vision implementors—supervisors and individuals.

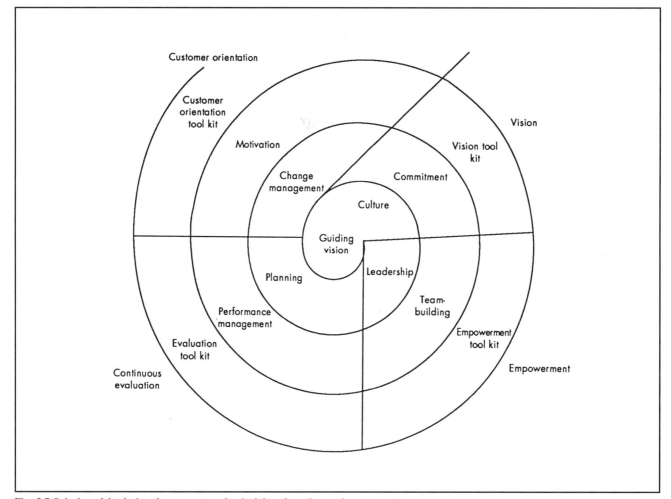

Fig. 5-7 Spiral model relating the concepts and principles of continuous improvement.

CHAPTER 5

IMPLEMENTATION OF CI

There are four "slices" that correspond to the four key principles of continuous improvement:

- Vision.
- Empowerment.
- Continuous evaluation.
- Customer orientation.

Some suggested tools are listed in Fig. 5-6. The model is generic enough to be adapted by any organization, and can be made specific quite readily.

IMPLEMENTATION ISSUES AND TRAPS

Management must not react to problems with a "How in the world did that happen?" but with a helping attitude. After all, should not the job of management be to help with exceptions to the routine? If things always worked smoothly, with no unusual circumstances or problems, why bother with managers?

Douglas Patterson points out a number of traps to avoid when implementing continuous improvement:

Trap 1: Delegating continuous improvement authority. Continuous improvement must be the responsibility of top management—first. Then it must become everyone's responsibility. As Deming, Juran, and others point out, the management culture must change in order for continuous improvement to be properly instituted.

Trap 2: Continuous improvement is a new name for existing programs. Statistical process control, analysis of variance, quality circles, and cross-functional teams are all part of continuous improvement, but do not constitute its entirety. Continuous improvement requires a vision and principles, as well as tools.

Trap 3: Do it right the first time. Too vigorous and literal an implementation of this concept will stifle creativity. Failures are more than OK; they are opportunities for learning.

Trap 4: Continuous improvement is statistical quality control and quality circles. Tools are only a part of the picture. Humanistic management principles, discussed in Management Dynamics, and the principles discussed in this section are necessary for achieving the goals of continuous improvement.

Barry Sheehy has developed some guidelines on surviving the inevitable the first crisis of a CI program:

1. Acknowledge the crisis.
2. Consider it a hidden opportunity.
3. Make sure everyone understands that there is no going back.
4. Give voice to fears and concerns—but do not back down in the face of naysayers.
5. Recall accomplishments that have occurred so far.
6. Get counsel from workers and suppliers about problems; do not base action solely on one's own perceptions.
7. Ask for advice. No voice is too small.
8. Revise the plan and inform everyone of the changes.

Sheehy also recommends the following preventive or mitigating factors:

1. Build backsliding and setbacks into the plan.
2. Underpromise and overdeliver.
3. Plan a renewal at about month 12.
4. Review goals for their attainability.
5. Record all accomplishments.

While a crisis is unavoidable, preparing for it can considerably lessen any damage to the credibility of the program. Some additional lessons learned:

- Any change effort which requires changes in individual behavior, regardless of initial focus, must include means for ensuring that such changes will in fact occur. Saying that it will be done does not make it so.
- Change is best accomplished when persons likely to be affected by the change are brought into the process as soon as possible.
- Successful change is not likely to occur following the single application of any technique.
- No single technique or approach is optimal for all organizational problems, contexts, and objectives. Diagnosis is essential.
- Not everyone will be a winner. Sometimes implementing CI requires that some people lose prestige, status, control, or even position entirely. Terminations brought about by change should be done as quickly as possible. Nobody said continuous improvement was easy, or that implementing discipline in a modern environment was simple.
- Do not forget what was done right previously. Change can be exciting, but do not lose sight of procedures that worked well before. Are those doing things the "old way" left unsupported?
- Cultures are changed and transformed, not removed. If a culture is destroyed, and not replaced, an undesirable culture may form.

THE OMNIPRESENT QUESTION

Continuous improvement is quite complex. Unfortunately it is difficult to recall all the scores of principles and techniques without referring to a text. Yet it is still pleasant to have a phrase that will serve someone like a Swiss Army knife—easy to remember, yet useful for almost any of the situations that may be encountered. For continuous improvement advocates, this phrase is: What is the value added? This can be the omnipresent question with which old bureaucracies, work rules, and product designs can be pruned down. It does not demand a complex answer, requiring consultants and reports and spreadsheets. The answer does not have to be numeric or measurable in a traditional sense. This question—what is the value added—gives insight into continuous improvement while maintaining a return on investment strategy.

AVOIDING MISTAKES IN IMPLEMENTATION

For many organizations the biggest hurdle to overcome is a traditional corporate culture which is hindering success in today's marketplace. The traditional culture is characterized by an emphasis on short-term profit, a highly autocratic management style, little or no empowerment of employees, an environment where access to information is jealously guarded, a lack of investment in training and development of human resources, and no objectives or systems for measuring performance in areas critical to cus-

CHAPTER 5

AVOIDING MISTAKES IN IMPLEMENTATION

tomer satisfaction. Overcoming this hurdle requires a significant company-wide change effort, and recognizing that there is a problem is the first step.

The key is to have that recognition occur at the top of the organization. The majority of successful organizational transformations start at the top, and are driven top-down. This was the case at Xerox, Motorola, and Harley Davidson; all are excellent examples of successful transformations. If the leaders of the organization do not perceive a need to change, long-term success is almost impossible. But even when companies get beyond recognizing the need for change, they often make critical errors when attempting to implement company-wide strategies focused on producing continuous improvement; consequently the strategies, while well intended, never live up to their original promises. A recent study conducted jointly by Ernst & Young and the American Quality Foundation suggests that "A lot of companies read lots of books, did lots of training, formed teams and tried to implement lots of practices simultaneously, . . ." but did not achieve the results they wanted.

One common mistake organizations make is to underestimate the nature and scope of the change required to sustain an environment of continuous improvement. It is one thing to create some *ad hoc* teams, train them, and have them make positive contributions to the organization; it is quite another to have a team approach and the empowerment of employees in achieving business objectives as integral parts of the way the organization is managed "day-in and day-out." The latter requires significant, fundamental changes in the way an organization is managed, which for the most part means changing the culture of the organization.

The purpose of this section is to explore some of the reasons why continuous improvement efforts sometimes fail, identify critical success factors, and present an approach which has been used in producing continuous improvement in scores of organizations.

First, everyone must agree on what is meant by culture. For this discussion culture will refer to the important understandings that members of an organization share. It is the set of shared beliefs, attitudes, and values which influences how people in an organization behave. The strength of the culture is dependent upon the degree to which these beliefs, attitudes, and values are shared. Culture is a liability when the beliefs, attitudes, and values are not in line with the objectives of the organization, its members, and its customers.

So how does an organization go about creating a cultural change focused on continuous improvement? Unfortunately, there is no one model which has universal acceptance as the correct way to change culture, and this represents a serious dilemma for organizations. There is currently a proliferation of quality consultants, each selling a different approach, and many with thousands of dollars per day price tags. Indeed, because of the interest in total quality inspired by the Malcolm Baldrige National Quality Award, the award has been called "the full employment act for quality consultants." The aforementioned American Quality Foundation study cited "a total of 945 different quality management tactics" being applied by companies which participated in the study. In assessing the total quality movement, the overall variability introduced just by the sheer number of different approaches has contributed to mixed results.

FLAWED APPROACHES

It is important to discuss two common, but somewhat flawed approaches to implementing a total quality culture. The first can be called the "special project" approach. In this approach an organization attempts to introduce total quality management exclusively via project teams or task teams. These teams are created and chartered to address specific problems or issues. Team members are taught traditional quality techniques and processes (that is, problem solving, statistical tools, effective meeting skills, etc.) which are then applied during regular (often weekly) meetings until the problem or issue is resolved. While these teams may be successful, the long-term impact of this approach in transforming the organizational culture is usually minimal. That is because the project teams themselves take on the identity of total quality within the organization, and they typically coexist with a predominant company culture which may, in fact, represent the antithesis of total quality. In other words, the perception within the organization is that total quality is something to be practiced only within the confines of the special project teams; outside of the teams it is "business-as-usual." The crux of the problem is that senior management has not planned for the integration of the total quality approach with the day-to-day management of the business. There is no plan to address the factors which are critical to supporting and sustaining a true cultural change. Without this, the special project approach is constantly "butting heads" with the predominant culture, thus requiring an ongoing life-support system of its own to maintain it. If and when the resources which feed the life-support system dry up, and they usually do, the organization is back where it started.

A second flawed approach is the "training only" approach. This approach treats the cultural change effort strictly as an exercise in training. Perhaps this is because many total quality consultants are recycled management training specialists. For whatever reason, many organizations erroneously believe that if every employee is trained in total quality tools and processes, the organization's culture will be magically transformed. Training in this context may reveal management's underlying attitude toward the worker. The message seems to be that the worker is the source of the organization's quality problems, and management hopes that training will provide the necessary attitude adjustment and skills improvement in employees, with improved quality as a natural consequence. Unfortunately, training by itself has little impact on organizational culture and performance. That is because the trainee goes back to the job, perhaps with every intention of applying newly learned skills and techniques, only to find that the work environment does not encourage or reinforce this, and in fact may be hostile to it. Training by itself often raises employee expectations and creates an impending sense of empowerment often followed by disappointment and a loss of management credibility because no plan for follow-through exists. What management must realize is that most problems with quality, and the lack of continuous improvement, have a systemic basis and are not due to employee deficiencies.

CULTURAL CHANGE MODEL

Management interested in producing and sustaining real cultural change must understand each of the basic factors that influence the perpetuation of culture. These factors are the key to change and must be addressed in a carefully developed change strategy which is created by the senior executives of the organization. Consider each of the basic factors with an understanding that no single factor, taken alone, can be said to be responsible for defining and sustaining an existing culture. It is the collective impact that these factors have on the beliefs, attitudes, and values of an organization that makes them important. Similarly, a change strategy, if it is to be successful, must not focus on only one or two

CHAPTER 5
AVOIDING MISTAKES IN IMPLEMENTATION

of the cultural change factors. This is the inherent flaw of the "training only" approach discussed previously. All the factors need to be addressed to ensure that together they support the desired change (see Fig. 5-8). The discussion presents a pre- and post-change description of each factor.

TRAINING

While training by itself will not create lasting change, it is an important factor in perpetuating an existing culture, and it plays a key role in any cultural change strategy. Many companies offer little or no opportunity for employees to develop skills. That training which does exist is often on-the-job, poorly designed, with little or no reinforcement or follow-up to ensure that the skills have been acquired and are being applied. Most successful change strategies which are focused on continuous improvement provide basic training in problem solving, effective meetings, teamwork, and statistical tools to all employees as part of the implementation of the new strategy. This may take six months to a year or longer, depending on the size of the organization. Beyond fundamental training, additional training needs should be addressed on an ongoing basis. This is especially true if the emphasis is on continuous improvement. Realizing that training is key to tapping the potential that this vast resource represents, many companies are establishing annual training budgets that represent five to eight percent of labor costs.

RECOGNITION AND REWARD

There is an old saying that "what gets rewarded gets done," and there is a lot of truth in this. Many organizations claim to place a priority on quality and continuous improvement, but the sole basis of that claim is the number of times these terms are used in their marketing literature. In reality, they have recognition and reward systems which reinforce the attainment of financially-based objectives. They usually lack quality-specific recognition and rewards which reinforce continuous improvement. In fact, many organizations leave recognition to chance, dependent on the individual styles of its managers. There is an inconsistent approach to what gets recognized and how it gets recognized. Thus, there are numerous missed opportunities to reinforce—and thereby increase and perpetuate—continuous improvement efforts. Unfortunately, many decisions and actions which then result have negative consequences for quality and customer satisfaction. The irony is that if organizations do put a high priority on quality-based objectives which are reflected in recognition and reward practices, the more traditional financially based objectives will be achieved in the long run. What needs to be in place is a well defined approach to recognition and reward which identifies the types of behaviors which should be reinforced (recognized) and results which should be rewarded. Recognition and reward alternatives should be part of a plan, and managers should be held accountable for implementation.

Recognition and reward systems, it should be noted, are not without controversy. Some people oppose recognition and reward specifically targeted to reinforce the transition to total quality on the grounds that applying extrinsic forms of reward ultimately demeans the overall quality initiative. They argue that the intrinsic rewards (that is, personal satisfaction) derived from activities of collaboration, team accomplishment, and continuous improvement are both adequate and appropriate as a means of reinforcing and thus sustaining continuous improvement efforts. While intrinsic rewards may appear to be an inexpensive way to reinforce quality efforts, this should not be used as an excuse for failing to develop an appropriate recognition and reward system. Human needs and desires for recognition, appreciation, fairness, and social esteem require explicit and tangible managerial attention. It should also be noted that getting people to learn and apply new concepts, principles, and techniques is not as intrinsically rewarding as one would hope. It often is a frustrating and time-consuming effort which requires ongoing reinforcement to sustain, until the ideas are assimilated by the new culture and are applied as a matter of routine.

COMMUNICATION

Many organizations do not spend much time or effort promoting communication internally. Management seems to be guided by what can be called the "mushroom approach." Employees are generally "kept in the dark and/or fed a line." A similar approach, used to determine the need for communications, is the "need to know" gage. If management feels an employee has a need to know in order to carry out the job, then the employee is given access to information. Unfortunately, there is usually an overall perception that, below the management level, employees do not have a need to know.

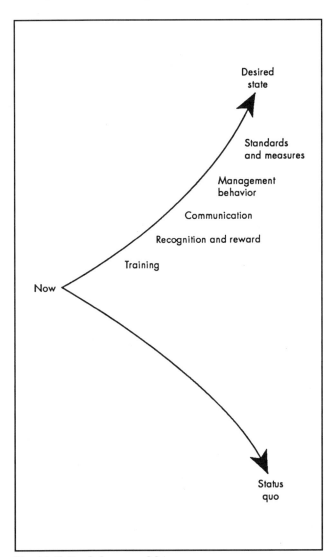

Fig. 5-8 Cultural change model.

CHAPTER 5

CONTINUOUS IMPROVEMENT USER GROUPS

Communication up, down, and across the organization is another key to the effectiveness of a continuous improvement, total quality strategy. If employees are to be effectively empowered to improve quality and improve processes continuously, they need to have access to timely, relevant, accurate information. Employees cannot be expected to make sound decisions if they are not properly informed. Communication must occur not only top-down, but horizontally and bottom-up. Management must be properly informed about issues and problems which are occurring at all levels of the organization. Horizontal communication is important because it often involves internal supplier-customer relationships, for which communication and feedback are critical to quality.

Communication should also be a cornerstone of any change process. The new culture should be communicated via formal and informal vehicles. Formal vehicles include announcements, pronouncements, memos, meetings, and other explicit communications. Informal vehicles include rituals, stories, metaphors, and other symbolic forms of communication. A comprehensive communications strategy must ensure that all available communication vehicles are being used to communicate the new beliefs and values, celebrate successes, identify problems, and track progress.

MANAGEMENT BEHAVIOR

Many organizations have tended to promote people in the organization who demonstrate technical expertise. They are high performers as individual contributors, so they will make good managers, right? Maybe. Technical competence does not assure good "people skills." At the same time, many organizations do little to prepare managers or provide ongoing support so they can acquire effective management skills. This is unfortunate because management behavior is the single most important determinant of an organization's culture. This may appear to be a contradiction because culture is defined as the set of shared beliefs, attitudes, and values which influence behavior. Yet, management behavior tells people in the organization which beliefs, attitudes, and values are important. That is why management behavior is such an important cultural element. Consequently, management must lead the change effort, not by words alone, but by actions as well. Total quality is not something that "the top tells the middle to do to the bottom." It is a process led by the top and cascaded throughout the organization. An important part of any change strategy, then, is a definition of the explicit roles and responsibilities of managers in the implementation of a culture reflecting continuous improvement. An understanding that these roles and responsibilities are expected behaviors for managers must also be reflected in training programs, communications, and approaches to recognition and reward. In fact, many organizations have developed survey instruments that provide managers with feedback from the people they manage as to how well their management style meets the stated roles and responsibilities. These are used to identify development opportunities for managers. The ultimate goal is that all managers are driving or facilitating the change process, and in fact many organizations send a strong message that in the long run they will not settle for anything less.

STANDARDS AND MEASURES

The previously cited axiom of "what gets rewarded, gets done" can be augmented with "what gets measured, gets done." In fact, organizations tend to measure performance that gets rewarded, and many companies, while claiming to assign a high priority to quality, do very little measuring and tracking of quality-specific performance. What does get measured relative to quality comes out of the quality control or quality assurance function, and is primarily focused on defect levels for products and services. This information is often not tracked over time to assess trends, and is not reviewed regularly by senior executives, with the possible exception of heads of operations. What does get measured, tracked, and reviewed regularly by senior executives are numerous indices of the financial health of the organization. There often is a total lack of data which indicates the "quality health" of the organization. If it is readily accepted that quality is "meeting customer requirements," then quality measures need to be developed for all areas which are critical to customer satisfaction. This is true for internal as well as external customers. Measurable standards of performance have to be defined for each of these areas, so that quality is quantified for employees. Next, improvement targets and objectives must be established. One common criticism of company-wide cultural change efforts is that they lack clear direction. Targets provide direction; they give definition to the change effort and provide standards against which progress can be measured. This is especially important to continuous improvement efforts—setting measurable targets which represent ongoing improvement. Knowledge of customer requirements, historical performance, process capability, and competitive position is important to goal setting. Finally, and in keeping with the previous discussion on communication, performance measurements in areas critical to customers and progress against targets should be disseminated and reviewed throughout the organization. The organization as a whole must be focused on improving performance and meeting goals so that all will participate in enhancing progress.

SUMMARY

While improving quality is a critical imperative that all organizations face if they are to survive into the next century, many organizations' attempts to bring about a lasting cultural change, focused on continuous improvement, have floundered. This section has examined why some of these attempts have failed by identifying the cultural elements which are important to perpetuating an existing culture and are key to effecting real change. Finally, a model for change has been presented. While not without risks, it does present a clear-cut approach with a proven track record.

CONTINUOUS IMPROVEMENT USER GROUPS

Continuous improvement user groups (CIUG) are a new forum for groups of small manufacturers (40 to 200 employees) to see and learn practical methods of implementing continuous improvement techniques—by working directly in manufacturing shops.

Two CIUG organizers (Rick Fleming and Mike Rother) form groups of six small manufacturers that meet at one another's plants for a full day every six weeks.

The host site for CIUG meetings rotates among the group's

CHAPTER 5
CONTINUOUS IMPROVEMENT USER GROUPS

members. At each meeting every member company has three representatives, from the shop floor and from management. Participants share experiences and ideas, get input from experienced facilitators, work on a specific focus topic (using the host company's shop floor as an example), and track progress on improvement projects under way at each member's home plant. Figure 5-9 lists the typical CIUG focus topics a member would see in the first year. Figure 5-10 illustrates a typical storyboard.

Employee involvement and team building are topics that are addressed at every CIUG meeting, in conjunction with each focus topic, to ensure that members understand the key role these topics play within continuous improvement. Over time, the group builds up a basic understanding of a model for continuous improvement. The individual technologies, tools, and techniques that are promoted by hundreds of seminars, books, articles, consultants, and equipment vendors can certainly be useful—*if* one has a clear picture of how they fit into the overall scheme of an effective manufacturing operation. The model of effective practice in continuous improvement provides a picture of pieces that make up CI, and how they fit together.

The focus throughout is on hands-on implementation, not on theory. The groups have two main goals: to foster management commitment to continuous improvement, and to begin implementing a continuous improvement philosophy grounded in quality improvement, just-in-time manufacturing, total productive maintenance, and employee involvement. Membership is on a fee basis, which supports the involvement of experienced facilitators who prepare and run the meetings.

CIUGs may also go on field trips to visit an exemplary manufacturer and see the effects of CI firsthand. This has proved to be an eye-opening experience for many group members.

CIUGs provide a forum for small manufacturers to learn and discuss specific issues in CI, using a real-world environment (the shop floor) to apply new skills and knowledge. CIUGs also offer ongoing feedback and guidance. Because they involve multiple firms (which may or may not be in the same industry), CIUGs concentrate on elements basic to the improvement process, not on techniques that are unique to specific companies. That is not a problem; group members find that 80% of continuous improvement deals with basic processes, while only 20% involves proprietary techniques. Figure 5-11 lists various reasons why CIUGs are unique.

WHY IS CI IMPORTANT FOR SMALL FIRMS?

The philosophy and practice of continuous improvement are important to small firms because of the low implementation cost and high return on investment. Small firms usually lack the cash flow or capital to make large investments in new technology. This is a blessing in disguise. Continuous improvement efforts free up capital by eliminating waste or non-value-added activity in their operations and give them productivity gains with minimal investment.

In their search for higher performance levels, firms often turn to technology fixes or other "magic dust" solutions. Unfortunately, these quick fixes often mean that automation is spread over complex and inefficient processes that should have been cleaned up first. The exceptionally high (and costly) failure rate for Manufacturing Resource Planning systems is a good example. Continuous improvement efforts lay the much-needed groundwork for successful implementation of new productivity-enhancing technology. Streamlining production processes through continu-

---

Topic 1: Finding waste.

Topic 2: Housekeeping and workplace organization (3-S).

Topic 3: Systems improvement and policy deployment.

Topic 4: Problem solving and improvement stories.

Other topics
- Improving flow.
- Reducing setup time.
- Quality at the source.
- Ingredients for effective teams.
- Machine maintenance.

Fig. 5-9 Typical first year CIUG focus topics.

---

ous improvement eases future technology implementation by reducing process complexity and increasing employee knowledge of the manufacturing operation.

Small companies have a distinct advantage in implementing continuous improvement due to their size. Most have a family-oriented atmosphere and direct communication channels, both of which are vital to effective continuous improvement efforts. Small firms also benefit from the discipline that continuous improvement requires, which helps companies avoid the "sawtooth" effect of making improvements and then later reverting to old habits. An example of this "sawtooth" effect is a manufacturing company that reduced setups to an average of 30 minutes, but three months later reverted to two hours. The problem? A failure to establish easy-to-follow setup procedures and reduce lot sizes. As with any company, small firms desperately need to tap all the capabilities of their people and limited resources, and maintain a strong customer focus. Continuous improvement helps in both these areas.

BENEFITS

The CIUG format fosters a close-knit team approach to productivity improvement. Members learn how to work as teams, assess opportunities, and identify and implement improvements. Lecture or seminar-style training becomes tedious after a short while, reducing the effectiveness of the learning experience for participants. The CIUG overcomes this drawback by making use of different training methods—lectures, discussions, and hands-on projects—retaining the best of each approach.

Continuous improvement requires a change in the way companies operate. The cultural change that must take place to achieve substantial improvements is very difficult to implement. Patience, support, recognition, and constancy of purpose are some of the key words. CIUGs smooth this difficult cultural change by involving both management and shop floor people from each member company. CIUGs emphasize the long-term nature of continuous improvement techniques, and foster a strong commitment through year-long membership. At the end of one year, member firms can decide to continue for another year, perhaps with even greater emphasis on implementation.

The CIUG serves as an effective forum to provide continual guidance through the winding road of CI and to gain experienced advice on how to avoid middle-of-the-road performance. As one visitor from Texas remarked at a CIUG meeting, "The only thing in the middle of the road is yellow paint and dead armadillos."

Managers are learning that CI is an ongoing process—a new way to run a company—instead of a program that can be turned on like a switch. Members are moving away from a preference for

CHAPTER 5

CONTINUOUS IMPROVEMENT USER GROUPS

CIUG Improvement Story
Theme: _____

Team Information	1. The Problem (Plan) How we measure it: Improvement goal: By when:
2. Current Situation (Plan)	3. Analysis of Causes (Plan)
4. Solutions (Do)	5. Results (Check)
6. Standardization (Act)	7. Future Improvement Plans
8. Celebrate Successes	

Fig. 5-10 An improvement "storyboard" used by CIUG members to present their progress on improvement projects.

CHAPTER 5
CONTINUOUS IMPROVEMENT USER GROUPS

> **Why CIUGs Are Unique**
>
> - They are designed for small manufacturers.
> - Membership requires a year-long commitment, with one-day meetings occurring every six weeks.
> - Both shop and management employees are involved. For continuity, the same employees come to every meeting.
> - Meetings are held at member companies, and members apply lessons learned directly in the shop.
> - The focus is on implementation, not theory.
> - Members get personalized input from experienced facilitators.
> - Project tracking stimulates improvement efforts and helps maintain progress.
> - Members and facilitators share experiences and information.

Fig. 5-11 Why continuous improvement user groups are unique.

"magic dust" technology fixes, and are making improvements in the basic efficiency of their processes.

Companies that had a pattern of reverting to the old way of doing things are now maintaining and charting the improvement gains. Members learn to Plan, Do, Check the results and then Act to refine or further improve what they did; instead of the haphazard Plan-Do-Plan-Do cycle that does not build upon past experience.

Member companies have become sensitized to waste, and are able to recognize non-value-added activity in their operations. Finally, while "employee involvement" has become something of a cliche, CIUG members are recognizing the need to listen to and support the ideas of their employees.

CIUGs IN ACTION

Early on the morning of December 7, 1990, sixteen people from seven different manufacturing firms across southeastern Michigan met for the second time to discuss the progress of their CIUG.

The group gathered in a small meeting area created out of extra space in the engineering department of the host company. Over coffee and donuts, they chatted about employee involvement issues back at their respective home plants. The value of the last meeting's topic, eliminating waste, was discussed and stories exchanged on how each had used the information to solve some of their own company's recurring problems. Several people related how the project tracking portion of the meeting provided the motivation required to successfully move forward on improvement projects. The group settled into their seats as the facilitators focused on one of the fundamental topics in continuous improvement: housekeeping and workplace organization.

After a presentation and discussion on this focus topic, the group was split into three teams, and headed for a tour of the host-company's shop floor. Each team analyzed shop floor operations, to discover good practices and uncover opportunities for improvement—all related to the focus topic for that day.

Each team then compiled their findings and presented them to the host stamping plant. The result: identification of 16 good practice ideas that everyone could apply at their home plants, and over 40 suggestions for improvements in workplace organization. The suggestions covered several areas including layout and location of die storage racks and coils, safety, visual control, posting visible data, quality improvement, and employee involvement.

In the days after the visit, the stamping company took the list of improvement suggestions, prioritized it, and began implementation. Die racks were moved, dies were color-coded, coil storage changed, staging areas added, buzzers and lights installed, digital read-outs for presses established, several areas of the shop cleaned and reorganized, and problem-solving teams were formed.

The result? Setup times were reduced by 40% on one press. A new cell arrangement resulted in greater throughput with higher quality. Employee morale was enhanced through a cleaner and more organized shop that the employees themselves designed. More importantly, a basic continuous improvement process had been established. All with minimal cost to the company.

References

1. Robert Kaplan, "Must CIM be Justified by Faith Alone?" *Harvard Business Review* (March-April, 1986), p. 87.

Bibliography

Berry, Thomas H. *Managing the Total Quality Transformation.* New York, NY: McGraw-Hill, 1990.

Brocka, Bruce, and M. Suzanne Brocka. *Quality Management: Implementing the Best Ideas of the Masters.* Homewood, IL: Richard D. Irwin, 1992.

Hunt, Daniel V. *Quality in America: How to Implement a Competitive Quality Program.* Homewood, IL: Richard D. Irwin, 1991.

Patterson, Douglas O. "Saying Is One Thing, Doing Is Another!" *Journal of the Institute of Environmental Sciences.* January/February 1991, pp. 17-20.

Scholtes, Peter R., and Hacquebord, Heero. "Six Strategies for Beginning the Quality Transformation, Part II." *Quality Progress.* August 1988.

Sheehy, Barry. "Hitting the Wall: How to Survive Your Quality Program's First Crisis." *National Productivity Review*, Vol. 9, no. 3. Summer 1990, pp. 329-35.

Weaver, Charles N. *Total Quality Management: A Step-by-Step Guide to Implementation.* Milwaukee, WI: ASQC Quality Press, 1991.

CHAPTER 6

SUPPLIER INVOLVEMENT IN CONTINUOUS PROCESS IMPROVEMENT EFFORTS

CHAPTER CONTENTS:

SUPPLIER CERTIFICATION 6-1

QUALITY FUNCTION DEPLOYMENT 6-7

BENCHMARKING 6-7

Continuous process improvement starts with supplier involvement. Process inputs (raw materials, information, and other resources and services required before an organization can perform the in-process tasks or operations) often represent more than half of the manufacturing cost[1] and are responsible for a significant amount of variation in the quality characteristics of the process outputs. The inputs usually are obtained from outside suppliers and can be 50 to 70% of product cost. (generally, this chapter addresses outside or external suppliers, as opposed to those internal to the organization; however, the concepts and ideas can be extended to include internal suppliers as well). For a company to be competitive, these costs must be controlled and reduced. Dr. W. Edwards Deming's Point Four,[2] "End the practice of awarding business on the basis of price tag," provides the philosophical foundation for involving suppliers in an organization's continuous process improvement efforts.

What is required first is a method which provides the customer a uniform assessment or evaluation of the quality and process capability of each supplier. This is usually called "supplier certification." Second, the customer's requirements must be clearly defined and communicated to the supplier(s); the method of determining these customer requirements and expectations has become known as "quality function deployment." The third relevant method is "benchmarking"—obtaining information about the processes and products of competitors and best-in-class ("world-class") organizations—which provides a target for excellence and identifies opportunities for technology transfer by allowing the emulation of high quality processes. The focus of this chapter is on the first of these three methods; the other two are briefly discussed. Figure 6-1 provides an overview of the general certification process.

SUPPLIER CERTIFICATION

The importance of each supplier's involvement is directly related to the criticality of the impact of the supplied product or service to the variation of the output product or service. (Note: for brevity, "product" will be used throughout the rest of this chapter in place of "product or service.") Consideration must be given to all controlling functions that will ensure final product compliance to the customer's specifications.

Effective supplier involvement requires establishing a customer-supplier partnership. This partnership begins with establishing a positive, cooperative, trusting environment with effective communication channels between the two organizations. In the partnership, the supplier has the responsibility for continuous process improvement, resulting in improved quality of supplied products to the customer.

Achieving proven supplier process capability (often documented by awarding the supplier a "preferred status" or "certification") allows the customer to eliminate incoming inspection, since defects are minimized and rarely produced. Audit requirements, however, must still be in place, to provide assurance to both the customer and supplier that the supplier's process maintains its capability, and to measure continuing improvement effectiveness.

The quality management system used by the supplier plays a critical role in documenting the quality of the supplier's products and processes. The customer's requirements must be reflected at all stages of the process, from guaranteeing good designs to ensuring quality production—leading to reliable performance, prompt delivery, and efficient service.

The Contributors of this chapter are: **Dr. Mary A. Hartz**, *Statistical Advisor and Total Quality Management Group Leader, IIT Research Institute;* **Ted T. Crosier**, *Total Quality Management Senior Management Specialist, IIT Research Institute.*

The Reviewers of this chapter are: **Dr. Sarah Brooks**, *Professor Emeritus, Mathematics and Computer Science, Mohawk Valley Community College;* **John Farrell**, *Senior Research Engineer, IIT Research Institute;* **Steven J. Flint**, *Director of Research, IIT Research Institute;* **Robert Giammaria**, *Director, Manufacturing Support Services, New York State Industrial Technology Service of the Mohawk Valley Applied Technology Commission;* **Paul MacEnroe**, *Executive Director, Mohawk Valley Applied Technology Commission.*

CHAPTER 6

SUPPLIER CERTIFICATION

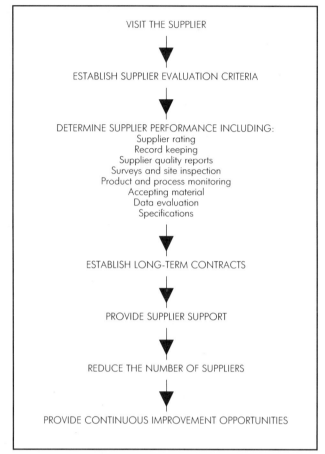

Fig. 6-1 General supplier certification process.

Supplier performance measurements need to be defined; these metrics are often based on delivery, quality, and total life-cycle cost because contracts are no longer awarded on lowest unit cost. The best *value* is sought; consideration is given to nonconformance costs due to poor quality as well as to quantity errors, missed delivery schedules (both too early which requires costly storage and too late which translates to missed production), transportation costs, and other customer concerns such as reliability, availability, maintainability, testability, and so on. Single point-in-time unit cost data are not sufficient; trends over time are examined to assess the vendor's ability to maintain specified values as well as to document improvements resulting in lower life-cycle cost. Rockwell, for example, uses both supplier quality performance and delivery performance combined in a single dollarized measure to reflect the cost of supplier nonquality.[3]

The resultant long-term partnership requires a mutual commitment to continuous process improvement, which is advantageous to both parties. The customer is assured of quality products, on time, and at a cost that represents value, while the supplier reduces costs, improves his or her competitive position, and receives a long-term commitment (contract) from the customer. Major savings occur from the identification of non-value added process steps which can be totally eliminated from the supplier's processes. Since both parties benefit (a "win-win" situation), the partnership is strengthened. The suppliers that survive the 1990s will be those that design their entire organization around satisfying the customer. As an example, specific results achieved by Westinghouse (as documented in its winning application for the Malcolm Baldrige National Quality Award) from its Supplier Rating System include:

- Reducing incoming inspection by 70%.
- Improving delivery schedules by 50%.
- Reducing delivery cycle time by 30%.
- Improving part quality levels by 30%.
- Reducing total costs by 25%.
- Reducing inventory by 25%.

Supplier certification can be organization-specific, as defined by the requirements of the Ford Motor Company's Q1 Preferred Quality Supplier Program; or the certification requirements can be applicable to a wider audience, such as those used by the United States Department of Defense (through documents[4] specified as part of the procurement such as MIL-Q-9858). These certification requirements define the minimally acceptable content and attributes of the supplier's quality system(s) and offer a consistent, structured, uniform, proven approach to assure that the supplier's product meets the specified customer needs.

Supplier process certification, rather than product certification, provides additional benefits and cost savings. For example, the United States Government currently is taking action to supersede the Department of Defense General Specification for Microcircuits (MIL-M-3851Q) that uses Qualified Products Lists (QPL) to identify parts acceptable for use in military systems. The replacement document, General Specification for Integrated Circuit (Microcircuit) Manufacturing (MIL-M-38535), is process (not product) based. It uses a Qualified Manufacturers List (QML) to identify approved vendors and technologies. This QML approach defines a procedure that certifies manufacturing processes, and not the individual certification/qualification of a product. Hence, all devices produced and tested using the QML certified/qualified technology flow are considered acceptable.

As the use of the quality system approach grew internationally, the need to establish a common basis was recognized. A supplier should not have to be audited and then certified (approved) by each customer in each country (the cost alone would be prohibitive). Instead, an internationally recognized equivalent certification (registration) process was developed. In the United States, the leadership is being provided by quality professionals primarily associated with the American Society for Quality Control (ASQC) in conjunction with the American National Standards Institute (ANSI), the official U.S. member of the International Organization for Standardization (ISO).[5] The ISO 9000 series has been endorsed without change as the national standard for supplier certification in over 90 countries; in the United States they are called the ANSI/ASQC Q90 series.[6] Auditors are now being certified to one of several levels using standardized criteria. The ISO 9000 standards only address the quality system, not the product. These standards are now being considered for use in the QML systems application.

Many consultants have developed various supplier rating systems. Numerous software packages are available to assist in the mechanics of tracking and evaluating the supplier information. Most of these are similar in that each includes steps related to planning, execution, and follow-up.

This overview of the general approach, without the details of any one specific product or endorsement of any specific quality expert, expands upon the general introduction to supplier certification, and provides specifics for each phase of the general certification process.

CHAPTER 6
SUPPLIER CERTIFICATION

VISITING THE SUPPLIER

The first step in obtaining supplier involvement is to make current suppliers aware of the customer's emphasis on obtaining quality product. This can be accomplished by holding a supplier's conference where the new emphasis is explained. The suppliers need to be told how they will be evaluated. The customer's expectations in terms of cost, delivery, packaging, testing, inspection, auditing, data, reports, process improvement, and corrective actions will be defined at this conference. A tour of the facility will give suppliers an understanding of how their material is used. Feedback is obtained from the suppliers regarding their understanding of the customer's requirements and specifications. Any required clarification is provided. Emphasis on the cost of poor quality product received from suppliers is important to gain supplier buy-in.

Often this group visit is followed by a preassessment visit to each supplier's facilities. Open meetings are conducted where representatives of both the customer and supplier discuss issues of mutual concern. An effort should be made to deal with vendors who are experienced in continuous process improvement; then the discussion phase, the associated consensus making, and, indeed, the entire certification task, will be much easier.

Individuals involved in certification maintain professional objectivity, are open to discussion and input from others, work well together as a team, and remain approachable throughout certification. Often the involved individuals are already part of in-house teams. For example, purchasing material, a process in itself, is a team effort. The "purchasing material" certification team (probably including members from procurement, supply, engineering, and the end-user) will need appropriate training in the philosophy and tools of TQM.[7] When this training is lacking or when suitable people are not available, problems often result. Often it is for these "people" reasons that some organizations hire outside, neutral third party organizations to perform the audits and establish the certification of their suppliers.

ESTABLISHING SUPPLIER EVALUATION CRITERIA

Supplier evaluation criteria are often developed by an in-house (customer organization) team that includes, as a minimum, the buyer, a technical expert, and one end user. This team develops quality improvement goals, quality audit procedures, corrective action procedures, cost standards (cost per lot, life-cycle cost, etc.), and delivery standards (percent on time, days late, etc.), expressing them with operational definitions. The ultimate standard is zero defects, minimum cost, and 100% on-time delivery. The criteria will vary widely by material complexity. For more complex material there may be a requirement for submitting a sample for evaluation. Additional "pre-control samples" may be required for a short pre-control run, before long-term production is approved.

A broader approach would establish a long-term partnership with each supplier. This requires both communication and cooperation. A cross-organization team would then work to develop the supplier evaluation criteria. The supplier is responsible for developing processes which have the ability to detect and, ideally, correct processing conditions or process output criteria that exceed specified statistical process control criteria. (These criteria are established by the supplier, by studying the process and documenting process stability and capability characteristics.) Placing the responsibility for quality and lack-of-quality detection on the supplier organization addresses the complex but important issue of supplier accountability. The cross-organization team also can address surveillance and noncompliance penalty issues. Shared responsibility implies less control—something that management may not be ready to accept. Usually both organizations include a feedback/surveillance/verification loop.

DETERMINING SUPPLIER PERFORMANCE

Supplier certification establishes a formal agreement between customer and supplier regarding quality, cost, and delivery requirements; the customer eliminates incoming testing and inspection and relies on the supplier's systems and procedures to deliver a product which meets specifications.

Certification requires extensive planning and a strong partnership between the customer and the supplier. The required, detailed product and process audits replace inspection. Supplier documentation is very important. Certification is normally reserved for those suppliers with an excellent quality history and a demonstrated willingness to apply continuous process improvement to their processes.

An important consideration in the certification process is the priority assigned to quality by the supplier's senior management. Their quality philosophy should be investigated to see if it is consistent with TQM. Corporate quality history and quality assurance systems, including the placement of the quality organization, must be considered.

Certification is reserved for top-notch suppliers. For example, it is Ford Motor Company's policy to deal with suppliers that continually strive to increase quality and productivity. To recognize outstanding quality suppliers, Ford developed the Q1 Preferred Quality Supplier Program. Ford's criteria are shown in Fig. 6-2. Ishikawa provides the guidelines shown in Fig. 6-3 for selecting a certified supplier. Others, such as Delco Electronics and Chrysler Motors, have similarly defined certified supplier programs.

Motorola is another organization which has defined requirements for supplier quality assurance, using ANSI Z1.8 (ASQC Standard C-1) as the basis.[10] The objective is to continually improve the quality of Motorola's products and services with the goal of achieving zero defects and the highest possible customer

---

The nomination of criteria and the factors in Ford's supplier quality rating system include:

- Adequacy of the supplier's system to control product quality including the use of statistical process control on the selected product characteristics or process parameters that are significant to part function, fit, durability, or appearance (for appearance items only).

- High success rate on initial sample and first production shipment evaluations.

- Outstanding ongoing quality performance at the using location.

- An absence of significant field problems.
- The commitment and active support of the supplier's management to pursuing never-ending improvement in quality.

- The ability and willingness to conduct manufacturing feasibility studies during the design process for parts within the supplier's area of expertise.

Ford Product Engineering, Purchasing, and Supplier Quality Assurance must agree that a supplier location meets these criteria for a commodity class. This supplier location will then be designated as a Q1 Preferred Quality Supplier. The Q1 designation is awarded separately by each division of Ford Motor Company.

Fig. 6-2 Q1 Preferred Quality Supplier Program Criteria.[8]

CHAPTER 6

SUPPLIER CERTIFICATION

> 1. The supplier knows the management philosophy of the pruchaser and continuously and actively maintains contact with the purchaser. He or she is also cooperative.
> 2. The supplier has a stable management system that is well respected by others.
> 3. The supplier maintains high technical standards and has the capability of dealing with future technologies.
> 4. The supplier can supply precisely those raw materials and parts required by the purchaser, and these meet the latter's quality specifications. The supplier also possesses process capabilities for that purpose or has the ability to enhance such process capabilities.
> 5. The supplier has the ability to control the amount of production or has the ability to invest in such a way to ensure its ability to meet the amount of production.
> 6. There is no danger of the supplier breaching corporate secrets.
> 7. The price is right and the date of delivery is met precisely. In addition, the supplier is easily accessible in terms of transportation and communication.
> 8. The supplier is sincere in implementing contract provisions.

Fig. 6-3 Ishikawa's Supplier Certification guidelines.[9]

satisfaction. Supplier certification may be granted on a selected part number basis to those suppliers who demonstrate the ability to meet the prerequisites and requirements. The primary objective of certification is to ensure that purchased material fully conforms to specification upon receipt and is ready for use without incoming inspection. It is Motorola's intent to establish a long-term partnership relationship with "best-in-class" suppliers for current and new products.

The major issues in determining supplier performance are further discussed in the supplier certification documentation of Ford, Chrysler, and other individual certification programs, as well as more generally by the American Society for Quality Control, (ASQC), in its book, *Procurement Quality Control*. The ASQC offers several books on performing quality audits (see Bibliography).

Supplier Ratings

Suppliers are rated on their ability to demonstrate trends in performance and to identify any necessary corrective action. A supplier information system must be established and performance measures must be defined. These measures typically include quality, delivery, cost, service, and compliance. The quality measure may consist of percent defectives (items or lots), internal losses (scrap, rework, delays) or external failures. Defects may be measured in parts per million. Delivery can consider percent on time, percent early or late, and percent over or under order. Cost may be based on the customer's cost-of-quality system. The rating can also include subjective elements such as cooperation, technical assistance, responsiveness to problems, compliance to instructions, and customer service.

The rating system identifies opportunities for improvement which become the focus of the supplier's continuous process improvement effort. The measurement criteria also play an important part in the supplier's training program. Most ratings systems use a weighted average of these measures.

Any rating system, however, may have potential problems. ASQC[11] identifies some of these:

- Lack of documentation.
- Cluttered reports.
- Inadequate database management.
- Lack of timeliness.
- Lack of supplier discrimination.
- Operational problems.
- Lack of management support.

Proper planning and coordination meetings will help overcome many of these potential problems. Effective rating systems are:

- Efficient, in terms of cost/benefit ratio.
- Documented, to support consistency and ensure fairness.
- Discriminating, so that good suppliers are rewarded.
- Timely, so gains are achieved immediately.
- Enforced, so that suppliers know that the customer means business.

Record Keeping

The importance of a documented, proactive system (sometimes called the "documentation pyramid") cannot be overstated. The customer must define for the supplier the data, reports, and quality system information that must be provided. The content, format, frequency, and use of each report must be specified. Data required may include variables-type information (such as physical measurements), attributes-type information (such as counts or rates), configuration management information (such as plot traceability) and/or certificates of analysis from independent testing laboratories.

Documentation deficiencies are a common reason for failing to become a certified supplier. Rapidly improving computerized tools have become available to assist organizations with this large task. Data definition, responsibility for collecting and recording, data integrity, and security of information all are topics to be considered when establishing a record keeping system. People across the supplier's organization must have timely access to, retrieval, and use of the large volume of information collected, in an efficient, cost effective manner. A methodology for rapidly enhancing the data collection system must also be in place. The computerized data system should focus on increasing the productivity of the supplier's organization while providing speed and ease of analysis for decision making.

Care must be taken not to establish a record keeping system just for its own sake. Data collected without subsequent analysis are meaningless. Decisions based upon collected data are valuable, while useless data collection and documentation can be very costly.

Supplier Quality Reports

Many organizations require that the identified information (as displayed using SPC charts, for example) accompany material delivery. Reports may be required which demonstrate statistical evidence of process control or material acceptability. A monthly report from suppliers on their continuous process improvement efforts is recommended for tracking progress.

Supplier rating report formats need to be flexible to meet the specific customer's needs; often the format includes several summary levels. The format should be planned with supplier input. Consideration should be given to using reporting by exception—that is, only reporting lots of material with problems.

Feedback is required if the supplier is expected to react to the customer's analysis of the delivered material and any problems it is causing. This information feedback must be accurate, clear, prompt, and consistent, and it must go to those who can initiate corrective action. The use of cost-of-quality data, including the dollar impact on the organization, is informative and provides an

CHAPTER 6
SUPPLIER CERTIFICATION

effective method of communicating the results of poor quality. The objective is to focus on areas requiring corrective action that will yield the highest payback potential.

Surveys and Site Inspection

Customer surveys and site inspections will determine if the supplier is effectively accomplishing the intended improvement effort. Further conformance to standards, such as ANSI/ASQC Standard 90-94, can be documented. A team composed of the buyer, technical expert, and quality assurance specialist is responsible for conducting the site survey and/or inspection. Both are usually required of all potential suppliers. Characteristics such as objective listening and questioning techniques are taught to the team as well as methods to help overcome aggressive or maladaptive behavior.

The potential supplier's business and technical capability should be reviewed during the site survey. Some site surveys are done using a checklist that can be completed in a relatively short time. However, enough time should be spent looking at actual evidence of process control and process capability — control charts and process capability studies — to help prepare for the site inspection. Effort is required to review the process control system: the training, discipline, equipment and tooling, staffing, control charts — in fact, everything being used to assure error-free output.

Evaluation techniques for the more detailed site inspection include questionnaires, checklists, survey forms, and rating systems. Observations of normal operations are preferred. Decisions and recommendations should be based on data including actual observations. The emphasis is on reviewing the process control system. Actual control charts should be reviewed with the operators at the process being controlled. Corrective action should be followed for timeliness and effectiveness.

Historical data on current suppliers should be reviewed prior to initiating the site inspection. An analysis of previous corrective action taken (timeliness and effectiveness) is the most important part of the review. The "Whys" behind both good and bad findings need to be investigated.

Routine issues can be handled at monthly meetings, usually held at the supplier's facility. Joint agendas should be exchanged before these meetings. Problems must be clearly identified, understood, and resolved as soon as they occur. Agreed-upon corrective action must be documented with milestones. The objective, as before, is to establish trust and confidence.

Product and Process Monitoring

Supplier products can be examined (audited) at either the customer's or the supplier's facility. Surveillance visits to audit either (or both) products or processes are typically scheduled at regular intervals; acceptable results are required for continued certification. Larger scale audits are typically scheduled at longer intervals but are more comprehensive in nature. Total reassessments typically occur every three years.

The objective of these audits is to assure that defects are not produced. The emphasis is on reviewing those procedures that are in place to prevent defects and to continuously improve processes. One way to identify these procedures is to review the use of the "usual" quality improvement tools, such as flow charts, checklists, and run charts. Reviewing the use of flow charts, for example, can lead to the following types of questions:

- Have flowcharts been developed, both at the "big picture" level and at the lower, more detailed levels of operation?
- Are these flow charts readily available in the workplace for reference?
- Has the workforce been trained to use the flow chart?
- Can selected employees explain it to the outside auditor?
- Is the flow chart used when new employees are trained in the specific process operation?
- Do employees and management consult the flow chart when an issue regarding process operating consistency arises?
- Does the actual process conform to that documented on the flow chart, or have individuals invented "work around" ways to by-pass the standard process flow?
- Do the standard operating procedures and related work instructions exactly mirror the actual process flow chart?

Other questions might focus on the documentation system. For example:

- Are the quality manuals, calibration instructions, etc., available for operator use?
- Are they used?
- Is the process control system well documented?

More fundamental questions which take a broader look include:

- Does the supplier really understand the customer's requirements?
- Does the supplier know the difference between process control (everything required to achieve a state of statistical control) and statistical process control (tools and techniques used to monitor the process)?
- Are corrective actions timely and effective?
- Is root cause investigation methodology used to discover the real reason for process variation?

Product monitoring is required of noncertified suppliers and may include routine and special testing as well as product audits. Routine testing consists of receiving inspection and first article inspection. Special testing is required when conformance to specifications cannot be assured through routine inspection methods. The identification of nonconforming material is the objective of product monitoring. The procedure to follow when nonconforming material is found must be understood by both parties.

Product audits simply mean reinspecting samples of the product. The purpose is to establish the acceptability of the product, not to focus on the supplier's inspection system (although the latter is an additional benefit). Product audits may encompass elaborate, formal reviews and laboratory testing. When performed on-site, time should be spent reviewing the quality system and procedures as well as inspecting samples of material.

Process audits determine the conformity of the design, specifications, and other documents to the requirements. The supplier is given the audit plan which explains how the process audit is to be conducted. A process audit examines the details of process control as part of the manufacturing cycle. Records of process parameters are evaluated for out-of-control situations and corresponding corrective action. Weaknesses in record keeping should be recorded. The absence of safeguards is significant. Specific controls for the process under review are of special interest. Summary findings should be reviewed with the supplier. Process improvement action may be required prior to certification. Continued surveillance will be scheduled following certification to assure the maintenance of the supplier's quality system.

Accepting Material

Each supplier must know how his or her products (or services) will be measured and evaluated. The initial inspection plan and all

CHAPTER 6

SUPPLIER CERTIFICATION

product/process audits must be defined. Communications play a key part. When an inspection plan is used, it must be able to identify problems. It should describe the method that the organization will follow to help the supplier with improvements. A standard sampling plan, as opposed to an *ad hoc* approach, should be used. However, the limitation of the plan (since it is based on an "acceptable quality level" such as specified in MIL-STD-105D) should be recognized since it allows acceptance of a lot that has been found to contain defects. This philosophy is unacceptable with today's emphasis on quality. The long-term objective is to make incoming inspection unnecessary as a result of continuous process improvement.

The supplier needs to know the type of testing to be performed. Testing is generally of two types, functional (performance) and dimensional. The amount and kind of functional testing, whether dynamic and/or environmental, is a major consideration. The supplier will be interested in how the testing will be performed and the pass/fail criteria. Dimensional inspection results determine how well the material conforms to drawings and specifications. Before conducting dimensional testing, the organization needs to ensure that its measuring capability is known. Studies must be conducted on all gages, and capability studies must be performed on all equipment to be used for testing. The supplier will want to know the details of the test plan.

Properly based, realistic specifications are critical to success. Ishikawa[12] states, "Specifications must be determined statistically, after the companies engage in quality analysis and process analysis (including a survey of process capabilities) and consider their economic feasibility." He further points out the need for customer and supplier to constantly work toward improvement of the specifications.

Data Evaluation

All supplier data must be statistically evaluated. Variables, not attribute, data should be required whenever possible. Histograms and control charts may be required from the supplier as statistical evidence that the material meets requirements. The type of analysis required will depend on the circumstances. The statisticians from the customer and supplier will meet to make these decisions. The customer should require statistical evidence of quality with the delivery of the supplied material.

Specifications

Specifications commonly describe the product or service, its required characteristics, acceptance limits, and packaging requirements. Specifications should define what is required to satisfy the customer. They must provide mutual understanding between customer and supplier by answering the fundamental questions:

- Who is responsible for design control?
- Are the tolerances and specifications reasonable and within the supplier's capability?
- Are the test requirements adequate to guarantee form, fit, and function?
- Are specifications operationally defined?

Both objective and subjective requirements such as cleanliness must be clearly understood. Any misunderstandings must be clarified by the customer.

ESTABLISHING LONG-TERM CONTRACTS

A long-term relationship must be established with the preferred suppliers. The basis for this relationship must be mutual trust and loyalty. Open communication drives out fear by eliminating suspicion. This long-term partnership will eventually lead to design teaming, where the supplier contributes to new product designs. This partnership with suppliers provides both organizations with the ability to forecast the quality, cost, and availability of future material. The long-term result will be better material at lower cost. Profits can increase for both partners.

For example, consider an end-user item which has a "fixed" final market price. The organization purchasing raw materials may negotiate automatic price reductions for these materials based on continuous process improvement results. As the cost of materials decreases, the profit increases for the purchasing organization. If the supplier is able to consistently provide high quality material, both the supplier and purchaser organizations can reduce testing, eliminate buffer stock, and move to a just-in-time inventory system, thereby reducing testing and inventory costs as well as reducing floor space requirements; both the purchaser and supplier organization improve profits. The supplier can negotiate a long-term contract with a favorable pricing structure. Further, a long-term relationship allows the supplier to take risks such as investments in technology (process automation), equipment (process monitoring), training, and facilities.

PROVIDING SUPPLIER SUPPORT

The management of suppliers is important to maintaining the flow of quality material. The motto to follow is "Our suppliers meet our expectations." Supplier support involves monitoring and responding to supplier problems and supplier improvement goals. An active role should be given to the procurement and supply functions. Set qualitative goals for both the suppliers and the procurement and supply functions. Develop a rating system and let the suppliers know how they will be measured. Improvement goals should be established and corrective action initiated when problems are found. A proactive program, with high expectations, is required. Suppliers must be held accountable, but the customer must also offer assistance. Nurture suppliers until they become certified. Nurturing may include helping to strengthen their quality management, sponsoring quality seminars, providing TQM training, and providing general guidance. Discussions about implementation planning, procedure preparation, recommended data and report formats, and workshops on analyzing data or on comparing data to requirements might be offered. It may take two to three years to develop a Certified Supplier.

REDUCING THE NUMBER OF SUPPLIERS

The goal is to have one supplier for each item of input material. When this is accomplished, further reductions in the number of suppliers can be achieved by combining the materials to be provided by the preferred suppliers. The number of suppliers should be reduced for several reasons. Foremost is the need to reduce variation. Multiple suppliers for the same material increase the possibility of variation. A proliferation of suppliers encourages buyers to jump from firm to firm shopping for price, not quality. Having one supplier for each item results in reduced material variation for that item. Fewer suppliers reduces contract administration time, allowing more time to be applied to continuous process improvement. There will be fewer contracts to support with fewer technical problems to solve. The number of procurement employees may even be reduced, lowering contract administration costs. Involving suppliers in TQM may increase the need to provide them with technical support; having fewer suppliers can reduce the demand on the organization's limited technical resources.

CHAPTER 6
BENCHMARKING

PROVIDING CONTINUOUS PROCESS IMPROVEMENT OPPORTUNITIES

Once the certification has been established (the processes have been documented to have a specific level of process stability and a specific level of process capability), these levels provide a baseline for continuous improvement. However, the improvement of supplier processes will require an intensive long-term partnership. Gap assessment and analysis will be required to enable the reduction of variation, with the resultant capability value increased. In a world of ever-tightening specifications, the challenge is to close the gap faster than the competition.

In addition to process improvements, "people" improvements are required. Training and education are essential in all aspects of the work improvement process. Both group and individual follow-up workshops are required. This balance of emphasis and resources between process and people improvement requires careful planning as a result of business strategy definition and development. Leadership by the senior management team is essential here, and throughout the certification process, to define the priorities of the organization in support of this strategic plan.

As with all aspects of process improvement, certification is, itself, a process. The certifying organization has the responsibility to evaluate the certification process at timely points along the process. Root cause analysis of corrective action situations is required. Change must be documented, and potential benefits verified against actual improvements in the certification process. The ISO 9000 series, for example, was reviewed in 1992 with minor refinements and upgrades; planning is under way for the 1996 upgrade.

SUMMARIZING THE SUPPLIER CERTIFICATION PROCESS

In the past, suppliers often were selected strictly on a price basis. If the unit purchase price measure included total life cycle cost (including the cost of quality), then using the lowest bidder could be the ideal approach. However, since unit purchase price is usually the only measure, low bidders typically are not cheap; they usually increase total cost by decreasing the quality of the material they supply and by increasing rework, scrap, and inventory. Schedule delays may also result.

Suppliers can also be selected based on past history with similar supplies, test results with similar supplies, or published experience of other customers. If at all possible, a supplier that specializes (in the particular material being purchased) should be selected. From the selected suppliers, those with a record of the highest quality will be considered for certification.

Establishing long-term partnerships with a small number of suppliers reduces variation, improves quality, reduces costs, and builds a sound technical team. As a result of the supplier certification process, both the supplier and the certifying organization will deliver higher quality products with greater consistency and efficiency.

QUALITY FUNCTION DEPLOYMENT

Quality function deployment (QFD) is a methodology and set of tools which enables the user to determine customer needs and then communicate this information throughout the user's organization. QFD focuses on designing quality into the products at the start, rather than by changing the process once it is in place.

Benefits of using the QFD approach usually include:

- Establishment of a common purpose across the organization.
- Development of a common language across the organization.
- Earlier determination of key product characteristics.
- Documentation of actual customers' needs rather than decisions based on opinions.
- Reduction in product development costs.
- Reduction in time required to bring a new product to market.
- Greater customer satisfaction due to lower costs and improved responsiveness.
- Increased productivity and market share.
- Reduction in number of engineering changes across the product's life cycle.

BENCHMARKING

Benchmarking is the method of continually comparing processes against the best in the industry (direct competitors) and the best in class (function). The knowledge of these "best" processes is used to set goals and improve the organization's processes.

Benchmarking allows the organization to break from the arbitrary, self-limiting improvement goals set in the past. The new knowledge of best processes is used to establish more challenging improvement goals.

Benchmarking is the link between planning for and achieving continuous process improvement. Benchmarking answers the questions:

- What are the critical success factors?
- Who are the best (competitors or by class)?
- How much better are they?
- Why are they better?
- How can the company improve to compete and excel?

Benchmarking helps create a culture that values and expects continuous process improvement. It increases knowledge about the competition's abilities and customer expectations. Benchmarking helps focus resources on the highest priority, largest return on investment process improvement efforts. It is critical to the long-term success of the organization.

CHAPTER 6

BENCHMARKING

References

1. As reported by engineers from Ingersoll in a 1982 presentation.
2. Dr. W. Edwards Deming's Point Four refers to Deming's Fourteen Points, a set of management obligations which cannot be delegated. These are presented in his book, *Out of the Crisis*, on pages 23-24, and explained throughout the rest of that manuscript.
3. This was discussed by William A. Kirsanoff, Rockwell International Quality Specialist, at the 1990 ASQC Quality Congress in San Francisco.
4. There are numerous documents published by the United States Government that present department defined methodology, specifications, and standards. Those from the Department of Defense are numbered "MIL-#-####." Others are issued by the Department of Energy (DOE), the Federal Aviation Administration (FAA), etc.
5. The ISO Committee on Quality Management and Quality was originally established in 1979 with 20 ISO member countries participating. In early 1993, 43 countries are active in working sessions and another 23 hold observer status.
6. For details see the ISO 9000 Handbook or Cottman's book, both published by the American Society for Quality Control.
7. A detailed discussion on recommended TQM training is discussed in *A Guide to Implementing Total Quality Management* by T. Crosier and M. A. Hartz.
8. Ford Motor Company, *Q1 Preferred Quality Supplier Program Criteria*, January 1986, p. 14.
9. Kaoru Ishikawa, *What is Total Quality Control? The Japanese Way* (Englewood Cliffs, NJ: Prentice-Hall, Inc., 1985), p. 162.
10. Motorola Purchased Material and Supplier Quality Certification Program.
11. An extensive definition of terms can be found in the *Glossary and Tables for Statistical Quality*, published by the American Society for Quality Control Statistics Division.
12. Ishikawa, ibid, p. 160.

Bibliography

American Society for Quality Control Statistics Division. *Glossary and Tables for Statistical Quality Control*. Second Edition, 1983.

Arter, Dennis R. *Quality Audits for Improved Performance*. Milwaukee, WI: American Society for Quality Control, 1989.

Bossert, James L., ed. *Procurement Quality Control*. Milwaukee, WI: American Society for Quality Control, 1988.

Bureau of Business Practice. *ISO 9000: Handbook of Quality Standards and Compliance*. Milwaukee, WI: American Society for Quality Control, 1992.

Cottman, Ronald J. *A Guidebook to ISO 9000 and ANSI/ASQC Q90*. Milwaukee, WI: American Society for Quality Control, 1993.

Crosier, T., and M. A. Hartz. *A Guide to Implementing Total Quality Management*. Reliability Analysis Center, 1990.

Deming, Dr. W. Edwards. *Out of the Crisis*. Massachusetts Institute of Technology Press, 1982.

Ford Motor Company. *Q1 Preferred Quality Supplier Program Criteria*. January 1986, p. 14.

Hartz, M. A. and T. Crosier. *Process Action Team Handbook*. Reliability Analysis Center, 1991.

Ingersoll Engineering, "It's Not So Much What the Japanese Do—It's What We Don't Do," Presentation at the 1982 ASQC Quality Congress.

Ishikawa, Kaoru. *What is Total Quality Control? The Japanese Way*. Englewood Cliffs, NJ: Prentice-Hall, Inc., 1985, p. 162.

Johnson, Ross H., and Richard T. Weber. *Buying Quality*. Milwaukee, WI: American Society for Quality Control, 1985.

Kirsanoff, William A. "Supplier Rating—A Total Quality Approach." Presentation at the 1990 ASQC Quality Congress.

Maass, Richard A., John O. Brown, and James L. Bossert. *Supplier Certification: A Continuous Improvement Strategy*. Milwaukee, WI: American Society for Quality Control, 1990.

Mills, Charles A. *The Quality Audit: A Management Evaluation Tool*. Milwaukee, WI: American Society for Quality Control, 1989.

Newkirk, Roger. *Motorola Purchased Material and Supplier Quality Certification Program*. Motorola Inc. Engineering Standards, 12S11062A, 10/13/88.

Robinson, Charles B. *Auditing a Quality System for the Defense Industry*. Milwaukee, WI: American Society for Quality Control, 1990.

Robinson, Charles B. *How to Make the Most of Every Audit: An Etiquette Handbook for Auditing*. Milwaukee, WI: American Society for Quality Control, 1992.

Talley, Dorsey, J. *Management Audits for Excellence*. Milwaukee, WI: American Society for Quality Control, 1988.

CHAPTER 7

BENCHMARKING

THE SEARCH FOR "BEST" PRACTICES

Benchmarking is not a new concept. It shares many elements with a multitude of activities ranging from competitive analysis and total quality management to ancient warfare. Since its inception as a formalized process at Xerox in 1979, it has evolved into a technique used by the majority of Fortune 500 companies. Not all companies that try it are pleased with their success, but many are. Some companies have embraced the practice and want to share their experiences by offering training to others. Notable among this group are AT&T, IBM, and Motorola. What began as a process to improve the design and manufacture of copiers is now used in a variety of industries and functions. Today, this diversity of applications has gone well beyond manufacturing industries and now includes studies in government, education, agriculture, health care, and financial institutions.

The purpose of this section is to answer some of the questions about benchmarking and to help understand why so much attention is being focused on the process. More important, it provides guidelines to help teams get started and describes pitfalls to avoid for teams that have begun the journey. Figure 7-1 lists the common pitfalls experienced in benchmarking.

WAKE-UP CALL FOR XEROX

With patent protection for the xerographic process, Xerox had prospered with little or no serious competition. Unfortunately, they had created a large, bureaucratic, complacent organization. The expiration of patent protection was Xerox's wake-up call. In the late 1970s, Xerox's domestic market share declined to 22% (unit shipments) and their revenues declined from 82% to 42% of the copier market. Their low-volume copier market was essentially lost (1% share) to the Japanese, who were selling copiers through wholesale dealers for prices below the Xerox manufacturing costs.

Xerox met the challenge with a strategic action in 1979 called Competitive Benchmarking, which focused on Japanese design and manufacturing techniques, particularly those of its joint venture partner Fuji-Xerox. The quality transformation at Xerox began in 1983, and benchmarking was one of three key processes. The change at Xerox was dramatic: in 1986 Xerox launched a family of mid-volume copiers that were designed in half the time and cost essentially half as much to design and manufacture as the previous generation introduced four years earlier. In 1987, a DataQuest newsletter stated that the Benchmarking program deserves the lion's share of credit for the company's turnaround in recent years.

During the 1980s, Xerox applied benchmarking to many functional areas of the company. The 1982 study with L.L. Bean, which focused on logistic and distribution issues, proved that benchmarking of functional areas in unrelated industries could be beneficial. Xerox enjoyed a 10% improvement in logistics and distribution productivity in the early 1980s, with 30% to 50% of the gain attributed to the L.L. Bean study.

By the end of the decade, Xerox's low-volume market share grew to 20% in the United States, with product produced in both the United States and Japan. The quality transformation at Xerox paid off and benchmarking played a key role in that rebirth.

BENCHMARKING AND COMPETITIVE ANALYSIS

Benchmarking can be confused with competitive analysis, which is a related but different exercise. Benchmarking shares many elements with competitive analysis. It often seeks information in the public domain and it looks for creative ways to obtain and analyze data, but there are significant differences. Two key, distinguishing characteristics of benchmarking are that the organization being studied is cooperating as a partner in the study and the focus of the study is on processes and practices, not just performance measures. Table 7-1 summarizes the major differences for organizations looking at the external environment.

The current literature on benchmarking offers a wide array of benchmarking categories. Many of these distinctions are quite arbitrary and have nothing to do with the process itself but more with the partners chosen and area of the business being analyzed. The benchmarking process does not vary significantly with the three categories of partners selected: internal organizations, direct competitors, or noncompetitors. This section assumes that there are two forms of benchmarking: strategic and functional or operational. It will address the latter form of benchmarking, which focuses on the operational processes and practices, and services offered by an

CHAPTER CONTENTS:

THE SEARCH FOR "BEST" PRACTICES   7-1

PROBLEMS AND PITFALLS IN BENCHMARKING   7-10

The Contributors of this chapter are: Roger Swanson, President, Competitive Dynamics, Inc., Culver City, CA; Seymour M. Zivan, Retired—Xerox Corporation.
The Reviewers of this chapter are: Fred Bowers, Manager of Benchmarking, DEC; Jim Henderson, President, Intellenet, Inc., Asheville, NC.

CHAPTER 7

THE SEARCH FOR "BEST" PRACTICES

TABLE 7-1

Characteristic	Competitive Analysis	Benchmarking
Approach	Independent	Partner
Performed by	Individual	Team
Target	Competitor	"Best Practices"
Focus	Performance measures	Processes and practices
Objective	Competitive intelligence	Process improvements

TABLE 7-2

Common Categories	Revised Definition
Strategic benchmarking	Strategic benchmarking
Process, service, and functional benchmarking	Functional/operational benchmarking
Performance, competitive, and industry benchmarking	Industry analysis

organization. Strategic benchmarking focuses on strategic marketing, financial, organizational, and technological issues facing an organization. Although it is important to benchmark how other organizations select and deploy the vision, goals, and policies to work units (that is, organizational goal alignment/policy deployment), it is not the subject of this section.

Table 7-2 summarizes benchmarking categories using this revised definition.

Industry analyses performed by professional and trade associations often have the look and feel of cooperative, competitive analysis, but on a broader scale. Often, they go beyond the mere capture of performance measures and include outcomes (that is, measures of business or process effectiveness and customer satisfaction) to help identify best practice companies. Generally, these studies include organizations of various sizes in a variety of industries, providing desirable diversity, but the identity of the participants is usually protected. This seriously handicaps the studies for use in functional/operational benchmarking, however. These studies do provide valuable information on data not normally available in the public domain and when these studies are repeated, they provide important trend information.

BENEFITS OF BENCHMARKING

The United States is no longer competitive in many global industries. As with other quality improvement efforts, improved methodologies result in higher productivity and lower costs. This can only occur if organizations understand the need to change, are willing to change, and have an idea of the outcome after changes. Benchmarking is particularly helpful in validating proposals for change. Although circumstances are different in other industries,

- Lack of management commitment and involvement.
- Not applied to critical areas first.
- Inadequate resources.
- No line organization involvement (process owners and stakeholders); staff exercise only.
- Too many subjects; scope not well defined.
- Too many performance measures.
- Critical success factors and performance drivers not understood or identified.
- Potential partners ignored: internal organizations, industry leaders, or friendly competitors.
- Poorly designed questionnaires.
- Inappropriate data collection method.
- Too much data; inconsistent data.
- Analysis paralysis; excess precision.
- Communication of findings without recommendations for projects to close gaps.
- Management resistance to change.
- No repeat benchmarking.

Fig. 7-1 Common pitfalls in benchmarking. (Copyright Competitive Dynamics, Inc.)

benchmarking often results in creative imitation and the adoption of new practices that overcome previous industry barriers. This search for diversity and for innovative breakthroughs applied elsewhere is at the core of benchmarking benefits. By sharing of information, all parties benefit since it is difficult to excel in all activities. Sharing of information and data is often the first hurdle to be overcome in the benchmarking process. Do not, however, attempt benchmarking in areas in which trade secrets or sensitive information determines the outcome of the process.

The benefits of benchmarking, as exemplified by Xerox, are fairly well documented by many organizations in a wide variety of industries. Benchmarking, used in conjunction with other quality techniques, or used alone, can influence how an organization operates. If the search for best, or just better, practices is performed correctly, then the likelihood of successful outcomes is quite high. This, however, assumes that pitfalls are avoided and prerequisites have been met before benchmarking is initiated.

PREREQUISITES FOR SUCCESS

Management commitment and support can overcome many of the barriers to successful benchmarking. The key requisite for success is the organization's readiness to accept change, given that it comprehends the need for change. Rather than resting on their laurels and previous success, organizations must become receptive to new ideas. Additionally, sufficient resources must be allocated and both awareness and skills training must be available.

Benchmarking shares success factors with other quality management processes. Teamwork, analysis of data, decisions based on facts, focus on processes, and continuous improvements are shared characteristics. The need for leadership, a customer focus, and empowered employees are equally important. For many organizations, success in quality management or benchmarking requires that reorganizations and culture barriers (such as NIH—not invented here—that exists in many professional disciplines) be addressed before beginning the journey.

As in other quality management activities, there must be someone to manage the introduction and application of benchmarking in the organization. The tasks to help internalize the benchmarking process within the organization include a wide variety of actions, such as:

- Identifying a benchmarking champion.
- Cascading just-in-time skills training to create centers of competency.
- Coordinating projects and interaction with consultants.
- Formalizing gateways to outside organizations.
- Establishing networks, information databases, and newsletters.

Not all of these tasks need to be in place before the first benchmarking project, but they should receive consideration in implementation planning.

CHAPTER 7
THE SEARCH FOR "BEST" PRACTICES

A final prelude to benchmarking is often referred to as step zero. In this preliminary step, a strategic and competitive assessment is performed to establish primary goals, objectives, and performance measures of success. The strategic assessment leads to the organization's vision of the future. This vision then is translated into organizational goals, policies, and, finally, operating objectives that are both measurable and actionable. Benchmarking teams need to have some idea of the strategic direction of the organization so that their efforts are aligned with the organizational goals.

Additionally, benchmarking teams need a clear understanding of both internal and external customer needs and expectations. Without this, they will have difficulty selecting important subjects first since they lack knowledge of critical success factors necessary to satisfy those customer expectations. An understanding of competitive strengths and weaknesses provides additional background that aids the selection process. Strategic and competitive assessments tell them what is important to focus on, and also give them an idea of how effectively they are currently achieving important goals. Finally in step zero, documentation of key processes (for example, flow charts and baseline performance indicators) and practices completes the desirable preconditions before starting the benchmarking process.

To illustrate the key points of this section, an example will be used: it will assume the case of a firm that designs, manufactures, sells, and services industrial/commercial equipment (any number of products could be used, and it will also assume equipment is HVAC, heating-ventilation-air conditioning) primarily in North America. The firm specializes in equipment for small office and industrial buildings, and they have only two direct competitors in their market niche. The firm, Clear Air, Inc., has just completed a strategic and competitive assessment. They discovered that their customers are very satisfied with low cost, new products that can be designed and manufactured with very short lead times. Clear Air assesses its critical success factors to be: innovative new products; fast response for design, manufacture, and installation; and low cost, reliable products.

EIGHT-STEP BENCHMARKING PROCESS

The benchmarking process consists of three general activities: planning, analysis, and integration/action (see Fig. 7-2). Overall, the process follows the Plan-Do-Study-Act cycle of all quality processes. Within each step, this logical cycle is present at a lower level of detail. This eight-step process can be easily linked to other benchmarking processes (such as the Xerox 10-step, the Alcoa 6-step, and the AT&T 9-step). This is important since the reader will

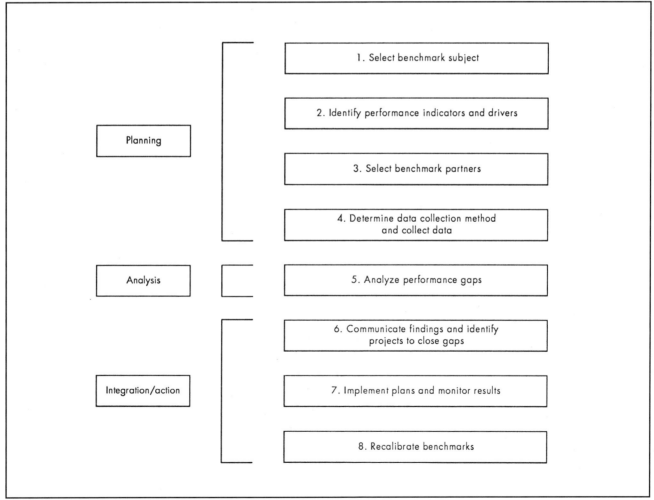

Fig. 7-2 The three general categories of benchmarking.

CHAPTER 7

THE SEARCH FOR "BEST" PRACTICES

have benchmarking partners that will likely select one of these models and also will need a common language for effective communication. Each step discusses the numerous pitfalls that might be encountered; benchmarking is like other quality processes in that learning is often enhanced by the failures along the way.

PLANNING

Selecting the Subject

The first step in the planning phase is to select a subject. Quality steering committees, senior management, functional department heads, or quality teams generally perform this function. The first benchmarking project may not be the most critical problem, but it may serve an important educational purpose. Additionally, it may help to convince management if they are skeptical of the value of benchmarking. Subsequent projects should be ones that are critical to the success of the organization. Key questions that can help to focus on the subject are:

- Where are competitive pressures and current problems occurring?
- Where is the most time and money spent?
- Where are the production or design bottlenecks?
- What areas are critical to the success of the organization?
- Where can the greatest improvement occur?

The selection of a benchmarking subject is similar to the methodology used to select processes for improvement by internal quality teams. The primary difference is the external focus of benchmarking and the added cost, which limits the processes addressed. Since organizational outputs generally depend on cross-functional processes and the practices associated with them, the selection task depends on identifying key processes and deciding which ones are the most important, which ones need improvement, and which ones will yield the greatest return if improved.

One technique that is quite useful in this ranking is the prioritization matrix (generally the analytical hierarchy process) coupled with assessments of process effectiveness. The disadvantage of this technique is that it requires significant effort by those with subject matter expertise. The primary advantage is that it is done only once (or easily repeated if necessary) and it provides a prioritized order of all major processes needing improvement. This approach is often used by senior management and steering committees, for example, when a more global view of the organization is necessary. Its strength lies in the focus on outcomes or goals, the identification of success factors needed to reach the goals, and the identification of key processes that drive performance.

A less rigorous approach can also be used, which relies on the typical problem solving/process improvement process: identify potential problem areas/process improvement opportunities, identify customers and outputs, conduct preliminary analysis of potential causes, and select key causes. Both of these techniques help to steer teams past the first serious pitfall of benchmarking: attempting several difficult projects simultaneously.

Another common tool used to prioritize projects and decide which ones to pursue is the cost-benefit analysis. Here, the potential benefits from a changed process or practice are compared with the cost of the study and costs to implement potential recommendations. Estimates of the cost of performing a benchmarking study, however, vary significantly depending on the scope, logistics, and maturity of the teams. Typical projects (five to ten partners, performed by a mature team) might require six team members working 25% of the time for six to ten months, or three quarters to one and a quarter man-years.

One final note on selecting a subject: avoid selecting subjects that involve processes or activities that involve trade secrets or sensitive information. Even though one might stipulate off-limit areas, the risk of divulging sensitive information may not be one that management will be willing to take.

Forming the Team and the Project Road Map

Next, a team needs to be formed (or modified if it is an existing quality team) and a project road map developed. These are integral activities in the subject selection step. The team should include from four to eight members, who are subject matter experts representing the various functions affected by the project. Teams composed of staff personnel who are not owners or stakeholders of the process being studied often make recommendations that meet significant resistance during implementation. This pitfall can easily be avoided by including members from all functions affected.

Teams often include several people in an advisory capacity: the project sponsor or customer, a benchmarking coach, or an internal or external consultant. This is a way of keeping the customer informed of the team progress and the findings. Both the customer and coach should provide advice—not dictate content.

The road map serves several purposes: it keeps the team on track and it documents activities and decisions. This helps to educate others and to document the team process for later use. Another advantage is that clearly defined projects help teams to ensure that the project scope is achievable. Basic elements of the road map include:

- Project scope and objectives.
- List of activities.
- List of deliverables.
- Individual roles and responsibilities.
- Meeting and review schedules, and
- Issues.

During the last year Clear Air has lost several contracts to a competitor that was able to deliver against a very tight delivery schedule. In the past, this competitor was unable to perform that quickly. Examples of projects that might be considered are manufacturing cycle times and product development cycles. They might want to explore the impact of reducing the work in process inventory with the application of just-in-time inventory management and the impact of shortening the design cycle with concurrent engineering techniques. These preliminary subjects are quite broad and the team will use top down and then deployment flow charts of key processes to determine non-value steps, sequential flows that can be parallel, and sources of process variation. Next, the team will collect additional data to confirm problem areas with a potential for significant improvement. They will then be able to do a preliminary cost-benefit assessment to help prioritize and select benchmarking subjects. An alternate approach would be to use a prioritization matrix with assessments of current levels of process effectiveness for a more global perspective.

Performance Indicators and Drivers

This step begins with the documentation of processes and practices associated with the subject. The elements of a process will be assumed to include customers, inputs, outputs, activities,

CHAPTER 7
THE SEARCH FOR "BEST" PRACTICES

decisions, sequence, and responsibilities for adding value. Practices can be process specific or have an impact on multiple processes, and generally define how activities and decisions are performed. For example, product assembly can be defined by manual or mechanical practices and communication can be either verbal or written. Documentation activities include the development of process flow charts and the collection of baseline data from internal operations, as well as from relevant public domain sources (for example, trade and professional association studies, newsletters, and trade publications). The primary goals are to identify the vital few performance indicators that confirm superior performance and to identify those processes and practices that "drive" performance. This search for "cause" and "effect" will be followed by the identification and documentation of internal process variables and attributes. A table of data and information that is both quantitative and qualitative, as seen in Fig. 7-3, facilitates the development and analysis of indicators and drivers. It will also be used to compare inputs from benchmarking partners.

The key to finding the vital few measures starts with the identification of customer satisfaction measures and measures of business effectiveness. Finding this measurement of results, or outcomes, is a difficult yet necessary task in successful benchmarking. The difficulty arises because effectiveness depends on selecting the right things to do and doing things right—or judgment and execution. Judgment creates the system of performance drivers (that is, processes, practices, and structural factors), while execution determines how efficiently the system performs. Poor results usually occur when either or both of these elements fail. Superior execution can seldom overcome wrong or poorly chosen drivers, however. The search for key drivers (that is, causes) is simplified if the failure can be identified as one of judgment or execution.

This dilemma will remain when selecting the vital few measures, in linking these measures with performance drivers, and later on, in selecting partners and in analyzing data. As would be expected, unfavorable outcomes often mask excellent processes that are executed poorly. This is the halo effect, where an organization that performs well is often assumed to do everything well, and conversely, an organization that performs poorly is assumed to do everything poorly. In benchmarking, organizations with poor outcomes are generally ignored because of credibility issues.

It is important to select performance indicators that cannot be manipulated or result in suboptimization (win the battle, but lose the war). Often the best measure may be one that is not currently in use, or is felt to be too difficult to measure on a continuous basis. The ideal is to pick ratios that can be continuously measured, but a good measure monitored periodically is better than a poor one measured continuously. A cost accounting labor overhead rate that is monitored regularly, is a good example of a poor indicator since there are so many factors that influence the

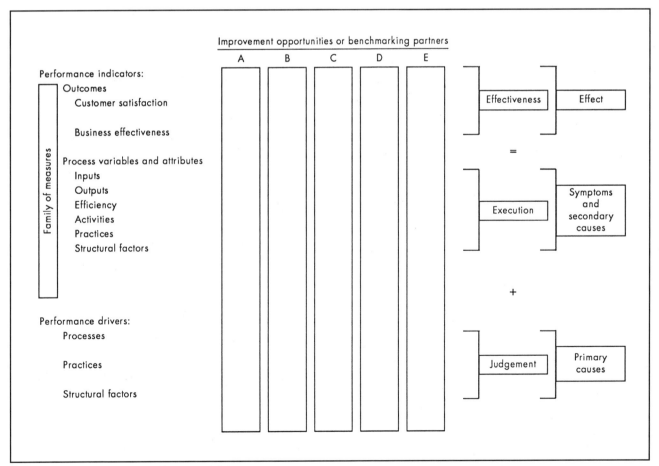

Fig. 7-3 Table of performance indicators and drivers.

CHAPTER 7

THE SEARCH FOR "BEST" PRACTICES

ratio. Since accounting systems vary significantly, it is often difficult to use financial data without major adjustments to ensure comparability. In most cases, measures of time (for example, machine hours, man-hours, and elapsed time) and various dimensions of quality (such as failure rates, acceptance levels, defects, and yields) are better than financial measures.

Indicators are only a means to an end, not an end in themselves. They track change resulting from the implementation of improved processes and practices; they are expected to improve over time as process variation is reduced and as process velocity is improved (for example, by eliminating nonvalue steps or by changing work and information flows). Process flow charts are particularly useful in identifying where process variation occurs, where failure data needs to be collected, where value is added, and where parallel flows can replace sequential flows.

After identifying the vital few measures, the team needs to understand the causal link between the performance drivers and these indicators. This critical step requires subject matter expertise using various techniques such as fishbone diagrams and interrelationship digraphs. The team should, however, remain open-minded regarding this relationship since the study might uncover other performance drivers and a better perspective of the cause and effect relationship. During the discovery of these relationships, the team should not lose sight of the customer, customer expectations, and success factors defined by the customer.

Finally, the team needs to address a special category of performance drivers: structural factors. These are the culture, organization structure, technology, environment, and any other administered factor (such as contractual agreements) that might have an influence on the performance indicator. Structural factors create barriers that must be either corrected or accepted but should never be ignored. Acknowledging these barriers helps to reconcile differences in performance between benchmarking partners during the analysis stage.

The team would then define the performance indicators and some preliminary performance drivers depending on the specific subject selected. The outcome measures of short manufacturing lead time and short design cycle time will be used as measures of business effectiveness with an assumed direct relationship to customer satisfaction. In addressing the manufacturing lead time problem, the team might want to look at several indicators such as: work in process, inventory turnover, rework costs as a percentage of unit manufacturing costs, number of suppliers, incoming raw material quality, final inspection yields, number of quality defects, raw material lead times, and supplier delivery frequency. An exploration of the drivers of manufacturing cycle times might include a review of some of the following: material requirements planning (MRP) systems and practices, MRP software, purchasing policies and practices, discrepant material handling, supplier alliances, supplier quality certification process, order entry, and factory scheduling systems and practices.

On the design side, the appropriate indicators might include: new product design cycle time, variant product (that is, second generation) cycle time, engineering change order frequency, unique drawings per engineer, approvals per drawing, organization layers in engineering, and percentage of drawings using computer-aided design (CAD). Performance drivers in the design area might include: extent of CAD design, product development process and practices (including customer focus groups, early supplier involvement, pilot production, and manufacturing involvement), engineering approval and review process, and robust design practices.

The primary focus should be on those areas where major variations in performance occur, process steps do not add value, and work flows are not optimized.

Selecting Partners

The selection of partners often involves the use of external data and information sources. The secondary research performed while establishing baseline information and data will often uncover articles and sources helpful in identifying organizations who perform well in the company's subject area. If these sources do not uncover potential partners, continue the review of periodical and trade sources, and begin a search focused on direct contact with trade associations, consultants, customers, suppliers, employees, Baldrige award companies, and benchmarking clearinghouses.

Most organizations performing benchmarking designate a champion to facilitate training and project management, and to provide a gateway for initiating projects. These champions form an informal network (sometimes formalized by their joint membership in various associations) that can be instrumental in locating partners. Tapping into the network is like finding the Mother Lode ore deposit.

Partners fall into two general categories: other internal units and external organizations. If there are other units within the organization that are performing similar functions, they should always be included. External organizations include:

- Direct competitors.
- Industry peers that serve a different market.
- Companies that serve another's market in a related, but different industry.
- Organizations outside a company's industry who perform similar functions.

The first category is often ignored because of the fear of divulging sensitive information, but they are often willing to participate in nonsensitive areas or when the anonymity of participants is protected by an outside third party, such as consultants or trade associations. The last two categories are often the source of greatest innovation since they are not generally influenced by the paradigms of the company's industry.

Once a preliminary list has been developed, partners are selected using the criteria below. Here, the selection process will be based on data and information available in the public domain, which often is general and not specific to the subject selected, so be careful of the halo effect.

- Profitability.
- Industry stature and potential for best practices.
- Functional and process expertise and potential for best practices.
- Ease of obtaining information (indicated by prior participation in earlier studies).
- Relevancy.
- Potential for innovation.

The benefit of selecting organizations outside of one's industry can be lost if issues of relevancy and credibility of the organization are not addressed in the selection process. The relevancy issue should be discussed immediately with partners that agree to participate, to avoid wasting each other's time. A brief discussion of the subject is generally sufficient to help both parties conclude that they have the potential of learning and applying innovations from each other. The search for the best organization is generally

CHAPTER 7

THE SEARCH FOR "BEST" PRACTICES

a waste of time because of the law of diminishing return. The objective is continuous improvement and future studies, if warranted, might find the best. Finally, do not place the ease of collecting data above the potential for best practices in the selection criteria weighting.

Most benchmarking studies include five to ten partners. Exceeding ten partners complicates project management and should be avoided for most studies. (The primary exception is industry studies funded by a trade association or by participants in an association study. In these studies, the scope and time frame are extended beyond those recommended for functional/operational benchmarking within an organization.)

To obtain the participation of partners, one must be willing to share information and data about one's own organization and provide a copy of the final report. In many cases, receipt of the final report is sufficient inducement to participate. The benefits of benchmarking should not be oversold, however.

The search for partners in the Clear Air example would use the same sources discussed above. The trade publications and organizations would be ones addressing HVAC design, manufacturing, and marketing. In addition, trade sources serving the same customers (for example, elevator manufacturers serving small office and industrial building contractors) or related industries (such as residential HVAC and major appliance manufacturers) could be explored. Given the conditions described above, Clear Air's direct competitors should be excluded, but industry peers serving other markets (that is, companies producing HVAC equipment outside their current geographic market, or different HVAC products and customers) should be considered.

Data Collection

Data collection can be the most difficult step in the benchmarking process. The primary objective is to gather information and data to confirm superior performance and to uncover best practices without burdening the partners with long, time-consuming data collection methods. Although data and information are collected in establishing baseline measures and in selecting partners, the majority of the data for most projects is collected in this step.

The types of data and information required often define the method of obtaining it. For most benchmarking projects primary research is required, since the level of detail goes far beyond the level available in secondary, published sources. This type of primary research generally involves a questionnaire or survey document, which will be used in combination with other activities. These documents address both the quantitative performance indicators ("whats") and the qualitative performance drivers ("hows"). Most organizations prefer to collect all information with one document, with follow-up questions to validate the data, while other teams prefer two documents: the first to collect the quantitative data, and the second for the qualitative information.

The typical sequence of data collection begins with a questionnaire mailed to each partner, which is followed by telephone interviews to clarify key points after the written response is received. An alternative approach of faxing the document and then conducting a telephone interview is a common technique. This works well for short questionnaires requiring minimal data collection. Be sure to schedule all telephone interviews in advance. Facility visits and meetings to discuss preliminary findings generally follow the preliminary data collection activity. It is recommended that two or three members of the benchmarking team conduct visits and that they adhere to the scheduled time and agenda. Each member should have a clearly defined role in the visit (such as note-taking, asking questions, process guidance) and the team should conduct a debrief as soon after the visit as possible (less than two days). Most projects include a final meeting of partners where the study findings are reviewed and discussed.

The actual methodology depends on data and information requirements (amount and accuracy of data) and other criteria such as costs, project schedule, and logistics issues. Facility visits and meetings to discuss preliminary findings are not a prerequisite for successful benchmarking, however. Visits to partners in other industries are generally recommended since the need to discuss normalization of industry differences often requires more time and interaction than is possible over the telephone. Visits to partners who will help to pretest the questionnaire are also recommended, but trips that add little value to the process, also known as industrial tourism, should be avoided. Be prepared to answer questions about the organization and its answer to the questionnaire, but make sure the responses do not influence the partner's answers (that is, by providing one's answers first).

The first step in developing a questionnaire or survey is to list all questions that need to be answered to validate the existence of superior performance. These questions are followed by ones that ensure that everyone is measuring the same thing and that a common denominator can be identified. The data capture mechanism (questionnaire, survey form, computer input form, etc.) should be logical and easy to use. Limiting the data captured to a few vital indicators will minimize the the respondent's time and facilitate data input and analysis. Detailed, statistically valid data is mandatory for some analyses, but a lower level of detail is often acceptable for most benchmarking projects. Open-ended questions about relevant processes and practices and requests for process flowcharts usually follow, to uncover the performance drivers.

Next, the team should answer all questions, and if possible, pretest the questionnaire with an internal unit and one external partner. Questions that are difficult to answer or not essential must be revised or eliminated. Add appropriate questions suggested during the pretest. Perhaps the most difficult task in developing a questionnaire is to ensure that it is focused on a clearly defined subject and scope. This will help to avoid long, detailed questionnaires that potential partners will either ignore or answer incompletely.

Since benchmarking requires interaction with other organizations, ethical and legal issues must be addressed. Always consult legal counsel when in doubt. Exchanging cost and marketing information, for example, can be construed as an action leading to restraint of trade, which must be avoided. Acquiring or disclosing trade secrets or sensitive information can likewise create legal problems that can be easily avoided. In many cases an outside consultant, who protects the anonymity of the participants of a study, can help to avoid problems resulting from face-to-face interaction between partners.

From an ethical standpoint the key cautions are for participants not to misrepresent themselves, to respect the confidentiality of information, and to use the information for its intended purpose. Results should not be shared with third parties until all partners have approved this disclosure. During the project, partners should be prepared for each activity, share data and information, honor commitments, and work through designated channels within partner organizations.

The initial stages of collecting performance data can rely on mailed questionnaires and telephone follow-up. The questionnaire needs to address the specific questions related to the subject plus

CHAPTER 7

THE SEARCH FOR "BEST" PRACTICES

more general, demographic questions that are helpful in analyzing organizations not in the same industry. Requests for process flow charts and descriptions of various practices can also be handled the same way. To interpret flow charts and to fully understand processes and practices, it is likely that the questionnaires would be followed with plant tours and personal contact with partners. This is particularly important when partners are not in the same industry and a thorough understanding is necessary to normalize the data (that is, for apples to apples comparisons) for analysis, the next step.

ANALYSIS

The objective of the analysis step is to identify the best performing organization and to determine the reasons for the superior performance. Since the Table of Performance Indicators and Drivers highlights the key elements needed to compare partners, it becomes a useful tool in analysis. The performance indicators define the benchmark standard and the gaps in performance for each participant. The processes and practices of the best organization are the benchmark performance drivers that each partner will try to imitate creatively.

Identifying this cause and effect relationship between performance drivers and the resultant measures is the most challenging part of benchmarking. Here, the impact of structural factors on performance is an important consideration, since many of these factors relate to qualitative, rather than quantitative information and are often more difficult to analyze. Analytical problems frequently occur when diverse organizations are being compared or when it is difficult to isolate organizational efficiency (that is, doing things right) from organization effectiveness (that is, doing the right things right). A good process that is executed poorly can easily be overlooked.

The primary concerns in data analysis are the accuracy and validity of the data. Difficulties can arise if there is too much, too little, or inconsistent data; this often results in analysis paralysis—an endless search for trends or relationships. The desire for excess precision can lead to a similar trap. The purpose of keeping the scope of the study focused on critical success factors and the vital few performance indicators is to avoid these problems. Aside from the volume of data collected, the next most challenging task is to normalize the data from organizations that are not identical. This problem is particularly acute when financial data is used, particularly if the partners reflect the recommended diversity. The caution about using financial data becomes evident in the analysis step.

Since benchmark performance is a moving target, it is necessary to forecast performance into the future based on current trends for both the searching organization and the best organization. The Xerox Z-chart, as shown in Fig. 7-4, is particularly useful in graphically portraying the competitive gap for key indicators and the extent of improvement required to close the gap. This chart highlights the need to take strategic actions as well as to continue or improve the current level of process improvement.

Progress in achieving performance improvements is then monitored relative to the forecasted trends. Keep in mind that indicators are only a means to an end, not an end in themselves. They track improvements resulting from changes in performance

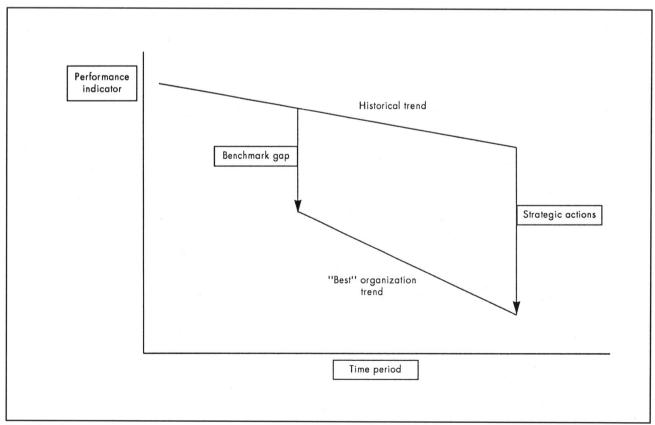

Fig. 7-4 Z-chart.

CHAPTER 7

THE SEARCH FOR "BEST" PRACTICES

drivers; they track change from both the historical trend and the strategic actions taken as result of the benchmarking study. The aim should be to measure, compare, and improve the processes, not mandate a fixed standard and hope that performance improves.

For most projects it is advisable to distribute preliminary conclusions, along with supporting data, to participants to have them validate their data inputs and to comment on the preliminary reasons for superior performance. Large, apparent performance gaps need to be treated with great sensitivity to ensure that partners have the opportunity to explain the differences and to make corrections in possibly erroneous data.

The Clear Air example does not require special discussion here and would be analyzed in the manner discussed above. The main differences between internally focused problem solving/process improvement and the benchmarking process usually merge in the analysis stage, which assumes that data from different industries has been normalized for comparison purposes. The primary reason for conducting a benchmarking study is to learn about and observe innovative processes and practices. Once this search is complete, gaps defined, and reasons for the gaps understood, the organization can then creatively adapt these innovations in the next phase where it takes strategic actions to change.

INTEGRATION AND ACTION

The objectives of the integration and action steps are to:

- Obtain organization buy-in.
- Initiate projects to close gaps, and
- Implement plans developed in these projects.

Benchmarking often becomes a catalyst for change by adding credibility to proposals for change, by helping to break out of industry paradigms, and by creatively imitating the innovative best practices of others. Before the team attempts to obtain organization buy-in, it must organize and present its findings. The team must reach consensus about the findings and its recommendations for action.

The key to getting favorable responses to findings depends on how the team was developed (for example, did it include stakeholders of the process) and how the project was managed (for example, was management involved in a periodic, formal or informal review process). Unfortunately, benchmarking projects can easily become sidetracked in the integration and action steps. Resistance to change, the primary barrier, can be minimized if the process is managed properly, and if the key prerequisite to success has been satisfied (that is, management support and willingness to change).

If the team has focused on critical subjects, recommendations will probably require strategic actions and a new project to plan and implement the proposed change. Often the benchmarking team is not well suited for both communicating findings and implementing plans for change. To prevent this, the team needs to be modified as it transitions into this phase. The team needs to be committed to conducting the benchmarking study, and also to following the implementation of recommendations.

Since the recommended actions will likely have an impact on various strategic and tactical plans of the organization, the team must consider the integration of recommendations within the planning process. Finally, the recommendations should address the responsibility for change, the timetable, and the estimated cost and benefits of the change. Realistic schedules and financial estimates are necessary to ensure the team's credibility.

Perhaps the most serious mishap in benchmarking occurs when implementation is made in large, diverse organizations without pilot programs to test the impact. This error is common to all change actions within an organization. Failure here should not be associated with the benchmarking process, but with the change process within the organization. An effective sequence for pilot activities might begin with a small-scale pilot that is closely monitored. Other factors that have an impact on the process must be isolated during the monitoring to ensure that changes result in the desired outcomes. Once success is obvious, results are communicated and, if needed, training curriculum developed to facilitate the cascade of change. Training then becomes the vehicle for standardizing the change throughout the organization.

When the first project is completed satisfactorily, the organization is ready to tackle the next most critical subject. Before leaving the first project, the team needs to address one final issue: when it will repeat the benchmarking of the new standard. Subsequent recalibrations are often triggered by a reasonable period of time (say, two to three years) or some specific change in performance indicators that are measured periodically.

KEY LESSONS AND NEXT STEPS

Benchmarking, like other quality techniques, works in organizations that have met certain prerequisites and that do not have major impediments to success. It is not a panacea, a fad, or a quick fix that is performed once, but a continuous process that is internalized within the organization. To be effective, benchmarking works best in an environment in which teamwork and management support are evident. Teams need to focus on results as measured by performance indicators, but the organization must change the performance drivers to achieve benchmark standards. Target-setting alone, without changes in the organization culture and other structural factors, processes, and practices, can only bring temporary relief.

Benchmarking often starts in an organization because a few key people have educated themselves on the process and its benefits. Members of the first team generally have read books or articles on benchmarking, have attended training seminars or workshops, and have problems that lend themselves to an externally focused problem solving effort. If this team is successful and communicates its success within the organization, the technique will be applied by others.

The first team members form a benchmarking competency center, and they are often relied on as coaches in subsequent benchmarking studies. Their educational efforts will be enhanced when the organization designates a champion who will promote benchmarking and help manage the implementation of the technique within the organization. The management of benchmarking (from the organizational level, not the project level) includes the cascade of just-in-time skills training, coordination of benchmarking projects and contacts with benchmarking partners, and the establishment of networks, newsletters, and project file databases.

For large organizations, the implementation process will take years; initial doubts will be eliminated as the organization continuously improves, as it strives to be the best and sets the standard for others to follow.

CHAPTER 7

PROBLEMS AND PITFALLS IN BENCHMARKING

PROBLEMS AND PITFALLS IN BENCHMARKING

Benchmarking has been adopted by many institutions to assure achievement at the highest possible levels of performance. By comparing an entire organization or its processes to those thought to be world class, appropriate targets are developed and subsequently used to stimulate needed change. The benchmarking concept was initiated as a response to the lethargy that was in the way of change in so many business enterprises. During the 1970s and 1980s there were massive inroads by new market entries, mostly from overseas, taking market share, revenues, and profits from those who did not see the necessity to change. They believed or at least acted as if there were no possible ways to improve performance. Benchmarking would awaken them to improvement opportunities by revealing the comparative successes of others. It is often practiced as part of an overall quality initiative bent on meeting the needs of customers and stakeholder. Xerox, Motorola, Digital Equipment, and Ford, among other companies, speak of the benefits achieved through benchmarking.

Benchmarking is a management tool. It is not a "silver bullet" whose mere adoption will assure success. Success will only result from using this tool correctly. Barriers and pitfalls to attaining maximum results from benchmarking must be recognized and overcome. A reasonable expectation of the benefits of the proper use of benchmarking is that in some time in the future, perhaps three to five years from its application within a specific process environment, hindsight will reveal that the organization's performance was directly enhanced toward goal fulfillment by its use.

To achieve this expectation, a strategy must be developed to use benchmarking correctly so that barriers and pitfalls to its success will be addressed. There are ten essential elements of such a strategy:

- Define benchmarking as a means of uncovering best practices.
- Employ benchmarking as an element of strategic planning or process re-engineering.
- Select processes to benchmark whose improvement will have a substantive effect on organizational performance.
- Assure that those who benchmark are responsible for implementing findings.
- Develop measurements that describe the performance of processes to be benchmarked.
- Design instruments to collect information from benchmarking partners.
- Know how the company's own process performs.
- Identify qualified benchmarking partners.
- Perform productive partner visits.
- Implement findings.

Addressing each of these strategy elements will provide the means to overcome a set of barriers and pitfalls capable of denying any benchmarking initiative success in stimulating needed change. This section will discuss each strategy element along with the pitfalls it is to overcome.

DEFINE BENCHMARKING AS A MEANS OF UNCOVERING BEST PRACTICES

Pitfall. Those embarking on benchmarking view it as comparing numerical results with their competition.

Given the scenario that spawned benchmarking, it can be easily seen how one can run into this pitfall. The competition is either catching up or has caught up or there is a fear they may catch up. Where is the company in regards to its competition? Is it still better than its competition?

Two major problems exist within this pitfall. First, the comparison of numbers may register bad or good feelings, according to where the company stacks up. However, simply comparing numbers does not make the company perform any better. Second, limiting comparisons to competition denies the benefits that can be derived from examining like activities across all industry as opposed to just the industry within which a company markets. In benchmarking customer service, an international bank chose overnight parcel delivery services as partners because of their excellence in customer service, not because they were in banking. In benchmarking distribution center operations, Xerox chose L.L. Bean because of its expertise in the subject area, not because it was a Xerox competitor.

Benchmarking must be defined as a search for best practices among organizations recognized as leaders in performing the process under study. When these best practices are implemented, performance will improve.

A search for best practices may begin by locating leaders through their measured performance. It must go beyond measured performance and seek out the root causes that lie in the practices employed to achieve the numbers. When these practices are emulated, levels of measured performance should improve. A person cannot emulate another's results—only emulate practices that in turn lead to results.

The benchmarking process (see Fig. 7-5) is designed to uncover best practices. In step 1, the area to be benchmarked is defined. In step 2, the team to benchmark this process is selected. In step 3, measurements that describe the performance of the selected area are identified. In step 4, partners recognized as leaders in this area are selected. In step 5, teams identify their own practices and performance prior to identifying their partners' practices and performance. In step 6, the team proposes programs and actions, corresponding to practices used by partners, that would improve the company's performance. Step 7 is implementation of these proposed practices, and in step 8 the performance is measured.

EMPLOY BENCHMARKING AS AN ELEMENT OF STRATEGIC PLANNING

Pitfall. Benchmarking is considered an additional process or task that has no relation to strategic planning or management control. Considered as one more of a continuous stream of initiatives introduced to improve performance, benchmarking gains the position of a "flavor of the month." One more fad, like so many others, thrust upon the organization and discarded when it did not provide immediate return. Its potential benefits are lost.

The planning process (see Fig. 7-6) describes how an organization's priorities are transformed into strategies and plans that, when implemented, provide the actions that lead to improved performance. People throughout the organization seek opportunity areas to exploit so that goals may be better achieved. But how do they locate these opportunities that can be transformed into

CHAPTER 7

PROBLEMS AND PITFALLS IN BENCHMARKING

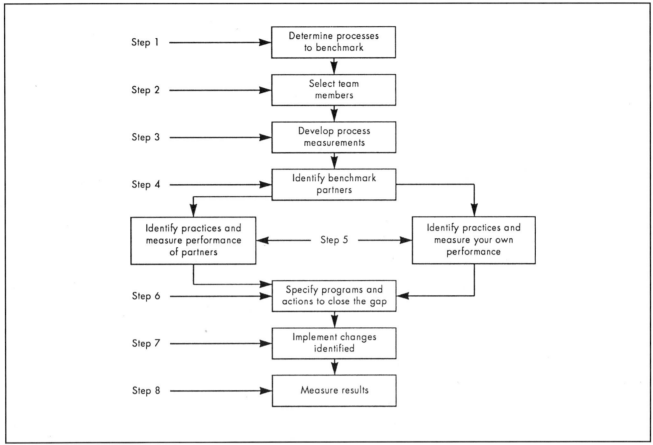

Fig. 7-5 The benchmarking process.

strategies and plans for improvement? Utilizing brainstorming techniques, people explore the means to the better way and change. How much more efficient this process of discovery would be if it was augmented by a structured comparison of performance with those who are considered leaders in the area where opportunity is sought. This is the place for benchmarking to add value. Planning performed without the benefit of an external perspective is denied the advantages provided by the experience of others. Benchmarking, performed outside the confines of the planning process, is easily taken on and as easily discarded when its relevance is lost. Benchmarking performed as an element of planning exposes to those responsible for each business process how others have gained improvements, so that plans will lead to the most aggressive targets. Benchmarking performed as an element of planning cannot be as easily discarded as a flavor of the month.

SELECT PROCESSES TO BENCHMARK WHOSE IMPROVEMENT WILL HAVE A SUBSTANTIVE AFFECT ON ORGANIZATIONAL PERFORMANCE

Pitfall. Everyone is encouraged to be trained in the use of benchmarking and to adopt it as a means of performance improvement. Without regard to the impact on the overall performance of the organization, areas are chosen to be benchmarked for the sake of adopting benchmarking. Use of the tool becomes more urgent than where the tool is used. Each organizational element is asked to perform one or two benchmarking initiatives by some specific date, say year end. Each scurries about to show it understands the benchmarking process. Given the propensity of some higher-level executives not to walk the talk, areas chosen are more often subprocesses or tasks within a subprocess than higher-level processes. A logistics organization (see Fig. 7-7) is responsible for inventory management. Its design for managing inventories dictates the form of its distribution network, which in turn dictates the form of each distribution center. Distribution center operations are chosen for benchmarking even though a much greater "bang for the buck" would have resulted from targeting the higher-level process. Also, benchmarking the distribution network first will assure the need for distribution centers before their practices are compared to others.

The planning process that benchmarking supports should dictate the areas to be benchmarked. What are the priorities of the organization? Where is improvement needed? Where is it possible? Most important, benchmarking should be seen as a top-down process with major processes compared before examining their subprocesses. In Fig. 7-7, applying benchmarking first to distribution center operations may lead to major improvements in centers that would have been shown not to have been needed if the process was performed top-down. The processes chosen for benchmarking should be the exact processes selected for improvement in planning. Selection should not be random.

CHAPTER 7

PROBLEMS AND PITFALLS IN BENCHMARKING

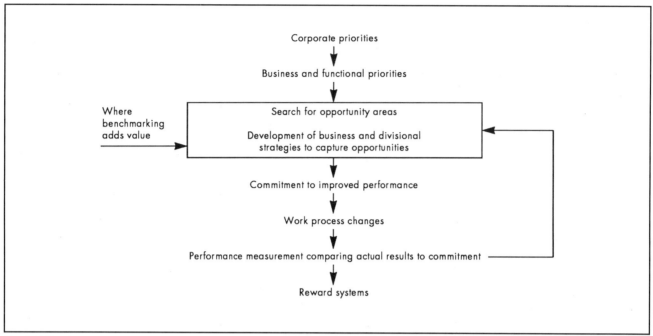

Fig. 7-6 Planning process.

ASSURE THAT THOSE WHO PERFORM BENCHMARKING ARE RESPONSIBLE FOR IMPLEMENTING FINDINGS

Pitfall. Organizations wishing to use benchmarking to improve performance commission a team of surrogates who are available to perform the actual search for best practices. The surrogate team is most often expert in benchmarking but not in the process to be researched. Best practice findings are presented to management as viewed through the eyes of others. Little happens as a result of this work. The benchmarking study becomes one more shelf item gathering dust. Change does not take place.

For benchmarking to succeed in bringing about needed change, the team performing the comparisons must be carefully crafted.

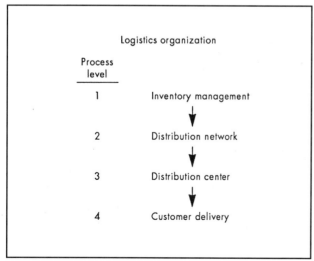

Fig. 7-7 Selected processes to benchmark.

Representatives from top management and process owners must be involved. Line management, not staff, must be the drivers. The best representatives of the organization should be involved and not just those available for assignment. Wherever possible, internal and external customers of the process should be involved.

A leading international corporation benchmarking inventory management involved vice presidents of manufacturing and a key business unit. Process customers were represented by a marketing manager and process ownership came from plant and logistics managers from across the United States.

DEVELOP MEASUREMENTS THAT DESCRIBE THE PERFORMANCE OF PROCESSES TO BE BENCHMARKED

Pitfall. Processes are benchmarked, seeking best practices, without establishing quantitative measures against which progress may be measured. Usually associated with a corporate culture that does not emphasize the need for measurements, organizations embark on a search for best practices without being able to trace their performance: where they have been, where they are, and where they will be if the practices of others are indeed implemented. In essence, nothing is provided to assure that the benefits associated with change actually are realized. Benchmarking may result in describing needed changes, but there is no assurance that it was directed at the correct targets or if goals were realized.

Though benchmarking is a search for best practices as opposed to a comparison of measurements between organizations, measurements must be developed to describe the performance that is sought from the adoption of best practices. It is not enough to seek to improve customer service. Customer service must be defined in terms that the customer both understands and needs. Figure 7-8 describes how such a measurement of process performance is identified. The output of a logistics process is the fulfillment of orders to customers. The customers of this process are distribu-

CHAPTER 7
PROBLEMS AND PITFALLS IN BENCHMARKING

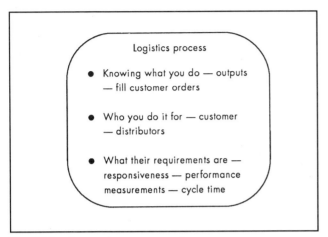

Fig. 7-8 Develop process measurements.

tors. Their needs involve responsiveness when orders are placed. Order-to-delivery cycle time is defined as the measure of process performance for this process. It is one of the key measures that benchmarking should be tasked to improve.

DESIGN INSTRUMENTS TO COLLECT INFORMATION FROM BENCHMARKING PARTNERS

Pitfall. Those attempting to benchmark seek "nice to know" answers or approach partners with an unstructured set of questions. With little experience in identifying the root causes or practices that result in achieving improvement in the areas studied, visits to partners are made seeking answers to a seemingly random set of questions. When the results of this benchmarking are surveyed, little is found that can be used to improve performance.

To achieve what is needed from each meeting with a benchmarking partner, a visit guide must be designed that structures the search for information. The visit guide identifies what must be learned to improve performance in the process area selected for benchmarking.

It should be based on overcoming the perceived barriers to improvement. Figure 7-9 shows how a cause-and-effect diagram or "fishbone" may be used for structuring the need for information that will lead to writing the questions to be addressed to benchmarking partners. One objective of the logistics process is the reduction of order-to-delivery cycle time. Barriers to achievement are brainstormed and placed on each of four major "bones". For personnel, barriers include:

- Not enough trained people.
- High turnover.
- No empowerment at the first level.

Corresponding to each barrier, questions would include:

- How is the number of people—needed to effectively operate a logistics network—determined?
- How is the rate of employee turnover controlled?
- How are those within an organization empowered to take command of the area of the logistics process they work within?

A thorough compilation of such questions will result in a productive visit guide. Starting each question with the words "How do you" will provide the discipline to seek partner practices.

KNOW HOW THE COMPANY'S OWN PROCESS PERFORMS

Pitfall. Those who benchmark seek the best practices of others without first developing a baseline for comparison with their own practices and results. With an imperfect understanding of practices employed within their own organizations, teams fail to understand how others execute the same process. Performance measurement is poor or nonexistent. Comparisons between the benchmarking organization and its partners cannot be made.

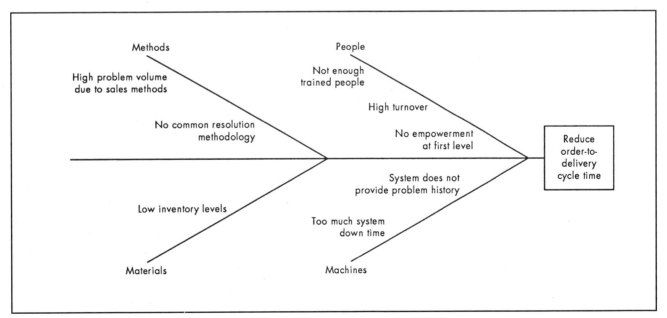

Fig. 7-9 Designing the visit guide.

CHAPTER 7

PROBLEMS AND PITFALLS IN BENCHMARKING

After defining the measures of performance for the process under study, the benchmarking organization should assure that its own process is measured. After developing the visit guide, the benchmarking organization should subject itself to each and every question. Baselines are developed that establish the only real means of comparing the practices and performance of partners. Without this baseline, findings from partners stand by themselves, lacking any meaningful means of evaluation.

Other benefits are involved through the work of first understanding one's own process. First, subjecting the internal organization to the same questions and measures as future partners will provide experience in performing the benchmarking process as well as test the operational readiness of the visit guide and measurement package. Second, understanding one's own process allows for common-sense changes and improvements prior to comparison with others. Third, those who benchmark often subject several of their internal organizations to the baselining process as a means of internal comparison and improvement prior to moving to external partners.

IDENTIFY QUALIFIED BENCHMARKING PARTNERS

Pitfall. Rushing into the visit portion of benchmarking, willing partners are chosen who do not possess unusual excellence in the area where improvement is sought.

It is difficult if not impossible to assure that partners chosen for comparisons are indeed best-in-class. A careful selection process will assure that partners present a high probability of providing major opportunities for improvement. A suitable benchmarking partner is an organization that excels in meeting the same set of customer needs the process under study is attempting to meet. Usually, such an organization is in an industry setting where there is an urgency to excel in a particular area.

- If benchmarking the product development cycle to improve time to market, partners in industries with short product life cycles may present the greatest opportunities. The fashion industry may be an example. A partner in this industry would have to excel in bringing new products to market fast to survive.
- If benchmarking safety practices to reduce lost-time accidents, partners in industries where employees may be more prone to work-related dangers may present the greatest opportunities. The chemical, mining, or construction industries would have to excel in their safety practices given this scenario.
- If benchmarking inventory management to increase turns, partners in industries with small profit margins but meaningful profits should be sought. The only way these firms can be profitable is by turning inventory fast.

Understanding the industry setting that produces the particular excellence sought, potential partners may now be listed. These are potential partners whose appropriateness will yet be tested. The list of potential partners can be generated through any combination of the following sources:

- Team brainstorming.
- References in journal articles uncovered through library databases.
- Presenters at trade associations.
- University experts.
- Customers.
- Vendors.
- Consultants.

The list of potential partners may number 25 to 30 organizations. Criteria must now be developed to rank them and select those most desired for comparison. Such criteria may be derived from the visit guide, or may involve defining readily available metrics that are thought to characterize excellence in the area under study. For inventory management, characterization would involve the level and movement of inventory turns, the concurrent market or sales performance, and the long-term levels of writeoffs for obsolescence.

Those selected as desired partners must now be called and quizzed by team members to expose their excellence and define their ranking on the final list of partners. Those most desired must be courted or sold on their participation in the study.

PERFORM PRODUCTIVE PARTNER VISITS

Pitfall. Team members do not stick to the visit guide, and leave without obtaining what they set out to learn.

The actual interview time for a benchmarking visit should be no longer than six hours. All of the foregoing preparation has been pointed toward making this time period as productive as possible, gaining valuable information to permit performance improvement. The following guidelines will assist team members to assure the most productive visits:

- Provide the partner with the visit guide questions prior to the visit, so that they may have people capable of sharing at the meeting.
- Organize the visit guide questions so that partner attendees need not be in attendance for the full duration of the meeting but only when their knowledge is needed.
- Maintain the benchmarking visit as a no-nonsense business meeting. Minimize small talk. Meet for dinner the night before to exchange pleasantries if necessary.
- Stick to the script. Get answers to each visit guide question.
- Do not tell war stories of your own. Do not share process facts at this meeting. Invite the partner to your offices to learn about your process.
- Organize the visiting team by task. There should be an appointed questioner, note taker, and time keeper.
- Agree on what was learned immediately after the visit.

IMPLEMENT FINDINGS

Pitfall. Nothing is implemented as a result of benchmarking.

Every element of the benchmarking strategy was developed to assure implementation of best practices when they were located. Unfortunately, the same organizational resistance to change that spawned benchmarking may stand as a barrier to the implementation of the findings. Reinforcing essential elements of this strategy will emphasize what it takes to best assure benefiting from benchmarking.

- Make sure those who are assigned to locate best practices are those who are going to implement what is learned.
- View benchmarking as part of the planning process. The best practices obtained through benchmarking should be the foundations for plans, programs, and strategies for the future. No long-range plans should be approved that do not result from the findings of benchmarking.
- Use the benchmarking gap—or difference between today's measured performance and the performance levels targeted after all best practices of partners are implemented—to describe the opportunity for improvement in major processes (see Fig. 7-10). Long-range plan strategic actions should include the implementations of located best practices.

CHAPTER 7
PROBLEMS AND PITFALLS IN BENCHMARKING

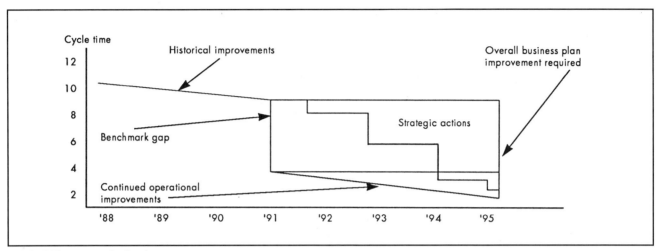

Fig. 7-10 Five-year cycle time improvement.

SUMMARY

Benchmarking permits an organization to target the most aggressive levels of performance by seeking best practices from outstanding operatives in all areas studied. These best practices define the levels of performance that should be attained when strategic actions correspond to uncovered practices.

A formal benchmarking process will best assure that benefits are realized. Numerous pitfalls and barriers exist to the realization of these benefits; most of these exist in the improper execution of benchmarking. The ten-step strategic approach to benchmarking should aid practitioners overcome these pitfalls.

CHAPTER 8

ACTIVITY-BASED COSTING

When non-accounting people, like operations personnel or engineers or managers, are asked how well their organizations' accounting information serves them and their coworkers, the answer is rarely flattering. Existing cost systems were simply not designed to deliver managerial information. Typically, accounting systems adequately satisfy external reporting requirements and some executive management tastes, but they usually fall short in helping managers detect problems or their solutions. Traditional accounting systems were not designed to do so.

Activity-based information significantly boosts the value and utility of financial data for decision-makers and empowered employees. People can relate to activity-based data. Users of this data routinely remark that it is common sense.

There is a lack of awareness of what activity-based costing (ABC) is and is not. This lack of awareness is an impediment to change. The remedies are either to attend seminars, buy lengthy textbooks, or read articles on the subject. Too often, the books never get read or attendees return from seminars enlightened but without the capability to act or get started.

This chapter is meant to present an overview of ABC. Everyone's time is valuable. The chapter is written with short, to-the-point sentences and assumes the reader is moderately informed of an organization's existing (and flawed) costing practices. Ideally, the reader is a member of a project team or is the pilot target of an activity-based costing or an activity analysis assignment.

CHAPTER CONTENTS:

THE RISE OF ACTIVITY-BASED COSTING   8-1

HOW ABC SYSTEMS ARE BUILT   8-4

ACTIVITY-BASED MANAGEMENT   8-13

THE THIRD DIMENSION— THE POWER OF ATTRIBUTES   8-17

ABM BENCHMARKING AND BUSINESS PROCESS REDESIGN AND REENGINEERING   8-24

UNDERSTANDING COST VARIABILITY FOR ACCURATE COST ESTIMATING AND BUDGETING   8-25

UNDERSTANDING CUSTOMER DEMANDS AND PROFITABILITY   8-30

THE UNIFICATION OF QUALITY, TIME AND COST DATE   8-31

THE RISE OF ACTIVITY-BASED COSTING

ABC is a simple concept, yet implementors and users find it complicated. Perhaps there have been too many articles and seminars stressing ABC concepts or benefits and not enough information explaining ABC. This guide is intended to eliminate the mystique surrounding ABC.

First, it must be understood that ABC is only data. If someone effectively and creatively uses ABC data, it will serve a purpose and prevent mismanagement of time and resources. Although ABC is only data, it can be very powerful and spark project teams or decision makers to take new steps or draw innovative conclusions. In this sense, ABC is part of change management. ABC is also an enabler for continuous improvement and decision support, which makes tools such as just-in-time (JIT), total quality management (TQM), and business process reengineering (BPR) more focused, effective, and profitable.

ABC is not a financial reporting system designed to serve regulatory agencies such as the IRS or SEC, but it may be in the future. ABC provides managerial information in financial metric form. Financial denominations, such as dollars, yen, marks, or pounds, serve as measures for the language of business. ABC communicates dollars to nonfinancial managers in a superior manner because ABC physically mirrors the activities of people and equipment. ABC communicates to people how many resources are used by activities, as well as why the resources are used. When people use common sense to implement ABC's clear, relevant data, it enhances their understanding of the data; thus, ABC creates benefits.

WHY ABC IS BECOMING POPULAR

Similar to JIT, ABC was described by accountants in the 1800s and early 1900s. Today, it is commonly heard that the new managerial techniques are simply a repackaging of old techniques. This is generally true regarding ABC. Although the basic mechanics of ABC are simple, the nature of what is costed—namely products and processes—is changing. As shown in Fig. 8-1, complexity, variety, and diversity in businesses have dramatically escalated. An increasingly complex business environment leads to higher overhead costs—simply put, complexity breeds overhead.

The manner of conducting business has shifted from the past, but costing practices have not correspondingly shifted enough. Consequently, there are grossly distorted valuations of product costs and inadequate managerial accounting information. ABC simply brings cost information to the level needed to make decisions and gain a competitive edge in the current business environment.

The Contributors of this chapter are: Gary Cokins, CPIM, Principal Consultant, Electronic Data Systems; Jack Helbling, Logistics and Cost Manager, Soap Sector, Procter & Gamble; Alan Stratton, Vice President, Cost Technology, Inc., Portland, OR.
   The Reviewers of this chapter are: John P. Campi, President, Genesis Consulting Group, Inc., Racine, WI; Robert C. Creese, Professor, Industrial Engineering Department, West Virginia University; Tom Pryor, President, ICMS, Inc., Arlington, TX; Julie Sahm, Manager of Client Services, ICMS, Inc.

CHAPTER 8

THE RISE OF ACTIVITY-BASED COSTING

In addition to increasingly complex businesses, there has also been a mix-shift in the elements of cost, as shown in Fig. 8-2. The direct costs of touch-laborers and purchased materials are being displaced by overhead costs. Overhead costs are, in reality, technology and people who sustain productivity gains and manage the complexity.

The overhead or indirect costs are typically controlled today by using responsibility cost center budgeting. Budget controls are becoming less effective for managing businesses because of growing dependencies among and between departments and functions. Departments and functions are now being ridiculed as organizational silos and stovepipes. Budget variance management at the account line-item level has always been controversial as a method of control. Good performers are either good negotiators at budget creation time or they know how to shift charges and credits between accounts. There is gamesmanship to budgeting.

In the last ten years, there has been much more recognition of and appreciation for the cross-functional behavior within businesses. Department walls are coming down, and there is less throwing-the-order-over-the-wall behavior. At the same time, managers are collectively rising above their walls and viewing the interconnectivity and mutual dependencies between their departments. Responsibility cost center budgets lose effectiveness as the internal supplier-customer chain is followed. For example, when a purchasing agent saves a dollar buying a substandard product, other departments will make up for it in rework, overtime, or quality, or in another nonvalue-added activity.

ABC is not old wine (data) in a new bottle (revolutionary accounting). It is new wine from an old bottle. What has changed is the introduction of the computer. Relational databases and fourth generation languages now allow the rapid reorganization of data. As previously noted, the proliferation in mix, variety, complexity, and diversity, as well as the displacement of direct labor and material costs by overhead, has overwhelmed traditional cost accounting practices. Without ABC, executives and managers are forced into a guessing game about what things really cost.

THE MULTIPLE USES FOR ABC

From a historical perspective, ABC was first used to describe improved product costing. Professors Robert S. Kaplan and Robin Cooper of the Harvard Business School became leading spokesmen by articulating in business periodicals how grotesque misallocations of overhead could distort the true costs of products. In conventional systems, direct labor stated in hours or dollars has traditionally been relied on as the basis to assign overhead costs to products. Kaplan and Cooper noted, as did engineers and product managers, that significant amounts of different overhead activities, from testing to material handling, are disproportionately consumed by certain parts, products, and product families. Traditional burden-averaging and labor-based cost systems do not capture the disproportion. Today's cost accounting systems are not mirroring the true economics of physical production and resource cost consumption. ABC provides a much closer match between costs and outputs.

ABC corrects the distortions so that people can know what processes, services, and products truly cost. In addition, ABC includes costs well beyond those used to compute inventory costs such as selling and distribution expenses. ABC has no readily identifiable boundaries because it is a managerial system and therefore depends on how it will be used. ABC can provide "total delivered cost" information.

Pioneering companies, such as Hewlett-Packard, became early ABC implementors by launching experimental pilot ABC models.

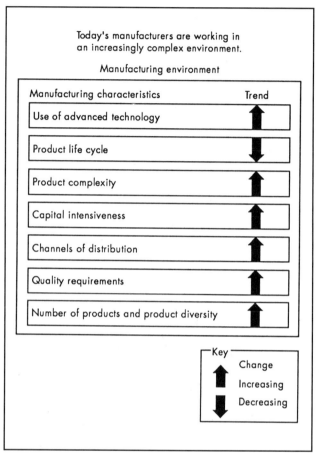

Fig. 8-1 Current business environment.

As time passed, practitioners learned ABC's capabilities, gained ABC experience, and expanded ABC's applications. ABC has evolved into activity-based management (ABM), which is a more encompassing term and includes the managing of costs, as well as a more proper assigning of costs to processes and products. In this chapter, ABC is the tool that identifies and computes costs for activities, processes, and the outputs of activities, such as products or services. Refer to Fig. 8-3 for CAM-I's (Computer-Aided Manufacturing-International) definition of ABC.

ABM, on the other hand, provides information that focuses on the management of activities using ABC data and other tools to achieve continuous improvement. Refer to Fig. 8-4 for the definition of ABM.

In the following pages, ABM is introduced as a method for reporting costs, also for managing them. However, do not equate managing with controlling. ABC-ABM data is used far more for predictive modeling than for control. Today, cost data for purposes of control has been eclipsed by faster feedback from total quality management, such as statistical process control practices, or from real-time, integrated information systems.

WHY ACTIVITY-BASED MANAGEMENT?

Why use activity-based management in addition to activity-based costing? ABM provides data to empower teams to reengineer business processes, to identify waste, to reduce cycle times, and to accomplish these tasks profitably. These improvements are achieved by providing activity-based metrics that traditional

CHAPTER 8

THE RISE OF ACTIVITY-BASED COSTING

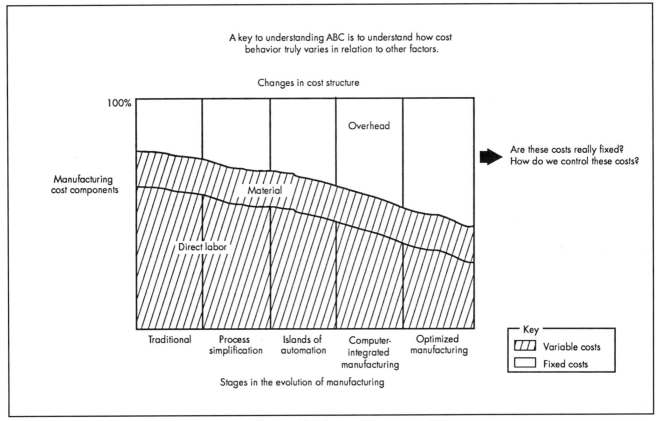

Fig. 8-2 Key principles of ABC. (*Courtesy The CAM-I Glossary of Activity-Based Management, 1990*)

Fig. 8-3 What is ABC? (*Courtesy The CAM-I Glossary of Activity-Based Management, 1990*)

Fig. 8-4 Definition of activity-based management.

accounting cannot provide. Figure 8-5 shows that, for data users, ABM provides the opportunity to receive and use relevant information. For data producers, ABM provides the opportunity to add significant value to the management process.

Companies are walking on thin ice if they believe their cost numbers from a traditional system. Generally, managers are under an illusion that if the accountants can produce the numbers, the system must be working. They are confusing information systems with costing practices.

ABC-ABM can be implemented in service businesses as well as discrete, job shop, and process manufacturing businesses. To date, most implementations have been in manufacturing environments. However, service businesses can also realize significant

CHAPTER 8

HOW ABC SYSTEMS ARE BUILT

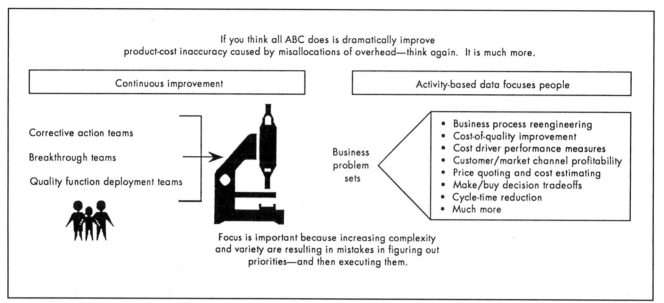

Fig. 8-5 Why use ABM?

benefits by understanding how cost behavior relates to business requirements. Service providers, such as banks and insurance companies, are deploying ABC-ABM.

How costly is it to implement and maintain an ABC-ABM system? Do the benefits exceed the costs? Is the climb worth the view? It depends. Since there is so much freedom and flexibility to design ABC-ABM systems, it makes sense to initially keep it simple and minimize the time and effort in collecting and calculating the data. Use shortcuts and workarounds. Estimates in ABC-ABM stand the credibility test because cost materiality and relevance are considered in the design phase. Further, Pareto's Law where "few account for many" is prevalent in understanding cost behavior.

The following section, "How ABC Systems are Built," relates how the assignment of costs is distorted and flawed. Understanding how to correct for errors of "misallocation," exposes the framework used in the section, "Activity-Based Management," and throughout the rest of this chapter. From that section to "Understanding Cost Variability for Accurate Cost Estimating and Budgeting," the focus is on performance improvement.

HOW ABC SYSTEMS ARE BUILT

THE IMPORTANCE OF UNDERSTANDING ABC DESIGN

Make no mistake about this next point. When ABC pilots are successful, the primary reason is that the pilots were designed for operational personnel to receive utility from the ABM data. The product costing data falls out as extra bonus information.

The explanation for this is behavioral. When operational people are initially involved in revamping an enterprise's cost practices and cost system, the new system reflects their needs. They buy-in and sustain its use. They like the new cost data to be reliably reported at timely intervals. If the accounting personnel revamp the system, operations personnel will perceive the event as just another meaningless financial exercise. Remember, it is easier to implement ABC than it is to sustain it.

Regardless of who utilizes ABC-ABM data, it is critical to understand how an ABC-ABM system is constructed. One does not *buy* an ABC-ABM system and plug it in. Like other programs, it is a process. But it is more tangible with hard data resulting from its design.

It is easier to learn how ABM works by first learning how ABC works. Consequently, the remainder of this section explains what action-oriented, operations personnel might call an instruction manual for constructing product costs. But this section is much more. It explains the underlying principles of costing that universally apply for both valuing and managing products and processes.

ABC-ABM is not strictly about costs. It is about resource use and consumption. The following section shows how costs are traced and assigned based on cause-and-effect behavior and on relationships with cost drivers. Granted, this section simply reslices overhead costs without affecting the total cost. The following section shows how companies can reduce overhead by managing resource consumption.

THE ABC-ABM FRAMEWORK

ABC design first focuses on activities. Activities are what people and equipment do to satisfy customer needs. Activities are what consume business resources. The lack of an effective activity-based focus would make ABC just another cost accounting system. By focusing on activities, instead of on departments or functions, ABC makes it possible for ABM to be a powerful tool for managing, understanding, and, most importantly, profitably improving a business.

CHAPTER 8

HOW ABC SYSTEMS ARE BUILT

The ABC model in Fig. 8-6 is similar to a pump, valve, and pipe system—costs are pumped to activities using special and sensitive valves. All resource consumption, represented by general ledger costs, is first accumulated in the general ledger using conventional business systems such as payroll, accounts payable, journal entries, and so forth. In the process of determining activity costs, all resource costs are transferred through the pipes into activity costs.

Resource costs represent people, computers, technology, equipment, machines, supplies, tooling, and other factors. These factors allow productive activity and the serving of customers, whether internal or external.

Because ABC-ABM focuses on activities, activity-based reports are more informative than traditional month-end department or cost center statements produced by the general ledger. As shown in Fig. 8-7, activity costs accurately mirror what an organization does—for better or for worse. Activities are custom-defined when a team designs the ABC system.

Activities are best defined using an active verb and object grammar convention. For example, create labor routings. Managers and employees relate to costs that are described in this manner, because they know they can change or affect an activity performed by a person or a machine. In some cases, they may choose to fully eliminate the activity (determined to be "non-value" adding).

After enough practice, custom-defined systems within a company may be supported by standard activity definitions using an activity dictionary. However, the initial learning experience (build from scratch) is important and valuable for project team members.

The weakness of traditional general ledger reports is that expenses are reported by department and spending account. General ledger reports only describe what is spent, while activities describe how expenses are spent. When managers review their monthly budget variance reports from the general ledger, they are either happy or sad, but not necessarily smarter! ABC corrects this.

After defining activities and computing activity costs, cost assignment is repeated. This time, activity costs are assigned to the cost object that uses the activity. Cost objects are usually parts, services, ingredients, products, customers, or distribution channels. Activity cost drivers recognize the proportionate discharge of each activity cost into its cost objects. Well-designed ABC systems remove skewed cost distributions by minimizing overhead averaging that is prevalent in today's allocation-based designs.

As shown in Fig. 8-8, activity-based costing is a two-stage cost distribution system. Products and customers consume activities. Activities consume resource costs.

Advanced ABC-ABM implementors eventually recognize that multiple steps can comprise the first-stage cost assignment, the resource drivers (the second-stage cost assignment uses activity drivers). After general ledger resource costs are unbundled and accumulated into activities, additional optional steps are taken to redistribute the activity costs into macro activities or processes. For example, the activity "unscheduled machine repair" may draw resources from multiple departments. This activity cost might be an intermediate step to be combined with similar activities to feed the macro activity of the machine, such as "drill holes." Experienced cost accountants will recognize this as the "step-down allocation," but with ABC it is accomplished at the activity level, not percent-of-department. This is a critical distinction, because it allows for assignment based upon activity driven relationships which are cause-and-effect in nature.

Ideally, all costs could be directly charged to activities and then directly assigned to end products. However, the cost and complexity of data collection can exceed the value of the improvements from information. In addition, most overhead costs are difficult, if not impossible, to directly charge to cost objects (for example, products, services, and customers). Traditional cost accounting arbitrarily allocates nondirect overhead costs, which corrupts cost integrity. To ABC-ABM, allocation is a dirty word. ABC-ABM resolves misallocations by using resource and activity drivers that reflect unique consumption patterns and link cause-and-effect to the cost-assignment process (see Fig. 8-9).

Misallocation occurs because the variability of indirect and overhead costs is not always proportionate with the allocation base. For example, distributing the cost of buyers in the purchasing department based on labor or machine hours makes little sense. Instead, ABC-ABM would use the number of purchase orders or another measure that more directly links the consumption of the costs of procurement activities to those parts and products that place demands on those activities.

Figure 8-10 shows that in ABC, direct material costs are attached to products using a bill of materials (BOM) or a formula. Similarly, direct labor costs may be attached using a routing bill of labor. Despite the direct attachment of material and labor costs, a large portion of resource costs remain as indirect or overhead costs.

ABC attaches indirect costs in a logical way to the consumers of these costs using cause-and-effect reasoning. First, ABC links resource costs to activities based on effort expended or material consumed. ABC then uses activity drivers to attach indirect activity costs to cost objects in proportion to the activity cost's consumption by the cost objects. Using a bill of activities makes it easier to understand the true cost of the cost object.

In job shop or custom manufacturing businesses that have less repetitive end-items, laborers and engineers might complete a work order or project accounting input form that effectively serves as a direct labor cost tool. ABC's two-stage distribution scheme is used to trace the remaining indirect costs to where they are consumed.

ABC's additional bill of activities serves, in effect, as a proxy direct-bill charge. An ABC bill of activity charge is less precise than a direct charge, but it is far superior to using allocations that have little or no correlation to how costs are consumed.

The defect of traditional cost systems is allocations that do not correlate with the resources they are supposedly consuming. As shown in Fig. 8-11, ABC removes such distortions.

The task for ABC project teams is to shift the cost assignment path from right to left on Fig. 8-11.

To date, many companies use ABC only to perform product costing. Some companies use ABC to determine costs consumed to serve different customers. In this case, the mix of customer services, not products, causes a significant portion of overhead costs. These costs are not proportionately consumed by activities for different customers in relation to sales or margins. Customers with large sales are not necessarily profitable to a supplier. ABC makes it easy to align and explain customer costs with profits or losses.

First-time observers of ABC fear a nightmare of astronomical amounts of data that must be routinely collected and edited. In the next sections, these fears are eliminated. Remember, with ABC, closeness is better than precision.

TOOLS FOR INFORMATION COLLECTION

Resource cost drivers are an innovative concept in ABC modeling. In traditional systems, financial controllers usually

CHAPTER 8

HOW ABC SYSTEMS ARE BUILT

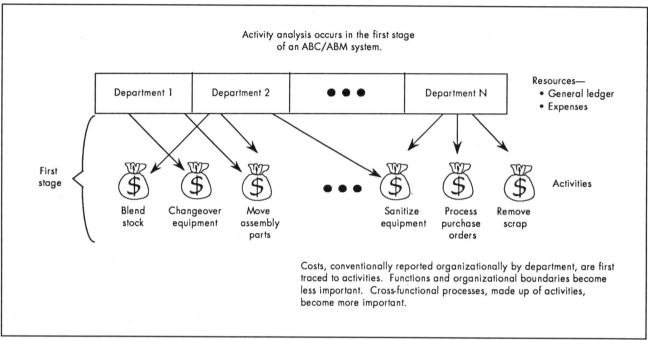

Fig. 8-6 ABC model.

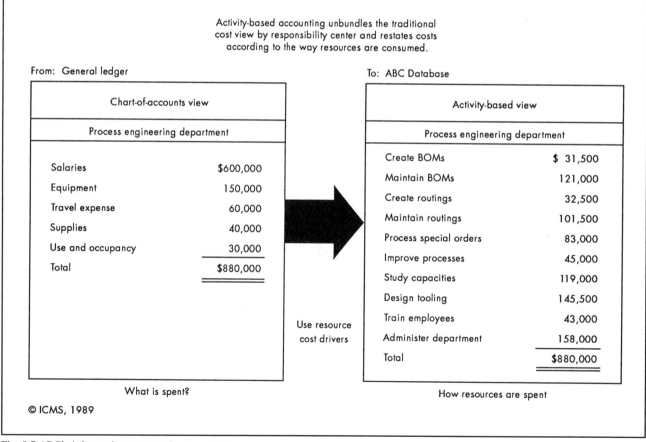

Fig. 8-7 ABC's informative report. (*Courtesy ICMS*)

CHAPTER 8
HOW ABC SYSTEMS ARE BUILT

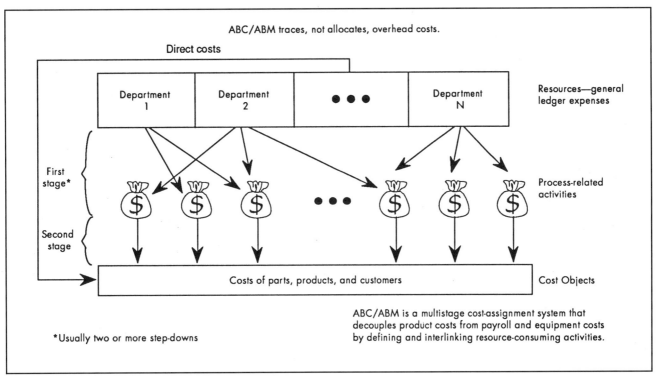

Fig. 8-8 ABC—the two-stage cost distribution system.

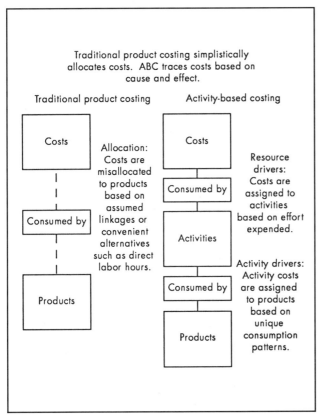

Fig. 8-9 ABC improves upon traditional costing. (*Courtesy Hal Thimony, The Clorox Company*)

create elaborate, step-down, sequential allocations that distribute service department cost to production work centers based on percentage estimates. In addition, these allocations are often flawed because they assume arbitrary relationships, for example, using square feet or head count. These measures do not reflect disproportionate resource consumption. Such burden-averaging techniques are convenient for accountants, but not for users of information.

With ABC, resource cost drivers replace the step-down allocations with cause-and-effect relationships at the activity level, not at the departmental level. Figure 8-12 shows that resource drivers are reasonable estimates of time, effort, or cost, and are often obtained from interviews. In some cases, labor reporting systems and work order systems, for example those used by maintenance departments, are additional sources of data.

Interviews and survey forms are not the only source for estimating resource drivers, but they are the most common source. Additional tools of data collection include:

- Observation.
- Time keeping systems.
- Questionnaires.
- Storyboards.

ABC activities are defined in physical terms, such as moving scrap. ABC users like these terms because they help them to relate to an action and to understand what receives or benefits from the stated action.

Today, few companies use ABC to generate a monthly accounting statement or variance report. Many companies use a representative time period under normal operating conditions, such as a fiscal quarter or one year. In these cases, ABC is a static snapshot of a time period with the time duration analogous to a camera's film exposure. If the mix of activities and their content changes quickly, ABC snapshots should be taken at more frequent intervals.

CHAPTER 8

HOW ABC SYSTEMS ARE BUILT

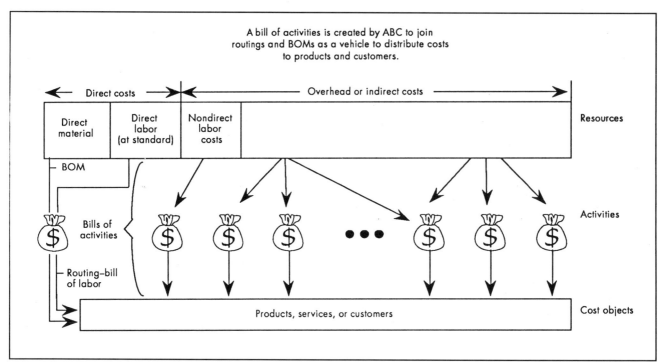

Fig. 8-10 ABC-ABM model—cost-assignment view.

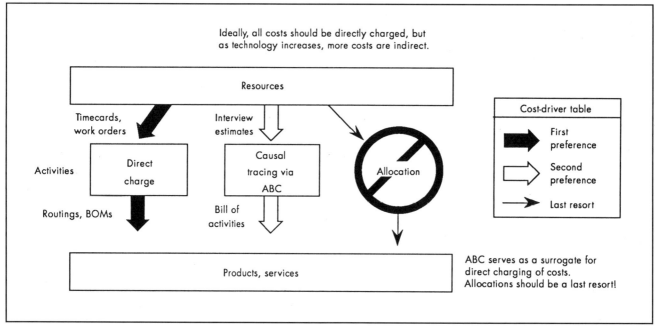

Fig. 8-11 ABC removes distortions.

If a dynamic picture of actual costs is desired for controlling activities and evaluating performance, managers should not refer only to accounting data. Figure 8-13 reveals how companies can be "better, faster, and cheaper" by leveraging financial and nonfinancial data. Managers should use metrics that encompass and combine measures for quality, time, and cost. Managers should never examine any one of these metrics in isolation.

Ultimately, performance measures combined with an enforced accountability for them is what really matters. One gets what one measures, and what gets measured, gets done. ABC-ABM assists in pruning dysfunctional measures and aligning the reformed measures with goals and strategies.

Timely and meaningful feedback to employees is critical to manage activities and processes. If necessary, ABC-ABM can

CHAPTER 8
HOW ABC SYSTEMS ARE BUILT

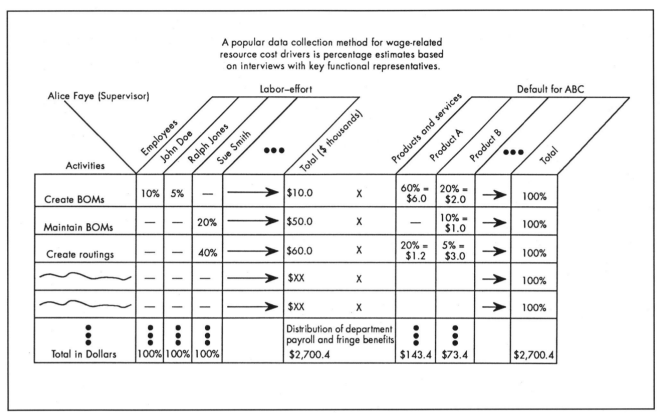

Fig. 8-12 ABC-ABM data collection.

generate a monthly income statement analysis. However, this task should occur well after the off-line ABC-ABM model is first accepted as a tool. It is important for employees to view ABC-ABM as a predictive cost planning tool, not as a control tool.

The next topic of discussion is how to define and structure the cost-assignment parameters and features. These concentrate on achieving more accurate product costing. Process costing, although involved, will be covered in detail in the section "Activity-Based Management."

DEFINING ABC-ABM TO SATISFY REQUIREMENTS

The first step in determining activity cost is to use the resource cost drivers to distribute factory overhead costs, and other product logistics and selling costs, into activities. A simple way to eventually distribute overhead resource costs to products is to imagine two separate broad routes. One route traces resource costs to equipment-related activities. The other route traces nonequipment-related, or people, activities. Figure 8-14 shows the importance of dividing activities according to what machines do and what people do.

Equipment-related activities are usually associated with technology and usually include unrecoverable (sunk) costs, as well as utility and energy costs. The maintenance and repair cost of engineers are often included in this category, as well as specialized, indirect support costs. Equipment-related activities represent technology-related resource consumption, such as depreciation expenses. An exception is when specialized, indirect people support an equipment-related activity.

Equipment-related activity costs are traced to cost objects, such as parts or products, using activity drivers that have the same cost behavior characteristics as the activity. These are usually unit-volume based, using activity drivers such as labor hours, machine hours, or units produced.

Nonequipment-related activities are usually indirect and people-effort intensive, such as material handling and order processing. These costs are traced to cost objects using activity drivers that likely have a step function, such as material handling batch loads or equipment setups. Most of these costs do not vary one-for-one with unit-volume outputs or machine hours. These costs are, rather, the result of quantity-insensitive batches, for example, the number of products, number of vendors, or number of engineering change notices (ECNs) of the cost object. Many of these overhead costs, such as a product family manager's salary, are exclusive to only a subset of the cost objects. Traditional cost systems spread these costs across all cost objects—ABC does not. In general, product complexity and variety causes nonequipment-related overhead in the form of people, the most flexible resource.

In addition to direct product-conversion work, the laborers classified as direct labor perform a variety of indirect tasks, such as material handling, for a surprisingly large portion of their work. These costs are more likely to be nonequipment-related. Direct laborers' indirect activities should be distributed to cost objects in the same manner that activities of employees who are classified as indirect laborers are distributed to cost objects.

ABC does not recognize any manner of class distinction—employees are employees, and their resources are consumed. ABC does not recognize employees as being either direct or indirect. Now that overhead costs have been recast as activity costs, ABC

CHAPTER 8

HOW ABC SYSTEMS ARE BUILT

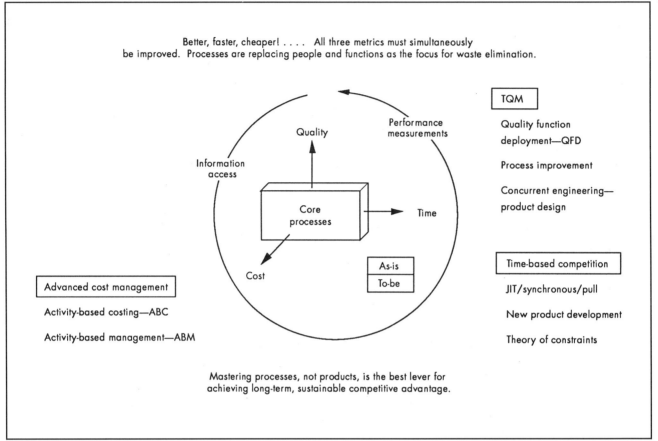

Fig. 8-13 Processes have three measures.

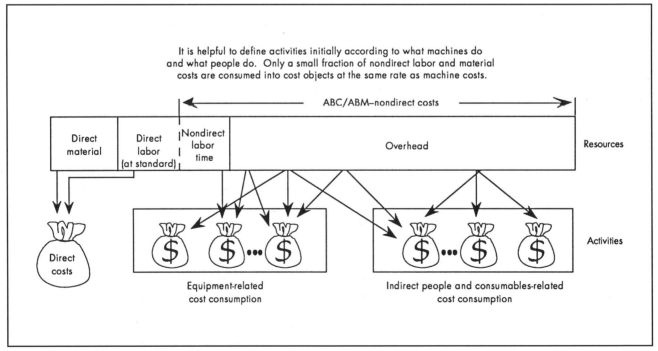

Fig. 8-14 ABC-ABM activity mechanics.

CHAPTER 8

HOW ABC SYSTEMS ARE BUILT

must additionally recast activity costs into those cost objects, such as products, which uniquely consume them.

The bottom of Fig. 8-15 shows how the ABC method uniquely calculates the distribution of each activity cost for each activity cost driver's measurable consumption by each cost object. In the final step, these elements of overhead cost for each cost object are summed across. For assembled items, these component part costs are rolled up through the product structure to give the total product cost. Powerful ABC software is currently commercially available. This software uses relational database technology to provide a reverse audit trail that reports the sources of all costs. In addition, database technology allows the attribute data that will be described in the section "The Third Dimension—The Power of Attributes" to accompany the activity cost information.

In reality, dividing resource costs into two categories is artificial. There are many ways to organize activity costs. However, the division of resource costs aids the design team in keeping the model accurate and relevant.

Companies with low direct-labor content and high equipment costs can substitute part features (for example, number of holes) for the traditional time-based activity cost driver. This data represents design for manufacturability (DFM) information that helps design engineers achieve target costs.

It is now understandable how traditional allocation practices actually misallocate costs. Why the misallocation occurs is explained in the next section, which is based on published research by Robert Kaplan and Robin Cooper.

MAINTAINING CONGRUENCY

As Fig. 8-16 shows, traditional cost systems view costs as a dichotomy of either fixed or variable costs and are often criticized for heavily spreading the fixed, indirect costs across products. Traditional cost systems allocate large overhead cost pools assuming the allocation base, such as direct labor hours, exactly varies with or reflects the degree to which those resources are consumed. ABC expands an assumed homogeneously behaving cost pool into multiple smaller heterogeneous cost pools, each with its own unique activity cost driver (for example, allocation base). With ABC, the heterogeneous cost pools and their activity drivers have similar parallel behavior.

As Fig. 8-17 shows, an ABC design team must recognize activity differences and activity-consumption differences. The team must also avoid combining activities from different levels of the variable and cost-intensive activity hierarchy, for example, combining unit-level activities and batch-level activities. If the team traces an activity to a cost object using an incongruent activity driver, the team is still spreading costs the traditional way and is continuing to introduce distortion error.

Note that the lowest category of the hierarchy—facility-sustaining activities—is extremely insensitive to changes in unit or batch variability or in the mix of cost-intensive activities. In effect, cost objects do not cause these costs to occur. It is recognized that these are fixed costs (or highly discretionary costs), which are inevitable and ultimately to be recovered with volume, any volume. There is a subtle difference between support costs and staff costs. Support departments serve other departments including the direct line-based departments. Staff departments perform administrative duties and would be classified as facility-sustaining. The section titled "The Third Dimension—The Power of Attributes" covers the controversies from full-absorption costing.

PROFIT DISTORTIONS FROM TRADITIONAL COST SYSTEMS

Markets are increasingly being segmented into smaller units, which are separately addressed with their own desired product and service variations. Although production managers would like a trend reversal back toward standard one-size-fits-all products, the growth of the consumer movement will not let it happen. The days of large consumer segments buying standard products have gone. The new market reality places competitive importance on the ability to manage a diversity of product offerings. Traditional cost systems mask the impacts from both diversity and variety and are providing highly inaccurate data.

Figure 8-18 graphically describes the resulting difference between traditional product costing and ABC product costing. The dashed horizontal line shows zero percent deviation from traditionally calculated product, the baseline cost. The vertical axis is the percent error deviation between the traditional and new ABC calculation. On the horizontal axis, individual part or product numbers (or services) are ranked left-to-right, from the most overcosted to the most undercosted. Although the shift in total overhead costs among products exactly nets to zero, the shaded areas will not equal because the percent deviation on the vertical axis does not reflect volume; it is on a per-unit basis.

Note how large the differences are. Errors are not five to ten percent. They are 50% to over 10,000%. Logarithmic-scaled graphs are needed to fit the measured error on a single page! This result has given rise to a popular observation: it is better to be approximately correct than precisely inaccurate.[1] Diversity, complexity, and product variety are satisfactorily handled in ABC-ABM. The design team strives to show diversity by examining the amount and extent of cost behavior that each activity contains. Cost behavior is linked to cost objects based on unique distinctions and features in the cost objects that consume each separate activity.

ABC VERSUS TRADITIONAL COSTING

Generally, products on the left side of the graph in Fig. 8-18 have high unit volumes, large lot sizes, limited engineering and technical support, minimal technology, and low complexity. Products on the right side of the graph have low unit volumes, small lot sizes, considerable engineering and technical support, and a large technology investment. The horizontal S-curve's height represents the degree of diversity. The ABC S-curve correctly reveals the distortions from misallocations of the traditional cost system. It shows a dramatically more accurate cost of products or services.

The shift in product costs illustrated by the S-curve in Fig. 8-18 is only part of the big picture. An alarming picture of ABC-computed profit is revealed when ABC costs are matched with total volume and sales dollars for each product or service sold.

Figure 8-19 shows cumulative ABC dollar profit, by product, based on highest-to-lowest ranked product margins. The $1.8 million profit at the right is correct and auditable; the steep incline and descent are, in effect, unrealized profits and losses simply displayed by ABC. The conclusion is that companies have the potential to make a greater profit on individual items than managers ever imagined. However, this profit can also be simultaneously lost. Financial profit and loss statements only report the correct total profit. ABC reveals how products comprise that profit (or loss), product by product.

Management's initial reaction to the right side of the graph shown in Fig. 8-19 is to consider dropping unprofitable products. However, some semi-fixed costs, such as the maintenance or

CHAPTER 8
HOW ABC SYSTEMS ARE BUILT

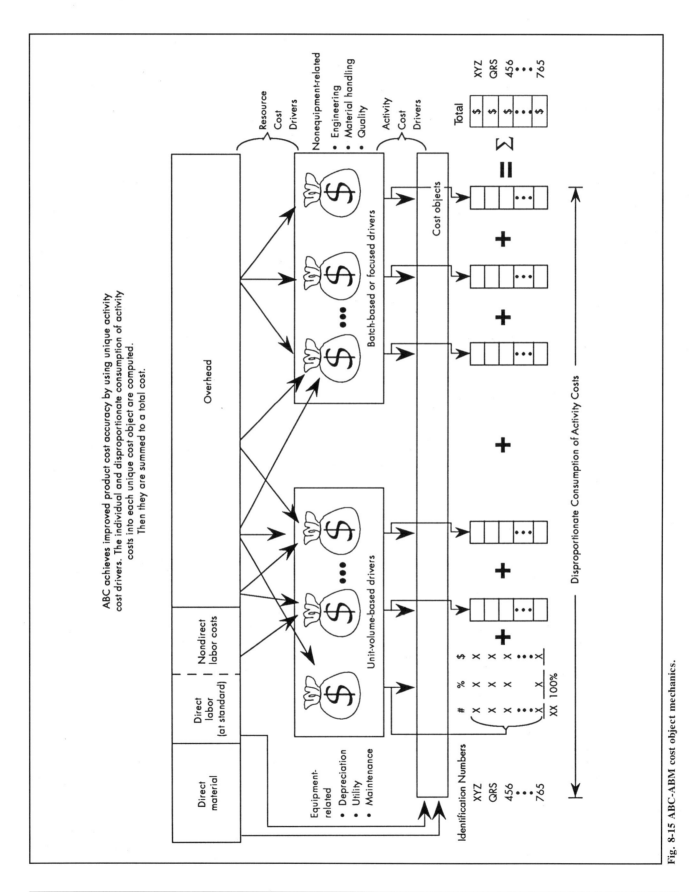

Fig. 8-15 ABC-ABM cost object mechanics.

CHAPTER 8

HOW ABC SYSTEMS ARE BUILT

inspection functions, may remain. These semi-fixed costs may cause an overhead death spiral when applied to the surviving product's costs, since the unprofitable products may go away but not all of the costs. Additional analysis of each data point along the graph leads managers to consider other options, such as:

- Repricing.
- Rationalizing product variety (consolidation and elimination).
- Changing minimum order policies.
- Purchasing some loss products from a clueless competitor and repackaging them for resale.
- Increasing specific product sales volume by creating incentives for the sales force.
- Redesigning the product or service (for example, design for manufacturability).
- Eliminating low value-added activities to reduce cost.
- Improving activity efficiency and reengineering processes.
- Reexamining site strategies.
- Reconsidering sourcing.
- Accepting the situation with understanding of the impact (do nothing).

Regardless of management's reaction, having a product profitability profile as shown in Fig. 8-19 is better than acting blindly without one. The response that management makes with this profile is, at least, an informed response. Companies with a product profitability profile have a competitive advantage.

Parts and products do not cause all overhead and activities to exist. Customers, markets, and channels of distribution also cause costs. Suppliers acknowledge that the same manufactured product can cost significantly more to sell to one customer compared to another. The section "Understanding Customer Demands and Profitability" explains how the ABC framework can also measure customer costs and their profitability as viewed from the supplier. The ABC model simply adds customers as cost objects. The calculation method for products and customers is identical and effectively seamless.

ABC is not a silver bullet. Here is a caveat because ABC for costing cost objects, like products, does have shortcomings. ABC reports a cost-time slice with full period expensing and no consideration to amortize long-term payback expenses, except those formally accounted for such as depreciation. For example, in research and development, a large amount of expenses for an abbreviated time period will be traced to a product thus overinflating the product's cost. Similar distortions can occur for seasonal products. All products have a life cycle, with front-end cost loading before the care and attention to them eventually settle down and costs stabilize. ABC only measures cost consumption for the period duration. Life cycle product costing, though not popular today, will likely be launched from insights gained via ABC.

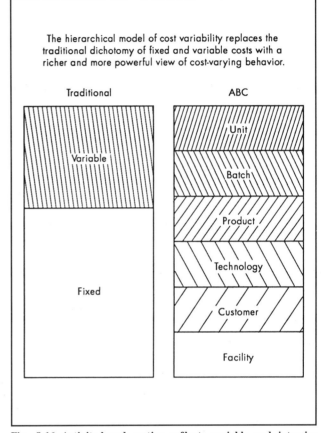

Fig. 8-16 Activity-based costing reflects variable and intensive behavior. (*Courtesy Robin Cooper and Robert S. Kaplan*)

In addition, target costing and simultaneous engineering, which occur at the front end of this lifecycle, are substantially enhanced with ABC data. Design engineers are availed of quantified downstream costs which are currently hidden from them and are nonvalue-adding.

ABC is regularly called a model as opposed to a system, because it is an "as-is" snapshot of the current state and conditions. ABM allows one to use this data for gain. Predictive modeling with activity-based data and relationships can generate proposed "to-be" pictures of future costs.

Now that the cost-assignment model behind ABC is understood, the section "Activity-Based Management" will present the more critical lessons involving the managing of costs using the same principles of cost consumption via drivers.

ACTIVITY-BASED MANAGEMENT

ABC-ABM'S DUAL AXES

Activity-based accounting serves as both an accurate product-costing tool and a performance-improvement tool (see Fig. 8-20). The cost-assignment view on the vertical axis (ABC) assigns resource costs to activities and activity costs to cost objects, such as products and customers. The process view on the horizontal axis (ABM) concentrates on managing processes and their constituent activities plus evaluating activity performance. ABM provides activity-based information to focus employee efforts to continuously improve quality, time, service, cost, flexibility, and profitability.

CHAPTER 8

ACTIVITY-BASED MANAGEMENT

Concentrating on the cost-assignment view provides an increase in product cost accuracy. This leads to better strategic decisions for pricing, product mix, sourcing, and product design. The process view provides operational and tactical data to improve performance.

Regardless of activity accounting's use for product costing or process management, the best use for ABC-ABM is to help managers to manage. ABC-ABM brings focus to the more important priorities and drives employee and management behavior in a business environment that is shifting toward increasing managerial complexity and an increased mix of products and markets. Figure 8-21 shows how resources are converted into output as explained in the previous section.

Figure 8-22 illustrates that an operational cost driver causes activities to use resources for accomplishing work and yielding output. These kinds of data are used for diagnostics and deeper analysis.

DEFINING ACTIVITIES AND THEIR LEVEL OF DETAIL

The capability to value products, services, and customer costs using ABC, and to analyze activities and processes using ABM, introduces a problem caused by the increased flexibility of using activities as a fundamental building block. This dual capability may confuse ABM system designers when they attempt to specify an activity's level of detail.

Detailed definitions of activities are not absolutely necessary to improve product cost accuracy. The key is in defining activities, which can be broadly aggregated into summaries. As described in "How ABC Systems Are Built," using appropriate activity drivers, mainly nonunit-volume drivers, corrects the majority of misallocations caused by spreading overhead costs based only on labor or machine hours. Cross-subsidies of under- and overcosted products are sufficiently corrected by a limited number of activity and activity driver definitions. Extensive detail on activities is not worthwhile for improving the accuracy of product or customer costing because of diminishing returns from incremental expansion and subdivision of activities. The extent of detail should be no more refined than that required by uses or decisions made with the data.

More detailed information is necessary, however, on activities that assist and benefit operational managers to effectively influence activities at a tactical level. At the tactical level, the trade-off is added detail versus the added cost to collect more data (see Fig. 8-23).

The most difficult part of ABC model designing is to define activities at an appropriate level of detail to satisfy the primary and (hopefully) predetermined purpose for the system. ABC designs for strategic purposes, such as product-mix profitability, use high-level summarized activities that are infrequently updated. Designs for operational purposes, such as cost reduction or process improvement, use a low level of detail and refinement of activities that are more frequently updated.

Do not mistake initial ABC pilots as a replacement for the financial accounting system that serves investors and regulators. ABC is a managerial system and the interval between updates depends on user requirements and the rapidity of change (for example, new equipment, new products, changed processes, different employee jobs, and so forth). Eventually, ABC will directly interface with the general ledger, but, today, this inter-

---

Activity-based costing employs both unit and nonunit volume-based cost drivers.

Driver trait		Level of activity	Description	Examples of activities
Variable		Unit volume	Performed every time a unit is produced	• Drilling a hole • Supplying electricity
Nonunit variable		Batch-related	Performed every time a batch is produced	• Setting up a machine • Moving a batch • Ordering a purchased part
Disproportionate	To products	Product-sustaining	Performed to enable a product to be produced	• Engineering a product • Marketing a product
	To processes	Technology-sustaining	Performed to enable a technology to produce a product	• Maintaining a machine • Attending a course
	To markets	Customer-sustaining	Performed to service customers and prospects	• Delivering a product • Resolving complaints
Fixed; discretionary		Facility-sustaining	Performed to enable production to occur	• Lighting the factory • Using janitorial services • Paying rent, insurance, taxes

Fig. 8-17 The hierarchical model of cost variability. (*Courtesy Robin Cooper, Harvard Business School*)

CHAPTER 8
ACTIVITY-BASED MANAGEMENT

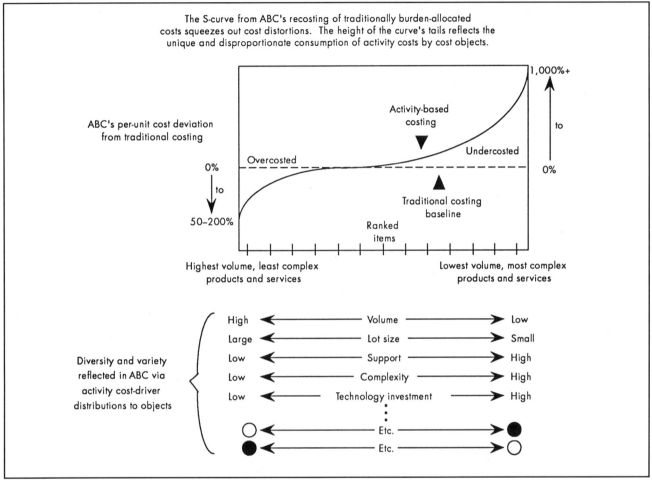

Fig. 8-18 Percent over and under product cost.

facing would be a distraction to a project team's objective—to achieve acceptance of ABC-ABM from managers and employees.

The next section deals with the roles of activities and drivers and puts them in the context of managing or altering processes for improvement.

THE ROLE OF ACTIVITIES

Activities are central and critical to an ABM system. If they are initially well-defined, the implementation project will overcome any implementation problems.

Businesses are now asking questions that can help determine the level of detail and intentions that are required to define activities. Business and operations managers rarely begin by asking questions such as:

- How much is being spent to move materials?
- How much is being spent to order parts?
- How much is being spent to fix breakdowns?

Instead, managers are asking:

- What is the cost of complexity?
- What is the cost of waste in the system?
- How much is a 10% improvement in reliability worth?
- What is the cost of idle capacity and where is it?
- How can throughput be increased?
- What should be focused on?
- What should be changed?

The answer to these questions is in knowing what the costs and value-added content of activities within processes are and how costs change with changes in the mix of activities and redesign of processes. Therefore, the initial ABC-ABM model design should begin with questions that a business is asking to solve its problems. Activities must be defined so that these business questions are answered. If the definition and assignment of dollars to activities are poorly planned in pilot models, the entire model will be viewed as useless by its users and will probably delay future upgrades to an enhanced and advanced cost management system.

Since there are many correct model designs, it is advantageous to focus the model design on a specific objective or problem set to be solved. Focusing the model design places emphasis on certain activities by using key words or phrases to define an activity, or by directing an activity definition toward influencing a certain behavior.

For example, if a company is undergoing a TQM campaign, the model should emphasize quality through waste elimination and conformance to expectations. If a company is pursuing a customer satisfaction, cycle time reduction, or time-to-market campaign,

CHAPTER 8

ACTIVITY-BASED MANAGEMENT

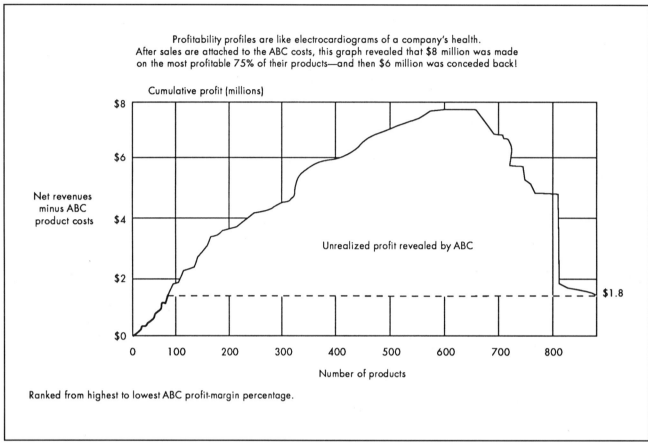

Fig. 8-19 Product profitability profile.

the model should be accordingly flavored with relevant campaign grammar. For a model to succeed, it must incorporate the values that management endorses.

The likelihood of an ABC-ABM project success will increase with the project team's attention to understanding problem sets of high interest to potential users. The team should strive to know in advance how users will use this new cost data to make decisions and effect changes (the section, "ABM Enables Continuous Improvement" expands on this important idea).

THE ROLE OF ACTIVITY AND OPERATIONAL COST DRIVERS

An activity driver measures how much of an activity is used by a cost object. It is a measure of output and is integral to ABC product costing. An activity driver provides a bridge among ABC's informational elements. This important bridge distributes activity dollars into cost objects. However, if the activity driver does not redistribute activity dollars into each cost object in some proportion to some diversity that is unique to that cost object, the design team has missed the objective and continues to introduce error into product and cost object data.

Activity drivers may not be the true drivers of cost in the sense of triggering or being the root cause that impacts an activity. These are called operational cost drivers. Activity drivers are consequences of what happened, whereas operational cost drivers reveal what is making it happen. Costs tend to be incurred at the process level, not at the activity level. Operational cost drivers are factors which influence a change in the cost of several related activities, whereas activity cost drivers measure the frequency and intensity of the demands placed on activities by output-oriented cost objects.

Since activity drivers are integral to product (or cost object) costing, they must be measurable—for example, the number of setups or material moves. In contrast, operational cost drivers may be less measurable, but more insightful or directional as a cause, such as inventory levels or machine schedule imbalances. Operational cost measures are more useful for measuring performance since they are highlighted at a causal point, but the activity driver is usually easier to measure performance as an output.

As shown in Fig. 8-24, activity drivers and the operational cost driver's work units for performance measures use similar data that may have different purposes. An activity driver mirrors the consumption of an activity by its cost objects. The operational cost driver mirrors how efficiently an activity (or group of activities) is performed.

In Fig. 8-24, an activity driver for issuing purchase orders could be the number of parts on the product's bill of materials. This consumption measure, however, does not reveal the efficiency of this activity. The output of issuing purchase orders is the operational cost driver—the number of purchase orders issued. A performance measure for issuing purchase orders could be the cost-per-purchase-order issued. The next section discusses how additional managerial information can be attached to each activity in the form of attributes.

CHAPTER 8

THE THIRD DIMENSION—THE POWER OF ATTRIBUTES

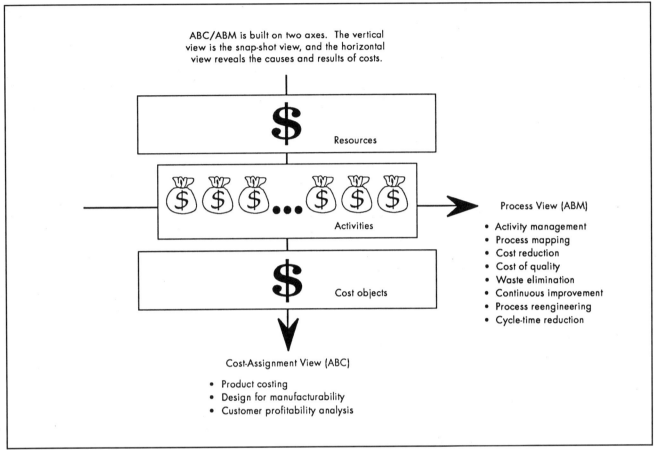

Fig. 8-20 The cost-assignment and process axes.

THE THIRD DIMENSION—THE POWER OF ATTRIBUTES

Benchmarking, business process reengineering, and performance measurement are important to managers. ABC-ABM can facilitate an increasingly fair computation of measures that aid comparisons, such as plant-to-plant, without the apples-to-oranges problems that managers currently confront.

Activity-based data can have additional managerial information, such as quality or time-speed, included in the form of attributes. This more robust activity-based data then becomes a truly powerful tool not only for strategic and tactical use, but also for behavioral change management.

HOW ATTRIBUTES TURN HUNCHES INTO FACTS

Attributes are descriptive labels given to activities. A popular attribute is one for nonvalue-added activities. Attributes make activities robust. As shown in Fig. 8-25, attributes are the third dimension of the ABC model. Activity attributes are an orderly way to accumulate data for making business decisions. Attributes quantify many different aspects of business processes, and they provide multiple concurrent views with which to focus, prioritize, and analyze detail and post measure.

Often companies intent on computing more accurate product costs using ABC pause after the first stage of collecting activity data. They redirect their efforts to use this fresh and valuable cost data to change processes before continuing with the next stage to cost products.

Activity analysis promotes creative ways to associate activities with attributes. Commonly used attributes include value-added content and cost-of-quality attributes. No limit exists for the type of attributes that companies may invent. By using attributes, ABC supports emerging management improvement programs such as business process reengineering and benchmarking. ABC also supports related value-engineering and process-focused programs.

Much is being said and written about shifting paradigms. The cost management discipline is experiencing changes that are making it support the new thinking on achieving customer satisfaction and effecting continuous improvement, as well as innovative breakthrough change for improvement. Traditional cost management systems were intended to control costs using

CHAPTER 8

THE THIRD DIMENSION—THE POWER OF ATTRIBUTES

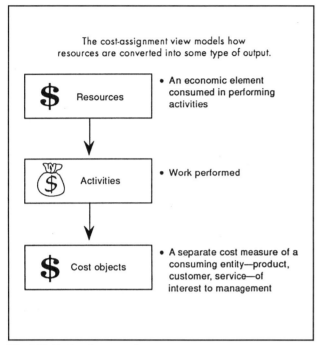

Fig. 8-21 Cost-assignment view.

cost-based budgets, standards, variances, and measures narrowly reported for each responsibility cost center.

There has been a shift away from this budgeted cost variance paradigm for several reasons. First, financial accounting data for control purposes has been superseded by TQM-based quick-feedback systems, such as statistical process control. Second, as businesses downsize and become more cross-functional for increased flexibility and customer satisfaction, the focus is toward managing processes, of which many involve several different departments.

A business process is a string of sequential and related activities with a purpose to achieve a specific goal. Costs may be measured as activities, but they are incurred at the process level. Examples of popular core business processes include fulfilling a customer's order, developing a new product, purchasing items, or enacting engineering changes. The new cost management paradigm focuses on understanding processes. Activity-based cost management data is being deployed for prospective forward-planning diagnostic and analytical uses more than for control-intended after-the-fact score keeping.

ABC-ABM AS A CHANGE MANAGEMENT TOOL

ABC-ABM should be thought of as a change-management tool. It is critical to understand processes. Once processes are understood, companies can achieve customer satisfaction, total quality management, time-based competition, and all of the other popular strategies, campaigns, and programs used to initiate change within the enterprise.

Managers require answers to many questions so that they can understand their business environment. Attribute analysis is a way to answer these and other questions:

- Why are there so many nonvalue-added costs?
- Why do costs of nonconformance exceed costs of conformance?
- What portion of costs can actually be locally controlled with responsibility?
- What portion of overhead costs vary with unit volume, or with batches, or are specifically product-sustaining, technology-sustaining, or customer-sustaining costs?

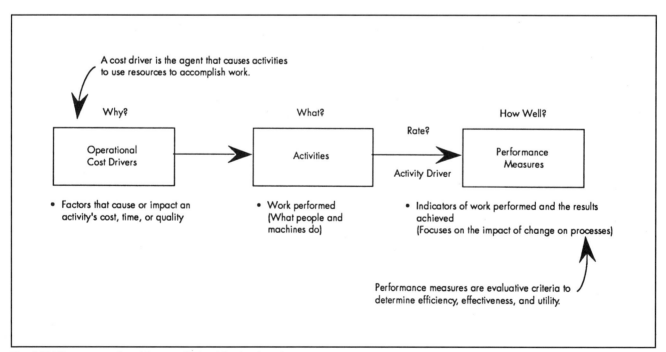

Fig. 8-22 The process view. (*Courtesy National Semiconductor*)

CHAPTER 8

THE THIRD DIMENSION—THE POWER OF ATTRIBUTES

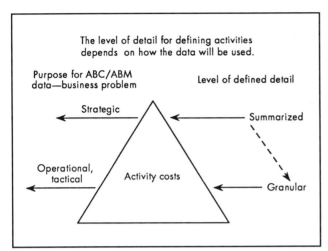

Fig. 8-23 Defining activity detail.

The success of installing an ABC-ABM model greatly depends on the level of acceptance from its potential users. Do not expect users to immediately feel comfortable with ABM. Initially, ABM may be viewed as an accounting gimmick, but attributes gain user attention. The insights that managers gain by using attributes tend to create interest for continuously using and refining the ABM model.

Figure 8-26 gives activities a three-dimensional appearance when attributes are added to each individual activity. Building ABC-ABM models and systems has been made easier using commercially available ABC software.

Various ABC-pioneering companies have imaginatively defined different types of attributes. Because of team creativity, it can be reasoned that new types of attributes will continue to appear.

Activity costs are what attributes are attached to. The initial use for activity costs is to rank activities by dollar amount and show them to employees and managers. People are either shocked at how they really spend money or they have their expectations confirmed. Either way, activity data provides useful feedback. Some day, ABC may become nicknamed "amnesty-based costing," because it brings out so much more of the previously hidden costs that were masked by misallocation practices. Examples of popular attributes, including a brief description, are listed below.

Value-added versus Nonvalue-added Costs

This classification technique is popular because it allows the prioritizing of eventual action steps that are important for gaining ABC-ABM acceptance (see Fig. 8-27). Assigning grades on a scale is straightforward, but defining rules to assign these grades can be a challenge. Many companies define key value-added activities from a customer's viewpoint. One asks "Given a choice, would a customer pay for this activity?" or "If you quit performing this activity, will the customer notice or care?"

An external customer's viewpoint is usually used to define the scale for value-added content, but an internal customer's viewpoint can also work in a complex enterprise. Some activities are mandated by government legislation, such as from the Food and Drug Administration; these may be specially coded outside of the value-added scale.

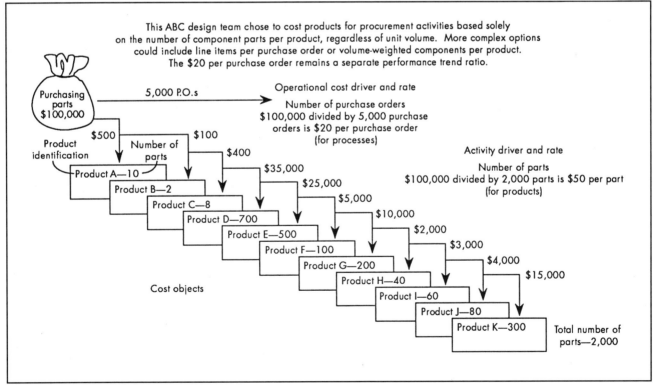

Fig. 8-24 Activity drivers are costed out. *(Courtesy Alan Stratton, National Semiconductor)*

CHAPTER 8

THE THIRD DIMENSION—THE POWER OF ATTRIBUTES

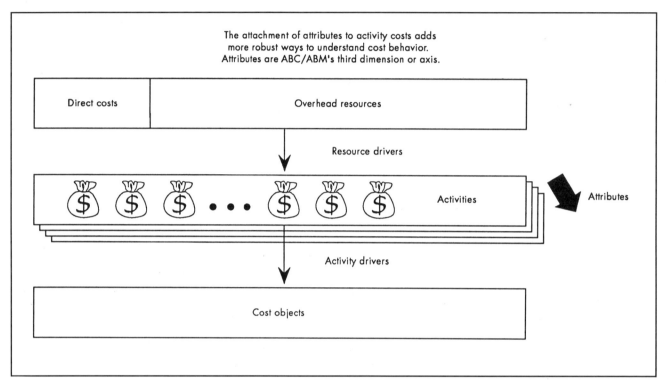

Fig. 8-25 Using attributes to understand cost behavior.

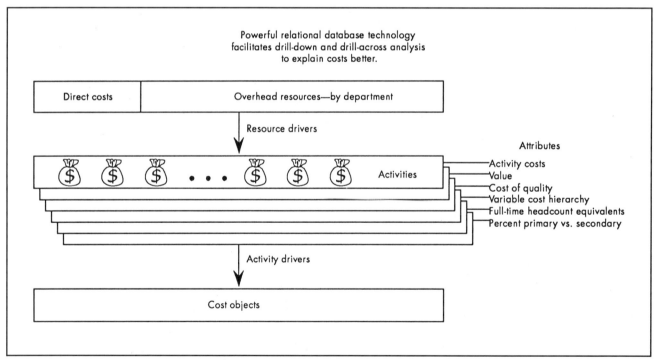

Fig. 8-26 Activity attributes.

Business Process Reengineering

Processes, such as administering engineering change orders, can be evaluated by combining two or more activities which have a common purpose. Figure 8-28 shows how collections of activities can be ranked by gross dollar cost. By including the value-added attribute, processes can be compared and evaluated in terms of their size as well as value-content.

The use of process value analysis using ABM data provides

CHAPTER 8
THE THIRD DIMENSION—THE POWER OF ATTRIBUTES

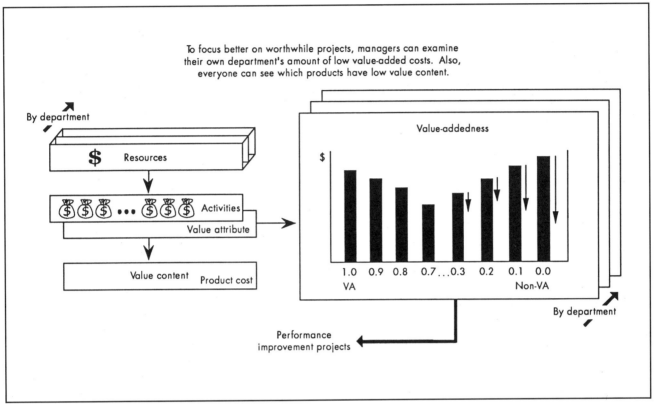

Fig. 8-27 Value-added versus nonvalue-added costs.

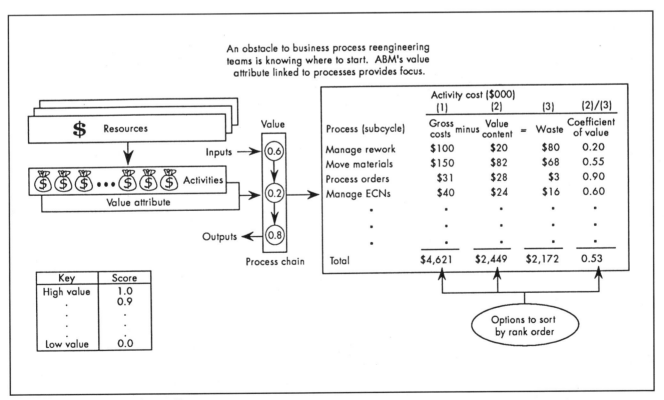

Fig. 8-28 Activity analysis using attributes.

CHAPTER 8

THE THIRD DIMENSION—THE POWER OF ATTRIBUTES

managers with a framework and systematic approach for planning, predicting, and influencing cost. This approach focuses management attention on the interdependency between departments and functional activities. Also, by analyzing the cost drivers of activities within business processes, managers can understand and act upon the causes of cost, not their symptoms.

When activities that do not add much value are reduced or eliminated, both cost and time are improved, as shown in Fig. 8-29.

Cost of Quality

This classification technique is consistent with the popular TQM leaders, such as Crosby, Juran, and Deming. As shown in Fig. 8-30, the technique places overhead activities into the following categories:

1. Cost of conformance.
   - Prevention activities.
   - Appraisal activities.
2. Cost of nonconformance.
   - Internal defect or failure activities.
   - External defect or failure activities.

Touch costs are necessary to convert materials and add value. Since touch costs are direct costs, they may be excluded from the overhead cost-of-quality categories mentioned above. Some companies estimate that their costs of conformance and nonconformance exceed 80% of their overhead costs. These costs can be reduced by employing TQM techniques such as root-cause analysis or employing process value analysis. Reducing the cost of nonconformance is important, but cost of conformance activities should also be considered as targets for improvement. All TQM activities have a cost that can be reduced.

Some companies are disappointed by the lack of actions following their TQM training programs. They claim difficulty in getting started. ABC-ABM cost-of-quality data, which is relatively easy to score, provides a spark to jump-start a quality program that has lost direction.

Cost-of-quality metrics are controversial in TQM, mainly because, in the past, users uncontrollably reclassified costs, sometimes including or excluding different cost types at each measure point in time. ABC-ABM introduces a closed system where changes must occur either inside activities, or as changes in activity-driver rates. ABC-ABM also substantiates any final cost-of-quality report with a list of detailed activities. In the past, the task to compute a cost-of-quality number was usually to satisfy an executive request, not to provide employees with a yardstick.

In summary, activity attributes add more to activity analysis than just the costs of activities. Attributes make activities more understandable, usable, and meaningful. Attributes also provide leverage for decision making.

ABC-ABM is fundamentally a database measuring the current conditions and rates of resource cost consumption. ABC-ABM's potential is unleashed when users manipulate the data by sorting and reorganizing activity costs, activity cost drivers, and outputs to create different reports with various attributes.

If the project team does not make changes or make decisions any differently with the new data, the team has basically failed with the implementation—it becomes a non-event. The use of ABC-ABM data to make changes should be considered as the test to determine the project team's success of implementation. The following points or tips should be considered to determine whether the goal is ABC or ABM.

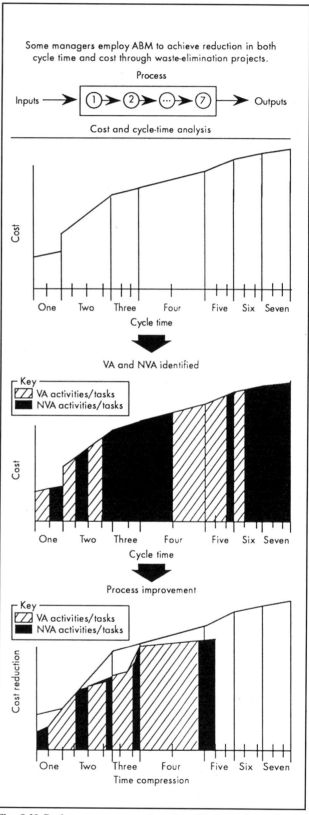

Fig. 8-29 Business process reengineering. (*Courtesy Coors Brewing Company*)

CHAPTER 8
THE THIRD DIMENSION—THE POWER OF ATTRIBUTES

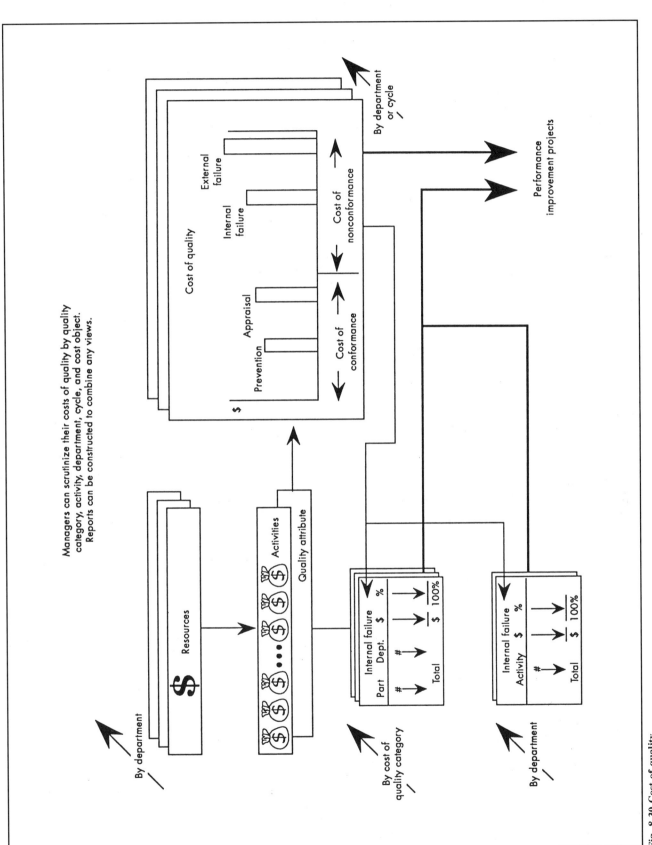

Fig. 8-30 Cost of quality.

CHAPTER 8

ABM BENCHMARKING

DESIGNING FOR ACTIVITY-BASED PRODUCT COST AND ACTIVITY-BASED MANAGEMENT

An activity-based product cost model, or macro model, is generally less complex than other options. This model is also typical of early, first-generation ABC efforts. In the past, the users of the new data were usually financially oriented. Product decisions from the ABC data usually involve product pruning, and rationalizing and optimizing the product mix. The design team will define activities in general terms and sensibly limit the number and diversity of activities. The outcome of the model is significantly improved product costs. The model will require updating as business conditions change, perhaps annually. Changes that lead to the occurrence of different activities or swings in their size require updating the model. Such changes include introducing new products, abandoning old products, or introducing new equipment, technology, skilled positions, or services.

Activities are used to capture and interpret data. When constructing a macro model, the level of detail for defining activities should not be more refined than the level of insight needed for decision making. The challenge is to keep this balance during design. The temptation to overdesign is strong and ABC software does not act as a deterrent.

An activity-based management model, or micro model, is generally more complex and refined than the macro model. Figure 8-31 shows the contrast between macro and micro model design. Note how the micro activity data is captured at the lower levels of an indented bill of activities. It is this ability to roll-up and consolidate micro activities into macro activities for product costing (ABC) that allows multiple uses of activity data in the same ABC-ABM system.

Two approaches exist for the sequencing of the ABC-ABM system design. Some companies focus first on product cost (ABC) and then on performance management. Other companies focus first on performance management (ABM) and then on product cost. For companies that focus on continuous improvement using ABM, the primary goal is process improvement.

As an example, National Semiconductor Corporation, an advanced user, recognized that the two views—cost-assignment (ABC) and process view (ABM)—serve two different management groups. The design team took advantage of this knowledge by defining micro activities to meet the needs of the managers who are directly responsible for performance. Defining micro activities allowed a manager to begin process improvements immediately. The micro activities were later consolidated into relevant product costing-oriented macro activities. The ABC-ABM team addressed both management groups without creating the massive data of two models, or sacrificing the needs of one group for the needs of the other group.

Management efforts for TQM or total cycle time reduction may initially yield results, but frequently are not effective in the long run because of a lack of reliable activity cost information. TQM projects suffer from an absence of financial metrics. ABM provides these useful metrics. The micro model addresses more business issues than product cost. The users of the ABM model include engineering and operations management, where the goal to improve the process is constant.

Some design rules for combining micro activities into macro activities using an indented bill of activities are:

Activities should have the same level of variability. For example, a batch-level activity should not be combined with a unit-level activity. A product-sustaining activity should not be combined with a batch-level activity. A facility-sustaining activity should not be combined with other nonfacility-sustaining activities.

Activities with similar variability should have the same activity driver. This ensures that the activity driver used for each macro activity reflects the activity consumption in a level, linear, and nondistorting manner.

Activities should be similar in function, which ensures that activities requiring separate visibility receive it, even if they use the same activity driver.

In summary, if companies understand the two views used to design the ABC-ABM system before beginning a project, they can initially capture activity data at a level of detail sufficient to serve tactical and operational purposes. For strategic purposes, individual activities can be combined and rolled-up into summary, or macro, activities, which are attached to products using a single activity driver. Individual activities can also be combined and rolled up into different macro activities (not used for product costing), which can equate to processes.

ABM BENCHMARKING AND BUSINESS PROCESS REDESIGN AND REENGINEERING

Benchmarking is a 1990s form of copying from the masters, similar to what apprentices do. Benchmarking involves sharing and full disclosure between partners. By predefining and standardizing in advance both activity and activity driver definitions, multiple sites can compare themselves on a level playing field.

Figure 8-32 shows how eight internal plant sites are able to rank themselves from worst to best for individual processes using ABM data. The activity driver per unit of process output is the measuring rod. Substandard performers now have data to analyze what better practices are required to be the best. More important, they have partners who will disclose how their best-in-class process achieves superior measurable results.

Although multiple activities comprise a process, for benchmarking, a process is considered as synonymous with an activity.

Why is business process redesign and reengineering becoming a popular management campaign? First, leading-edge companies coined the phrase "bleeding edge" when they observed their competitors rapidly copying their products without incurring the expenses of research and development. The lesson learned is that superior business and manufacturing processes offer greater and more sustainable advantages than products. Superior processes provide long-term advantages and are more difficult for competitors to copy.

Second, it is no longer sufficient to endorse continuous productivity improvement as an incremental and gradual journey. Outright breakthrough innovation must be combined with continuous improvement. Companies must first do the right things, and then do the right things well. Business process redesign and reengi-

CHAPTER 8

UNDERSTANDING COST VARIABILITY

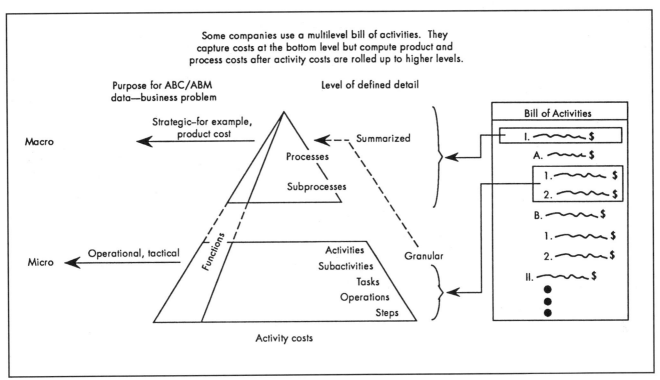

Fig. 8-31 Macro versus micro design.

neering encourages this behavior; its war cry, coined by productivity consultant Michael Hammer, is "don't automate, obliterate!" In effect, companies are realizing they cannot afford their existing overhead and infrastructure, and they must act more quickly.

Finally, downsizing and work force reductions, together with mission-critical cross-functional coordination, is resulting in a different employee base. Businesses are retaining more flexible workers to focus on and leverage the enterprise's core competencies, regardless of organizational structure and pecking orders. Beyond some level of employee cutbacks, lean and stretched organizations can no longer simply remove bodies to improve profits. Instead, they must extract waste out of the system's business processes.

Business process reengineering dramatically affects cost, quality, service, and speed. It is distinguished by results that are quantum leaps and order-of-magnitude improvements, rather than just day-to-day continuous improvements preached by quality management devotees. To accomplish innovation and breakthrough changes requires creative, step-out-of-the-box thinking and solutions. It further requires the cross-functional redesign of major processes. This becomes an expansion of scope for stove-pipe managers who are today limited to only office-talk about what is upstream and downstream from them—but in which they must now get involved. Business process reengineering is business as usual.

Figure 8-33 shows how ABM data is linked to processes. A business process is a sequence of two or more activities with a common purpose. Costs are incurred at the process level and are measurable at the activity level. The link of activity costs and their attributes (for example, value-added content, cost-of-quality, and so forth) with time-sequenced process steps provides the power for new insights.

At the bottom of Fig. 8-33, a traditional flow chart reveals how items are worked on and passed across workstations such as departments or functions. Note that the resource drivers from the input form exactly equate to the activities in the flow chart. This is not an accident. With advanced planning, activity costs with their attributes can be mapped to "as-is" processes.

The next step is to develop and achieve a "to-be" process that will better serve the customers of the process—hopefully faster and cheaper. ABC-ABM data joins time-based and quality-based metrics for the brainstorming needed to make the migration from the "as-is" to the "to-be" state happen.

The following section covers fresh uses of ABC-ABM data to gage the consequences of changes in activities.

UNDERSTANDING COST VARIABILITY FOR ACCURATE COST ESTIMATING AND BUDGETING

The "as-is" ABC-ABM model of cost consumption, including all of its cost pools, rates, and parameters, is used to predictively model outcomes to solve problems. First, however, the thorny problems involved with truly fixed costs (for example, facility-sustaining costs) must be dealt with.

CHAPTER 8

UNDERSTANDING COST VARIABILITY

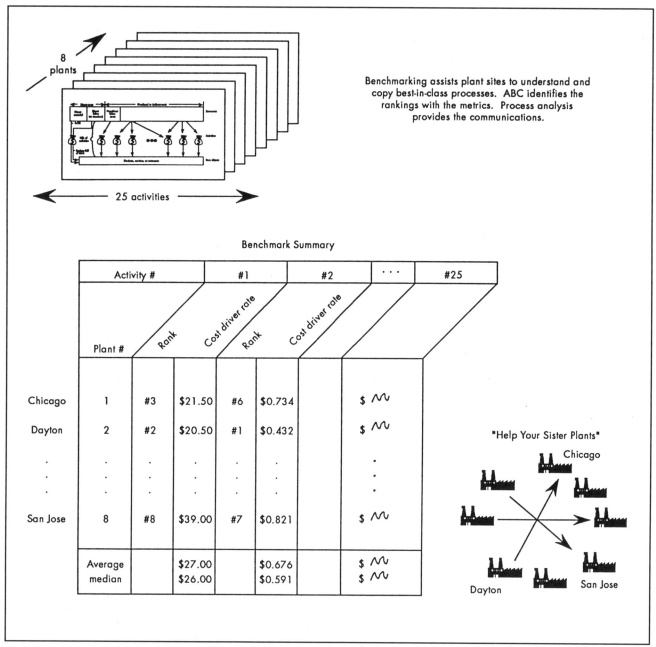

Fig. 8-32 ABC benchmarking.

THE MYTHS OF FULL-ABSORPTION COSTING

Variable cost behavior analysis shows how the magnitude of activities varies with volume and mix changes of cost objects. By classifying activities, they can be systematically matched to activity drivers with similar variable behavior. This analysis is critical for simulating what-if scenarios that gage the impact of proposed changes in sales and production on cost. These are the basic analyses used for justifying investments, for cost estimating new business and orders, and for cost budgeting.

ABC provides greater insight into the behavior of costs than traditional fixed-versus-variable cost analysis. There are facility-sustaining costs, such as recruiting, janitorial services, security officers, and paying rent and insurance. These costs are depicted as nontraceable overhead to the right in Fig. 8-34. Such costs are not caused by products or customers, which makes it difficult, if not impossible, to attach them to products or customers.

Facility-sustaining costs do not have direct cause-and-effect relationships on primary business functions and processes. ABC model designers are encouraged to minimize the size of facility-sustaining costs. They can accomplish this by aggressively examining initially classified facility-sustaining costs for potential causality by any activity. For example, fractions of the time spent by executive officers can be linked to some of the same activities

CHAPTER 8
UNDERSTANDING COST VARIABILITY

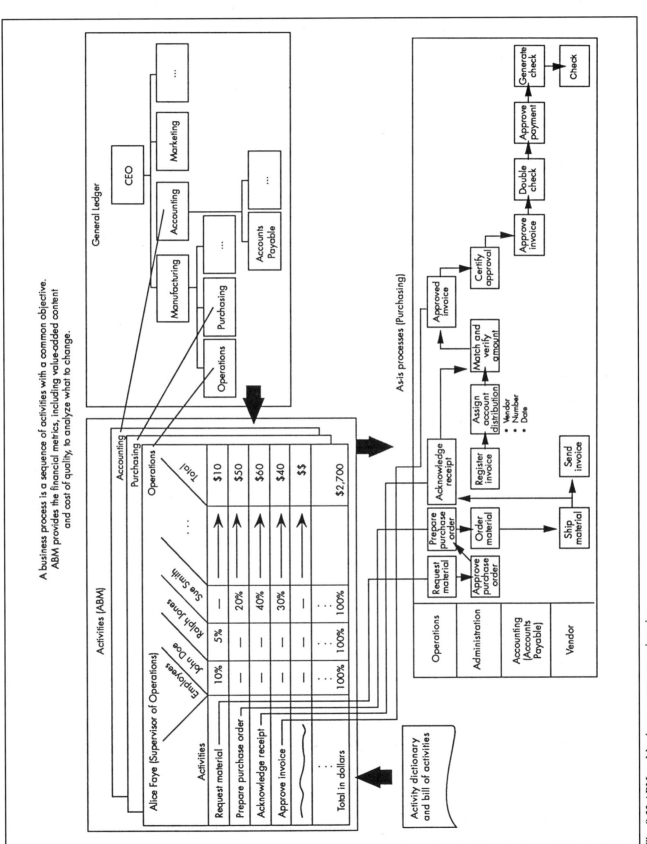

Fig. 8-33 ABM and business process reengineering.

CHAPTER 8

UNDERSTANDING COST VARIABILITY

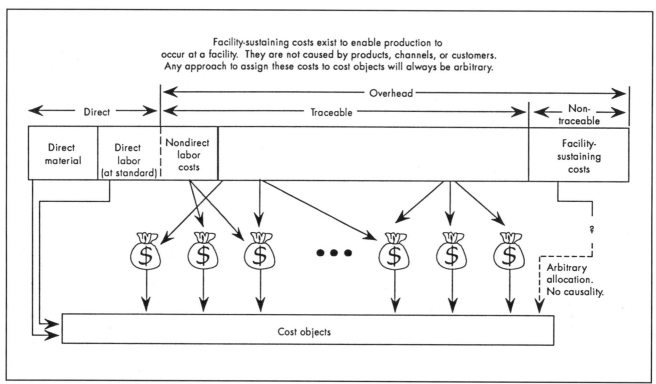

Fig. 8-34 Facility-sustaining costs.

their subordinates spend time on. Eventually, costs with only the highest fixed or extremely discretionary content will remain as a residual—the truly facility-sustaining costs.

When companies perform traditional cost estimating to bid for orders or new business, they add a tax-like cost-plus surcharge to recover the so-called fixed costs. Companies using the substantially expanded cost elements from ABC data are learning to selectively assign the recovery of these truly fixed costs to specific products or customers.

This is not accomplished by evenly spreading costs based on sales dollars or any other assumed link. Rather, the disconnect between these costs and any cost object is first recognized and then associated with those customers (or products) that can absorb the costs. The purpose for associating costs when there is no causality is for inclusion in quoting prices for new businesses.

Figure 8-35 displays how this practice gives the sales function a greater range to price, which may allow winning an order or customer. The primary caution, however, is to not repetitively price too many orders in this manner because the older base business will eventually expire without sufficient price coverage for the truly fixed costs.

Companies eventually begin to understand that products do not cause facility-sustaining costs. Despite this understanding, a company may insist that products absorb all costs. This insistence occurs because companies like to analyze profit margins that show the recovery or absorption of all costs. If a company insists on absorption of all costs, they can spread facility-sustaining costs after all direct and activity-based costs have been traced to cost objects. A reasonable allocation base could be the cumulative ABC costs for total product cost, because, at that point, all products will have been costed based on logic—a cost profile not to be distorted.

ACTIVITY-BASED BUDGETING OR ACTIVITY-REQUIREMENT PLANNING (ARP)

What follows are powerful ways to use ABC-ABM data for predictive modeling. Activity-based budgeting is the reverse engineering of the ABC cost-assignment framework. By estimating the quantities and volumes over a future time period for all of the cost objects and activity cost drivers, the level of resource costs can be computed. This technique is effectively an elaborate flexible budget; however, the flexing uses more factors than simply estimated units of output or labor hours, which are traditionally used.

To readers who are familiar with the inventory production method called material requirements planning (MRP), the reverse explosion math for placing purchase orders is strikingly similar. Therefore, ARP places demands on activities, similar to the way MRP places demands on suppliers. Activity costs are then translated into the mix and magnitude of resource costs.

Refer to Fig. 8-15 and imagine reverse computing the ABC model. Assume that the activity rates are calibrated from those used to initially construct the model. This new predictive model is computing the amount of activities based on the estimated mix, quantities, and volumes of cost objects and activity drivers, and must also equal the level of resources predictably consumed. Of course, this newly computed level of resources will never equal the enterprise's point-in-time payroll and indirect spending. In fact, the difference is actually a computation of a capacity difference. ABC effectively brings visibility to where excess (or deficient) capacity exists.

Activity-based costing measures consumption, not spending. There will always be a lag in the response time between when changes occur in the mix and quantity of cost object demand (for example, a drop in orders) and subsequent responding changes in

CHAPTER 8

UNDERSTANDING COST VARIABILITY

spending (for example, overtime, hiring, or downsizing).

Activity-based budgeting is appealing because it is consistent with the notion of continuous improvement. The estimated costs of resources can be immediately computed as soon as estimates for cost objects and activity volumes are collected. The focus is on identifying and managing excess capacity. Activity-based budgeting is also appealing as a replacement to the traditional budget cycle because of criticisms to current budget practices. Criticisms include:

- There is enormous time and cost by employees to produce the annual budget with questionable benefits.
- Data becomes stale relative to rapid product changes, rapid competitor events, and continuous internal organization changes during and after preparation of the budget.
- The budget format is excessively organizational (for example, departmental), while structure becomes less important and cross-functional integration becomes more relevant.
- Spending control can be accomplished with alternatives, such as with signature approval policies.
- After-the-fact budget variances usually reflect negotiating skills of supervisors at the beginning of the budget year and do not measure their performance in reacting to the variables that realistically happen.

ACTIVITY-BASED COST ESTIMATING FOR QUOTING

The same logic behind activity-based budgeting and activity-requirements planning is currently being used for quoting bids. Traditional quotes are constructed from rigorous estimates of the direct labor and material costs. Next, a spread percent factor is used, which presumes all new orders will consume overhead activities at an enterprise-wide average rate. This is nonsense.

An ABC cost estimate uses a checklist activity by activity. The quoted order is tested for its impact to place demands on activities. For those activities impacted, the activity driver is multiplied by the quoter's estimate of driver volumes.

For example, will a 200,000 piece order require four setups of 50,000 units, or 20 setups of 10,000 units? The batch-related costs will differ.

Similar to the S-curve in product cost subsidizing, ABC logic accurately predicts how quoted orders will disproportionately place demands on all activities and processes. Then it costs them out, using its calibrated driver rates. ABC can tolerate some error without much adverse effect.

CONTROLLING ACCEPTABLE IMPRECISION

Design teams often consist of accountants and engineers who have a tendency to create detailed models. ABC-ABM software does not deter teams from over-engineering their models; rather, it enables and encourages them to do the opposite. These teams can unknowingly use time and effort to develop elaborate activity driver profiles that affect only a small fraction of total activity costs. Such detailed models may use activity costs with less-than-imaginative activity drivers.

In addition, design teams tend to define inappropriate activity drivers for cost objects, usually at a level that is too detailed. In

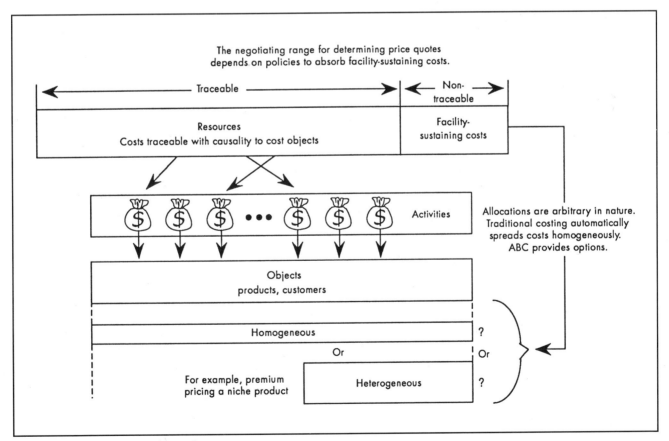

Fig. 8-35 Arbitrary full absorption costing.

CHAPTER 8

UNDERSTANDING CUSTOMER DEMANDS AND PROFITABILITY

this use, it is more acceptable to trace activities to a product brand or product family than to a specific product. It is also acceptable to trace activities to a market segment instead of to a particular customer. From this less detailed level, cost can be spread by annual unit volume or, more simply, the number of products or customers. At this stage, the model has separated most of the significant cost diversity. This makes it acceptable to allocate, which has been doggedly resisted. Developing the model from a less detailed level minimizes distortion from misallocation. Figure 8-36 shows why acceptable levels of imprecision are acceptable.

With ABC-ABM, closeness is much better and more practical than precision. Further, it is simpler, more economical, and produces less headaches.

Nonmanufacturing activity costs, often white-collar in nature, can be assigned directly to customers. This facilitates understanding profitability by customer.

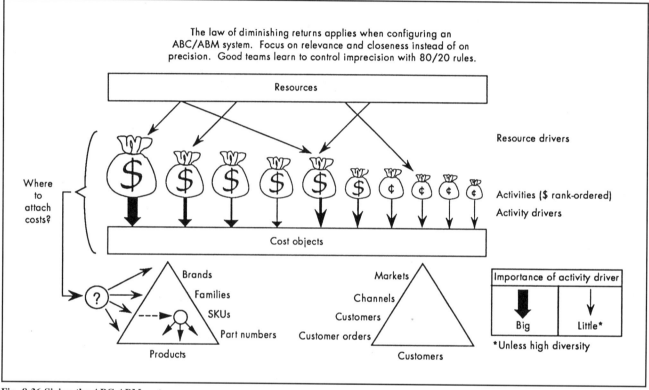

Fig. 8-36 Sizing the ABC-ABM system.

UNDERSTANDING CUSTOMER DEMANDS AND PROFITABILITY

Products do not cause all resource consumption, but they cause most manufacturing resource consumption. There are also many other types of resource consumption. Being further removed from the manufacturing floor and moving toward the production planning, engineering, and sales offices means that current production products have less impact on these types of white-collar resources.

Moving toward the sales offices means that customers and channels of distribution have a greater impact on consumption of administrative resources. These customer and marketing costs of conducting business deserve the same amount of attention that product costs traditionally receive.

As shown in Fig. 8-37, a parallel cost continuum, with business function coinciding with associated cost objects, can be matched process by process. As this resource-cost continuum moves (left to right) from production to administrative functions, the cost-object continuum also moves from parts to customers and markets served.

Note that the two main cost-object families, products and customers, can be expressed in pyramid hierarchies. This occurs because certain resources, such as a product engineer, cannot be attached at a part number level, but they can be charged higher in the hierarchy and then fanned or allocated to products within their product family.

As Fig. 8-38 shows, companies are beginning to recognize that customers with the highest sales volumes do not necessarily generate high profits. Figure 8-39 shows that the use of ABC principles provides companies with a more realistic view of the sources of profit. This view occurs because sales and marketing

CHAPTER 8

THE UNIFICATION OF QUALITY, TIME AND COST DATA

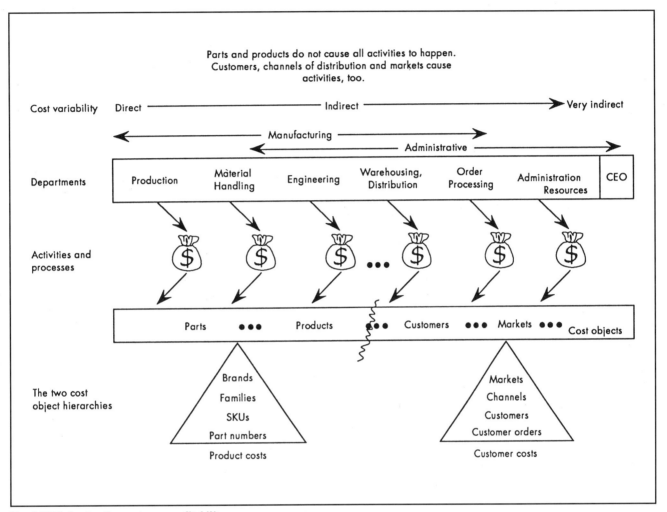

Fig. 8-37 Understanding customer profitability.

costs consumed by specific customers or markets are combined with the ABC manufacturing stock-to-dock product costs.

As increasing segmentation of consumer feature and function preferences occurs without any relaxation of management edicts to maximize customer satisfaction, ABC will assist in making tough decisions. Consumer demands for affordable prices with prompt and reliable service will keep cost as a highly visible element of any evaluation equation, despite increasing attention on the elements of service, quality, and flexibility.

THE UNIFICATION OF QUALITY, TIME AND COST DATA

Just as the falling of the Berlin Wall initiated substantial changes around the globe, a similar monumental event is occurring in the business world. Organizations that were always vertically structured are also tearing something down. Businesses are now recognizing that their stovepipe-like functional departments are impediments, and they are being tipped over sideways to better manage cross-functional processes.

The popular business buzzword for this is business process redesign and reengineering. The jolt that has tipped over organizational silos is the emergence of the concept that customer satisfaction is the only critical success factor; everything else is secondary. With this fresh thinking, manager-desired programs "of-the-month" are giving way to customer-benefiting programs.

Although business process redesign or reengineering may sound like another this-too-shall-pass management program, this one is different because there is so much *common sense* involved. Consider:

1. A *process* is the integrating theme for the organization of work.

CHAPTER 8

THE UNIFICATION OF QUALITY, TIME AND COST DATA

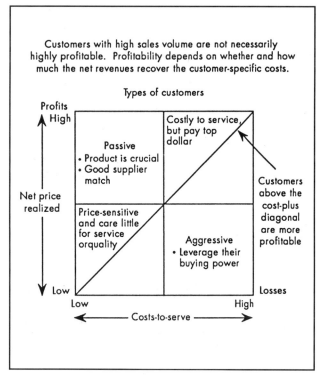

Fig. 8-38 Customer profitability matrix.

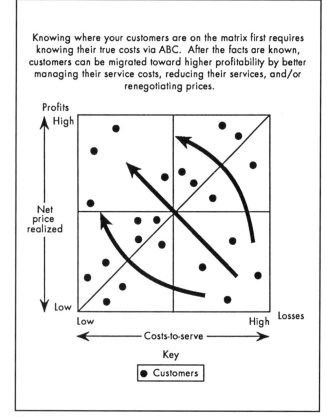

Fig. 8-39 Customer migration matrix.

2. A process consists of two or more *activities* having a common purpose, which usually involves a *customer* of some sort.
   - Activities are the foundation and building blocks for improvement programs.
     a. Total quality management is performing activities without error.
     b. Just-in-time flow control is doing activities without waste.
     c. Simultaneous engineering is incorporating into design not only what customers want but also the features that will minimize the low-value-adding *activities downstream*, such as assembling, inspecting, reworking, or repairing.
   - In effect, *business processes reengineering* synchronizes activities across functional boundaries. It forces customer-desired, not department manager-desired changes.
3. A horizontal, cross-functional view of processes is essential because the *current* vertical organizational structure repeatedly loses sight of the customer between department hand-offs.
   - Superior execution to achieve speed requires superior process design.
   - When people collectively better understand their core processes and those processes' purpose, they *change* things more *quickly* and more *radically*.
   - The *rate of organizational learning* is a critical differentiator between competitors, particularly in high-tech companies—which eventually is everybody.
4. Investing in processes and process innovation offers *sustainable advantages* longer than innovations in products or services, which can always be copied by competitors.

Business process redesign and reengineering involves both processes and organizational structure. Both must change.

Business process reengineering is the radical redesigning of processes to speed the flow of materials and to document communications and decisions. Only by viewing processes side-by-side can one understand the obstacle course and hurdles that slow down the movements of what it takes to make customers happy. Actions taken from these understandings result in companies that can finally make progress at getting a grip on how to accomplish things faster, better, and cheaper...all three simultaneously!

Quality programs focus on how to improve what people are doing. Reengineering focuses on *eliminating* what people are doing.

Business process reengineering is not blind head-count reduction. It is not tinkering with incremental improvements. It is not cutting fat or just reorganizing people. It is not brute force automation to cement outdated procedures or to simply do the same archaic practices faster. What it does is realign resources with the customers they are intended to serve.

As the pace of change accelerates due to fierce competition and emerging technologies, tools and techniques are not leading the charge for performance improvement as much as is the integration of all the tools. ABC and JIT were in practice fifty years ago, but under different names such as material flow control. Today, there is a convergence of all these managerial programs and philosophies. Figure 8-40 presents a genealogy chart that depicts agile manufacturing as the descendant of the many separately dedicated, but currently merging, advances in manufacturing.

CHAPTER 8

THE UNIFICATION OF QUALITY, TIME AND COST DATA

Agile manufacturing is the polar opposite of Henry Ford's mass production route to riches. In the past, the consumer had no choice—exemplified by the black Model T—and volume-based economies of scale dominated decisions. Today, widely segmented consumer preferences are being courted under the banner of maximizing customer satisfaction. Agile manufacturing emphasizes flexibility so consumers can order customized, one-of-a-kind products with quick delivery at affordable prices.

The bottom line in Fig. 8-40 is accounting's ancestry. Historically, accounting's mission emphasized precision and control based on hindsight. Although activity-based costing and management (ABC-ABM) is introducing new and potentially powerful data for strategic use and for predictive planning utility for managers, its reputation is tainted by accounting's old generation's fixation on numbers without relevance.

Most activity-based cost accounting systems are successful; but when there is trouble, it often occurs when individuals either are uninformed of the system's function or have not experienced a system implementation. Because ABC-ABM projects compete with other initiatives such as TQM, JIT, computer-integrated manufacturing (CIM), DFM, and so forth, managers may be reluctant to initiate such a project. To gain a growing acceptance for advanced costing techniques based on activity-based measurements, managers must understand two principles:

1. ABC-ABM does not require two financial sets of books for internal managerial and external regulatory purposes.

2. It is imperative that managers realize how ABC-ABM fits together with other business improvement programs (such as TQM, DFM, and so forth), which use time-based and quality-based data (see Fig. 8-41).

Misunderstandings about ABC occur because questions still remain from the first pilot ABC projects of the late 1980s. The debate revolves around two questions:

1. Does activity-based accounting simply correct the distortions from the traditional allocation-based method and compute revised product costs with more accurate product profits?
2. Or does ABC-ABM clarify the cost and degree of value-added activities and processes by restating general ledger-based reports into better information so managers can relate and respond regardless of what products cost?

The answers to these questions are that ABC-ABM accomplishes both of these and more. Based on recent activity-based accounting projects in companies that use ABC software, a majority of project teams begin with a strong sense of the business problems that must be solved and then learn from trial and error. This discovery process is currently the best way to apply ABC-ABM and to use the ABC-ABM data.

Figure 8-42 catalogs the various activity-based management applications that ABC-ABM project teams pursue. When cost-based data is combined with quality-based and time-based data,

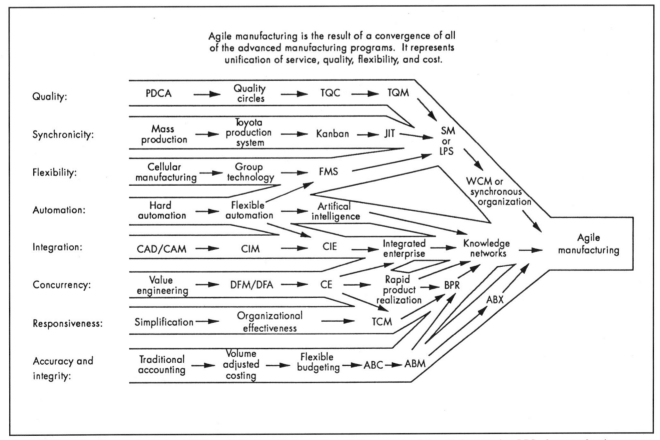

Fig. 8-40 Toward agile manufacturing (ABX—activity-based information; CIE—computer-integrated enterprise; LPS—lean production system; SM—synchronous manufacturing; TCM—total change management; WCM—world-class manufacturing).

CHAPTER 8

THE UNIFICATION OF QUALITY, TIME AND COST DATA

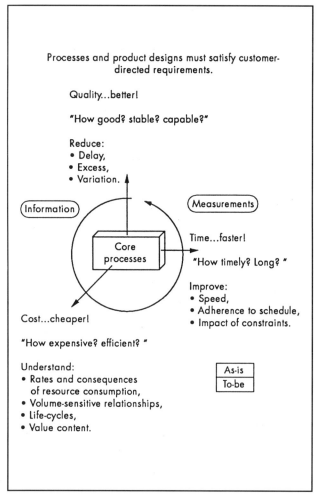

Fig. 8-41 Better, faster, and cheaper.

results get even better. The extent of uses is broad because activity-based data is an enabler that empowers teams and task forces. The strength of this management tool is its ability to measure and organize data at an extremely fundamental level where resource-consuming activities are performed. ABC-ABM makes it easier to understand variable cost behavior information. It dramatically supersedes simplistic categorizing of costs as either fixed or variable with a superior continuous view of cost viscosity—within processes. ABC-ABM also brings visibility to the degree of value-added cost content and cost-of-quality for activities and processes. This information can be used to enhance continuous improvement and breakthrough innovation projects.

Natural laws describe how resources are consumed by activities. Activity-based management systems have four conditions that impact cost consumption:

1. The mix of products, services, customers, and channels of distribution.
2. The physical design features of and the specifications for products.
3. The effects of processes that are often influenced by formal or informal policies.
4. The presence of waste due to neglect or nonconforming errors.

These conditions cause costs to occur. As shown in Fig. 8-43, insight gained from activity-based analysis will favorably impact resource consumption by reducing or eliminating activities, or by replacing an inefficient activity or unnecessary process with a more economical alternative. Today's popular corrective-action programs, intended to suppress these conditions, are product rationalization, design for manufacturability, and quality-managed waste elimination by reengineering processes with time-compression to better serve customers.

ABC-ABM will eventually be universally adapted because traditional allocation-based costing methods actually misallocate and mask the causes of costs. With ABC-ABM, cost drivers link causes to cost consumption and remove this masking of cost behavior. Once product-cost distortions are corrected using activity-based tracing methods, management realizes that more profits are generated by fewer products than previously recognized. Managers also realize that substantial erosion of profits comes from offering too many unnecessary products and services. When managers have a better understanding of the relationship between processes, products, and profits, corrective action can occur, including abandoning product lines, repricing, and targeting process improvements as candidates for reengineering.

Strategic applications for ABC-ABM are more obvious than tactical and operational applications. Today, a controversy surrounds cost reporting as a control tool. For example, Eli Goldratt, a theory-of-constraints advocate, believes that cost accounting is the "number one enemy of productivity." Other professors, including Thomas Johnson of Pacific Lutheran University, argue that cost control and reporting are not effective for improving process controls and efficiency. These controversies disappear as ABC information is used in conjunction with other improvement philosophies to profitably focus improvement efforts. The improved awareness that comes from the ABC-ABM process view will evolve with experience. ABC-ABM will emerge much more as a planning and decision-making tool than as an after-the-fact report on control performance.

Companies can reduce difficulty getting started with ABC-ABM and TQM programs by initiating the discovery process using a repetitive plan-do-check-act (PDCA) approach. ABC-ABM systems should be concurrently designed while quality and process redesign teams construct root-cause diagrams, map processes, and so forth. All project teams will eventually unite and reconcile the different ways they define and account for activities within processes. ABC-ABM data initially serves as financial metrics to aid in identifying project opportunities (ABC's diagnostic phase) and, later, to stimulate ideas for options and alternative processes (ABC's analytical phase). After improvement projects are prioritized and completed, ABC-ABM is used to measure the cost impact and consequences of change (measurement phase).

ABC's strategic value is its ability to correct product-cost distortion. ABM's operational value is in linking the fresh financial metrics of resource-cost consumption to improvement programs. Managers who understand ABC-ABM realize that accurate, relevant cost data enables identification for and accomplishment of innovation and continuous improvement.

References

1. H. Thomas Johnson, and Robert S. Kaplan, *Relevance Lost: The Rise and Fall of Management Accounting* (Boston: Harvard Business School Press, 1987).

CHAPTER 8
THE UNIFICATION OF QUALITY, TIME AND COST DATA

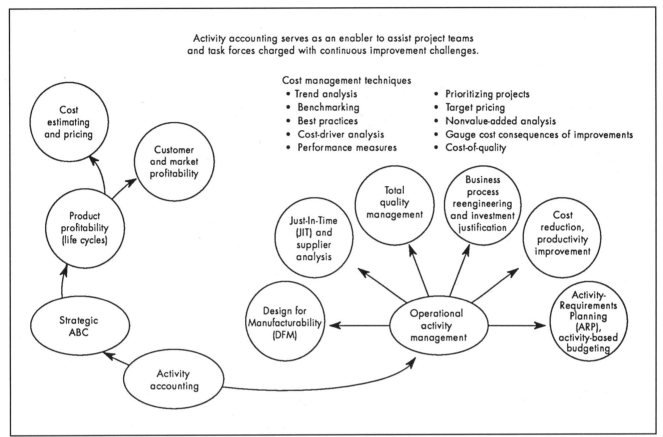

Fig. 8-42 Activity accounting as a catalyst for change.

Bibliography

Brimson, James. *Activity Accounting: An Activity-Based Costing Approach.* New York: John Wiley & Sons, 1991.

Brinker, Barry J., ed. *Emerging Practices in Cost Management* (a compilation of 50 articles from the *The Journal of Cost Management*, Spring 1987 to Fall 1990). Boston: Warren, Gorham & Lamont, 1990.

Cooper, Robin and Kaplan, Robert. *Design of Cost Management Systems: Text, Cases and Readings.* Englewood Cliffs, NJ: Prentice-Hall, 1991.

Cooper, Robin et al. *Implementing Activity-Based Cost Management: Moving from Analysis to Action.* Montvale, NJ: Institute of Management Accountants, 1992.

Johnson, H. Thomas. "It's Time to Stop Overselling Activity-Based Cost Management." *Management Accounting,* September 1992, pp. 26-35.

Johnson, H. Thomas and Kaplan, Robert. *Relevance Lost: The Rise and Fall of Management Accounting.* Boston: Harvard Business School Press, 1987.

Kaplan, Robert S. "In Defense of Activity-Based Cost Management," *Management Accounting.* November 1992, pp. 58-63.

Pryor, Tom. *Activity Dictionary.* Arlington, TX: ICMS, Inc., 1992.

Rummler, Geary A. and Brache, Alan P. *Improving Performance: How to Manage the White Space on the Organization Chart.* San Francisco: Jossey-Bass Publishers, 1990.

Turney, Peter B.B. *Common Cents: The ABC Performance Breakthrough.* Hillsboro, OR: Cost Technology, 1992.

Turney, Peter B.B. and Alan J. Stratton. "Using ABC to Support Continuous Improvement." *Management Accounting*, September 1992, pp. 46-50.

CHAPTER 8
THE UNIFICATION OF QUALITY, TIME AND COST DATA

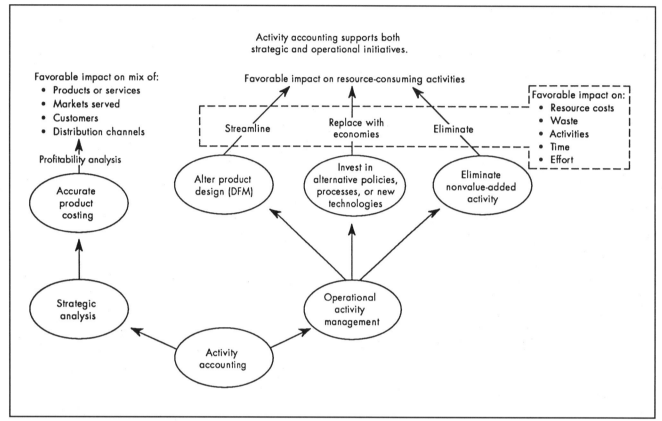

Fig. 8-43 The favorable impact of activity-based analysis.

CHAPTER 9

CONTINUOUS IMPROVEMENT AND JUST-IN-TIME

Just-in-time (JIT) is a significant factor in pursuing continuous improvement (CI). JIT is commonly thought of as delivering parts "just in time" for them to be used at an automotive assembly line. Some will argue that JIT has changed manufacturing more than anything since Eli Whitney invented mass production or Henry Ford introduced the assembly line. Others will state that JIT is a methodology which embraces the continuous improvement philosophy for a mass assembly process. Either way, JIT has made and will continue to make a profound impact on industry.

This chapter will begin by developing a common foundation so that readers will understand the terminology. Then it will explore the similarities in the JIT definitions culled from recognized experts in the field. One of these similarities is that JIT is pursuing the elimination of waste. Continuous improvement also seeks out the elimination of waste or non-value added activities. Since the elimination of waste is a key ingredient, the categories of waste will be explored in more detail. If one is to eliminate waste, one needs to know how to find it.

Next, the elements of a JIT system will be introduced. These elements will be grouped into sections of Technical, People, and Organizational/System issues. Each element will be briefly introduced. The intent is not to provide a full discussion on each element but to begin to show how the elements are interrelated.

After an introduction of the elements, an implementation strategy will be discussed. This implementation strategy is not unique to JIT. The same steps of the strategy can be used to implement other initiatives (for example, total quality management, statistical process control, and project management).

JIT is an approach to achieving excellence in a manufacturing company, based on the continuing elimination of waste and consistent improvement in productivity. Waste is then defined as those activities that do not add value to the product.[1]

JIT is a management philosophy aimed at eliminating waste from every aspect of manufacturing and its related activities. The term JIT refers to producing only what is needed, when it is needed, in just the amount needed.[2]

Just-in-time is a strategy for achieving significant continuous improvement in performance through the elimination of waste of time and resources in the total business process.[3]

In the strict sense, being just-in-time means having only the correct part in the correct place at the correct time. The ideal is to compress production lead time as much as possible. JIT production has come to mean, in the broadest sense, a philosophy of manufacturing which strives to produce with the shortest possible lead time and the fewest possible mistakes.[4]

Just-in-time is actually no more than a strategy to achieve nonstock or stockless production.[5]

The idea of producing the necessary units in the necessary quantities at the necessary time is described by the short term just-in-time. Just-in-time means, for example, that in the process of assembling the parts to build a car, the necessary kinds of subassemblies of the preceding processes should arrive at the product line at the time needed in the necessary quantities.[6]

In the broad sense, an approach to achieving excellence in a manufacturing company based on the continuing elimination of waste (waste being considered as those things which do not add value to the product). In the narrow sense, just-in-time refers to the movement of material at the necessary place at the necessary time. The implication is that each operation is closely synchronized with the subsequent ones to make that possible.[7]

A philosophy of manufacturing based on planned elimination of all waste and continuous improvement of productivity. It encompasses the successful execution of all manufacturing activities required to produce a final product, from design engineering to delivery and including all stages of conversion from raw material onward. The primary elements of zero inventories are to have only the required inventory when needed; to improve quality to zero defects; to reduce lead times by reducing setup times, queue lengths, and lot sizes; to incrementally revise the operations themselves; and to accomplish these things at minimum cost. In the broad sense it applies to all forms of manufacturing, job shop and process as well as repetitive.[8]

There are many similarities in these definitions. One is the general recognition that JIT primarily applies to manufacturing organizations. Although JIT principles can be applied to administrative pro-

CHAPTER CONTENTS:

CATEGORIES OF WASTE 9-2

ELEMENTS OF JUST-IN-TIME 9-2

IMPLEMENTATION STRATEGY 9-7

The Contributor of this chapter is: **Roger A. Reeves**, *Manager, Manufacturing Services Division, GPS Technologies, Troy, MI.*
The Reviewers of this chapter are: **Robert A. Carringer, CMfgE**, *Senior Vice President, George Group, Inc.;* **Richard Clements**, *President, Solution Specialists, Alto, MI;* **Dr. Neal P. Jeffries**, *President, Center for Manufacturing Technology, Cincinnati, OH.*

CHAPTER 9

CATEGORIES OF WASTE

cesses, the most common application is in manufacturing repetitive, discrete components. More frequently, JIT has been applied to job shop operations and to process industries.

Another similarity is that JIT is a management strategy aimed at changing the culture of an organization to improve the flow of any process. The ultimate goal is to increase customer satisfaction.

CATEGORIES OF WASTE

Elimination of waste is another common thread in these definitions. Waste is simply defined as any non-value-added activity. Customers do not want to pay for anything that does not add value. Value-added tasks or activities in manufacturing typically change the form, function, or features of a product or service. To eliminate waste, one needs to know how to identify it. Once it is known where the non-value-added activities are, then workers can work to reduce and eventually get rid of all cost added actions. Most authors of just-in-time material agree on the seven categories of waste:

- Overproduction.
- Queue time (wait).
- Transportation time (move).
- Manufacturing process.
- Inventory.
- Motion.
- Scrap, rework, inspection.

Overproduction is as undesirable as underproduction because it causes many of the other wasteful activities. Since many managers have been taught that their objective is to keep the machines running and making parts in the name of efficiency, it is difficult to convince them that overproduction is wasteful. But if the parts are not needed by the customer, cost will be added to the process by moving and storing the additional parts. In addition, overproduction pre-emptively uses materials and machine resources to build product that is not needed currently, while overdue or late orders suffer.

Queue time is the time a part sits and waits for the next machine or operation. This wasted time includes storage or delay time that is spent waiting in inventory or work-in-process.

Transportation or move time is the time spent in moving the product from one activity to the next. In some cases, products are moved from one storage or queue area to another instead of moving from one value-added operation to the next. Transportation should be evaluated for the time spent and the distance moved. The prime objective is not to reduce the movement but to eliminate it. Many manufacturing sites use yarn pinned to a plant layout drawing to illustrate the flow of parts. This visual technique effectively shows complex flows.

Manufacturing process waste is confusing to some people since manufacturing operations are typically value-added. While most effort should be spent reducing and eliminating the non-value-added activities, the manufacturing processes should also be reviewed. These two questions should be asked:

1. Why does this operation need to be performed?
2. How can this operation be improved?

For example, in a drilling operation, is the hole even necessary? Can it be engineered out of the product using techniques such as design for manufacturability (DFM)? Can the drilling operation be improved by designing automated loading/unloading fixtures or by drilling more than one part at a time?

Figure 9-1 shows how manufacturing process waste can be reduced in a drilling operation.[9]

Inventory was traditionally seen only as an asset on the financial balance sheet. Now managers are beginning to understand the real cost of carrying additional inventory "just-in-case." By reducing inventory, a company's cash flow can be dramatically improved. This additional cash flow can then be used for other technical improvements.

Motion can be analyzed by the traditional Industrial Engineering technique of ergonomics, which evaluates the man-machine interface and determines the proper distance, height, and weight of movements. This area is getting more attention since the Occupational Safety and Health Administration (OSHA) is issuing guidelines for cumulative trauma disorders caused by repetitive motions. The two aspects of ergonomics to be considered are the health and safety of the worker and the economy of motion itself.

The final category of waste is scrap, rework, and inspection. Most people can see the waste in the first two items, but it was not until the advent of the total quality philosophy that inspection has also been considered to be a waste. Much effort has been put into education and training to implement Statistical Process Control (SPC). If a company does in fact control the process, then a natural benefit should be the reduction of inspectors and auditors.

One of the highly acclaimed books in the field of just-in-time is Suzaki's *The New Manufacturing Challenge: Techniques for Continuous Improvement*.[10] In addition to the seven recognized wastes, he adds an eighth to the list: "the waste of underutilized people's skills and capabilities." Without people's skills, the other seven wastes would not even be identified.

ELEMENTS OF JUST-IN-TIME

Just-in-time is a strategy for the elimination of waste, and is composed of several interrelated elements. One method of grouping divides the elements into three sections:

- Technical Issues.
- Human/People Issues.
- Organizational/System Issues.

CHAPTER 9

ELEMENTS OF JUST-IN-TIME

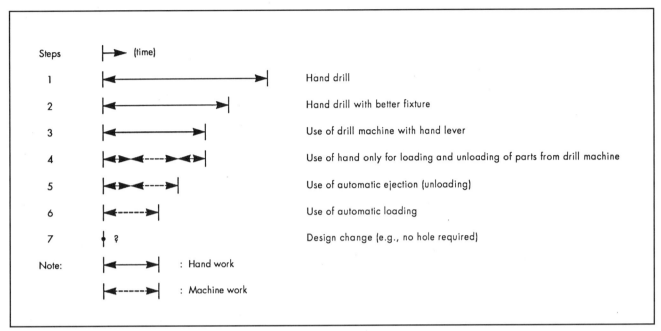

Fig. 9-1 Elimination of processing waste in a drilling operation.

All three categories are vital to any change process. If technical or "hard" issues are addressed without considering the other two areas, then the improvement effort will be less successful. An emphasis on ensuring that people have been properly trained regarding the technical issues is a step in the right direction. However, by neglecting the organizational or system issues, the effort will still not be totally successful. The organizational issues hold the other two areas together. People must be properly encouraged and expected to use the new skills and knowledge in their jobs in order for the full benefit of just-in-time to be realized.

TECHNICAL ISSUES

There are four elements within the technical section of just-in-time:

- Structured flow manufacturing.
- Small lot production.
- Setup reduction.
- Autonomation.

Structured Flow Manufacturing

Structured flow manufacturing is the organization and physical layout of manufacturing resources to produce products in a continuous flow. This technique can be applied equally well to an office environment. In a structured flow environment, the focus is on evaluating the entire process, not just one operation or machine.

The objective is to maximize the efficiency of the entire process or to reduce the throughput time. Throughput or lead time is defined as the time required to produce a product from receipt of raw material to shipment of the final product. In the past, most organizations focused their efforts on maximizing a department's efficiency instead of evaluating the total lead time. A quick technique to evaluate the organization is to look at the inventory of either parts or paperwork in process. A structured flow environment will have minimal inventory.

The layout of the facility will be based upon products as opposed to grouping similar processes together. In traditional plants, similar equipment is all placed into one department. For example, all grinding is done in one area. In a structured flow environment, all equipment necessary to produce a part (or family of parts) is grouped together so that products travel only very short distances.

The benefits of structured flow manufacturing include reduction of inventory, simplification of the process, quicker quality feedback, reduced material handling, and better communications.

Structured flow manufacturing is also known as U-cells, C-cells, factory within a factory, or group technology. Figure 9-2 shows a simple example of product-oriented flow.

Small Lot Production

Small lot production is producing products or paperwork in smaller and smaller quantities. There are many opportunities for improvement associated with producing in smaller lots. The ultimate goal is to produce in lot sizes of one, if setup time can be eliminated.

The best way to implement small lot production is to begin by cutting lot sizes in half. This may appear to be impossible, but in the spirit and culture of continuous improvement, organizations must break out of past practices. Cutting lot sizes in half is usually more simple than moving equipment to achieve a structured flow environment, because it often means only a change in scheduling practices. Caution however, cutting lot sizes in half by itself will have a significant impact on operations. This strategy is best used in combination with reducing setup time and scrap/rework.

Another facet of small lot production is to move parts between operations before the entire lot is completed. This concept is known as a transfer batch. For example, if the production order is for 100 parts, lots of 20 can be transferred, as opposed to waiting for the entire lot of 100 to be completed before moving the parts to the next operation. Although it may increase material handling costs, throughput time will be reduced—thereby completing the

CHAPTER 9

ELEMENTS OF JUST-IN-TIME

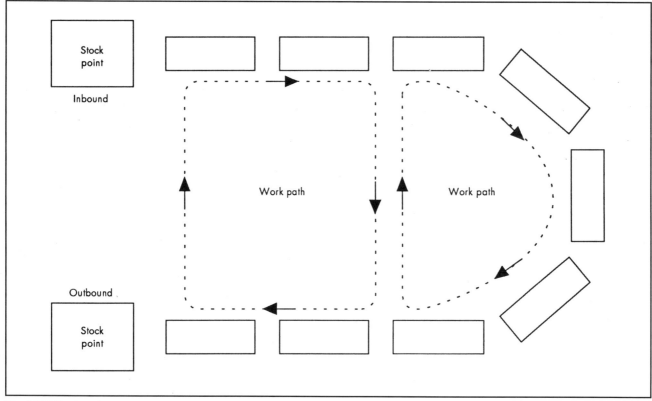

Fig. 9-2 U-line layout.

total order in less time. Remember, the goal is to maximize the whole process as opposed to sub-optimizing on reducing material handling costs.

Benefits of small lot production include reduced cycle times, reduced inventories, improved visibility of problems, and greater flexibility in responding to schedule changes.

Setup Reduction

Setup time is defined as the time from the last good piece of the previous job to the first good piece of the new job. It is the absolute minimum amount of time needed to change over from one activity to another. This element also applies in the office environment; how much time does it take to go from one application software package to another or to switch from one project assignment to another?

Setup time can be reduced using three steps:

1. Identify external and internal activities. Complete external activities while the machine is still running on the current job.
2. Convert as many internal activities as possible to external activities.
3. Eliminate as many adjustments as possible.

External activities are those tasks that can be accomplished while the machine is still running. Internal activities are those that must be done when the machine is shut off. Setup time can be significantly reduced by preparing and organizing the external activities.

Benefits of setup reduction include less machine downtime, more production capacity, reduced operating costs, improved flexibility to respond to customer orders, and the ability to implement a structured flow environment.

Autonomation

In order to fully realize just-in-time, good units must flow to subsequent operations 100% of the time. Autonomation prevents defects by building error-detecting mechanisms into equipment. Some writers call this "automation with a human touch."[11] The Japanese term for autonomation is *jidoka*. The concept allows each employee to run multiple pieces of equipment without having to check up on the machines that are in process. Another writer calls this process "Zero Monitoring Manufacturing."[12]

Autonomation is not automation. To automate is to have the equipment run by itself. Autonomation includes the additional process of checking for abnormalities or defects. To be fully automated, a machine must be able to detect and correct its own operating problems. Equipment is being designed today to detect problems through error-proofing techniques (*poka-yoke* or *baka-yoke*) but it still relies on human intervention to correct the operating problem.

Having operators watch machines doing the work does not add value to the product and should therefore be considered waste. Since the goal of just-in-time is the elimination of waste, autonomation will allow the workers to handle more work by separating operators from machines. The goal is to put man's intelligence into the machines. Hence, a human touch is given to the equipment.

Benefits of autonomation include increased productivity of the workforce, reduction in scrap and rework, expansion of worker responsibilities, and cost reduction.

CHAPTER 9

ELEMENTS OF JUST-IN-TIME

HUMAN/PEOPLE ISSUES

There are four elements with the human/people section of just-in-time:

- Employee involvement.
- Housekeeping.
- Visual controls.
- Quality focus/fitness for use.

Employee Involvement

People run companies; without people, both technology and systems are useless. Everyone, from the president to the shop floor employees, from the engineers to the secretaries, must participate and cooperate for continuous improvement to occur. Top management cannot solve all of the problems, although they have all of the company resources available to dispense to CI efforts.

However, total employee involvement does not mean that management abdicates responsibility for the running of the company. Top management must communicate where the company is heading and why. Middle management must figure out what needs to be done to accomplish top management's direction. Lower management and the workers need to determine the hows and then follow-through with the implementation. Figure 9-3 depicts each group's role.

Organizing teams to work on common objectives is one way to begin employee involvement. Work groups must have specific goals, objectives, time frames, and resources. Before the teams can be effective, they must be given the proper "where" and "why." Top management must stay involved in following up on the team's progress. The success of total employee involvement depends on several factors:

- Highly visible management support and recognition.
- Goals that are easily understood and effectively communicated to group members.
- Education and training to empower the groups to take action.
- Results that are publicized and groups that are recognized for their accomplishments.
- Clearly defined roles and expectations of group members.
- Supervisors who are actively involved as team leaders.

Housekeeping

Housekeeping is the organization of the work place. The popular saying is, "A place for everything, and everything in its place." This certainly applies on the shop floor or in an office. An effective housekeeping system will quickly identify problems because they can no longer be hidden.

There are five steps to creating an effective housekeeping system:

1. Simplification.
2. Organization.
3. Discipline.
4. Cleanliness.
5. Participation.

The first step is to clear the area of all unnecessary equipment and materials.

The next step is to organize the remaining required materials and equipment so that everything has assigned locations. Materials and equipment should be organized so they are easy to locate at the time they are needed. When a tool is missing, a quick scan of the area should locate the tool.

The third step is the discipline required to ensure that the work area stays clean. Cleaning up should be done throughout the work day, rather than being reserved for the last five minutes of the shift. Discipline demonstrates that management cares about the safety of the work force by creating a safe and efficient workplace. Any housekeeping policy should be endorsed and enforced by management.

The fourth step is to keep the work area clean. Housekeeping should not just be done the day before the "big tour." It must be part of the daily work schedule. Clear directions and expectations should be communicated to all employees on their responsibilities in the housekeeping program.

The last step is the active participation of all employees. Each employee is responsible for his/her own work area. Management can set a good example by cleaning up the areas on their walks through the plants and offices and by consistently communicating the housekeeping standards.

If a company cannot handle implementing and maintaining a good housekeeping program, it will never succeed on tougher projects like structured flow manufacturing, preventive maintenance, or a pull system.

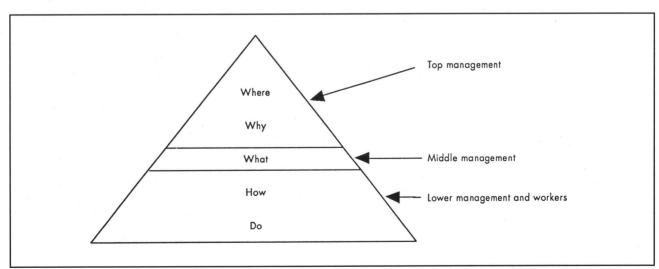

Fig. 9-3 Pyramid of responsibilities.

CHAPTER 9

ELEMENTS OF JUST-IN-TIME

Visual Controls

Visual controls provide instant status of information. They provide a fast and effective means of evaluating any operation, either on the manufacturing shop floor or in an office. Control through visibility is a prerequisite for a pull scheduling system that will be discussed later in this chapter.

Visual controls can take on many forms. One of the simplest is to mark squares or lanes on the floor to identify specific material. Anyone walking by can see when material is missing and should be replenished. An office example might be to identify where to put the different colors of markers used for flip charts or white boards. This can be done by putting a colored mark or strip of tape where the marker should be placed. When the marker is missing, one can "see" that something is missing from that location.

However, regardless of the actual form, control through visibility is important for continuous improvement because it:

- Provides effective feedback about the system.
- Is easy to use, and easy to understand.
- Is simple; not many things can go wrong.
- Is cost-effective to implement.

Quality Focus/Fitness for Use

Employees should have a quality focus as part of their work ethic. What does "quality focus" mean? Each company should have an ongoing educational process that defines what quality is. For example, some may define quality as "conformance to requirements" or "consistent conformance to customers' expectations."

Quality is improved through the actions of people. People should be expected to evaluate and improve the process, not inspect the product.

Much has already been written on the subject of quality improvement. One point to emphasize is that quality is improved one project at a time. Once a problem (or opportunity) has been identified, each company can use problem-solving methodology to eliminate the root cause of the problem. An important key is the consistent use of the problem-solving methodology.

Fitness for use is delivering a product or service that exactly meets the customer's specifications. Fitness for use includes communicating with customers, both internal and external, and finding out what their expectations are. Expectations include both the product or service specifications and also the intangible "wants" of the customer. These wants may include prompt response to requests or delivery of the part in proper orientation or alignment.

Fitness for use is a culture in which all personnel understand who their customers are and have talked with those customers to get a better understanding of their expectations. Communications between departments must be encouraged, and communication must extend to suppliers as well.

To effectively implement a structured flow environment, the fitness for use concept must be understood. Less than acceptable presentation of material to the next operation cannot be tolerated.

ORGANIZATIONAL/SYSTEM ISSUES

The four just-in-time elements associated with organization and systems are:

- Level load and balanced flow.
- Preventive maintenance.
- Supplier partnerships.
- Pull systems.

The systems elements are generally harder to work on than the elements of technical or personnel management.

Level Load/Balanced Flow

Level load is establishing work schedules that are equal to or close to the rate of customer demand. Balanced flow is adjusting the production rate to the sales rate. Inventory is no longer used as a buffer between manufacturing and the customer.

A perfectly level schedule is difficult to achieve since the customer does not always require the same amount every day. However, a company can take its customers' needs into account and level its internal production schedule each day. The time frame for leveling should be daily or weekly, not monthly.

An example of mixed model scheduling is shown in Fig. 9-4.

If the customer requires an A, B, and C each day, the company should strive to produce each model each day. A more common approach would be to build all of the A's first. This approach might work if all three models could be built in a day's time. A better approach would be to produce the parts required using the least common denominator approach. In the example above, there is 1 A needed for 1 B and 1 C.

Level load and balanced flow require:

- Focus on cycle or lead times.
- Removal or correction of bottleneck areas, to create linearity.
- Utilization of one piece of equipment or department is of lesser importance than the flow of the entire process.
- Reliability of equipment.
- High level of product quality.

To effectively level and balance the flow, flexibility and shorter manufacturing cycles are required. Mixed model scheduling will be common practice, building every part every day. This requires quick setups, small lot production, high quality, and reliability of equipment. The goal is to shorten the manufacturing lead time.

Preventive Maintenance

Preventive maintenance preserves the process capability of equipment and tooling within the defined specifications. The equipment, whether a press, assembly line, or copy machine, must be available when the customer or process requires it. Required maintenance hours should be calculated into the production schedule just as customer requirements are.

Preventive maintenance should be performed on a regular, predetermined schedule, for example every three or six months, or every 10,000 cycles. Establishing the time frames can be done by reading the equipment manufacturer's operating manual or using experience on similar types of equipment. It is better in this case to err on the side of more frequent maintenance than to wait too long and have the equipment break down.

Easy preventive maintenance can be done by the actual production workers themselves (if the union contract allows). Since the people who run the machines know them best, it makes sense to make them responsible for their work areas. And, in linking this concept to total quality focus, the workers who are

```
AAAAA – BBBBB – CCCCC
ABC – ABC – ABC – ABC – ABC
```

Fig. 9-4 Mixed model scheduling.

CHAPTER 9
IMPLEMENTATION STRATEGY

responsible for product quality must be able to control the process quality as well.

The basic philosophy of just-in-time is to continuously improve the process and to eliminate waste. Machine downtime and out-of-tolerance equipment is extremely wasteful in terms of time and money.

As a point of caution on preventive maintenance, do not think that just buying a computer software program means that a preventive maintenance program is in place. Maintenance management is a comprehensive business system with many interdependent variables.

Supplier Partnerships

Supplier partnerships are cooperative business agreements with suppliers of parts and equipment. In a just-in-time environment, the goal is to effectively manage the entire supply chain from raw materials to components to subassemblies to finished products to the ultimate consumer.

The overriding factor in establishing supplier partnerships is total cost, not just the lowest price. Total cost includes the price as well as other factors, such as quality, delivery, and working relationships. Quality includes costs of sorting, rework, reordering, disposition of nonconforming material, and problem-solving efforts. Delivery includes the right quantity of the right part at the right time in the right place. Working relationships include attitude and willingness to satisfy the customer.

Supplier partnerships then should be a cross-functional working relationship with purchasing, manufacturing, quality, engineering, and materials management. Purchasing can be the leader or first point of contact, but effective communication throughout the entire company is required.

Reducing the total number of suppliers is critical to effectively managing supplier partnerships. Supplier partnerships are characterized by single (not necessarily sole) sourcing, longer term contracts (three to five years or life of the part), ongoing regular communications, and sharing of long-term business forecasts and objectives.

Pull Systems

Pull systems should be the last JIT element implemented. An effective pull system requires that all of the other 11 elements are implemented and effectively getting the desired results. An alternative to implementing pull systems last is to begin with implementing pull systems. This will cause problems and issues to come to the forefront. These issues are the same that will have to be addressed in the prior 11 elements.

The essence of a pull system, as opposed to a push system, is that material is only produced when it is needed. Note, the key word is "produced" and not just shipped or delivered when it is needed. Producing the part just-in-time requires short manufacturing leadtimes, flexibility, and quick communication channels. Hence, visual controls are vital for quick communications.

Pull systems do not require zero inventories. However, inventory is strategically placed in order to balance the production flow. Empty baskets, *kanban* cards, or other visible control signals such as lights or an empty marked square on the floor can signal the supplier to start producing. When the buffer or container is full, the supplier stops producing.

Pull systems require a brand new way of thinking about and running manufacturing processes. Be sure to start on the other elements before rushing to implement a pull system.

IMPLEMENTATION STRATEGY

Knowing the elements is the first step to implementing just-in-time environment. This education should extend beyond just the top management of the company. People need to know what they are trying to do and how it relates to what they already know.

The next step in creating a JIT environment is to have a picture in mind—a vision of what the end result will look like. Can a description be made about how the organization will be operating in its future state? The details will probably not all be in, but the key is to keep communicating the future state frequently to all employees so that they too can "see" it and keep striving for it.

In order to create the vision, an organization should do its homework by studying the JIT elements and by visiting and talking to organizations who are currently working on JIT. The purpose is not to duplicate the steps but to discuss the lessons learned by that organization in its implementation. What did they do right that they would recommend to others? If they had to do it over again, what would they do differently?

A comparison of the organization's current practices against the JIT elements will generally give some direction as to which element to concentrate on first. It is important to keep the whole picture in mind; otherwise, suboptimizing may occur. Organizations should take an active role in performing their own assessments as opposed to purchasing this service from another firm.

Developing a project plan for the JIT implementation should be the next step. This plan should indicate what the action is, who is responsible, and when it should be accomplished. More formal project management techniques such as statements of work, work breakdown structures, network logic diagrams, and responsibility matrices may be useful for large-scale implementation projects.

A pilot location in the organization should be selected to begin the implementation. Do not get caught in analysis paralysis by spending a lot of time in the planning step. Start in a small area of the organization where success can be easily attained. This small, isolated pilot will enable the company to keep things in control while maintaining the visibility on doing something new. Communicate the results of the pilot (both good and bad) to the organization in a timely manner.

Finally, keep the project plan updated as the implementation of JIT progresses throughout the organization. Implementation never stops with the continuous improvement mindset, but results should be forthcoming in six months to a year. Do not expect quick results overnight. Two to three years is a good time frame (depending on the size of the organization) for planning the organization-wide implementation of all JIT elements.

SUMMARY

Top management commitment and support to just-in-time is absolutely critical to success. The first step is education for top management. They must know what they are trying to achieve, and can then begin educating others in the organization of their

CHAPTER 9

IMPLEMENTATION STRATEGY

vision. They must express not only the "what" but perhaps also the "how," and, most importantly, the "why." Effective communication takes a lot of time. Managers should not just do the talking, they must also spend at least equal time listening actively to ensure that the intended message was understood.

The foundation of just-in-time implementations is the belief in people. All the textbooks and consultants in the world cannot implement JIT within the organization; it must come from the people within the organization. The people must be given the proper education and training in what is to be implemented, the goals, objectives, and timeframes expected, and the necessary resources (time, money, and support) to do the work.

An important concept to remember in any JIT implementation is understanding the interrelationships between elements. For instance, setup reduction should occur before cutting lot sizes in half. Employee involvement and housekeeping are generally good places to begin, and pull systems should be near the bottom. However, each organization is different and should develop its own implementation plan.

The benefits of just-in-time are many, but in order to achieve them, one must begin the journey. Remember, there are two ways of getting information about JIT: "pulling" the information as needed, or being supplied with it "just-in-time."

References

1. Walter E. Goddard, *Just-in-Time: Surviving by Breaking Tradition* (Essex Junction, VT: Oliver Wight Limited Publications, Inc., 1986), p. 9.
2. Kiyoshi Suzaki, *The New Manufacturing Challenge: Techniques for Continuous Improvement* (New York, NY: The Free Press, 1987), p. 6.
3. *Just-in-Time Manufacturing VIDEOPLUS Manufacturing Excellence Series* (David W. Buker, Inc. and Associates, 1989), Unit 1 Lesson 1 p. 1.
4. James H. Greene, *Production & Inventory Control Handbook, Second Edition* (New York, NY: McGraw-Hill Book Company, 1987), p. 24.10-11.
5. Shigeo Shingo, *A Study of the Toyota Production System from an Industrial Engineering Viewpoint* (Cambridge, MA and Norwalk, CT: Productivity Press, 1989), p. 97.
6. Yasuhiro Monden, *Toyota Production System: Practical Approach to Production Management* (Industrial Engineering and Management Press, 1983), p. 4.
7. Thomas F. Wallace and John R. Dougherty, *APICS Dictionary, 6th Edition* (Falls Church, VA: American Production and Inventory Control Society, Inc., 1987), p. 16.
8. James Cox III, John Blackstone, Jr., and Michael S. Spencer, *APICS Dictionary, 7th Edition* (Falls Church, VA: American Production & Inventory Control Society, Inc., 1992), p. 24.
9. Suzaki, p. 15.
10. *Ibid*, p. 208.
11. Shingo, p. 58.
12. Iwao Kobayashi, *20 Keys to Workplace Improvement* (Cambridge, MA and Norwalk, CT:Productivity Press, 1988), p. 81.

Bibliography

David W. Buker, Inc. & Associates. *Just-in-Time Manufacturing VIDEOPLUS Manufacturing Excellence Series*. Antioch, IL, 1989.

Cox III, Dr. James F., Dr. John H. Blackstone, Jr., and Michael S. Spencer. *APICS Dictionary 7th Edition*. Falls Church, VA: American Production & Inventory Control Society, Inc., 1992.

Goddard, Walter E. *Just-in-Time: Surviving by Breaking Tradition*. Essex Junction, VT: Oliver Wight Limited Publications, Inc., 1986.

Goldratt, Eliyahu M. and Jeff Cox. *The Goal: Excellence in Manufacturing*. Croton-on-Hudson, NY: North River Press, Inc., 1984.

Greene, James H. *Production & Inventory Control Handbook, 2nd Edition*. New York, NY: McGraw-Hill Book Company, 1987.

Hall, Robert W. *Attaining Manufacturing Excellence*. The Dow Jones-Irwin/APICS Series in Production. Homewood, IL, 1987.

Imai, Masaaki. *Kaizen: The Key to Japan's Competitive Success*. New York, NY: Random House Business Division, 1986.

Just-in-Time Reprints. Falls Church, VA: American Production & Inventory Control Society, Inc., 1987.

Kobayashi, Iwao. *20 Keys to Workplace Improvement*. Cambridge, MA and Norwalk, CT: Productivity Press, 1990.

Monden, Yasuhiro. *Toyota Production System: Practical Approach to Production Management*. Industrial Engineering and Management Press and Institute of Industrial Engineers, 1983.

Ohno, Taiichi. *Toyota Production System: Beyond Large-Scale Production*. Cambridge, MA and Norwalk, CT: Productivity Press, 1988.

Ohno, Taiichi and Setsuo Mito. *Just-in-Time: For Today and Tomorrow*. Cambridge, MA and Norwalk, CT: Productivity Press, 1988.

Schonberger, Richard J. *Japanese Manufacturing Techniques: Nine Hidden Lessons in Simplicity*. New York, NY: The Free Press, 1982.

Schonberger. *World Class Manufacturing: The Lessons of Simplicity Applied*. New York, NY: The Free Press, 1986.

Sepehri, Mehran. *Just-in-Time, Not Just in Japan: Case Studies of American Pioneers in JIT Implementation*. Falls Church, VA: American Production & Inventory Control Society, Inc., 1986.

Shingo, Shigeo. *A Study of the Toyota Production System: From an Industrial Engineering Viewpoint*. Cambridge, MA and Norwalk, CT: Productivity Press, 1989.

Shingo, Shigeo. *Zero Quality Control: Source Inspection and the Poka-yoke System*. Cambridge, MA and Norwalk, CT: Productivity Press, 1986.

Suzaki, Kiyoshi. *The New Manufacturing Challenge: Techniques for Continuous Improvement*. New York, NY: The Free Press, 1987.

Vollman, Thomas E., William Lee Berry, and D. Clay Whybark. *Manufacturing Planning and Control Systems*, 2nd ed. Homewood, IL: Dow Jones-Irwin, 1988.

Wallace, Thomas F. and John R. Dougherty. *APICS Dictionary, 6th Edition*. Falls Church, VA: American Production & Inventory Control Society, Inc., 1987.

CHAPTER 10

DEMING, JURAN AND TAGUCHI

Since the mid-1970s, there has been an increasing emphasis on quality by the consumer and consequently the media (or conversely). One element of this movement is the recognition of the present quality of Japanese products and an acknowledgment of the almost miraculous improvement the Japanese have made in quality since the end of World War II. This complete turnaround of the reputation of the quality of Japanese manufactured products was accomplished with the assistance of several notable consultants in the quality and management sciences. This assistance was in support of the perseverance, innovation, and cooperation on the parts of Japanese industry, academia, and government. A primary element of the Japanese success has been the effective integration and utilization of statistical methodology within the industrial organization and structure. The consultant singularly instrumental in initiating and nurturing this movement is Dr. W. Edwards Deming. Consequently it is of no surprise that Dr. Deming has received special attention from American industry and media. Although many industries and people are interested in, and much has been said about Dr. Deming, there is still an aura of mystery surrounding Dr. W. Edwards Deming and what he advocates.

CHAPTER CONTENTS:

DR. DEMING'S PHILOSOPHY 10-1

DR. JURAN'S PHILOSOPHY 10-11

TAGUCHI'S CONTRIBUTIONS AND THE REDUCTION OF VARIABILITY 10-14

DR. DEMING'S PHILOSOPHY

During the fuel embargo of the early 1970s, many consumers purchased imported vehicles because of their good fuel efficiency. But the consumers also received quality vehicles. The American automotive industry has never been the same. Similar statements can be made for many other commodities, for example: electronics, cameras, clothing, color televisions, electric motors. By 1980 the Japanese had captured portions of the American market that America has yet to win back. Many Americans first became aware of how the Japanese were achieving their success when NBC aired its television white paper "If Japan Can, Why Can't We." Much of the Japanese success is attributed to an American named Dr. W. Edwards Deming; an American whose teachings fell mostly on deaf ears at home at that time. Subsequently, Dr. Deming has become identified with the need for a transformation of management.

Will slogans and exhortations for the work force improve quality? According to Dr. Deming, "Do it right the first time. A lofty ring it has. But how could a man make it right the first time when the incoming material is off-gage, off-color, or otherwise defective, or if his machine is not in good order, or the measuring instruments not trustworthy? This is just another meaningless slogan, a cousin of zero defects."[1] That is, everybody doing their best is not going to improve quality.

Dr. Deming has become known for the transformation of management through his "14 points for Management." These 14 points are criteria for a company to become successful, with the primary agent and focus of change being management. Management must provide the leadership and resources in order for industry to regain its competitive position.

Management must:

- Understand special and common causes of variation.
- Remove the barriers that are robbing workers of their pride in workmanship.
- Recognize that it is their obligation to optimize the total system.

DEMING THE MAN

Dr. Deming describes himself as a statistician who sees the statistical methods as a system of services and science to industry. This man, who has had tremendous impact on the development of industry worldwide, was born in Sioux City, Iowa on October 14, 1900. After graduating from the University of Wyoming, he began his career in the 1920s by filling several academic positions, teaching engineering and physics while pursuing his Master's and Doctor's degrees. In 1927, Deming became a mathematical physicist with the U. S. Department of Agriculture and in 1928 he was awarded a PhD by Yale University. He continued working for the Department of Agriculture until 1939 when he became an advisor in sampling for the Bureau of the Census.

The Contributors of this chapter are: **Linda Beach, Ph.D.**, *Director, Quality Assurance, NYMA, Inc.;* **Forrest W. Breyfogle, III, P.E., CQE, CRE**, *President, Smarter Solutions, Austin, TX;* **Gregory F. Gruska**, *President, The Third Generation, Orchard Lake, MI.*

The Reviewers of this chapter are: **Michael L. De Pree**, *Project Engineer, Product Development, Herman Miller Inc.;* **Maureen Heaphy**, *Principal Consultant, The Transformation Network, Inc.*

CHAPTER 10

DR. DEMING'S PHILOSOPHY

In 1946, he established an office in Washington D. C. and pursued his career as a Consultant in Statistical Studies.

From July to August 1950, Dr. Deming taught the elementary theory of random variation and basic techniques such as control charts to several hundred Japanese engineers and statisticians. The start of the Japanese 'revolution in quality and economics' began in this time period but not through these classes. Not wanting to repeat the unsatisfactory American experiences in promulgating and perpetuating statistical techniques in industry, Dr. Deming held several conferences with top management during the summer of 1950. This was accomplished through the offices of Mr. Ichuro Ishikawa, the president of JUSE and of the great *Kaidenren* (Federated Economics Societies). Additional management meetings were held during two trips to Japan in 1951 and some 15 trips in subsequent years. The subject of these conferences has evolved into what is referred to as the Deming Philosophy.

To commemorate Dr. Deming, the JUSE created the Deming Prize in 1951. This prize is awarded annually to the Japanese company with the most outstanding achievements in Quality Control—in the utilization of statistical theory in organization, consumer research, design of product, and production. (Recently, the award process was expanded to allow for overseas companies.) Today, Dr. Deming's name is well known within Japan's industrial community.

DEMING THE PHILOSOPHY

In today's economic environment, the major strategy to achieve customer satisfaction and economic success is continual improvement. This is reflected in the Malcolm Baldrige National Quality Award criteria as well as industry-specific requirements (Total Quality Management (DoD), Six Sigma (Motorola), Targets for Excellence (General Motors), Total Quality Excellence (Ford), etc.). The focus on continual improvement is not only in manufacturing operations but all activities—from basic design and supplier relations to employee relations and customer service. The transformation of management necessary to use this strategy has been fostered and advanced by Dr. W. Edwards Deming. Even though Dr. Deming takes much pride in being a statistician, he realized early that the advantages of statistical thinking would be lost unless it was thoroughly integrated within the operating philosophy of a company. Dr. Deming has identified and taught management the key elements (14 points) for this transformation, and the need for these elements. But he also teaches that training in techniques and skills (even statistical methods) is not enough. It must be integrated with subject matter expertise (product and process knowledge) throughout the organization, throughout the total system. Further, efforts and investments to integrate continual improvement into the everyday operating activities must be guided by profound knowledge.

Dr. Deming has identified, and is constantly refining, a system of profound knowledge for industry, education, and government as consisting of four parts, all relating to each other:

    A. Appreciation for a system.
- Optimize the total system.
- Interdependence between customer and supplier throughout the system.
- Need for communication and cooperation.

    B. Theory of variation.
- Common cause and special cause.
- Descriptive and predictive analysis.

    C. Theory of knowledge.
- Without theory, examples teach nothing.

    D. Knowledge of psychology.
- Intrinsic and extrinsic motivation.

This system of profound knowledge is not "The System of Profound Knowledge." That is, it does not include all knowledge that is profound or necessary for successful enterprise, but only that which relates to the transformation of management. Further, it is not intended to supersede or replace the 14 points for management but provide a basis for a better understanding of them. Indeed, "the 14 points...follow naturally as application of the system of profound knowledge, for transformation from the prevailing style of Western management to one of optimization."[2]

But "A System of Profound Knowledge" and the "14 points" are not the magical keys sought after by many in the western world. There are no such keys. True, a person attending Dr. Deming's seminars learns of tools like the control chart and strategies like the plan-do-study-act (PDSA) cycle. But these are presented as applications and in examples. The main theme of Dr. Deming's seminars is the need for a Transformation of Management to a {personal and organizational} philosophy of continual improvement.

A review of any successful integration of the continual improvement philosophy into the normal operating strategy shows that it is comprised of interrelated people, process, and product strategies. Dr. Deming encompasses these in his 14 points. But strategies are not enough. Fundamental change must occur. In all successes, this requires both philosophical and psychological struggles within the organization, as shown in Table 10-1.[1]

Of these philosophical struggles, the one which causes the most concern is Target versus Tolerance. American workers (and managers) do want to satisfy their customers. They will get upset or angry if anyone tries to say otherwise. Discussions on the need for a Target Philosophy can become emotional. Yet this point is key. The primary objective of establishing a continual cycle of improvement, as taught by Dr. Deming, is doomed to decay and eventual failure in an organization that holds the classical tolerance philosophy rather than the customer-focused target philosophy.

But what are the major differences between the two philosophies? The definitions may seem similar (see Table 10-2). Yet, there are distinct differences in attitude and approach and subsequently results. With the Tolerance Philosophy the goal is Customer Satisfaction; that is, to assure oneself and the customer that all products will be within the specification, or close enough. On the other hand, the Target Philosophy focuses on Customer Delight.

TABLE 10-1
Philosophical Struggles

	Continual Improvement	"Traditional"
Orientation:	Target.	Tolerance.
Initiative for improvement:	Prevention.	Detection.
Approach:	Statistical Process Control.	Statistical Product Control.
SPC knowledge resource:	Workforce.	"Experts."
Production emphasis:	Stable and acceptable process.	Volume.
Economic view:	Long term.	Short term.

CHAPTER 10
DR. DEMING'S PHILOSOPHY

TABLE 10-2
Tolerance versus Target

Philosophy:	Tolerance	Target
Definition:	Make the product to print.	Products are provided from a stable process that is centered on a customer designated target and is continually improved to reduce variability and increase confidence of all customers.
Consequences:	All parts within specification are equally acceptable.	Parts closer to the target are more desirable.
	Each activity is considered separately.	The process must be considered as a whole.
Emphasis:	Product control.	Process control and continual improvement.
Result:	Customer satisfaction.	Customer delight.

The other philosophical points become easier to accept if the concept of "customer delight" is combined with the realization that in "optimizing the total system" the customer is not only the individual who purchases the final product or service but all the individuals involved in the system.

But a philosophy consistent with continual improvement is not enough. Successful organizations also had to undergo psychological struggles. Besides the normal human reluctance to any change, these struggles include the need for:

- Factual and analytical decision making.
- Decision making at the lowest level.
- Focused cross-organizational teamwork.
- Clear performance standards.
- Suppliers as extensions of the planning and production system.
- Customer focus.

The changes that individuals and organizations have to undergo in the transformation of an organization to continual improvement are neither easy nor rapid nor ever complete.

SYSTEMS INTEGRATION

Dr. Deming's 14 points "...provide a yardstick by which any one in the company, stockholders, and the bank may measure the performance of the management."[3] These points can be itemized into the following areas:

The primary requirement of quality is that it be an attitude which permeates the entire organization. There must be a quality consciousness—a total commitment to consistently provide a product or service of maximum value to all (internal and external) customers. Further, there must be a responsibility for quality, a confidence in all of a company's products and services—an enthusiasm for quality.

Besides "getting the faith," an organization must review its entire operating philosophy and change those ideas, areas, and procedures which are not conducive to the production of economical quality products and services.

1. Management must innovate and allocate resources to fulfill the long-range needs of the company and all of the customers, rather than consider only short-term profitability.
2. The operational strategy of accepting defective products (or equipment, procedures, etc.) within the process and repairing after final production is counterproductive and uneconomical. "Acceptance of defective materials and poor workmanship as a way of life is one of the most effective roadblocks to better quality and productivity."[4]

It is not possible to inspect quality into a product. That is too late, the quality is already there. Consequently, management must eliminate the dependence on mass inspection to assure quality. Product and process design and control must have the goal of the production of quality that the customer wants. Management and worker must acknowledge the natural variation of the systems and be focused on continual improvement. That is, management must stress prevention, not appraisal. Further, these systems and controls must continually be monitored to assure that they are still effective and economical. They should be modified or eliminated when their usefulness is over.

In order to achieve and maintain the desired system, management must provide the proper environment and tools. The tools promoted most enthusiastically by Dr. Deming are statistical methods. Unaided by statistical techniques, management's "normal" reaction to trouble, according to Deming, is to blame the workers. This gives rise to the myth that there would be no problems in production if only workers would do their job correctly. Yet, of the sources of waste, 85% are faults of the system or common causes and thus are management's responsibility. Only 15% of the problems are local faults or special causes—attributable to the worker. Confusions between these two types of problems leads to frustration at all levels and produces still greater variability and higher costs. Besides identifying the sources of waste, statistical techniques should also be used to maintain acceptable and stable processes. "Statistical control of the process provides the only way...to build quality in and the only way to provide...evidence of uniform repeatable quality and cost of production."[5] But Dr. Deming cautions against gathering numbers without doing something with them. It is not enough to have an elaborate information system that crunches hordes of numbers but does not report the whole story. Quality must also be designed into the information system to tell the users what they need to know.

In order to reap the benefits of continual improvement, communications and willingness to work together toward a common goal must exist across all departments, plants, and divisions. Management must "drive out fear throughout the organization. The economic loss resulting from fear to ask questions or report trouble is appalling."[6] Quality cannot be owned by a single staff or organization. Ownership must be broad based. Employees must stop thinking in terms of departmental or plant parochialism and start thinking of themselves as a group of people who "own" common processes. Teams should be established to work on sources of waste. This requires communication and feedback.

CHAPTER 10

DR. DEMING'S PHILOSOPHY

Further, management should examine closely the impact of work standards and supervision on the production of a quality product or service. The aim of both should be to help people do a better job. Supervisors must have more time to help the people on the job, know and utilize statistical techniques to identify and solve problems, and be responsible for the overall costs of their (sub)system. Quality, not quantity, must be king. Increased productivity and lower cost are the natural result of a quality controlled environment. Dr. Deming suggests that as a substitute, the total (system) cost of a product or service be established as an appropriate performance measure. But even this does not quantify customer dissatisfaction, which may be "unknown or unknowable."

Unfortunately, the American educational system has not yet provided adequate training in statistical skills. Consequently management must institute elementary statistical training on a broad scale across all levels and disciplines. This should be part of a vigorous program for (re)training people in new skills to keep up with changes in the product and process technologies. "One bottleneck to this endeavor though is the shortage of competent statisticians for teachers."[7] To offset this problem, management must identify and make maximum use of existing statistical knowledge and talent within the organization. These individuals can be assisted in part by competent statistical consultants.

In retrospect, it is quite evident that the term "management" appears quite often in this section. The reason is obvious—quality is management's responsibility. Dr. Deming states that "only ...management can bring about the changes required. Failure of...management to act on even one of the 14 points will bring failure to reach the maximum quality and reliability that could otherwise be achieved."[8] The ways of doing business with suppliers and with customers that were good enough in the past must now be revised. Drastic revision is required.

But who is "management?" The answer is anyone who can affect the variation of the process. Since "top" management holds the economic keys of the process, they have the largest piece of the responsibility pie. However, continual improvement will not provide its maximum benefits unless everyone is involved.

IMPACT ON MANUFACTURING ENGINEERING[9]

At the engineering and operations levels, the emphasis under continual improvement shifts from problem detection to prevention: from finding problems and solving them to preventing situations where problems can exist. A necessary part of this change is the integration of factual decision making, using statistical methods, into the everyday operating activities. The move to a continual improvement approach to manufacturing integrates ongoing quality into normal process considerations. This dramatically changes the traditional scope of the manufacturing engineer's responsibility.

For a process to be considered acceptable, all operations are now required to be running in statistical control with acceptable variation, on target, and meeting all customer requirements (including productivity). Furthermore, the process team (which includes the manufacturing engineer) must agree with the conclusion.

When this approach is initially introduced, many if not most engineers can be "reluctant" to participate since it can be viewed as:

- An infringement on the engineers' inalienable rights; that is, stepping on their toes.
- An added workload; "Not only do I have to do my normal job, etc."
- A loss of trust; "How come they (production) get to say whether the process is OK?"
- Another attempt to force statistics on them.

The requirements and duties of the manufacturing engineers are changed. They have to consider not only the basic machine but also the measurement system which will be used to monitor the process and the impact of support activities. The main question asked of them is not

Will it work?

but

What are the sources of variability and how can their effects be minimized?

And they must answer these questions as part of a process team. After the initial reluctance, the engineers begin to learn that this approach has its positive points too—in fact, more positive than negative. The requirement for statistical verification provides them with ammunition to get machine builders and maintenance to modify and correct unacceptable and marginal machines. Since they are able to quantify what the process is doing, process improvement—including reprocessing—is not restricted to only "making it work" but also to achieve improved quality and productivity.

The process team becomes a means of exchanging and expanding ideas, not restricting them. Instead of a burden, the team becomes a resource to the manufacturing engineer. Process quantification and improvement activities become a team activity, not an engineer's activity.

Finally, the continual improvement approach broadens the manufacturing engineer's scope of "who is the customer."

Processing for Control

The main benefits of continual improvement are not with existing processes but with future ones. The manufacturing engineer becomes aware of external as well as internal sources of variation. With this in mind the engineer will look at the design of the process not as

Can it be made (to nominal)?

but

What will it take to maximize customer satisfaction throughout the production cycle while minimizing the economic impact?

The design of the process no longer takes place in a vacuum. The engineer truly becomes responsible for the marriage of form (economy of design) and function (ease of operation).

Continual improvement provides the bridge between all participants of the process—product engineering, manufacturing engineering, production management, and the machine operator. Everyone becomes part of the system to satisfy not only the product engineer but all customers. This approach makes each participant think about the others. It fosters teamwork and makes all parties involved smarter about the big picture—from product conception to customer satisfaction.

SUCCESS AND FAILURE

The most visible and touted example of the economic potentials of the concepts and teachings of Dr. Deming is the Japanese industry. Unfortunately the media coverage of their successes has completely overshadowed the accomplishments of other industries, for example: Xerox, Motorola, Federal Express, and Harley-Davidson. And the list can be expanded. A key feature of such a list is that not all entries are Japanese companies. There are quality leaders—companies embracing the underlying concepts of

CHAPTER 10

DR. DEMING'S PHILOSOPHY

the teachings of Dr. Deming—on every continent. It would be intellectually and emotionally satisfying to believe that these companies have initiated and maintained such programs for purely altruistic reasons. But alas, their motivations have been tainted by that crass element—profit. Clear data exist which indicate that market share is improved by having a high quality product or service in the market place—as viewed by a company's customers. Further, quality has a demonstrated close relationship with financial performance. The bottom line (return on investment, ROI) is improved by quality systems over the long term. But perhaps it will be more instructive to examine those companies which did not succeed and determine why.

The reasons for lack of success fall into one or more of the following areas:

Equating the Transformation of Management with SQC

Many individuals, when they are initially exposed to Dr. Deming's teachings, come to believe that it is the same as the traditional Statistical Quality Control movement. Although his approach encompasses the techniques of SQC, it is much more. It considers and impacts the total organization philosophy and structure. It instructs and expects results from management as well as the technical and operational elements.

Stressing Techniques, Not Communications

When learning a new language (communications skill), a person starts by building a vocabulary and some basic rules. Only after a person is comfortable with this level is the study of structure, poetry, or other linguistic skills attempted. Yet many companies attempt to teach the hard skills of statistics to all levels without any preparation in the soft skills of basic understanding, terminology, and uses. The primary reason for this is that it is far easier to find someone to teach the hard skills (that is, the typical academic or short course) than it is to get a person experienced and skilled enough to successfully cover the simpler (?) soft skills. Further, the emphasis should be on the inter- and intracommunications among operating levels and departments. Statistical techniques are the means of communications, not an end in themselves.

Becoming Statistical Fanatics

Statistical techniques can help management achieve and maintain an economically optimal system. But this not achievable overnight. Management often attempts to speed things along by using management by objectives (MBO) techniques. The focus becomes the tool, not the proper application. Numbers become paramount: "I have more charts (designed experiments, studies, etc.) than you, so I must have a better process." Because of a lack of full understanding, people will require a "statistical solution" and often bypass the often more economical control plan solution. This leads only to frustration and dissatisfaction.

Demanding the "Quick Fix;" "Bringing in the Miracle Men"

These two characteristics will inevitably be seen together and can accentuate the impact of the other characteristics above. Because of the emphasis on short-term return, many managers are looking for "solutions" that can be implemented and have results during their tenure. Consequently they are fast to jump on the bandwagon and hire a high paid consultant to "install" the system which is then in vogue. This can lead to many of the characteristics above. As a result, the installed system will most certainly fade away.

Dr. Deming's approach is concerned with the long-term economic health of a company. Certainly the upfront investment will yield short-term economic and/or quality gains, but the advantage of this strategy is the continual improvement of quality and therefore of ROI. Consequently, a company's experience with the teachings of Dr. Deming will be dissatisfying and uneconomical if at any level there is a lack of knowledge or a lack of commitment.

IMPLEMENTATION

The primary element in the successful integration of continual improvement into an organization's operating philosophy is the full understanding by top management of what they are getting into. Continual Improvement is not just a collection of techniques that can be disseminated via classes or books. It requires a conscious change in the way people are doing their jobs and interacting with others. "Failure ...to act on any one of the 14 points...will bring failure to reach the maximum quality and productivity that could otherwise be achieved." Only management can bring about the changes that are required. but they need the knowledge as well as the desire. The need for new knowledge and skills is not restricted to top management. Communication and feedback via statistical methods are primary to the teachings of Dr. Deming. Consequently all levels of management and worker must become familiar with the "language" of statistics. Care must be taken not to become "overexuberant" in this area. Although everyone should understand the "language," everyone need not be an expert "linguist." That is, the training should be geared to assist individuals with their own jobs—not to provide everyone with the same level of expertise. Recognition and satisfaction of these distinct levels of need will be more rewarding psychologically and economically. Coupled with this training should be a review of work standards and performance measures. These should be modified to be consistent with the new strategy. With the integration of quality attitudes and statistical methods people will accomplish more by working smarter, not harder. Both productivity and quality will increase. People also need to be trained when to ask for assistance. This requires the availability of statistical expertise as well as the confidence of the employees in the system. Dr. Deming states "(a) close second for quick results would be to start to drive out fear, to help people to feel secure to find out about the job and about the product, and unafraid to report trouble with equipment and with incoming material. A close third, and a winner, would be to break down barriers between departments."

Provisos/Concerns

Implementation of continuous improvement requires the assistance and guidance of competent statistical and management consultants, with emphasis on *competent*. Whether internal or external resources are used, care must be taken that these individuals do not focus on the techniques of statistics. This may yield some initial and isolated successes, but advances in the uses of statistical techniques come not by searching a plant for a chance to apply this or that technique but by providing the mechanism to identify problems and apply the appropriate method. Finally, although small gains will be visible shortly after a company mobilizes for continual improvement, it must be realized that sweeping improvements over the whole company will take a long time—but they will continue forever.

CHAPTER 10

DR. DEMING'S PHILOSOPHY

DEMING'S 14 POINTS (ANNOTATED)

1. Create Constancy of Purpose

"The business process starts with the customer. In fact, if it is not started with the customer, it all too many times abruptly ends with the customer. ... One of the bigger problems in this process is to define operationally the customer's needs throughout the organization. ... Only top management can establish the constancy of purpose necessary to know and then meet customers' needs and expectations. Only they can make policy, establish the set of core values, or set the long-term course for the corporation."[10]

2. Adopt the New Philosophy

"American style of management rode along unchallenged between 1950 and 1968, when American-manufactured products held the market. Anyone anywhere in the world was lucky for the privilege to buy an American product. By 1968, forces of competition could no longer be ignored. What happened in Japan could have happened in America, but did not."[11] Learn the new philosophy. One can no longer accept defective material, material unsuited to the job, defective workmanship, defective product, equipment out of order. The philosophy "I can make it work, (no matter what they throw at me)" should no longer be encouraged.

3. Cease Dependence on Mass Inspection

"Routine 100% inspection to improve quality is equivalent to planning for defects, acknowledgment that the process has not the capability required for the specifications. Inspection to improve quality is too late, ineffective, costly. When product leaves the door of a supplier, it is too late to do anything about its quality. Quality comes not from inspection, but from improvement of the production process. Inspection, scrap, downgrading, and rework are not corrective actions on the process."[12] Statistical control of the process provides the only way to assure that quality is built in, and the only way to provide evidence of quality, uniformity, and cost. The use of statistical tools to control a process is required, but remember: A statistical chart detects the existence of a cause of variation that lies outside the system. It does not find the cause.

4. End the Practice of Awarding Business on the Basis of Price Tag Alone

"We can no longer leave quality, service, and price to the forces of competition for price alone—not in today's requirements for uniformity and reliability."[13]

"If you are going to meet your customers' needs at a price they are willing to pay, you must begin to establish long-term relationships with suppliers, encouraging them to adopt the philosophy of continuing improvement. The reason for the long-term relationships is obvious. Our suppliers can invest in the future and be secure that someone isn't going to replace them next year by underbidding on the price tag. Of course you would only want to enter into a long-term relationship with an organization that could consistently meet your needs and expectations and would improve its ability to do so over the duration of the relationship. Dr. Deming states that when you try to find an organization that meets these criteria, you will be lucky to find even one. Do not worry about finding more than one. If you follow these criteria, however, the number of multiple source items will dramatically decrease along with the associated high cost of waste."[14]

5. Improve Constantly and Forever the System of Production and Services

Use statistical techniques to identify the two sources of waste: faults of the system due to common causes (85%), controllable by management; and local faults due to special causes (15%). Strive constantly to reduce this waste. "The Process of Continuous Improvement spirals toward a customer target. Improvement is possible because integral to the process is the Deming Cycle. (Deming calls it the Shewhart Cycle; some know it as the Plan, Do, Study, Act or PDSA Cycle.)"[15]

6. Institute Training

"Training must be totally reconstructed. Management needs training to learn about the company, all the way from incoming material to customer. A central problem is need for appreciation of variation. ... Japanese management has by nature important advantages over management in America. A man in Japanese management starts his career with a long internship (4 to 12 years) on the factory floor and in other duties in the company. He knows the problems of production. He works in procurement, accounting, distribution, sales. ... Money and time spent for training will be ineffective unless inhibitors to good work are removed."[16]

7. Adopt and Institute Leadership

"The job of management is not supervision, but leadership. Management must work on sources of improvement, the intent of quality of products and of services, and on the translation of the intent into design and actual product. The required transformation of the Western style of management requires that managers be leaders. Focus on outcome (management by numbers, MBO, work standards, meet specifications, zero defects, appraisal of performance) must be abolished and leadership put in place."[17] The aim of leadership should be to help people to do a better job.

8. Drive Out Fear

Fear exists in many forms. A familiar example is provided here by William Scherkenbach. "The vice president for sales is presiding at a meeting of his sales managers. They know full well that if any of them predicts an increase of less than 10% for the next year, he won't have a job there. It makes no difference whether he makes the predicted increase. They all must go through this wasteful ritual."

This is similar to the left-handed hockey stick syndrome where an indicator of progress shown over time is flat but the presenter shows a prediction of progress that goes upward and to the right. Unfortunately, often there is no plan to implement improvements, just wishful thinking.

9. Break Down Barriers Between Departments

"A new president came in; talked with the heads of sales, design, manufacturing, consumer research, etc. Everybody was doing a superb job, and had been doing so for years. Nobody had any problems. Yet somehow or other the company was going down the tube. Why? The answer was simple. Each staff area was optimizing its own work, but not working as a team for the company. It was the new president's job to coordinate the talents of these men for the best good of the company."[18]

10. Eliminate Slogans, Exhortations, and Targets for the Work Force

"Slogans, in the absence of quality control, will be interpreted correctly by the work force as management's hope for a lazy way

CHAPTER 10
DR. DEMING'S PHILOSOPHY

out, and an indication that the management has abandoned the job, acknowledging their desperation and total inadequacy."[19]

11. Eliminate Numerical Quotas for the Work Force

"A quota is a fortress against improvement of quality and productivity. I have yet to see a quota that includes any trace of a system by which to help anyone to do a better job. A quota is totally incompatible with never ending improvement. There are better ways."[20]

12. Remove Barriers that Rob People of Pride of Workmanship

"...Top management asks: 'What do you want me to do?' You tell them to first learn about variability, and then to begin to change company systems. They won't be expecting that answer. I have found that they are more willing (well, some more willing than others) to spend money on training their people, or to commission the development of a case study, or to buy some new technology. But it is always for someone else. Even though it is for their organization, it is not personally for them. ... They need to personally understand that this is for them."[21] Management must understand variation so that they can ask the right questions and make the necessary changes in the system.

13. Institute a Vigorous Program of Education and Self-Improvement

"What an organization needs is not just good people; it needs people that are improving with education. ... Everybody has responsibilities in the reconstruction of Western industry, and needs new education. Management must go through new learning."[22]

14. Put Everyone in the Organization to Work to Accomplish the Transformation

"Optimization (of the total system) is management's job." But who is management? In broad terms, anyone who can affect the variation of the process is a manager. Although the areas of responsibility will vary greatly, everyone involved in the process is a manager since they all impact the process at some level. Further, in his *System of Profound Knowledge*, Dr. Deming discusses the need to "nurture and preserve the positive innate attributes of people" by involving them in the change process. "Everyone wins with optimization."[23]

PLANNING FOR QUALITY

In simple terms, Dr. Deming's approach is to plan for quality—at every stage of the product cycle and by every activity—and then follow through. For ease of discussion, it is convenient to consider the specific elements in four major groupings:

1. Design of the product.
2. Design and control of the process.
3. Supplier relations.
4. Management philosophy.

The following will discuss each of these groups, identifying which of Deming's 14 points (above) are associated with each group.

Design of the Product

Refer to Deming's points 1, 2, 5, 8, and 9. Quality must be designed into the product. This is the first step in assuring quality. Management, and the design element, must be aware of and satisfy the customer's actual needs and perceived desires. However, the design element must realize that the customer is not just the end user. Each segment of the product cycle is also a customer of the design element's product. The design element cannot exist in isolation, expecting manufacturing to guarantee quality, marketing to sell the design, and service to cover any deficiencies. The design element must work together with the other functions to continually evaluate the product and process for buildability and reliability as well as function. Dr. Deming reflects on the fact that not all defects cause problems. Products and processes must be continually reexamined and redefined to yield an economic product which will satisfy all the customers.

Design and Control of the Process

Refer to Deming's points 1, 2, 3, and 5 through 13. Although the product design cycle has the initial and primary responsibility for product quality, it is the responsibility of manufacturing to assure conformance to requirements on a daily basis. This should be accomplished through initial process design and daily ongoing process control. Communication and feedback are necessary between the design and manufacturing elements at all stages of the product cycle in order to assure that the system allows "Building it right the first time." The design element must understand the capabilities of the process in order to produce a quality product which is buildable at an acceptable economic level. In return, the manufacturing elements must, in time, provide the machines, tooling, material, gaging, and training necessary to support the design. Further, work standards must be developed and installed to assist, not impede quality. Communication and feedback is not restricted to the design process. Maintenance of the integrity and effectiveness of the process requires an information system to report how the system is operating compared to the requirements. According to Dr. Deming, it is here that statistical techniques excel. In fact, Deming states that these techniques are prerequisite for a successful system. "The statistical control of quality is application of statistical principles and design, maintenance, and service, directed toward the economic satisfaction of demand." "The statistical control of quality is a system of applications that embraces all formal quantitative aspects of planning, design, purchase, production, service, marketing, and redesign of product ....It provides a plan, a road map, that leads to a better competitive position."[23]

Supplier Relations

Refer to Deming's points 2, 3, 4, and 9. Outgoing quality is directly related to the quality of materials received from suppliers. Dr. Deming has definite comments about vendor-vendee interactions. He states "Sporadic variation in the quality of incoming material is a sore disease everywhere." "Vendors...must learn and use process control. There is no other way for them and you to know the quality that they are delivering and no other way to achieve the best economy and productivity."[25] To minimize incoming variation, Dr. Deming recommends the manufacturer must:

1. Reduce the number of suppliers for the same item.
2. Demand and expect suppliers to use Statistical Process Control.

Dr. Deming also warns about being victimized by the "paradox of thrift;" purchasing will buy from the low priced supplier and the rest of the organization will pay for the poor quality

CHAPTER 10

DR. DEMING'S PHILOSOPHY

supplies through scrap, rework, internal and field repairs, etc. The purchasing elements must understand that price has no meaning without evidence for quality. The ability of potential suppliers to provide the required quality at the stated costs must be assessed early in the product cycle. Further, sufficient time must be made available to interact with the selected suppliers in order to set up the communications to coordinate specifications, measuring, and control practices. The Japanese have developed this point to their economic advantage. They utilize a materials management strategy called "kanban hoshiki" in which suppliers effectively deliver their product to the line itself. This offers the manufacturers advantages in terms of space and inventory savings. But for a just-in-time material scheduling system to succeed, incoming material must satisfy all the process (and customer) requirements. If the supplier has sent nonconforming products, this strategy will be ill-fated. In general, the weakest link of a chain determines the overall strength. Consequently, the supplier must be recognized and considered as an integral member of the product cycle. The suppliers must be required to economically establish and maintain their processes to produce quality products and services.

Management Philosophy

Covers all of Deming's 14 points. Leaders are the glue that holds all the pieces together. They are the primary element in the establishment, implementation, and maintenance of a system to produce good quality. But management must recognize that "good" quality is not necessarily perfection. Good quality, Dr. Deming says, means a predictable degree of uniformity and dependability at a low cost, with quality suited to market. Further he states that "it is good management to reduce variation of any quality characteristic even when no or few defectives are produced." Management must recognize that quality is a business parameter. "When uniform, dependable quality is reached through statistical methods, costs of production decrease, productivity improves, and the manufacturer has a solid basis for the prediction of the quality of his product and his costs."[26]

A SYSTEM OF PROFOUND KNOWLEDGE[27]

One need not be eminent in any part of this system of profound knowledge in order to understand and apply it. However, it is necessary to comprehend that the various segments of the system cannot be separated. They interact with each other.

Appreciation for a System

Profound Knowledge is a system of interacting elements, not isolated concepts. A system is a series of activities with an aim and is managed for optimum output. For example:

- An orchestra—individual musicians working together to produce a symphony. No single musician is "better" than the other. All are needed if the music is to be beautiful and harmonious.
- The synergistic effect of people working together.

To understand the system of profound knowledge, one must consider all segments taken together, not as isolated entities.

Profound knowledge initially comes from outside the organization and by invitation. It cannot be forced onto anybody. People work hard under stress, but they do not generate knowledge. In order to improve a process, one must seek those who have knowledge about the process. Enlargement of a committee is not an effective way to acquire profound knowledge—especially if all participants are from the same background or experience. If the system cannot experience profound knowledge, it cannot produce profound knowledge. Profound knowledge initially comes from outside the system. But if the system wants and accepts this knowledge, the external resource will not be needed forever.

Role of Management

Management's obligation is to optimize the system, whether it be industrial, education, or government. To do this, management must understand and know:

- The boundaries of the system.
- The target (aim) of the system.
- That in a stable system, the only way to achieve something outside the system is to change the system.
- How to make predictions (that is, what will happen in the future).
- The change necessary for continual improvement will not occur if management retains a short term mentality.

Additionally, management must include temporal spread (longevity) to change focus.

Optimization requires the ability to make predictions. This is the responsibility of management, even though most managers fail to realize it. Most decisions made by management are impacted by future considerations. A manager who decides to ship "marginal" products is predicting that the customer will not detect the inconsistencies or will not react to them. Management will promote an employee with an evaluation of "outstanding:" that is, management expects future work to be equally "outstanding."

THEORY OF VARIATION (STATISTICAL THEORY)

Knowledge for Study of Variation

There will always be variation, between people, in output, in service, and of product. What does it mean?

Appreciation of a Stable System

Management must understand the difference between special and common causes of variation. People do not recognize that a process has two states: stable and unstable. They see the world as a single state, a continuous stream of problems, or things they do not like reacting with each other. This is the "I can make it work, regardless of what they throw at me" syndrome. Prediction is folly if special causes are active within the process; that is, the process is out of statistical control. In a stable system, all 'products' are subject to the same statistical variation (common causes). If the process is unstable, something is changing. The variation seen today may not be the same that will be seen tomorrow, because of these special causes of variation. Knowledge of specific sources of variation do not determine stability. Just because a special cause is known or allowed to exist in a process, the process does not become in statistical control.

Leadership of people is entirely different in the two states, stable and unstable. Leadership principals are constant, but specific actions will be different for the two states.

Leadership requires the identification of "what should be done" (that is, the right thing to do) and the establishment of a process, to be able to do things right. Management maintains the common cause system by providing constant and relevant feedback. To improve the system, management must change the system. In a sense, this is causing the system to become unstable

CHAPTER 10

DR. DEMING'S PHILOSOPHY

until the new level of performance is achieved. Consequently, during the improvement phase, increased control and monitoring activities are prudent.

- In a stable system, the role of management is to improve the system.
- In an unstable system, the role of management is to determine what has changed.

There are two mistakes in attempts to improve a process:

1. Tampering. Treating a common cause as if it were a special cause. This mistake may be made by an operator trying to "help the system" by making unnecessary adjustments to "fine tune" the system. It also occurs with quality-related discipline, that is, penalizing the worker for the variation of the system.
2. Treating a special cause as if it were a common cause. Management of a process requires understanding of that process—one must look at the total system, the total process. The measurement system must be stable and acceptable. Measurement systems, like all other systems, have variation. To be usable, the data (results) produced by a measurement system must be stable, in statistical control. Further, the measurements must be based on an operational definition which is related to the requirements of the customer (operator, manager, end user, etc.)

Only a process that is stable has a definable capability. The capability of a process is the variation due only to the common causes of variation. This is its "inherent variation." Consequently, the capability can be determined only when the process is in statistical control. The acceptability of the capability of a process is independent of its stability. A process can be in statistical control and still produce unacceptable results. On the other hand, a process with all acceptable results may be out of statistical control.

To improve the system, management needs to understand:

1. What the process should be doing.
2. What can go wrong.
3. What the process has been doing.

To be able to identify what is changing, management must have:

- Knowledge of procedures that minimize the net economic loss of mistakes. (Shewhart control charts). This knowledge includes the understanding of the distinction between enumerative studies (dealing with past and present activities) and analytic problems (relating to the results of future processes.)
- Knowledge of the interaction of forces.
  —The effect of the system on the performance of people.
  —The new economics of cooperation.
  —The win-win philosophy.
  —The interdependence among people, groups, divisions, companies, countries.
- Knowledge about loss functions, and which customer requirement is most sensitive to process variation.
- Knowledge about the production of chaos and the loss that results from successive application of individually important, random forces. This includes the losses due to:
  —Worker training worker.
  —Executives meeting together without guidance of profound knowledge.
  —Committees and government agencies working without guidance of profound knowledge. (Enlargement of a committee does not necessarily improve results, and is *not* necessarily a way to acquire profound knowledge.)
  —Competition for market share; win-lose philosophy.
  —Barriers to trade.
  —Demands that lie beyond the capacity of the system.

THEORY OF KNOWLEDGE

Any rational plan, however simple, requires prediction. A statement devoid of prediction conveys no knowledge.

- There is no knowledge, no theory, without prediction.
- There is no observation without theory.

Experience teach nothing unless studied with the aid of theory. An example teaches nothing unless studied with the aid of theory.

- No number of examples establish a theory, yet a single unexplained failure of a theory requires modification or even abandonment of the theory.

There is no true value of any characteristic, state, or condition that is defined in terms of measurement or observation.

- There is no such thing as a fact concerning an empirical observation. Any two people have different ideas about what is important about any event.
- Operational definitions are necessary for consistent observations and communications.
- Interpretation of data from a test or experiment is prediction. This prediction will depend on knowledge of the subject matter.

It is only when a process is in a state of statistical control that statistical theory aids prediction.

KNOWLEDGE OF PSYCHOLOGY

Psychology helps in understanding people, interactions between people and circumstances, interactions between teacher and pupil, interactions between leaders and followers, and any system of management.

Management's role is to optimize the total system. But the total system has a mechanical portion and a people portion. Mechanical systems are one side of the picture; living systems are on the other. They have two separate focuses. Physical science is basic for understanding the mechanical side; psychology is necessary for the human side; statistics is necessary to quantify (and understand) both.

Psychology can assist in using statistical tools effectively. For example, if a company is interested in setting up a brainstorming (cause and effect diagram) session, its group should consist of eight or more people. In order to get the synergy necessary for effective results, the members of the group should have different backgrounds. Brainstorming may not be possible when all participants have the same background. Diverse backgrounds assist individuals to shift or expand their framework.

Psychology comes in different "flavors." It is not one concise field. Psychological theories can be divided into two major approaches based on their view of causation:

- Physical causation. This is the behavioral-response school. Attribution (the why something happens) is the result of an

CHAPTER 10
DR. DEMING'S PHILOSOPHY

inference from observed behavior. This gives rise to classes of "reinforcers"—different quantity and timing of the response to the behavior (interval, ratio, fixed, and variable). These reinforcers (responses) can provide positive, negative, or neutral reinforcement. In this school, all employees require supervision by a person who has "supervision" powers. This fits in well with the Management Science approach to management, but not Dr. Deming's approach.
- Personal causation. Personal causation means doing something intentionally to produce a change. This always and invariably raises the question of intention when viewing behavior. This goes beyond the concept of a reinforcer. Rigorously defined, a reinforcer is an observable event that increases (or decreases) the probability of the observable response that preceded it. Yet, failure often increases attempts to succeed and success reduces the necessity to attempt further. In this school, not only is the specific effect desired (for example, solution to a puzzle), but people desire to produce it themselves.

People are different from one another. A leader must be aware of these differences, and use them for optimization of everybody's abilities and inclinations.

Management of industry, education, and government operate today under the supposition that all people are alike. "People" processes, like every other process, have variation. If the different sources of variation acting on these systems are not recognized and understood, then optimization of the total process is impossible. An example of the lack of understanding of this concept is the following:

- People learn in different ways, and at different speeds:
  1. Theoretically—through words and numbers: visual, auditory.
  2. Qualitatively—through sensory experiences: auditory, visual, tactile, etc.
- People must be taught in different ways. The present system of education and training (academic and industrial) tacitly assumes that all people learn in the same manner. It is the factory model for learning; the mass production school. All the raw materials (students) are subject to the same course of instruction and are expected to ingest and comprehend the material at the same levels. And yet studies have shown that this is not true, regardless of the rewards provided.
- A manager, by virtue of his or her authority, has the obligation to make changes in the system of management that will bring improvement. If management does not understand psychology and statistical theory, then the effect of any changes in the system of management to bring improvement to the total system will be haphazard.
- There are many people that follow the behavioral-response school of psychology. But, because they do not understand a stable system, they try to reinforce behavior which may be due to a special cause. Further, this physical causation approach bases its actions on the result of an inference from observed behavior—that is, the analysis of an Enumerative study (about the past). Yet, the improvement of the total system is an Analytic problem (about the future). It is this lack of understanding that causes much confusion and disagreement. An example of these last points is the merit system (appraisal systems, grades, awards, etc.). Management "knows" through experiences gained during the first half of this century that rewards have increased productivity and quality. Consequently, they are amazed when today's employees do not respond "correctly," that is, in the same manner. In the old days a grade of D may have motivated a student to try harder. Today, that same grade may discourage the student from continuing to learn.

Motivation can be:
- Intrinsic:
  1. For improvement and innovation.
  2. For "joy" in work and learning.
- Extrinsic:
  1. A day's work for a day's pay.
  2. Drudgery or avoidance of "punishment." Necessary for some activities.
  3. Overjustified (for example, excessive reward). This will take the 'joy' out of work.

Motivation is affected by demands placed on people beyond the capacity of the system.

Dr. Deming recognizes the significant impact of intrinsic motivation. People are born with a need for relationships with other people, and with a need to be loved and esteemed by others. There is an innate need for self-esteem and respect.

- Circumstances provide some people with dignity and self-esteem. Circumstances deny other people these advantages.
- Management that denies their employees dignity and self-esteem will smother intrinsic motivation.
- Some extrinsic motivators rob employees of dignity and of self-esteem. If for higher pay, or for higher rating, someone does something known to be wrong, that person loses dignity and self-esteem.
- No one, child or other, can enjoy learning if he or she must constantly be concerned about grading and gold stars for performance, or about rating on the job. The educational system would be improved immeasurably by the abolishment of grading.

Everyone is born with a natural inclination to learn and be innovative. They inherit a right to enjoy their work. Psychology helps to nurture and preserve these positive innate attributes of people. Extrinsic motivation is submission to external forces that neutralize intrinsic motivation. For example, "pay" is not a motivator. Learning and joy in learning in school are submerged in order to capture top grades. Under extrinsic motivation, people are ruled by external forces. They try to protect what they have. They try to avoid punishment. They know no joy in learning. Extrinsic motivation is a zero-defect mentality versus a continual improvement mentality. Removal of a demotivator does not create motivation.

Overjustification comes from faulty systems of reward. Overjustification is resignation to outside forces. It could be monetary reward to somebody, or a prize, for an act or achievement that he or she did for sheer pleasure and self-satisfaction. The result of reward under these conditions is to throttle repetition: he or she will lose interest in such pursuits. Monetary reward under such conditions is a way out for managers that do not understand how to manage intrinsic motivation.

Many psychologists say that all motivation is internal and that reinforcers are external. This is the physical causation approach. *Intrinsic* and *extrinsic* relate to the center of causality (personal causation)—the "why we do what we do." Extrinsic motivation has an external focal point of causality. This yields the pawn experience—the feeling of being externally pushed around. Intrinsic motivation, with its internal focal point, provides the *origin experience* because of the strong sense of originating one's own actions. This discussion assumes that the "hygiene factors" (Herzberg) have been taken care of. If the physiological (hunger, thirst, sex, and shelter) and security needs are not satisfied, then the primary locus of causality will be external.

There is a necessity for a transformation (government, industry, education) to a leadership style which eliminates competition, ranking people, grades in school, and prizes for athletics in school.

Competition in a win-lose relationship:

- Competition is an artificial construct: situations created by people, puts people into "slots," creates artificial scarcity.
- The focus is short-term gratification.
- Examples include: market share, sports activities, academic grades, employee evaluations.
- Fallacies:
  1. Competition breeds character (remember, history is written by the winners).
  2. Customers want a competitive market (the "winner" is not the customer).
  3. Growth requires "survival of the fittest" (short-term definition of "fit").
- The focus of trying to beat your competitor is different (and less desirable) than satisfying your customer.

Management must obtain and use knowledge about the psychology of change.

SUMMARY

Of the many reasons given to explain why the Japanese industries have achieved their present market strength, quality of products and services is an undisputed element. A person instrumental in initiating this quality revolution is Dr. W. Edwards Deming. The concepts and philosophy advocated and taught by Dr. Deming are the major building blocks of Japanese quality systems. It is a systems engineering approach to the objective of developing, implementing, and maintaining a system which economically produces products and services which delight all customers. It has two highly visible characteristics: the primacy of statistical techniques and the habit of improvement. Statistical methods are used as the main communication and feedback mechanisms. These tools are utilized to maintain and improve the system as well as to assist in problem solving. The system remains viable over time by a discontent with the status quo. There must be continual activity to improve quality and minimize costs. The teachings of Dr. Deming impact and help all departments and all levels. But it also has its demands and requirements. There must be total commitment by all employees to quality. This means an active involvement and visible participation in the quality system. Knowledge of statistics and statistical techniques is also required. To some, this may mean understanding basic terminology—what tools are available, and their advantages and disadvantages. To others, this requires the skills to select and apply the proper method. Finally, "to be successful in any company, there must be a learning process, year by year, from top to bottom with accumulation of knowledge and experience, and under competent tutelage."[28] It is the key element of competitive economic strategy in the closing of the twentieth century, but there are no shortcuts.

DR. JURAN'S PHILOSOPHY

Total quality management (TQM) is an integrated approach to continuous improvement in performance and customer satisfaction. The totality of the philosophy is in the principle that everyone is responsible for quality. Customer satisfaction is concerned with both external and internal customers. The total integration of effort is aimed at managing to continuously improve performance at all levels. Though many subscribe to the TQM philosophy, most companies struggle with the implementation of the concept and the management of the program.

TQM is a philosophy that embraces the works of many pioneers of quality. Perhaps the most overlooked expert in the area of planning for quality is J.M. Juran. The Juran quality trilogy provides a basic structure for viewing manufacturing activities as a total system. The quality trilogy—quality planning, quality control, and quality improvement—provides a basic structure for viewing manufacturing activities as a total system.

An important concept in quality planning is identifying internal and external customers. Juran provides three customer categories: suppliers, customers, and the processor team. Suppliers provide the input to a process. Customers are receivers of the output. A processor team is any individual or group responsible for a particular process.

To integrate Juran's framework into quality planning for manufacturing systems, several facts must be accepted. These facts are:

- The entire work force is, in some form or other, a customer.
- The processor teams are the individuals with the knowledge and expertise to provide input to continuous improvement.
- The structure to allow teams to communicate both inter- and intraprocess is provided and supported by management.
- Solicitation from all processor teams is vital to any quality planning effort.

QUALITY PLANNING

Juran defines quality planning as defining the products and processes required to meet customers' needs. In manufacturing, internal and external customers have different needs. Juran suggests preparing a flow diagram of the product's processes as a means of determining a company's customers. Figure 10-1 is an example of a hypothetical flow diagram. Advantages to using flow diagrams are:

- The flow diagram will graphically depict an entire system.
- The requirements and the boundaries of the work flow are defined.
- Opportunities to improve the manufacturing processes are easier to identify.

For example, Fig. 10-1 depicts each box as a process. Each box defines the boundaries of one processor team. Every team is dependent on input from another team. The output from the last

CHAPTER 10

DR. JURAN'S PHILOSOPHY

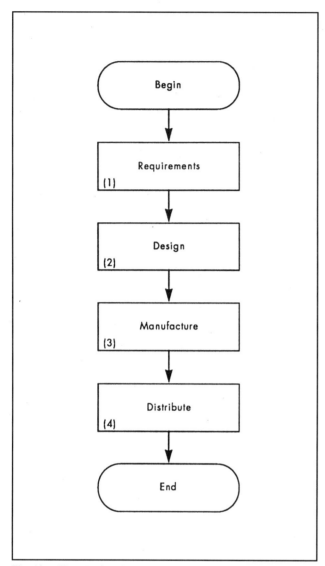

Fig. 10-1 Hypothetical flow diagram.

processor team goes to the external customer. If each processor team does not perform their job well, meet all requirements, and ensure that their output is of the highest quality, the external customer will be dissatisfied. In such events, the manufacturer suffers through rework and waste, increasing costs, and declining customer satisfaction, which ultimately leads to lower sales and profit. Figure 10-2 applies customer identification concepts to the hypothetical flow diagram.

A *manufacturing system* can be defined as a set of processes that are related, and bound together to accomplish a specified goal (see Fig. 10-3).

Processes are a series of actions or functions that bring about an end result that meets a company's goals. Processes are defined according to people, equipment, procedures, and the organization's external and internal customers. As shown in Fig. 10-4, all systems have three basic components: input, processing, and output.

Processes show the relationship of activities and the necessary communication. This communication is important because each process must share certain input and output.

In summary, identifying the manufacturing system components aids managers in defining:

- Required functions.
- Performance standards.
- Expected customer behavior.
- Required processor team responses.

Together, they define the system that produces the product. Manufacturing managers needs to focus on improving the processes performed to meet defined goals.

QUALITY CONTROL

By expanding the flowchart in Fig. 10-1, the activities of each processor team can be defined. For example, one can take Process Number 3 "Manufacture," and define the necessary activities needed to complete this process. Such an expanded flow diagram may look like Fig. 10-5.

Management knows that output Process 2 will only be as good as the input from Process 1. That is, if the boards produced from Process 1 are of poor quality, then drilling holes would be a waste of time and resources. If Process 3 solders the boards incorrectly, then Process 4 will be a complete waste of time. The chain of events continues until the final product is of such poor quality that the entire product is scrapped or completely reworked.

Quality control activities are centered around error prevention. Mechanism are needed to build quality into the products that are produced by each processor team. To accomplish this, quality control processes are built into each activity. This allows for error detection and correction mechanisms early in the manufacturing cycle. Figure 10-6 is an example of building quality checkpoints into a system using a flow diagram.

Quality control involves ways to reduce errors in the manufacturing process and involves statistical process control (SPC) techniques. SPC techniques allow for identification of variances so errors can be identified and corrected as they occur. Even with such measurement systems in place, employees should know how to interpret and process the information. A flow diagram provides employees with an understanding of the entire system and the interfaces involved in the communication process.

QUALITY IMPROVEMENT

Quality improvement is a formalized method for continual improvement of processes. It involves process measurements and controls that allow management to identify variances from predetermined standards. More important, quality improvement focuses on getting the right people together to solve problems. Understanding the work flow allows management to resolve problems by identifying employees knowledgeable in each area. A team formed and controlled by the employees can then brainstorm the problems and recommend solutions. The important point is that management must provide the infrastructure for teaming to be productive.

CHAPTER 10
DR. JURAN'S PHILOSOPHY

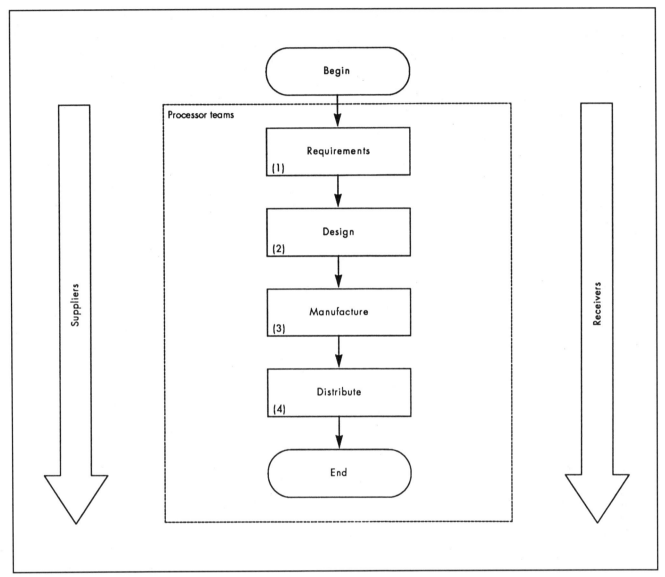

Fig. 10-2 Identifying customers.

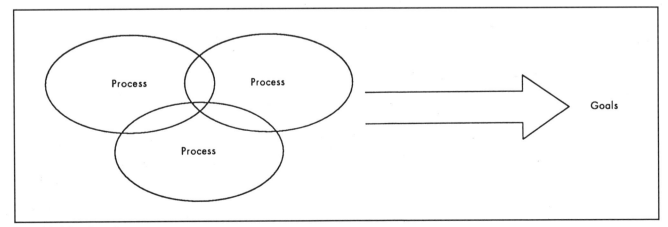

Fig. 10-3 Manufacturing system.

CHAPTER 10

TAGUCHI'S CONTRIBUTIONS AND THE REDUCTION OF VARIABILITY

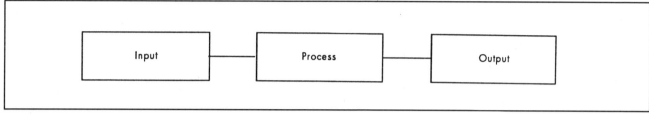

Fig. 10-4 System components.

TAGUCHI'S CONTRIBUTIONS AND THE REDUCTION OF VARIABILITY

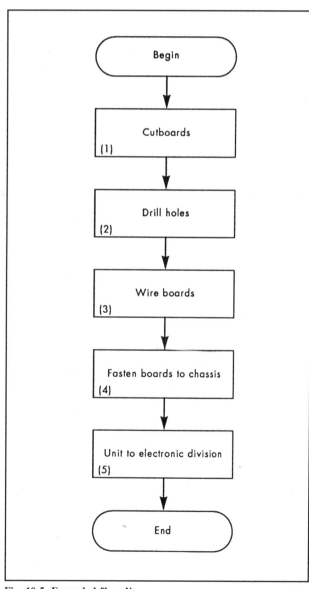

Fig. 10-5 Expanded flow diagram.

Why, in the 1980s, did the doors of Japanese automobiles often sound better when they were closed than the doors of automobiles manufactured in America? Upon examination one might find that the two types of doors have a similar basic design; however, there are noted differences when several doors of each type are examined. The component parts of the better sounding door typically measure closer to the nominal specification requirements and exhibit less part-to-part variability. A higher quality sound is then possible since the clearances between mating parts is less, resulting in a tighter fit and better sound when the door is closed. This basic "reduction in variability" and "striving for the best dimension" strategy is consistent with the philosophy of Genichi Taguchi, which differs from "traditional" industrial practices often found in American industry.[29]

The experimentation procedures proposed by Taguchi have brought both acclaim and criticism. Some claim that the procedures are easier to use than classical statistical techniques, while statisticians have noted problems that can lead to erroneous conclusions. However, most statisticians and engineers will probably agree with Taguchi on the issue that more emphasis should have been given in the past to the reduction of process variability within the product design and manufacturing processes.

Many articles have been written on the positive and negative aspects of the mechanics used by Taguchi.[30-34] Rather than dwell upon all the specific mechanics proposed by Taguchi, an overview of a strategy is given here that illustrates the application of Taguchi philosophy with classical statistical tools, at times noting differences between the mechanics.

To illustrate the strategy, consider a specification of 75 ± 3 (that is, a measurement between 72 and 78 is within specification), where a sample of parts measured the following relative to this specification:

75.7   75.9   76.1   76.3   76.7   76.8   76.9   77.1   77.4   77.7

A "traditional" American industry evaluation would be to look at each component part from this type of data as either passing or failing specification. With this line of thinking there often persists either the naive expectation that "all" parts from the population will be within specification or the statement that "out of specification" parts will be discarded or fixed after inspection.

This type of thinking does not typically lead to the consideration of what could be done to improve the process for the purpose of better meeting the needs and desires of the customer. If an industry were to use this alternative line of reasoning, there would be typically fewer rejected components and often less money would be spent on inspection. Instead of evaluating the measurements as

CHAPTER 10
TAGUCHI'S CONTRIBUTIONS AND THE REDUCTION OF VARIABILITY

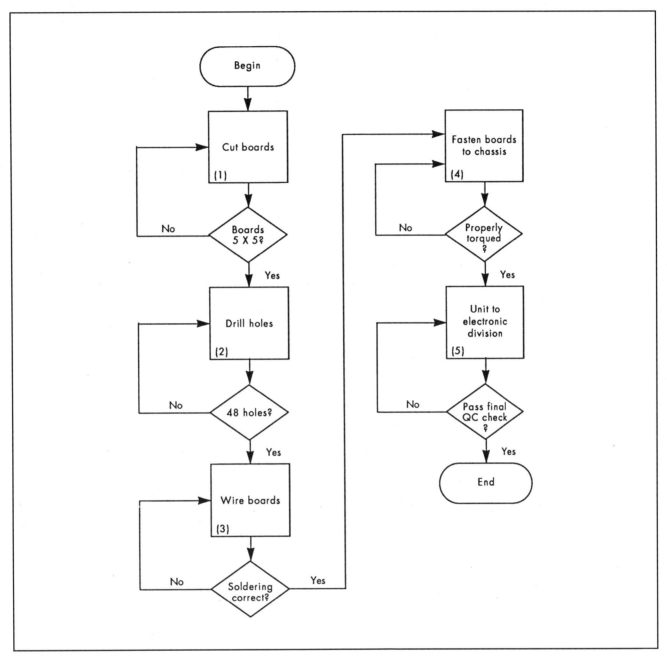

Fig. 10-6 Quality control processes.

pass/fail, the process is better understood if the actual measurement values are examined. One form of pictorially looking at the data is to present the information in the histogram form shown in Fig. 10-7 with an estimate for the shape of the probability density function (for example, the "bell-shaped" PDF shown). From this type of plot, it can now be seen that the data is skewed toward the upper specification limit and that a noticeable portion of the PDF curve extends beyond the upper specification limit.

The percentage of the total area under the PDF curve beyond the specification limits is an estimate of the total percentage of parts that can be expected to be outside these limits. By examining the data from this point of view, the previous "no defect" statement could be changed to a percentage failure rate estimate of approximately 3%. However, the accuracy of this estimate is questionable, since the data may not be normally distributed, which is an important assumption that can dramatically affect an estimated PDF. Also, "eyeballing" this area percentage relative to the total percentage is subject to much error and inconsistencies between people. A better approach to determine this estimate is to make a probability plot of the data.

If data are plotted on normal probability paper and follow a straight line, then the data are presumed to be from a normal distribution and a percentage of population estimate is obtainable directly from the probability paper. Because of this more direct

CHAPTER 10

TAGUCHI'S CONTRIBUTIONS AND THE REDUCTION OF VARIABILITY

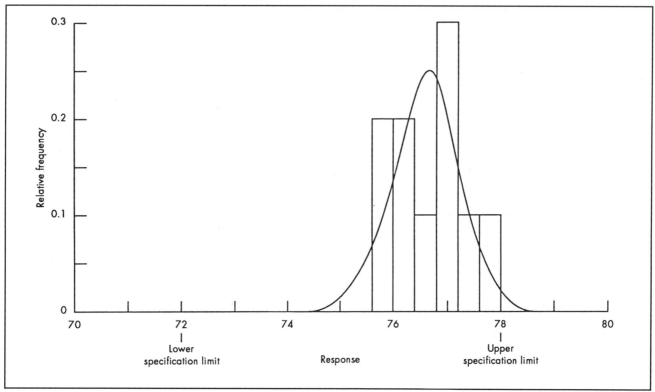

Fig. 10-7 Histogram and probability density function (PDF) of data.

approach, the inaccuracies from the two previously discussed problems are reduced, yielding a more precise estimate. Some computer programs can create a normal probability plot for the previous data, such as the one shown in Fig. 10-8. This type of plot can also be created manually by using special probability paper, where the ranked data values are matched with plot positions that can be determined from tables or equations.[35, 36]

The probability plot shown in Fig. 10-8 for the above set of data indicates (if the sample are random) that approximately 2% (that is, 100 − 98 = 2) of the population would be outside the upper bound of the specification limit of 78. One should also note that the median or mean is skewed by approximately 1.7 from the nominal specification criterion (that is, 76.7 − 75.0 = 1.7).

By looking at the actual data values in lieu of making pass/fail component judgments, one is now able to make an assessment with a relatively small sample size that the process mean should be adjusted lower to decrease the amount of unsatisfactory production parts. It should also note the amount of variability that the process is producing relative to specification (that is, 98% of the part-to-part variability consumes approximately 50% of the specification range). In some situations this amount of variability could lead to noticeable inconsistency in part-to-part performance; even though all components may be within specification. A Taguchi philosophy stresses the importance of reducing variability; however, management is often more interested in explanations that relate to monetary units, as opposed to part tolerances and data variability. To make this translation Taguchi suggests using a loss function.

To illustrate the Taguchi loss function, Fig. 10-9 shows how component loss is typically viewed by American industries. From this figure, it seems reasonable to question the logic of consider-

ing that there is no "loss" if a part is at a specification limit, while another part has a loss value equal to its scrap value if it is barely outside its specification criteria. An alternative is to consider a "loss to society" as expressed in Taguchi's Loss Function. Figure 10-10 shows a quadratic loss function where scrap loss at the specification limits equates to that shown in Fig. 10-9. However, with this strategy a component part has an increasing amount of "loss" as it deviates from its nominal specification criterion, even though it may be within its specification limits. Unlike Fig. 10-9, the curve in Fig. 10-10 does not have the illogical dramatic shift in loss when a part is slightly beyond the specification limits.

A Taguchi loss function strategy emphasizes reducing variability and striving for a process mean that equates to the nominal specification. Companies using the basic philosophy, of examining key specifications and striving for a mean of measurements equating to nominal specification values along with a reduction in data variability, can expect to produce products that are perceived by customers to have higher, more consistent quality. In the television industry, for example, the quality of a television picture would then be expected to be consistently good from television to television (as opposed to a picture on many televisions that "is good enough" when they barely met specification).

The next question of concern is how to efficiently improve the processes so that responses have less variability and are closer to nominal criteria. To illustrate a strategy, consider that in the previous example it was desired to improve the process so that the mean is closer to the nominal specification of 75.0.

A brainstorming session with people who are familiar with the process could create a list of many factors or variables that might affect the magnitude of the response, which is currently measuring on the average 76.7. A "natural" experimentation approach is

CHAPTER 10

TAGUCHI'S CONTRIBUTIONS AND THE REDUCTION OF VARIABILITY

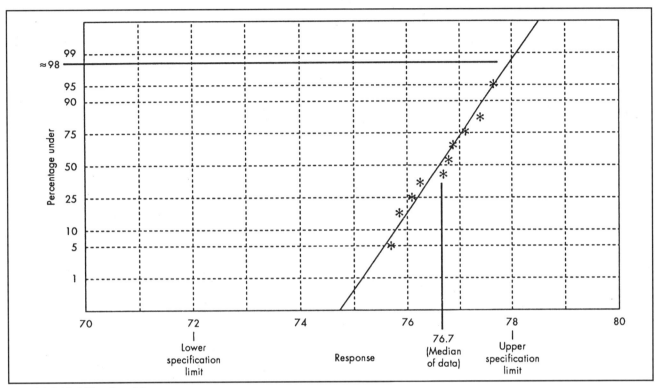

Fig. 10-8 Probability plot of data.

to assess each factor for its importance by changing its level individually while noting the change in a response. However, this one-at-a-time approach is very inefficient and can lead to erroneous conclusions, since factor interactions are not considered. A one-at-a-time experiment, for example, would not typically detect the situation where the two factors *temperature* and *pressure* do not individually affect a process; however, high temperatures in conjunction with low pressure causes a significant change in a response (that is, the two factors interact).

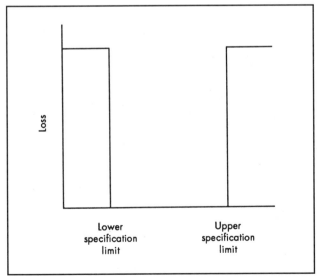

Fig. 10-9 Traditional method of interpreting manufacturing limits.

A structured design of experiments (DOE) strategy assessing many factors within one experiment is an effective alternative to the one-at-a-time strategy.[37] A properly executed DOE can efficiently give direction to determining important changes (that is, the significant factors and their associated effects) to make when improving the process.

To get the maximum benefit from a DOE, the experiment needs to utilize an effective basic setup/analysis strategy for the given situation. A Taguchi-based strategy encourages the use of three-level factors (for example, three temperature settings), one large experiment, and two-factor interactions that are to be identified before the experiment is begun. In a Taguchi strategy, a confirmation experiment after a DOE is especially important because erroneous conclusions can result when, for example, a two-factor interaction unknowingly made another factor appear significant. For example, a temperature/pressure two-factor interaction existed; however, from the Taguchi analysis a humidity factor was instead thought to be significant.

One DOE approach, often considered to be a more effective strategy, is to initially plan the use of multiple experiments, where the first experiment utilizes roughly one fourth of the total time and money allotted for the experiment. Also within this strategy only two levels of the factors (for example, an experiment trial has temperature either at 100°C or 110°C) are initially considered and two-factor interactions are managed by the experiment resolution.[38]

If there are many factors that could affect a response (for example, 15), a reasonable length initial experiment (for example, 16 trials) could be chosen, where two-factor interactions are aliased with the main effects in the analysis (that is, a resolution III design matrix). A resolution IV experiment could next be executed with a lesser number of factors (that is, the significant

CHAPTER 10

TAGUCHI'S CONTRIBUTIONS AND THE REDUCTION OF VARIABILITY

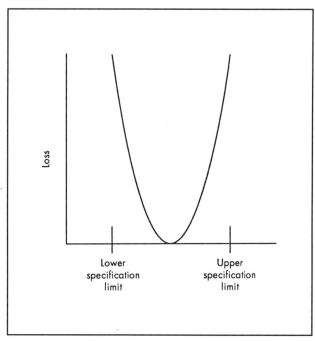

Fig. 10-10 Taguchi loss function.

factors from the initial resolution III experiment analysis) for the purpose of assessing the factor main effects without two-factor interaction aliasing. Within a resolution IV design, two-factor interactions are aliased; however, if a particular contrast of two-factor interactions are found significant (for example, temperature/humidity, pressure/conveyor speed, or voltage/design level), a technical assessment could give needed insight to the two-factor interaction(s) that most likely are causing the contrast to appear significant. A resolution V experiment can then be conducted with a lesser number of factors at appropriate levels (that were previously thought important through main and/or two-factor interaction considerations) such that each two-factor interaction effect can be assessed separately from the main effects and other two-factor interaction effects.

Some additional experiment design considerations using this strategy are:

1. Two-level, 16-trial experiment designs are commonly used. They can have the properties:
   - Resolution III: 9 to 15 factors.
   - Resolution IV: 6 to 8 factors.
   - Resolution V: 5 factors.
2. A resolution IV experiment design can be a better starting point than a resolution III design if the number of factors are not excessive.
3. Because of the additional knowledge that is gained during a multi-experiment strategy, it may be appropriate to add factors and/or change the levels of factors as the follow-up experiments progress.
4. Within an experiment, center points can be added to evaluate for nonlinearity between factor levels. If nonlinearity exists, a response surface design can then be used to estimate the nonlinearity effects for the purpose of better determining factor levels that optimize the process.
5. A confirmation experiment is useful to validate the conclusions from the DOEs.

There are several approaches to determine the combination of factor levels to use within a DOE (that is, the DOE design matrix). Some computer programs give experiment design matrices that are a function of the number of factors and the resolution that is desired for information about two-factor interactions (for example, the temperature and pressure two-factor interaction previously noted). Taguchi and Konishi give orthogonal arrays and linear graphs, where many of these experiment designs are similar to those described in other sources.[39] Bisgaard has tables for 8 and 16 trial designs with blocking,[40] and Breyfogle has 8, 16, 32, and 64 two-level full and fractional factorial tabular designs along with associated two-factor interaction aliasing for resolution V, IV, and III designs.[41]

There are several methods to analyze the results from a DOE. In the analysis, each effect (for example, main or two-factor interaction) is evaluated to see if it is significantly large relative to an estimate for experimental error. Analysis of Variance (ANOVA), t-tests, or a simple probability plot of the effects can be used to make this assessment. However, in a typical DOE the effects determined from these analyses relate only to the mean and do not address how a factor level can affect the variability of a response (for example, a process yields a more consistent day-to-day output at low pressure since it is more robust to typical "noise variations" in ambient temperature and humidity).

To address the effect of factor level on the variability of a process, Taguchi suggests using an inner and outer array (that is, a combination of two DOE matrices). For each inner array trial all outer array trials are executed (that is, a 16-trial inner array with a 4-trial outer array would have a total of 64 trials). The inner array is a design matrix with factors and levels that can be controlled (for example, part tolerances). The outer array design matrix has "noise" factors that typically vary within a process (for example, ambient temperature and humidity).

To analyze an inner/outer array experiment, Taguchi devised a signal-to-noise ratio technique, which some research shows can yield debatable results.[42] However, Box states: "It can be shown that use of this signal-to-noise ratio is equivalent to an analysis of the standard deviation of the logarithm of the data."[43] In addition to determining the significant factors and the estimate of their mean effects, this type of experiment could yield insight as to how the inner array factors could best be set to reduce the variability in a response as a function of changes in the outer array factor levels.

A variation to the above basic inner/outer array strategy that some practitioners might find more useful and easier to implement is to construct an experiment design where each trial is replicated and the standard deviation between replications is also considered a trial response to analyze (with the above noted logarithm transformation, if needed).

For companies to survive and prosper in the future they must continually work to do things "smarter." They must look at things differently and direct their efforts during both development and manufacturing toward the goal of better meeting the needs of their customers. Wisely applied statistical techniques often need to be an integral part of this change activity. Manufacturers commonly need to break a paradigm of just assessing whether component parts are "good enough" relative to specification limits. More effort should be given to identifying specifications that are important to better meet the needs/desires of customers. Effort should then be given to improve the underlying processes so that these criteria components are more closely aligned to nominal specifications and have less part-to-part variability. As Edwards Deming notes: "Inspection to improve quality is too late, ineffective, costly."[44]

CHAPTER 10

TAGUCHI'S CONTRIBUTIONS AND THE REDUCTION OF VARIABILITY

References

1. Dr. W. Edwards Deming, *Out of the Crisis* (Cambridge, MA: MIT Press, 1982, 1986).
2. Dr. W. Edwards Deming, Class Notes—4 day seminar—*A System of Profound Knowledge for Industry, Education, and Government*, 1992.
3. Dr. W. Edwards Deming, *Quality, Productivity, and Competitive Position* (Cambridge, MA: MIT Press, 1982).
4. *Ibid.*
5. *Ibid.*
6. *Ibid.*
7. *Ibid.*
8. *Ibid.*
9. Adapted from J. Evans, J. Szuba, and M. Schultz, "How Process Control Plans Affect Manufacturing and Process Engineering," *Transactions—TMI: Innovations in Quality* (Detroit, Engineering Society of Detroit, 1987).
10. W. W. Scherkenbach, *The Deming Route to Quality and Productivity* (CEEPress Books, George Washington University, 1987).
11. Deming, *Out of the Crisis*.
12. *Ibid.*
13. James Bakken of the Ford Motor Company, January 1981.
14. Scherkenbach, *The Deming Route to Quality and Productivity*.
15. *Ibid.*
16. Deming, *Out of the Crisis*.
17. *Ibid.*
18. Deming, *Quality, Productivity, and Competitive Position*.
19. *Ibid.*
20. Deming, *Out of the Crisis*.
21. Scherkenbach, *The Deming Route to Quality and Productivity*.
22. Deming, *Out of the Crisis*.
23. Dr. W. Edwards Deming, *The New Economics for Industry, Education, Government* (to be developed, 1993.)
24. Deming, *Quality, Productivity, and Competitive Position*.
25. *Ibid.*
26. *Ibid.*
27. Adapted from the "Report of the Deming Study Group of Greater Detroit," c/o G. F. Gruska, *AQC Transactions* (Milwaukee: ASQC, 1990-1992).
28. Deming, *Quality, Productivity, and Competitive Position*.
29. Philip J. Ross, *Taguchi Techniques for Quality Engineering* (New York: McGraw-Hill, 1988).
30. G. E. P. Box, "Signal to Noise Ratios, Performance Criteria and Transformations," *Technometrics* 30(1): 1-40 (with discussion), 1988.
31. George, Box, Soren Bisgaard, and Conrad Fung, "An Explanation and Critique of Taguchi's Contributions to Quality Engineering," *Quality and Reliability Engineering International*, 4(2), 123-131, 1988.
32. Joseph J. Pignatiello Jr. and John S. Ramberg, "Top Ten Triumphs and Tragedies of Genichi Taguchi," *Quality Engineering* 4(2), 211-225, 1991-92.
33. Robert Lochner, "Pros and Cons of Taguchi," *Quality Engineering* 3(4), 537-549 (1991).
34. Eugene E. Sprow, "What Hath Taguchi Wrought?," *Manufacturing Engineering* (April, 92), pg. 57-60.
35. TEAM, Technical and Engineering Aids for Management, Box 25, Tamworth, NH 03886.
36. Forrest W. Breyfogle III, *Statistical Methods for Testing, Development, and Manufacturing* (New York: Wiley, 1992).
37. *Ibid.*
38. *Ibid.*
39. Geniichi Taguchi and S. Konishi, *Taguchi Methods Orthogonal Arrays and Linear Graphs* (Dearborn, MI: American Supplier Institute Inc., Center for Taguchi Methods, 1987).
40. Soren Bisgaard, *A Practical Aid for Experimenters* (Madison, WI: Starlight Press, 1988).
41. Breyfogle.
42. Box.
43. *Ibid.*
44. W. E. Deming, *Out of the Crisis* (Cambridge, MA: Massachusetts Institute of Technology, 1986).

Bibliography

Bridgman, P. W. *Reflections of a Physicist*. New York: Arno Press, 1980.

Carlisle, J. A. and Parker, R. C. *Beyond Negotiation*. New York: Wiley & Sons, 1989.

DeCharms, R. "Personal Causation and Perceived Control." *Choice and Perceived Control*. Hillsdale, New Jersey: Halsted Press, 1979.

DeCharms, R. "The Origins of Competence and Achievement Motivation in Personal Causation." *Achievement Motivation*, New York: Plenum Press, 1980.

DeCharms, R. *Enhancing Motivation*. Irvington Press, 1976.

Deming Study Group of Greater Detroit. "Report of the Deming Study Group." *Transactions, AQC 1990*, ASQC.

Deming Study Group of Greater Detroit. "Report of the Deming Study Group II." *Transactions, AQC 1991*, ASQC.

Deming Study Group of Greater Detroit. "Report of the Deming Study Group III." *Transactions, AQC 1992*, ASQC.

Gruska, G. F. "Enumerative vs Analytic Studies." *Quality Concepts '90*, ASQC/ESD 1990.

Gruska, G. F. *Comprehensive Process Control Plans*. Orchard Lake, MI: The Third Generation, Inc., 1986.

Heaphy, M. S. *Theory D—An Overview of Dr. Deming's Philosophy*. The transformation Network, West Bloomfield, MI 1989.

Henderson, D. et al. "Total Quality Requires Total Involvement." *Transactions, AQC 1990*, ASQC.

Kohn, A. *No Contest*. Boston: Houghton Mifflin Co., 1986

Mann, N. R. *The Keys to Excellence The Story of the Deming Philosophy*. Prestwick Books, 1985.

CHAPTER 11

PROCESS APPRAISAL

THE PROCESS VIEWPOINT

A proven methodology for achieving improvement is to look at each operation as a *process* with inputs, actions, outputs, and feedback loops. A model of a universal process is shown in Fig. 11-1. Material inputs to a unit operation/process come from outside suppliers of parts and materials, from a prior manufacturing process, or from other internal sources. Action is taken by people or equipment (or both) following written procedures; using input materials and subassemblies; using resources from engineering design, sales, or production scheduling; and working in an environment unique to the process. The output goes to inspection and test, to a subsequent manufacturing process, or to the external customer. There may be several feedback or feedforward loops for controlling and smoothing the process.

Quality output requires quality input, and this demands that all personnel know exactly what the process needs for excellent performance and therefore that specifications for suppliers are unambiguously defined, and that purchases are made only from qualified suppliers. For technical specifications (dimensions, weights, compositions, voltages, etc.) tolerances are the most important parameters. Too often they are set arbitrarily, pulled from a vendor's catalog, or based upon similar products. One basis for a robust manufacturing process and a robust product design is insensitivity to variations in the characteristics of parts and materials. Therefore, the tolerance range on specifications should be set as narrow as possible for the least trouble in manufacturing or in the product's reliability, and as wide as possible for minimum cost. For critical material and part characteristics it is often valuable to experimentally determine how the manufacturing process or product reliability is affected by variations in each characteristic, so that a tolerance specification can be optimized for quality performance and acceptable cost. The realm of supplies specifications and tolerances can be valuable territory, full of opportunities for improvement.

Turning attention to the action of an operation, one might first look at written procedures. A written procedure should describe the operation to which it applies, for example, ''welding truss members to the frame;'' the material and equipment to be used, for example, specific welding equipment and type of rods; the procedure for using the equipment and carrying out the operation; reference to applicable technical documents; and any special instructions where the procedure differs from standard practice. It is especially important to note aspects of the work which have a critical effect on the quality of the result, for example, the strength or freedom from warpage of a welded junction.

Examine the equipment (machinery, tooling, part movers, computers, etc.) to see if it is really adequate for the work to be done. Talk with the people doing the work (whether ''operators,'' ''engineers,'' ''schedulers,'' etc.) to discover if they have the skills, training, and understanding necessary to do outstanding work. Consider all the resources needed to do the work, especially information resources, and see if they are described in the written procedure. Ask people doing the work if they get the resources they need, when they need them. This simple line of exploration can open many vistas for improvements. Also notice if any aspects of the environment (humidity, temperature, noise, cleanliness, disruptions, etc.) could adversely affect the quality of the work. Closely observe the operation and discern whether the work actually done corresponds with the written procedure. Investigate the reasons for any differences, choose the best way, and modify the procedure accordingly. Mundane work perhaps, but an overlooked area for improvements.

Feedback (and anticipatory feedforward) loops are important elements for the quality of a process, especially its responsiveness, stability, and ability to adjust to unanticipated changes in conditions. Feedback loops to control machinery may be automatically or manually executed. Some of the important feedback/feedforward loops consist of intermittently transmitted information between operators of connected processes, concerning marginal output or an alert about an unavoidable change of input to the next process. Search for these loops, describe what they are supposed to accomplish, then do a reality check to determine if the feedback/feedforward actually accomplishes what it should do. If it does not, analyze the situation to discover what is wrong and why. After the analysis is completed and suitable corrective actions are known, an engineering change order (ECO) can be initiated to correct the situation. To enhance the effectiveness of an ECO the situational analysis must include a discussion with the operators.

CHAPTER CONTENTS:

THE PROCESS
VIEWPOINT   11-1

COST OF
QUALITY   11-11

THE ROOTS OF
QUALITY
COSTS   11-12

GETTING STARTED
WITH QUALITY
COSTS   11-13

STRUCTURING
THE QUALITY
COST
SYSTEM   11-13

CONCLUSION 11-15

The Contributors of this chapter are: Don Hoernschemeyer, Ph.D., President, Manufacturing and Development Technology, Soquel, CA; J. Stephen Sarazen, President, EXL Group, Nashua, NH.
The Reviewers of sections of this chapter are: Jack Campanella, Corporate Quality and Reliability, Underwriters Laboratories, Inc.; Miki Magyar, M.A., Technical Editor, Technical Communications, Colorado Data Systems, Inc.; James F. Rafferty, President, Eagle Engineering, Inc., Troy, MI.

CHAPTER 11

THE PROCESS VIEWPOINT

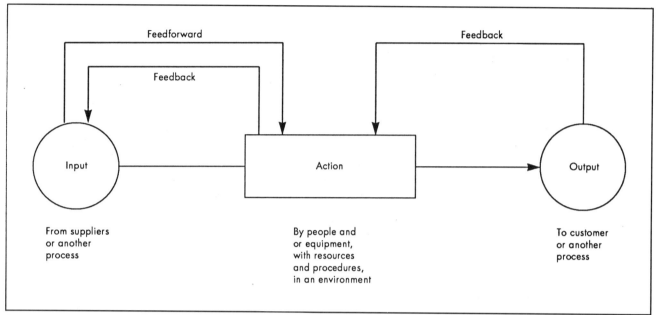

Fig. 11-1 Model of a universal process.

As feedback inherently involves information comparing actual with desired results, one can improve manual feedback loops by looking for quality deficiencies in the information—errors, incompleteness, lateness, and extraneous data which represents waste and noise. Deficiencies in people-mediated loops represent another untapped lode of riches for continuous improvement. Communication links between producers and vendors, between producers and customers, and between people in internal operations are an excellent place to look for process improvements. Such interfaces represent the "soft" feedback loops which generally receive far less attention than the "hard" feedback loops in manufacturing equipment. Adding new feedback loops where they could reduce variation is another excellent way to improve a process. A common situation which can be improved by adding feedback is where a part is difficult to fabricate because of wide variations in one dimension of a component part (or other characteristic), and to compensate the operators work around as best they can. Another area frequently missing a feedback loop is the interface between the external customer's receiving inspection/test and the producer's final testing. A customer may find that an important product characteristic (for example, amplifier gain) is consistently off the specification target, but still within the specification range, and uses the unit without communicating the variance to the producer. The corresponding producer's tests show the amplifier gain to be right on target. Neither knows that their tests are giving different results—until larger deviations push the gain outside the specification's range, according to the customer's tests, and he or she returns the goods because they are nonconforming. Meanwhile, the producer, believing that the units are perfectly good, is surprised and annoyed by the returns.

UNDERSTANDING PROCESSES

A fundamental requirement for improving a process is a thorough understanding of its technical basis, its ergonomics, and its idiosyncrasies. This is certainly an obvious truth, but key details of a process are often not perceived or understood, and therefore not mentioned in written procedures nor paid attention to. The domain of understanding holds diamonds in the rough for mining and polishing.

Attempting to answer the question, "How well do we really understand this process?" can be a provocative stimulus to action. An excellent starting point to macro understanding is a flowchart of unit manufacturing processes and of the overall manufacturing operation. An example of the latter is given in Fig. 11-2. Overall flowcharting is an important step because it makes explicitly visible the interfaces and feedback loops between operations. Interdepartmental discussion of an overall production flowchart is a leveraged improvement activity, because a small investment in time can disclose surprises and sources of errors which were not previously known to the participants. By the time consensus is reached regarding what actions do or do not occur, and who gets what from whom, many sources of problems and aggravations will have disappeared and there will be a better understanding of each other's work. In the right environment the participants will naturally conceive and implement improvements in overall manufacturing operations.

Another place to look for better understanding of a process is in the "universal process factors," which are the various actions taken by people and equipment, using resources, supplies, and procedures in an environment, to achieve the desired result or output. With these generic factors, and examples from the printed circuit board industry (see Figs. 11-3 and 11-4) the significance of most of the items in Fig. 11-3 is probably self-evident, but a few comments may be in order.

Concerning the action factors, it is natural to initially analyze and plan the work of a process, yet seldom review it. People seem to concentrate on the execution of work while neglecting conscious attention to analysis of the results and planning for improvements. Beware of the mindset, "We're action oriented." It is a deadly platitude! CI requires continuous cycles of measurement, analysis, correction and improvement plans, execution of modified work, and checking on the results to see what has been achieved.

CHAPTER 11

THE PROCESS VIEWPOINT

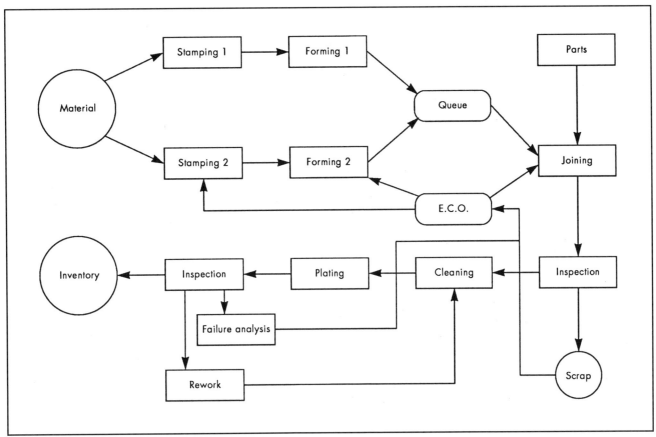

Fig. 11-2 Production of a sheet metal cabinet for radar electronics.

Regarding the people factors, skills requirements and development can easily be neglected. For example, in using personal computers it is common to find people making only minimal use of the capabilities of the software because they have not had training in its use, or had not been given time to learn the requisite skills. For engineers there is always the issue of keeping skills current with evolving technology and computer capabilities. How many would say that their performance could not be enhanced by learning new technical skills? Very few! It is easy to neglect people's continuing education. The importance of relationships between people in different departments cannot be overemphasized. Multifunctional *ad hoc* teams, formed to deal with a single issue, can make outstanding improvements. Habits and biases act invisibly, and their influence is subtle, but seldom weak. Insight into the presence and roles of these factors can disclose limitations to the effectiveness of processes. Some common limiting biases are: perceiving quality and productivity as competing goals; not-invented-here mindset in engineering groups; and the unconscious view that people are directly responsible for errors and defects. Common debilitating habits are firefighting and band-aid fixes.

Making equipment easier to use is a simple way to increase throughput and reduce errors and defects. What is required is that the users think about improving it, rather than passively accepting it the way it is, and that those who can change the equipment are receptive to suggestions and act to get something done.

Two resources always ripe for improvement are information and time. Ask people if they have all the information necessary to do the best job. Notice if they have the time—are allowed the time—to pay attention to improvement work and to do it. The inherent difficulty here is the habit of thinking only of material resources. People become demotivated when they do not receive the nonmaterial resources they believe they need.

Examine whether the people doing the work have contributed to the formal procedures and whether they've had an opportunity to review them before they are finalized. If they have not, make it so, with the double goals of improving the procedures and having people feel ownership for the way they work. Changes to procedures and drawings clearly need to go through a formal review and approval process. In most manufacturing companies this is handled by an Engineering Change Control Board. But it is an activity (process) that, in the rush of daily business, is easily executed in a less than optimal way. Not only may engineering changes make manufacturing process more difficult, but they can also be the cause of a product problem—for example, lowered reliability. In the realm of environment and culture, factors such as standards of performance, people's attitudes about the possibility and importance of continuous improvement, and the constancy of purpose and support by management are crucial to long term success. Otherwise people will view CI as just another passing fad, with obvious consequences. The environmental factors of an organization are usually invisible to its inhabitants. To perceive them it is useful to solicit the impressions and insights of an outsider who is attuned to such nuances. Neglect of the physical working conditions is a certain prescription for low motivation. To make an improvement, simply notice a need and do something about it.

CHAPTER 11
THE PROCESS VIEWPOINT

Another powerful route to improving a process through greater understanding is in modeling its behavior on mechanical, thermodynamic, kinetic, or other technical principles. Such an analysis can disclose a minimum set of technical process parameters which must be specified and controlled, and disclose their interrelationships. Technical modeling can also indicate whether the output of the process depends in a linear, exponential, logarithmic, or other way upon the input and process parameters. It may also indicate a maximal or minimal response behavior. There is certainly no need to experimentally search for this information if it is known from technical principles, although in a complex system it may be useful to check on the expected relationships between variables. A technical model of electroplating holes in printed circuit boards would disclose cleanness of drilled holes (freedom from resin smear) as a mechanical factor, composition of the plating bath as a thermodynamic factor, and electric current as a kinetic factor. An example of a much more complex technical process where thermodynamic and kinetic modeling has been useful, is the chemical vapor deposition of metals, oxides, and nitrides onto silicon wafers in the fabrication of integrated circuits.[1] For assembly operations, a model may be generated from the work of the best and worst operators, and used to supplement a mechanical analysis.

Finally, make a concerted effort to discover all the influential factors in a process. They can be found among the "universal process factors" (see Fig. 11-3), in the technical model analysis, by asking people doing the operations, and by using DOE. To make effective use of the universal process factors it is necessary to replace the general factors by specific factors for each unit process. Such customized tables (as shown in Fig. 11-4) can provide a solid basis for understanding by those directly involved. DOE is a statistical methodology designed to disclose which parameters (among those examined) are influential, and which interact or have synergistic effect. It gives a quantitative estimate of the relative power of the parameters to affect the output of the process. When using DOE it is imperative to consider all parameters in the analysis, without *a priori* exclusions. Failure in this step is responsible for many failures or limited successes in technical process improvement. There is extensive literature and seminars on the important subject of DOE. It is a very powerful technical tool.

MONITORING PERFORMANCE

To know how a process is performing, whether a process needs improvement, or if an intended improvement action has been successful, one obviously needs measures of that performance. In fact, a comprehensive performance monitoring system is a fundamental requirement for manufacturing excellence. Companies typically measure overall performance—for example, the cost of production, manufacturing yields (but not always the more telling

---

Actions
- Analysis.
- Planning.
- Execution.
- Measurement.

People
- Skills.
- Motivation.
- Habits and biases.
- Relationships.

Equipment
- Technical factors.
- Kinds of controls.
- Ease of use.
- Reliability.

Resources
- Materials and supplies.
- Information.
- Time.
- Space.

Procedures
- Approved.
- Written and understood.
- Simplicity.
- Verification and changes.

Environment
- Expectations and standards.
- Attitudes about quality and productivity.
- Constancy of purpose and support.
- Physical working conditions.

Fig. 11-3 Universal process factors.

Actions
- Forming a conducting surface by electroless metal deposition.
- Figuring the proper current, time, and geometry.
- Doing the plating correctly, with changing mixes of boards and plating conditions.

People
- Operators understand general plating principles.
- Old habits about "tweaking" are explored and corrected if necessary.
- Platers have a good relationship with drillers, and discuss common issues.

Equipment
- Manual control of timing is replaced by automatic control.
- Out-of-limit bath chemistry is signaled by an alarm.
- Board and electrode placement for each board type is easily set.

Resources
- Vendor's information on the use of specific chemicals.
- Time for operators to make changes between different types of boards.
- Lab results on plating quality are quickly fed back to the operators.

Procedures
- The vendor's recommendations are checked for actual performance.
- Current, time, and board placements are listed for each type of board.
- Proposed changes in drilling conditions are checked for effect on plating quality.

Environment
- Production management does not keep changing priorities of what is most important.
- Operators get support from maintenance and engineering.
- Physical working conditions are safe and conducive to quality work.

Fig. 11-4 Examples of process factors in electroplating of holes in printed circuit boards.

CHAPTER 11

THE PROCESS VIEWPOINT

first-pass yields), meeting delivery schedules, etc. But as necessary as these measures are, they give no clues to the performance of unit operations, and hence no clues to where improvements could be most beneficial. What is needed are real-time measures of key performance parameters for each manufacturing process. Three of the most fundamental generic measures are Quality, Cycle Time, and Cost. Their relationship to a process model is shown in Fig. 11-5. Cycle Time is measured by the time from initiation to completion of work, and includes machine set-up, waiting for input materials, and delivery to the next step. Cost is measured by the cost of materials and labor, and the cost of quality failures (for example, scrap and rework), although amortized equipment cost can be included. Quality of the output can be measured by yield, number of defects or errors, and deviations from target specs. An example of performance measures for a sheet metal joining operation is shown in Fig. 11-6. With existing software and networked computers it is relatively easy to monitor real-time process performance and compare these measures with the minimum possible cycle time, with a 100% yield cost, and with complete conformance to output specifications.

But it does not suffice to just measure and examine trends in average values of Quality, Cost, and Cycle Time indicators; one must also look at the variability of these indicators. According to G. Taguchi, any deviation from a target value represents a loss to society. That is, there will be a cost to pay somewhere through the birth and lifetime of the product. The cost to society for a field failure is borne by the owner or the manufacturer, depending upon whether the failure occurred during the warranty period. Products in which components are within engineering tolerances will experience breakdowns at a rate roughly proportional to the magnitude of deviations from the drawing. Deviations which propagate through a multiple step manufacturing operation, causing larger or additional deviations, will invariably lead to low yields or low product reliability—all expensive. Hence, it is most important to have measures of the variability of process output. As the output of processes varies with time, it is essential to have a statistically valid way of interpreting the performance data.

Statistical Process Control (SPC) is a well known means for obtaining such information.[2] It provides a rigorous way of monitoring process performance, as well as monitoring variation in key input and process control parameters. It is also used to measure the ability of the process to meet the output specifications. The so-called Capability Index, C_p, is defined as the specification width divided by the statistical process width (equal to three times the standard deviation of the process average). The larger C_p, the more capable the process. A value of at least 1.0 is generally sought. SPC-type charting provides an easy way to notice trends, in particular whether a change made to improve a process did or did not in fact improve it. Charting displayed on walls should be easy to read, color-coded, and without the numerical data. As real process data always contains noise and sporadic jumps, statistical analysis is needed to make valid conclusions. SPC methods automatically provide this guidance in a simple to interpret form.

Performance measurements will however do no good unless they are used as an opportunity for improvement, and acted on. One of the benefits of posted SPC charts is their announcement of performance trends or the maintenance of steady state control. They display performance information in a universal format which everyone can read and interpret, which facilitates (but does not ensure) rational discussion and decisions about action.

CORRECTING AND PREVENTING PROBLEMS

A back-door way of systematically raising performance is by eliminating those problems that drag down performance. The first step in correcting existing problems is to notice them—which is not as easy or as trivial as it sounds. Why? Because of a natural human inclination not to admit the existence of problems, particularly when they are self-caused, and to ignore them when they are

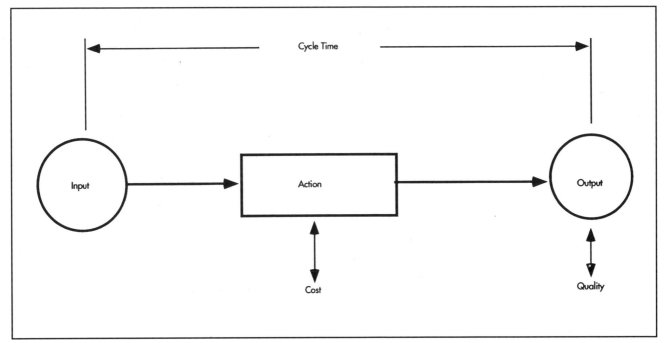

Fig. 11-5 Measurement of process performance.

CHAPTER 11

THE PROCESS VIEWPOINT

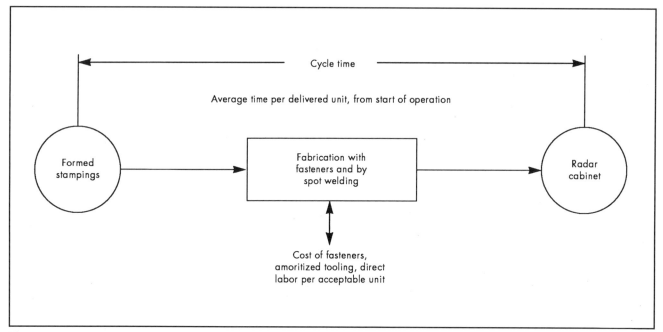

Fig. 11-6 Measurement of joining performance.

recognized. As Albert Einstein said, "Problems cannot be solved at the level of thinking that created them." This is particularly applicable to problems caused by organizational structure, or by management decisions and actions a few levels above where the problem is occurring. Really tough problems which have resisted all improvement efforts typically require the participation and cooperation of management. Even "purely technical" problems can have their roots in a management decision.

An important tactic of the unrelenting war on waste and poor quality processes is to force problems into the open by systematically detecting, measuring, and reporting them. Therefore, a comprehensive performance monitoring system is mandatory, not only because it exposes problems, but also because it quantifies them for cost analysis and prioritizing. If the problem cost is high enough it will get action. Three specific tactics for forcing corrective action are:

1. Publicly displaying performance charts.
2. Pushing supplies and finished inventories down to minimal levels (this is one method of JIT).
3. Stopping the production line for serious or repetitive problems.

Any unwelcome action to expose serious problems can be hazardous. The support and encouragement of management is essential for successful improvement work and for the wellbeing of the participants. It is wise to elicit the support of management for the exposure and reporting of sensitive performance measures.

Prominent display of overall performance charts (for example, yields, warranty returns, or delivery times) is likely to get the attention of managers and spark discussions in staff meetings. However, such charts do not indicate the source of the problems, nor directly translate to responsibilities for unit manufacturing operations. Hence, they do not usually invoke corrective action. What is needed are charts at every level of operations so that problem tracking can be established. The use of small inventories forces any production problem into the open because there is no buffer stock to fall back on. This is particularly true for work-in-progress, that is, work that is standing idle in queues. With line-stoppage, each operator has the authority to sound an alarm and/or stop the production line if there is a serious difficulty or a repetitive problem in his or her area. This gets immediate attention! Protocol then invokes the assistance of whatever people are necessary to quickly solve the problem, or to initiate longer term diagnosis and correction work when the causes of the difficulty are not readily apparent. A related and powerful technique for information transactions (for example, about supplies in inventory) is to notify the responsible person of a serious error as soon as it is noticed, and request or demand corrective action, depending upon the frequency or severity of the problem. Obviously this technique requires good communication skills and agreements, and often a formal procedure.

One guiding principle for corrective action is to detect and act as close to the difficulty as possible. A way of accomplishing this while using SPC charts is to have the operator responsible for the construction, updating, and interpretation of the control chart, and responsible for initiating action whenever the data points signal an out-of-control, or approaching out-of-control condition. A corollary is that staff analysis and assistance is needed only when the problem cannot be solved by those closest to it. Operators ask for help when they need it.

A cardinal principle of good corrective action work is to make every attempt to discover the root cause(s) of the problem, rather than being satisfied with "band-aid solutions" which merely patch up symptoms. This principle can be illustrated with another example from printed circuit board manufacturing. Suppose that poor plating is observed in the component mounting holes. This problem could be due to a poor electroplating process, or to deficient electroless metal deposition in the prior operation, or to incomplete cleaning of drilled debris from the holes, or to improper drilling (for example, speed, drill sharpness, etc.), or to incorrect lamination conditions causing incomplete resin cure, or to a defective resin formulation in the purchased laminating

CHAPTER 11
THE PROCESS VIEWPOINT

material. Compensation for tool wear can be built into the software on CNC equipment, thus the machine checks and corrects itself automatically. But when it fails—and it will fail—humans still have to do the troubleshooting. And stopping short of the root cause is a sure recipe for recurrence of the problem.

An aid that can be used to push a problem analysis down to root causes is a Cause and Effect (C&E) Diagram, with an example shown in Fig. 11-7. Notice the strong similarity between the elements of this diagram and that of the universal process model in Figs. 11-1 and 11-2. "Output" is replaced by "Effect," and the Input and Action factors all become Cause factors. The items listed under each of the six cause categories are just a few of the many typical causes which may affect the output of an operation.

When using a Cause and Effect diagram one would replace the generic factors by whatever specific factors are applicable. For example, "Information" may be replaced by "Welding Bulletin AP 4.2" or "Production Schedule," or "Job Traveler No. PC-5," and "Measurement" replaced by "Tensile Tester K2," or "Ultrasonic Imaging Unit." A cascade of C&E diagrams is used to analytically get down to the root causes. The observed problem is taken as the "Effect" in the first diagram and possible causes are listed as branches in each of the six categories. After an initial investigation these possible causes are pared down to a smaller set of probable causes. Next, each probable cause from the first diagram is made an "Effect" in a second tier of diagrams. Again using the printed circuit board example, "poor plating of holes" would be the Effect in the first diagram and "unclean holes" would be one of the cause factors under the "Material Resources" category. In the second diagram, "unclean holes" becomes the Effect and possible cause factors might be dull drills ("Equipment") or incorrect drill speed ("Method" and "People"). By generating successive diagrams until no further causes can be conceived, one will get to the root causes. People find this technique easy to use and very effective. Another power of the C & E diagram is that it tends to force an examination of all categories of causes which could produce the effect.

Not surprisingly there are several other formal methods which have been developed for solving problems, for example, Kepner-Tregoe, FMEA (failure mode effects analysis) and Morphological Analysis. All are extensively described in the literature. Another effective method is to combine several SPC control charts (or simple trend charts) and look for a correlation between out-of-control points in a process output parameter and out-of-control points in process parameters (for example, temperature or speed) or input parameters (for example, composition of a material). Charting all key process parameters, back to purchased supplies, can in many instances readily trace cause and effect back to the root causes. As long as the input, process, or output data is measured and stored in a spreadsheet or database it can be readily displayed as a chart. Several SPC-type software programs make it easy to simultaneously view several charts in time alignment.

For effective corrective action, four other items need attention:

1. Follow up to insure that action was taken.
2. Determine if the action did indeed achieve what it was supposed to.
3. Notice whether the action produced any unwanted or unintended consequences. It is not uncommon for an engineering change, made to correct one problem, to cause another. This possibility must be blocked by rigorous Engineering Change reviews.
4. Any incomplete corrective action should be brought to the attention of the operations manager and discussed in staff meetings. Corrective action items must be completed and closed.

In addition to correcting existing problems, the aim is to prevent their occurrence in the first place. Many of the techniques of safety work—where it is essential to prevent accidents—can be used. A "what if, then" analysis can be applied to the universal

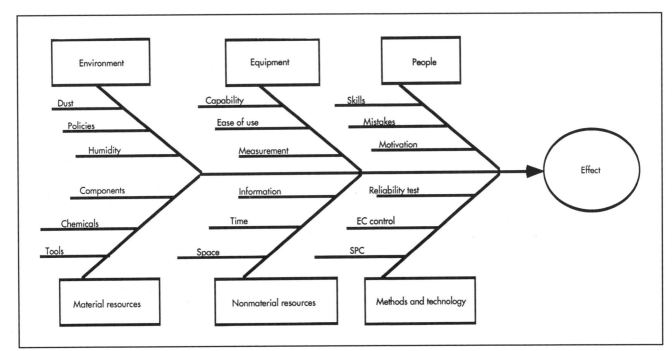

Fig. 11-7 Cause and effect diagram (ECC = Engineering change control).

CHAPTER 11
THE PROCESS VIEWPOINT

cause factors of Resources, Methods, Equipment, People, and Environment to anticipate what could occur if there was a change in any key parameter. This analysis clearly requires good understanding of the process; allowing the process opeators to do it is convenient and empowering. They have the requisite experience, often knowing "if the drill speed is too high on this kind of laminated board, the guys in plating are going to have real difficulties." The technical staff can apply the same analysis to thermodynamic, kinetic, mechanical, or electromagnetic process factors. Potential problem analysis has a way of forcing ignorance about the process into the open.

An excellent place to look for improvements is at interfaces between operations, where there is a hand-off of parts, drawings, information, etc., because this is an area where many problems originate. Some key interfaces are: customer-to-marketing; marketing-to-design engineering; design engineering-to-manufacturing; engineering-to-purchasing; and purchasing-to-suppliers. These interfaces involve technical specifications, customer expectations and desires, process specs, material, and part specs. Many things can and do go wrong during informational transactions between functions or departments. A powerful and contemporary way of systematically handling interface issues in a manufacturing environment is with "Concurrent Engineering," wherein marketing, design engineering, application engineering, test engineering, and manufacturing people work as a team. A simple way to initiate improvements in interface transactions is to deal with unsatisfactory hand-offs as they occur and get the two sides of the transaction together to discuss the issue. To provide a solid basis for such discussions, it is essential to have a record of errors which is creditable to all persons.

WASTE AND COMPLEXITY

The most obvious way of improving an operation is to eliminate wasted effort, time, and materials. For a perspective on this question consider a quote from Dr. S. Shingo.[3] He tells of a visit to a plant in which the slogan "Eliminate Waste" was prominently displayed on a large sign. "When I saw this sign, I asked the firm's president if all his employees were idiots." When the president reacted with chagrin and puzzlement, Dr. Shingo inquired whether this sign had been posted because "some workers would not get rid of waste even if they saw it." "It seems to me," he said, "that as long as someone knows that something constitutes waste he will get rid of it. The big problem is not noticing that something is really wasteful." He told the president that the slogan on the sign ought to be "Find Waste!" Dr. Shingo stresses, "We must always keep in mind that the greatest waste is waste we don't see."

Some forms of waste are quite visible, but other forms are less obvious. A guide to finding obscure waste is to look for work that does not add value to the product or service. Examples are: machine setup time, rework of defective products, and inspection (unless demanded by the customer). Can inspection or testing really be eliminated, and if so, how? It can be eliminated when the probability of nonconformance with specs is very low: specifically, when it is so low that the cost of shipping a defective item is less than the cost of inspecting or testing it. (The cost of shipping a defective unit is incurred when the customer returns the unit under warranty.) A low probability of nonconformance occurs when a manufacturing processes has a high capability, for example, a $C_{pk}>1$. (C_{pk} is defined like C_p except that it accounts for deviation of the average output value from its target value.[2])

With a C_{pk} of 1.0 the probability of a defective unit is about 2600 ppm; with a C_{pk} of 1.17 the chance of a defect is only 465 ppm; and with a C_{pk} of 1.33 it is only 65 ppm. When the probability of a defective unit is that low it is impossible to judge the quality of products by testing or inspecting a sample, and the cost of 100% testing/inspection can be much higher than the cost consequences of using the product or part which may be defective. Even if one cannot eliminate outgoing inspection/testing, one can often eliminate receiving inspection/testing. All that is required is buying from vendors with demonstrated high process capabilities. It is certainly a waste for both producer and customer to test the same items. It all comes down to probability and trust.

Another way to eliminate waste is by eliminating unnecessary complexity: strive to simplify every input, action, and output. With the proliferation of computers a common tendency is to "computerize" all data, irrespective of its value or extent of use. Examine the output of a process and attempt to simplify it by asking the recipient whether any part is not wanted. This is particularly true for information outputs. It also shows up in product features which have little or no value to the customer, and whose presence is due to the producer's infatuation with it or to historically valid reasons. Quality Function Deployment is a good technique for not only insuring that the external customers get the design features they want, but also that they do not get what they do not want or are unwilling to pay for. An example of unnecessary information complexity is massive quality assurance procedures—which merely gather dust and are not used. It is reported that when Rene McPherson took over leadership of the Dana Corporation in 1973, one of his first acts was to destroy 22 inches of policy manuals and substitute a simple one-page statement of the philosophy of the company.[4]

Complexity is a major enemy of consistency and a source of errors. A complex manufacturing process, unless tightly controlled, is likely to exhibit a low consistency (low C_{pk}), and complex products, unless very well designed, are likely to be difficult to use or prone to use-induced failure. A household example of a complex and difficult to use item is a VCR—which few persons can program correctly. An example of a product engineered for simplicity of use, although it is more complex than its predecessors, is the highly automated "point-and-shoot" camera. With these cameras there is no need for the user to make complex decisions about focus, aperture, or speed settings—and photo quality has risen considerably. Another example is an automatic bread-making machine: it is simple to operate, self-cleaning, and makes delicious bread.

INNOVATION

When the simpler means for making improvements have been exhausted, one must look to somewhat more complex or difficult or creative methods, or be innovative. Although innovation usually carries the connotation of new products, it also applies to new processes, services, and new ways of doing things. In the present context this could be innovation for new process technology (that is, manufacturing processes not described in the engineering or patent literature) or better methods of employing the existing technology. A key to innovation is the recognition of an opportunity or a possibility: "we might be able to adapt one of those XYZ machines to replace our WXY machine," or, "that seems like a difficult and unreliable way of producing that part: there must be better ways." The task then is to convert the recognition of an opportunity into a useful and profitable result. The "how" of doing this has naturally received considerable attention.[5,6,7]

CHAPTER 11

THE PROCESS VIEWPOINT

Some key features of the innovation process are shown in Fig. 11-8. A new perspective can be an optical wedge to expand the vision of what is possible. But how to get those new viewpoints and perspectives? Meditation, group gestation sessions, or a hike in the mountains may work. New perspectives can also be obtained by listening to the views and ideas of operators, new employees, operators in adjacent work areas, and consultants. A new perspective can be immensely valuable in opening up a whole new approach to the issue. The perspective that transistors could be fabricated on a single silicon chip, and connected to each other without wires, made possible the microelectronics industry.

A fertile field for sprouting new technical ideas is the patent literature—a source often neglected by companies which do not apply for patents. The best ideas frequently come from technology outside of one's primary area of interest. For example, a technique used in the printing industry has been adapted to manufacturing slow-release drug products and a new form of high energy battery. The technology of integrated circuit fabrication has been applied to the manufacture of very small pressure sensors fashioned from silicon.[8] Cross-fertilization is a powerful way to generate new and useful technology, and patents are an excellent source of pollen.

Another systematic and effective technique for getting new ideas is called "benchmarking." (See Reference 9 for an overview.) The reason benchmarking is so effective is that proven ways of excellence are copied.

Benchmarking can be defined as measuring performance against a best-in-the-class company, determining how the best in that class achieves those performance levels, and using the information as a basis for the surveying company's own targets, strategies, and implementation. Although "copying" may not seem innovative, some creative modifications are often needed to adapt the borrowed method to one's own situation. According to Robert Camp of Xerox Corporation (a pioneer in developing successful methods of benchmarking), "Benchmarking is a positive, proactive process to change operations in a structured fashion to achieve superior performance."[10] In benchmarking the manager's goal is to identify those superior companies, regardless of industry, so that their practices and methods can be studied and documented. One key is to look beyond one's own industry because the leading practitioner of what is being sought (for example, shortened cycle time) may be in another industry. As Camp expressed it, "Benchmarking should not be aimed solely at direct product competitors. In fact it would be a mistake to do so since they may have practices that are less than desirable."

So benchmarking is about learning from the best-in-the-class and doing the necessary self study, to clearly discern the distinctions—especially fundamental distinctions—between the practice in one's own organization and in the best that has been identified. Thereafter it is straightforward planning to improve processes by incorporating the best methods. As may be expected, getting good benchmarking information from another organization requires a close relationship and mutual trust. The International Benchmarking Clearinghouse has recently been established to help companies with the process.[11]

Innovation can also be done right on the factory floor. Although the testing of new processes is usually done outside the manufacturing area, ultimately the innovation has to be tried in the manufacturing arena. Innovation is often required when transferring a new process into the production plant, because the conditions and requirements of the production environment differ from those in the development environment. As Marcie Tyre has written,[12] "to implement new technology successfully, (production) plants need to schedule time for testing and experimentation just as diligently as they would schedule time for production. They need to create what I call 'Forums for Change'—to set aside times and places in which users and experts can gather and reflect on data, formulate new questions, and develop new solutions without disrupting normal production...What I discovered was that where managers set up forums for change, they saved months of start-up time, and the resulting processes were both more reliable and more efficient."

OPTIMIZATION AND CONTROL

After major problems have been corrected, manufacturing operations simplified, bottlenecks and waste eliminated, and better process technology installed, it is appropriate to consider

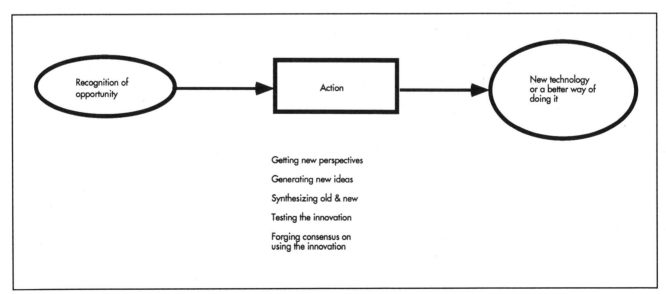

Fig. 11-8 Features of an innovative process.

CHAPTER 11

THE PROCESS VIEWPOINT

technical means for optimizing and controlling the manufacturing processes. At a minimum, control of a manufacturing process means maintaining the output parameters within specification limits. At a higher level, control also means keeping the variation of output parameters about their target values below some set values. Together this means maintaining a specified minimal value of process capability (C_{pk}). This results in a robust manufacturing process, one which is insensitive to small random variations of the input and action parameters. Optimization means maximizing or minimizing the product performance characteristics (for example, amplifier gain, signal-to-noise ratio, strength, impurity concentrations, mean time between failures), or enhancing the throughput (minimizing the cycle time or maximizing the yield). The yield is, in turn, a function of specifications and variation from the target value; yield is a direct function of C_{pk}.

There are three major tasks to optimizing and controlling: acquiring fundamental process information, improving basic quality factors, and tuning the technical parameters. The first was covered earlier in this section. Some basic quality factors to consider are: accurate and reliable test methods, good written process procedures, operator understanding and skills, well maintained equipment, and supplies that meet all requirements. The major subtasks of technical tuning are:

- Bringing a process into statistical control by removing disturbing causes.
- Determining process capability.
- Adding or modifying automatic control devices.
- Setting key factors for highest performance and stable operation.
- Implementing SPC.

First, direct attention to achieving control of a process, because it is generally a waste of effort to try to improve a process that is not stable. A process is said to be in "statistical control" when the output variations are stable in time, that is, balanced around the center point in a random manner. To bring an out-of-control process into control, it is essential to understand the types of disturbing causes which may act on the process. There are four types of causes which produce variation: Common Causes, which are the myriad of ever-present factors contributing in small and random ways; Special Causes, which sporadically induce variation over and above that inherent in the system due to common causes; Tampering Causes, which are unnecessary adjustments made in an attempt to compensate for the other sources of variation; and Structural Causes, which are regular or long-term changes due to environmental factors such as humidity, temperature, holiday periods, etc. A process is brought into (statistical) control by eliminating special causes and tampering causes.

Tampering is a faulty way of trying to control a manufacturing process.[13] It is faulty because it involves making a change when no change is called for (so-called "tweaking"), that is, when there is no statistically valid reason for concluding that the process is out of control. Tampering causes can be eliminated by prohibiting adjustments to a process when a deviation in the measured value of an output parameter does not represent a statistically significant deviation from the norm. Adjustments are allowed only when the measured values show an "unnatural" pattern in a SPC-type chart[2] and when there is a sound linkage between a possible cause factor and the observed unnatural variation.

Sources of special disturbing causes can be sought in the elements of the universal process model: equipment, procedures, people, resources, and environment; for example, in changes in product or process specs, component specs, supplies and materials, testing accuracy, etc. If the culprit special cause is not apparent after a search of such factors, it is highly advisable to initiate formal corrective action. Technical sources of special causes can often be found by looking for correlated changes in output parameters and one or more of the input or process control parameters (for example, composition or temperature).

After a process is stable, its capability (C_{pk}) is determined by taking enough measurements over a time period sufficient for all normal operation conditions to be included. If the process capability does not meet the goal, it must be improved further—either by shifting the average value toward the target value; by lowering variation through eliminating special or common causes; or by changing the output specifications. At this point it is always wise to re-examine the validity of the specs: perhaps a subsequent manufacturing process is now robust enough to accommodate the input variance, or perhaps customers do not really need such tight specs.

In optimization the aim is often to change the average value of an output parameter or the cycle time—for example, raising the color saturation value of a photographic film by changing the film composition and manufacturing process conditions. While raising the performance it is essential to maintain stability of the process by locating the new process settings on a plateau of the response surface. The technical approach to optimization must involve statistical design and statistical analysis applied to realistic physical models of the process. The heart of optimization work is Design of Experiments (DOE).

DOE is a statistical methodology for determining which parameters are affecting a process output, whether they are acting separately or in concert, and the quantitative response of output variables to the major process parameters. DOE can highlight significant special and common causes; it is not only an optimization tool, but also an efficient tool for increasing process capability. In the words of Bhote (reference 14), "DOE is the key to the magic kingdom of quality." There are many versions of DOE, including methods such as factorial designs, screening designs, self-directing optimization, evolutionary optimization (EVOP), Taguchi orthogonal arrays, and Shainin methods. Many of these are discussed in other chapters.[14]

When a process has been improved through reducing in the variation and bringing the average output into correspondence with target values (or raising it), it may be necessary to implement new controls to maintain the process in this optimal state of operation. Some simple ways of maintaining the achieved improvements are by changing applicable written procedures, by regularly auditing the operation, or by auditing the conformance of supplies and materials to specifications. Another way to lock in performance gains is to install new or different automatic control devices on the equipment. Automatic controls must be designed to react only to statistically significant deviations from the set point. That is, the control device must distinguish the signal, as an indicator of the process condition, from the noise of other sources, so that adjustments are appropriate and do not amount to automatic tampering. When a process is in statistical control, variations from the target value should not be reduced by a feedback compensation method which is based simply on the deviation of the last outcome from the target value. Attempting to control in this manner will appreciably increase the variation and lower the C_{pk}. (See Reference 13 for an illustration.) One can also use SPC control charts to manually maintain control. A trained operator using SPC charts knows when it is appropriate to make an adjustment, when it is not, and when to confer with the engineering staff to find the cause(s).

CHAPTER 11

THE PROCESS VIEWPOINT

INTEGRATING AND HARMONIZING

The remaining action for improving manufacturing operations consists of harmonizing the collection of unit manufacturing operations, integrating them into a smooth overall operation. This requires congruent inputs and outputs of sequential processes; that is, actual output values must match input specs. Although input/output specs may have been congruent when originally designed, in the course of time engineering changes or changes in actual practice can shift them out of alignment. Moreover, optimal settings of input parameters may not match the original input design specs, and the actual range of an output parameter may be too broad to allow good control of variance in a subsequent manufacturing process. For robust manufacturing operations, the tolerance range of the input parameters of a process must be larger than the actual achievable range of the output parameters of a prior process. Accordingly, each process must have a high capability index with respect to the following process.

A second way to harmonize operations is to look for leveraged action, where one action affects several other processes and, hence, produces an effect out of proportion to the original action. In Peter Senge's description,[15] "Leverage comes from new ways of thinking, new insights, and from an awareness of the structure in which the system unconsciously operates." One potent area for leveraged action is matching the input and output rates of the various processes, and simultaneously searching for material or paper bottlenecks due to mismatches in the rates of linked processes. Production scheduling is generally figured on the basis of average throughputs of the separate operations, but common causes of variation continually distort the actual rates, resulting in work pile-up in queues between processes. Analysis of the variance of actual manufacturing rates can provide a more rational basis for scheduling, and for determining whether or not flow variances are normal. Perturbations of flow rates by rework and scrap largely result from special causes.

Another area for leveraged action lies in dealing with the propagation of errors and deviations through successive manufacturing operations. In the printed board example described previously, a deviation in the prepreg resin composition can cause a poorly cured board, causing problems of resin-smear during drilling, which can cause poor plating of the holes. It is obvious that leveraged action lies in obtaining quality prepreg. This is similar to the situation in corrective action work, where leveraged action lies in eliminating the root cause(s). Design target specs will never be met exactly all the time, hence it is necessary to know the effects of variances all the way down the production line. The basic question is, "If this deviation occurs, even though it is within the natural control limits and within the tolerance range (or specification limits), what will be the consequences in every subsequent operation and in the final product?"

Another area for leveraged action is in the dynamic complexity of the whole manufacturing system, that is, in direct manufacturing operations plus all supporting functions. Peter Senge defines "Dynamic Complexity" as "the situation in a system of interconnected processes where causes and effects are separated in space and time, so that persons associated with the causes receive no feedback, or incomplete or delayed feedback, on the effects. Hence, well-intentioned efforts to improve the results often produce unexpected and undesirable effects." Engineering changes are one example of dynamic complexity in manufacturing operations. An engineering change designed to correct a problem or make an improvement in a product's performance, or in the ease of manufacturing, may unintentionally produce undesirable side effects. Common detrimental effects are a greater difficulty in manufacturing, or a decreased product reliability. For complex products the engineering change process offers vast room for leveraged improvement.

MANAGEMENT SUPPORT

The last type of action to be considered is that taken by management in support of continuous improvement. This topic has been eloquently addressed by P. Crosby, W. Edwards Deming and J.M. Juran in connection with quality improvement. For continuous manufacturing improvement at least four aspects of management support should be considered: Context, Communication, Empowerment, and Obstacles.[16]

First and foremost, management must create a suitable context in which continuous improvement can flourish. Attitudes of just getting by, complacency with the present status, and acquiescence to mediocre conditions must be eradicated and replaced with positive attitudes. Managers need to believe in continuous improvement and promote improvement in all processes. Second, management has to communicate the guiding policy and action principles of continuous improvement, and facilitate honest discussions about how CI is, or is not, occurring. Cross-functional meetings can be a good communication vehicle if they promote dialogue and agreements about cooperative actions. Third, people have to be empowered with information, time, resources, and emotional support. No one will work on improvement if they do not have time for it, or if they perceive that everything else is of higher priority. Managers need to encourage people to work in new ways, and to acknowledge people's accomplishments. This is a no-cost action that is easily ignored or shorted. Fourth, managers need to discover and dissolve obstacles to people's performance. Common obstacles are human resistance to change (especially to being changed) and organizational attitudes which stifle creativity and contribution. One way to discover obstacles is to use Peter Drucker's dictum: (paraphrased) "What gets in the way of your working and accomplishing? What do I do—and what does the organization do—that hinders you in doing what you're paid to do?" Resistance to change can be ameliorated by promoting a culture of innovation and learning. Ultimately an organization on the path of continuous improvement becomes a learning organization.

COST OF QUALITY

Over the years, there has been a paradigm established which holds cost as a driver and quality as a result. The first point to be made in this section is that the inverse is more likely to be true. Quality is a critical driver and cost is a result (see Fig. 11-9). Another way to view this is, the better the quality of processes, products, and services, the lower the ultimate cost. The primary purpose for establishing a quality cost system is measurement. As organizations implement quality they must measure to determine

CHAPTER 11

THE ROOTS OF QUALITY COSTS

> From: Cost as the driver, quality as the result.
>
> To: Quality as the driver, cost as the result.

Fig. 11-9 Paradigm shift #1.

opportunities, and measure to determine if improvements are real. A quality cost system, if structured and managed properly, is an excellent measurement vehicle.

There is a second paradigm that many organizations suffer from as well—the perception that products and services may be improved as separate from the processes through which the products and services were designed, developed, and delivered (see Fig. 11-10). This paradigm too, must be shattered. The bottom line is that all products and services are the result of a process and the quality of the process determines the quality of the output. This, in turn, determines the cost.

The shift in these two paradigms must be kept in mind throughout this discussion of quality costs. This section outlines the basic definition of quality cost; the various elements that comprise a quality cost system; and most importantly ways in which to understand the often confusing process of structuring and managing a quality cost system.

To provide a better perspective on quality costs a case example is used, to underscore that organizations need not begin the measurement process with all elements at once. The example is an international manufacturing business. The case further demonstrates the value of forming Quality/Finance/Operations partnerships and operationally defining the elements of the system.

> From: Improving products and services directly.
>
> To: Improving the processes through which the products and services are designed, developed, and delivered.

Fig. 11-10 Paradigm shift #2.

THE ROOTS OF QUALITY COSTS

In Dr. J. M. Juran's first "Quality Control Handbook" published in 1951, the first chapter is entitled "The Economics of Quality." Later, Phil Crosby wrote "Quality Is Free." The implication in the two titles is clear: sound quality yields economic benefits to the organization. In the early 60s, the costs associated with quality were broken into four categories: prevention, appraisal, internal failure, and external failure. These remain widely recognized today. Table 11-1 depicts examples of the elements of cost that comprise each category.

It should be clear to everyone that improved quality ultimately translates into decreased cost, higher returns, more satisfied customers, and so on. The advent of total quality has helped many organizations see that a sound quality cost measurement can provide guidance to the quality improvement process in much the same way that cost accounting does for general management. Quality costs are simply a measure of the costs associated with the achievement of quality. These costs relate not only to "waste" costs such as scrap and rework, but also to investment costs such as education and training.

Quality costs are those costs incurred by preventing nonconformances to products and services; appraising products and services to determine conformance; and failing to meet requirements.

The four elements of quality cost mentioned in Table 11-1 require further definition prior to discussing setting up a quality cost system. Figure 11-11 contains a brief definition of each of the four elements.

Organizations have wrestled with the definition and visualization of "cost of quality" for many years. Some contend all that is needed is to capture the four types of cost. It is not that simple. While most managers understand the relationships of cost to profit and to productivity, they have a very difficult time making the transition to quality costs. The term "cost of quality" causes confusion. They see it as the money being spent building quality into products and services rather than the cost of nonquality or waste. This is an educational issue that may be best handled through the formation of cross-functional partnerships, and a calculated, ever improving approach to capturing, analyzing, and eliminating unwanted cost.

TABLE 11-1
Four Categories of Quality Cost and Examples of Each

Prevention	Appraisal	Internal Failure	External Failure
Process control	Inspection	Scrap	Lost sales
Quality improvement programs	Customer surveys	Meetings not starting on time	Penalties
Inspection	Audits	Overnight letters due to poor planning or execution	Field change orders
Education and training	Data collection efforts	Administrative costs associated with recommunication	Complaint handling
Quality planning	Lab supports	Lost time due to safety issues	Legal actions
	Field reports	Some employee morale issues	Accounts receivable
		Supplier training	Allowances

CHAPTER 11

GETTING STARTED WITH QUALITY COSTS

> Prevention costs: The costs associated with the prevention of defects in products and services, such as inspection for the purpose of data collection which will drive process improvement; education of people in customer-driven design techniques; and general quality training.
>
> Appraisal costs: The costs associated with auditing, measuring, and analyzing products and services to assure conformance to requirements (customer requirements, specifications, standards). These costs include inspection or testing of purchased materials and in-process inspection; field tests; and depreciation allowances for test equipment.
>
> Internal Failure costs: Costs incurred in the correction of nonconforming products and services prior to shipment to the customer or delivery of the service. These costs include redesign, rework, sorting, and retest and diagnosis.
>
> External Failure costs: Costs incurred in the correction of nonconforming products and services after shipment to the customer or delivery of the service. These costs include customer complaint handling, returns and credits, and past due accounts receivable management.
>
> All of these costs should include salaries and fringe benefits in the cost calculation.

Fig. 11-11 Definitions of four quality cost elements.

GETTING STARTED WITH QUALITY COSTS

Every product or service is designed, developed, and delivered through a process. Each process consists of a few or many steps. In fact, a simple definition of process might be "a series of steps that yield some outcome." If various processes are isolated from the overall business system, it becomes relatively simple to calculate the costs associated with those processes. All that is required is to break each process down to its various steps and elements of cost within each step. For example, if a manufacturing facility has an area where all rework is performed, the cost associated with that area can be calculated. This requires that the costs for labor (including fringe benefits), equipment depreciation, overhead, materials, etc., be calculated. In a total quality process there should be no need for a rework area, or for rework at all. In the real world however, there is and likely will continue to be some rework. It then becomes the challenge of the organization to measure the cost of rework, to establish a baseline for improvement, set goals, and measure the continuous improvement effort using the quality cost process. It is important to note that quality costs are a measurement tool, not a quality program. Linkage must be made to an overall quality process.

One of the best ways to test the theory out is to select a simple process and construct a flow diagram. Be sure to set clear beginning and end points for the process and be sure to document the process as it is actually functioning, not as people believe it should be functioning. Once the diagram is complete, work with someone in finance to assign estimated costs to each step of the process. This can be accomplished by determining the time required for one product to pass through the process. Time should be calculated for each step, to establish the cycle time for the product. This may be calculated for service delivery and administrative processes as well. With this complete, the next step is to review the process, as documented, looking for those process steps that do not add value. Any step that is associated with rework, inspection, repair, scrap, retest, or other work that should not be required under a total quality system is highlighted. By adding up the associated costs for each of the nonvalue adding steps a cost of quality, for that process, is determined. Further, the percentage of this cost to the overall cycle cost may be determined. Either the absolute cost or the percentage of "waste" to the overall cycle cost may be used as a baseline for improvement.

Case Example: Business Overview

The business is high-tech electronics manufacturing. The company is headquartered in the U.S. but has operations worldwide. In one division there are eight manufacturing sites in seven locations outside the U.S.: Puerto Rico, Canada, Mexico, Hong Kong, Singapore, Taiwan, and China. Leading the list of results over the three years since implementation of a continuous improvement effort is a cumulative cost reduction of eight figures.

The business has undertaken a continuous improvement strategy with four primary elements:

- Education and training in quality.
- Improved quality information systems (QIS).
- The application of several techniques such as statistical process control, process certification, formation of employee involvement teams.
- The use of quality costs as the primary measure of success.

Note: Not every plant is using the same methodologies for continuous improvement. It was decided early on that while each plant must follow a common strategy, individual sites could choose the methods that worked best for them. The Far East plants used quality circles, while several of the plants chose more conventional approaches.

STRUCTURING THE QUALITY COST SYSTEM

Some people become quite passionate discussing quality cost systems. They are emphatic that such a system must take each of the cost types into account. Many quality cost efforts are doomed to failure because quality professionals and engineers attempt to lead by exhortation or by trying to include all cost elements, in every area of the organization, from the beginning. In some companies, there are those who argue that no quality cost system is complete unless it captures every detail of the four cost types.

CHAPTER 11

STRUCTURING THE QUALITY COST SYSTEM

There are numerous examples of organizations where groups responsible for setting up the quality cost system spent days debating whether or not to include a cost that, in the aggregate, would not matter. This time itself is non-value-added activity.

The trouble with a "do everything" approach is that it takes longer to plan, define, sell, and implement. While this is being debated, valuable time passes. One way to address this issue is to "eat" the quality cost "elephant" one bite at a time. By focusing on the key areas of waste, such as rework or excess inventory, it is easier to gain support for the effort. The most successful quality cost efforts begin small and remain focused on those elements of cost that are the best measure of the overall business system.

Case Example: The Approach to Quality Costs

The business has a strategy built on continuous improvement. The person responsible for overseeing the continuous improvement effort is the Quality Engineering Manager, who must also ensure that the quality cost system is in place and that it works. The first step in the process is to form a partnership. Recognizing that the one organization in any business that has the most congruent goals with Quality is Finance, the QE manager set about convincing them that a full partnership was the best means for driving this important goal and performance measurement process. The senior quality manager for the business and the senior finance manager teamed up. Partnerships were then formed at each subsequent level. Once agreement in principle was reached, the task of selecting the elements began. At this point in the process the partners determined that additional expertise would be helpful, so a cross-functional steering committee was formed of representatives from the various functional areas such as Materials, Administration, and Operations, as well as Quality and Finance.

The steering committee selected six key elements: scrap, rework and repair, reinspection, excess inventory carrying costs, obsolescence write-offs, and retest/diagnosis. These were selected for several reasons. First, they were readily captured through the existing reporting process. There was little work involved to design the algorithms for compiling the data. Second, all of management clearly understood the impact of these elements on the bottom line. Third, the business did not try to eat the whole elephant in one sitting. One bite at a time was sufficient.

GOAL SETTING

The goal setting process for the first year is a difficult one. Each entity, whether it is a plant or other business center, must identify a baseline for each element from the previous year. This may require some data restructuring. In most manufacturing facilities, the elements of cost are tracked. For example, scrap, rework and repair, excess inventory carrying costs, etc., are usually contained in the financial reports. If they are not, then the first year will simply be one of establishing the baseline. Some organizations collect the data for a quarter and then extrapolate the costs for a year.

The use of year over year trends is an excellent measure. Further, the cost of quality as a percentage of total spending is perhaps the best measure of cost. Table 11-2 demonstrates this.

Looking at the column FY1, if overall spending for fiscal year one (FY1) was $100M and the cost of waste, as defined by the selected elements, was $20M, then waste was 20% of total spending. For FY2, spending increases to $200M. If $30M is spent on waste, there is an increase in spending of 50%, or 15% of total spending. On the surface this appears to be bad news, but the year-to-year comparison as a percent of total spending shows a 25% reduction. Obviously, there would still be room for continuous improvement, although the quality cost has improved significantly.

Once the goals are set, a process must be put in place for collecting, analyzing, and reporting the information on an ongoing basis. These goals must become an integral part of the business' overall goal set, and be viewed with the same regularity and urgency as profit and loss, headcount, productivity metrics, sales, or other key metrics are reviewed and ultimately drive organizational behavior.

THE REPORTING PROCESS

As with any key business metric, the performance metric must be accompanied by a well planned, complete reporting process. In most large organizations the accounting system is well established. The quality cost measurement and reporting process should be compatible with the financial reporting process. In fact, the process should align with existing financial reporting processes. In smaller organizations where the financial reporting process may be less mature, or in some cases, virtually *ad hoc*, it is more difficult to establish the necessary reporting structure.

In organizations where quality cost reporting has been established and is successful, the results are reviewed monthly, quarterly, and annually. These reports include commentary relative to future issues and trends. There is also an annual review which coincides with the annual goal setting process. Throughout this process, discrepancies are investigated and resolved. Trends are plotted and analyzed each month.

IMPLEMENTATION

There are numerous approaches to implementing a quality cost system. If the organization chooses to structure a steering committee as in the case example, then the committee should begin the communications process early, pulling in ideas from other parts of the organization and sharing progress. Once the approach and process are established, it is critical to ensure that all the people who will be participating understand what is expected of them and the time frames involved. One approach is to hold a series of meetings where the process is outlined and people have the opportunity to ask questions.

TABLE 11-2
Cost of Quality as a Percentage of Total Spending

	FY1	FY2	% Change
Total Spending	$100 M	$200 M	+ 100%
Cost of Quality Spending	$20 M	$30 M	+ 50%
Percentage	20	15	- 25

CHAPTER 11
CONCLUSION

Perhaps the most important issue to resolve is definitions. It is most important to ensure consistency of definition for each element. To do this, operational definitions for each element may be established. A good operational definition will remove points of potential confusion and serve to clarify key points. This ensures that questions and interpretations are kept to a minimum. As the organization becomes more sophisticated, the definitions are continually reviewed and fine-tuned.

To ensure that the process stays on track, there should be constant follow-up and ongoing communications. The process must be kept "alive." A good indicator of a successful process is when the cost data and information from various parts of the operation are used in the routine running of the business.

Case Example: Implementation

The steering committee decided to track waste reduction as a percent of overall spending rather than as an absolute number. This eliminated wide swings in the data as the business changed. They determined that the prime indicator of success was the trend rather than the numbers themselves. The important point is that they were positioned to see results while at the same time, beginning to educate management and create the behavior required to drive continuous improvement.

It was determined that each plant would submit numbers on a monthly basis, broken down by element. To ensure the right level of attention, the numbers were presented in aggregate at the monthly senior management operational review. The intent was not to compare plants but to highlight progress and opportunities. This was viewed as a primary means of educating the senior staff. On a quarterly basis, a report was distributed to senior management, showing group level goals, performance against those goals, and future trends. The plants were asked to provide commentary each month explaining the various results. If a cost rose, there was discussion as to why. If costs were reduced more than anticipated, there was analysis that explained the reasons. This approach was used as a means to get the plants to better understand the concept of cause and effect as it related to process performance and costs.

Operational definitions were developed for each element. Even so, there were occasional glitches. An example of misinterpretation occurred when a plant in the Pacific Rim decided to amend the scrap definition to read "net of salvage value." In their mind, they had recovered some of the cost of waste. The operational definition addressed this by clearly spelling out that scrap included "...all scrap resulting from process, handling, human as well as machine errors,... and is reported in full, regardless of recoveries."

The steering committee took on the task of determining who would be responsible in each plant for the data collection and reporting process. A decision was made to make the plant manager, the quality manager, and the plant finance manager jointly responsible. They were asked to form a partnership in their respective plants and jointly drive the process. The steering committee continuously reinforced the partnership by addressing all correspondence relating to waste elimination to these three managers for each operation. It was also determined that every operation must set goals and forecast spending against those goals on a quarterly basis. The primary emphasis is on the continuous improvement strategy, the elimination of waste as a performance measurement, and full support for the operations through the three key components of the strategy: education, information management, and process certification.

To maintain focus on the longer term and further drive the concept of cause and effect, each plant was asked to identify the top three programs they would be driving for the current year which would have a significant impact on the reduction of waste. The various projects, and the successes and lessons learned, were then shared between plants. This provided a unique opportunity for internal benchmarking and additional leverage for the overall process.

CONCLUSION

As the quality cost process becomes integrated with the business, any business changes there may call for modifications to the process. The operational definitions may no longer be complete; the reporting format may change; business requirements or products may change. The performance metric must be continually upgraded and improved to keep pace with the business.

It is important to link quality costs to an overall quality strategy. It is critical that the components of the strategy are linked architecturally. This means that each strategic element must be clearly related to the others. For example: education supports and integrates with a quality information system, and the quality information system supports and integrates with education and process certification; process certification integrates with quality education and quality information systems. In this example there is a "three legged stool" of sorts, with the cost of quality measure binding all three legs together. This performance measurement becomes the principal success indicator in driving the strategy. The components of strategy must integrate with the various cultures and unique business applications from country to country, as well as from plant to plant.

Keep in mind when developing and implementing a quality cost measurement:

1. Link the performance measurement to the continuous improvement process. If it is not inherently clear to management *why* they should be doing this, they may not do it.
2. Form a strong partnership with Finance and the operations. Once they are part of the team, they will help drive the education process.
3. Drive toward process control and breakthrough. Once gains are realized, have a system in place to hold the gains and to improve further.
4. Focus on a few key areas of major cost reduction opportunity. One bite at a time will ensure immediate success—then continue to add elements over time.
5. Ensure that the definitions, data collection, and reporting processes are well defined. A simple definition, even though incomplete, is better than no definition at all.
6. Set clear goals and assist the operating entities in achieving their goals. Mentor as well as monitor.
7. Watch for ways to improve the improvement process. Look for unique "cultural" approaches that may be transportable. Continuously update and clarify the process to keep pace with changing times.

CHAPTER 11
CONCLUSION

Case Example: Results

After three years of driving continuous improvement throughout the business there were numerous breakthroughs:

- The cumulative cost reduction across all the plants was over $20M.
- The cumulative cost avoidance was an additional $60M+.
- The organization enhanced the goal setting process, setting the next year's goals earlier in the current year.
- New elements were added each year.
- One plant attained a yield of 17 parts per million in a highly complex and critical process.
- The Quality-Finance-Operations partnership in waste reduction resulted in other "joint ventures." A task team was formed to link quality systems to internal business controls. Quality and Information Services formed a partnership to drive the development, implementation, and ongoing support of a quality information system for process monitoring.

References

1. D.W. Hess, K.F. Jensen and T.J. Anderson, *Chemical Vapor Deposition*, Reviews in *Chemical Engineering*, 3, 97, 1985.
2. *Statistical Process Control*, The Western Electric Company, 1956.
3. Dr. S. Shingo, quoted by J.H. Sheridan in *Lessons from the Gurus, Industry Week*, 35 (August 6, 1990).
4. T.J. Peters and R.H. Waterman, *In Search of Excellence*, Harper & Row, 1982.
5. E.A. Gee and C. Tyler, *Managing Innovation*, John Wiley & Sons, 1976.
6. P.F. Drucker, *Innovation and Entrepreneurship*, Harper & Row, 1985.
7. G.M. Prince, *The Practice of Creativity*, Harper and Row, 1970.
8. M. Leonard, *IC Fabrication Techniques Sculpt Silicon Sensors, Electronic Design*, 39, October 26, 1989.
9. K. Bemowski, *The Benchmarking Bandwagon, Quality Progress* (January, 1991).
10. R.C. Camp, *Benchmarking: The Search for the Best Practices that Lead to Superior Performance*, ASQC Press, 1989.
11. American Productivity and Quality Center, 123 N. Post Oak Lane, Houston, Texas, 77024-7797.
12. Marcie Tyre, *Managing Innovation of the Factory Floor, Technology Review*, 59 (October 1991).
13. T.J. and E.C. Boardman, *Don't Touch that Funnel, Quality Progress* (December, 1990).
14. K.R. Bhote, *World Class Quality*, AMACOM Publishing, 1991.
15. P.M. Senge, *The Fifth Discipline*, Doubleday, 1990.
16. D. Hoernschemeyer, *Creating Conditions for Excellence, Manufacturing Systems* (December 1986).

Bibliography

ASQC Quality Costs Committee. *Principles of Quality Costs*. 2nd ed. 1990.

J. Stephen Sarazen, *How to Eat a Quality Cost Elephant*, 1990.

CHAPTER 12

THE ROLE OF ISO 9000 IN CONTINUOUS IMPROVEMENT

INTRODUCTION

The International Organization for Standardization (ISO), with membership from over 90 countries, develops and promotes standards that facilitate international exchange of goods and services. ISO/Technical Committee 176 (ISO/TC 176), formed in 1979, harmonizes international activities in quality management and quality assurance standards. ISO/TC 176 created and has ongoing responsibility for the ISO 9000 family of quality standards.

The ISO 9000 family currently includes more than 21 standards either already published or under development. Table 12-1 lists the standards that are formally published. It should be noted that some of them are assigned numbers other than "9000."

The standards can be grouped into three major categories:

- Quality assurance standards providing requirements for effective and efficient quality systems. ISO 9001, ISO 9002, and ISO 9003 fall into this category.
- Quality management standards providing guidance on the scope, structure, and elements of effective quality systems. ISO 9000-1 through ISO 9000-4 and ISO 9004-1 through ISO 9004-4 fall into this category.
- Quality technology standards providing detailed guidance on elements within a quality system, such as equipment calibration and auditing. ISO 8402, ISO 10011-1 through ISO 10011-3 and ISO 10012-1 fall into this category.

Over 60 countries have adopted standards in the ISO 9000 family as national standards, including virtually all industrialized countries and many developing nations. To date, the United States has adopted ISO 9000-1, ISO 9001, ISO 9002, ISO 9003, and ISO 9004-1, collectively, as the American National Standards Institute/American Society for Quality Control Q90 Series (ANSI/ASQC Q90 Series). When these five standards are revised (see footnote to Table 12-1) and reissued by ISO/TC 176, the United States national standards will carry the same numbers as the respective International Standards. The United States is also adopting other standards in the ISO 9000 family, and those standards will carry the same numbers as the respective International Standards.

USES OF THE STANDARDS

Guidance and registration have emerged as the two major uses for the standards. The guidance standards provide recommendations for the design, implementation, and maintenance of effective quality systems and quality system elements. Their primary use is within a company.

The quality assurance standards—ISO 9001, ISO 9002, and ISO 9003—are generating the most interest, however. Demonstrated compliance to one of these standards is an expected part of doing business in Europe, and it is fast becoming a customer requirement for doing business in the United States as well.

"Demonstrated compliance" in the United States usually means registration of a company's quality system by an accredited, independent third party. However, second-party registration by customers may become more important in some industry sectors, such as the automotive industry. Registration involves a rigorous, on-site quality system audit, performed by certified auditors, against the requirements of the relevant quality assurance standard. Registration is usually applied on a facility basis, and the facility, once registered, is subject to ongoing surveillance and requalification. By the end of 1993, it is predicted that more than 2000 facilities in the United States will have sought registration.

APPLICATION OF THE STANDARDS

The standards are generic. They describe what elements should be part of a quality system, but not how they should be designed and implemented. They also do not indicate who should be responsible for elements in the quality system, with the significant exception of clearly delineating the requirements for

CHAPTER CONTENTS:

INTRODUCTION 12-1

CONCEPTUAL FOUNDATIONS FOR CONTINUOUS IMPROVEMENT 12-3

IMPROVEMENT FROM INHERENT DISCIPLINE 12-4

EXPLICIT MECHANISMS FOR CONTINUOUS IMPROVEMENT 12-5

PROBLEM PREVENTION 12-7

SUPPLEMENTAL STANDARDS 12-10

SUMMARY 12-11

---

*The Contributors of this chapter are: **April Cormaci**, Executive Vice President, Service Process Consulting, Inc., Edison, NJ; **Ian Durand**, President, Service Process Consulting, Inc., Edison, NJ; **August B. Mundel, P.E.**, White Plains, NY.*
*The Reviewers of this chapter are: **Dan Burns**, Certified Lead Assessor, ISO 9000, Thomas Conrad Corp.; **Robert J. Craig**, Director of Quality, Berg Electronics; **Barbara Fossum, Ph.D.**, Director, Quality Management Consortia Program, University of Texas at Austin, College of Business Admininstration; **Bob Hammill**, ISO Services Manager, Perry Johnson, Inc.; **Kent Hemingson**, Program Manager, ISO 9000, IBM; **Brian Kitka**, ISO Services Asst. Mgr., Perry Johnson, Inc.*

12-1

CHAPTER 12

INTRODUCTION

TABLE 12-1
ISO 9000 Family

Standard No.	Title
	Quality systems—Models for quality assurance
ISO 9001*	In design/development, production installation and servicing
ISO 9002*	In production, installation and servicing
ISO 9003	In final inspection and test
	Quality management and quality assurance
ISO 9000-1*	Part 1: Guidelines for selection and use
ISO 9000-2	Part 2: Generic guidelines for application of ISO 9001, ISO 9002, and ISO 9003
ISO 9000-3	Part 3: Guidelines for the application of ISO 9001 to the development, supply, and maintenance of software
ISO 9000-4	Part 4: Application for dependability management
	Quality management and quality system elements
ISO 9004-1*	Part 1: Guidelines
ISO 9004-2	Part 2: Guidelines for services
ISO 9004-3	Part 3: Guidelines for processed materials
ISO 9004-4	Part 4: Guidelines for quality improvement
	Guidelines for auditing quality systems
ISO 10011-1	Part 1: Auditing
ISO 10011-2	Part 2: Qualification criteria for quality system auditors
ISO 10011-3	Part 3: Management of audit programs.
	Quality assurance requirements for measuring equipment
ISO 10012-1	Part 1: Metrological qualification system for measuring equipment
ISO 8402	Quality management and quality assurance—Vocabulary

*Following ISO directives, these standards (originally published in 1987) have been reviewed and are being revised. The Draft International Standard (DIS) versions are being balloted in 1993 by ISO/TC 176 member countries. If approved, they are expected to be published in 1994 as full International Standards. A new standard, ISO/DIS 10013, Guidelines for developing quality manuals, is also out for ballot.

management with executive responsibility for quality. This approach recognizes that the needs of companies vary. The design and implementation of a quality system must be influenced by a company's size, objectives, market offerings, processes, and practices. For example, no particular organization structure is described, and the existence of a "Quality Department" is not required or necessarily recommended.

All market offerings (hardware, software, processed materials, services, and combinations of these) are included under the generic term "product." This is possible because the subject of the standards is the quality system itself, not the products a company produces. The recommendations and requirements for a quality system are distinct and separate from the "technical requirements" or "specifications" for a particular product.

The generic nature of the standards, however, sometimes poses difficulties with interpretation. To address this, ISO/TC 176 developed several supplemental standards. For example, ISO 9000-3 helps companies apply ISO 9001 to the development, supply, and maintenance of software; ISO 9004-3 guides companies producing processed materials in the application of the quality management concepts in ISO 9004-1; and ISO 9004-2 guides service companies in the application of the concepts in ISO 9004-1.

THE QUALITY ASSURANCE STANDARDS: ISO 9001, ISO 9002, AND ISO 9003

Coverage

The generic product realization cycle can be used to illustrate the coverage of the requirements in the quality assurance standards. As shown in Fig. 12-1, the cycle begins with a definition of products needed by customers in the marketplace. It continues through the design of products and processes; purchase of raw materials, goods, and services; and operation of production and delivery processes. Where appropriate, the cycle may also include installation and maintenance processes.

The quality assurance model in ISO 9001 includes only contract or order review in the marketing function, and then follows the cycle through maintenance (referred to as "servicing" in the standards). ISO 9001 is the most comprehensive of the quality assurance standards. ISO 9002 does not include design or servicing, but it covers everything else; so it is still a comprehensive model. ISO 9003 covers only final inspection of a delivered product. In practice, this minimal coverage makes ISO 9003 a feasible choice for very few businesses, and provides significantly reduced confidence to customers.

Scope

ISO 9001, ISO 9002, and ISO 9003 each have four main clauses. Clauses 0 through 3 contain introductory material, including the scope of the standard, definition of terms, and references. Clause 4, the core of the quality assurance standards, is divided into subclauses, each of which describes one of the elements of an effective quality system. Figure 12-2 shows the contents of clause 4 in ISO 9001. Subclause references in the rest of this chapter are to ISO 9001.

Some of the subclauses describe elements that are clearly related to functions in the product realization cycle (for example, subclause 4.4, Design control). Other subclauses describe elements that span many or all functions (for example, subclause 4.1, Management responsibility and subclause 4.18, Training).

ISO 9000 FAMILY SUPPORT FOR CONTINUOUS IMPROVEMENT

A continuous improvement philosophy is embedded in the ISO 9000 family guidance and requirements. This is evidenced in several ways:

- The conceptual foundations for continuous improvement are integrated into the ISO 9000 family.
- The discipline necessary to implement a quality system that meets the requirements of the quality assurance standards inherently results in improvement, especially in the process of fulfilling requirements related to documentation and quality records.
- Explicit mechanisms for continuous improvement of products, processes, and quality systems are described as elements in several standards, including the quality assurance standards.

CHAPTER 12

INTRODUCTION

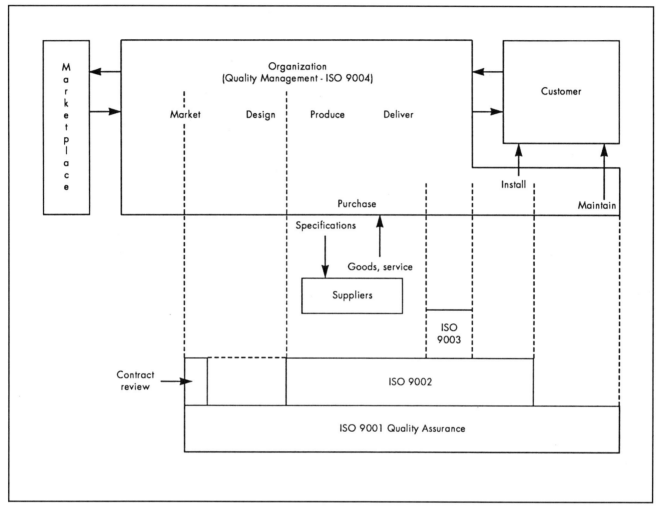

Fig. 12-1 Product realization cycle.

- Requirements throughout the quality assurance standards and recommendations in the guidance standards support problem prevention.
- Supplemental standards that focus specifically on quality improvement and managing processes have been developed.

This chapter focuses on the discipline, explicit mechanisms, and preventive elements found in clause 4 of ISO 9001. However, the last section of this chapter provides a brief summary of two important supplemental standards: ISO 9004-4, Quality management and quality system elements—Part 4: Guidelines for quality improvement; and ISO 9004-2, Quality management and quality system elements—Part 2: Guidelines for services.

TERMINOLOGY

Some differences in terminology exist among the standards in the ISO 9000 family, many of which will be addressed in revisions to the standards. Table 12-2 compares terms used for customers, vendors, and the implementing organization in this chapter with those in ISO 9001, ISO 9002, and ISO 9003.

CONCEPTUAL FOUNDATIONS FOR CONTINUOUS IMPROVEMENT

While the current versions of the quality assurance standards and most quality management standards do not explicitly advocate or require continuous improvement, they implicitly reflect it through several specific beliefs:

- Management of the quality system should have higher priority than quality control of products.
- The quality system should be integral to the business.
- Senior management is responsible for the quality system.

CHAPTER 12

IMPROVEMENT FROM INHERENT DISCIPLINE

- Preventing problems costs less and results in higher customer and employee satisfaction than correcting problems.
- Effective, two-way communication—vertically through management levels and horizontally across functions—is critical.
- Appropriate levels of documentation are necessary for communication, consistency, and internal self-discipline.
- Management by fact is essential.

**TABLE 12-2
Terminology**

Standard	Vendors to a company or organization	Implementing company or organization	External customer
ISO 9001, ISO 9002, ISO 9003	Sub-contractor	Supplier	Purchaser
This chapter	Sub-supplier	Company	Customer

```
4.1   Management responsibility
      4.1.1 Quality policy
      4.1.2 Organization
      4.1.3 Management review
4.2   Quality system
4.3   Contract review
4.4   Design control
4.5   Document control
4.6   Purchasing
4.7   Purchaser supplied product
4.8   Product identification and traceability
4.9   Process control
4.10  Inspection and testing
4.11  Inspection, measuring, and test equipment
4.12  Inspection and test status
4.13  Control of nonconforming product
4.14  Corrective action
4.15  Handling, storage, packaging, and delivery
4.16  Quality records
4.17  Internal quality audits
4.18  Training
4.19  Servicing
4.20  Statistical techniques
```

Fig. 12-2 ISO 9001 Clause 4 quality system requirements.

IMPROVEMENT FROM INHERENT DISCIPLINE

The quality assurance standards explicitly require discipline in: definition of the quality system based on documented procedures, consistent adherence to these procedures, and collection and analysis of relevant data. Every acceptable quality system must demonstrate five characteristics:

- Communicate. Unless quality system policies, structure, responsibilities, and procedures are captured formally, they are not likely to be thought through well, communicated, understood, or consistently implemented.
- Perform. Consistent adherence in daily practice to the quality system defined by the documentation reduces variability, prevents known problems, and benefits from proven procedures.
- Record. Without a record of what occurred, made at the time of occurrence, companies are not likely to assess the level of quality effectively, and are less likely to be able to analyze performance, identify unforeseen problems, or anticipate unsatisfactory results.
- Check on the Results and Act on the Difference. Recording results is useless unless the results are analyzed and checked against specifications or criteria. And, the quality system loses meaning if action is not taken when results are unsatisfactory or problems are indicated. Remedial action must be taken when a problem is detected. Critical problems must be investigated and root causes identified and removed.

Two explicit requirements pervade ISO 9001 and ISO 9002 and reflect this structured discipline: documentation and records. In addition, many requirements support prevention.

DOCUMENTATION

The quality assurance standards require that a company document procedures for all elements of a quality system (subclause 4.2, Quality system). The scope of documentation may appear onerous at first, but the intent is to document only to the extent that is effective and appropriate for a company.

Documentation describes what must be done and how it will be done. Many companies choose a three-level documentation structure to fulfill the requirements of the standards:

- A high-level document (often in the form of a company quality manual) includes, for example, the quality policy, the company's quality objectives, responsibilities related to quality, and an overview of company-wide processes.
- Procedures (usually at the department level) describe how each functional organization will implement the relevant quality system requirements, for example, engineering standards, design rules, and a purchasing handbook for agents.
- Work instructions describe specific, detailed steps that must be followed to assure the quality of the output. Examples include the steps in crucial laboratory tests, or activities to process applications in an insurance company where completeness and consistent assessment of factors is critical.

As an alternative or an adjunct to work instructions, a company may require demonstration of defined skill levels by employees performing some jobs, and/or provide training before an employee performs certain tasks (subclause 4.18, Training). In this case, the procedures or work instructions may be minimal. However, the company must be able to produce records indicating that an assessment of the required skill level/training has been performed, and that employees have the qualifications, demonstrated the skills, or completed the training.

Document control, subclause 4.5, requires that all documents describing or used in the quality system be controlled to preclude unauthorized changes or use of obsolete versions. Typically, this means that documents are assigned a unique number, reviewed and approved prior to release in accordance with documented

CHAPTER 12

IMPROVEMENT FROM INHERENT DISCIPLINE

procedures, and bear a revision number and effective date. Moreover, controlled documents must be correct, clear, and available, and obsolete documents must be removed from all places of use and distribution. A readily accessible master list of all controlled documents, including document number, title, date, and latest revision, must be maintained.

Preparing documentation to meet the standards should stimulate a company to identify its operational processes and understand what activities actually occur. (Registrars' auditors emphasize consistency between documentation and practice.) To do this effectively, a company must involve employees at all levels and in all functions, which provides opportunities for communication within and across functions.

As a practical result of documenting processes and procedures, companies frequently recognize areas where breakdowns in internal communication adversely affect process operations, as well as inefficient and redundant activities within and among processes. Opportunities for immediate improvement often become obvious, and provide input for the problem identification stage of continuous improvement.

Documentation supports continuous improvement in other ways as well. When procedures are documented and implemented consistently, a company can be assured of its operations and reliably measure performance. Reliable measurements are necessary to assess the effect of attempted improvements. In addition, up-to-date documentation formally captures validated improvements to help ensure that the gains are maintained.

RECORDS

Quality records are required in almost all elements of the quality assurance models in ISO 9001 and ISO 9002, with documented retention periods determined by the company. Records capture the results of following the documented procedures. Examples include the results of:

- Internal audits of the quality system (subclause 4.17, Internal quality audits).
- Management evaluation of the company's capabilities and performance (subclause 4.1.3, Management review).
- Inspections and testing (subclause 4.10, Inspection and testing).
- Calibration and measurement assurance (subclause 4.11 Inspection, measuring, and test equipment).
- Review and disposition of products that do not conform to specifications (subclause 4.13, Control of nonconforming product).

Quality records, subclause 4.16, requires documented procedures for the identification, storage, collection, indexing, filing, maintenance, and disposition of records. The intent is to ensure that records are accessible and legible for the retention period specified.

Quality records provide objective information on how elements in the quality system are operating, which becomes input to all stages of continuous improvement. Once improvements have been implemented, records help a company verify the immediate and continuing effectiveness of those improvements.

PROBLEM PREVENTION

Requirements throughout the quality assurance standards are intended to ensure that actions are taken to prevent potential causes of nonconformities. Several subclauses contain requirements that promote accurate and complete translation of customer needs throughout phases in the product realization cycle (for example, subclause 4.3, Contract review; and subclause 4.4, Design control). In addition, the data collected in records can be applied to problem prevention as well as problem correction. This chapter discusses problem prevention more fully in a later section.

THE EFFECT OF PREPARING FOR REGISTRATION

While ISO 9001 and ISO 9002 provide requirements for quality systems, it is preparing for the formal registration assessment by outside auditors that produces results. Objective assessment and ongoing surveillance provide a company with a realistic view of the effectiveness of its quality system deployment, and identify areas of nonconformance to the standards.

Less than 5% of companies in the United States currently meet the requirements set out in ISO 9001 or ISO 9002. Enhancing, or designing and implementing, a quality system based on these standards is a major effort for most companies. It often requires 1-1/2 to 2 years and the commitment of substantial resources. However, companies that successfully prepare for registration report significant benefits.

One extensive study of companies in the United Kingdom indicated that manufacturing companies experienced cost reductions of 15% to 35%, and some service companies saw reductions as high as 60%. Related studies and individual experiences in the United States validate these results.

In general, a quality system based on the requirements of ISO 9001 or ISO 9002 leads to lower variable costs by reducing error, scrap, and rework. (Many studies indicate that unnecessary work caused by inattention to quality and poor process design accounts for up to one-half of the total work effort in a typical company.) Reducing waste, especially through process improvement and simplification, results in reduced cycle times and increased process capacity that was previously dedicated to rework.

Effective relationships with sub-suppliers that result from purchasing requirements in subclause 4.6 inevitably lead to lower life cycle costs for raw materials and off-the-shelf goods and services.

Last, but perhaps most important, companies experience higher customer satisfaction as a result of fewer quality problems (including service aspects such as better on-time delivery).

EXPLICIT MECHANISMS FOR CONTINUOUS IMPROVEMENT

While all of the elements required in ISO 9001 and ISO 9002 work together to create an effective quality system, four requirements explicitly drive continuous improvement:

- Process control (subclause 4.9)
- Corrective action (subclause 4.14)
- Internal quality system audits (subclause 4.17)

CHAPTER 12

EXPLICIT MECHANISMS FOR CONTINUOUS IMPROVEMENT

- Management reviews (subclause 4.1.3).

Figure 12-3 illustrates their interaction, as well as indicating the integration of documentation and records. These four requirements are also key elements in the Malcolm Baldrige National Quality Award criteria.

PROCESS CONTROL (SUBCLAUSE 4.9)

The intent of this requirement is to ensure that processes produce consistent quality. Specifically, this subclause requires a company to identify and plan all product realization processes that directly affect quality, and to ensure that these processes are carried out under controlled conditions. Controlled conditions include: documented procedures, work instructions where necessary, approval of process design and equipment and qualification of employees, and process measurement and control (using statistical techniques if and when they are appropriate). The type and frequency of process measures are not defined in the standards, but they must provide a reliable basis for process control.

Subclause 4.9 also recognizes that subsequent inspections and testing cannot verify the outcome of some processes, notably in material processing industries and services. To ensure consistent quality in these cases, processes must be carried out by qualified personnel and/or must be continuously monitored and controlled.

The results of process measurements provide input for corrective action procedures, internal quality system audits, and management reviews.

CORRECTIVE ACTION (SUBCLAUSE 4.14)

Corrective action requirements are intended to ensure that significant nonconformities, defects, or other undesirable situations are analyzed to resolve existing problems, and to uncover and remove underlying causes.

Embedded in the requirements is the necessity for each corrective action to investigate four time periods: the past, the present, the near term, and the future. Immediate, or present, remedial action must be taken as soon as a problem is identified.

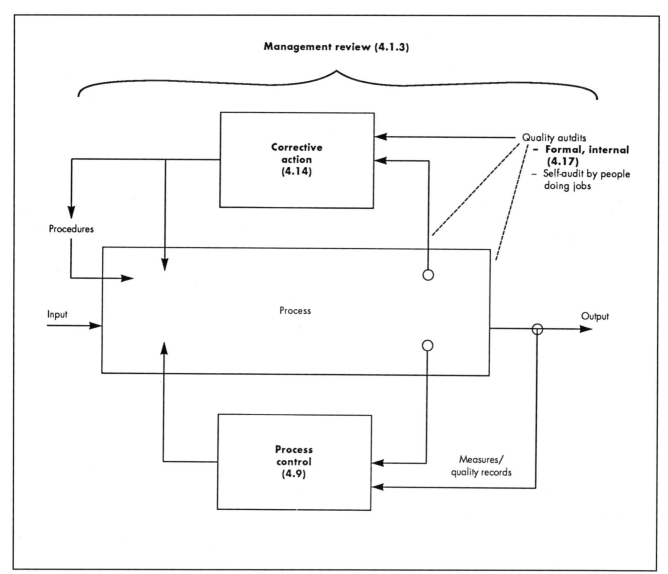

Fig. 12-3 Explicit mechanisms for continuous improvement.

CHAPTER 12

EXPLICIT MECHANISMS FOR CONTINOUS IMPROVEMENT

But this is rarely sufficient. Action is necessary to assure that the problem does not recur in the immediate future (the near term) while investigation of root causes takes place. As important is attention to the past—determining what defective products may have been produced or delivered between the time a problem began to occur and the time the problem was detected. (Subclause 4.8, Product identification and traceability, and subclause 4.12, Inspection and test status, support this need.) And finally, there is a requirement to investigate root causes and take action necessary to prevent recurrence in the future.

While the standards do not define a particular structure for the corrective action procedures, they do identify information to be analyzed (for example, process operation, procedures, quality records, service reports, customer complaints) and define what must be accomplished by the corrective action activities.

A key insight in subclause 4.14, indicated by listing "customer complaints" among the information to be analyzed, is the necessity for a company to view itself from the outside as well as the inside.

The requirements of subclause 4.14 stimulate:

- Data collection and analysis that support decision-making based on fact.
- Development and use of a systematic approach to problem analysis and solution that is based in a continuous improvement cycle: identify and define symptoms, diagnose root causes (which may be product-, process-, or quality system-related), identify and evaluate alternative solutions, and implement the chosen solution.
- Validation that the chosen solution is effective.
- Standardization of effective corrective actions and updating of any relevant documentation.

INTERNAL QUALITY SYSTEM AUDITS (SUBCLAUSE 4.17)

The intent of this requirement is to:

- Assure adequacy of quality system documentation.
- Verify that quality system activities comply with the documented quality system procedures.
- Assess the effectiveness of the quality system.

This requires periodic audits of documentation and deployment of the system (against the requirements of ISO 9001 or ISO 9002) and quality system results, usually as indicated by records. Subclause 4.17 specifically requires that:

- Audits be conducted on a regular basis (with the frequency determined by the company) depending on the criticality of a quality system element.
- Audits be conducted by personnel who are independent of the area being audited.
- Results be recorded and brought to the attention of people who are responsible for areas audited.
- Responsible personnel address findings.
- The effectiveness of changes be assessed.
- The followup be documented as part of a company's quality records.

Some companies choose a quarterly audit of critical elements of the quality system. Others perform monthly audits of different elements to complete a thorough audit within 6 to 12 months.

While the intent of these internal audits is to confirm satisfactory implementation and operation of the quality system, a frequent outcome is identification of areas that do not fully meet quality system requirements. Internal audits, in particular, provide input for decision making on a strategic level (through management reviews) as well as on a local level (using corrective action procedures).

Internal audits can make two additional, significant contributions: 1) to assess operations prior to designing a quality system based on ISO 9001 or ISO 9002, and 2) to perform a formal internal assessment to confirm compliance before the actual registration assessment. This "pre-assessment" can identify areas of nonconformance to be addressed through corrective action procedures, or give confidence that the system does conform to the requirements prior to the formal assessment by a registrar. Without some form of formal pre-assessment, a high percentage of facilities fail their initial registration assessment.

MANAGEMENT REVIEW (SUBCLAUSE 4.1.3)

Although this subclause is quite short, it implies a substantial responsibility for the management with executive responsibility for quality. Those responsible must periodically review the suitability and effectiveness of the quality system with regard to: 1) the requirements of ISO 9001 or ISO 9002, and 2) the company's quality policy and objectives. Two common inputs to management review are the results of internal audits, and analyses of all relevant quality records. While the interval between reviews is determined by the company, many choose an annual or semi-annual review once registration is successfully achieved.

The advantages of a regular management review include: ability to adjust the structure and focus of the quality system based on changes in strategic direction; more informed resource allocation decisions; increased understanding of actual operations; and, of course, identification and implementation of improvements in business processes.

PROBLEM PREVENTION

The cost and time to resolve quality problems increases exponentially at each stage in the product realization cycle (see Fig. 12-1). The standards require activities throughout the product realization cycle that can prevent or minimize problems, with particular emphasis on the early phases.

Seven subclauses, in particular, facilitate prevention:

- Contract review (subclause 4.3)
- Design control (subclause 4.4)
- Purchasing (subclause 4.6)
- Inspection and test status (subclause 4.12)
- Control of nonconforming product (subclause 4.13)
- Handling, storage, packaging, and delivery (subclause 4.15)
- Training (subclause 4.18).

CONTRACT REVIEW (SUBCLAUSE 4.3)

For registration purposes, a formal contract with a customer is not required for this subclause to apply; a customer order of any

CHAPTER 12

PROBLEM PREVENTION

kind is intended. This subclause specifically requires that contracts, purchase orders, or other indications of customer requirements be formally reviewed to ensure that the requirements are complete, understood, and documented. In addition, a company must have mechanisms to confirm that it can meet requirements, including delivery time. These mechanisms might confirm, for example, availability of parts, raw materials, and processing capability. Differences or ambiguities must be resolved before actually accepting the order. And a mechanism must exist to amend contracts and transfer the modified requirements quickly and accurately to relevant functions in the company. Records of the activities and results of contract review must be maintained.

Contract review is an early opportunity to take action that will eliminate changes later in the production and delivery processes. Misunderstanding of a customer's requirement causes rework, increased cost, processing delays, decreased productivity, and dissatisfied customers. Miscommunication between the functional areas in a company may lead to inability to meet schedules or fulfill specifications.

Data collected from contract reviews and captured in quality records support efforts to improve the communications with customers and within a company.

DESIGN CONTROL (SUBCLAUSE 4.4)

Design control is the longest subclause in ISO 9001 and ISO 9002. This emphasis reflects the premise that most of the problems in output quality can be traced to either faulty specifications, incomplete specifications, problems with the design of the product, or problems with the design of the processes that produce and deliver the product.

The emphasis on processes underscores that every company does design. A company may not have a design department, an engineering organization, or a development organization, but it still must design its work processes. And those processes have profound effects on the effectiveness and efficiency of its quality system.

The general intent of subclause 4.4 is to control and verify design activities so that input specifications are effectively translated into process and product designs, and that these designs can be realized with high quality and low cost. To achieve this result, procedures to implement several prevention-oriented measures are specified. They fall into three categories:

- Planning (integrated design and development plans for each project that must be generated and updated as necessary; assignment of activities to qualified personnel with adequate resources available; organizational and technical interfaces identified and mechanisms in place to manage them).
- Assurance (input specifications identified, documented, reviewed, and ambiguities resolved; design output documentation, written in terms of input requirements, includes acceptance criteria and identifies characteristics critical for safe operation of the product).
- Verification (periodic verification to ensure that design continues to meet requirements throughout the design process; may include, for example, formal design review, qualification testing, and comparisons with proven designs).

In addition to these activities that minimize miscommunication and misinterpretation of customer requirements, subclause 4.4 requires that design changes and modifications be identified, documented, reviewed, and approved before incorporation. The records that result from planning, assurance, and verification can provide input to corrective action as well.

PURCHASING (SUBCLAUSE 4.6)

ISO 9001 and ISO 9002 consider production (not the finance organization) to be the internal customer of purchasing. Therefore, the primary job of purchasing is to ensure that production processes have a high degree of confidence in the suitability and timeliness of incoming goods and services. (The implied focus is on lowest life cycle cost, not lowest first cost.)

To that end, the requirements in subclause 4.6 facilitate relationships between a company and its sub-suppliers that provide high confidence. They require that a company: identify critical purchases; develop and maintain an approved sub-supplier list for critical purchases based on evaluation of sub-suppliers' ability to meet requirements (which may include quality system requirements); and purchase critical items only from approved sub-suppliers. Criteria for approval and disapproval of sub-suppliers and a mechanism to address problems with sub-supplier performance must be established and documented, and records maintained.

Requirements related to purchase orders cover creation, review, and approval, with the objective of ensuring that sub-suppliers receive complete and accurate specifications. Purchasing is another early point in the product realization cycle where problems can be avoided by issuing complete, accurate, and precise product or service specifications to sub-suppliers, and assuring timely receipt of purchases.

A company with high confidence in a sub-supplier's ability to meet quality and timeliness requirements can realize significant reductions in operating expenses. The company needs less inventory; less work will be in process awaiting input; and cycle time is reduced.

INSPECTION AND TEST STATUS (SUBCLAUSE 4.12), AND CONTROL OF NONCONFORMING PRODUCT (SUBCLAUSE 4.13)

The intent of these subclauses is to prevent inadvertent use or installation of nonconforming or untested product at all stages in the product realization cycle. Subclause 4.12 requires that the inspection and test status of a product at any stage in its production be identified by suitable means (for example, tags, labels, stamps, routing cards, physical location). Subclause 4.13 provides additional requirements for recording the decisions on review and disposition of nonconforming product, including the identification of those authorizing the actions.

HANDLING, STORAGE, PACKAGING, AND DELIVERY (SUBCLAUSE 4.15)

Subclause 4.15 requires procedures to preclude damage, deterioration, or misuse of raw material, work in process, or finished product. The coverage of this subclause extends from receipt of incoming material through acknowledged receipt of product by customer or distributor. In addition to the possibility of physical damage, the procedures should address adequacy of identification, labeling, and shipping destination information to avoid losses and misdirected transfers.

TRAINING (SUBCLAUSE 4.18)

The intent of subclause 4.18 is to ensure that all personnel have the skills, knowledge, and competencies to do their jobs. This subclause might be better entitled "Job Qualifications." Specif-

CHAPTER 12

ACHIEVING REGISTRATION

ically, the standards require systematic methods to identify education and skills needed for different jobs, and to assure that the people in the jobs have them. This subclause refers to all job requirements that affect the quality of work output, not just those that relate specifically to quality techniques. A company is required to provide training or qualify personnel on the basis of appropriate education, training, or experience, and maintain records of training and qualification.

The prevention-related aspect of this subclause is clear: qualified employees have the required skills and know the appropriate procedures, resulting in less rework (and, consequently, higher customer satisfaction). The requirements in subclause 4.18 also support continuous improvement in two ways: they result in data for corrective action, and employees who know their jobs may be freer to identify opportunities for improvement and be more valuable in the improvement process.

ACHIEVING REGISTRATION

There are three major steps in achieving registration: preparation, review, and audit.

Preparation requires a complete internal review of all the operations and procedures, and a careful review of the International Standard, noting whether each of the items required by the appropriate ISO 9000 category are included; and then making the Quality program and the documentation conform to the requirements of the 9000 standard.

Let it be assumed that the organization wants to qualify for registration in accordance with ISO 9001. If any of the listed categories does not apply to the operation, a special note should be made, in order that the registration organization will be advised that there is a logical reason for not including the listed category. It is the organization's responsibility to develop and install an orderly system of controlling the quality of the product. Having developed a satisfactory system, the manual should then be written to reflect what is done in the plant. All operations that are performed should be included and operations that are not performed should not be included. The operations and the manual must be in agreement.

Having decided to apply for and secure a registration, the next step will be to investigate the advantages that various registrars provide, and how well they will operate within the particular industry. For example, there is one registrar who specializes in steel operations. Most organizations may be adequately served by several registrars.

The registrar chosen should also be acceptable to the customer. As agreements between countries become firmer, all properly accredited registrars should be equally acceptable.

The registrar will request a copy of operations procedures and manuals for what is called a Desk Audit. A Desk Audit is a review of the documented system to determine if it agrees with the concepts of the ISO 9000 standard to which the organization wishes to be registered. It is also the responsibility of the Registrar to determine whether the documented procedure can properly control the quality of the process and product.

The auditor will either agree that the operation is prepared for a facility audit or request changes before going further.

Once the desk audit is completed, and found to be acceptable, the registrar will make an appointment to audit the operations. The audit team will consist of a lead auditor and one or more auditors working with him.

Sometimes arrangements are made for the accreditation agency to observe the audit. The purpose of the accreditation auditors is to check on the accuracy of the work of the registrar's auditors.

In the early days of the registration process, particularly with some of the European audits, the insistence on compliance with all the details of the ISO 9000 documents was perfunctory. Some early grants of registration were much less rigorous than the standards being exercised in the United States today. In fact, it is suspected that the audits being performed in the U.S. are more rigorous than some performed in Europe. Although this may make it slightly more difficult to obtain a registration in the United States, it will (if true) result in a level of control and excellence of product of U. S. producers being superior to that of the Europeans, and should result in more business for U.S. companies.

The auditors, after reviewing the operations, will have a closing interview. They will report any findings. A "finding" is the auditor's term for a deficiency or discrepancy between the operations and the documentation, or any other shortcoming they discover. If there are findings, they must be cleared up before a registration is issued, unless the deficiencies are extremely minor. Many discrepancies are due to the failure of operating personnel to conform to the documentation furnished, and to the failure of the documentation to agree with the procedures in use. In some instances a review of the operations may indicate that there is really insufficient control to warrant registration.

When the organization becomes registered, it will be listed by the registrar and the accreditation board. At present these listings are revised monthly.

Having become registered as conforming to the ISO 9000 standard, the organization will be subjected to supplementary audits, of a lesser magnitude, and after a period of several years a re-audit to maintain the qualification.

There are some instances where a company has many plants, all operating with similar equipment and procedures, and each delivering the same product to different customers. The company needs each plant to be qualified as an ISO registered supplier. The registrars and the accrediting agencies may not require a thorough review of each plant.

It is unethical for an auditor to use information picked up during an audit for his or her own benefit, or to tell one organization about procedures of another.

One of the restrictions in the United States is that the registered operations are prohibited from marking the product to identify it as a product that comes from a plant whose quality system is registered. There is an important distinction between product certification and quality system registration. Product that is registered or listed under a product certification scheme is often marked or tagged appropriately. The organization with a quality system that is registered as conforming to the ISO 9000 system can advertise and use suitable insignia on its letterheads to inform the public and customers that its quality system has been registered. The product and product packaging must not be so marked.

CHAPTER 12

SUPPLEMENTAL STANDARDS

SUPPLEMENTAL STANDARDS

While extensive relative to common industry practice, the quality assurance models in ISO 9001 and ISO 9002 provide minimal requirements. Member countries of ISO/TC 176 are well aware that some aspects affecting continuous improvement are not covered in these models, or need further explanation. The committee has created supplemental guidance standards that address these aspects. Standards with particular relevance to continuous improvement are: ISO 9004.4—Quality management and quality system elements—Part 4: Guidelines for quality improvement; and ISO 9004-2, Quality management and quality system elements—Part 2: Guidelines for services.

GUIDELINES FOR QUALITY IMPROVEMENT (ISO 9004-4)

ISO 9004-4 presents a set of management guidelines for implementing continuous improvement within a company. The standard describes fundamental concepts and principles, management guidelines, and a methodology (including tools and techniques) for quality improvement.

Figure 12-4 lists the contents of ISO 9004-4. It is clear that the guidelines cover the major elements of a successful continuous improvement system:

- Process orientation; that is, the belief that quality improvement is achieved through process improvement, including administrative, production, and delivery processes.

---

1. Scope.
2. Definitions (process, supply chain, quality improvement, quality losses, preventive and corrective action to improve processes).
3. Fundamental concepts.
    3.1 Principles of quality improvement.
    3.2 Environment for quality improvement (management responsibility and leadership; values, attitudes, and behaviors; quality improvement goals; communication and teamwork; recognition; education and training; quality losses).
4. Managing for quality improvement.
    4.1 Organizing for quality improvement.
    4.2 Planning for quality improvement.
    4.3 Measuring quality improvement.
    4.4 Reviewing of quality improvement activities.
5. Methodology for quality improvement.
    5.1 Involving the whole organization.
    5.2 Initiating quality improvement projects or activities.
    5.3 Investigating possible causes.
    5.4 Establishing cause and effect relationships.
    5.5 Taking preventive or corrective action.
    5.6 Confirming the improvement.
    5.7 Sustaining the gains.
    5.8 Continuing the improvement.
6. Supporting tools and techniques.
    6.1 Tools for numerical data (control chart, histogram, Pareto diagram, scatter diagram).
    6.2 Tools for non-numerical data (affinity diagram, benchmarking, brainstorming, cause and effect diagram, flowchart, tree diagram).
    6.3 Training in applying tools and techniques.

Fig. 12-4 Contents of ISO 9004-4.

- Communication horizontally and vertically.
- Focus on internal and external customers.
- Attention to workplace issues (management's responsibility for creating an atmosphere that enables and empowers employees, and improving their own work processes; setting goals; teamwork, training; recognition).
- Involvement of the whole company.
- A systematic methodology for identifying opportunities for improvement, investigating root causes, making and validating changes, and sustaining improvements.
- Measurement (for example, customer satisfaction, process efficiency, societal losses) and decisions based on facts.
- Use of tools and techniques.

The standard also recognizes that continuous improvement efforts can only realize their full potential if applied and coordinated within a structured framework. This framework requires organization, planning, measurement, and review of activities.

GUIDELINES FOR SERVICES (ISO 9004-2)

ISO 9004-2 recognizes that some characteristics of service offerings differ from those of other product offerings. In particular, services are often produced in real time as part of the delivery process. While ISO 9004-2 is intended to apply the principles of quality management described in ISO 9004-1 (Quality management and quality system elements—Part 1: Guidelines) to services, this guidance standard also provides useful insights into effective responses to the requirements in ISO 9001 and ISO 9002.

Three main points require consideration:

- Final inspection and test is usually not an effective alternative for service and service delivery processes because customer consumption of the outcome of the process is often immediate. In these cases, corrective action at the time of occurrence may not be possible or effective. Therefore, control of service and service delivery characteristics must primarily be achieved by controlling the processes that deliver the service. Major activities in process control are to monitor and change the process, as necessary, to ensure that outputs continuously meet customer requirements.
- ISO 9004-2 explicitly places the customer at the focus of the quality system. It includes substantial discussion of subjective aspects of a quality system, such as: the quality of the human interaction between a company's personnel and its customers and the need to manage it; the motivation of personnel to meet customers' expectations and continuously improve; emphasis on the communication and team skills needed by personnel who interface with customers; and the importance of a customer's perception of a company's performance and credibility. These aspects must also be monitored, results recorded, and steps taken to improve them.
- Customer assessment is the ultimate measure of quality, and a company must proactively, continuously, and systematically seek feedback from customers as a basis for improvement.

CHAPTER 12
SUMMARY

SUMMARY

People familiar with industry-specific quality standards often find them prescriptive in approach. While this promotes consistency of practice and, frequently, reduces variability of product characteristics, it tends to inhibit improvements in the quality system itself.

In contrast, the content, use, and descriptive nature of the quality standards in the ISO 9000 family stimulate continuous improvement of products, processes, and the quality system through multiple vehicles. The commonly accepted conceptual foundations for continuous improvement are integrated into the quality assurance and quality management standards. Improvements result from the discipline built into the quality assurance standards related to documenting procedures, consistent adherence to the procedures, and collection and analysis of quality records. This is especially true during implementation and use of the standards because few companies presently have an operating quality system that complies with the requirements. Once a quality system based on ISO 9001 or ISO 9002 is well established, explicit mechanisms built into the standards promote continuing improvement. In practice, these explicit mechanisms reinforce the emphasis on problem prevention that pervades many of the quality system elements.

CHAPTER 13

THE BALDRIGE CRITERIA AS A SELF-ASSESSMENT TOOL

Self-assessment is a structured approach to establish the true operational state of a business. It is the act of evaluating operations and organizations using a pre-established set of criteria. For the purposes of this chapter, the Malcolm Baldrige National Quality Award Criteria (Baldrige Criteria) will be used. A complete set of the Baldrige Criteria can be obtained from:

National Institute of Standards and Technology
Office of Quality Programs
Route 270 and Quince Orchard Road
Administration Building, Room A537
Gaithersburg, MD 20899

Based on the evaluation against the Baldrige Criteria, action plans can be prepared to close the gaps and achieve "world class" competitive capabilities.

CHAPTER CONTENTS:

THE IMPORTANCE OF SELF-ASSESSMENT 13-1

THE ASSESSMENT TOOL 13-1

BALDRIGE CRITERIA FRAMEWORK 13-2

TRAINING MALCOLM BALDRIGE EXAMINERS 13-3

MALCOLM BALDRIGE SCORING PROCESS 13-4

MANAGING EXPECTATIONS 13-10

SELF-ASSESSMENT VERSUS CONTINUOUS IMPROVEMENT 13-12

HOW COMPANIES HAVE USED SELF-ASSESSMENT 13-13

SELF-ASSESSMENT LESSONS LEARNED 13-15

PERSPECTIVES BEYOND SELF-ASSESSMENT 13-17

THE IMPORTANCE OF SELF-ASSESSMENT

Self-assessment provides a structured, benchmark process for improvement. This benchmarking capability can be against a company's previous accomplishments (the assessment from the previous year), or against other companies. Sources for comparison include:

- Companies who will share information. This includes the Malcolm Baldrige National Quality Award (MBNQA) winners, who are ready to share their methodologies and lessons learned from applying the Baldrige process. This sharing formally starts at the Quest For Excellence Conference which is held during the first quarter of the year. At this conference the previous year's winners describe the approach they took, and are open to discuss all seven Baldrige Categories.
- The range of scores from the previous Baldrige applicants (this information is shown for a three-year period later in this chapter).
- Companies that have a business relationship with another company, which will allow the sharing of information.

Self-assessment enables people at all levels of an organization to better understand what is needed to be a world class competitor. It provides a road map to fulfill their vision for the organization. This is particularly true when the vision of the company includes the core values and concepts endorsed by Baldrige:

- Customer-driven quality.
- Leadership.
- Continuous improvement.
- Employee participation and development.
- Fast response.
- Design quality.
- Long-range outlook.
- Management by fact.
- Partnership development.
- Corporate responsibility and citizenship.

THE ASSESSMENT TOOL

A company can apply for the Baldrige Award as an opportunity to display its world class status, or as a vehicle for improvement. This assessment of progress can be particularly effective when potential improvements are identified from the assessment, and the company views the "areas for improvement" as valuable opportunities.

---

*The Contributor of this chapter is: John Vinyard, President & CEO, Bekaert Associates, Inc., Marietta, GA.
The Reviewer of this chapter is: Robert E. Chapman Ph.D., Senior Economist, Malcolm Baldrige National Quality Award Office, National Institute of Standards and Technology.*

CHAPTER 13

BALDRIGE CRITERIA FRAMEWORK

It is unfortunate, however, when the award is viewed as an objective rather than a milestone along a continuous process. Companies who are merely trying to win an award have missed the point, and may not understand the true power of using the Baldrige process.

The power of the award is centered in its nonprescriptive nature. The criteria will talk about issues (such as empowerment of employees) but not specify precisely how a company should implement the concepts discussed in the criteria. That fact has elicited criticisms from "gurus" who feel they have the prescription which can be implemented in a cookbook fashion for all companies and cultures.

BALDRIGE CRITERIA FRAMEWORK

The core values and concepts of Malcolm Baldrige are embodied in seven categories, as follows:

1. Leadership.
2. Information and analysis.
3. Strategic quality planning.
4. Human resource development and management.
5. Management of process quality.
6. Quality and operational results.
7. Customer focus and satisfaction.

A brief description of each of these categories is given in Fig. 13-1.

The framework of Baldrige has four basic elements:

- Driver. Senior executive leadership creates the values, goals and systems, and guides the sustained pursuit of quality and performance objectives.
- System. The system comprises the set of well defined and well designed processes for meeting the company's quality and performance requirements.
- Measures of progress. Measures of progress provide a results-oriented basis for channeling actions to deliver ever-improving customer value and company performance.
- Goal. The basic aim of the quality process is the delivery of ever-improving value to customers. The interrelationship of this framework is shown in Fig. 13-2.

For each of the categories there are subcategories (called items), and each category and item is weighted with a point value. The examination items and their point values from 1993 are shown in Fig. 13-3. A more detailed listing of the examination items and point values is shown in Fig. 13-4.

During a Baldrige scoring process, the examiners review applications individually (First Stage Review) and score them based on comparing the application to the criteria. If the application scores high enough to go on to the second stage (determined by the judges) a Second Stage Review is conducted by a group of examiners, led by a senior examiner. They discuss the original scoring of the application and reach consensus regarding the scoring and key issues around each subcategory.

From this Consensus Review a limited number of applicants are selected for the Third Stage Review—a site visit. Based on the findings of the site visit, the judges will finally recommend winners. The final winners are selected by the Secretary of Commerce. This overall flow is shown in Fig. 13-5.

The **Leadership** Category examines senior executives' personal leadership and involvement in creating and sustaining a customer focus and clear and visible quality values. Also examined is how the quality values are integrated into the company's management system and reflected in the manner in which the company addresses its public responsibilities and corporate citizenship.

The **Information and Analysis** Category examines the scope, validity, analysis, management, and use of data and information to drive quality excellence and improve operational competitive performance. Also examined is the adequacy of the company's data, information, and analysis system to support improvement of the company's customer focus, products, services, and internal operations.

The **Strategic Quality Planning** Category examines the company's planning process and how all key quality requirements are integrated into overall business planning. Also examined are the company's short- and long-term plans and how quality and operational performance requirements are deployed to all work units.

The **Human Resource Development and Management** Category examines the key elements of how the work force is enabled to develop its full potential to pursue the company's quality and operational performance objectives. Also examined are the company's efforts to build and maintain an environment for quality excellence conducive to full participation and personal and organizational growth.

The **Management of Process Quality** Category examines the systematic processes the company uses to pursue ever-higher quality and company operational performance. Examined are the key elements of process management, including research and development, design, management of process quality for all work units and suppliers, systematic quality improvement, and quality assessment.

The **Quality and Operational Results** Category examines the company's quality levels and improvement trends in quality, company operational performance, and supplier quality. Also examined are current quality and operational performance levels relative to those of competitors.

The **Customer Focus and Satisfaction** Category examines the company's relationships with customers and its knowledge of customer requirements and of the key quality factors that drive marketplace competitiveness. Also examined are the company's methods to determine customer satisfaction, current trends and levels of customer satisfaction, and these results relative to competitors.

Fig. 13-1 Malcolm Baldrige National Quality Award category descriptions (1993 criteria).

CHAPTER 13
BALDRIGE CRITERIA FRAMEWORK

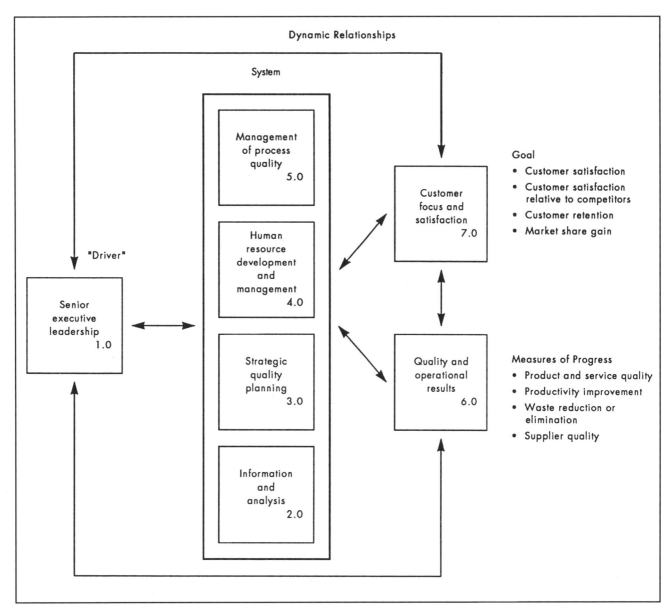

Fig. 13-2 Baldrige award criteria framework.

TRAINING MALCOLM BALDRIGE EXAMINERS

Although this chapter is not intended to recap the entire Malcolm Baldrige process, an understanding of the training for examiners might be helpful as a company starts the self-assessment process. Examiners are, primarily, trained on case studies, and the scoring of those case studies. Group work on the consensus process emphasizes facts and interpretation, rather than a mathematical averaging of the scores. Unlike traditional team meetings, where participants may be judged on how close the final output is to their original input, the consensus teams are aiming at the best group output from their discussions.

Other examiner training activities include determining key business factors for a company, and determining site visit issues.

Site visit issues are what the examiners would like to see upon visiting a company to verify and clarify the information in the application. Much of the material used to train Malcolm Baldrige examiners is publicly available from:

National Institute of Standards and Technology
Office of Quality Programs
Route 270 and Quince Orchard Road
Administration Building, Room A537
Gaithersburg, MD 20899
Telephone: (301) 975-2036
Fax: (301) 975-2128

CHAPTER 13

MALCOLM BALDRIGE SCORING PROCESS

MALCOLM BALDRIGE SCORING PROCESS

The system for scoring Examination Items is based upon three evaluation dimensions: (1) Approach, (2) Deployment, and (3) Results. All Examination Items require applicants to furnish information relating to one or more of these dimensions. Specific factors associated with the evaluation dimensions are described below. Scoring Guidelines are outlined on Table 13-1.

APPROACH

"Approach" refers to the methods the company uses to achieve the requirements addressed in the Examination Items. The factors used to evaluate approaches include, as appropriate:

- The appropriateness of the methods, tools, and techniques to the requirements.
- The effectiveness of methods, tools, and techniques.
- The degree to which the approach is systematic, integrated, and consistently applied.
- The degree to which the approach embodies effective evaluation/improvement cycles.
- The degree to which the approach is based upon quantitative information that is objective and reliable.
- The degree to which the approach is prevention-based.
- The indicators of unique and innovative approaches, including significant and effective new adaptations of tools and techniques used in other applications or types of businesses.

DEPLOYMENT

"Deployment" refers to the extent to which the approaches are applied to all relevant areas and activities addressed and implied in the Examination Items. The factors used to evaluate deployment include, as appropriate:

- The appropriate and effective application of the stated approach by all work units to all processes and activities.
- The appropriate and effective application of the stated approach to all transactions and interactions with customers, suppliers of goods and services, and the public.

RESULTS

"Results" refers to outcomes and effects in achieving the purposes addressed and implied in the Examination Items. The factors used to evaluate results include:

- The performance levels.
- The quality and performance levels relative to appropriate comparisons and/or benchmarks.
- The rate of performance improvement.
- The breadth and importance of performance improvements.
- The demonstration of sustained improvement or sustained high-level performance.

Traditional educational scoring systems have a range of 30%. The scores normally range from 70% to 100%. Any one of those scores could be a passing grade, and the overall bell-shaped curve is significantly compressed. The Malcolm Baldrige scoring process is very different from that used in educational systems. Unlike the process most people used in school which requires above 90% for an excellent score, the Malcolm Baldrige process takes the range of excellence and spreads it out significantly.

Baldrige has intentionally spread the bell-shaped curve over a much wider range, with 50% as the "anchor point." Entries will score in the 50% range (in the 1993 criteria the scoring guidelines show a 40% to 60% scoring range for these characteristics) if they:

1. Have an approach which has:
   - A sound systematic approach responsive to the primary purposes of the item.
   - A fact-based improvement process in place in key areas addressed by the item.
   - More emphasis on problem prevention than on reaction to problems.
2. Have deployment which has no major gaps, though some areas may be in the early stages of deployment.

1993 Examination categories/items	Point values
1.0 Leadership	95
1.1 Senior executive leadership	.45
1.2 Management for quality	.25
1.3 Public responsibility and corporate citizenship	.25
2.0 Information and analysis	75
2.1 Scope and management of quality and performance data and information	.15
2.2 Competitive comparisons and benchmarking	.20
2.3 Analysis and uses of company level data	.40
3.0 Strategic quality planning	60
3.1 Strategic quality and company performance planning process	.35
3.2 Quality and performance plans	.25
4.0 Human resource development and management	150
4.1 Human resource planning and management	.20
4.2 Employee involvement	.40
4.3 Employee education and training	.40
4.4 Employee performance and recognition	.25
4.5 Employee wellbeing and satisfaction	.25
5.0 Management of process quality	140
5.1 Design and introduction of quality products and services	.40
5.2 Process management: product and service production and delivery processes	.35
5.3 Process management: business processes and support services	.30
5.4 Supplier quality	.20
5.5 Quality assessment	.15
6.0 Quality and operational results	180
6.1 Product and service quality results	.70
6.2 Company operational results	.50
6.3 Business process and support service results	.25
6.4 Supplier quality results	.35
7.0 Customer focus and satisfaction	300
7.1 Customer expectations: current and future	.35
7.2 Customer relationship management	.65
7.3 Commitment to customers	.15
7.4 Customer satisfaction determination	.30
7.5 Customer satisfaction results	.85
7.6 Customer satisfaction comparison	.70
TOTAL POINTS	1000

Fig. 13-3 Examination items and point values (1993).

CHAPTER 13

MALCOLM BALDRIGE SCORING PROCESS

The Leadership Category examines senior executives' personal leadership and involvement in creating and sustaining a customer focus and clear and visible quality values. Also examined is how the quality values are integrated into the company's management system and reflected in the manner in which the company addresses its public responsibilities and corporate citizenship.

1.1 Senior Executive Leadership (45 pts.). Describe the senior executives' leadership, personal involvement, and visibility in developing and maintaining an environment for quality excellence.

- ☑ Approach
- ☑ Deployment
- ☐ Results

AREAS TO ADDRESS

a. Senior executives' leadership, personal involvement, and visibility in quality related activities of the company. Include: (1) reinforcing a customer focus; (2) creating quality values and setting expectations (3) planning and reviewing progress toward quality and operational performance objectives; (4) recognizing employee contributions; and (5) communicating quality values outside the company.
b. Brief summary of the company's customer focus and quality values and how they serve as a basis for consistent communication within and outside the company.
c. How senior executives regularly communicate and reinforce the company's customer focus and quality values with managers and supervisors.
d. How senior executives evaluate and improve the effectiveness of their personal leadership and involvement.

1.2 Management for Quality (25 pts.). Describe how the company's customer focus and quality values are integrated into day-to-day leadership, management, and supervision of all company units.

- ☑ Approach
- ☑ Deployment
- ☐ Results

AREAS TO ADDRESS

a. How the company's customer focus and quality values are translated into requirements for all managers and supervisors. Summarize: (1) their principal roles and responsibilities within their units, and (2) their roles and responsibilities in fostering cooperation with other units.
b. How the company's customer focus and quality values (1.1 b) are communicated and reinforced throughout the company with all employees.
c. How company and work unit quality and operational performance plans are reviewed. Describe: (1) types, frequency, content, and use of reviews and who conducts them; and (2) how the company assists units that are not performing according to plans.
d. Key methods and key indicators the company uses to evaluate and improve awareness and integration of quality values among managers and supervisors.

1.3 Public Responsibility and Corporate Citizenship (25 pts.). Describe how the company includes its responsibilities to the public in its quality policies and improvement practices. Describe also how the company leads as a corporate citizen in its key communities.

- ☑ Approach
- ☑ Deployment
- ☐ Results

AREAS TO ADDRESS

a. How the company integrates its public responsibilities into its quality policies and practices. Include: (1) how the company determines or sets operational requirements and goals taking into account risks, regulatory, and other legal requirements: (2) a summary of the principal public responsibility areas addressed within the company's quality policies and/or practices and how key operational requirements are communicated throughout the company; and (3) how and how often progress in meeting operational requirements and/or goals is reviewed.
b. How the company looks ahead to anticipate public concerns and to assess possible impacts on society that may derive from its products, services, and operations. Describe briefly how this assessment is used in planning.
c. How the company leads as a corporate citizen in its key communities. Include: (1) a brief summary of the types and extent of leadership and involvement in key communities; (2) how the company promotes quality awareness and sharing of quality related information; (3) how the company seeks opportunities to enhance its leadership; and (4) how the company promotes legal and ethical conduct in all that it does.
d. Trends in key indicators of improvement in addressing public responsibilities and corporate citizenship. Include responses to any sanctions the company has received under law, regulation, or contract.

2.0 Information and Analysis (75 pts.). The Information and Analysis Category examines the scope, validity, analysis, management, and use of data and information to drive quality excellence and to improve operational and competitive performance. Also examined is the adequacy of the company's data, information, and analysis system to support improvement of the company's customer focus, products, services, and internal operations.

2.1 Scope and Management of Quality and Performance Data and Information (15 pts.). Describe the company's data and information used for planning, day-to-day management, and evaluation of quality and operational performance. Describe also how data and information are managed to ensure reliability, timeliness, and rapid access.

- ☑ Approach
- ☑ Deployment
- ☐ Results

AREAS TO ADDRESS

a. Criteria for selecting data and information for use in quality and operational performance improvement. List key types of data and information used and briefly outline the principal roles of each type in improving quality and company operational performance. Include: (1) customer-related; (2) product and service performance; (3) internal operations and performance, including business processes, support services, and employee-related; (4) supplier performance; and (5) cost and financial management and evaluation.
b. How the company assures reliability, consistency, and rapid access to data of throughout the company. If applicable, describe how software quality is assured.
c. Key methods and key indicators used to evaluate and improve the scope and management of data and information. Include: (1) review and update; (2) shortening the cycle from data gathering to access; (3) broadening access to all those requiring data for day-to-day management and improvement; and (4) alignment of data and information with process improvement plans and needs.

2.2 Competitive Comparisons and Benchmarking (20 pts.). Describe the company's processes, current sources and scope, and uses of competitive comparisons and benchmarking information and data to support improvement of quality and overall company operational performance.

- ☑ Approach
- ☑ Deployment
- ☐ Results

AREAS TO ADDRESS

a. How the company uses competitive comparisons and benchmarking information and data to help drive improvement of quality and company operational performance. Describe: (1) how needs are determined; and (2) criteria for seeking appropriate comparison and benchmarking information—from inside and outside the company's industry.
b. Brief summary of current scope, sources, and principal uses of each type of competitive and benchmark information and data. Include: (1) customer-related; (2) product and service quality; (3) internal operations and performance, including business processes, support services, and employee-related; and (4) supplier performance, company operational performance.

Fig. 13-4 Detailed listing of examination items and point values.

CHAPTER 13

MALCOLM BALDRIGE SCORING PROCESS

c. How competitive and benchmarking information and data are used to improve understanding of processes, to encourage breakthrough approaches, and to set "stretch" objectives.
d. How the company evaluates and improves its overall processes for selecting and using competitive comparisons and benchmarking information and data to improve planning and company operations.

2.3 Analysis and Uses of Company-level Data (40 pts.). Describe how data related to quality, customers, and operational performance, together with relevant financial data, are analyzed to support company-level review, action, and planning.

☑ Approach
☑ Deployment
☐ Results

AREAS TO ADDRESS
a. How customer-related data and results (from Category 7.0) are aggregated with other key data and analyses, analyzed, and translated into actionable information to support: (1) developing priorities for prompt solutions to customer-related problems; and (2) determining key customer-related trends and correlations to support status review, decision making, and longer-term planning.
b. How operational performance data and results (from Category 6.0) are aggregated with other key data and analyses, analyzed, and translated into actionable information to support: (1) developing priorities for short-term improvements in company operations, including cycle time, productivity, and waste reduction; and (2) determining key operations-related trends and correlations to support status reviews, decision making, and longer-term planning.
c. How the company relates overall improvements in product service quality and operational performance to changes in overall financial performance.
d. How the company evaluates and improves its analysis as a key management tool. Include: (1) how analysis supports improved data selection and use; (2) how the analysis-access cycle is shortened; and (3) how analysis strengthens the integration of overall data for improved decision making and planning.

3.0 Strategic Quality Planning (60 pts.). The Strategic Quality Planning Category examines the company's planning process and how all key quality requirements are integrated into overall business planning. Also examined are the company's short- and long-term plans and how quality and operational performance requirements are deployed to all work units.

3.1 Strategic Quality and Company Performance Planning Process (35 pts.). Describe the company's strategic planning process for the short term (1 to 2 years) and longer term (3 years or more) for customer satisfaction leadership and overall operational performance improvement. Include how this process integrates quality and company operational performance requirements and how plans are deployed.

☑ Approach
☑ Deployment
☐ Results

AREAS TO ADDRESS
a. How the company develops strategies, goals, and business plans to address quality and customer satisfaction leadership for the short term and longer term. Describe how business plans consider: (1) customer requirements and the expected evolution of these requirements, (2) projections of the competitive environment; (3) risks: financial, market, and societal; (4) company capabilities, including human resource development, and research and development to address key new requirements or technology leadership opportunities; and (5) supplier capabilities.
b. How the company develops strategies and plans to address overall operational performance improvement. Describe how the following are considered: (1) realigning work processes ("re-engineering") to improve operational performance; and (2) productivity improvement and reduction in waste.

c. How plans are deployed. Describe: (1) the method the company uses to deploy overall plan requirements to all work units and to suppliers, and how it ensures alignment of work unit plans and activities; and (2) how resources are committed to meet the plan requirements.
d. How the company evaluates and improves its planning process, including improvements in: (1) determining company quality and overall operational performance requirements; (2) deploying requirements to work units; and (3) receiving planning input from company work units.

3.2 Quality and Performance Plans (25 pts.). Summarize the company's quality and operational performance goals and plans for the short term (1 to 2 years) and the longer term (3 years or more).

☑ Approach
☑ Deployment
☐ Results

AREAS TO ADDRESS
a. For the company's chosen directions, including planned products and services, markets, or market segments, summarize: (1) key quality factors and quality requirements to achieve leadership; and (2) key company operational performance requirements.
b. Outline of the company's principal short-term quality and company operational performance goals and plans. Include: (1) a summary of key requirements and key operational performance indicators deployed to work units and suppliers; and (2) a brief description of resources committed for key needs such as capital equipment, facilities, education and training, and personnel.
c. Principal longer-term (3 years or more) quality and company operational performance goals and plans, including key requirements and how they will be addressed.
d. Two-to-five-year projection of improvements using the most important indicators of quality and company operational performance. Describe how quality and company operational performance might be expected to compare with competitors and key benchmarks over this time period. Briefly explain the comparisons, including any estimates or assumptions made regarding the projected quality and operational performance of competitors or changes in benchmarks.

4.0 Human Resource Development and Management (150 pts.). The Human Resource Development and Management Category examines the key elements of how the work force is enabled to develop its full potential to pursue the company's quality and operational performance objectives. Also examined are the company's effort to build and maintain an environment for quality excellence conducive to full participation and personal and organizational growth.

4.1 Human Resource Planning and Management (20 pts.). Describe how the company's overall human resource plans and practices are integrated with its overall quality and operational performance goals and plans, and address fully the needs and development of the entire work force.

☑ Approach
☑ Deployment
☐ Results

AREAS TO ADDRESS
a. Brief outline of the most important human resource plans (derived from Category 3.0). Address: (1) development, including education, training and empowerment; (2) mobility, flexibility, and changes in work organization, processes, or work schedules; (3) reward, recognition, benefits, and compensation; and (4) recruitment, including possible changes in diversity of the work force. Distinguish between the short term (1 to 2 years) and the longer term (3 years or more), as appropriate.
b. How the company improves its human resource operations and practices. Describe key goals and methods for processes/practices such as recruitment, hiring, personnel actions, and services to employees. Describe key performance indicators, including cycle time, and how these indicators are used in improvement.

Fig. 13-4 Detailed listing of examination items and point values. (cont.)

CHAPTER 13

MALCOLM BALDRIGE SCORING PROCESS

c. How the company evaluates and uses all employee-related data to improve the development and effectiveness of the entire work force and to provide key input to overall company planning and to human resource management and planning. Describe: (1) how this improvement process addresses all types of employees; and (2) how employee satisfaction factors (Item 4.5) are used to reduce adverse indicators such as absenteeism, turnover, grievances, and accidents.

4.2 Employee Involvement (40 pts.). Describe the means a available for all employees to contribute effectively to meeting the company's quality and operational performance goals and plans; summarize trends in effectiveness and extent of involvement.

☑ Approach

☑ Deployment

☑ Results

AREAS TO ADDRESS
a. Principal mechanisms the company uses to promote ongoing employee contributions, individually and in groups, to quality and operational performance goals and plans. Describe how and how quickly the company gives feedback to contributors.
b. How the company increases employee empowerment, responsibility, and innovation. Briefly summarize principal goals for all categories of employees, based upon the most important requirements for each category.
c. Key methods and key indicators the company uses to evaluate and improve the effectiveness, extent, and type of involvement of all categories and all types of employees. Include how effectiveness, extent, and types of involvement are linked to key quality and operational performance improvement results.
d. Trends in the most important indicators of the effectiveness and extent of employee involvement for each category of employee.

4.3 Employee Education and Training (40 pts.). Describe how the company determines quality and related education and training needs for all employees. Show how this determination addresses company plans and needs and supports employee growth. Outline how such education and training are evaluated, and summarize key trends demonstrating improvement in both the effectiveness and extent of education and training.

☑ Approach

☑ Deployment

☑ Results

AREAS TO ADDRESS
a. How the company determines needs for the types and amounts of quality and related education and training for all employees, taking into account their differing needs. Include: (1) linkage to short- and long-term plans, including company-wide access to skills in problem solving, waste reduction, and process simplification; (2) growth and career opportunities for employees; and (3) how employees' input is sought and used in the needs determination.
b. Summary of how quality and related education and training are delivered and reinforced. Include: (1) outline of methods for education and training delivery for all categories of employees; (2) on-the-job application of knowledge and skills; and (3) quality-related orientation for new employees.
c. How the company evaluates and improves its quality and related education and training. Include how the evaluation supports improved needs determination, taking into account: (1) relating on-the-job performance improvement to key quality and operational performance improvement goals and results; and (2) growth and progress of all categories and types of employees.
d. Trends in the effectiveness and extent of quality and related training and education based upon key indicators of each.

4.4 Employee Performance and Recognition (25 pts.). Describe how the company's employee performance, recognition, promotion, compensation, reward, and feedback approaches support the attainment of the company's quality and performance plans and goals.

☑ Approach

☑ Deployment

☑ Results

AREAS TO ADDRESS
a. How the company's employee performance, recognition, promotion, compensation, reward, and feedback approaches for individuals and groups, including managers, support the company's quality and operational performance goals and plans. Address: (1) how the approaches ensure that quality is reinforced relative to short-term financial considerations; and (2) how employees contribute to the company's employee performance and recognition approaches.
b. Key methods and key indicators the company uses to evaluate and improve its employee performance and recognition approaches. Include how the evaluation takes into account: (1) effective participation by all categories and types of employees; (2) employee satisfaction information (Item 4.5); and (3) key indicators of improved quality and operational performance results.
c. Trends in key indicators of the effectiveness and extent of employee reward and recognition, by employee category.

4.5 Employee Wellbeing and Satisfaction (25 pts.). Describe how the company maintains a work environment conducive to the wellbeing and growth of all employees; summarize trends in key indicators of wellbeing and satisfaction.

☑ Approach

☑ Deployment

☑ Results

AREAS TO ADDRESS
a. How wellbeing factors such as health, safety, and ergonomics are included in quality improvement activities. Include principal improvement goals, methods, and indicators for each factor relevant and important to the company's employee work environment. For accidents and work-related health problems, describe how root causes are determined and how adverse conditions are prevented.
b. Special services, facilities, and opportunities the company makes available to employees.
c. How the company determines employee satisfaction. Include a brief description of methods, frequency, and the specific factors for which satisfaction is determined. Segment by employee category or type, as appropriate.
d. Trends in key indicators of wellbeing and satisfaction. This should address, as appropriate: satisfaction, safety, absenteeism, turnover, turnover rate for customer-contact personnel, grievances, strikes, and worker compensation. Explain important adverse results, if any. For such adverse results, describe how root causes were determined and corrected, and/or give current status. Compare results on the most significant indicators with those of industry averages, industry leaders, key benchmarks, and local/regional averages, as appropriate.

5.0 Management of Process Quality (140 pts.). The Management of Process Quality Category examines the systematic processes the company uses to pursue ever-higher quality and company operational performance. Examined are the key elements of process management, including research and development design, management of process quality for all work units and suppliers, systematic quality improvement, and quality assessment.

5.1 Design and Introduction of Quality Products and Services (40 pts.). Describe how new and/or improved products and services are designed and introduced and how processes are designed to meet key product and service quality requirements and company operational performance requirements.

☑ Approach

☑ Deployment

☐ Results

Fig. 13-4 Detailed listing of examination items and point values. (cont.)

CHAPTER 13

MALCOLM BALDRIGE SCORING PROCESS

AREAS TO ADDRESS
a. How designs of products, services, and processes are developed so that: (1) customer requirements are translated into product and service design requirements; (2) all product and service quality requirements are addressed early in the overall design process by appropriate company units; (3) designs are coordinated and integrated to include all phases of production and delivery; and (4) key process performance characteristics are selected based on customer requirements, appropriate performance levels are determined, and measurement systems are developed to track performance for each of these characteristics.
b. How designs are reviewed and validated, taking into account key factors: (1) product and service performance; (2) process capability and future requirements; and (3) supplier capability and future requirements.
c. How the company improves its designs and design processes so that new product and service introductions and product and service modifications progressively improve in quality and cycle time.

5.2 Process Management: Product and Service Production and Delivery Processes (35 pts.). Describe how the company's key product and service production and delivery processes are managed to ensure that design requirements are met and that both quality and operational performance are continuously improved.

☑ Approach

☑ Deployment

☐ Results

AREAS TO ADDRESS
a. How the company maintains the quality of production and delivery processes in accord with the product and service design requirements (Item 5.1). Include: (1) the key processes and their requirements; (2) key indicators of quality and operational performance; and (3) how quality and operational performance are determined and maintained, including types and frequencies of in-process and end-of-process measurements used.
b. For significant (out-of-control) variations in processes or outputs, how root causes are determined, and corrections made and verified.
c. How the process is improved to achieve better quality, cycle time, and overall operational performance. Include how each of the following is used or considered: (1) process analysis/simplification; (2) benchmarking information; (3) process research and testing; (4) use of alternative technology; (5) information from customers of the processes—inside and outside of the company; and (6) challenge goals.

5.3 Process Management: Business Processes and Support Services (30 pts.). Describe how the company's key business processes and support services are managed so that current requirement are met and that quality and operational performance are continuously improved.

☑ Approach

☑ Deployment

☐ Results

AREAS TO ADDRESS
a. How key business processes and support services are designed to meet customer and/or company quality and operational performance requirements. Include: (1) the key processes and their requirements; (2) key indicators of quality and performance; and (3) how quality and performance are determined and maintained, including types and frequencies of in-process and end-of-process measurements used.
b. For significant (out-of-control) variations in processes or outputs, how root causes are determined, and corrections made and verified
c. How the process is improved to achieve better quality, cycle time, and overall operational performance. Describe how each of the following are used or considered: (1) process analysis/simplification; (2) benchmarking information; (3) process research and testing; (4) use of alternative technology; (5) information from customers of the business processes and support services—inside and outside the company; and (6) challenge goals.

5.4 Supplier Quality (20 pts.) Describe how the company assures the quality of materials, components, and services furnished by other businesses. Describe also the company's plans and actions to improve supplier quality.

☑ Approach

☑ Deployment

☐ Results

AREAS TO ADDRESS
a. How the company defines and communicates its quality requirements to suppliers. Include: (1) a brief summary of the principal quality requirements for key suppliers; and (2) the key indicators the company uses to evaluate supplier quality.
b. Methods the company uses to assure that its quality requirements are met by suppliers. Describe how the results of these methods and other relevant performance information are communicated to suppliers.
c. How the company evaluates and improves its own procurement activities. Describe the feedback sought from suppliers and how it is used in improvement.
d. Current plans and actions to improve suppliers' abilities to meet key quality and response time requirements.

5.5 Quality Assessment (15 pts.) Describe how the company assesses the quality and performance of its systems, processes, and practices and the quality of its products and services.

☑ Approach

☑ Deployment

☐ Results

AREAS TO ADDRESS
a. Approaches the company uses to assess: (1) systems, processes, and practices; and (2) products and services. For (1) and (2), describe: (a) what is assessed; (b) how often assessments are made and by whom; and (c) how measurement quality and adequacy of documentation of processes and practices are assured.
b. How assessment findings are used to improve: products and services; systems; processes; practices; and supplier requirements. Describe how the company verifies that assessment findings lead to action and that the actions are effective.

6.0 Quality and Operational Results (180 pts.) This category examines the company's quality levels and improvement trends in quality, company operational performance, and supplier quality. Also examined are current quality and operational performance levels relative to those of competitors.

6.1 Product and Service Quality Results (70 pts.) Summarize trends in quality an current quality levels for key product and service features; compare the company's current quality levels with those of competitors and/or appropriate benchmarks.

☐ Approach

☐ Deployment

☑ Results

AREAS TO ADDRESS
a. Trends and current levels for all key measures of product and service quality.
b. Current quality level comparisons with principal competitors in the company's key markets, industry averages, industry leaders, and appropriate benchmarks.

6.2 Company Operational Results (50 pts.) Summarize trends and levels in overall company operational performance and provide a comparison of this operational performance with competitors and/or appropriate benchmarks.

☐ Approach

☐ Deployment

☑ Results

Fig. 13-4 Detailed listing of examination items and point values. (cont.)

CHAPTER 13
MALCOLM BALDRIGE SCORING PROCESS

AREAS TO ADDRESS
a. Trends and current levels for key measures of company operational performance.
b. Comparison of performance with that of competitors, industry averages, industry leaders, and key benchmarks.

6.3 Business Process and Support Service Results (25 pts.). Summarize trends and current levels in quality and operational performance improvement for business process and support services; compare results with competitors and/or appropriate benchmarks.

☐ Approach
☐ Deployment
☑ Results

AREAS TO ADDRESS
a. Trends and current levels for key measures of quality and operational performance of business processes and support services.
b. Comparison of performance with appropriately selected companies and benchmarks.

6.4 Supplier Quality Results (35 pts.). Summarize trends in quality and current quality levels of suppliers; compare the company's supplier quality with that of competitors and/or with appropriate benchmarks.

☐ Approach
☐ Deployment
☑ Results

AREAS TO ADDRESS
a. Trends and current levels for the most important indicators of supplier quality.
b. Comparison of the company's supplier quality levels with those of appropriately selected companies and/or benchmarks.

7.0 The Customer Focus and Satisfaction (300 pts.). The Customer Focus and Satisfaction Category examines the company's relationships with customers and its knowledge of customer requirements and of the key quality factors that drive marketplace competitiveness. Also examined are the company's methods to determine customer satisfaction and retention, and these results relative to competitors.

7.1 Customer Expectations: Current and Future (35 pts.). Describe how the company determines near-term and long-term requirements and expectations of customers.

☑ Approach
☑ Deployment
☐ Results

AREAS TO ADDRESS
a. How the company determines current and near-term requirements and expectations of customers. Describe: (1) how customer groups and/or market segments are determined, including how customers of competitors and other potential customers are considered; (2) the process for collecting information, including what information is sought, frequency and methods of collection, and how objectivity and validity are assured; (3) the process for determining specific product and service features and the relative importance of these features to customer groups or segments; and (4) how other information such as complaints, gains and losses of customers, and product service performance are cross-compared to support the determination.
b. How the company addresses future requirements and expectations of customers. Describe: (1) the time horizon for the determination; (2) how important technological, competitive, societal, economic, and demographic factors that may bear upon customer requirements, expectations, or alternatives are considered; (3) how customers of competitors and other potential customers are considered; (4) how key product and service features and the relative importance of these features are projected; and (5) how changing or emerging market segments are addressed and their implications on new product/service lines as well as on current products and services are considered.
c. How the company evaluates and improves its processes for determining customer requirements and expectations. Describe how the improvement process considers: (1) new market opportunities; and (2) extension of the time horizon for the determination.

7.2 Customer Relationship Management (65 pts.). Describe how the company provides effective management of its relationships with its customers and uses information gained from customers to improve customer relationship management strategies and practices.

☑ Approach
☑ Deployment
☐ Results

AREAS TO ADDRESS
a. For the company's most important processes, and transactions that bring its employees into contact with customers, summarize the key requirements for maintaining and building relationships. Describe key quality indicators derived from these requirements and how they were determined.
b. How service standards that address the key quality indicators (7.2a) are set. Include: (1) how service standards requirements are deployed to customer contact employees and to other company units that provide support for customer-contact employees; and (2) how the overall service standards system is tracked.
c. How the company provides information and easy access to enable customers to seek assistance, to comment, and to complain. Describe the main types of contact and how easy access is maintained for each type.
d. How the company follows up with customers on products, services, and recent transactions to seek feedback and to help build relationships.
e. How the following are addressed for customer-contact employees: (1) selection factors; (2) career path; (3) deployment of special training to include: knowledge of products and services; listening to customers; soliciting comments from customers; how to anticipate and handle problems or failures ("recovery") in customer retention; and how to manage expectations; (4) empowerment and decision making; (5) satisfaction determination; (6) recognition and reward; and (7) turnover.
f. How the company ensures that formal and informal complaints and feedback received by all company units are aggregated for overall evaluation and used throughout the company. Describe: (1) how the company ensures that complaints and problems are resolved promptly and effectively; and (2) how the company sets priorities for improvement projects based upon analysis of complaints, including types and frequencies of complaints and relationships to customers' repurchase intentions.
g. How the company evaluates and improves its customer relationship management strategies and practices. Include: (1) how the company seeks opportunities to enhance relationships with all customers or with key customers; and (2) how evaluations lead to improvements in service standards, access, customer-contact employee training, and technology support. Describe how customer information is used in the improvement process.

7.3 Commitment to Customers (15 pts.). Describe the company's commitments to customers regarding its products/services and how these commitments are evaluated and improved.

☑ Approach
☑ Deployment
☐ Results

AREAS TO ADDRESS
a. Types of commitments the company makes to promote trust and confidence in its products/services and to satisfy customers when product service failures occur. Describe these commitments and how they: (1) address the principal concerns of customers; (2) are free from conditions that might weaken customers' trust and confidence; and (3) are communicated to customers clearly and simply.

Fig. 13-4 Detailed listing of examination items and point values. (cont.)

CHAPTER 13

MALCOM BALDRIGE SCORING PROCESS

---

b. How the company evaluates and improves its commitments, and the customers' understanding of them, to avoid gaps between expectations and delivery. Include: (1) how information/feedback from customers is used; (2) how product service performance improvement data are used; and (3) how competitors' commitments are considered.

7.4 Customer Satisfaction Determination (30 pts.). Describe the company's methods for determining customer satisfaction, customer repurchase intentions, and customer satisfaction relative to competitors; describe how these methods are evaluated and improved.

☑ Approach
☑ Deployment
☐ Results

AREAS TO ADDRESS
a. How the company determines customer satisfaction. Include: (1) a brief description of methods, processes, and measurement scales used; frequency of determination; and how objectivity and validity are assured. Indicate significant differences, if any, in these satisfaction methods, processes, and measurement scales for different customer groups or segments; and (2) how customer satisfaction measurements capture key information that reflects customers' likely market behavior, such as repurchase intentions.
b. How customer satisfaction relative to that for competitors is determined. Describe: (1) company-based comparative studies; and (2) comparative studies or evaluations made by independent organizations and/or customers. For (1) and (2), describe how objectivity and validity of studies are addressed.
c. How the company evaluates and improves its overall processes, measurement, and measurement scales for determining customer satisfaction and customer satisfaction relative to that for competitors. Include how other indicators (such as gains and losses of customers) and customer dissatisfaction indicators (such as complaints) are used in this improvement process.

7.5 Customer Satisfaction Results (85 pts.). Summarize trends in the company's customer satisfaction and trends in key indicators of customer dissatisfaction.

☐ Approach
☐ Deployment
☑ Results

AREAS TO ADDRESS
a. Trends in indicators of customer satisfaction. Segment by customer group, as appropriate. Trends may be supported by objective information and/or data from customers demonstrating current or recent (past 3 years) satisfaction with the company's products/services.
b. Trends in indicators of customer dissatisfaction. Address the most relevant indicators for the company's products/services.

7.6 Customer Satisfaction Comparison (70 pts.). Compare the company's customer satisfaction results with those of competitors.

☐ Approach
☐ Deployment
☑ Results

AREAS TO ADDRESS
a. Trends in indicators of customer satisfaction relative to competitors. Segment by customer group, as appropriate. Trends may be supported by objective information and/or data from independent organizations, including customers. This information and/or data may include survey results, competitive awards, recognition, and ratings.
b. Trends in gaining and losing customers, or customer accounts, to competitors.
c. Trends in gaining or losing market share to competitors.

Fig. 13-4 Detailed listing of examination items and point values. (cont.)

3. Have results which show:
   - Improvement or good performance trends reported in key areas of importance to the item requirements and to the company's key performance-related business factors.
   - Some trends and/or current performance can be evaluated against relevant comparisons, benchmarks, or levels.
   - No significant adverse trends or poor performance in key areas of importance to the item requirements and to the company's key performance-related business factors.

MANAGING EXPECTATIONS

Before any company starts the self-evaluation process, one of the key issues is the expectations of management. As described in the previous paragraphs, the scoring process is not the traditional "70 is a C" activity. Under the Baldrige process 70 would be a very strong score, and might warrant a site visit.

As previously discussed, if the objective of self-assessment is to win the award, a limited approach is being taken. The strength of self-assessment is the ability to continually improve the capabilities of the company and to continually strengthen its competitiveness in the marketplace. For these reasons, and due to the scoring system, managing management expectations is critical if the mid-level supporters of the process do not want to be victims of "killing the messenger."

One way to develop a set of management expectations is to go through a planning exercise with management where their expectations of the self-assessment process are developed. This planning activity should not include discussion of the "Baldrige scores expected," but should focus on a specific level of management understanding and continuous improvement expected out of the process over the years. Clearly, a self-assessment process should be a long-term commitment, and not a one-time activity to develop a numeric score. Companies who have not focused on the Baldrige process must be ready for scores of 100 to 300. In a number of instances management has gone through self-assessment activities for the first time, not understanding the Baldrige scoring process, and has been upset by a score in this range. In some instances the management teams have attempted to cover up the score, and discontinue the self-assessment activity.

One step to help manage expectations is to spend approximately four hours in training the management team in the Baldrige process, including some preliminary consensus review activities. Any management team should be aware of what the scoring categories mean, and the previous Baldrige applicants' distribution of written scores. This summary is shown on Table 13-2.

CHAPTER 13
MANAGING EXPECTATIONS

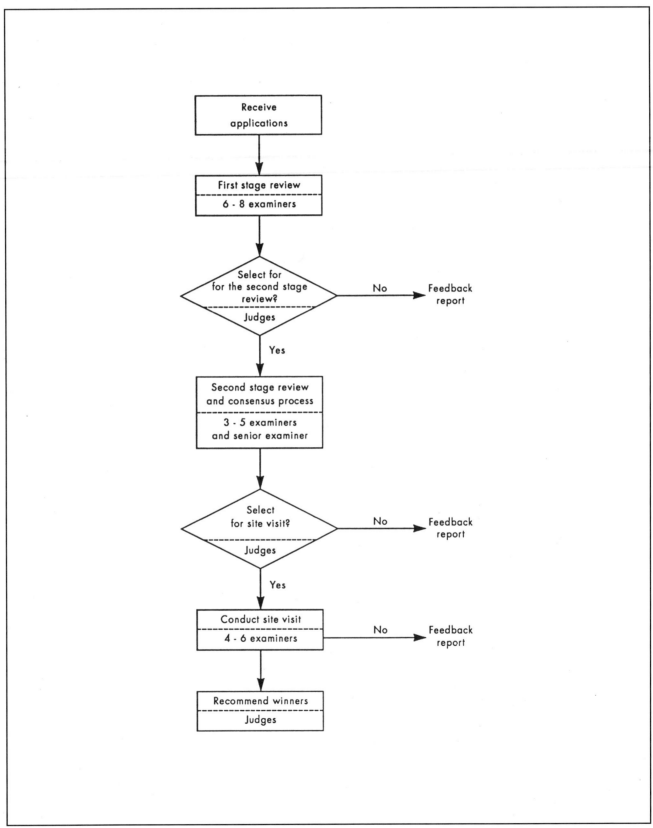

Fig. 13-5 The evaluation process.

CHAPTER 13

MANAGING EXPECTATIONS

TABLE 13-1

Score	Approach/Deployment	Score	Results
0%	Anecdotal information; no system evident information presented.	0%	No data reported or anecdotal data only. Data not responsive to major requirements of the Item.
10% to 30%	Beginning of a systematic approach to addressing the primary purposes of the Item. Significant gaps still exist in deployment that would inhibit progress in achieving the major purposes of the Item. Early stages of a transition from reacting to problems to preventing problems.	10% to 30%	Early stages of developing trend data. Some improvement trend data or early good performance reported. Data are not reported for many to most areas of importance to the Item requirements and to the company's key performance-related business factors.
40% to 60%	A sound, systematic approach responsive to the primary purposes of the Item. A fact-based improvement process in place in key areas addressed by the Item. No major gaps in deployment, though some areas may be in early stages of deployment. Approach places more emphasis on problem prevention than on reaction to problems.	40% to 60%	Improvement or good performance trends reported in key areas of importance to the Item requirements and to the company's key performance-related business factors. Some trends and/or current performance can be evaluated against relevant comparisons, benchmarks, or levels. No significant adverse trends or poor current performance in key areas of importance to the Item requirements and to the company's key performance-related business factors.
70% to 90%	A sound, systematic approach responsive to the overall purposes of the Item. A fact-based improvement process is a key management tool; clear evidence of refinement and improved integration as a result of improvement cycles and analysis. Approach is well deployed with no significant gaps, although refinement, deployment, and integration may vary among work units or system activities.	70% to 90%	Good to excellent improvement trends in most key areas of importance to the Item requirements and to the company's key performance-related business factors, or sustained good to excellent performance in those areas. Many to most trends and current performance can be evaluated against relevant comparisons, benchmarks, or levels. Current performance is good to excellent in most areas of importance to the Item requirements and to the company's key performance-related business factors.
100%	A sound, systematic approach, fully responsive to all the requirements of the Item. Approach is fully deployed without weaknesses or gaps in any areas. Very strong refinement and integration-backed by excellent analysis.	100%	Excellent improvement trends in most to all key areas of importance to the Item requirements and to the company's key performance-related business factors, or sustained excellent performance in those areas. Most to all trends and current performance can be evaluated against relevant comparisons, benchmarks, or levels. Current performance is excellent in most areas of importance to the Item requirements and to the company's key performance-related business factors. Strong evidence of industry and benchmark leadership demonstrated.

SELF-ASSESSMENT VERSUS CONTINUOUS IMPROVEMENT

Companies who have used self-assessment to their advantage have included it as a continuous part of the management process. In many cases, this has been included in the compensation and incentivization of key management. This step has been shown to continually focus on improving a company's score, which means a step toward continuous improvement activity throughout the year.

Additionally, companies must think through the steps necessary to institutionalize this scoring process in a way that it becomes ongoing.

CHAPTER 13

SELF-ASSESSMENT VERSUS CONTINUOUS IMPROVEMENT

TABLE 13-2

Range of score	Range number	% Applicants in Range 1990	% Applicants in Range 1991	% Applicants in Range 1992	Comments
0 to 125	1	0	2.8	0	No evidence of effort in any category. Virtually no attention given to quality.
126 to 250	2	7.2	13.2	12.0	Only slight evidence of effective effort in any category. Early recognition of quality issues.
251 to 400	3	18.6	35.8	30.0	Some evidence of effective effort in a few categories, but not outstanding in any. Limited integration of efforts. Largely based on reactions to problems, with little preventive effort.
401 to 600	4	52.6	34.0	40.0	Evidence of effective efforts in many categories, and outstanding in some. A good prevention-based process. Many areas lack maturity. Further deployment and results needed to demonstrate continuity.
601 to 750	5	19.6	14.2	18.0	Evidence of effective efforts in most categories, and outstanding in several. Deployment and results show strength, but some efforts may lack maturity. Clear areas for further attention.
751 to 875	6	2	0	0	Effective efforts in all categories, and outstanding in many. Good integration and good to excellent results in all areas. Full deployment. Many industry leaders.
876 to 1000	7	0	0	0	Outstanding effort and results in all categories. Effective integration and sustained results. National and world leaders.
		100%	100%	100%	
Number of Applicants		97	106	90	
Winners		4	3	5	

HOW COMPANIES HAVE USED SELF-ASSESSMENT

REVIEW OF THE MALCOLM BALDRIGE NATIONAL QUALITY AWARD (MBNQA) PROCESS WITH THE MANAGEMENT TEAM

One of the key aspects of any self-assessment process is involving management. It is key that they understand what is being undertaken, have been able to impart their own expectations into the effort, and feel they have a participative role. This first step includes briefing the Management Team on the Malcolm Baldrige process, the criteria, and how First Stage (Application Scoring), Second Stage (Consensus), and Third Stage (Site Visit) Reviews are conducted. This step includes a number of aspects:

- Formal briefing on the MBNQA at an executive level.
- A more detailed briefing of the Baldrige process.
- A detailed discussion (including questions and answers) of the Criteria.
- An open question and answer session with management. This could be conducted in a group and/or one-on-one, as appropriate.

DEVELOP AND DEFINE MANAGEMENT EXPECTATIONS

After the initial step of briefing management in the Malcolm Baldrige process, the company needs to clearly understand management expectations. These expectations should be defined for key management groups, as well as for "decision influencer" groups within the organization. These expectations will be determined again, through group and one-on-one meetings within the organization. Defining these expectations early is critical to subsequent success. The Management Team must feel their concerns have been addressed, and the overall effort is headed in a direction which they support.

TRAINING IN THE MALCOLM BALDRIGE EVALUATION PROCESS

This step is oriented toward educating groups (each from 16 to 36 employees) in the Malcolm Baldrige Criteria, the scoring process, and the process of consensus evaluation of applications.

CHAPTER 13

HOW COMPANIES HAVE USED SELF-ASSESSMENT

Malcolm Baldrige case studies are used. Both manufacturing and service case studies are available, for a wide range of companies. It should be noted that the training includes two days of classroom work preceded by approximately 16 to 20 hours of homework for each attendee. Additionally, homework is given during the two-day session. Essentially, the training is the same as that conducted for the Malcolm Baldrige examiners. Attendees learn:

- The criteria.
- How the criteria can be interpreted.
- The scoring system.
- The consensus process.
- The stages of a Malcolm Baldrige review.
- How to determine the weight of particular items included (or omissions).
- How to determine site visit issues.

This two-day session is an excellent opportunity for the attendees to meet each other and discuss common total quality issues. Normally these sessions include significant question and answer periods where the attendees can get many of their Total Quality/Malcolm Baldrige questions answered or discussed by a knowledgeable group.

These types of meetings are also a significant bellwether to the organization, and symbolize a change in the Total Quality activities. For example, the organizations which send attendees should clearly understand that this is the early stage of a major focus of the company on Malcolm Baldrige assessment activities.

DRAFTING THE MALCOLM BALDRIGE APPLICATION

The attendees from the training session are given the responsibility to draft a Malcolm Baldrige type application. It does not make sense to try to hold the application to "85 pages" which is the Baldrige standard for large companies. Nevertheless, if the application expands to 150 to 200 pages it becomes much more difficult for the examiners to score. Therefore, a page range might be used for each one of the sections which approximates the value of the section to the overall score. Notable exceptions to the page limitations might be Section 1.0 (Leadership) and Section 4.0 (Human Resource Development and Management) since these two sections drive the entire thrust of the other sections of the application.

SCORING THE APPLICATION

Once the application is prepared, Baldrige Examiners score it independently. This step (frequently called the Stage One Review) is done by each examiner without input from the company or from other examiners. The individual scores will then be summarized by the 28 items, and by the seven major categories.

CONSENSUS REVIEW

After the application is scored, the examiners will conduct a consensus review (probably over the telephone). The examiners will discuss each one of the categories if there is a significant difference in scoring. The general guidelines (by item) are:

- If the range of scores is within 20 points—average the scores.
- If the range of scores is 30 points—either average or consensus.
- If the range is 40 points or more, a consensus conversation must be held.

From this review several key documents will be prepared:

- A summary of the scores from the consensus activity, and the highlights of strengths and areas for improvement.
- Identification of site visit issues.
- Highlighting areas of particular strengths or areas for improvement.

One of the advantages of this type of internal review, as compared to the Baldrige process, is that many of the comments which may not strictly adhere to the Baldrige criteria can be captured. In many instances these are valuable comments regarding improvements which may be viewed favorably by an examination team even if they are not specifically mentioned in the Baldrige criteria.

SITE VISIT

Once the consensus score is reviewed by management, a site visit will be conducted. A reasonably limited site visit is recommended (that is, approximately two days on-site) where the specific areas are discussed in an open manner. Unlike a true Baldrige site visit, this can be viewed as a cooperative effort to get at the details behind the statements in the document.

Prior to the site visit, discussion is held with management regarding the level of conversations they would like to have regarding the findings of the consensus review. Obviously, a real Baldrige site visit does not include any conversation regarding the internal teams' discussions. Nevertheless, this activity is being conducted to augment many ongoing improvement activities. As such, the spin-off of ideas can be very beneficial.

COMPUTERIZED BALDRIGE SURVEY

A number of companies offer the ability to assess Baldrige position based on a survey of the company's employees. Although this is not a technique which is as accurate as a formal self-assessment, it can be an important bellwether and point out areas which can be opportunities for improvement.

This survey (typically conducted with several hundred employees) can give management guidance regarding the areas for improvement based on employee perceptions. This can be an important secondary data point when combined with the Malcolm Baldrige type of assessment discussed elsewhere in this chapter.

DEVELOP FEEDBACK

When the site visit and/or survey is (are) completed, feedback is prepared for the Management Team. It is envisioned that the company's management is involved in the preparation of this document, since this is viewed more as a step toward continuing to improve rather than an outsider's "cold assessment."

FEEDBACK TO MANAGEMENT

When the survey is completed, the assessment is finished, and the feedback document is developed, the findings will be discussed with the appropriate Management Teams. This can be tailored as a presentation only to a selected group, or could be expanded and given to any number of employees.

One of the most important aspects of this feedback will be managing audience expectations. The Baldrige scoring process is clearly not the "70% is a C" scale many people used in college. Helping the audience to understand the meaning of a Baldrige score is critical.

Particular attention is paid to areas where improvements can be made, while springboarding off achievements already established or under way. It is critical for any outsider to recognize the

CHAPTER 13

SELF-ASSESSMENT LESSONS LEARNED

accomplishments of a company, and not try to take a Greenfield approach to recommendations.

IDENTIFY OPPORTUNITIES FOR IMPROVEMENT IN THE ASSESSMENT PROCESS

Since this entire proposal is centered around establishing an internal self-assessment capability, this step can be particularly important. It is envisioned that this is a discussion with the examiners who went through the assessment process, as well as many of the individuals who wrote the application and/or were involved in the site visit.

This can result in valuable "lessons learned" documentation activity which can improve subsequent steps toward an ongoing self-assessment process.

DEVELOP NEXT STEPS

These steps are merely preparing for subsequent activities which can continue to add value to the company. As such, it is envisioned that the participants will work with the company to develop a set of next steps which can be presented to management to show them a plan for institutionalizing the self-assessment process, and ongoing improvement.

SELF-ASSESSMENT LESSONS LEARNED[1]

Obviously, companies who have dedicated themselves to the self-assessment process have learned a number of lessons along the way. This section attempts to capture some of those lessons. It should be strongly emphasized that the Baldrige process is nonprescriptive, as are the lessons learned. Each company should evaluate its own internal culture, and whether the lessons are applicable to the culture, products, locations, services, customers, and people.

Lessons learned include:

1. The applicant does not have to follow the rigor of the formal Baldrige process. On an internal self-assessment there are a number of shortcuts that can be developed which can ease the time and effort required. The assessment process can also be prescriptive in an attempt to help the applicant in a way the Baldrige Office cannot do without hurting the impartiality and integrity of the national award.

2. Do not underestimate the time required. Many companies who have started a self-assessment process have initially underestimated the time required within the organization. This has included indicating that some of the internal examiners involved in the process should do it part time while retaining most of their normal job responsibilities. It becomes quickly apparent, however, that the time required for this process is significant.

   The old axiom "if it's worth doing, it's worth doing well" certainly applies to this process, but this is an area most frequently overlooked.

3. Develop a plan and resource requirements like those for any other major project. Like any other major effort, a self-assessment process needs to be planned in detail. The plan should be communicated to all key employees, particularly those who will be directly impacted by the process. This includes the application writers, assessors (examiners), the managers of those employees, and the senior executive team.

   Like other major projects, someone needs to be in charge. This includes a person with the primary responsibility for the activity, and another group with responsibility for the various aspects (or categories) of the assessment.

   Where additional resources are required, they should be planned, and approved by the appropriate management level. At a minimum the plans should include the steps required, responsibility, timing, resources required, and potential barriers.

4. Examiner training is the key to the calibration of scores. In most instances, new examiners need a period of time to adjust to the Baldrige scoring system. One of the first things new examiners look for in the training exercise is how their scores compare to others. In some instances these scores can be skewed by their own criteria of excellence, by their own company's level of achievement, or by their desire to continue to the next stage to verify or clarify what is briefly described in the application. In any of these instances, the calibration of the new examiner may be higher or lower than would be expected after the examiner has had a chance to go through the cycle several times.

   Some companies have studied the differences between the scores of examiners who have not been trained, and examiners who have been through the training. These studies have shown that examiners who have not been through the training can score applications 100% higher than examiners who have been through the training! For example, a trained examiner may score an application at 350 points when an examiner without training could score that same application at 700 points.

   In some instances the profile of the scores developed by untrained examiners may be the same as that of the trained examiners (that is, the examiners will both recognize which items are stronger and which have the most areas for improvement), but the level of the scores can be higher or lower. Part of the examiner training emphasizes the ability to "see the same things" in the application, and to carefully relate those items to the criteria. The risk in inflating the scores is two-fold. First, with an inflated sense of wellbeing, the company team may not approach improvement with as aggressive an attitude as they might if they knew the opportunities for improvement were larger. Second, if the scores are not calibrated, the company will not know if it is genuinely improving the next year, or how it compares to other companies. There are probably many other reasons that it is important to be calibrated, but the two listed here are most often noted.

5. Continually strive for improvement. If companies are going to dedicate the resources required for self-

CHAPTER 13

SELF-ASSESSMENT LESSONS LEARNED[1]

assessment, there must be some type of return. That return should be in strengthening the competitiveness of the company, which can only be accomplished if genuine improvements are continually emphasized.

One of the lessons learned is to continually emphasize this fact to management, and to continually ask what improvements are being made as a result of using and leveraging the knowledge from the assessment.

6. It is a never ending journey. Clearly, the self-assessment process must become a way of life and cease to be a program. If it is a program there will certainly be some groups within any company who wait for the program to die, or feel it does not apply to them. If it is an integral part of the management structure and process, self-assessment is a stronger tool to ensure that competitiveness is continually improved.

7. Stay current with the annual criteria changes. As the Baldrige criteria change each year companies can benefit by staying current with those changes.

When using the self-assessment process, one of the primary objectives is continuous improvement. Changing the criteria annually helps a company emphasize internally that no improvement can be the last.

8. Stay the course. Companies who start this process normally receive pressure from within to stop assessments. The true benefit is achieved over a number of years, when the company recognizes that this is not a program or a transient activity, but is a fundamental change in the way the company measures progress and achievement.

9. Do not let divisions of the company compete in a win/lose manner. If the groups within the company are competing for a fixed prize, or if there can only be a limited number of "Winners," the competition may help one division more than the others, but it will not maximize the benefit to the entire company. In short, do not incentivize negative competition where the groups will not share lessons learned, or improvement techniques. It does not do a corporation any good if its divisions are improving at each other's expense.

This may sound like a simple concept to implement, but there are any number of traps which one can fall into, such as:

- Saying "whoever improves the most."
- Saying "whoever gets the highest score."
- Saying "only one division can win the top company award."

Clearly goodness should not be measured in terms of a raw score. Some companies have been able to avoid this negative competition by measuring whether each division improves its division score from year to year.

10. Top management support is mandatory. Although this is a truism for almost anything a company does, self-assessment is no exception. As noted previously in the Stay the Course lesson learned, once self-assessment is started there will be pressure to stop the assessments.

One interesting note from companies who have been through a number of annual self-assessment cycles: the most difficult year appears to be the second year. This hurdle is toughest since the process has not been in place long enough to show the true potential, and everyone is complaining about the amount of effort required.

11. Do not get too detailed on the scores given to the applicants. A number of companies have found that there is a benefit to giving the applicants feedback by a scoring range, and staying away from specific item scores.

12. Communicate realistic expectations at the beginning. For the long-term support of a self-assessment process, it is much better to let management (and the participants) know how much effort will be required at the beginning. Many self-assessment processes have been criticized when it becomes apparent that the initial estimates of the workload fall short of the actual amount of work required.

13. Writing the application is a learning process. The initial complaint from a self-assessment process is the amount of work required to write the application. Nevertheless, the people who have been through this admit that it is a significant learning exercise. Very few times in their careers will they be required to learn as much in as short a period of time.

14. Take steps to make the writing process easier. For internal self-assessed applications, do not make it an exercise in writing skills, wordsmanship, etc. Some companies who have been through this exercise several times want their divisions to spend their time on improvement activities, and not on writing. Some of them even have the saying, "Wordsmanship does not count."

15. Tying the process to compensation can have a big reward. As previously discussed, companies want to stay away from making this a numbers game. Applicants, however, take the process much more seriously when they have their performance evaluations, salary, or incentives tied to their ability to improve their Baldrige scores.

16. The quality of the examiners is key to internal credibility. The examiners, and their knowledge of the criteria and the scoring process, are key to the management team taking the process seriously. This means that the best training must be made available, and the best people must be used. One company made the comment "If the person you want is available, then you have chosen the wrong person."

To emphasize the importance of training, one company made the comment "As much training as you do up-front, people will still misinterpret the criteria."

17. Give timely feedback. The perspective of what is timely feedback can be a matter of interpretation, but the definition which should be used is that of the customer. In this instance the customer is the applicant. Most applicants want feedback at the end of the site visit. That might not be practical, but it should be the goal.

In addition to being timely, the feedback should be two-way. This is one of the advantages over a formal Baldrige application, where feedback is only one-way and written. This two-way conversation can be a powerful technique in helping the applicant to improve, and to understand the perspective of the examiner team.

18. All corrective actions need a process owner, a focus, and a delivery date. Each time an Area for Improvement is accepted by the applicant, the applicant must develop a plan for how those items will be improved. Although the corrective actions may be short, medium, or long-term, a process must be in place which verifies that the appropriate steps are in place to determine the root cause, and to ensure that actions are progressing as planned.

Additionally, the process controls must include a verification that the corrective action did, in fact, correct the Area for Improvement. This means that there must be an audit trail, and appropriate responsibilities for reporting.

CHAPTER 13

PERSPECTIVES BEYOND SELF-ASSESSMENT

Self-assessment is an important tool in a company's ability to improve continuously. One of the things which can make the assessment process easier is an evaluation/site visit cycle which is streamlined. This is particularly helpful when the streamlining cuts the time required, and tailors the events to the company's needs. Since every company and management team is different, it is difficult to propose one approach which will fit every company. Proposed here, however, are two approaches which may be considered to shorten the cycle:

1. One week cycle:
   - Day 1: Score applications. This can be an abbreviated scoring process, and could include specific examiners taking primary responsibility for specific categories.
   - Day 2: Review scores (consensus process).
   - Day 3: Site Visit preparation.
   - Day 4: Site Visit—NOTE: Unlike an impartial site visit team, an internal team can tell the applicant what they are looking for, and shorten the process.
   - Day 5: Final adjustments of scores, and preparation of the feedback comments.
2. Two week cycle:
   - Day 1 and 2: Score application.
   - Day 3: Consensus.
   - Day 4: Preparation for site visit.
   - Day 5 and 6: Site visit.
   - Day 7: Review site visit findings, and develop open issues.
   - Day 8: Return for an additional day of site visit.
   - Day 9: Prepare feedback.
   - Day 10: Debriefing the applicant.

References

1. These lessons learned are based on John Vinyard's experience, as well as the experience of Patrick Norausky, Vice President, Total Quality, Dresser-Rand and Steven Detter, Director, Total Quality, McDonnell Douglas Corporation.

CHAPTER 14

GENERAL PRODUCTIVITY IMPROVEMENT

USE OF THE ANSI Y14.5M STANDARD IN PRODUCT DEFINITION

While searching for a competitive edge in manufacturing, it is easy to forget that the product must be defined with real numbers rather than through a simple conceptual definition. In the dynamic atmosphere of product development, management's path of least resistance is to apply the latest techniques in a quest for product success. What suffers most from this approach are the details associated with applying the product concept to the design and manufacture of a real entity.

Past practices, that would allow for success even with inefficient design cycles, no longer yield acceptable results. Following the traditional product cycle may now lead to the failure of the product and, possibly, the business. Global competitors, time-based competition, shortage of highly trained employees, and changing employee desires require the elimination of inefficiency in bringing a product to market. The product cycle is no longer forgiving.

In the quest for better product performance, the one key element that must be controlled is the quantitative (geometric) definition of the product. For success to be obtained in the marketplace, this definition must be fleshed out in a way that provides the customer with value. A successful product eventually is fully described in a precise form. This can be done using a structured process or by trial and error. The latter approach requires a less competitive market, to allow the time needed to go through the design/build cycle repeatedly until the job is done correctly. Unfortunately, the market conditions that once allowed this approach no longer exist.

This section moves into the bowels of the business where the work of actually bringing a product to market is done—and where decisions are made that form the major influences on product cost and time-to-market. While the emphasis in business education is on the idea that general management concepts can be applied successfully by anyone with the correct background, at some point in the life cycle of a product decisions unique to the product will be made that determine its ultimate success or failure. As pointed out in the TOOL AND MANUFACTURING ENGINEERS HANDBOOK, Volume 6, an inadequate "script" will prevent the most elaborate production plans from succeeding. For a manufactured product, the product definition, which includes specification of the manufacturing process, is a script. To be built successfully, the product must be fully described prior to implementing production.

The intent here is to take structured methods and show how they both complement and form a foundation for the continuous improvement process. Placing concurrent engineering methodologies within the framework of a continuous improvement program reinforces their power and provides an economic argument for their implementation and continued use.

Powerful techniques exist that can integrate the effects of continuous improvement into the technological and human resource realms. The ultimate goal of these techniques is to develop the individuals that populate the organization, linking their unique technological knowledge with the organization that is responsible for bringing the product to market. Even the best time-tested management techniques do not provide the depth of structure needed to accomplish this. The combined use of concurrent engineering, based on geometric control, and continuous process improvement provides such a structure. For an organization that must survive and prosper in a competitive global environment, the methods provide the details—the numbers—necessary to give definition to the product and allied processes at the time and level appropriate to the economic goals of the firm.

The one major theme is the use of *product*

CHAPTER CONTENTS:

USE OF THE ANSI Y14.5M STANDARD IN PRODUCT DEFINITION 14-1

SHOP FLOOR MANAGEMENT 14-18

AUTOMATIC IDENTIFICATION 14-22

SIMULATION AND WASTE ELIMINATION 14-24

SIMULATION AS A DECISION MAKING TOOL 14-31

FAILURE MODE AND EFFECTS ANALYSIS (FMEA) 14-35

A VERSATILE TOOL TO AID IN DOWNTIME REDUCTION 14-39

The Contributors of this chapter are: **Richard J. Caldwell, Jr.**, Engineer, Diamond Star Motors Corp.; **Robert G. Campbell, PE**, President, Advanced Manufacturing Technology, Inc., and Professor of Engineering, Harper College; **James Higley**, Associate Professor, Purdue–Calumet; **Morey Kays**, Professor, Purdue–Calumet; **Edward S. Roth, P.E., CMfgE**, President, Productivity Services, Inc.; **Ken Sharma**, Senior Partner, Intellection, Inc.; **Dr. Dean Stamatis, CQE, CMfgE**, President, Contemporary Consultants, Southgate, MI; **David M. Wilke**, Synchronous Manufacturing Implementer, General Motors.

The Reviewers of sections of this chapter are: **Gloria L. Campbell**, Associate Professor of Business Administration, Waltburg College; **Paul A. Carr**, Senior Sales Engineer, Automatic Inspection Devices, Inc.; **Arthur D. Harmala**, V.P., Sales and Marketing, Automatic Inspection Devices; **Alvin G. Neumann**, President, Technical Consultants, Longboat Key, FL; **Frank Rack**, Managing Change, Inc.; **Arthur Thomson, MSE, PE**, Principal, Thomson Associates, Cleveland, OH.

CHAPTER 14

USE OF THE ANSI Y14.5M STANDARD IN PRODUCT DEFINITION

definition to drive a continuous improvement program. There are two distinct areas of explanation: the product definition and the continuous improvement program.

WHAT IS PRODUCT DEFINITION?

The process and the allied techniques are built around the concept of product definition. To create the sequence of steps that are involved, and to show how they can be used to support a continuous improvement process, it is necessary to explain what the product definition is. Once the extent of the definition is understood, managers will be able to relate its elements to the continuous improvement process and concurrent engineering concepts.

Working Drawings

In designing and manufacturing a product, an enormous amount of information is necessary. The working drawings are the first concrete data. These normally are composed of a design layout (see Fig. 14-1) of the complete product, and the details (see Fig. 14-2) of the individual components.

The design layout, implicitly containing the product's bill of material, provides the necessary information showing how the individual components interrelate with each other and form the final assembly that the customer identifies as the product. (While design layouts are used to illustrate preliminary design alternatives, the emphasis here is on the final concept that survives the design process and is to be implemented in hardware.) This layout provides a baseline that is used for the various stages of the design, development, and improvement process.

The design baseline provides a benchmark, to which any variations or alterations of the elements of the design are to be compared during the continuous improvement cycle. It provides a decision making tool that adds rationality to the discussions and eliminates conjecture as to the physical nature of the results.

The ideal course of events in the design office should be to create the details of the components after the assembly layout is completed. Once complete definition is provided by the baseline assembly, the individual components can be identified and designed. It must be emphasized that the detailed level of design requires a continuing reference to the assembly. All of the

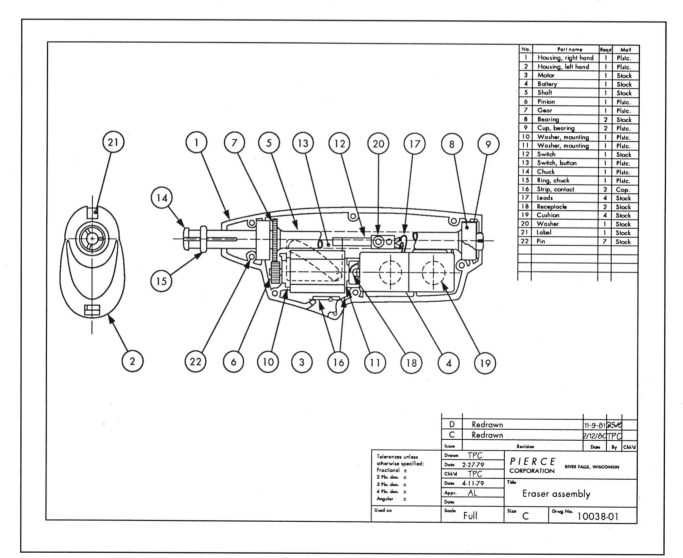

Fig. 14-1 Example of the final form of the design layout, the assembly drawing. (*Courtesy Pierce Business Products*)

CHAPTER 14

USE OF THE ANSI Y14.5M STANDARD IN PRODUCT DEFINITION

individual decisions relating to a particular component are based on the interrelationship and function of that part as described in the assembly drawing. Adequate design of the individual parts is not possible in isolation. A completed baseline assembly must be in hand prior to component design, and ongoing reference to the layout and mating components must be made as this design phase progresses.

Process Description

Included in the definition of the product is the information necessary to produce and verify it. Much of the product's success depends on the ability of manufacturing personnel to execute the design in an economic fashion; the need to define and document the process of fabrication is crucial.

The relevant set of documents incorporated in the definition includes the process plans for the individual components and the various assembly levels. These plans provide a description of the concrete steps through which the components will pass in production.

Once the manufacturing sequence is determined, information can be developed on the various tooling components envisioned in the plan. These tooling packages range from specification of commercial components to the design and fabrication of custom tooling. They are part of the process definition and encompass all gage requirements along with the more traditionally defined tooling. The key elements, the process plans and the tooling requirements, are as critical to the final success of the product as the baseline design.

The last element in the process description is the quality assurance procedures that support the delivery of a product and assure the desired level of customer satisfaction. These techniques can be either qualitative or quantitative in nature. They can be sophisticated techniques that are organizationally based and embedded in the design process, or as commonplace as the use of statistical process control methods. In any event, the quality concerns are an integral part of the product definition and a distinctive part of the process documentation.

The Geometric Description

To understand how the concept of geometric control can drive the design process and subsequent improvement efforts, it will help to identify the key geometric characteristics of the product

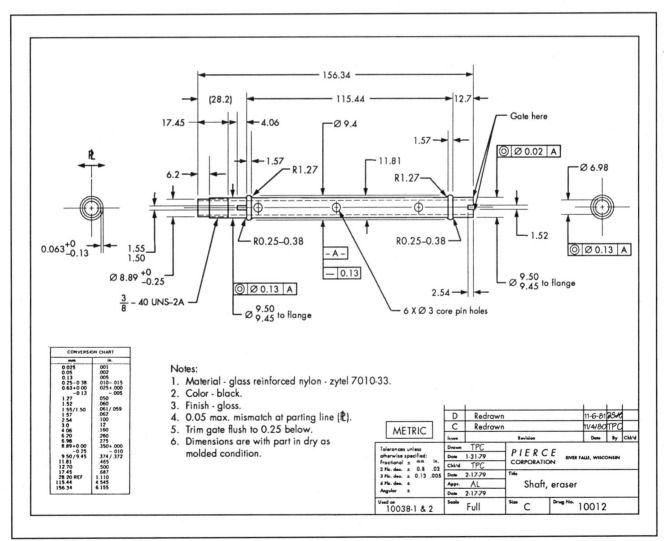

Fig. 14-2 Example of detail drawing extracted from design layout. (*Courtesy Pierce Business Products*)

CHAPTER 14

USE OF THE ANSI Y14.5M STANDARD IN PRODUCT DEFINITION

description. As already mentioned, the product's geometric elements are its primary attributes. Without these specific features the product would not exist.

In some order of priority relating to the design process, the product must initially take on both size and shape. The size and shape information provides the scale of the product and the topology of its geometric features. The shape of the object—the topology—is three-dimensional. The size information currently specified on the detail drawings is not truly three-dimensional, since no standard exists that relates these size specifications to three-dimensional topology. This point becomes important when individual part variations (tolerances) are described.

The next element of the product definition relates to the kinematic control of the component. This is one of the more crucial pieces of information that will be used to initiate the process definition sequence and the improvement techniques. This is also the information that is most likely *not* to appear when the product definition is developed under more traditional design methods; any formal documentation of this type of control is usually done in the manufacturing area. However, from an economic standpoint, early specification of kinematic control is a great advantage since it allows greater understanding of product variation.

Relating this to the American National Standards Institute (ANSI) Y14.5M standard, kinematic control is achieved by identifying the datum features that compose what is called the datum reference frame. This frame is specified in a geometric control applied to the component's features on the detail drawing (see Fig. 14-3). In simplest terms, the specified controls are derived from the effects introduced by mating parts interacting with the component in its assembled position. The essence of this is shown in Fig. 14-4 that illustrates the six possible motions that are to be constrained by the assembly—three translations and three rotations. Inadequate control of these six motions introduces product variation in either manufacture, inspection, or assembly.

The last major element of the geometric definition involves specification of the level of variation that can be tolerated in the product. This may take two forms. The first deals with the variations that occur when the size specifications vary during production. This variation is dealt with primarily at the two-dimensional level, even though the objects produced are three-dimensional by nature. Inherent in the standard is some three-dimensional form control (see *TMEH* Vol. 4, Chapter 4), but this is not sufficient to satisfy the demands of most precision products.

To overcome the limitation of size tolerances, the ANSI Y14.5M standard provides for the addition of explicit control of form, orientation, and position. Even though the addition of this information creates more complexity in the product definition, there is still an advantage: a true three-dimensional tolerance zone is identified that allows for direct analysis of the possible effects of variation in the product features. The conjecture as to what may result from the size variations is eliminated by reference to a standard that defines the size and shape of the tolerance zones and allows for a logical analysis of the interaction of these tolerances. This eliminates fuzziness in the initial design and ensuing improvement projects.

WHY FOCUS ON THE DEFINITION?

The next order of business is to demonstrate the relationship of the product definition to the continuous improvement process. The rationale behind the emphasis on the definition falls into two major categories: the human resource or organizational area, and the technical area. Of these two, the organizational concern will have the most positive impact upon the continuous improvement process.

Management Tool

Starting with the premise that continuous improvement is both a desirable process and necessary in the competitive world of manufacturing, management must have a variety of techniques to focus and direct these improvement efforts. Concurrent engineering lends structure to the initial design of the product, which may be extended to the ongoing efforts for improvement. A common thread, woven through the product life-cycle, achieves the continuity lacking in the project management of most products. It is a rationalized management tool providing a standardized definition of the product and organizing the necessary baseline information to be utilized in the improvement process.

There are two overarching themes in the use of these techniques. First is the need to integrate the *design* and *build* portions of the product cycle to eliminate the inefficiencies that occur in the traditional sequence. In particular, problems with the traditional

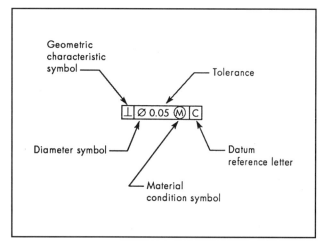

Fig. 14-3 Feature control frame that provides specification of geometric control. (*Courtesy American Society of Mechanical Engineers*)

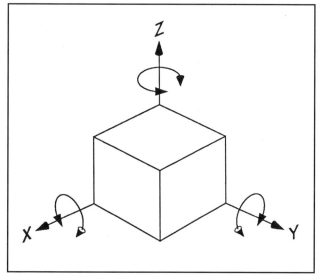

Fig. 14-4 Six degrees of freedom for kinematic control.

CHAPTER 14
USE OF THE ANSI Y14.5M STANDARD IN PRODUCT DEFINITION

technique are organizational in nature. This method, with its clear demarcation between the design and manufacturing functions, is a tremendously inefficient way for an organization to operate, and results in more trial-and-error approaches to product design than can be tolerated.

The second theme involves the desire to move product design and its allied processes away from a creative art toward something that is more structured and scientific in nature. The more rationality that is contained in the design cycle, the more predictable will be the results.

Design Methodology

Another area closely allied to management control is the use of concurrent engineering methodologies to structure the continuous improvement process as it relates to the design of individual components. The Six Step Methodologies that were explained in Volume 6 of the *TMEH* (pages 10-53 and 10-56) are applicable to the review of existing components and subsequent alteration to improve either the product or the process.

The methodologies, in effect, formalize asking the "right questions." Since the design of a modern product involves the interaction of large numbers of highly specialized individuals dispersed over both time and distance, anything that ensures consideration of the appropriate information at the correct time in the design cycle has major positive benefits on project economics. Because of the nature of assembled, interchangeable products, there are intricate information flows that have large impacts on the product's cost structure. Without a methodology, it is inconceivable that these flows would yield the desired results by happenstance.

Communication Device

The use of the Y14.5M standard provides a common language to discuss the geometric characteristics of a product. Since this language is used by the design, manufacturing, and quality organizations within the company, the standard provides a medium in which the concerns and needed decisions relating to the project can be communicated.

A more important point must be made. The use of this common language is intricately bound to the design and improvement methodologies. The correct application of these methods requires precise use of the language of geometric control. However, to correctly use this language necessitates the implementation of the design methodologies. The two elements cannot be separated. If one is used, the other follows by necessity.

As an element of the continuous improvement process, the Y14.5M language forms an indispensable means to ensure that the identification of candidates for improvement and subsequent project efforts are not affected by the use of fuzzy language. The hazy thought processes that result from imprecise communication reduce the opportunities that can be identified and restrict the results that should occur.

Process Definition

When structured design techniques are used by the team, they provide conceptual form to initial elements of the production process. To provide the detailed level of product definition that results from the Six Step Methodology, the designer must provide specific information relating to geometric control on the drawing; this places constraints on the production process. In effect, the product definition begins to describe an idealized production process that would bring the design to physical reality. This is shown in a simplified example in Figs. 14-5, 14-6, and 14-7. In Fig. 14-5, the component is specified using elements of the structured design methods. Figure 14-6 illustrates how this part functions as a member of an assembly. The datum features that are identified document the part's relationship with its mating component. Figure 14-7 is an example of the required tooling and shows how the tooling mimics the mating part. Much of the essence of the final tool design is incorporated in the definition of the part.

Of particular concern is the product definition's ability to identify tolerance levels providing recognition of the variation inherent in all production processes. This early consideration of variability is another positive gain. The sensitivity of the product to particular types of variation provides insight into the areas that may be explored for purposes of product improvement. Furthermore, should unexpected levels of sensitivity surface, the design methodology provides a rationale for investigation and identification of the means to improve the product and the process.

Deviations or changes that occur in designing and implementing production can be compared to the idealized process implicitly contained in the product definition, to see if the revised goals can be achieved by altering existing processes and techniques. The incremental improvement of existing components can be facilitated by the recognition that geometric controls, incorporated in the original product definition, provide an idealized process.

Educational Function

One of the organizational benefits that result from the improvement process is educational. With the rapid changes that take place in the general business environment and the continuous change that is instituted by the improvement process, a major undertaking is training and development. While formal training sessions can teach general principles and concepts, much of the information and experience that is unique to a given company is either difficult or impossible to impart in this fashion. This information may well be the final arbiter in determining the market success of the product.

The use of a concurrent engineering team is an excellent mechanism to deliver the specific training that should be provided as an ongoing investment in the firm's personnel. Overlaying structured design methods—based upon geometric control on actual projects—allows the education of individual team members, in the context of situations with sufficient complexity to bring product designers to optimum design efficiency. Additionally, immediate economic benefits are provided by working on real projects, rather than generalized academic exercises that cannot be applied to the company's situation.

Another aspect of the process is the development of a proprietary body of knowledge, derived from the project team, which develops from the dynamics of the group. Methods, procedures, and informal organizational links are created that help bring the project to a successful conclusion. Properly attended to, these results are not transient but long-term assets to the organization. The leverage obtained through this type of organizational development may be one of the major factors in justifying these specific improvement methods.

Organizational Acceptance

The team approach that is required to implement concurrent engineering concepts has a psychological benefit. The organization is made up of individuals, who are the people who must accept and institute the prescribed changes. The creation of the team establishes its members as the agents of change and provides

CHAPTER 14

USE OF THE ANSI Y14.5M STANDARD IN PRODUCT DEFINITION

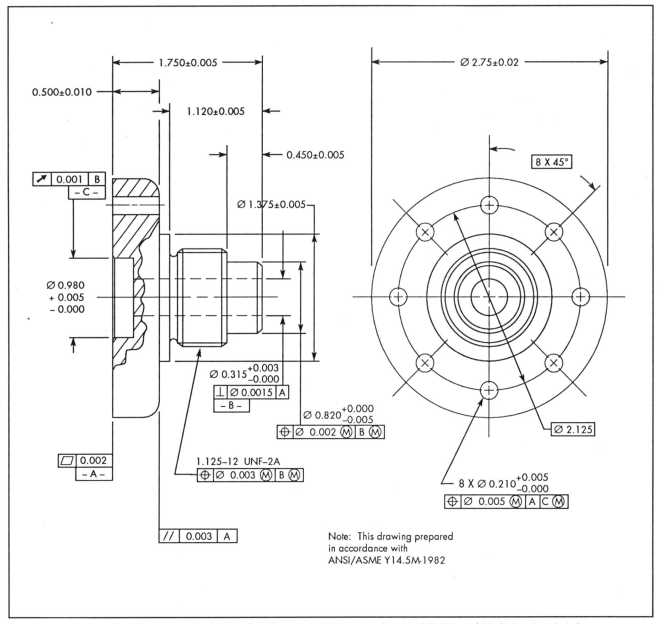

Fig. 14-5 Component detail showing geometric controls. (*Courtesy Lowell Foster and National Tooling and Machining Association*)

an opportunity for direct influence and action in the decision making process. They become stakeholders, with significant psychic capital invested in the project; ultimately they become advocates for continuation of the underlying process. The benefits that accrue to the firm result from the employees' identifying with the success of the project and, ultimately, with the success of the process.

Further benefit will result, as the team influences other employees to become involved with the structured process and these methods are adopted by additional areas of the firm. The technique has its greatest benefit through the wider dispersal of both responsibility and authority within the company. It must be recognized that such a result is not automatically achieved. Use of the methods will expand over a period of time, but only after they become widely accepted as integral parts of the team's activities. When done properly, structured design's need for information, interaction, and communication provides contact with personnel outside the team. These contacts are a mechanism that will draw ever widening circles of people into the processes.

Problem Solving

Enhanced problem solving is the result of the improvement and concurrent engineering processes. The most visible reason for this relates to the product definition. Without a complete understanding of the physical reality of the product, much of the subsequent activity would lack direction and become one large experiment, whose ends are reached by trial and error—an approach which is no longer economically viable in a time of increased competition.

CHAPTER 14

USE OF THE ANSI Y14.5M STANDARD IN PRODUCT DEFINITION

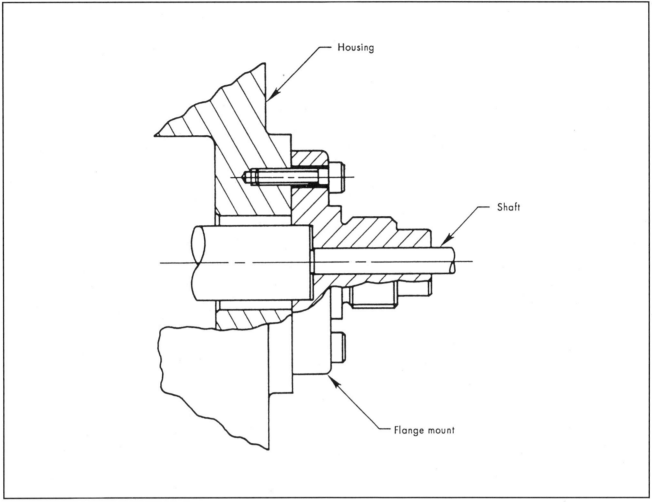

Fig. 14-6 Assembly drawing showing functional relationships necessary to define component detail in Fig. 14-5. (*Courtesy Lowell Foster and National Tooling and Machining Association*)

The specific types of information and decisions that are related to the product and process definition require recognition of many of the problems that classical management techniques left undefined to be solved in the course of full production. The improvement process, coupled with concurrent design concepts, forces early recognition of these concerns at points in time and at levels in the organization that can provide acceptable and economical solutions. Production is not disrupted by the need to sort out problems that occur due to incomplete product definition.

From a management standpoint, the use of these techniques provides the means to initiate and control problem solving. The structured design approach can be an entry point into the early stages of design and process improvement, formalizing various general concerns and reducing them to concepts that can be quantified. The most important concept from which all these concerns follow is *variability*. The techniques provide a well ordered set of controls and design procedures that place reasonable boundaries around the areas to be considered as candidates for the improvement process. The methods create the mechanisms to institute consideration of specific projects: the internal logic of the techniques simultaneously institutes a way to control the project. It is all wrapped up in a neat package.

HOW IS PRODUCT DEFINITION USED IN CONTINUOUS IMPROVEMENT?

The product definition has been established as an important component of the improvement process. The next thing to be gained is an understanding of its use. The following applications of the definition again fall into two groups: technological and organizational. Of these two areas, the organizational provides the greatest payback to the company implementing a continuous improvement process. The productivity gains that can be achieved result more from the logical structure added to the work than from any clever implementation of technological solutions. It cannot be overemphasized that what is being offered here is *not* basically technological in nature. The human element far surpasses the technical in importance to the success of these techniques.

A Benchmark

One of the tenets of continuous improvement requires the existence of a standard. In manufacturing a product, the product definition—which in its broadest sense includes information about both product and process definition—provides a standard from which the improvement process begins.

CHAPTER 14

USE OF THE ANSI Y14.5M STANDARD IN PRODUCT DEFINITION

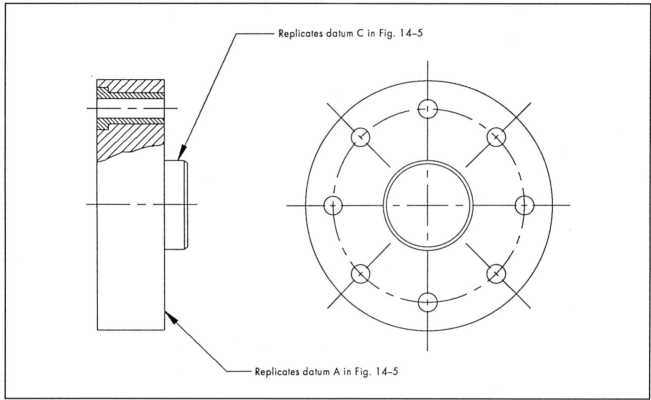

Fig. 14-7 Drill jig design based on data reference frame in Fig. 14-5.

The baseline design layout, containing all of the geometric information provided by dimensioning and tolerancing techniques, provides a benchmark from which an idealized manufacturing process is defined. The actual implementation of production is usually the result of many compromises, associated with existing processes and practices that have evolved over the years. A clear definition is provided for the various elements contained in the optimum process by using the structured design approach based on concurrent engineering and geometric control principles. This establishes a goal for the concurrent engineering team and provides a basis for formulating a continuous improvement program.

While benchmarking in current usage implies finding the best product or process available to use as the standard, the product/process definition that results from concurrent engineering activities allows the product to be self-referencing. Thus, many of the problems that occur (in trying to benchmark a product which has no similar competition in the marketplace) are eliminated. Additionally, benchmarking the process elements of the definition is facilitated by using the design concepts to determine the idealized process sequence. The problems, encountered in trying to get externally generated information based on what is probably proprietary knowledge, are overcome. In the event that outside sources of benchmarking information are available, these can be folded into the improvement process while still using the internally generated product definition.

Quantifying the Product

In order to generate a standard, the product must be quantified. Many market-related characteristics of a product design are initially difficult to define in a quantitative manner. However, once the product moves from a marketing concept to physical implementation, the qualitative notions of product satisfaction must be reduced to something that can be brought to physical reality. The product definition is the device that assigns the numbers that allow production to happen.

The structured design approach requires specific types of information that can only be provided in quantitative form. The functional definition that is assumed to be the desired result of the design process is, by its very nature, a precise entity that can only be described in these terms. The metrics that are used in the improvement process are contained in the definition. Problem identification and prescription of solutions require a precise statement of the conditions as they exist and concise targets to be achieved during the improvement process. Traditional design techniques lack the precision to provide these.

A Common Definition

The use of geometric control in the design process provides a common vocabulary that has concise meanings and precise effects on the production and delivery of a satisfactory product. One of the major difficulties that structured design seeks to eliminate is the classical confrontation between the designers and the manufacturing personnel. The vocabulary that underlies geometric control fosters the development of a design language that is governed by the functional outcomes of the product delivery process, and coincides with many of manufacturing's concerns.

It is not possible to produce a product in the tight timeframes that are becoming the norm if an inadequate or incomplete design definition is provided to the operations staff. The goal of the designer should be a unique specification of the product that is not open to alternative interpretations. The drawing should tell one

CHAPTER 14
USE OF THE ANSI Y14.5M STANDARD IN PRODUCT DEFINITION

unique story. Achievement of this result requires close interaction and communication between the various disciplines responsible for the product, communication that does not readily occur in this age of specialization. Such communication will not happen if individuals use terms and phrases that are only decipherable within the context of their own discipline.

Geometric dimensioning and tolerancing techniques used in conjunction with concurrent engineering force interaction and communication through a common language base. Assessment of the success of the process is easy, since the product definition provides the desired concrete outcome. Individual signoff on the resulting definition indicates acceptance and understanding of the interpretation. With this information in hand, much of the trial and error design that occurs in classical manufacturing can be eliminated.

The disciplined use of geometrically descriptive language also enhances the improvement process by providing the quantitative definitions upon which to analyze functional outcomes and to plan and implement subsequent improvements. Since traditional design techniques involve only two-dimensional analysis of variation, the importance of the concepts from the Y14.5M standard cannot be overstated. Without them, it is not possible to perform a three-dimensional tolerance analysis that can be easily communicated across disciplinary boundaries. More to the point, even when some consensus is reached on how to quantify product variation at the two-dimensional level, the precision upon which the consensus is based is so low that it prevents acceptable results from being fully achieved. There must be a common language.

Sources of Variation

The reality of manufacturing is that the idealized concept contained in the designer's mind is never achieved in physical terms. Each step of the manufacturing process introduces some variation. If the cumulative effects are too large, the resulting product is not acceptable. The ability to identify and quantify this variation is an important step in differentiating a product from others in the marketplace.

The use of geometric control provides the mechanism to anticipate likely sources of variation; to discuss their ramifications relative to the product design; and to provide alternative process definitions to keep the variation within acceptable levels. All of these concerns apply to both the initial design stages and subsequent improvement of the design and production processes.

Elements of the Y14.5M standard are built around concepts that reinforce the need to make distinctions between what the designer sees as the idealized product and what the physical process of production will yield. To refer back to the datum reference frames, there are three levels of physical realization of the features forming these frames which emphasize the existence of variation. In the standard, the term "datum" is used to identify the feature that the designer conjures up in the process of designing the product. It is a perfect geometric feature that can be described on the drawing but cannot be created in the physical world. The "datum feature" is the actual feature that can be generated on a real part. The associated equipment required in the production and verification process contains features called "simulated datums" which contact the datum features. The sources of variation manifest themselves in the datum features and the simulated datums and work against the designer's concept. Key concerns become the identification, quantification, and control of these variations.

The use of structured methods raises the level of sophistication of the design by allowing tolerance analysis at the three-dimensional level, which is not possible with size and shape descriptions. The logical method allows assessment of tolerance interactions that cannot be computed from the standard two-dimensional information contained in the orthographic views traditionally used in product definition. Even when it is possible to do some degree of two-dimensional analysis, the results are so ambiguous as to be misleading and, ultimately, useless.

Manufacturable Design

The manufacturability of a product is directly influenced by the techniques described here. By following the structured design approach in a collaborative environment (the concurrent engineering team), it is possible to incorporate information within the design definition that will have impact upon the manufacturing process. Early consideration of the design's influence on subsequent manufacturing procedures allows for considerable improvement of the manufacturability of the product compared to products designed in isolation.

The structured approach using the Y14.5M standard provides a checklist of the common geometric variations that are encountered in production of three-dimensional components, and can be used to anticipate their effects on the completed product. This aids in wedding the art of design to the science of design. For the individual with less mastery of design and manufacturing procedures, this is a confidence-building tool that leads to consideration of the most important aspects of the product: those that affect its actual fabrication and function. The concept of process improvement is pushed upstream into the design phase.

The generation of the specific information needed for a product definition moves decisions upstream in the product life cycle. Thus, as a decision making tool, the geometric design process can be used to facilitate the making of timely decisions and the achievement of consensus much earlier in the process than is the normal situation under traditional product management. This ties the design and process definition together as integral rather than separate entities.

In the context of the continuous improvement process, this linkage between the design and the process must be stated explicitly. The interweaving of the various elements of the product delivery system makes it impossible to alter any of the elements of the system without causing an impact on other elements. Any possible improvement must be judged on the basis of its impact on the system, and not just on the basis of convenience for a particular department or functional group.

STAGES OF IMPLEMENTATION

This section is based on the assumption that an existing traditional design structure is already in place. The techniques are as readily applied to the improvement of existing products as to the design of new products developed under concurrent engineering techniques. The most complicated implementation of these ideas is in an organization already producing a large variety of products using traditional design techniques. The explanation of the stages of implementation assumes that a core group of individuals will embrace the process and become its advocates. Without their fervor, the techniques may become one more fad that never lived up to expectations.

Audit Existing Design Process

Whether attempting to apply the techniques to a new or existing product, the starting point is to audit the current design practices and learn where the important elements of the design are

CHAPTER 14

USE OF THE ANSI Y14.5M STANDARD IN PRODUCT DEFINITION

actually being specified. At one level this audit can be performed using the notion of design refinement that was introduced in *TMEH* Vol. 6, pages 10-47 to 10-48 which advocates the use of geometric control in concurrent engineering applications.

Within the specialized fields of engineering, the long term trend has been to have engineers move toward the area of analytical design; support groups provide the creative design elements associated with a particular product. As engineers and allied designers have become more involved with the analytical elements of product design, more of the information that is required for a successful product implementation has been left to an informal design organization to provide. This informal organization includes elements of purchasing, manufacturing, quality assurance, suppliers, and other groups. The relevant point is that the current product definition found in the engineering drawings (or databases) is not sufficient for successful completion of the project and must be fleshed out as the design moves downstream.

The mechanism to audit this level of the design process is the concept of design refinement. The audit looks at the point in the life of the product where various elements of definition refinement occur. This audit can be organized on the basis of form deviation that is contained in the German standard, DIN 4760. The levels of deviation or variation that are shown in Fig. 14-8 have been modified for use here and, as a basis for an audit, provide some idea of who creates this information, when it is created, and where it is documented. From the standpoint of the economic success of a particular product, these are all critical questions.

The essence of the audit is to determine whether a complete product definition is ever assembled and documented, at any time in the design and delivery cycle, and how this comes about. If the definition is assembled in piecemeal fashion and dispersed throughout the organization, there is much economic inefficiency built into the process. This should be addressed through a more organized design procedure. The audit proceeds from the level of size and shape information (two-dimensional) and continues through the levels that are shown in Fig. 14-8. Certain products might require movement to the material structure, but the majority of products can achieve significant benefits by proceeding to the surface texture level.

The audit has three ultimate goals. There is an attempt to learn if the design intent—the functional requirements—is carried throughout the creation of the product and process design. With dispersed or segregated decision making, this may not always be the case. Second, the need to design manufacturability into the product is assessed by finding out if the realities of the production processes are recognized in achieving product function. A designer's request for features or levels of control that are unrealistic does much to discredit the integrity of the product definition and leads to subversion of the design intent. Lastly, the audits give quick readings on whether the product definition takes the physical realities into account by anticipating the effects of individual tolerance interaction on total product variation.

Education in Documentation Principles

Should the audit reveal that the design definition is put in place over time by a disconnected group of individuals and functional areas, then the first step in implementation would be to provide an overview of product documentation techniques (see Fig. 14-9). The logical starting point for this would be the concepts that are contained in and allied with the Y14 graphics standards which include the geometric controls. Additional areas of exposure should include the B4 standards on limits and fits, the B89 standards that deal with applied dimensional metrology, and the B46 standards on surface texture. In total, these standards provide the basis for the structured design techniques that are advocated, and are well established methods that should be incorporated in any professional design process.

As the move to integrated systems—computer-based or not—occurs, it is also important that the design organization include elements of configuration management that emphasize the interdependence of the information and the individuals responsible for its generation and documentation. This awareness of interdependence will lay the groundwork for training project team members in small group dynamics and team building at a later time. The first step is to use this awareness to create discipline in the product design cycle which can be used in subsequent improvement processes. Team building activities should not be introduced at this stage without having the structured design techniques in place. The design methods will be easier to institute, since they build on existing activities that the individuals will see as their primary responsibility. The team responsibilities will require more radical behavioral changes. Acceptance of these behavioral changes may be fostered by delaying their imposition.

Senior Management Support

The preceding stage is one that can stand scrutiny on its own. The suggested tools, contained in the standards that were mentioned, should be implemented in every engineering effort no matter what its level of sophistication. Without a standardized level of documentation and communication, much of the proprietary knowledge that results from putting a product into produc-

---

Size.

Limits of Size (Taylor's Principle).

Datum Reference Frame.

Geometric controls.
- Form.
- Orientation.
- Location.

Surface texture.
- Waviness.*
- Roughness (systematic).*
- Roughness (random).*

Surface integrity.
- Crystalline structure.*
- Lattice structure.*

*Adapted from DIN 4760.

Fig. 14-8 Levels of tolerance refinement.

B4	Standards	Limits and fits
Y14	Standards specifically	Engineering graphic standards Y14.5M, Y14.8M, and Y14.36
B46	Standard	Surface texture
B89	Standards specifically	Applied dimensional metrology B89.1.12M, B89.3.1, and 89.6.2

Fig. 14-9 Domestic US standards applicable to structured product documentation.

CHAPTER 14

USE OF THE ANSI Y14.5M STANDARD IN PRODUCT DEFINITION

tion is lost; as a result the system keeps "reinventing the wheel." Thus, management support for standardized documentation at this level can be acquired on a supportable basis.

The difficulty in getting management approval to start the continuous improvement program is likely to arise as movement is made through the levels of refinement beyond the size and shape information found on all engineering drawings. As the information density is increased, the costs incurred in initiating the design process are also increased. It naturally takes longer to provide the additional information. It also requires more highly trained engineers and designers. Furthermore, since this information is based on part function, it needs considerably more thought and calculation to arrive at the desired controls. These activities take time, which inflates the cost of the early project stages.

Senior management must be convinced that the expenditure of additional resources at the front end of the cycle will have economic benefits that reduce total project costs. One suggestion to garner management support is to show that the information being recommended for inclusion in the initial product definition is generated within all existing projects at some point in the cycle. The earlier this information is incorporated into the definition, the less likely it becomes that the costs of changes will be incurred. With engineering change costs running from an estimated $3600 per change in the electronics industry to $12,000 in the aircraft and aerospace industry,[1] it would not take long to convince managers that the trial and error process of product design is economically inefficient at best and suicidal in an extreme case.

To get the complete design definition early in the cycle requires the use of more highly skilled workers. Reviewing what is known about the productivity of these "knowledge-oriented workers," it can be concluded that their work cannot be automated as easily as that of direct labor. The knowledge that individual employees have accumulated through education and experience becomes one of the most important assets available to the company, and any improvement in their productivity will have large positive effects on cost. The goal of management is to get access to this wealth of knowledge in a manner and at a time appropriate to the needs of the product development cycle. Senior management must be convinced that the most efficient way to tap into the technical knowledge of its people is through the use of concurrent engineering concepts that are based on geometric control principles. Once this is implemented, the continuous improvement process can be laid on top of the concurrent engineering procedures and teams.

A helpful technique that can be used to get management's attention for the program involves total project costing and activity-based accounting. While these approaches are still in development, the more traditional cost accounting techniques have become less useful as the structure of manufacturing has changed. As fewer activities are involved with direct labor and more with the knowledge workers, significant problems have arisen in correctly allocating costs to specific products, and to the various stages of development when considering a specific project. The existence of these questions and concerns helps support the argument that new techniques and methods are needed to ensure that the business continues to thrive. When viewed from this perspective, the productivity of these individuals becomes a significant part of the total resources used to bring the product to fruition. It becomes imperative that some method be developed to first identify their contributions and then to improve their productivity. Dependence on direct-labor hours to allocate costs leads to decisions based on misleading information. More realistic ways of matching the revenue stream with the expense stream may be found in the new accounting methods.

Require a Structured Design

The actual design process should take advantage of the structured techniques that are the normal evolution of geometric control concepts. Once having obtained top management acquiescence for the use of geometric control and the concurrent engineering teams, the use of the structured baseline design must be implemented and enforced. Corruption of the database, upon which much of the subsequent improvement efforts will be based, is not the best way to instill confidence in the use of these techniques.

Previous design approaches in industry relied on the simultaneous design of individual components and the assembly of which they were members. The typical argument that justified this technique was that it expedited the design process and got the product into production more rapidly. With the increasing emphasis on time as a competitive weapon, the pressure to design components prior to complete definition of the assembly is intensified. Unfortunately, what is really expedited is the production of faulty designs due to a lack of complete product definition.

The basic element of the design process, that management must insist upon, is that the assembly shall be completely defined prior to starting any of the detailed level design of specific components. The complete conceptual definition of the product must be reduced to an engineering design, with sufficient definition that the quantitative elements of subsequent component designs can be identified and analyzed. This layout of the product concept forms the baseline design from which all further product decisions will emanate. No alterations should be made to this definition as part of the improvement efforts unless all parties to the original agreement are involved and consensus is again reached. The impact on the design is the touchstone of all decisions that must be made to implement production.

Following from this baseline layout are all of the individual decisions required to define the specific components comprising the completed product. Management concern should be that all of the design decisions for these details must be based on part function. In its simplest terms, the geometric design of the part is derived directly from the interrelationship of the component with the next level of assembly in which it is an element. This shows why the baseline layout must be completed before the detailing stage of the project is entered. Without complete definition of the assembly, it is not possible to define the part at the level of control that is necessary in the current competitive environment. All decisions that are made relating to the individual components prior to definition of the assembly are pure conjecture and loaded with risk. Management may take these risks for prudent business reasons, but the risks should be made explicit. Do not hide behind an elaborate component design that is based on faulty premises or on none at all.

Core Implementation Group

To obtain the necessary definition of a product, the concurrent engineering team must be composed of the appropriate members. These individuals must cover the full range of interests and functional areas that have impacts on the product cycle. They must also have complete authority to represent their constituencies. The ideal project structure would be to have all of the various functions incorporated under a project manager who has direct lines of authority to these individuals such that they are bound to the success of the project. An example of this type of management structure is found in the Japanese automotive firms and has led to quite successful product management outcomes.

CHAPTER 14

USE OF THE ANSI Y14.5M STANDARD IN PRODUCT DEFINITION

In any event, the team that is assembled, be it a matrix structure or a cohesive and formal team, must contain members of all the critical elements of the product structure.

Training Issues

In establishing the improvement team after identifying the core group, the next important undertaking is the creation of a common perspective to approach the issue of continuous improvement. The use of the concepts contained in the Y14.5M standard provide the common language in which to conduct discussions of a project's merits. This is a precise language with enough complexity in it to require formal training efforts that bring all of the participants to the same level of facility.

Three technical areas can be viewed as foundation concepts to explain the issues that must be highlighted by the training effort: datum references, geometric controls, and control verification.

Datum reference frames. The idea of datum reference frames (DRF) has been around for a very long time. The old machinist's dictum to take hold of a part once and machine it all over is a example of this. Each time a component is taken out of a machine tool and placed into another setup (which requires another piece of tooling), variation—or error—is introduced into the process. Each additional method of locating the part establishes a new set of reference surfaces, that add to the variation of the part features and lead to tolerance stackups that play havoc with the functionality of the assembled product.

The DRFs that are specified as part of the product definition establish a unique location in space for the part, based upon its function in the assembly. For relatively simple parts such as cylinders or prismatic solids, the DRF can be explained readily in terms of the Cartesian coordinate system. As shown in Fig. 14-4, the object has six degrees of freedom in three-dimensional space. Three of these are the motions that it can have along any one of the coordinate axes. These linear motions are usually referred to as translations and, in creating or assembling the component, must usually be constrained. Additionally, there are three further motions, called rotations, that can occur about the coordinate axes. Again, in the fabrication or assembly of the product, these motions must be prevented.

The significance of the DRFs goes back to the old machinist's rule-of-thumb. The minimum level of variation inherent in the design occurs when a single DRF is identified from the baseline layout and used to manufacture the product. Additional methods (DRFs) that are used to locate the component during manufacture or inspection will introduce feature variations that can cause problems in achieving desired part function. With increasing demands on part quality and significant increases in machine precision, the fits that were previously designed into a product to account for such variation are no longer acceptable. The use of the single setup (a single DRF) is now sufficiently accepted to have become a marketing tool to sell machinery. While the single DRF is not always obtainable, it does provide a benchmark toward which the improvement team can work.

Geometric controls. Having identified the DRF that defines the component's position in the assembly, the functional features that form the elements of the component must be related to this DRF. This provides a description of the desired end product. The traditional shape and size descriptions do not provide the necessary three-dimensional description required to guide the product to physical reality. Only through refinement of the end product definition can the three-dimensional functionality be achieved; the only way to do this is by the use of the geometric controls contained in ANSI Y14.5M.

The implementation of these controls includes three distinct stages (see Fig. 14-10). The first stage requires identification of the DRF that places the component in its functional position in the next higher level of assembly. Having determined the functional datum features, the next step is to qualify and relate these datum features to each other. The qualification involves the specification of the necessary form and orientation controls that will limit variation due to positioning inaccuracies. Relating the features to each other will most likely involve the specification of orientation or positional controls that will limit possible kinematic variation when the part is located by the datum features.

The remaining step is to take each of the functional features contained in the product definition and relate them to the DRF. This is accomplished by using the controls that are allowed by the standard. It should be reinforced at this point that there are additional levels of refinement possible (the DIN 4760 levels of control). All of the decisions that are used to determine the precision with which the product must be defined should be based on part function. It is an axiom of structured design that none of the component definitions can be created until the baseline layout is completed.

Verification of controls. Implicit in the quantification of the product design is the presumption that the features will be verified. If the features are to be inspected, then consideration must be given to the design of the inspection process.

The first consideration in this verification is the feasibility of checking for the control that is specified in the product definition. It serves no purpose to specify controls or control levels that cannot be checked using available technology. Yet, it is not unusual to see controls called out on a drawing that are related to nonfunctional features, that currently available technology cannot verify, or that are specified at levels that obscure the results with large errors inherent in the measurement process.

To elaborate on these last items, the first and most egregious problem involves the specification of controls relating to nonfunctional features. The obvious solution to this is training of design personnel and implementation of the structured design methodology. The regimen introduced by these design procedures will provide adequate definition of the product based only on those features explicitly required to define the end product.

Another concern deals with the level of control that is specified. The numerical value that appears in the control is determined initially by the functional requirements of the design. Having made the appropriate determination of the control level, a significant element of the process definition includes the specification of the inspection techniques.

The measurement system is a process that needs definition and control just as the fabrication process does. Of particular concern is sufficient understanding of the system, so that the variability introduced by the measurement process is also identified and placed within the appropriate bounds. Most of the opportunities that exist—and much of the improvement that can be achieved—will be missed if measurements cannot be made with the desired level of accuracy.

To tolerance a part

- Establish the DRF
- Relate the Datum Features
- Relate features to DRF

Fig. 14-10 Stages to implement geometric controls. (*Courtesy A. Neumann, Technical Consultants, Inc.*)

CHAPTER 14

USE OF THE ANSI Y14.5M STANDARD IN PRODUCT DEFINITION

Identification of an Advocate

An impressive number of techniques—magic bullets—are vying for the attention of management. If these ideas are imported rather than generated internally, many of the methods that catch management attention will eventually fall on unfertile ground. The cultivation of these ideas and the resources that bring them to fruition need careful attention that can only be provided by a true believer. In the situation where the implementation of such methods is assigned randomly, there is little probability that anything positive will come of the effort. Too many negative experiences along these lines will create an unresponsive workforce.

To implement structured design and improvement methods, there must be a careful selection of the resources used, to introduce and integrate them into the existing product design cycle. These may be the most crucial decisions necessary to reap the rewards of this endeavor.

Choose an advocate. Since the use of these methods is not the norm in industry, the efficacy of applying them may not be self-evident. In extreme cases, there may be individuals with 30 years or more of design experience—probably with considerable success—using traditional design methodology where these "new" concepts were not used. The inertia preventing implementation of these changes can be overpowering. Sometimes, if an edict comes down from upper-level management, people will go through the motions for some period of time—until management becomes frustrated and gives up, or sufficient negative experience accrues to prove that the method is "no good."

At this point the realization is reached that a *behavioral* change is being asked for along with a change in design technique. This behavioral change will not occur by simply ordering it to happen. It must be fostered by example; to set the example, an advocate must be appointed.

The selection of an advocate is an extremely important, and probably difficult, decision. The individual chosen must be both a highly structured designer and a master politician with a high energy level. In the area of design expertise, the advocate must already have operable knowledge of the requisite techniques—the structured design techniques—that are being recommended. The benefits will not accrue if a technical novice is placed in charge of implementing changes in design procedures.

On another plane, the advocate must have the interpersonal skills necessary to interact with senior designers who must be convinced that what they have been doing throughout their professional lives is no longer acceptable in the new competitive world. This person must genuinely like people and enjoy educating them. Organizational and interpersonal skills are imperative because the whole process is highly structured. Individual islands of acceptance will not provide the benefits that are expected, and may well cause failure of the entire project.

Provide Advanced Training. Advanced training opportunities must be made available to the advocate, both in areas of small group dynamics and in advanced applications of geometric control. A team is being assembled to deal with the product development cycle from start to finish, from conception to delivery. The ultimate success of the product depends heavily on the cohesiveness of the team. Thus, the advocate must have working knowledge of small group dynamics in order to facilitate the desired change. In a small firm, the advocate may well be both project manager and facilitator all rolled up in one person. A larger organization may have a sufficiently large number of product teams that the advocate functions only as a facilitator, a change agent. Without the direct authority of the project manager to influence group behavior, many of the skills needed to accomplish the desired ends will probably have to be acquired through formal training; they must also be based on some inherent ability in this area.

Addressing the training needs in the technical area, there have been attempts to create a certification procedure covering knowledge of geometric control. One of the thoughts associated with this is to provide different levels of certification. The first level would possibly involve the ability to interpret the controls that have been placed in the product definition. After acquiring this experience, the individual would move to the next level involving the application of the controls to a design. The final step would be the ability to teach the material, a trainer's level. Choice of the advocate should involve someone with the application ability and experience who has the desire to reach the trainer's level. It would be impossible to facilitate the team's discussions and negotiations without the requisite technical knowledge to implement the concurrent engineering methods. The prime mover in this instance is the technical information necessary to provide the geometric description defining the product.

Management support. It should be obvious that the structured design and improvement process should be undertaken only with active management support. The fact that the change in timing within the expense stream now moves costs to the initial stages of design makes management tolerance of this front-loading of expenses imperative. Without tangible comprehensive management support the techniques should not be implemented. Without such a champion on the senior staff, the program implementation will take the form of a management edict that will not foster the desired behavioral and technical changes.

Mechanisms must be put in place to indicate the high level of management support that is given to these techniques, and also to make their implementation highly visible. One of the most obvious mechanisms would be to place the advocate in a separate position reporting to upper management. The most logical situation would be to vest either the project manager or a senior member of the project team with the authority to require the design technique changes from which the behavioral changes will follow. Both the resources that are devoted to filling such a position and the access that the position has to management would go far toward demonstrating the high level of commitment needed for success.

Other ways of supporting the effort, while less visible, include management participation in design reviews and audits, with the insistence that a complete product definition is required early in the design cycle. In the improvement area, the use of the complete definition as the benchmark against which changes may be analyzed and judged will reinforce the need for structured design.

A more formalized way of supporting implementation would be to create external parallels to the company's in-house thrusts, by requiring all contractors to become educated in the techniques of geometric design and continuous improvement. This can be done most readily by inserting requirements in vendor audits and, ultimately, as terms in purchasing contracts. In some sense, much of this is already in line with the philosophy found in many national and international standards. ISO 9000 would be a good example of standard-driven impetus to foster these concepts.

Controlled Implementation

Like all new things in life, success comes more readily if things are learned gradually. In a similar vein, the use of the product definition to foster structured design and continuous improvement

CHAPTER 14

USE OF THE ANSI Y14.5M STANDARD IN PRODUCT DEFINITION

has the greatest probability of success if it is introduced in a controlled manner. This operates at two levels.

At the organizational level, the existing department structure will work against wholesale introduction of these techniques. The current reward structure helps to foster parochial interests related to individual disciplines and makes it difficult for consensus to evolve. Without consensus, the function of the team will be impaired.

On the technical level, it would be a rare organization that has many individuals with facility in the use of geometric controls. Because of the power of the method, the underlying language is somewhat complicated and requires some effort to master. Traditional product design practice, achieving robustness in the product through larger tolerances, did not encourage such learning. Of even more importance, acceptance of the results of this traditional process—which no longer yields competitive quality levels—did much to convince current designers that their design approaches were good methods needing no change. The product quality picture now leads to reassessment of this stance. However, the reevaluation has not proceeded far enough to provide the level of knowledge required for wholesale organizational change. Not many product designers understand the limitations of their methods, or are ready for the change.

To overcome problems in these two arenas, introduction of the structured methods should start with a project of controlled magnitude. One possibility would be to introduce the methods as part of a continuous improvement project where an existing product offers opportunities for productivity and quality improvements. Preferably, the product should be one of low assembly complexity or a subassembly. The rationale supporting this deals with the possible need to alter production methods. This might include incurring additional tooling charges to effect changes in production processes or sequences. The smaller scale of the prototype project places a boundary on the possible expenses incurred to implement the improvements.

A further reason to keep the project scale small relates to the need to achieve some visible and quantifiable results before questions begin to arise about the efficacy of the project. Even in the improvement mode, the front-loading of expenses may make management advocates of the process begin to second-guess their support of the test. A manageable scope for the project reduces the timeframe and, of more importance, may allow prediction of the outcome with a higher level of confidence. The initial implementation of these ideas should not be unsuccessful. Even marginal levels of success may cause reevaluation and possible cancellation of the attempt. Choose the first battles carefully. This is to be a quiet revolution, that seeks to avoid major risks associated with more radical change.

Upgrade Metrology/Inspection Capabilities

The need for a quantifiable definition of the product upon which to base the improvement cycle has been emphasized. This assumes that once all of this information is provided, there exists some means to verify it. Without a high level of precision in the ability to verify the controls—to measure things—the ability to manufacture interchangeable product is reduced. The result is a reduction in the level of precision in the measurement process, and a necessary increase in the level of employee skills. An extreme case of lack of precision reverts to the craft method of producing products: an individual craftsman produces an individual product through handfitting of each component. No interchangeability exists in this situation. Elements of this occur in the instance where a product is inadequately specified and the manufacturing people must provide fixes—more definitions—to bring the product into existence.

A key element in the improvement process is the realization that verification is a complicated undertaking requiring the same amount of attention and planning that goes into setting the process for the rest of the manufacturing sequence. Many routings show the inspection process identified as a single step in the overall operations sequence, with little more than the one-word description "inspection" and a gage number called out. As the general level of precision is improved and this ability is incorporated into the inspection sequence, the process specification must not be left to the inspectors. If an individual inspector redefines the measuring process each time it is implemented, or there is variation in the techniques applied by different inspectors, then the advantage of a more precise product definition is lost—the wrong things are being verified. The inspection process must have a direct link to the product definition.

An important change in the thought process occurs with the realization that the measurement process is subject to variation just as the more traditionally defined manufacturing operations are. Once statistical variation of the measurement process is accepted, careful specification of the parameters governing the inspection setup is a necessity. It is imperative to realize that a significant portion of the allowable product tolerance can be used up by the measurement error introduced by variability in the inspection setup.

The importance of specifying and monitoring the inspection process can be illustrated by the inclusion of requirements relating to gage repeatability and reproducibility (commonly referred to as GRR) found in purchasing contracts. Three particular concerns relate to the need for careful specification and control of this process.

The first concern involves the accuracy of the measurement which is affected by two major sources, the product definition and inadequate specification. Any ambiguity in the specification of what is desired leaves decisions to the discretion of manufacturing and inspection. The results that follow from this latitude will likely have wider variation than is acceptable. A good, simple example as to why a complete product definition is required can be taken from the ANSI Y14.5M standard. Figure 14-11 describes the dimension origin symbol, a version of the datum reference frame. The center figure shows how the product description provides a direction to the line of measurement and guarantees that the desired information is taken. The product detail firmly establishes the reference point and the measured point. There is no discretion left to the inspector. The lower figure shows how an inadequate specification of the functional surface allows a different feature to be used for setup by the inspector. This violates the designer's intent. The result is a major error in measurement. This source of error can be avoided by controlling the location of the part through complete specification of a datum reference frame and appropriate related geometric controls.

The second concern is the measurement process itself. Even with complete specification of the desired end product, elements of the verification process must be fleshed out. The individual instruments have error inherent in their design and manufacture. In conjunction with the inherent instrument error, the interaction of the elements of the setup may cause variation. Once the operator is added to the equation, observational and manipulative error also must be considered.

Illustration of this second source of error can be shown readily by considering the inspection of the part shown in Fig. 14-12. With complete specification of a datum reference frame, the order

CHAPTER 14
USE OF THE ANSI Y14.5M STANDARD IN PRODUCT DEFINITION

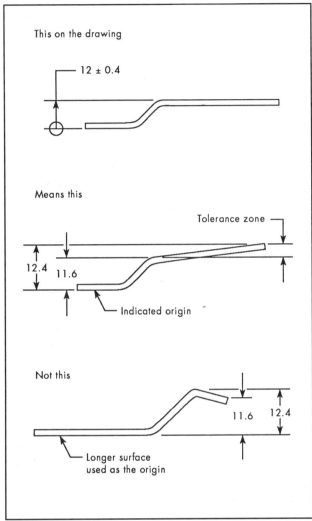

Fig. 14-11 Origin of dimension. (*Courtesy American Society of Mechanical Engineers*)

in which the datum features must be contacted to locate the part in space is rigidly specified. For a real part with surface variations rather than the designer's idea of the perfect surface feature, changes to this datum order (precedence) will change its location. This is shown in Fig. 14-13.

The third obvious, but critical, concern that affects the measurement process and the usefulness of its results includes such things as calibration technique and frequency, environmental controls for temperature and vibration, and training and development of inspection personnel. It cannot be overemphasized that the more complete the specification of the inspection process and the system within which it exists, the more the measurement error will be held to its minimum level—given the techniques that are being applied. Furthermore, this will enhance the stability of the system over time and allow the results of the improvement process to be more readily achieved and implemented.

Review and Critique

After completing the first project, time should be taken to review what has been accomplished. In a very real sense, this undertaking is an application of the continuous improvement process; this is the fourth stage of the Plan-Do-Check-Act (Shewhart) cycle. Because this was a project of limited scope used to prove the feasibility of the methodologies, no surprises should arise during the assessment. However, since each organization is unique, there is much that can be gleaned from the project to guide full implementation of the methods and procedures.

At the organizational level, the assessment will be used to justify continuing with the implementation. This is a business decision that will require expression of the project's results in monetary terms. Because many of the economic benefits that accrue from these techniques are measured in "dollars not spent," activity-based accounting may be one of the areas that needs to be developed simultaneously with instituting continuous improvement. It would be an unusual firm whose executives would rely solely on qualitative arguments to approve the expenditure of resources and allow the training and organizational restructuring necessary for full implementation.

Audits of the results should be performed on both the technical and the human resource aspects of the prototype process. In the technical area, the audit would focus on the actual documentation of the product and process definition. It can be readily seen whether or not the geometric controls that drive the design are being provided as integral elements of the definition. The source of the controls should be noted, and the stage of the design cycle when they are determined should be ascertained. A more detailed look at the definition would involve the level of deviation from the idealized design as it relates to the process description. In particular, it was noted that in a perfect world, a single DRF would be used in describing and fabricating each component. If this is not true for components selected for the audit, there should be documented justification for the departures and analysis of the effects that are to be expected in the downstream processes.

Another method of auditing the results of the product's technical specification would involve reverse engineering of a component that mates with a particular part. A significant amount of information concerning the functional characteristics of the mating component is necessary to describe any part. By assembling the information from the sample product definition, the auditor should be able to describe the functional design elements of the mate. If this can be done, then the original product definition was done adequately. If the mating part cannot be described functionally and, to some degree, geometrically, then the original definition does not contain sufficient description to produce it. This technique can also be used to identify subjects for continuous improvement efforts.

In the human resource area, the audit should focus on successful resolution of the issues of training and placement. Members of the concurrent engineering team are key players in the success or failure of the projects. The auditors should determine if the appropriate people were placed on these teams and if these individuals were provided with the necessary resources, including the appropriate training, to accomplish their assigned tasks. Not all individuals are able or willing to subordinate years of parochial interests to the needs of the product team. If such individuals are present, this should be recognized and alternate members brought on board.

One last major area to be audited is project management. Two distinct concerns are involved here. The first of these deals with the structure that management applies to the project. There is no clearly definitive way that this should be done. A variety of methods are suggested by management consultants, ranging from the traditional functional structure, to matrix management, to a

CHAPTER 14

USE OF THE ANSI Y14.5M STANDARD IN PRODUCT DEFINITION

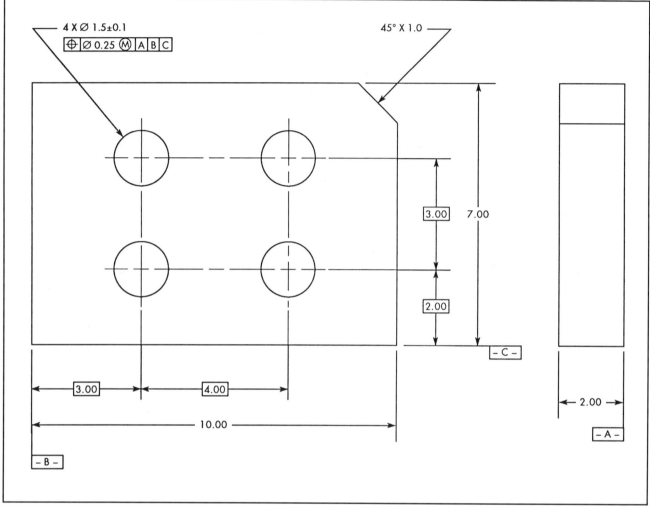

Fig. 14-12 Detail showing datums and datum precedence in position callout.

separate project team containing all the necessary functional personnel. The object of this audit is to determine if the specific structure that was applied to the prototype project has worked. The guidelines used in this determination are not whether they follow what purports to be current management tenets, but whether they *work*. Even if the techniques do not follow the latest fads, the functional results should be looked at by the audit team as the deciding factor. In one sense, the ends justify the means in this type of situation.

From the personnel standpoint, the team leader and the advocate should be given performance reviews. Assuming that these individuals were selected on the basis of their technical competence, the review should be directed toward the success of their efforts in coordinating the activities of the team's functional members and the interactions with other functional groups. Assessment items of interest would include the leader's ability to foster active involvement by team members. The review would also focus on the training aspects of the project cycle. This last concern is important because the team members are learning problem-solving skills that must be generalized for application to other products. The success of this enterprise is demonstrated by behavioral changes in the members of the team.

Expand Training

After successfully completing the audit, the next stage is to expand the use of concurrent engineering and continuous improvement. To institute expanded use, training must be provided in the use of the technical tools that support concurrent engineering and in the area of team building to facilitate the continuous improvement process.

Training personnel in the use of geometric control techniques should proceed in two phases. The first involves a basic introduction into the geometric control system, allowing the individuals to interpret existing product definitions. While the degree of application is at a low level, this is still a complicated undertaking involving significant interactions between the geometric language, underlying design concepts, and the manufacturing and verification processes. This introduction must create a sensitivity to the interdependence of all stages of the product cycle before an adequate understanding can be developed. The second level of training will provide necessary experience in using the control concepts to give definition in specific product applications. It is important to understand that the standard does not provide detailed guidelines that can be extracted directly from the printed document and thrown on the drawing. Each application is unique and

CHAPTER 14

USE OF THE ANSI Y14.5M STANDARD IN PRODUCT DEFINITION

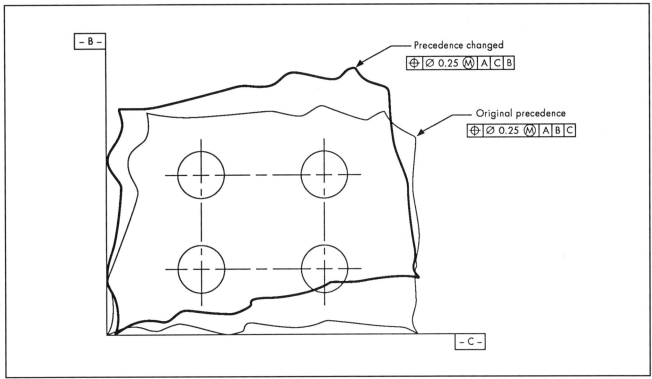

Fig. 14-13 Part from Fig. 14-12 showing effects of changes in datum precedence.

requires a degree of intellectual effort—thinking—that cannot be avoided. While it would be nice to buy a set of specific instructions with the standard, the designers must develop their own application guidelines. Thus, the second level of training is really an ongoing effort that continues as long as the firm develops new applications or products.

This last point contains additional import for continuing the application of structured design and improvement. The training efforts must take place within the context of actual projects. Little of value will be learned by formal training that utilizes artificial examples that are not relevant to a company's product line. The nature of the techniques requires learning by doing, since the design process is unique to each product. The methods cannot be taught solely by classroom communication, but must take place in the context of productive design projects. Concrete examples of the company's products are needed to create the specific techniques and supporting organization for successful product development.

A core group of trained individuals will be needed to expand the use of the methodologies beyond the experimental stage. These individuals must be qualified at the application level of structured design so that they may serve as facilitators as additional teams are created. Introduction to the techniques can be accomplished by including new individuals in the continuous improvement cycle. At this level, the new members can begin to contribute in a positive way without having complete knowledge of the methods that are used. This is an important point. The more rapidly that people can see the benefits of the techniques, the more likely it is that they will continue with the training efforts that will bring them to a high level of competence. The desired outcome of the whole exercise is the development of creative problem solving skills that can be applied to the design and improvement of a large variety of products. This cannot be achieved by management fiat. It must be reached by the individual as a personal goal.

Require Use

Under the guise of further training, the use of structured design concepts can be expanded to include additional projects and the standard procedures supporting the continuous improvement of existing product lines. The desired outcome of all of this effort is the required use of these techniques in every stage of the product design cycle.

Effective company-wide use of these techniques can be accomplished directly through implementation of the continuous improvement process. Any of the goals and behavioral changes that are necessary for successful implementation of continuous improvement are directly fostered by structured design based on the product definition. This emphasis on the definition fits in directly with the Shewhart cycle. The product definition provides the standard on which to base further improvement. What is now required is continuing motivation by management to keep these concepts and techniques alive. Constant attention to the activities and the results of the processes is needed to reach the level of mastery.

SUMMARY

While many products have attributes that are derived from creative design elements, most products have a geometric form that takes these design concepts and translates them into physical reality. The methods presented are useful in determining the functional elements of part geometry needed for efficient and economical production of the product. Furthermore, the methods seek to provide this information in a timely manner that assures efficient economical production. Consideration of this informa-

CHAPTER 14

SHOP FLOOR MANAGEMENT

tion at appropriate times in the product cycle will identify the causes of product variation. This allows prediction of the effects of variation on product function and correction of the design where appropriate. At a minimum, knowledge of this variation will quantify elements of risk in the design so that intelligent business decisions can be made.

The product definition provides a set of numbers that allow technical solutions to be found based on objective methods rather than visceral responses. Getting to the numbers requires both a team structure in the firm and a common language with which members of the team communicate. This language is necessary from an engineering standpoint to provide the basic dimensions required to achieve part function and also to determine how much variation can be tolerated in these values. The language and the methodology tie the design and manufacturing processes together, formally eliminating the artificial barriers that segregated these functional areas in the past.

The design methodologies work well in the context of teams. In fact, it is difficult if not impossible to enforce the structured design concept where concurrent engineering teams do not exist. The formal extension of the product definition and structured design into the area of continuous improvement is a logical next step, since improvement techniques require tight communication between large numbers of individuals in many functional areas.

Since the basis of continuous improvement is the establishment of a standard to start the process, the product definition provides a logical benchmark for these activities. In fact, much of the methodology in this article takes ideas from the improvement process and incorporates them into the early stages of the product cycle. Continuous improvement is effectively built into each phase of the design cycle. The emphasis on teams to create the product definition and to facilitate the improvement process provides an organizational structure that is mutually supporting. The elements found in continuous improvement activities are the same ones that are necessary to implement structured design.

The key to understanding structured design and improvement is the simple fact that all of the information that these techniques generate will eventually be identified and documented for a successful product. If the information is identified and documented early in the design cycle, the greatest economic benefits will be achieved. If the information is generated by trial-and-error during the later stages of the cycle, the price is high in lost market opportunities and avoidable expenses. The worst case would be where these components of the product definition are not identified. The resulting variability of the product will lead to its failure in the marketplace because the customer does not get the desired value. If it is not done correctly, the rewards of a successful product cannot be reaped.

SHOP FLOOR MANAGEMENT

If a manufacturing plan has been generated correctly, the execution of the plan should go smoothly. Three key issues should be considered in the planning phase:

1. Capacity.
2. Inventory.
3. Desired due-date performance.

If these issues are not addressed in planning, the execution phase will be nothing more than a continual series of fighting fires while productivity wanes. Productivity is defined as meeting or exceeding the operational parameters derived from the business goals. These operational parameters typically are throughput, inventory, operational expenses, lead time, and due-date performance.

Assuming even the most effective plan is in place, that plan is still subject to two primary agents of change:

1. The dynamics of the shop itself.
2. The volatility of demand.

These two forces are at work constantly and will unravel the best laid plans. The dynamics of the internal force means the shop may not behave the way the plan said it would, and the volatility of the external force means the customer will change his or her mind (probably with accelerating frequency and shortened order lead times).

In the execution stage or in managing the shop floor, both the internal and external forces must be factored into the equation for the end result of improving business performance and customer satisfaction. To reach optimal business performance, the shop floor must implement continuous improvement.

Basic operational parameters require the top-of-mind awareness of the professional charged with shop floor management. These parameters are:

1. The throughput of the production system.
2. Work-in-process (WIP) inventory on the shop floor.
3. Operating expense.
4. Production lead times and order lead times.
5. Due-date performance.

The shop floor manager must continually ask if improvements are being made in these areas.

GLOBAL VERSUS LOCAL

Many approaches attack the symptoms rather than the problem, which is inherently ineffective in finding a working solution and improving overall plant performance and productivity. Solving intermittent symptoms—such as a growing pile of material in front of one machine on the shop floor—will not address the reason the pile has accumulated. It may not be a constraint or bottleneck; it may be a result of the way that line is managed. Taking a step back and examining the problem globally instead of locally will not only solve the immediate problem of the overstocked pile, but it may solve it so that it does not occur again, with the exception of an unplanned event.

Too often, managers are concerned with optimizing the performance of all the departments in the plant. But by not recognizing overall constraints, the plant's performance will not be improved. The sum of local optimums is not the global optimum, especially in shop floor management. A global perspective must be implemented in order to see improvements in bottom-line performance.

CHAPTER 14

SHOP FLOOR MANAGEMENT

Suppose there are seven departments or areas in a plant: A, B, C, D, E, F, and G, and the constraint lies in area E. The shop floor manager can take two approaches:

1. All areas A, B, C, D, E, F, and G can be optimized, which usually will result in high inventories, long lead times, and increased cost; or
2. E alone can be optimized, and all others subordinately optimized in concert with E, which results in better performance, decreased lead times, and lower costs (see Fig. 14-14).

The latter option designs a solution from the point of view of the constraint, a global perspective. The former option assumes that if each part is optimized the whole will benefit, a local perspective. Such an assumption is simply not valid for anything other than a single stage manufacturing process or an environment where shop dynamics and market volatility do not exist.

CONSTRAINT-BASED MANAGEMENT

A global perspective looks at the system with constraints as its point of reference. The concept of constraint-based management produces continuous improvements in shop floor management. The first step for continuous improvement lies in identifying the constraints. The constraints most often impacting the shop floor are

1. Capacity constraints.
2. Sequence-dependent constraint.
3. Policy or measurement constraints.

Capacity constraints are resources (people, skills, money, machinery, tools, fixtures, or materials) that are limiting throughput of the production system. For example, in Fig. 14-15a, the primary constraint in Line 1 dictates the capacity of the entire production system even though Line 2, Line 3, and Assembly can outproduce Line 1.

Sequence-dependent constraints could become capacity constraints if ignored. For example, say a particular machine that makes gears is in Line 2 and while developing the schedule for the primary constraint in Line 1, the gear pitch is ignored. This could cause many more setups than originally planned and, as a result, the gear-making machine in Line 2 may become the primary constraint in the production system.

Policy and measurement constraints often dictate batch size and period requirements. To minimize setups, many systems use an approach that combines a certain number of future period requirements with current period requirements. Policy constraints also address procedures to combine or split orders. Measurements, particularly those applied on the shop floor, can generate constraints, build up inventory, and lengthen production lead times.

Every manufacturing environment has evolved over time and what may have once been valid may no longer be valid, with the dynamics of the shop floor in continuous flux and with today's volatile market demands.

For example, a particular manufacturer experienced a policy constraint relating to manufacturing lot size (see Fig. 14-16). The current month (February) had an assembly requirement of seven pieces for part number 73121-2. The policy of combining forward

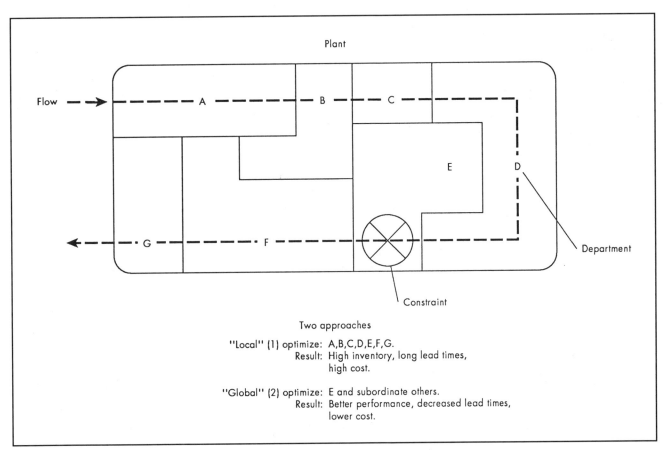

Fig. 14-14 Two approaches to flow optimization.

CHAPTER 14

SHOP FLOOR MANAGEMENT

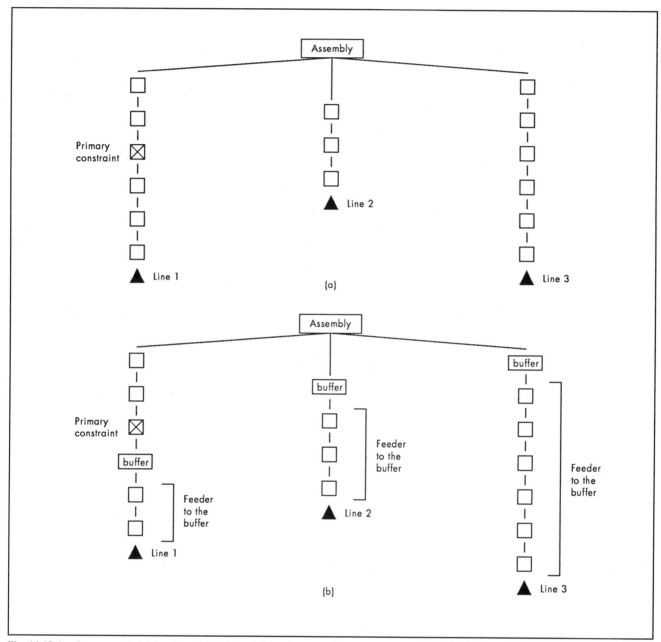

Fig. 14-15 A primary constraint in a line limits flow of the entire operation even when other lines can outproduce it, *a*. Addition of buffers, *b*, can alleviate this problem.

periods resulted in 210 pieces being made. The requirements were seven pieces for February, 80 pieces for March, and 123 pieces for April. Meanwhile, while this area was making 203 extra parts, work was being neglected for other part numbers that had February requirements on the same machine. The result of the policy combining forward periods (to reduce setups to reduce cost): the production line was five weeks past due.

MANAGING CONSTRAINTS

After the first step has been taken in continuous improvement by identifying the constraints, those constraints must then be managed to maximize effectiveness. Capacity problems can be addressed by off loading, working overtime, or sending part of the requirements to be produced elsewhere.

The sequence-dependent constraint can be addressed by developing schedules at the primary constraint that do not violate the sequence requirements of those constraints. If the system becomes too complex and has multiple, sequence-dependent constraints, the situation may require "decoupling" inventory or capacity.

And finally, policy or measurement constraints may be addressed by asking if the policy is valid for the chosen operating philosophy and methodology. For example, if the operating philosophy is "lean manufacturing," then policies and measurements that build inventory should be examined. Measurements

CHAPTER 14

SHOP FLOOR MANAGEMENT

> February assembly requirements: 7 pieces of part #73121-2
> Manufacturing work order quantity: 210 pieces for part #73121-2
> Actual requirements for part #73121-2: 7 pieces in February
> 80 pieces in March
> 123 pieces in April
> The manufacturer had a policy in place to combine forward periods, even if other needed part numbers were being bumped off the machine producing unnecessary (at the time) parts.
> Result: The production line was five weeks past due.

Fig. 14-16 Impact of policy constraint.

and policies must be consistent with the operating philosophy. Many policies and measurements can be changed without creating new constraints.

UNDERSTANDING CONSTRAINT BEHAVIOR

Constraints should dictate the behavior of all other parts of the system. The shop floor manager should run all resources with regard for and in concert with all known constraints. If the constraints are ignored and not made an integral part of the management analysis, the execution will suffer.

Continuous improvement tells the shop floor manager to carefully watch and analyze the behavior of constraints. Are they being managed properly? Are they being fully exploited? Constraint management is characterized by real-life situations. For example, the shop floor manager is charged with assuring that all resources are manned during lunch, that all lines continue to run during shift changes and, even more complex, that maintenance personnel are responding on a priority basis to the critical areas that dictate optimal productivity. Providing for quick changeover and improving overall machine reliability are just two prerequisites that result in shop floor productivity improvement. For every 1% improvement in output of the primary constraint resource, there is a 1% improvement in the throughput of the system. The potential benefits of even small, focused improvements, that is, constraint-based management, are substantial.

The identification and strategic management of constraints will show the shop floor manager where maximum benefit can be achieved. Another approach to providing for maximum productivity at the shop floor is to introduce "shock absorbers" so when the rubber meets the road, the ride is smooth no matter the pot holes or hazards.

AN IMPROVED APPROACH: BUFFER MANAGEMENT

Often, solutions are contrived that require expensive, ultra-automated equipment in an integrated, close-coupled system that experiences more downtime due to malfunction and breakage than productive uptime. These systems exhibit "brittleness" because they are closely coupled and have little or no shock absorbers. The high capital investment in these solutions almost always forces decisions that remove the elements of flexibility. There are solutions that require little capital investment and more rethinking of traditional manufacturing and shop floor management practices.

The manufacturing environment is inherently subject to internal dynamics and external volatility. In order to absorb the shock to the system from these forces, buffers that act as shock absorbers can be introduced at strategic points in the system. The primary forces impacting shop floor productivity are capacity and inventory. Traditional solutions have been to maintain a capacity level at the required load, and large inventories of finished goods and work-in-process.

Pure inventory and capacity solutions (that is, maintaining inventory and capacity substantially in excess of load) are not as effective as in the past. The market has changed in its actions and demands. This volatility directly impacts shop floor management and the way its operations are run. Maintaining large inventories was a viable option when the market was stable and large order lead times were acceptable to customers. The excess capacity approach worked well with market volatility, but at a cost of maintaining capacity far above load requirements.

The answer for today's manufacturing environment is a strategically managed, critical mix of capacity and inventory.

The concept of buffer management is to provide strategic shock absorbers. One buffer is maintenance of strategic inventory to protect system throughput. Buffers are placed at strategic points on the shop floor (see Fig. 14-15b). These strategic points may be in front of constraints, in front of assembly, or at other convergence points. Buffers can be designed in size to protect constraints even if the feeder areas into the constraint are experiencing disruptions, etc. Buffer management also aids in the identification of process improvements in the feeder areas. If certain parts are arriving late, the shop floor manager is alerted that a particular feeder area needs process improvements. Overall, buffers protect and enhance the throughput of the system.

In addition to buffers (which are strategic inventory), there needs to be "protective" capacity. This is capacity, in the feeders to the constraints and in the areas downstream from the constraints, that is above the load requirements to absorb the dynamics in those areas. This is a level of capacity that will prevent these resources from becoming constraints. The more constraints in the system, the more the system will be difficult to manage and the more it will show performance deterioration. The design of protective capacity is therefore important in determining how many constraints there will be in the production system.

Improvements relating to protective capacity and how much to maintain will depend upon the system itself. When evaluation and improvements of an unstable system first begin, the resident protective capacity needs to be high. As the system becomes more stable, throughput and due-date performance will improve. At this stage, the shop floor manager should consider "ratcheting" down the size of the buffers and the protective capacity. Before this, factors such as large batch sizes need to be eliminated or reduced to improve flow and reduce production lead times.

Indications that the system is becoming more reliable are that production shock waves subside and decrease in intensity. Also, the need for excess capacity will be lessened.

RE-CORRECTING TO STAY ON COURSE

Once the buffer inventories and protective capacities have been sufficiently reduced, there will be interruptions in throughput. At this stage, a second cycle of improvement must be implemented. Constraints may have changed, and strategic buffers may have to be moved or re-sized accordingly.

In the ongoing cycle of continuous improvement on the shop floor, the shop floor manager constantly will be reducing buffer inventory and resident protective capacity. As a system becomes more stable, these shock absorbers are reduced. A critical point exists in knowing the balance between an effectively buffered system and one that is too brittle. A brittle system will be indicated by missing dates on orders and lengthening response time to the

CHAPTER 14

SHOP FLOOR MANAGEMENT

market. The gap lengthens between the date of customer demand and delivery on that demand.

A reliable system will always have two basic characteristics:

1. It will behave the way it is predicted to behave.
2. Normal market volatility and shop dynamics will be absorbed.

Normal volatility does not account for unplanned peaks in demand, such as a major competitor going on strike or customers building a strike hedge. It is impossible to factor in every possible nuance and force that will impact the system. Even if it were possible, it would not be desirable because such a system would not be cost effective.

Two critical areas to watch in keeping tabs on system reliability are

1. Lead time to market, which applies to overall response time and is what the customer sees.
2. Production lead time, which is the time in the shop.

A good manufacturing plan should have certain parameters stated. These should include: throughput, inventory, operating cost, lead time, and due-date performance. For example, overtime should occur primarily at the constraints in a reliable system. If the initial manufacturing plan has been organized around constraint-based management with a global rather than local perspective, shop floor management will be inherently more productive.

How does the shop floor manager develop a process of continuous improvement for increased reliability, performance, and productivity? It requires a method of attacking the diseases instead of the symptoms. Attacking the disease through constraint-based management of a system has proven successful in today's manufacturing environments. It has proven even more successful when the concept is properly applied through software as a tool in the plant environment. Software solutions require a low capital investment and have a high rate of return which makes them a highly justifiable means to implement planning and scheduling technology. Software can identify constraints automatically and execute schedules by addressing the problem globally while comprehending shop dynamics and the market volatility impacting plant performance and productivity. Broader planning software solutions lay out the big picture and generate a more comprehensive, longer-term plan. Scheduling solutions are more attuned to the subtle, day-to-day or minute-to-minute changes that impact the flow of materials from beginning to end of production.

For optimal productivity the plan should be as close to reality as possible. Good planning and scheduling software aids in achieving this mechanically and should know intelligently where and when to make mid-course corrections.

PLANES, TRAINS AND AUTOMOBILES

The analogy of a voyage has been used to describe the manufacturing process. The plan is synonymous with a flight plan that a pilot must file before takeoff. The destination, load, type of aircraft, wind speed and direction, and other elements are all factored in developing the flight plan before takeoff. Once in the air, things may change from point A to point B. The scheduling or execution of the manufacturing plan reflects changes that must be made in the air in order reach point B at a particular time. Scheduling makes short-term corrections such as moving the joystick slightly to adjust for changing wind direction. Because a plan's granularity is much broader, it is not best equipped to detect such subtle changes. It is in this scheduling or execution stage that the shop floor manager and other personnel are critical. The human element in the manufacturing environment can be the greatest asset (or greatest liability). The mid-course corrections often depend on the human ability to think and reason. Humans are able to make decisions with incomplete and conflicting data. No existing technology is able to duplicate that with equal results.

But technology does exist to bring the intimate details of manufacturing planning and scheduling closer to the reality of daily and hourly shop floor management challenges. While philosophy precedes technology, implementing constraint-based management along with buffer management and protective capacity makes theory a reality on the shop floor resulting in increased performance and productivity.

AUTOMATIC IDENTIFICATION

A prerequisite for any continuous improvement program is availability of real time accurate information. In order to continuously improve any process, the decision maker or process controller must know the status of that process at all times. Automatic identification technologies provide a means of delivering fast, accurate information.

BAR CODING

Bar coding has been most widely recognized for its use in the retail grocery business. There it provides the retailer with fast, accurate information about sales. Manufacturers are learning that the same basic technology can provide them similar benefits with far greater rewards. The universal product code (UPC) is only universal to the retail grocery trade. Manufacturers have adopted different symbologies (arrangements of bars and spaces) that provide more information than UPC. Currently, the most popular symbology in industry is Code 39, primarily because the automobile industry and the Department of Defense have mandated that all parts shipped to them be labeled with bar codes, using Code 39.

While originally bar coding was dictated by customers, manufacturers were quick to realize the benefits of using bar codes internally. Manual entry of data, on average, will yield one error per 300 entries. This is true for both handwriting and keyboard entry. Bar coding, on the other hand, yields only one error per 3 million entries. In comparing speed, manual entry varies from one to two characters per second, compared to 10 to 30 characters per second for bar code readers.

Quality Control

One of the most obvious applications of bar coding as it relates to continuous improvement (CI) is in quality control. In order to control quality, there must be fast, accurate information about that quality. Using bar codes in quality control involves applying bar codes to parts (or containers) and providing bar code readers, along

CHAPTER 14

AUTOMATIC IDENTIFICATION

with menus listing defects, to workers who are monitoring quality. When a defect is noted, the part and the defect code are read by the worker. When the bar code reader is stationary, the data can be transmitted using "hard wire" to the host computer. In circumstances where the reader must be moved about, Radio Frequency Data Communications (RFDC) can be used to transmit the data in real time. A less expensive, but less timely method is to use portable data collection devices that are periodically connected to a terminal; the information is downloaded in a batch fashion. The accuracy of bar coding is maintained, but the advantage of real time information is lost.

Work In Progress

Another fertile application for bar coding related to continuous tracking is Work In Progress. In this application, specific products, containers, or orders are labeled with a bar code. As the product works its way through the system, each operator (human or automatic) is provided with a bar code reader. As the product comes into a work area, the operator "WIPS" it in by scanning the bar code. When the work is completed, the operator once more scans the bar code to "WIP" it out of that work area. The ability to find any order and its progress at any time at any place in the factory, instantly, can provide a tremendous competitive edge for a manufacturer. In addition, a continuous record is made of each order processed. This record includes the operators, the machines, the times, and any other needed information about any order. The potential to improve processes with this kind of information available is great.

VOICE RECOGNITION

In order to provide continuous improvement, inspection will still be required in some circumstances. To find the cause of defects, there must be some means of recording the time, place, and nature of the defects. Inspection in nearly all instances requires the use of the eyes and frequently, the use of the hands. Using manual data entry (keyboard, handwriting, or even bar code scanning) requires the operator to look away from the work to record necessary information. Voice recognition, while still limited in vocabulary, does very well when the data (words spoken) is finite and reasonably small. Recording of defects by an inspector is therefore a natural application for Voice Recognition to continuous improvement. The inspector typically speaks into a microphone on a headset, identifying the part number, the nature of the defect, the location of the defect, and any other information that may be required. The voice recognition system is connected to a computer that translates the data into computer-recognizable format. That information can be stored for later analysis, but more likely will be sorted and fed back to the process controller so that corrective action can take place immediately.

RADIO FREQUENCY IDENTIFICATION

Radio frequency identification is a relative newcomer to the manufacturing world. The technology involves the use of three basic components. The first is an integrated circuit chip with a tuned coil which acts as an antenna. This is referred to as the tag. The second component is a reader, a computer controlled device that sends, receives, and modulates data to and from the third component, the antenna. The antenna, when energized by the reader, sends out radio frequency (RF) waves which are picked up by the tuned coil on the tag. The coil uses the minute energy received to trigger the attached integrated circuit. The data, typically a number, recorded on the IC in turn, modifies the frequency of the tuned coil just enough so that the RF energy which is reflected back to the antenna can carry with it the data contained in the IC.

The RFID system is then similar to bar coding in its application to continuous improvement with two major exceptions. First, the data contained on the tag can be picked without "line of sight." This is important when the environment is dirty, the part is painted, walls or other obstructions lie between the part and the reader, or when reading distances of greater than 10' (3 m) are required. The second major difference is that RFID does not require operator intervention. In some very rigid environments this is possible with bar codes, however it is always possible with RFID. It is also possible to carry much more information in a small space using RFID as compared to bar codes.

RFID has application for continuous improvement beyond quality control and work in progress, where it is used in much the same way that bar codes are used. In order to control indirect cost such as materials handling, tags can be placed on fork lift trucks and other materials-moving devices. With readers strategically placed throughout the plant, there is a continuous record of activity. "Active" RFID tags which are battery powered and can be updated by the reader are now being placed on parts as they go through the manufacturing process. At every process, the time, operator, machine ID, and any other pertinent data can be added to the tag. At the end of all processing the tag is removed and all the information about how that specific part was made is permanently recorded. This sort of process tracking can be invaluable when field reports come in showing part failure. The continuous improvement system can have a database which contains all information about each part so that problems in processing can be quickly identified and isolated.

ELECTRONIC DATA INTERCHANGE

While Electronic Data Interchange (EDI) is technically not an Automatic Identification technology, it is closely related. Quite simply, EDI involves the transmission of data from one computer to another using hard wires or telephone lines. This transmission can take place between computers within one company or between computers at different companies or divisions. One major task to be accomplished in CI is the elimination of redundant, non value added activities. An example of that kind of activity is the use of a keyboard to enter data into a computer when that data already exists, somewhere, in computer-readable format. It has been estimated that up to 50% of data entered via keyboard is from a document that was generated by a computer.

Linking computers requires physical links as well as software compatibility. The latter issue, software compatibility, is not minor, particularly when the two computers belong to two companies. Several professional organizations and associations are beginning to address this issue. Once resolved however, the cost reduction and quality improvement opportunities are great. When data is entered into a computer once, it will not have to be entered again. This will reduce the expense of the data entry operators, but more importantly reduce the errors which are typically made in data entry.

Some major manufacturers are now requiring that their vendors be capable of receiving purchase orders and transmitting billing information using EDI. As organizations move towards JIT in order to meet continuous improvement goals, this elimination of two to three days from the order entry cycle is significant.

CHAPTER 14

SIMULATION AND WASTE ELIMINATION

RADIO FREQUENCY DATA COMMUNICATIONS

While not an Automatic Identification technology, Radio Frequency Data Communications is closely related. RFDC allows for the transmission of data from remote points using radio frequency waves. A major requirement for continuous improvement is having information in real time in order to facilitate decision making. Unfortunately, the products and materials about which information is required are frequently moving in an unpredictable pattern, or are not located near a computer. In these instances, some sort of portable data collection device such as a bar code reader must go to the product or process. Without RFDC the data is gathered as the operator moves about the facility and is stored in the collection device. Then, on some regular basis, the device is connected to a computer and data downloaded. This is not real time information. The information may be received too late to take timely action. Using RFDC, the portable data collection device can transmit the data as it is gathered to a central location. The information is available to the decision maker in real time so that timely decisions can be made.

SUMMARY

Automatic identification and its related technologies have been in use for several years and have proven to be reliable. Their major contribution to continuous improvement is to provide fast, accurate information to decision makers and process controllers. As companies implement continuous improvement, these important technologies must be considered.

SIMULATION AND WASTE ELIMINATION

The heightened awareness of potential benefits stemming from the use of just-in-time (JIT) production techniques for manufacturing has reached unprecedented levels. The JIT bandwagon has started its American journey and companies are going to extreme lengths to apply the principles and become involved. Those exposed to the approach are immediately confronted with the cornerstone of the concept—Waste Elimination. While the thought of waste elimination is quite appealing, there is often confusion as to what exactly are wasteful or non-value adding activities, or conversely value adding activities.

Even beyond that is the problem of implementing commonsense corrective actions which are labeled by cost accounting principles as "poor returns" or "nonjustifiable." Measurement conflicts abound in currently practiced conventional thinking, making it difficult for JIT to get a fair review or opportunity for application. At the very minimum, the focus must be turned:

- From a "static" operating cost approach which views manufacturing costs as relatively undynamic, and tacks on a desired profit margin to derive a selling price of the product for the marketplace—selling price = cost + profit.
- To a "dynamic" operating cost which acknowledges that a majority of manufacturing costs are extremely dynamic and can be better controlled, while the selling price of the product is determined by the marketplace and competition, causing profit to become a variable based on the level of control exerted to maintain or minimize manufacturing costs—profit = selling price−cost.

Additional concern is raised when a review of trends in product cost breakdown shows a strong shift from predominant labor costs (in past years) to a higher material cost content (in recent years). The question can be asked, "Has performance measurement been modified to reflect changes in cost structure, so that the proper incentive for cost control is achieved?"

While this section will not attempt to address the issues surrounding overdue cost accounting changes which must be made to allow JIT to prosper in America as it has abroad, it will focus on the use of simulation models which could demonstrate—either visually, graphically, mechanically, or otherwise—the need for significant changes to occur in the form of waste elimination from manufacturing operations.

To comprehend how simulation might be able to assist in explaining why change must occur, it is important to understand the approach. Simulation does not necessarily have to be computer driven. Numerous representations can be developed from the use of common everyday objects and a little creativity. For example, it might be desired to develop a scenario which depicts the variables in the manufacturing process involved with part movement through an interconnected series of operations. To get general points across, the items listed in Table 14-1 could work into such a simulation.

The point is that simulation represents a process or set of activities developed within a model, computerized or not. Certain assumptions surround the extent of change that can occur in input, processing, and output, which should reflect the scenario being represented. The success of the simulation depends on the realistic capturing of elements surrounding a given situation. Models that provide such capabilities can be extremely valuable to support and provide insight toward problem resolution, or in the case of manufacturing—waste elimination.

On the other hand, there may be a need to utilize computer driven simulations for a variety of reasons. The price range and capability of simulation packages vary almost as extensively as the scenarios they are simulating. The bottom line determination for choosing between computer driven simulation techniques or

TABLE 14-1
Common Objects for Use with Simulation

Common Object	Purpose
Dice	Variable output from production resources (parts per shift)
Altered Dice	Impact of reduced variability from preventive maintenance program (change 1→2, 6→5)
Poker Chips	Production inventory
Legos/Blocks	Represent subassemblies/components
Playing cards	Some form of variability—that is, customer orders, absenteeism, machine downtime, etc.

CHAPTER 14

SIMULATION AND WASTE ELIMINATION

reasonable facsimiles thereof, will probably be found within the following list of factors:

- Available budget: are funds available to cover the needed simulation application?
- Skills of simulators: are the personnel chosen to use the tool qualified, educated, and trained properly to use the simulation?
- Available computer hardware: what computers, printers, plotters, terminals, communications, etc., can be used and are available?
- Compatibility of chosen application to existing hardware: will it run on what exists or does more hardware have to be purchased?
- Sheer volume of the task: is a computer really needed, or will a calculator suffice for control of inputs, processing, and output?
- Need for iterative trials: how many times is the application to be run?
- Repeatability, predictability, consistency: the need to have these characteristics might lean toward a computer driven application.
- Ability to replicate a situation: will common objects suit the purpose or are more sophisticated techniques required?
- Quantity and type of data needed: is a computer needed to facilitate input to database, processing, and resulting output reports?
- Desired turnaround of results: speed of result delivery might lean toward use of computer, depending on the application.
- Ability to replicate activities: if this purpose is not served the results are meaningless—choose techniques accordingly.

WASTE/NON-VALUE ELIMINATION: MODEL OPPORTUNITIES FOR IMPROVEMENT

Perhaps the best way to initiate the process of looking at where or how waste can be eliminated, is by reviewing generic definitions which identify wasteful and non-value adding activities. Wasteful activities can be described as those involved with the manufacturing process that can be eliminated or improved upon immediately, with no noticeable disruptions or impairments to productivity in the areas of cost, quality, or delivery. They include:

- Stockpiling of material (just-in-case).
- Meaningless material movement.
- Unclean working conditions.
- Waiting with no "improvement oriented" use of idle time.

Non-value adding activities are those activities that add no value to the product but must be performed under current operating conditions. To obtain improvements in cost, quality, or delivery, modifications to the manufacturing process need to occur. Examples include:

- Receiving parts at a location other than the destination.
- Unnecessary material movement and handling activities, due to poor layout of facilities.
- Inspecting the part to scrap or rework rather than the process.
- Undoing excessive packaging from supplier, to make parts available for production.
- Breaking bulk packages down into usable productive lot sizes.
- Excessive operator motion, due to poor tooling and work station layout, such as adjusting through trial and error.

Value-adding activities can be defined as those activities that add value by moving the product functionally closer to customer delivery by ensuring required quality levels, while incurring minimal cost. Basic processing activities such as changing the shape of the product, strengthening, altering qualities in the product, assembling, and numerous others fall into this category. Examples include:

- Welding.
- Shaping.
- Assembling.
- Painting.
- Forming.
- Milling.
- Drilling.
- Grinding.
- Broaching.
- Casting.
- Stamping.

Even value-adding activities can be improved upon. There is virtually no limit to the opportunities that can be found within the standard operating facility. Every activity should be scrutinized to determine where improvements in cost, quality, and delivery can be better realized.

At this point, people who are not familiar with simulation techniques may be overwhelmed with potential opportunities. The reason for such a feeling stems from the fact that those who have not had to alter practices in the past, may not be ready to accept change as a necessity for future survival. It becomes necessary to use questions that start with, "Why...?," "How...?," and "What...?," as trigger mechanisms for simulation applications.

To strategically plan the best path to follow, the highest priority or need for improvement must be established. There is no generic starting point that would be best for all cases. It often requires that the process under study be evaluated, based upon where the greatest amount of time is spent in wasteful or non-value adding states. The following areas will be explored in the context of how cost, quality, and delivery can and should be improved—using representative simulation models to demonstrate the need for change—to eliminate waste:

- Quality designed into the product.
- Production scheduling.
- Impact of current processing limitations (bottlenecks).
- Unplanned/unscheduled downtime.
- Supplier/carrier management.
- Processing methods.
- Material flow and resource layout.

QUALITY BY DESIGN

Often the initial roadblock lies in the design of the product. Trying to build a product in manufacturing to a design that is burdened with critical dimensions that cannot be held when utilizing existing resources—or conversely using such wide tolerances that mating parts can be within specification, yet never match up—is an exercise in frustration and expense. It inevitably yields a communication loop that exercises deviation, many customer contacts, and numerous other wasteful activities such as waiting for permission to build "out of spec," sorting, rework, and potential warranty problems in the field. The challenge becomes to design the product in a manner which lends itself to the manufacturing process.

CHAPTER 14

SIMULATION AND WASTE ELIMINATION

Role of Simulation

Simulation through computer-aided design (CAD), or physically building prototype units that represent a model of expected product appearance, can offer a means of addressing issues that otherwise would be left to abstract discussions. From the start, there should be interaction between the designers and builders of the product, to ensure that concerns are addressed prior to build start-up. Even for products currently in existence, it is important to review ongoing improvements surrounding the product design since often the design sells the product—thus keeping the company in business. In this regard, it becomes vital to have continuing communication between the builders and designers of the product, to ensure that concerns surrounding cost, quality, and delivery are properly addressed. One of the best means to foster such communication is through use of simulation techniques, to visualize the concern.

Potential Cost Concerns

Cost improvements come with being able to support the design that the customer wants, to the communicated specifications, in a manner that best utilizes available resources. It is extremely important to understand the critical dimensions of the product, as well as the value adding activities to be performed by the process. When a product design forces use of machinery that was never intended to hold tight tolerances or intricacies, it can be expected that forecasted profits will rarely materialize. Extreme waste and non-value adding activities will be incurred from the start. If such incapabilities persist, it can also be expected that the customer will go elsewhere to find future builders of the required product. Simulation models that understand the limitations of manufacturing resources might prevent such oversights from happening.

Potential Quality Concerns

Objective quality improvements can be derived by monitoring the process that supports the customer design. If the process is expected to build quality into the product, there should be predictive monitoring controls in place, as well as "error-proofed" fixtures and tooling. Such techniques will be discussed in detail at a later point in this section.

Potential Delivery Concerns

Delivery, as it relates to product design, means the ability to support the customer specifications in the contracted manner. The customer expects the "agreed upon" quality of the design and has reflected a willingness to pay the established price. In order to stay in business, there must be control over process variability, so that consistent performance can allow for reliable delivery of the product to the customer. The primary objective becomes to minimize or to eliminate the extent of deviation and disruption, without jeopardizing customer satisfaction. Simulation can help to forecast where attention is needed most, by focusing on delivery that may fall short of reaching scheduled expectations and impair attainment of performance objectives. Delivery can mean many different things; however, there is one common binding element—customer satisfaction must be the undeniable, supreme goal.

SATISFY CUSTOMERS THROUGH PERFORMANCE TO SCHEDULE

Scenario

The production schedule should provide the target around which all manufacturing activities are centered. It should translate customer requests into meaningful actions geared toward satisfying those needs, at the lowest cost possible. The driving force should be that 100% customer satisfaction is the primary goal and fundamental reason for remaining in business.

Targets are developed to identify goals. The satisfaction or attainment of these goals is based on a variety of factors such as clarity of the objectives, definition of the goal, fixed nature of the target, consistency of the input, and means of measuring performance.

Does the schedule provide a target that is unambiguous, defined, fixed for a period of time, consistent or leveled, without conflict in measurement, and known by all personnel? If not, there could be a potentially large number of activities performed which are either wasteful or simply non-value adding.

The ability to provide exactly what the customers want—in the manner specified, on time, in the communicated quantity, combined with a world class price—seems to be the standard measure of how well a given function is satisfying customer needs and preserving the right to stay in business. Are the customer's needs being translated via the production schedule, in a cost-effective manner, that will guarantee delivery? Where this is not the case, there is typically a reliance on costly nonstandard modes of operation that are invoked to insure delivery such as overtime, premium freight, subcontracting, and numerous others, which invariably dig a hole that is hard to climb out of. The driving force should be that 100% customer satisfaction is the supreme goal. Where manufacturing activities stray from such a plainly stated objective, a number of actions usually occur—the majority of which are quite unpleasant. Customers stay satisfied through finding a supplier who can guarantee performance in world class price (through having the lowest cost), highest quality, and dependable delivery.

Role of Simulation

Simulation probably has its greatest potential in the area of scheduling. There are numerous applications available that can represent the manufacturing process in a model that utilizes animation to graphically show whether or not the customer is being satisfied, to whichever measure of performance is used. Since there are many variables to be considered when deriving a schedule, simulation can offer the luxury of being able to control or vary the direction and magnitude of almost any input, so that the expected performance can be reviewed prior to schedule release. Virtually any environment, policy, activity, or product can be simulated, so that the impact can be analyzed prior to "real-time" implementation.

Potential Cost Concerns

When the customer can capture customer needs in a manner that provides a target—that is fixed for a given period of time, is leveled to the extent of having a "pulse-like" beat of expected production, and is realistic for the resources of the manufacturing process—there is a great likelihood that costs to perform to the schedule will be minimized. Inputs to a simulation run can reflect the costs involved with manufacturing the product, so that the resulting output can be impacted based on how certain variables of cost are controlled. When trying to keep a localized (that is, departmental) cost at a minimum, there may be a much greater cost incurred through sacrificing global performance. Simulation has the versatility to draw conclusions regarding the impact of local decisions on global results.

CHAPTER 14

SIMULATION AND WASTE ELIMINATION

Potential Quality Concerns

An interesting dichotomy exists in manufacturing. It has become extremely important to be associated with a product that possesses a high level of quality. Yet on the other hand, it becomes equally as important to support customer demands in quantity, so that their production builds are not jeopardized. Quality versus Quantity. The key has been to find a formula that guarantees both. The role of production scheduling is to identify the process capability of manufacturing resources, with proper maintenance reflected, that can insure a quality output. Such consideration requires strong feedback from manufacturing personnel, process controls, and support functions, to update in real-time how performance is unfolding from both perspectives—quantity and quality.

An important awareness must exist within the manufacturing environment: changed circumstances do not necessarily require changing the schedule. There must be focused efforts toward permanent removal of the root causes of events that disrupt the attainment of the schedule. If issues are not addressed as they arise in the short term, their continued recurrence may be a contributing factor for making, or breaking, longer term survival. Simulation provides a means for developing scenarios which can depict areas of opportunity to address root cause quality concerns, or equally as important, a means of deriving greater throughput, without sacrificing quality.

Potential Delivery Concerns

Performance to a properly developed schedule is the primary means for ensuring delivery. There are two key factors that influence the ability to satisfy the customer. It is merely a matter of supply versus demand. The customer sets the stage by identifying demands that are expected to be met. Typically, for production purposes, the demand involves specifying the detailed product, quantity, and when the product is required for use. The manufacturing environment has to be able to identify the means of supplying the proper mix of products from available resources in the quantity needed, by the expected date.

Where schedules consider the realities of manufacturing operations and remain level, smooth and stable for a reasonable period of time, the only factor that is left becomes delivery. However, the required data needed to support reality plays a major role as to whether or not simulation can be used for scheduling, and can be followed. If the variables on the shop floor cannot be controlled to provide delivery to the customer, a simulation to help predict future direction and potential concerns will be useless. The environments where simulation works best are those that work on planning destiny, rather than reacting to circumstances beyond control.

UNDERSTANDING LIMITATIONS

Scenario

Typically, the individuals responsible for deriving and performing to production schedules are very aware of the current processing capability limitations of bottlenecks. Growing focus on the importance of such hindrances has coined the term *Constraint Management*, which centers around the fact that a small number of "bottlenecks" control global levels of production (throughput), material stock levels (inventory), and cost to support the manufacturing operation, both directly and indirectly (operating expense). When viewed from such a global perspective, it can be seen how any wasteful or non-value adding activities performed on these critical resources can generate extreme costs, and impact quality levels as well as delivery. All of these factors severely slash potential profit.

Role of Simulation

Simulation can offer means of reviewing constraint(s) in detail so that the variables impacting performance can be isolated and systematically controlled. The important feature which simulation provides is that potential loss of performance due to "real-time" experiments done on the shop floor can be offset by capturing the process in a model, trying similar logic prior to implementation. In essence, the learning curve of installation can be significantly trimmed and made less wasteful due to introduction and evaluation using simulation techniques.

Potential Cost Concerns

New projects or systems are invariably given a reasonable amount of latitude when first starting up, since some mistakes due to lack of experience are tolerated. Depending on the focus, a large amount of costs and wasteful activities can accumulate in a short amount of time. Simulation can be of great benefit in this area, since a model representing the new approach can be developed and can identify where the processing limitations or bottlenecks can be found. In a number of cases, unforeseen constraints enter the picture, which in operation could have paralyzed the process. There have been numerous instances where the use of simulation has paid for itself with just one application, through foresight in constraint awareness and management control.

Potential Quality Concerns

The importance of allowing the bottleneck resource(s) to work on, producing nothing but quality parts, cannot be stressed enough. Since subsequent processes are totally dependent on the limited processing capability of the constrained resource, defect-free part production is not only a must, it is a matter of survival. Equally as important is the understanding that until the product has fulfilled the customer need, it still carries the label of a bottleneck produced part. As such, there must be protection from scrapping, or allowing defects to occur on such parts, until they are physically in use at the customer location. Controls and customer feedback mechanisms, that continually check the process, are vital to ensuring the safe journey of parts following the bottleneck. Simulation can be of assistance by identifying areas of focus and expected frequencies of necessary audit control, when fed with the proper data. Use of these techniques can plan the itinerary of the product, with quality taken into account as a vital part of the productive process.

Potential Delivery Concerns

Awareness of the role that a bottleneck plays in product delivery truly determines whether or not the customer's demands are to be satisfied. Since it has been proven that relatively few resources control global levels of throughput, inventory, and operating expense, it becomes of utmost importance to realize the greatest levels of quality and quantity so as not to impair delivery to customer specifications. Once again, simulation plays a major role in assisting the user to identify areas of opportunity for improvement, through representing the impact of altered processing methods or capacity investments prior to their installation. Using simulation as a tool can establish whether or not seemingly excellent investments can be realized to the forecasted potential, by plugging in expected capabilities to a simulation run and evaluating the results.

CHAPTER 14

SIMULATION AND WASTE ELIMINATION

PREDICTIVE VERSUS REACTIVE RESOURCE DOWNTIME: PROCESS CONTROL, PREVENTIVE MAINTENANCE AND QUICK SETUPS

Scenario

Problems occur and people react. No one knows where, or when, the next catastrophe will strike. Personnel are instructed on who to contact when the resource goes down, yet there are no provisions in place for predicting when such an occurrence is about to happen. Parts are periodically inspected well after the quality has been built in, and the process has changed to another part. Resource changeovers occur not by plan, but by shortage of material. Personnel assigned to planned tasks are pulled out to address "fires." Are such actions value adding, or might there be some way to trim the prevalent wasteful and non-value adding elements? Some simple doctrines come into play. If one is in a constant mode of reaction, it is impossible to plan. If the causal factors for downtime are not properly identified and addressed for permanent removal or correction at the time of the occurrence, there is a great likelihood of recurrence. Customer satisfaction becomes extremely difficult when the resources required to produce the product are not available. The ability to predict and prevent extensive downtime can be a contributing factor in the formula for long-term survival. Predictive and preventive control must replace reactive firefighting.

Role of Simulation

Process controls can capture frequencies, history, and real-time performance to specifications, from properly equipped machinery. Feedback readings, taken from the resource, can flag approaching "out of control" conditions, rather than finding out through experiencing costly, wasteful defects to the part. Additional control comes through the development of engineered "error-proof" tooling. The use of such fixtures removes the need for skilled operator subjective adjustments, and provides consistent, repeatable, reliable processing from standard operations. A simulation model can demonstrate how specific process controls will interact with the machinery, as well as modeling prototypes of potential tooling changes. Through the use of simulation, varying levels of control and capability can be identified and evaluated for "best-fit" application.

Audit frequency of periodic resource capability checks can be established and represented within a detailed resource schedule, so that machine-specific preventive maintenance can be planned. A simulation can demonstrate the impact of expected lost productive time needed to perform such periodic maintenance procedures, if the required resource cannot be made available on the off-shift, or typically nonproductive time.

Another prevalent downtime activity that is receiving attention is the lost time of changeovers for a common piece of machinery. Currently, there is often no incentive to perform such changeovers, due to the mode of measuring productive labor, and the inefficiency involved with existing setup practices. The prevailing thought is that longer batch runs are much more economical than addressing all prohibiting factors involved with eliminating the "waste" of a changeover. How often do subsequent processing activities require "lot-splitting," forced setup from lack of material, and/or waiting until the "required" parts come through the pipeline? Are these activities not viewed as wasteful?

The need to employ quick setups on shared resources has been labeled as absolutely essential for providing the flexibility required to support small lot production and reduced manufacturing leadtime—two key factors for improving costs, quality, and delivery. Simulation can be of extreme value for demonstrating the ramifications of more frequent setups, through review of reduced inventory levels, and removal of quality-related issues that occur due to a "too-late" mode of detection—primarily as a result of large batch transfer.

Potential Cost Concerns

The pursuit of eliminating waste or non-value adding activities can be equated to extracting some degree of control over the events in the manufacturing environment. There can be enormous savings generated from being able to expect and plan activities to be performed. Simulation provides the capability for experimentation with a variety of approaches, some of which may be more applicable than others. At the very least it can enlighten the user as to whether or not a chosen direction is worth pursuing, or that an alternate path may warrant further attention.

Many times the initial process is designed expecting failures or defects to occur. Where this approach is practiced, the expectations are usually met. Considerable benefits can be derived from extracting higher levels of process control. Consider the savings from avoidance of costly rework or scrap related activities. Think of all the expense in overtime and premium transport that had to be invested to recover "good" parts from an incapable process. It would not be difficult to identify many more similar types of related savings. Simulation can represent varying levels of control, and establish the ongoing costs associated with each approach.

Preventive maintenance is viewed ambivalently. Those working in the manufacturing environment know that most machines require routine maintenance on a regular basis, such as oil or lubrication, just to keep operating. Periodic maintenance, on the other hand, is typically put off until a more opportune time. Unfortunately, such opportune times surface not by design, but by catastrophic failure. The costly waste of unplanned machine downtime is something that can be avoided, or at the very minimum, lessened. It is simply a matter of priority. A television commercial comes to mind, where there exists a choice between purchase of an oil filter or having to replace the entire engine at a not too distant date. The transmitted message strikes home as it says, "You can pay me now...or pay me later."

The cost carried to support inefficient changeovers is typically paid through having higher levels of inventory than required to support current customer demand. But there are hidden costs as well. The concept of keeping excess stock to mask wasteful or non-value adding activities is at the core of the elimination process. Inventory does not come cheap. It ties up valuable money that could be applied to needed process controls or other well directed tools that could further eliminate unnecessary activities. Beyond the evils of inventory are the factors that enter the picture due to the lack of flexibility. Considering all the corresponding premium expense that must be incurred due to so-called efficient batch runs, there may be some question as to whether a given environment can manage to survive under the current definition of efficiency. Simulation can shed some light on alternate practices and may provide insight for discovering the difference between utilizing resources to satisfy customer demand and activating for the purpose of providing parts.

Potential Quality Concerns

The ability to predict where problems will occur, rather than reacting to a problem once it arrives, is at the core of the quality

CHAPTER 14

SIMULATION AND WASTE ELIMINATION

issues. Predictive control, or planned actions, need to be established and adhered to, so that the corresponding functions can identify activities that are truly needed and those that exist to support the former reactive mode of operation. Process control capabilities must be utilized so that out of control resources are never used to produce parts for the customer.

Simulation can help to incorporate the planned events of maintenance and setup into the heart of a production schedule, so that expected, or planned downtime can be factored accordingly. Quality improvements come from being able to identify the variables in the process that restrict planning, so that these instabilities can be transformed from wasteful expense, into predetermined actions.

Potential Delivery Concerns

Delivery of products from the manufacturing environment to the customer relies on the capability and control of available resources. Where process controls are present to monitor performance and the presence of out of control situations, there is a greater likelihood of being able to produce and consistently deliver a quality product, utilizing cost-effective techniques. Simulation can forecast expected delivery using historical data; however, there is no known package available that guarantees manufacturing performance.

Preventive maintenance establishes a means for prolonging use from a given piece of machinery. Simply put, preventive maintenance strives to ensure that resources will not break down for prolonged periods of time. By not utilizing such techniques, the machine is virtually assured of eventual disaster. Resources need to be functional in order to deliver.

Changeovers are absolutely essential to accommodate a diversity of customer demands over shared equipment. The quicker the changeover, the higher the degree of flexibility, leading to greater customer responsiveness and delivery. The elements involved with a setup must be isolated to determine where improvements can be made to facilitate better coordination and use of resources while the machine is still running, rather than when the machine is down. In this manner, the resource can remain as productive as the process allows, utilizing internal cycle time, rather than costly and wasteful downtime. Simulation runs depict the differences in performance and flexibility, stemming from improved changeover capability.

SUPPLIERS/CARRIERS AS AN EXTENSION OF THE PROCESS

Scenario

The need to include suppliers and carriers in business decisions regarding product delivery and performance may be the key area for improvements in the future. Such consideration may be prompted by:

- Bulk dropoffs of material on a semi-weekly or monthly basis.
- Packaging that needs further handling for production use.
- Freedom of the week delivery.
- Needed parts tied up upon arrival for receiving inspection.

Clearly, much gain can be realized through eliminating or minimizing these wasteful or non-value adding activities.

When inventory campaigns are launched, one of the first places reviewed is supplier delivery—by quantity, frequency, and mode of packaging. If it is felt that too much is on hand, purposeful discussion with the supplier, purchaser, manufacturing personnel and others should take place, to determine the manner in which the materials are to arrive for running the business in a more cost-effective manner.

Role of Simulation

Simulation can identify scheduled arrivals of carriers, from specific supplier locations to the receiving docks, in a sequenced manner. Time windows can be set aside for the loading and unloading of material, so that the next arrival can be properly scheduled. The entire process can be represented, from the back door of the supplier to the shipment of parts into a given manufacturing facility. In a model, simulation can physically represent the supplier and carrier as an extension of the process.

Potential Cost Concerns

Numerous inefficiencies can be eliminated through detailed review of the current policies and practices of supplier deliveries. The first question is, "Do the parts arrive immediately for production use?" Most often the answer reveals the extent of opportunity available to make substantial cost improvements.

The start of all manufacturing activity, for every manufacturing facility, begins at the receiving area. Suppliers and carriers must be coordinated to the extent that what appears at the dock is in line with what is required for production. Where this is the case, by default, there are significant reductions of inventory-related waste throughout the entire process—the parts would not be available to misuse. Simulation can be extremely valuable in plotting a manageable course for attaining such a mode of operation.

Potential Quality Concerns

Many quality problems remain unaddressed, since existing inventory levels mask the need for resolution. By developing supplier schedules that make sure there is always enough stock available at the "raw" stage of production, there is a much greater likelihood of creating opportunity for misuse. Building parts before need exists is one of the worst kinds of waste. Simulation can assist in providing sequenced schedules to the supplier that reflect expected production build. Assuming that suppliers are capable of providing parts in line with stated needs, all that is left for the manufacturing facility to do, is to perform.

Potential Delivery Concerns

Delivery of parts for production use requires the manufacturing facility to communicate and perform to stated needs, the supplier to produce and ship in line with projected builds, and the carrier to transport the product in the manner to ensure the delivery at the needed time. Typically, the carrier is not viewed as playing a major role; however, the following activities may change that thinking:

- Planned receiving dock arrival (strictly enforced).
- Elapsed unloading time.
- Return trip designed to carry empty returnable containers.
- Delivery on a daily or by-shift basis (for close-by suppliers).

From the supplier side, work has to be done to identify the limiting capabilities from a production standpoint, so that realistic, fixed, leveled schedules can be derived. Wherever such limitations are exceeded, expected delivery from both the supplier and the manufacturing facility is highly unlikely. Thus it becomes

CHAPTER 14

SIMULATION AND WASTE ELIMINATION

extremely important to work with suppliers to identify targets for ongoing continuous improvement.

MAKE STANDARDIZED WORK THE STANDARD

Scenario

Processing methods have always been and still remain a primary target for waste elimination. The motions required to produce a product have long been studied under the disciplines of industrial engineering, process engineering, and numerous other areas, all focusing on the same intent—to develop the most efficient manner for producing a given product. Several critical concepts entered the picture in recent years and altered conventional thinking, such as the importance of bottleneck resource considerations, group technology or cellular manufacturing, quick changeover capability, small lot production, and numerous others. These approaches have centered around a more global theme, one that focuses on customer satisfaction. However, perhaps the most basic of all concerns is the need to standardize work, so that processing activities can be performed in a similar manner for all workers; this allows for potential rotation capability and increased flexibility should absenteeism, training needs, tardiness, sick leaves, and vacation coverages be concerns.

Role of Simulation

Activities can be represented in a model or through the use of representative facsimiles to physically capture how improvements can be made to the process. Most operations are dependent on the preceding department or resource as the supplier, while the customer becomes the subsequent set of activities along the fabrication route. If manufacturing operations are seen as a series of supplier-customer relationships, the viewpoint should be held that customer support, even internally, is of primary concern. Tools that enhance material flow for improved delivery, increase quality, and utilize value added activities should be considered for use in developing the standard mode of operating. Simulation can demonstrate the resource layout, material flow, operator motions, and numerous other considerations, so that the use of wasted or non-value adding motions is kept to an absolute minimum.

Potential Cost Concerns

Where operations or resources require an increased skill to process parts, there are raised levels of ongoing expenses incurred to support the function. The dependence on a heightened degree of knowledge or skill from a worker severely limits the potential for rotation or worker flexibility. Factors such as absenteeism, vacation, and others can inhibit global performance. The movement to standardized work is an effort to minimize dependence upon a designed-in limitation, and move toward a more simplistic, less demanding set of machinery and processing activities. An additional concern arises for resources with numerous bells and whistles: who can fix it when a problem arises? There is typically a greater risk of prolonged machine downtime due to a lack of "high-tech" problem solving knowledge. All these factors can be represented in a model to demonstrate alternatives and cost factors for dependence versus flexibility scenarios.

Potential Quality Concerns

A movement towards standardized work can be one of the greatest assets in a quality improvement campaign. Where variability can be restricted, control can be better attained. Having the machinery and operators available to provide consistent, reliable, defect-free parts is the major thrust of standardized work. It is important that this valuable consideration not be overlooked when identifying means of fostering quality improvements.

Potential Delivery Concerns

The ability to deliver product depends on the combined performance of operator and equipment. As such, dependence on a certain class of worker can limit the chances of attaining customer satisfaction. It is vital that some degree of standardization be factored into the operation of a given man-machine configuration, to insure that repeatability, capability, and other key factors of performance will consistently happen.

HURRY UP AND WAIT—THERE MUST BE A BETTER WAY

Scenario

The manner of measuring shop floor efficiency has long revolved around maximizing the output from each productive worker. Such focus encourages that wherever material physically exists in front of machinery, it is pushed through the process, regardless of need—the important consideration being that the worker not remain idle. One way to continue this practice is to encourage long batch runs, where material can be hoarded to get the most economical usage (on a per unit basis) from a given setup. From a local viewpoint, this generates very efficient usage of designated labor hours. What happens from a global perspective? Quite frequently, certain customers starve while waiting for stock to move through the process. Often there is a need to invoke changeovers to maximize the use of available workers in line with existing material. The phenomenon of "hurry up and wait" is an ongoing mode of operation. One department reaps the benefits of the performance measurement while another is pounded for factors beyond its control. The problem is complicated further by the fact that "specialty centers" such as milling, welding, painting, etc. are isolated functions rather than supporting material flow.

Role of Simulation

Simulation can be an invaluable tool for informing appropriate personnel of the need for change. One of the fastest growing applications of simulation technology is the modeling of the manufacturing process, so that the material flow can be represented. It is important to utilize the graphic capabilities of a simulation package, so that if the underlying concepts of a certain scenario are not being grasped, a picture can be developed to visualize the concept.

Smaller lot sizes, more flexible changeovers, cellular layouts, tighter process control through predictive tools, preventive maintenance, and key approaches all hinge on the fact that material must flow through the process to support customer needs. The most effective use of simulation can be in representing how the manufacturing process can transform current wasteful/non-value filled methods into an environment that fosters value added material flow, totally in line with customer demands.

Potential Cost Concerns

The cost of waiting due to unbalanced material flow can be seen in the expense for premium modes of operation. Premium use of labor for additional hours, shifts, or weekends; premium shipment of parts since the material was not available for the

CHAPTER 14

SIMULATION AND WASTE ELIMINATION

contracted mode of transport; subcontracting labor to add incoming flow; and numerous factors contribute to the practice. The costs that can be saved are monumental. It becomes a matter of focus; balance labor capacity or balance material flow. Simulation can be an invaluable tool providing the vision to see the benefits of moving to an operation which may look inefficient in labor usage (for the short term), but is quite effective in long-term customer satisfaction (and survival).

Potential Quality Concerns

Quality problems can arise when under pressure to perform in an expeditious mode of operation. Often the conventional quality practices are shortcut or abandoned, when there exists the possibility of customer shutdown due to lack of parts. As a result, it is not uncommon to have activities such as sorting, rework, scrap, and potential warranty problems, left as reminders of missing quality concern. Scarcity of parts from inadequate material flow is often the result of long batch runs or inefficient changeover capabilities. Flexibility through quick setups, and inherent process control and resource capability through preventive maintenance, allow for the "hurry up and wait" syndrome to become a thing of the past. Simulation can provide excellent insight on where effective implementation of needed direction should occur.

Potential Delivery Concerns

Where there are roadblocks in material flow, there are instances of impaired delivery. Improvements in material flow are important to respond to the diversity of customer demands. However, caution must be observed when designing activities for local improvements that are expected to impact global performance. Bottlenecks dictate the resulting throughput or material flow of a given operation. In order to best realize the improvements that are expected to occur, simulation can forecast whether or not projected benefits are in line with paper or predetermined calculations. Bear in mind the following considerations when looking for potential areas of opportunity:

- The operating components for material flow considerations from a bottleneck resource are either setup or processing time; for a nonbottleneck there is a third component—idle or opportunity time, which is the time left at the resources not involved with either setup or processing.
- Use of smaller lot transfer, which moves completed parts for subsequent processing in smaller increments.
- Increased flexibility from reduced setup time allows more setups to be done, providing capability to respond quicker to changed customer demands and provide a greater mix of production.
- Increased process control and capability allow for a greater yield from existing equipment.

Numerous other related capabilities should strive to increase throughput in line with customer needs, decrease "needed" levels of inventory, and reduce operating expense by having required material available at time of need, thus avoiding use of premium modes of operation.

SUMMARY

Perhaps the most rewarding use of simulation is that activities can be foreseen and forecasted prior to occurrence—as long as the inputs are representative of the environment. In this regard, numerous oversights can be avoided, which would greatly reduce the potential amounts of costly waste.

Simulation shows that where actions can be predictable, they can also have some level of control. It makes more sense to strive for planned destiny, rather than continually reacting to seemingly uncontrollable circumstances.

Wherever possible in today's globally competitive environment, there should be reminders of the need to extract control from the surrounding variables of the manufacturing process. It is only in this manner that companies can make sure that simulation and waste elimination actually do become partners for ongoing, continuous improvement.

SIMULATION AS A DECISION MAKING TOOL

Performing continuous improvement on a manufacturing system requires testing many ideas to find the ones that actually help. While a few promising changes can be tried directly on the system, some method for trying ideas out without disturbing production would be ideal. Many companies have found this method through the use of simulation modeling.

Simulation modeling has different meanings depending on the context. To understand what can be learned from a simulation, one should begin by looking at a plastic model of an automobile. The model shows the relationships between all the major physical parts, but does not indicate the engine or power train operation. Hence, this simple physical simulation represents some of the behavior of the much larger, more expensive real automobile.

Moving up a step in complexity, consider a cut-away model of the automobile's engine. Now, the model provides physical relationships as well as interference information because it can be seen how the parts move with respect to each other. While providing useful information, the model has faults, mostly because it cannot be changed easily when a different design is wanted.

To increase the flexibility of a model, the physical model is changed into a computer model. Using a computer, an engineer can design, assemble, and even move parts on a computer screen. The computer model delivers the same benefits as the physical model, but with added flexibility. The speed of this technique encourages many design iterations to help create the best design quickly.

Expanding the scope of the problem even more, imagine looking down at a manufacturing facility from a height that allows a complete top view. From here people, product, and machines can all be seen moving. With a little imagination, the movement is not unlike the movement of the engine model. Therefore, if it is desired to model the production facility, past modeling experiences are drawn upon and it is decided that a computer-generated model has the most benefits.

Problems are encountered when trying to model the production facility. While people, product, and machines are all in the same facility, they are not connected like the parts in an engine. The new factor in the model is time. The plant's efficiency depends on the timing and coordination of all processes. Manually choosing

CHAPTER 14

SIMULATION AS A DECISION MAKING TOOL

the number of people and machines for efficient plant operation becomes impossible with even the simplest facility. Simulation provides a quantitative method to measure the effects of changes. Hence, when an attempt is made to define simulation as it relates to the design and management of production facilities, it can be seen that simulation is the abstraction of a time-based flow system into a computer-based representation, or model.

With a good simulation system, a model of a system can be quickly built; tests can be run, modifications made, and meaningful results produced. This section briefly describes the development of simulation systems and goes on to give applications of simulation for decision making and operational improvements.

HISTORY OF SIMULATION

Numerical analysis, statistical methods, and simulation have evolved from the early part of this century. Until the 1960s, these three techniques remained interesting topics for academicians with little practical value due to the immense number of calculations required to obtain results. The general availability of mainframe computers in the 1960s brought these technologies out of textbooks and into the hands of engineers and analysts for practical applications.

Early simulationists used computer languages like FORTRAN to create models, a difficult task that led to the development of specialized simulation languages. One of the earliest languages, GPSS, came with the operating system on the IBM 360 computers. Hence, many people tried GPSS simply because it was a standard feature of the machine. Other special purpose languages were added, including SIMSCRIPT, GASP, SLAM, SIMAN, and many others. Most are still used today in improved forms.

The introduction of IBM's personal computer (PC) in 1982 marked an important year for computer users. The PC gave anybody who wanted access to a computer a good environment in which to work. Simulation languages became easier to use, more powerful, and less expensive. By the mid-1980s, graphical animation for simulation systems became available. In addition to being a powerful analysis tool, animation provides an excellent method for explaining system design and operation at all levels from sales to production to operator training.

As the twenty-first century dawns, graphical simulation systems are becoming a practical reality. To build a model, the simulationist simply draws a representation of the system on the screen and provides information about product flow. The simulation system then runs the animation without the need for any programming. More powerful computers and better software have brought simulation from a specialist's technique to a useful tool for engineers, production planners, marketing specialists, and managers.

Simulation Operation

Simulation for manufacturing generally takes two forms—continuous and discrete. In some cases, continuous and discrete models are linked together to create a combined model.

A continuous model represents an analog system with time as the independent variable. Examples include heating a furnace, cooling a casting, and chemical reaction rates. Typically, algebraic and differential equations describe continuous systems, so knowledge of advanced mathematics is required to build continuous models.

Discrete models represent most manufacturing systems with processes that start and stop at various times. A machining center, for instance, begins processing a part at a specific time and finishes the part at a specific, or discrete, time.

Combined models contain both continuous and discrete information linked together. In the production of canned liquids, for instance, continuous features are used to model the chemical reaction rates needed to create the liquid product, and discrete techniques model the product after it is placed in containers and palletized.

Most systems tend to be discrete in nature or can be approximated to be discrete, so this discussion concentrates on discrete models. To clarify terms for the rest of the discussion, simulation model, simulation, and model all refer to the computer-based representation of the system being described.

To illustrate the concepts of simulation operation, recall the machining center just discussed. Figure 14-17 shows a block diagram representation of the machining center, and Fig. 14-18 lists timing information for ten different parts arriving at the machining center. Note that the time units are not indicated. For now, assume minutes are used; the significance of time units will be explained shortly.

Figure 14-18 gives much useful information about the machining center's operation. For instance, the list shows how many parts are waiting at any given time, when the machine is busy, and when the machine sits idle. The data shows the machine is busy except for the first 2.6 minutes, the 1.1 minutes between part 2's finish time and part 3's start time, and the 1 minute between parts 6 and 7. This translates into 59.3/64.0 = 0.927 or nearly 93% busy

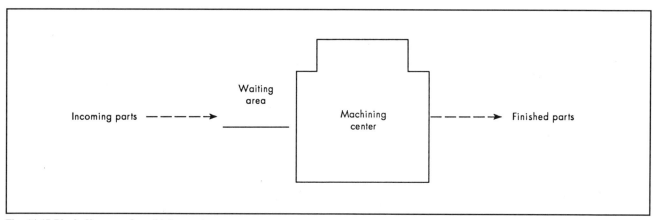

Fig. 14-17 Block diagram of machining center.

CHAPTER 14

SIMULATION AS A DECISION MAKING TOOL

time. Similarly, an average of 0.37 parts waiting for the machine can be computed, and part time in the system ranges from 4.6 to 11.5 minutes with an average of 8.3 minutes.

Figure 14-19 shows a portion of the timing information from Fig. 14-17, but in linear form. Figure 14-19 represents the actual operation of the machine in linear time, the same way everyone experiences time. This list of events is called an events chain. For the simple case of the machining center, the events chain can be manipulated manually. However, the addition of just one more machine complicates the system beyond a reasonable person's ability to keep track of the events chain. A simulation system, then, must manipulate the events chain as processes start and finish. The concepts of the events chain are simple, but the quantity of events in any real system demand a simulation system for an accurate representation of the system.

The simulationist translates the operation of the system into code, either graphically or through programming, that the simulation system understands. With the data manipulation power of a computer, the simulation system generates an events chain, adds and removes events as the state of the system changes, and collects data just as fast as the computer can execute. This means that a simulation of a system generally runs much faster than real time. In fact, months or years of a plant's operation may be simulated in just minutes, depending on the complexity of the simulation model. This is a key advantage of simulation, the ability to predict the behavior of a system in very little time. More advantages of simulation will be discussed under "Simulation Applications."

Recall the assumption that the machining center used minutes for time units. By looking at Fig. 14-19 though, it can be seen that time units are arbitrary and may be chosen in any convenient form. The key to choosing time units is consistency—all times must be represented in the same units. Some simulation systems allow the user to enter different units for different processes, but the simulation system then translates everything into the same time units.

SIMULATION APPLICATIONS

The prime advantage of simulation is the ability to predict the behavior of a system quickly. Given that, three general areas of application arise:

1. Facility design.
2. Facility improvement.
3. Scheduling.

Part Number	Arrival Time	Number in line	Process time	Start processing	Finish processing
1	2.6	0	6.2	2.6	8.8
2	7.9	1	5.8	8.8	14.6
3	15.7	0	6.0	15.7	21.7
4	19.9	1	9.1	21.7	30.8
5	23.5	1	3.7	30.8	34.5
6	28.2	2	2.1	34.5	36.6
7	37.6	0	4.6	37.6	42.2
8	40.0	1	8.5	42.2	50.7
9	48.1	1	4.2	50.7	54.9
10	52.5	1	9.1	54.9	64.0

Fig. 14-18 Machining center history.

Facility Design

The design of new facilities requires considerable planning in all areas from the types of processing equipment to the material handling methodology. Simulation plays an important role in facility design because of the ability to test different scenarios quickly. A simulation model should be constructed beginning with the initial planning stages of all new facilities.

Facility design models normally simulate long periods of time, are general in nature, and rely on statistical data; hence, output results must be analyzed carefully. Some simulation systems have output processors to assist with this task. By running many iterations with varying input data, the modeler develops a general feel for the proposed system on a gross scale and can predict potential problem areas.

From the initial simulation information, the facility design team then modifies the design to remove the potential problem areas and the modeler changes the simulation to follow. As the facility design becomes more detailed, the model becomes more detailed also and provides more information. After one or more iterations and careful data analysis, a good design emerges. At this point, the model can help the engineer predict the amount of equipment required, the number of operating personnel, and the locations and interrelationships of the major items in the facility.

The accuracy of the simulation model relies heavily on the accuracy of the data used. Timing information from equipment manufacturers and the amount of time required by operators should be carefully scrutinized for best results. When there is doubt about data, simulation runs at worst case, proposed normal, and best case may be made for comparison purposes.

Simulation as a facility design tool provides confidence in the system before ground has even been broken. The cost of the simulation study is usually negligible in comparison to the cost of a new facility, and the potential savings are tremendous. For example, if the simulation saves the cost of moving just one major piece of equipment, the model has more than paid for itself. Once the facility begins operation, the model can still be used as a scheduling tool.

System Improvement

Existing facilities may be improved with the use of simulation by careful modeling and analysis. Generally, existing systems have known data available that can be used to help create an accurate model. At times, the standard operating procedures can be changed just slightly with drastic improvements in productivity. Changes in such areas as operator break schedules and

Time	Operation
2.6	Part 1 arrives.
2.6	Part 1 starts processing.
7.9	Part 2 arrives, wait for machine.
8.8	Part 1 finishes processing.
8.8	Part 2 starts processing.
14.6	Part 2 finishes processing.
15.7	Part 3 arrives.
15.7	Part 3 starts processing.
19.9	Part 4 arrives, wait for machine.

Fig. 14-19 Linear machining center history.

CHAPTER 14

SIMULATION AS A DECISION MAKING TOOL

preventive maintenance may help remove bottlenecks and improve production. At other times, machines and operators may be added or removed. The prime benefit of simulation in facility improvement is the ability to try different scenarios without actually changing the real system and interrupting production.

An example of system improvement might be trying to reduce costs in a warehouse served by automated guided vehicles (AGVs). Obviously, when fewer AGVs are required to service the system, costs may be lower. By modeling the system, a simulationist can evaluate the effect of maintenance schedules, the location of charging skids, and store/retrieve schemes. Careful analysis may produce operating procedures that minimize the number of AGVs and the time spent storing and retrieving material.

Another example of system improvement might be trying to maximize the production of a facility. Normally, certain processes create bottlenecks that limit the output of a system. These bottlenecks are not always easy to find on the shop floor. A simulation model can help the analyst locate the bottlenecks and try to find solutions. At times, the addition of a few more people or another machine can increase production significantly.

Scheduling

A previous section pointed out the difficulty in trying to manage an events chain in a complicated system. Scheduling a real system is usually just as complicated, and a simulation can be a great help in creating acceptable schedules.

Models built for scheduling have varying levels of detail depending on the time frame under consideration. Long-term scheduling can normally be performed with a simple model, usually a modification of the facility design model. As the time frame to be scheduled shortens, the model must include more detail.

Short-term scheduling simulations require timely, accurate information about the current state of the system. These models use real system data and generally only predict the system behavior for a short period of time, a few hours to about a week. Figure 14-20 illustrates the use of a simulation for scheduling.

Note that the simulation does not optimize the schedule; it only predicts the schedule. An experienced planner views the results of the simulation and makes any changes required, and runs the model again. Given the speed of the simulation, the planner can run several iterations in a few minutes. The state of the system includes information about equipment and operator status, orders or parts in the system, and new orders. All this information can be read from data files or can be gathered from the real system via sensing devices.

Simulation for scheduling places the greatest demands on the modeler and the simulation system, but gives great benefits by improving operations on a day-to-day basis.

GETTING STARTED WITH SIMULATION

The method for getting started with simulation depends upon one's needs. For work of limited scope, the most economical approach is to hire a known, respected firm that does simulation on a consulting basis. This minimizes the time staff personnel must spend on the project, and the consultant can always be brought back in for any changes arising at a later date.

As the scope of work grows, in-house personnel may learn simulation as part of their responsibilities. The advantage to this approach is that the person doing the modeling learns the system operation well and keeps this knowledge in-house. When choosing a simulation system, several factors must be considered, including:

1. Ease of use.
2. Flexibility.
3. Availability of technical support.
4. Cost.

In the past, flexibility and ease of use often conflicted, but improved systems now combine these features. The vendor must have accessible technical support for the inevitable problems that occur with any computer system. Cost should not be the overriding consideration since any difference in cost will usually be quickly returned by a system that is easier to use.

SUMMARY

The key advantage of simulation is its ability to predict the behavior of a complex system without disturbing the system. This ability creates many opportunities to improve manufacturing facilities with simulation. When designing a new facility, simulation can help engineers determine a practical plant layout that reduces the number of surprises at startup. Simulation also has many applications in system improvement—from reducing costs to increasing output. And finally, simulation provides a fast, flexible tool for determining how to schedule a complicated facility. Modern simulation tools make the technology accessible to anyone interested in obtaining these benefits.

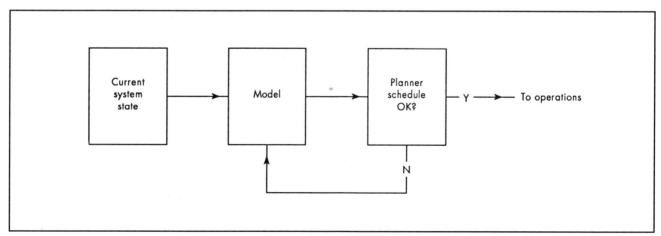

Fig. 14-20 Precision simulation for scheduling.

CHAPTER 14

FAILURE MODE AND EFFECTS ANALYSIS (FMEA)

FAILURE MODE AND EFFECTS ANALYSIS (FMEA)

Today when total quality management (TQM) is the byline of every industry's existence, one of the many concerns that management and engineers continuously face is the issue of improvement through prevention. One of the many tools to identify this improvement is an FMEA.

An FMEA is a hybrid of a reliability program that allows the engineer or anyone else to identify any known and/or potential problems and provide followup and corrective actions. It is used as a method to prioritize these problems, in both design and process (manufacturing) stages.

Is FMEA worth the effort? When problems can be assessed before they occur, or events can be evaluated quickly, the person doing the assessment or evaluation demonstrates a real understanding of the process (whether it is design or manufacturing) and a desire to improve it. Because it provides this understanding and improvement, FMEA is well worth using.

Second, in its most rigorous form, an FMEA summarizes the engineer's thoughts while developing a process and also documents the rationale for the manufacturing process involved.

To do an FMEA properly certain prerequisites must be recognized:[2]

1. Not all problems are important. This is very fundamental to the entire concept of FMEA, because unless it is internalized people are going to "chase fires" in the organization. Some problems have to be defined as being more important than others for whatever reason. The fact is that some problems indeed have higher priority than others. FMEA helps identify this priority.
2. The customer must be identified. The definition of "customer" normally is thought of as the "end user." However, a customer may also be defined as a subsequent or downstream operation as well as a service operation. When using the term from an FMEA perspective, the definition plays a major role in addressing problems. For example, as a general rule in the design FMEA "customer" is the end user, while in the process FMEA the "customer" is the next operation in line. This next operation may indeed be the end user, but does not have to be. Once the customer (internal or external) has been defined, it may not be changed—at least not for the problem at hand—unless it is realized that by changing it, the problem and/or the consequences may change.
3. The function must be known. It is imperative to know the function, purpose, and objective of what is to be accomplished, otherwise time and effort will be wasted in redefining the problem based on situations. If necessary, take extra time to make sure that the function or purpose of what is to be accomplished is understood.
4. The company must be prevention oriented. Unless it is realized that continuous improvement is in the company's best interest, the FMEA will be a static document to satisfy customer or market requirements. The push for continual improvement makes the FMEA a dynamic document changing as the design and/or process changes with the intent *always* to make a better design and/or process.

WHY DO FMEAs?

The propensity of managers and engineers to minimize the risk in a particular design and/or process has forced managers to look at reliability engineering, to minimize and define the risk. A pictorial perception of this risk is illustrated in Fig. 14-21.

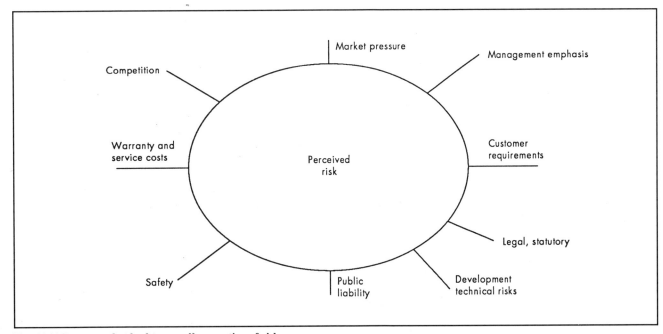

Fig. 14-21 Pressures that lead to overall perception of risk.

CHAPTER 14

FAILURE MODE AND EFFECTS ANALYSIS (FMEA)

These risks can be measured by reliability engineering and/or statistical analyses. However, instead of their complexity, the FMEA extracts the basic principles without the technical mathematics and provides a tool that anybody committed to continuous improvement can utilize.

Statistical process control (SPC) is another tool that provides the impetus for the implementation of an FMEA, especially for a process FMEA. SPC provides information about the process in regard to changes. These changes are called *common* and *special* causes. From an FMEA perspective, common causes can be viewed as failures that are the result of inherent failure mechanisms; as such, they can affect the entire population. This is a cause for examining the design.[3]

On the other hand, special causes are looked at as failures that result from part defects and/or manufacturing problems; they can affect a relatively small population. In this case, there is cause for examining the process.

Customer requisition is, of course, a very strong influence as to why an FMEA is being done. All major automobile companies in their supplier certification standards require an FMEA program from their suppliers.

Courts—through product liability—may require some substantiation regarding the level of reliability of products.

International standards such as the ISO-9000 series may define the program of documentation in a design. Product liability Directive of EC 1985 stipulates that manufacturers of a product will be held liable, regardless of fault or negligence, if a person is harmed or an object is damaged by a faulty or defective product (this includes exporters into the EC market). This liability directive essentially reverses the burden of proof of fault from the injured to the producer. Quality systems incorporating specific tools such as FMEA, fault tree analysis (FTA), or failure mode and critical analysis (FMCA) with safety prevention provisions will be particularly important in protecting a company from unfounded liability claims. Furthermore, proposed safety directives would oblige manufacturers to monitor the safety of their products throughout their foreseeable life.[4]

Even though all the above reasons are legitimate, the most important reason for writing an FMEA is the need to improve. Unless this need is part of the culture of the organization, the FMEA program will not be successful.

VOCABULARY OF FMEA

To understand the FMEA one must understand its language. There are several terms to understand.

Function. The task that a component, subsystem, or system must perform. This function is very important in understanding the entire FMEA process. It must be communicated in a way that is concise, exact, and easy to understand—no jargon. To facilitate this, it is recommended that an active verb be found to describe the function. The active verb by definition defines performance, and performance is what a function is. Examples of this may be found in the following words:

- Position.
- Support.
- Retain.
- Lubricate.

Failure. The inability of a component, subsystem, or system to perform to design intent. This inability can be defined as both known and potential. Of great importance here is that when potential failures in terms of *functional defectives* are identified, the FMEA is fulfilling its mission of prevention.

Functional defectives are failures that do not meet the customer's requirements but are shipped with the failures anyway because:

- The customer will never know the difference.
- The customer will never find out.
- The customer may find out, but whatever is delivered can be used.
- The customer may find out, but whatever is delivered has to be used because there are no alternatives.

Examples of failures are:

- Broken.
- Worn.
- Corrosion.
- Noise.

Causes of Failure. What is the "root cause" of the listed failure? Next to the function, cause of failure is perhaps the most important section of the FMEA. It is here that the way toward preventive and/or corrective action must be pointed out. The more focused people are on the "root cause," the more successful they will be in eliminating failures. In this regard, they must be careful not to be too eager for solutions, because they become victims of symptoms and short term remedies rather than eliminating the real problems.

Examples for Design:

- Wall thickness.
- Vibration.
- Torque specifications.
- Shockloads.

Examples for Process:

- Voltage surge.
- Dull tools.
- Improper setup.
- Worn bearings.

Effects of Failure. The outcome of the failure mode on the system and/or the product. In essence the effects of the failure have to do with the question of: what happens when a failure occurs? The effects of the failure must be addressed from two points of view. The first, Local, in which the failure is isolated and does not effect anything else; the second, Global, in which the failure can and does effect other functions and/or components. It has a domino effect. Generally speaking, the failure with a global effect is more serious than the one of a local nature.

The effect of the failure will also define the severity of a particular failure. Example:

Local—A parking light bulb failure.
Global—Power steering failure.

In the first case, one can identify a nuisance, where in the second case a catastrophic failure is imminent.

Product validation. Controls which exist to prevent the cause(s) of the failure from occurring and to validate repeatability for certain processes (especially with FDA).

Current controls. Controls which exist to prevent the cause(s) of the failure from occurring in the process phase. Examples include:

- Any of the SPC tools.
- Capability.
- Operator(s) training.

CHAPTER 14
FAILURE MODE AND EFFECTS ANALYSIS (FMEA)

Design verification. Controls which exist to prevent causes of the failure from occurring in the design phase. Examples include:

- Design guidelines.
- Design reviews.
- Specific specifications.

THE MECHANICS OF FMEA

Team. To do a complete job with the best results, an FMEA MUST be written by a team. The FMEA should be a catalyst to stimulate the interchange of ideas between the functions affected. A single engineer or any other individual cannot do it.

The team should be made of five to nine persons (preferably five). All team members must have some knowledge of group behavior, the task at hand, and the problem to be discussed; some kind of either direct or indirect ownership of the problem and above all, they *must all* be willing to contribute. Team members also must be cross-functional and multidisciplined. Furthermore, whenever possible and/or needed it is encouraged to have the customer and/or the supplier actively participate.

Design FMEA. A design FMEA is a systematic method to identify and correct any known or potential failure modes before the first production run. A first production run is viewed as the run that produces a product and/or service for a specific customer with the intent of getting paid. This definition is important because it excludes initial sample runs (ISR), trial runs, prototype run(s), etc. The threshold of the first production run is important, because up to that point modifying and/or changing the design is not a major thing. After that point, however, the customer gets involved through the letter of deviation, waiver of change, or some other kind of formal notification. Once these failures have been identified, they are ranked and prioritized.

The leader (the person responsible for a design FMEA) is the most knowledgeable about the design and can best "anticipate" failures. To facilitate the meeting, the Quality Engineer may be designated as the facilitator.

The minimum make-up of the team for the design is:

- Design Engineer.
- Process (Manufacturing) Engineer.

Anyone else who can contribute or who the design engineer feels is appropriate may also participate. A typical design team includes:

- Design engineer.
- Manufacturing engineer.
- Test/Development engineer.
- Reliability engineer.
- Material engineer.
- Field Service engineer.

Of great importance in the make-up of the team is that the team must be cross-functional and multidisciplined. Remember there is *no* such thing as "THE" team. Each organization must define their optimum team participation, recognizing that some individuals may indeed hold two different positions.

The focus of the design is to minimize failure effects on the system by identifying the key characteristics of the design. These key characteristics sometimes may be found as part of the a) customer requirements, b) engineering specifications, c) industrial standards, d) government regulations, or e) courts—through product liability.

The objective of the design FMEA is to maximize the system Quality, Reliability, Cost, and Maintainability. It is important here to recognize that in design there are only three possible areas to look for defects. They are:

- Components. The individual unit of the design.
- Subsystem or subassembly. Two or more combined components.
- System or assembly. A combination of components and subsystems for a particular function.

Regardless of what level the design is in, the intent is the same—no failures on the system.

To focus on these objectives, the team of the design must use consensus for their decision and they must have the commitment of their management.

The timing of the design FMEA is initiated during the early planning stages of the design and is continually updated as the program develops. A team must do the best it can, with what it has, rather than wait until all of the information is in. By then, it may be too late.

Process FMEA. A process FMEA is a systematic method to identify and correct any known or potential failure modes before the first production run. (The first production run was defined in the design section). Once these failures have been identified, they are ranked and prioritized.

The leader (the person responsible for the process FMEA) should be the Process/Manufacturing Engineer primarily because she/he is the most knowledgeable about the process structure and can best "anticipate" failures. To facilitate the meeting the Quality Engineer may be designated as the facilitator.

The minimum make-up of the team for the process is:

- Process/Manufacturing Engineer.
- Design Engineer.
- Operator(s).

Anyone else who can contribute or the process engineer feels is appropriate may also participate. A typical process team includes:

- Process/Manufacturing engineer.
- Design engineer.
- Quality engineer.
- Reliability engineer.
- Tooling engineer.
- Operator(s).

Of great importance in the make-up of the team is that the team must be cross-functional and multidisciplined. Remember there is *no* such thing as "THE" team. Each organization must define their optimum team participation, recognizing that some individuals may indeed hold two different positions.

The focus of the process FMEA is to minimize production failure effects on the system by identifying the key variables. These key variables are the key characteristics of the design, but now in the process, they have to be measured, controlled, monitored, etc. This is where the SPC comes alive.

The objective of the process FMEA is to maximize the system *quality, reliability, and productivity*. Of note here is that the objective is a continuation of the design FMEA—more or less—since the process FMEA assumes the objective to be as designed. Because of this, potential failures which can occur because of a design weakness are not included in a process FMEA. They are only mentioned if those weaknesses effect the process.

The process FMEA does not rely on product design changes to overcome weaknesses in the process, but it does take into consideration a product's design characteristics relative to the planned manufacturing or assembly process to assure that, to the

CHAPTER 14

FAILURE MODE AND EFFECTS ANALYSIS (FMEA)

extent possible, the resultant product meets customer needs, wants, and expectations.

Another important issue in a process FMEA is the fact that it is much more difficult and time consuming than the design FMEA. The reason for this is that in a design FMEA there are three possibilities of analysis; there are six in the process and each one of those may have even more. The six major possibilities for process are:

- Manpower.
- Machine.
- Method.
- Material.
- Measurement.
- Environment.

To show the complexity of each one of these possibilities, take the ''Machine'' for example. Some of the contributing failures may in fact be in one of the following categories. The list is by no means complete.

- Tools.
- Workstation.
- Production line.
- Process itself.
- Gages.
- Operator(s).

As in the design FMEA, the process team must use consensus for their decision and they must have the commitment of their management.

In the final analysis, regardless of what level the FMEA is being performed the intent is always the same—NO FAILURES ON THE SYSTEM.

The process FMEA is initiated during the early planning stages of the process before machines, tooling, facilities, etc., are purchased.

The process FMEA, just like the design FMEA, is continually updated as the process becomes more clearly defined.

FORMS

There are no standard forms available. Each organization has their own, with Automotive Industry Action Group (A.I.A.G.) trying to standardize some of the automotive forms by the end of the 1992 year.

GUIDELINES

The guidelines are numerical values based on certain statistical distributions that allow failures to be prioritized. They are usually of two forms, the Qualitative and the Quantitative. The qualitative form bases the meaning on theoretical distributions such as normal, skewed to the left, and discrete, for the occurrence, severity, and detection respectively.

The Quantitative form uses actual statistical and/or reliability data from the processes. This actual data may be historical and/or current.

In both cases the numerical values are from 1 to 10 and they denote set probabilities from low to high.

The first criteria in the guidelines is the *occurrence*. Occurrence looks at the frequency of the failure as a result of a specific cause. The ranking that is given to this occurrence has a meaning rather than a value; the higher the number the more frequent the failure.

The second criteria in the guidelines is the *severity*. In severity, the seriousness of the effect of the potential failure mode to the customer is assessed. Severity applies to the effect and to the effect *only*. The ranking that is given is typically 1 to 10, with 1 usually representing a nuisance and 10 representing a major noncompliance to government regulations or safety items.

The third criteria in the guidelines is the *detection*. In detection, the probability of a failure getting to the customer is assessed. The numbers 1 to 10 again represent meaning rather than value. The higher the number, the more likely the current controls or design verification failed to contain the failure within the organization. Therefore, the likelihood of the customer receiving a failure is increased.

A general complaint about these guidelines is frequently heard in relationship to consensus. For example, How can a group of people agree on everything? The answer is that they cannot, but that does not mean that a consensus must be reached in order to make a decision. One of the ways to use it is the following.

If the decision falls in an adjacent category the decision should be averaged out. If on the other hand, it falls in more than an adjacent category, a consensus should be reached. Someone on the team may not understand the problem, or some of the assumptions may have been overlooked—or maybe the focus of the group has drifted.

If no consensus can be reached with a reasonable discussion of all the team members, then traditional organizational development processes (group dynamics techniques) must be used to resolve the conflict. Under *no* circumstances should agreement or majority be pushed through, so that an early completion may take place. All FMEAs are time consuming, and the participants as well as their management must be patient for the proper results.

RISK PRIORITY NUMBER (RPN)

This number is the product of the occurrence, severity, and detection. The value should be used to rank the concerns of the design and/or process. In themselves, all RPNs have no other value or meaning.

A threshold of pursuing failures is a RPN equal to or greater than 50 (although there is nothing magical about 50). This threshold can be identified by a statistical confidence. If 99% of all failures must be addressed for a very critical design, what is the threshold? There are 1000 points available ($10 \times 10 \times 10$) from occurrence, severity, and detection; 99% is 990. So, anything equal to or over 10 as a RPN must be addressed. Please note, that this threshold is organization dependent and can change with the organization, the product, and/or the customer.

If there are more than two failures with the same RPN, severity, detection, and occurrence are used in that order of priority. Severity is used first because it deals with the effects of the failure. Detection is used over occurrence because it is customer dependent, which is more important than the frequencies of the failure.

RECOMMENDED ACTION

When the failure modes have been rank ordered by RPN, corrective action is first directed at the highest ranked concerns and critical items. The intent of any recommended action is to reduce the occurrence, detection, and/or severity rankings. Severity will change only with changes in design; otherwise, more often than not, the reductions are expected in either occurrence and/or detection. If no actions are recommended for a specific cause, then this should be indicated. On the other hand, if causes are not mutually exclusive, a DOE recommendation is in order.

CHAPTER 14

FAILURE MODE AND EFFECTS ANALYSIS (FMEA)

In all cases where the effect of an identified potential failure mode could be a hazard to manufacturing or assembly personnel, corrective actions should be taken to prevent the failure mode by eliminating or controlling the cause(s). Otherwise appropriate operator protection should be specified.

The need for taking specific, positive corrective actions with quantifiable benefits, recommending actions to other activities and following up all recommendations cannot be overemphasized. A thoroughly thought out and well developed FMEA will be of limited value without positive and effective corrective actions. It is the responsibility of all affected activities to implement effective followup programs to address all recommendations.[5]

THE PROCESS OF DOING AN FMEA

To do an FMEA effectively requires a systematic approach. The recommended approach is an eight-step method which facilitates both design and process FMEAs.

1. Brainstorm. Try to identify in what direction the group wants to go. Is it design or process? What kind of problems are being encountered in a particular situation? Is the customer involved, or is the company pursuing continual improvement on its own? If the customer has identified specific failures, then the job is much easier, because there is already direction.
2. Process flow chart. Make sure everyone in the team is on the same wavelength. Does everyone understand the same problem? The process flow chart will focus the discussion. If nothing else, it will create a base line of understanding for the problem at hand.
3. Prioritize. Once the problem is understood, the team must begin the analysis of the flow chart. What part is important? Where does the team's work begin?
4. Data collection. Begin to collect data of the failures and start filling out the appropriate form.
5. Analysis. Focus on the data and perform the appropriate analysis. Everything is fair provided it is appropriate. Here, available tools may be QFD, DOE, another FMEA, SPC, and anything else that the group may think suitable.
6. Results. Based on the analysis, results are derived. Make sure the results are data driven.
7. Confirm/evaluate/measure. Once the results have been recorded, it is the time to confirm, evaluate, and measure the success or failure. This evaluation takes the form of three basic questions. They are:
   - Is the situation better than before?
   - Is it worse than before?
   - Is it the same as before?
8. Do it all over again. Regardless of how question seven is answered, improvement must be pursued all over again because of the continuous improvement philosophy. The long-range goal is to eliminate all failures completely and the short term goal is to minimize failures if not eliminate them. Of course, the perseverance for those goals must be considered in relationship to the needs of the organization, customer, and competition.

SUMMARY

FMEA is a tool to identify the priority of failures in both design and process. It demands that a team be utilized to identify several points of view and a corrective action. Furthermore, it utilizes some qualitative and quantitative approaches to prioritize these failures.

Finally, it always focuses on the corrective action with the intent of eliminating failures from the design or the process.

A VERSATILE TOOL TO AID IN DOWNTIME REDUCTION

The Diamond-Star Motors (DSM) plant in Normal, IL, with more than 1,000 programmable logic controllers (PLCs) and 500 robots, continues to be regarded as the most automated automobile assembly plant in the world. DSM employs approximately 3100 people and has an annual production capacity of 240,000 vehicles. The vehicles Diamond-Star produces are the Mitsubishi Eclipse, Eagle Talon, and Plymouth Laser.

At this level of automation, downtime is costly. For a plant considered state of the art in most aspects, the means of determining the status in each shop was archaic. All pertinent data (downtime, cycle times, production status, etc.) was gathered manually. Discrepancies often existed in the data tallies, and it became difficult to determine what the real problems were. Therefore, a system that would locate and troubleshoot line problems as they occurred was devised in an effort to minimize down time and maximize production. The system also acquires and analyzes production data in real time which permits identification of problem areas, optimizes equipment waiting times, identifies trends, improves just-in-time processes, and produces shift reports.

BODY SHOP IS THE COMMUNICATION TARGET

Diamond-Star's Body Shop was targeted as a test bed for this data acquisition system. The system needed to be versatile in its ability to present data, able to change as the shop changed, and grow as the shop grew, and be easy to maintain and modify as necessary.

Shop status information and data is presented in four ways: computer screens, printed reports, light boards, and message displays. Three local computers on an ethernet and one remote unit, via high speed data modems, make available information from 200 computer screens, graphically representing the assembly lines in the shop. Information about cycle times on the line stations, waiting time periods on other equipment, and trends in cycle times are also available from the network and remote units.

Reports are automatically generated on these computers at each shift's end, to either a floppy disk or a printer. The reports show a log of all faults for that shift, with the on time, off time, and duration of each fault. The accumulated amount of time that a station caused the production line to wait—and the number of

CHAPTER 14

A VERSATILE TOOL TO AID IN DOWNTIME REDUCTION

these occurrences—is also recorded. All reports generated are also recorded on a hard drive as historical data for future reference.

To display shop conditions to production personnel on the floor, an incandescent annunciator light board (built by CEC Controls Company, Detroit, MI) was designed with white plexiglass windows arranged to appear as an aerial view of the Body Shop (see Fig. 14-22).

Each window represents an assembly line. Behind each window are green, yellow and red bulbs. If all conditions are good, the window lights green. If a line is waiting on another, either upstream or downstream, the window changes to yellow. If there is a fault at any station on the line, the line window changes to red. Three of these boards are placed in key areas of the Body Shop. One is a double-sided sign, the other two are single-sided.

A large Uticor PMD 3000 message display hangs beneath each board. When a line window on the annunciator board turns red, a text message scrolls across the message display identifying the station at fault and the nature of the fault.

The entire system, from the PLC programs created to access all faults to the computers and communication hardware, was designed with expansion in mind.

ACQUIRING 17,400 DATA POINTS

Adding a new fault at a particular station takes only a few keystrokes at the PLC and the computer. Each PLC program has several inactive lines of logic in place, configured as spare bit addresses. The software in each computer is configured to recognize these bits and has the text (spare 1, spare 2, etc.) attached to them. To add a fault, the correct fault bit is placed in the inactive PLC logic—thereby activating that address—and the text for the corresponding bit at the computer is changed from spare to the name of that fault.

To physically add a new line of equipment to the system, all that is needed in hardware is the proper PLC serial communication card and an appropriate length of RS422 twisted pair wire. There are numerous spare communication ports at the computer to handle significant future expansion.

To eliminate the need for a special person to maintain this system, large frame computers and less familiar operating systems were avoided. The entire system runs on personal computers (PCs), with Digital Research's DR DOS 6.0 and Novell's Netware Lite as the operating systems. Body Shop maintenance personnel are currently undergoing training that will provide them with the

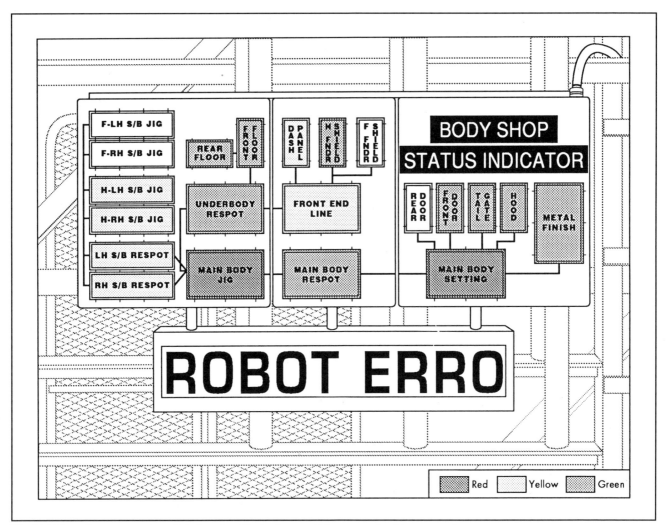

Fig. 14-22 Incandescent annunciator light board to display shop conditions to production personnel.

CHAPTER 14

A VERSATILE TOOL TO AID IN DOWNTIME REDUCTION

knowledge needed to take full responsibility for the maintenance of this system.

The complete system is comprised of 270 Mitsubishi MELSEC A series PLCs, and seven Pcs, with Xycom's industrial computing equipment in the shop at the heart of the network and standard IBM compatible Pcs elsewhere.

Running on four of these Pcs is a data acquisition and control software called Citect, with a fifth PC running a software called Scanner. Both of these software products are available from Control Instrumentation, Ltd. of Australia. The sixth PC runs a custom C program to control the Uticor message display units. A seventh PC was added as a supervisory computer, which can diagnose the network internally with a Citect utility called Probe, or call the network hub via modem and perform an external diagnosis. The remote Citect software can also be used at the supervisory computer to call the Citect 1 or Citect 2 computers and observe the status of the shop.

The Xycom 4190 computer was the choice for a shop display computer with its Nema 4 rating, dual keyboard, and 19" monitor. The system design required the data acquisition and control software to have multiline communication capabilities. It also had to handle very large numbers of I/O points (total points configured to date are 15388 digital and 2014 analog), and it needed to support MELSEC PLCs. To meet these requirements, the Scanner and Citect software packages were selected. The Scanner PC's sole function is to communicate with the PLCs, making information available to the Citect Pcs over the ethernet as required. With the work divided between the Scanner and Citect Pcs in this fashion, the system performs well.

At one end of this system lie 270 PLCs: 264 of these are on 24 individual data link networks (MELSECS proprietary network), six are stand-alone devices. Each data link network is comprised of a master PLC and various local PLCs. All local PLCs were programmed to share pertinent data with their master PLCs. The masters were then programmed to organize all local PLC data and their own data in a fashion easily read by the Scanner. This type of optimization was extremely important on a system of this size. The 24 data link masters and 6 stand-alone PLCs were then connected to the Scanner PC by means of a serial communication card at each of these 30 PLCs, 10,000 feet of RS422 twisted pair wire, and a DigiBoard 16-port serial communication card mounted in the Scanner.

Seven of the sixteen communication ports are currently being used. Each of the ports has three to six PLCs multi-dropped from it. By connecting these 30 PLCs to the Scanner, data from all 270 PLCs is obtainable. The Scanner PC is a passive backplane industrial PC with a Xycom 4100-AT4 processor card (25 MHZ 486) in place as the CPU. All Pcs in this system are either 25 or 33 MHZ 486 machines.

The DigiBoard is a PC/16i 12.5 MHZ serial communication card. An Intel Ether Express 16 ether adapter permits connection to the local area network (LAN) in order to share information with the Citect machines. All LAN products, twisted pair medium access units (MAUs), Fiber Hub, Fiber MAU, and twisted pair hub are produced by David Systems. Figures 14-23 and 14-24 show a System Overview and System Hardware Detail.

There are four Citect machines on this network. Each Citect machine is equipped with a DSI-32 processor card to run the Citect software.

This unit resides in the Body Shop Maintenance area. Citect 1 also transmits all faults serially as they occur to another machine, running a custom C program that sorts, prioritizes, and sends the faults to the Uticor message displays.

Citect 2 is at the other end of a 1,400 foot glass fiber line in the engineering office. This PC is a standard grade office machine and runs a database that is nearly identical to the database in Citect 1. Citect 4 resides in the executive offices and runs a streamlined version of the Citect 2 database. Citect 1, Citect 2, and Citect 4 generate reports, and each has the 200 graphic screens mentioned previously showing shop conditions in great detail. To provide ease of maneuverability, a simple system was devised that uses eight function keys generically throughout the majority of the screens. The first screen is an overview of the Body Shop, with representations of every assembly line in the shop. When on this screen, the function keys provide information on a shop-wide basis. By moving the cursor to an assembly line and clicking on the line or by typing the highlighted letters in the line name, the screen can be changed to a view of that line.

Once on a line screen, the same options are available—providing information about stations on that line only. Most keyboard options are shown on menus on each screen. All screens have a *help* page, and advanced keyboard options are visible there. This is an intuitive system that requires minimal training to use. For an example of screen keyboard options, see Fig. 14-25.

Citect 3 performs calculations needed by the other three Citect computers for reports. It also controls various PLC bits, including the outputs that drive the annunciator light boards. The Scanner, Citect 3, and the message control computer all reside in an enclosure beneath Citect 1. The remote Citect can run at any off-site location via high speed data modems. It allows full control over all functions configured in Citect 1 or Citect 2.

References

1. Joseph T. Vesey, "The New Competitors: Thinking in Terms of 'Speed-to-Market,'" *Manufacturing Systems* (June 1991), p. 20.
2. Dean Stamatis, *Failure Mode and Effect Analysis: Training Manual* (Southgate, MI: Contemporary Consultants Co., 1989, 1991).
3. D. Denson, "The Use of Failure Mode Distributions in Reliability Analysis," *MCNewsletter* (Spring 1992).
4. Dean Stamatis, "ISO-9000 Series: Are They Real?" *Technology* (Detroit Engineering Society, 1992).
5. Dean Stamatis, *Potential Failure Mode and Effect Analysis*, Ford Motor Co., 1989.

Bibliography

R. Adams, *Source Book of Automatic Identification and Data Collection*, Van Nostrand Reinhold, 1990.
D. Collins, and J. Whipple, *Using Bar Code: Why It's Taking Over*, Data Capture Institute, 1990.
C. Harmon, and R. Adams, *Reading Between the Lines*, Helmers Publishing, Inc., 1989.
R. Palamer, *The Bar Code Book*, Helmers Publishing, Inc., 1989.

CHAPTER 14

A VERSATILE TOOL TO AID IN DOWNTIME REDUCTION

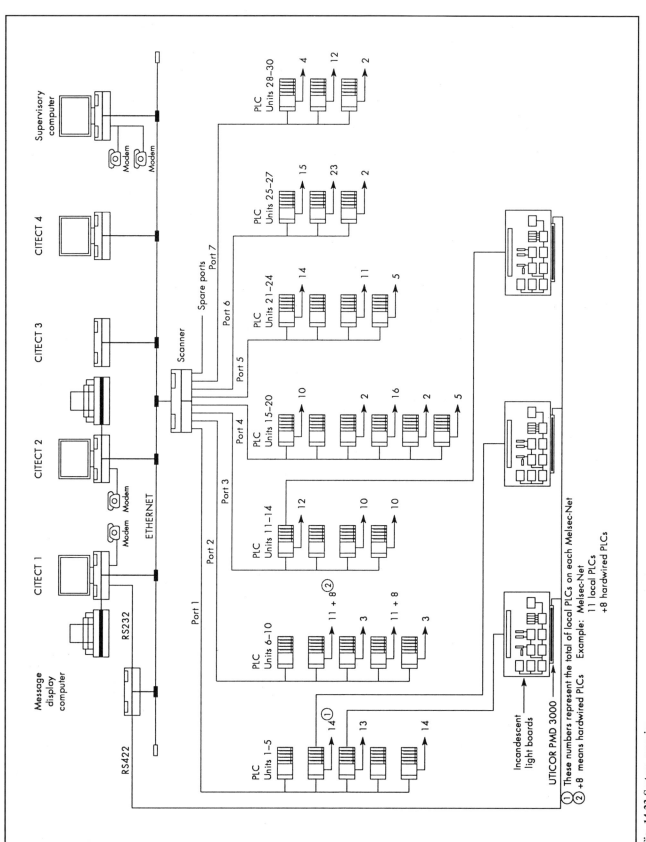

Fig. 14-23 System overview.

CHAPTER 14
A VERSATILE TOOL TO AID IN DOWNTIME REDUCTION

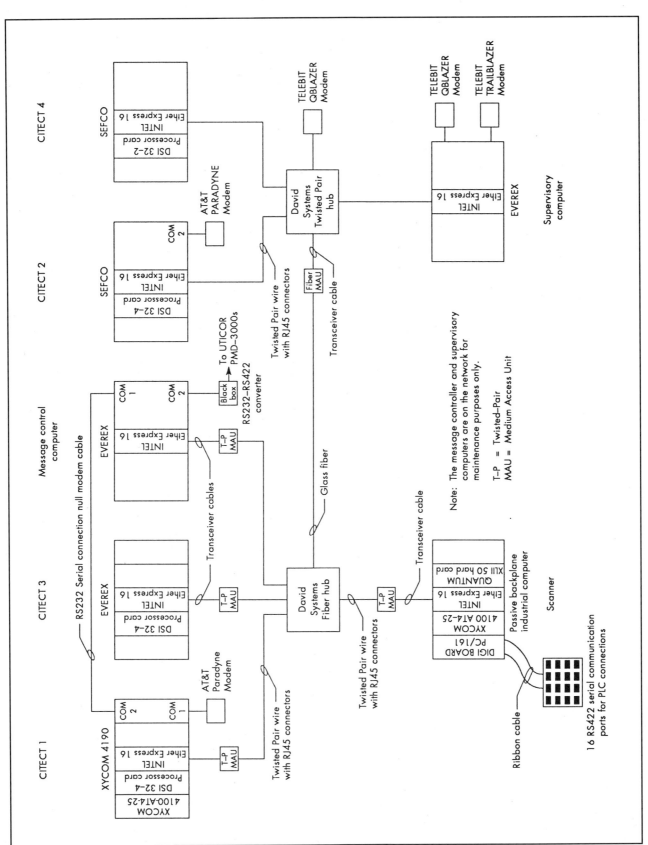

Fig. 14-24 System hardware detail.

CHAPTER 14

A VERSATILE TOOL TO AID IN DOWNTIME REDUCTION

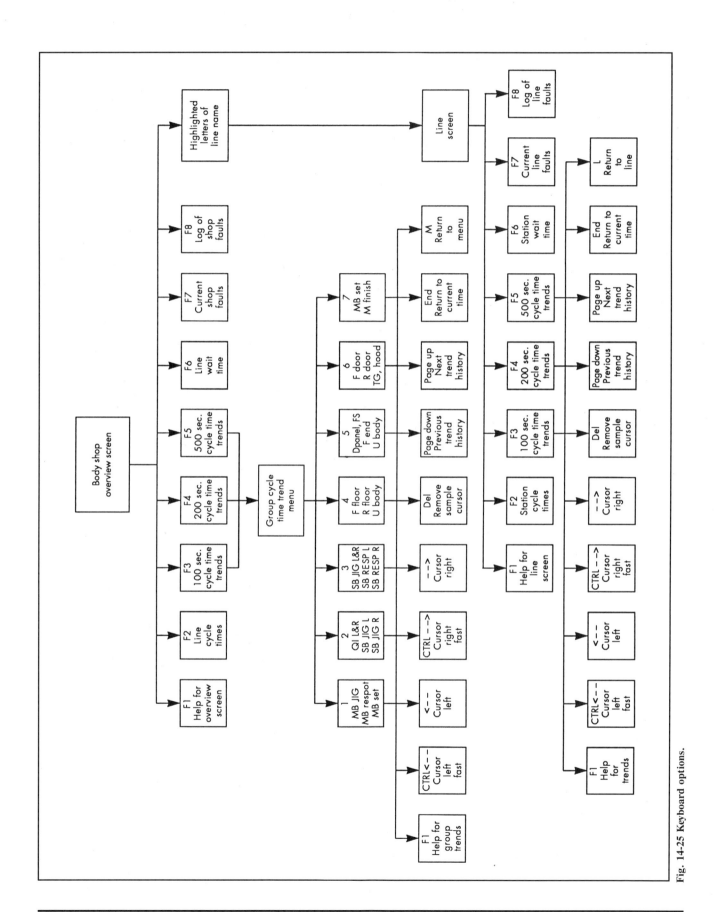

Fig. 14-25 Keyboard options.

CHAPTER 15

TOTAL PRODUCTIVE MAINTENANCE

Total productive maintenance (TPM) is a technique evolved by the Japanese in the 1960s from maintenance practices they learned from the United States during the 1950s. Total productive maintenance could possibly be a misnomer for the technique, since it goes beyond just maintenance. Total productive maintenance is a management technique that involves everyone in a plant or facility in equipment or asset utilization.

In most companies today, management is looking for every possible competitive advantage. There are companies focusing on total quality programs, just-in-time (JIT) programs, and total employee involvement (TEI) programs. These programs require complete management commitment and support if they are to be successful. However, no matter how management works to make these programs produce results, it is impossible to be successful without setting up a TPM program.

This may seem a bold statement, however, it is impossible to make the previously mentioned programs function without equipment or assets that are currently maintained in a World Class condition. Consider:

Is it possible to produce quality products on poorly maintained equipment?

Can quality products come from equipment that is consistently out of specification or worn to the point that it can not consistently hold tolerance? Can a JIT program work with equipment that is unreliable or has low availability?

Can employee involvement programs work for long if management ignores the pleas to fix the equipment or get better equipment so a "World Class" product can be delivered to the customer on a timely basis, thus satisfying the employee concerns and suggestions?

Many companies are beginning to realize that the management techniques and methods previously used to maintain equipment are no longer sufficient to compete in the world market. Attention is beginning to focus on the technique of TPM. However, the number of companies successfully implementing a TPM program is relatively small. The reason is that many companies try to use TPM to compensate for an immature or dysfunctional maintenance operation. They fail to realize that TPM is an evolutionary step, not a revolutionary one. In order to understand this statement it is necessary to consider the evolution of a typical quality program in a company.

In Fig. 15-1, the various stages of the maturity of a quality program are highlighted along the bottom of the arrow. In the early days, the company would ship almost anything to the customer. If the product did not meet their standards, nothing was done about it until the customer complained and shipped it back. However, this eventually became costly when competitors would ship products that the customer would accept, since there was no quality problem. In order to remain competitive, the company would have to make changes in the way they did business.

The second step in maturity was to begin inspecting the product in the final production stage, or in shipping just before it was loaded for delivery to the customer. While this was better than before (since it reduced the number of customer complaints) the company realized that it was expensive to produce a product only to reject it just before it was shipped. They realized that it would be more economical to find the defect earlier in the process, and eliminate running defective material through the rest of the production process.

This led to the third step in the path to quality maturity, the development of the quality department. This department's responsibility was to monitor, test, and report on the quality of the product as it passed through the plant. This seems much more effective than before, with the defects being found earlier, even to the point of statistical techniques being used to anticipate or predict when quality would be out of limits. However, there were still problems. The more samples the quality department was required to test, the longer it would take to get the results back to the Operations Department. It was still possible to produce minutes', hours', or even shifts' worth of product that was defective or out of tolerance, before anyone told the operator to stop.

Solving this problem led to the fourth step in the path to quality maturity, that is, training the operators in the statistical techniques necessary to monitor and trend their own quality. In this way, the term "quality at the source" was developed. This step enabled the operator to know down to the individual part, when it was out of tolerance—and no further defective components were produced. This eliminated the production of any more defects and prevented rework and expensive downstream scrap. However, there were still circumstances beyond the control of the operator that contributed to quality problems.

CHAPTER CONTENTS:

OVERALL EQUIPMENT EFFECTIVENESS 15-4

A SAMPLE OEE CALCULATION 15-5

HOW TO DEVELOP A SITE-SPECIFIC TPM PLAN 15-6

CONTINUOUS IMPROVEMENT IN A TPM ENVIRONMENT 15-8

A TYPICAL TPM TRAINING SESSION 15-9

The Contributors of this chapter are: **Terry Wireman**, *President, Wireman & Associates, Eaton, OH;* **Kathleen Whitcomb**, *TQC Manager, The Goodyear Tire & Rubber Company.*
The Reviewers of sections of this chapter are: **Fred Florian**, *TPM Resource, Fibers, DuPont;* **Robert K. Hall**, *Manager, Time-Based Study, American National Can;* **Jimmy Hendrix**, *Training Specialist, CPD Engineering, Sonoco Products Co.*

CHAPTER 15

TOTAL PRODUCTIVE MAINTENANCE

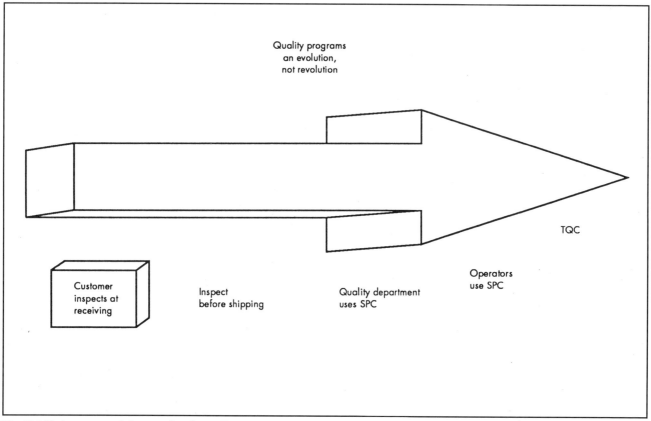

Fig. 15-1 Various stages of the maturity of a quality program.

This led to the next step, the involvement of all departments of the company in the quality program. This meant from the product design phase, through the purchasing of raw materials, to the final production and shipping of the product, everyone realized that producing a quality product for the lowest price, the highest quality, and the quickest delivery was the goal of the company. This meant that products were designed for producibility, the materials used to make the product had to be of the highest quality, and the production process had to be closely monitored to ensure that the final product was of perfect quality. The company had evolved to the stage of maturity necessary to be a "World Class" company.

How does this path to maturity compare with the path to maturity for asset or equipment maintenance? Figure 15-2 highlights the comparison. In Stage 1 of the path to TPM maturity, the equipment is not maintained or repaired unless the customer (operations, production, or facilities) complains that it is broken. Only then will the maintenance organization work (or in some cases be allowed to work) on the equipment. However, over time companies began to realize that when equipment breaks down it always costs more and takes longer to fix. This cost is compounded when the actual cost of the downtime is calculated. The company begins to question the policy, realizing that it is cost-effective to allow the equipment to be shut down for smaller amounts of time for minor service to reduce the frequency and duration of the breakdowns.

This leads to the second step on the road to TPM maturity, the establishment of a good preventive maintenance program. This step allows for the inspection and routine servicing of the equipment before it fails. This step results in fewer breakdowns and equipment failures. In effect, the product is inspected before the "customer" gets it. Some techniques used here will include routine lubrication and inspections for major defects.

The second step, while producing some results, is not sufficient to prevent many failures. The third step is to begin the utilization of predictive and statistical techniques for monitoring the equipment. At this point, the "hidden" problems are discovered before they develop into major problems. Some techniques used include:

- Vibration analysis.
- Spectrographic oil analysis.
- Thermographic or infrared temperature monitoring.
- Nondestructive testing.
- Sonics.

The information produced from proper utilization of these techniques reduces the number of breakdowns to a low level, with overall availability being over 90%. However, the quest for continuous improvement emphasizes the need to do better. This leads to the fourth step: involvement of the operators in maintenance activities.

This step does not mean that all maintenance activities are turned over to the operators. Only the basic tasks are included: some inspection, the basic lubrication, adjusting, and routine cleaning of the equipment. The reason it is more effective to have the operators involved in these activities is that they are the ones who know when something is not right with the equipment. In actual practice the tasks they take over are the ones that the

CHAPTER 15

TOTAL PRODUCTIVE MAINTENANCE

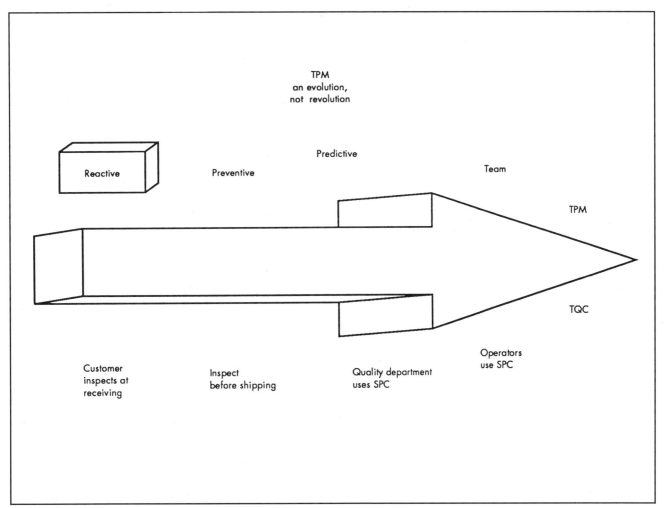

Fig. 15-2 Various stages of the maturity of a TPM program.

maintenance technicians have trouble finding the time to do. Freed of the burden of doing some of the more routine tasks, the maintenance technicians can concentrate on refining the predictive monitoring and trending of the equipment. They also will have more time to concentrate on equipment failure analysis, which will prevent future or repetitive problems on the equipment. This step increases the availability of the equipment, but also the reliability of the equipment over its useful life.

The last step of the evolution process is the involvement of cross-functional teams. This allows for problem solving for habitual or troublesome equipment problems. The cross-functional teams will utilize team members from various organizational disciplines to produce total solutions for these problems. The team members realize the need and importance of each team member. In a spirit of understanding and cooperation, the teams will allow for production and service to reach "World Class" standards.

What are the obstacles to achieving these goals? They can be divided into three main categories:

- Lack of management support and understanding.
- Lack of sufficient training.
- Failure to allow sufficient time for the evolution.

The lack of management support is related to management not completely understanding the true goal of the program. For example, if management begins the program by emphasizing its desire to eliminate maintenance technicians, they have failed to understand the true purpose of the program. The real goal is to increase the equipment's effectiveness, not reduce the labor head count. Without management understanding of the true goal of asset utilization, the program is doomed to failure.

Lack of training is another major obstacle, since the training must be given at two different levels. The first is the increased skills required for the maintenance technicians. The technicians will become skilled in advanced maintenance techniques, such as predictive maintenance. They also must have extensive training and guidance in advance data analysis to find and solve equipment failure and effectiveness problems.

The second area for training is the operator's training. The operators must be trained to do basic maintenance on their equipment. Subjects such as lubrication, torque requirements, proper cleaning techniques, all must be taught in detail to the operators. Also, if the operators are doing any repairs, they must receive training to be certified to do the assigned tasks. Without this level of training, the equipment's effectiveness will decrease instead of increase.

CHAPTER 15

TOTAL PRODUCTIVE MAINTENANCE

The last step is allowing enough time for the evolution to occur. The change from a reactive program to a proactive program will take time. By some estimates it may be a three- to five-year program to achieve a competitive position. By failing to understand this point, many managers condemn their programs to failure before they ever get started.

Successful TPM programs focus on specific goals and objectives. When the entire organization understands the goals and how they impact the competitiveness of the company, the company will be successful. There are five main objectives for any TPM program:

1. Insure equipment capacities.
2. Implement a program of maintenance for the entire life of the equipment.
3. Require support from all departments involved in the use of the equipment or facility.
4. Solicit input from all employees from all levels of the company.
5. Use consolidated teams for continuous improvement.

Ensuring equipment capacity ensures that the equipment performs to specifications. It operates at its design speed, produces at the design rate, and produces a quality product at these speeds and rates. The problem is that many companies do not even *know* the design speed or rate of production of the equipment. This allows management to set arbitrary production quotas. A second problem is that over time, small problems cause operators to change the rate at which they run equipment. As these problems continue to build, the equipment output may only be 50% of what it was designed to be. This will lead to the investment of additional capital in equipment, trying to meet the required production output. This point is further emphasized in the equipment effectiveness section that follows.

Implementing a program of maintenance for the life of the equipment is analogous to the popular preventive and predictive maintenance programs companies presently use to maintain their equipment. This step has a basic difference; it changes just as the equipment changes. All equipment requires different amounts of maintenance as it ages. A good preventive/predictive maintenance program takes these changing requirements into consideration. By monitoring failure records, trouble calls, and basic equipment conditions, the program is modified to meet the changing needs of the equipment.

A second difference is that this program involves all employees, from the operator to upper management. The operator may be required to perform basic inspecting, cleaning, and lubricating of the equipment, which is really the front-line defense against problems. The upper managers may be required to assure that maintenance gets enough time to properly finish any service or repairs required to keep the equipment in condition so that it can run at designed ratings.

Requiring the support of all departments involved in the use of the equipment or facility will insure full cooperation and understanding of affected departments. For example, including maintenance in equipment design/purchase decisions assures that equipment standardization will be considered. The issues surrounding this topic alone can contribute significant financial savings to the company. Standardization reduces inventory levels, training requirements, and startup times. Proper support from stores and purchasing can help reduce downtime, but more importantly it will aid in the optimization of inventory levels, thus reducing on-hand inventory.

Soliciting input from employees at all levels of the company provides employees with the ability to contribute to the process. In most companies this step takes the form of a suggestion program. However, it needs to go beyond that; it should include a management with "no doors." This indicates that managers, from the front line to the top, must be open and available to listen to and consider employee suggestions. A step further is the response that should be given to each discussion. It is no longer sufficient to say "That won't work" or "We are not considering that now." In order to keep communication flowing freely, reasons must be given. It is just a matter of developing and using good communication and management skills. Without these skills, employee input will be destroyed and the ability to capitalize on the greatest savings generator in the company will be lost.

The development of consolidated teams for continuous improvement is a development beginning with Step 4. The more open management is to the ideas of the work force, the easier it is for the teams to function. These teams can be formed by areas, departments, lines, process, or equipment. They will involve the operators, maintenance, and management personnel. They will, depending on the needs, involve other personnel on an "as-needed" basis, such as engineering, purchasing, or stores. These teams will provide answers to problems that some companies have tried years to solve independently. This team effort is one of the true indicators of a successful TPM program.

OVERALL EQUIPMENT EFFECTIVENESS

The overall equipment effectiveness (OEE) is the benchmark used for TPM programs. The OEE benchmark is established by measuring the equipment performance.

MEASURING EQUIPMENT EFFECTIVENESS

Measuring equipment effectiveness must go beyond just the availability or the uptime. It must factor in all issues related to equipment performance. The formula for equipment effectiveness must look at the availability, the rate of performance, and the quality rate. This allows all departments to be involved in determining equipment effectiveness. The formula could be expressed as:

$$\text{Availability} \times \text{Performance rate} \times \text{Quality rate} = \text{Equipment effectiveness} \qquad (1)$$

The availability is the required availability minus the downtime, divided by the required availability. Expressed as a formula this would be:

$$\frac{\text{Required availability} - \text{Downtime}}{\text{Required availability}} = \text{Availability}. \qquad (2)$$

The required availability is the time production is to operate the equipment, minus the miscellaneous planned downtime, such as breaks, scheduled lapses, meetings, etc. The downtime is the

CHAPTER 15

A SAMPLE OEE CALCULATION

actual time the equipment is down for repairs or changeover. This is also sometimes called breakdown downtime. The calculation gives the true availability of the equipment. It is this number that should be used in the effectiveness formula. The goal for most Japanese companies is a number greater than 90%.

The performance rate is the ideal or design cycle time to produce the product multiplied by the output and divided by the operating time. This will give a performance rate percentage. The formula is:

$$\frac{\text{Design cycle time} \times \text{Output}}{\text{Operating time}} = \text{Performance rate} \qquad (3)$$

The design cycle time (or production output) will be in a unit of production, such as parts per hour. The output will be the total output for the given time period. The operating time will be the availability from the previous formula. The result will be a percentage of performance. This formula is useful for spotting capacity reduction breakdowns. The goal for most Japanese companies is a number greater than 95%.

The quality rate is the production input into the process or equipment minus the volume or number of quality defects divided by the production input. The formula is:

$$\frac{\text{Production input} - \text{Quality defects}}{\text{Production input}} = \text{Quality rate} \qquad (4)$$

The product input is the unit of product being fed into the process or production cycle. The quality defects is the amount of product that is below quality standards (not rejected, and there is a difference) after the process or production cycle is finished. The formula is useful in spotting production quality problems, even when the poor quality product is accepted by the customer. The goal for the Japanese based companies is a number higher than 99%.

Combining the total for the Japanese goals, it is seen that:

$$90\% \times 95\% \times 99\% = 85\%$$

To be able to compete for the national TPM prize in Japan, the equipment effectiveness must be greater than 85%. It is sad to say that the equipment effectiveness in most U.S. companies barely breaks 50%. It is little wonder there is so much room for improvement in typical equipment maintenance management programs.

A SAMPLE OEE CALCULATION

A plastic injection molding plant had one press with the following information:

- The press was scheduled to operate 15 8-hour shifts per week.
- This gave a total possibility of 7200 minutes of run time per week.
- Planned downtime for breaks, lunches, and meetings totaled 250 minutes.
- The press was down for 500 minutes for maintenance for the week.
- The changeover time was 4140 minutes for the week.
- The total output for the operating time was 15,906 pieces.
- The design cycle time was 9.2 pieces per minute.
- There were 558 rejected pieces for the week.

What is the OEE for the press for the above week?

A form to collect and analyze OEE information is pictured in Fig. 15-3. The equipment availability is calculated in the first section of the form.

The Gross time available for the press is entered in line 1. The planned downtime, which involves activities that management sets a priority on and cannot be eliminated, is entered in line 2 (the 250 minutes for the week). The net available time for operation is entered in line 3 (This is actually line 1 minus line 2). The downtime losses, which are all unplanned delays, are entered in line 4. This would include maintenance delays, changeovers (which can be minimized), setups, adjustments, etc. The actual time the press operated is entered on line 5 (This is the difference between lines 3 and 4). The equipment availability (line 6) is line 5 divided by line 3 times 100%.

The equipment operating efficiency is calculated in the next section. The total output for the operating time is entered in line 7. The actual design cycle time (this number must be very accurate) is entered on line 8. The operational efficiency is

Overall equipment effectiveness	
1. Gross time available (8 X 60 = 480 minutes) X 15 turns	7,200 minutes
2. Planned downtime (for PM, lunch, breaks)	250 minutes
3. Net available run time (1 - 2)	6,950 minutes
4. Downtime losses (breakdowns, setups, adjustments)	4,640 minutes
5. Actual operating time (3 - 4)	2,310 minutes
6. Equipment availability (5 / 3 X 100)	33%
7. Total output for operating time (pieces, tons)	15,906 pieces
8. Design cycle time	0.109 minutes / piece
9. Operational efficiency (8 X 7 / 5 X 100)	75%
10. Rejects during shift	558 pieces
11. Rate of product quality (7 - 10 / 7 X 100)	96.8%
12. OEE (6 X 9 X 11)	23.96%

Fig. 15-3 Form to collect and analyze OEE information.

CHAPTER 15

HOW TO DEVELOP A SITE-SPECIFIC TPM PLAN

calculated and entered on line 9. The operational efficiency is line 7 (the total output) times line 8 (design cycle time) divided by line 5 (the actual operating time) times 100%. This number should be evaluated carefully to insure that the correct design capacity was used. If the percentage is high or exceeds 100% then the wrong design capacity was probably used.

The quality rate is determined by the total output for the operating time (line 7) minus the number of rejects for the measured period (line 10) divided by the total output (line 7) times 100%.

In the sample, the availability is 33%, the operational efficiency is 75%, and the quality rate is 96.8%. The OEE for the press for the week is 23.96%.

What do the present conditions mean? What do the indicators show the typical manufacturer? The answers are evident when a second model using the same press is examined. In Fig. 15-4, all of the parameters are set at "world class" standards to give an OEE of 85%. As can be quickly observed, the major improvement is in the total output for the operating time (line 7).

The press now will make 54,516 parts, compared to 15,348 with the 23.96% OEE. Since the resources to make the parts (labor and press time) are the same, it makes the company more product and ultimately more profits. With the press operating at an OEE of 85%, it is the same result as 3.5 presses running at the 23.96% OEE. The potential for increased profitability and ultimate competitiveness is staggering.

HOW TO DEVELOP A SITE-SPECIFIC TPM PLAN

The development of a TPM plan begins with establishing the current status of the organization. This evaluation will include more than just the maintenance organization; it should include all parts of the organization involved in the operation, maintenance, design, and purchasing of the assets for the company. The most important benchmark for TPM is the OEE. By establishing the current OEE for the plant assets, the organization is given a goal to work toward. The primary objective must be to optimize the equipment effectiveness. Any other objective fails to unite the various parts of the organization into a competitive, focused unit.

When considering the maintenance part of the organization, the evolution chart (shown previously) helps to determine the basic foundation. The first focus is on a preventive maintenance program. Without a good preventive maintenance program, the equipment is never maintained to a sufficient level to ensure that the organization has assets that are capable of producing a world class product or service.

The predictive maintenance program is further utilized to monitor equipment and ensure that wear trends are monitored. In this way, equipment can be overhauled, with worn components being changed, before a failure occurs. Figures 15-5 through 15-9 show typical inspection forms for hydraulic systems, pneumatic systems, belts, chains, and general equipment respectively.

The improvement in the organization's ability to service the equipment is followed by an increase in the amount of data utilized to perform failure and engineering analysis. This highlights the need for a computerized database for the tracking and trending of equipment histories, planned work, the preventive maintenance program, the maintenance spare parts, the training and skill levels of the maintenance employees, etc. The systems commonly used for this are called computerized maintenance management systems (CMMS). These systems are similar to computerized systems for other organizational disciplines in that they will be configured to track the data required to support the maintenance function—provided that all parts of the organization requiring service from maintenance actually use the system. With the current market (1992) showing over 300 packages commercially available, there should be a package suitable to every company's needs. These packages appear in annual listings in various trade publications such as *Engineer's Digest*, *Plant Engineering*, and *Maintenance Technology*.

The CMMS system will also be further utilized by the operations and engineering groups as the program matures. These groups will be required to assist in putting data they are tracking into the CMMS. For example, an engineer making a change in a piece of equipment would be responsible for entering the change in the equipment information section of the CMMS. Also, an

```
              Overall equipment effectiveness

 1. Gross time available                    7,200 minutes
    (8 X 60 = 480 minutes) X 15 turns
 2. Planned downtime                          250 minutes
    (for PM, lunch, breaks)
 3. Net available run time                  6,950 minutes
    (1 - 2)
 4. Downtime losses                           695 minutes
    (breakdowns, setups, adjustments)
 5. Actual operating time                   6,255 minutes
    (3 - 4)
 6. Equipment availability                           90%
    (5 / 3 X 100)
 7. Total output for operating time        54,516 pieces
    (pieces, tons)
 8. Design cycle time                  0.109 minutes / piece
 9. Operational efficiency                           95%
    (8 X 7 / 5 X 100)
10. Rejects during shift                      545 pieces
11. Rate of product quality                          99%
    (7 - 10 / 7 X 100)
12. OEE                                              85%
    (6 X 9 X 11)

 This produces 3.5 times as much product as Fig. 15-3
```

Fig. 15-4 Completed form from Fig. 15-3.

CHAPTER 15
HOW TO DEVELOP A SITE-SPECIFIC TPM PLAN

Hydraulic Inspection	Item	O.K.	Needs Repair
Hydraulic pump	Proper Oil Flow?		
	Proper Pressure?		
	Excessive Noise?		
	Vibration?		
	Proper Mounting?		
	Excessive Heat?		
Intake Filter	Clean?		
	Free Oil Flow?		
Directional Control Valves	Easy Movement?		
	Proper Oil Flow?		
Relief Valve	Proper Pressure?		
	Excessive Heat?		
Lines	Properly Mounted?		
	Oil Leaks?		
	Loose Fittings?		
	Damaged Piping?		

Fig. 15-5 A typical inspection form for hydraulic systems.

Pneumatic Inspection	Item	O.K.	Needs Repair
Compressor	Proper Air Flow?		
	Proper Pressure?		
	Excessive Noise?		
	Vibration?		
	Proper Lubrication?		
	Excessive Heat?		
Inlet Filter	Clean?		
	Free Air Flow?		
Directional Control Valves	Easy Movement?		
	Proper Air Flow?		
Muffler	Proper Air Flow?		
	Proper Noise Reduction?		
Lines	Properly Mounted?		
	Air Leaks?		
	Loose Fittings?		
	Damaged Piping?		

Fig. 15-6 A typical inspection form for pneumatic systems.

CHAPTER 15

A SAMPLE OEE CALCULATION

Belt Inspection	Item	O.K.	Needs Repair
Belt Sheave	Sidewall Wear?		
	Dirt in Sheave?		
	Alignment		
	Mounting		
	Support Bearings		
	Guards		
Belt	Stretch?		
	Narrow Spots?		
	Tension		
	Tracking of Belt		
	Contamination		

Fig. 15-7 A typical inspection form for belts.

Chain Inspection	Item	O.K.	Needs Repair
Chain Sprocket	Hooked Teeth?		
	Wear on Tooth Sides		
	Alignment		
	Mounting		
	Support Bearings		
	Guards		
Chain	Elongation		
	Side Plate Wear		
	Tension		
	Engagement of Sprocket		
	Proper Lubrication		

Fig. 15-8 A typical inspection form for chains.

operator performing a preventive maintenance task or service would be responsible for entering the necessary completion information into the CMMS. Effective use of the CMMS generates the database necessary to optimize the maintenance resources and help them achieve the equipment availability required to be World Class.

CONTINUOUS IMPROVEMENT IN A TPM ENVIRONMENT

As the TPM program matures, the company sees how much support maintenance gives to other World Class programs. The JIT program is enhanced, since now the equipment runs when it is scheduled and produces at the rate it was designed to operate. The total quality control program is enhanced, since the equipment is stable and produces a quality product every time it operates. The TEI program is matured, since most of the problems the employees are currently having relate to the equipment. This makes their

CHAPTER 15
A TYPICAL TPM TRAINING SESSION

Equipment Inspection		Item	O.K.	Needs Repair
Main Equipment Cleaning	Contamination	Dust		
		Dirt		
		Oil		
		Foreign Material		
Main Equipment Wear	Mechanical	Loose Nuts		
		Missing Nuts		
		Excessive Play		
Main Equipment Lubrication	Oil	Proper Level		
		Leakage?		
		Covers?		
Main Equipment Lubrication	Grease	Fittings Clean?		
		Proper Grease?		
		Proper Amount?		
5 Senses Inspection		Temperature?		
		Noise?		
		Vibration?		
		Visual?		
		Odors?		

Fig. 15-9 A typical inspection form for general equipment.

responsibilities easier to fulfill and leads to happier employees, concerned about maximizing their competitive strengths; TPM provides an advantage to companies trying to achieve the World Class competitive edge.

A TYPICAL TPM TRAINING SESSION

The following TPM training session of one company lasts two days and includes these elements:
1. Autonomous maintenance—involves the operator in the daily inspection and cleaning of the equipment.
2. Product design—considering ease of manufacturing when designing products.
3. Equipment effectiveness—method for gaging equipment performance.
4. Productive maintenance—a preventive approach to maintenance involving time-based analysis.
5. Equipment design—consider the ease of maintenance and operation when designing equipment.

The training session focuses on autonomous maintenance. After four hours in the classroom learning about the elements of TPM, the group proceeds to the factory floor to clean a designated piece of equipment. The cleaning of the machine is not for housekeeping purposes; rather it is to identify and place tags on items that can create waste. The tagging includes: defective parts, inaccessible areas, sources of contamination, and to question if a part is needed at all.

Participants are given two to four hours to clean and hang tags. On the second day the tagged items are compiled into a list. Team members then prioritize the items and assign responsibilities for completing items.

One of the major points of TPM is to prevent downtime by identifying potential problems before a piece of equipment actually has problems. Then the participants generate lists for the operators of items that must be checked daily, weekly, and monthly. Items on the daily sheet might include checking for fluid leaks, loose bolts, improper fluid levels or pressure, etc. These checkoff sheets are placed at the machinery, and as the operators check the items the sheets are initialed. Checklists are also generated for maintenance technicians on a daily, weekly, and monthly basis along with a lubrication schedule.

CHAPTER 15

A TYPICAL TPM TRAINING SESSION

By having production operators and maintenance technicians perform regular standardized inspections, the chance of having unscheduled downtime is greatly reduced. The goal is to be able to predict as well as prevent problems with equipment, and to have zero unscheduled downtime.

Ten to twenty participants are trained during each session. Whenever possible, the groups include associates from all shifts that operate that piece of equipment, along with the maintenance associates that are responsible for maintaining the equipment.

Follow-up is the most important element for TPM to work. The benefit of TPM requires many months, even years, of dedicated follow-up. Japanese companies have said it can take four to ten years to fully implement TPM in a manufacturing facility. One American manufacturer reported that it took four years to begin TPM in 50% of its operation.

Biweekly follow-up meetings are scheduled until all the tagged items are corrected, and periodic follow-up sessions are held to track the improvements.

Bibliography

Nachi-Fujikoshi Corporation, ed. *Training for TPM: A Manufacturing Success Story* (Cambridge, MA: Productivity Press, 1990).

Nakajima, Seiichi, ed. *TPM Development Program; Implementing Total Productive Maintenance* (Cambridge, MA: Productivity Press, 1982).

CHAPTER 16

MACHINING

DEVELOPING A STRATEGY FOR INNOVATIVE MACHINING

CHAPTER CONTENTS:

DEVELOPING A STRATEGY FOR INNOVATIVE MACHINING 16-1

FACTORS AFFECTING PROCESS QUALITY 16-6

PROCESS IMPROVEMENT POINTERS 16-7

WORKHOLDING 16-12

IN-PROCESS GAGING SYSTEMS 16-19

ADVANCED TOOLING 16-22

REFINED TOOLING MODULES 16-32

INTERFACES: MACHINE TOOL/CUTTING TOOL 16-36

OPTIMIZING PROCESSES AND PARAMETERS 16-38

STRINGENT FINISH REQUIREMENTS 16-42

THE NEED FOR FURTHER ADVANCEMENT 16-49

LASER PROCESSING 16-54

MACHINING COOLANTS 16-57

The globalization of the metalcutting industry has forced and will continue to force American manufacturing into developing and implementing sophisticated new technologies in order to remain competitive with foreign manufacturing. These new technologies could create dynamic change over the next decade which may very well exceed that of the past 25 years.

Trends in machining will include multi-axis and multi-function machining (simultaneous machining with both static and rotating tools of identical design for quick-change and modular flexibility) along with a variety of automation technologies that will change the face of manufacturing as it is known today. Following are some of the anticipated automation technologies:

- Semi-automatic and automatic quick-change tooling for flexibility between turning and rotating equipment.
- Machines with high-speed capability and improved rigidity, accuracy, and flexibility.
- Modular workholding devices.
- Manufacturing information systems which communicate with office and engineering systems.
- Machine controls which encourage innovative part processing, and programming with functions for tool management, gaging, and in-process tool sensing.
- Automatic tool storage, transportation, and change systems.
- In-process and post-process gaging for workpiece and cutting tool with feedback for tool compensation.
- Health and safety devices.
- Tool condition sensors to monitor in-process machining and provide protection for machine, workpiece, and cutting tool.
- Software for tool identification and tool management which will interface with gaging, storage, tool kitting, and production software.

Significant advancement can be expected in the area of software integration that coordinates the aforementioned products and technological developments. As a result, the major challenge of the decade will be in managing conflicts and change in the following areas:

- Capital investment versus people investments.
- Product features and quality versus product cost.
- Labor versus management.
- Marketing and engineering versus manufacturing.
- Technology versus knowledge.
- New philosophies versus old order.
- Quality versus productivity.
- Short-term goals versus long-term goals.
- Automation versus health and safety.

The need to apply present technology with an understanding of the total process will exceed the need to develop new technology. No new technological breakthrough will obsolete traditional machining methods for lathes, machining centers, and transfer lines over the next 10 to 20 years. However, an investment in human resource programs designed to support evolving manufacturing strategies and automation technologies will pay impressive dividends in the years to come. Many of these programs will work toward raising the standards for quality, productivity, innovation, and customer value.

THE MACHINING PROCESSES

The machining processes will be most affected by new technologies, just-in-time (JIT) inventory strategies, global competition, and quality control programs.

For the purpose of explanation, the manufacturing process can be divided into three distinct functional areas.

1. Inventory planning control.
2. Pre-production planning and setup reduction programs.
3. In-process manufacturing.

---

*The Contributors of this chapter are: **John Anderson**, Marquis-Bennett Associates, Inc.; **Ken Anderson**, General Automotive Team Leader, Ingersoll Cutting Tool Company; **Charles R. Brown**, Manager of Application Systems, Kennemetal, Inc.; **Subhajit Chatterjee, Ph.D., P.E., CMfgE**, Department of Industrial Engineering, The University of Tennessee—Knoxville; **Philip Z. Chrys**, Market Manager, Advanced Filtration Materials, Lydall, Inc.; **Randy Delenikos**, Marketing Manager, Lakos, Laval Corp.; **Charles S. Delonghi**, Marketing Manager, ITW Woodworth; **Werner K. Diehl**, Vice President and General Manager, ITW Woodworth; **Bert P. Erdel, Ph.D.**, President, Mapal, Inc.; **Arthur C. Fedrigon**, President, Beckart Environmental, Inc.; **Brian Sheehan**, Laser Operator, HGG Laser Fare, Inc.*

*The Reviewers of sections of this chapter are: **Brian Baker**, General Manager, Advanced Technology Center, GTE Valenite Corp.; **Timothy L. Cole**, Product Manager, Franklin Oil Corporation (Ohio); **Don Maurer**, Edmunds Gages; **Tim McCabe**, Sales Manager, Mid-State Machine, Inc.; **Mark Meyer, Ph.D., ME**, Professor, Manufacturing Technology, College of DuPage; **Sassan Moradian**, President, C&L Consultants, Antioch, TN; **Richard Walker**, Vice President, Marketing, Rofin-Sinar.*

CHAPTER 16

DEVELOPING A STRATEGY FOR INNOVATIVE MACHINING

INVENTORY PLANNING AND CONTROL

The primary objective of inventory planning and control is to maintain enough finished goods inventory to meet customer demands while keeping levels low enough to minimize costs.

By producing smaller lots more frequently, inventory carrying costs and shelf-life problems such as rust, contamination, and deterioration can be reduced. This technique is known as just-in-time inventory management.

For example, rather than producing one 8000-unit lot of goods to deplete over a four-month period, it is more cost effective to produce four lots of 2000 units each at one-month intervals over the course of four months (see Fig. 16-1). This reduces the average inventory level from 4000 units to 1000 units, and results in a corresponding reduction in carrying costs. In addition, the goods do not remain on the shelf as long, thereby reducing or eliminating typical shelf-life problems.

Although this type of inventory management saves money, it has placed additional burdens on the manufacturing process by requiring more frequent setups. In the preceding example average inventory was reduced by 75% which resulted in a significant savings. However, to achieve that rate of reduction, the manufacturing function will incur a 300% increase in setup time. As a result, setup reduction programs utilizing a variety of products and procedures have been developed to minimize machine downtime during tool and part change operations. This will drive manufacturing to improve pre-production planning and initiate setup reduction programs to improve machine and operator efficiencies, ultimately leading to increased profit potential and a stable market position (see Fig. 16-2).

PRE-PRODUCTION PLANNING AND SETUP REDUCTION PROGRAMS

Pre-production planning and setup reduction programs identify and organize all elements of a machining operation in advance of the production run. These programs can significantly reduce machine downtime and help maximize the productivity of machine and cutting tools.

One of the products which facilitate planning and setup reduction programs is quick-change tooling. These systems offer flexibility and standardization between turning and rotating equipment, enabling manufacturing to take a process which had been manual for 2000 years, and move it into the age of automation.

The modular quick-change tooling systems of today can be utilized on turning, rotating, and transfer machines in either manual, semi-automatic, or fully-automatic mode. This is a quantum leap forward for computer-integrated manufacturing. Quick-change tooling systems can facilitate machine utilization strategies which are designed to maximize utility from capital equipment and permit continous improvement.

A machine utilization strategy consists of several elements:

- Tool kitting.
- Pre-gaged tooling.
- Quick-change tooling.
- Advanced cutting tool materials.
- Continuous tool maintenance procedures.
- Tool management software.

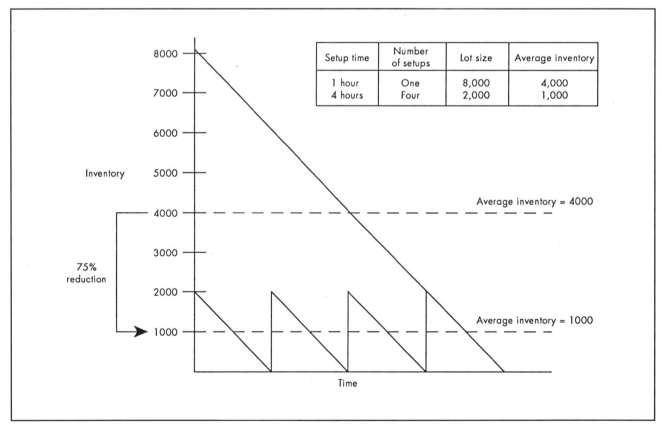

Fig. 16-1 Just-in-time inventory management.

CHAPTER 16

DEVELOPING A STRATEGY FOR INNOVATIVE MACHINING

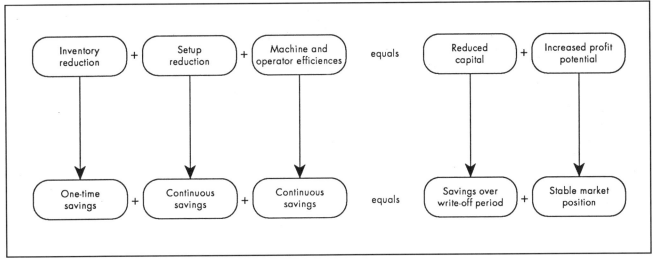

Fig. 16-2 The effects of inventory reduction, setup reduction, and machine and operator efficiency.

These products and procedures, when implemented together throughout a facility, have the potential to reduce downtime by as much as 60%. This will translate into improvements in productivity, quality, and profits for the entire organization.

Tool Kitting

Tool kitting pulls together all of the tooling (including backup tooling) and process documentation necessary to complete a production run or shift of operation. All tools, fitted with new cutting edges and pre-gaged, are placed in a tool taxi and kept near the machine. With all tooling readily available, no time is lost searching the shop floor for tools, changing inserts, making offset adjustments, or performing on-machine tool maintenance. When tool change is necessary the process is quick and efficient, permitting more machine run time and increasing productivity. A typical shift may require the tooling shown in Fig. 16-3.

Pre-gaged tooling

Once all of the necessary tools are fitted with new cutting edges, the "F" and "C" dimensions are measured and recorded for each tool. The data will be used to make offset adjustments following tool change operations. This saves time by eliminating the need to measure and run test cuts, compute offset deviations, and make offset adjustments during in-process machining (see Fig. 16-4).

Quick-change Tooling

Quick-change tooling is the central component of a machine utilization strategy. It is designed to reduce the time needed for tool change operations by enabling the operator to change an entire pre-gaged cutting unit, as opposed to changing an individual insert.

Quick-change tooling consists of two basic parts: the clamping unit and the cutting head. The clamping unit mounts to the machine and acts as a receptacle for the interchangeable cutting

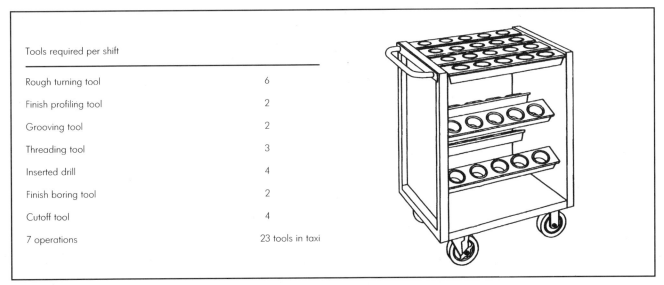

Tools required per shift	
Rough turning tool	6
Finish profiling tool	2
Grooving tool	2
Threading tool	3
Inserted drill	4
Finish boring tool	2
Cutoff tool	4
7 operations	23 tools in taxi

Fig. 16-3 A typical tool kitting arrangement.

16-3

CHAPTER 16

DEVELOPING A STRATEGY FOR INNOVATIVE MACHINING

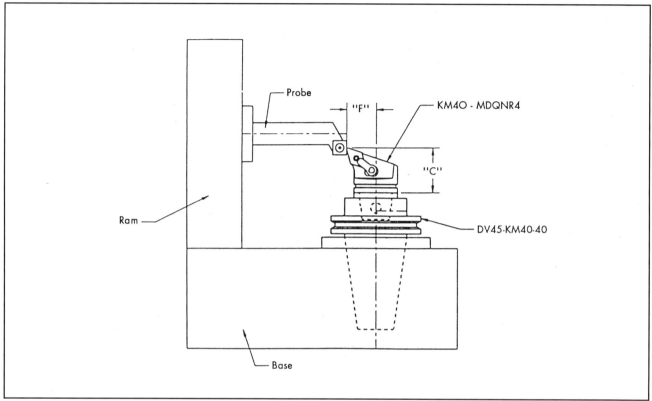

Fig. 16-4 Pre-gaging.

unit. It is the interchangeability of the cutting unit that makes this type of tooling "quick-change." To change tooling, the operator simply releases the locking system, changes the cutting unit, and locks the new tool in position. The operator then makes the offset adjustments according to the previously recorded data and continues machining the part. The total machine downtime is about thirty seconds.

Cutting Unit. The quick-change cutting unit is a compact, center-line system for boring, turning, and rotating applications. It is equipped with through-the-tool coolant capabilities, gripper finger grooves, and a tapered face contact shank for manual or semi-automatic actuation on either stationary or rotating equipment.

Quick-change Systems. Quick-change tooling systems should be compact, accurate, rigid, and cost effective. They should also be adaptable to lathes, machining centers, and transfer equipment with manual and/or automatic locking mechanisms. Quick-change systems that meet these criteria can be primary components of an effective machine utilization strategy.

Quick-change tooling can be adapted to existing machine tools at a relatively inexpensive cost. The increase in productivity that results from the conversion to quick-change tooling can pay back the initial investment in a very short time, usually about three to four months.

Manual Systems. New and existing machines that utilize conventional tooling can easily be converted to quick-change simply by replacing the current tooling with square shank, round shank, or builder-specific tool blocks which incorporate manual quick-change locking devices. Systems utilizing the European VDI adaptor can also be converted to quick-change tooling.

Semi-automatic and Automatic Systems. Semi-automatic quick-change tooling, which incorporates disc spring locking mechanisms, can be installed on new equipment providing a high level of automation at a moderate cost. The clamping system design enables tools to be locked or unlocked at the push of a button, or by depressing a foot switch allowing the operator to use both hands when changing tools. The semi-automatic system is quick-change, qualified, and wrenchless, which meets the criteria established to reduce setup and tool change time.

Fully automatic quick-change systems add tool storage and handling capabilities, and include locking mechanisms which are activated by a signal from the machine control. Fully automatic quick-change tooling, combined with other automated systems, such as part-loading and tool management software, can result in a highly productive manufacturing process.

Modular Systems. Modular machine tooling is comprised of rotating adapters in a variety of configurations. Each is designed with a compact, rigid, quick-change joint to accommodate rotating toolholders and boring tools. Some modular components can also be used on lathes and transfer machinery, to reduce perishable tooling inventories and tool setup time.

Advanced Cutting Tool Material

Major advancements in the near net shape of parts and cutting tool material will play an important role in determining the selection of workholding devices and metal removal rates over the next decade. Because near net shape implies less metal removal, increased rates of speed and new metalcutting materials will be needed for use in quick-change tooling systems. Advanced cutting tool materials will play a very important role in increasing

CHAPTER 16
DEVELOPING A STRATEGY FOR INNOVATIVE MACHINING

productivity. Materials such as cermets (ceramics with metallic binders), ceramics, polycrystalline diamond, and polycrystalline cubic boron nitride (CBN) enable machines to run faster and longer between tool changes, thereby utilizing the full performance potential of the machine and cutting tools by reducing the number of downtime tool change penalties. Figure 16-5 shows the relationship between tool speed and tool life for advanced cutting tools, coated carbides, and carbide tooling.

Continuous Tool Maintenance

Off-line continuous tool maintenance programs remove tool cleaning, repair, maintenance, pre-gaging, test cutting, and tool kitting from the in-process machining environment. They can have a significant impact on setup reduction programs and, in turn, improve machine uptime and productivity.

Tool Management Software

Tool management software automates tool crib transactions and helps the tool crib participate in setup reduction programs such as the kitting and pre-gaging of quick-change tools. This type of software helps the tool crib deliver the tools, fixtures, and gages required to implement the machine utilization strategy. This system provides process documentation which is easy to maintain, modify, and monitor for process changes both on the factory floor and in the engineering department.

Tool management software should:

- Be a computerized method of tracing tools, fixtures, gages, and supplies used in manufacturing environments.
- Help maintain an up-to-date tool inventory in the tool crib as well as on the shop floor.
- Report all tools charged to a given job, employee or location, and easily record inventory changes as tools are withdrawn, returned, transferred, reworked or scrapped.
- Automatically report tool shortages when an adjustable minimum stocking level is reached.

Tool management software should also:

- Provide continuous tool tracking.
- Minimize delays and expenses from lost tooling.
- Eliminate excess tool crib inventory.
- Maximize return on tooling investments.
- Reduce tool setup time.
- Provide process documentation.
- Organize tool kitting.

All of the products and procedures described here are designed to maximize manufacturing efficiencies and can be combined to create an extremely effective machine utilization strategy. The bottom line is a dramatic improvement in machine and cutting tool utilization which ultimately translates to improved productivity and increased profit opportunities.

Benefits of a machine utilization strategy include:

- Reduced downtime for tool search and tool change.
- Reduced setup time for part change-over.
- Reduced cut measuring time for obtaining offset adjustment data.
- Reduced cost.
- Improved tool maintenance.
- Improved tool utilization, management, and standardization.
- Reduced scrap.
- Improved part quality.
- Increased productivity.
- Existing machinery can evolve to a fully automatic system.

Heavy investments in office information systems and engineering support systems such as computer-aided design (CAD), manufacturing (CAM), and engineering (CAE) will begin to pay multiple dividends as automation technology is implemented on the factory floor. Most products will be applied to reduce physical labor at the point of metal removal and to improve productivity and quality while reducing cost. However, the greatest improvements will result from procedures and processes such as JIT inventory management, pre-production planning, and setup reduction program.

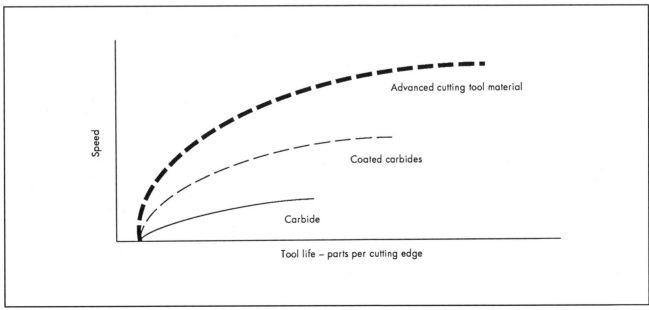

Fig. 16-5 Relationship between tool speed and tool life for advanced cutting tools, coated carbides, and carbide tooling.

CHAPTER 16

FACTORS AFFECTING PROCESS QUALITY

FACTORS AFFECTING PROCESS QUALITY

Continuous improvement of manufacturing processes is desirable for improved process output quality. To achieve it, the operator or the manufacturing engineer must constantly focus on many aspects of the process, the machines, and the environment. The operator and the manufacturing engineer can significantly improve the process output quality, and collectively contribute to the overall efficiency of the manufacturing facility. This section attempts to identify some key areas in machining and suggests, through operator or engineer efforts, possible improvement avenues for better process output in those areas.

Some of the factors affecting process output quality are:

1. Important machining parameters.
2. Machine tool aspects.
3. Setup and workholding aspects.
4. Process planning aspect.
5. Work material condition.
6. Value-added concept.

IMPORTANT MACHINING PARAMETERS

The three most important machining parameters are the cutting speed, the feedrate, and the depth of cut.[2] The cutting speed is the rate at which the work material flows past the cutting tool. The feed rate is the rate at which the cutting tool is advanced per revolution into or across the workpiece.

The output of a material removal process is very much dependent on the selection of these three parameters. Errors in selection of any one of them could have drastic effects on the quality of the finished product. For example, in machining, the temperature to which the tool, chip, and work are heated increases with an increase in cutting speed.[3] This effect may be deleterious for surface finish. High speeds, feeds, and depths of cut, singularly or in combination, may produce chatter which in turn results in a deteriorating surface finish, increases tool wear, and causes the emission of a harmful high-frequency sound.[4] By observing the machining process, a qualitative determination can be made regarding surface quality.

MACHINE TOOL ASPECTS

The condition of a machine tool is another very important factor. Process output quality deteriorates with the deteriorating condition of a machine tool. Thus an important aspect in monitoring process quality is the monitoring of the machine tool condition.

Ambient conditions are also very important, as fluctuations in ambient temperature affect the accuracy of the machine tool performance and thus directly affect the dimensions of the finished product.[5] Therefore, a knowledge of environmental conditions is extremely important in controlling process quality and in precision manufacturing.

External vibrations can be extremely detrimental the quality of the finished product. Such vibrations affect the machine tool structure and resultant form and size of the part. Thus, sources of such vibrations must be identified and controlled.

SETUP AND WORKHOLDING ASPECTS

The accuracy and efficiency of setups have a notable effect on the overall efficiency of any process. Accuracy of setups includes the accuracy of the individual components and of the procedure. For example, if an operator is setting up for a holemaking operation, how accurate is the positioning of the cutter? Also, how good is the workholding equipment? Workpiece location also plays a significant role in ensuring a successful operation and quality of finished product. Proper design of the locating device and workpiece location also ensures that each successive workpiece, when loaded and clamped, will occupy the same position in the workholding device.[6]

Efficiency of setup is crucial to improved manufacturing processes. An efficient setup process minimizes changeover time, thereby reducing downtime and increasing flexibility. Reducing setups, by using sound operation design principles, improves part quality through reduced changeover errors. Additionally, the quality of the workholding devices is very important for improved part quality. Worn or loose fitting fixtures, jigs, and other workholding devices are ineffective in countering the cutting forces and can result in defective products.

PROCESS PLANNING ASPECT

Process planning is defined as the task of developing the detailed series of steps necessary for the manufacture of the product. This function establishes the processes and the parameters necessary, the machine tools and the workholding devices required, and the cost, quality, and production rate.[7] The operator and the manufacturing engineer play a significant role in the creation of plans and in identifying weaknesses in them. A good process plan will streamline operations by eliminating redundant operations, reducing machine load imbalances, and removing interference or dependencies among operations, thereby improving plant efficiency and product quality.

WORK MATERIAL CONDITION

Large variations in incoming raw material quality can have a significant impact on the quality of the finished product. Such variations include variation in stock dimensions, variations in work material microstructures due to differences in manufacturing or heat treating, poor surface quality for cast or forged materials, and existence of subsurface inclusions or defects such as voids or cracks. Variations in stock dimensions affect the final dimensions of the part, whereas variations in work material microstructure and subsurface defects affect the machining process. For example, differences in microstructure affect the chip type produced;[8] the presence of inclusions affects cutting tool wear, and subsurface voids and cracks act as sites for crack initiation and material failure.

VALUE-ADDED CONCEPT

The lack of this concept usually results in a lack of industrial competitiveness through poor product quality and high manufacturing costs. The operator, the manufacturing engineer, and upper management should have (or embark on) a clear notion of which processes add value. Elimination of the non-value added processes will assist in product quality and plant efficiency improvements.

CHAPTER 16

PROCESS IMPROVEMENT POINTERS

PROCESS IMPROVEMENT POINTERS

This section identifies important parameters for process improvement, provides pointers for detecting deteriorating processes, and suggests remediations.

MACHINING ASPECTS

The primary parameters that govern the quality of a finished product are the correct selection of speeds, feeds, and depths of cut. Most operators, through years of experience, can select these parameters correctly for a range of work materials, machines, and cutting tools. However, when machining unfamiliar materials, such as newly developed high strength materials, it is a good idea to consult handbooks, prior research, or published experiences in the determination of these parameters. This is very important because there is as yet no established database for the machinability of some high strength materials, such as intermetallics. Interpolating data from the results of machining of other high strength materials may not be appropriate.

The color of chips and the sound of the machining process provide significant inputs regarding the quality of the process. Processes that develop high temperatures or excessive tool wear generate high-frequency sound and burnt chips. Both of these characteristics result in increased surface roughness of parts. Thus, operators should check the chip types and color, and listen to the process. Impending process deterioration may thus be avoided.

The effect of chatter should also be a "fingerprint" of process deterioration. A combination of high depth of cut, feeds, and speeds results in chatter; if chatter is to be avoided the proper parameters must be known for the particular work material and machine tool condition. This can only be determined through prior testing, and initially should be established through the acceptance testing of machine tools.[5] However, as machine tool conditions deteriorate with age, the initial parameters may be unsatisfactory and may require relaxation. Therefore, the operator and the manufacturing manager should be cognizant of the machine tool conditions and appropriately determine the machining parameters.

Proper selection of cutting tool materials and angles is vital to the success of the process. Operator experience is useful to determine the time between regrinds. Additionally, the correctness of the tool-grinding process and subsequent examination of the accuracy of the ground angles is equally important in ensuring that the quality of the tool is satisfactory. All tool angles, nose radius, and honing of cutting edges affect the forces evolved during the cutting process and the dimensional accuracy of the finished part.

In summary, the following points should be helpful in improving the quality of the machining process:

- Appropriate selection of cut, speeds, and feeds.
- Periodic examination of the color of the chips.
- Observation of sound emitted during machining.
- Appropriate selection of cutting tool material and angles.

MACHINE TOOL ASPECTS

The condition of a machine tool determines the finished part quality. Typically, acceptance testing of the machine tool is performed prior to and after its installation at the user's site. Various machine tool parameters can be measured during such testing: linear displacement accuracy, angular error about each axis due to motion along machine axes, spindle runout and growth in axial and radial directions, spindle tilt, squareness errors in machine axes, repeatability, etc.[5]

Machine tool performance degrades with age,[9] and manufacturing personnel should periodically check for parameter errors. The motivation for error detection lies in improving part quality through machine tool maintenance. Machine tool performance testing is expensive, but machining of parts to determine chatter limits,[5] positioning errors, or dimensional and form error can provide valuable information regarding the condition of the spindle, available horsepower, and wear in gibs and slideways.[10,11]

Operators should not run machines at loads exceeding designed specifications. For older machines such action may result in immediate chatter; however, such results may not be immediately evident in newer machines. Thus, occasional torque-testing of the machine spindle is a good idea.

Operators should periodically check for external sources of significant vibration in the vicinity of the machine, because such vibrations may induce additional vibrations during the cutting process.

The variation of the ambient temperature in the work place should be less than the temperature variation error constant of the machine tool; otherwise machine tool positional accuracy and the resultant part dimensions will be significantly affected.

The operator should also periodically examine the rigidity of the machine mounts. Looseness in machining mounts will result in increased vibration, with a deterioration of part quality.

In summary, the following points are significant in improving the quality of the finished parts:

- Acceptance testing of machine tools.
- Performance testing of machine tools over time. Machine operation below designed loads.
- Periodic checkup for the existence of significant external vibration sources.
- Periodic examination of the rigidity of the machine tool mounts.

SETUP AND WORKHOLDING ASPECTS

The importance of setup reduction for process improvement is an established fact. Setup reduction improves process efficiency by reducing setup time and setup positioning errors. The process flow time is thereby reduced and part quality improved.

The machine operator or the manufacturing engineer should detect areas where setup reduction is possible. They should examine part manufacturing and design similarity and determine which parts can be grouped together for processing, either using the same setup or with minimal changes in setup. This will also aid in the development of quick-change, modular setup and fixturing equipment. In modular fixturing, a fixture module can be set up externally and then loaded on the machine as required. This requires understanding the distinction between external and internal setups; the machine operator and manufacturing engineer should analyze all setups and segregate the external and internal procedures. External setups can be performed while a machine is operating, whereas internal setups require machine stoppage.

CHAPTER 16

PROCESS IMPROVEMENT POINTERS

External setup of tools, workholding, and workpieces to specific accurate dimensions enhances quality by reducing trial cuts and scrap.[12]

The use of modular tooling provides repeatability, improved rigidity, and tooling flexibility. Modular tooling principles utilize presetting of cutting tools into tool holders which can then be plugged into the required machine tools. Modular tooling also assists in tool standardization, reducing tool purchase, storage, and maintenance costs. The following pointers are beneficial in developing a modular tooling system:[12]

1. Definition of setup cost. This requires determining where and how the operators spend time locating barstock, castings, NC tapes, tool holders and tools, chuck jaws, gages, hand tools, etc.
2. Reduction of setup costs by presetting all tooling, providing backup tooling in racks near the machine, pickup and delivery of NC tapes at the machine, provisioning raw material availability at the machine, etc.
3. Development of a quick-change plan by:
   - Planning action for machines to be converted to modular tooling.
   - Grouping of parts into families.
   - Analyzing existing setups.
   - Reviewing yearly production.
   - Planning an implementation schedule.

Good workholding practices are also important. Sophisticated machinery will not produce quality parts if workholding devices do not impart adequate holding forces. Thus the operator should periodically examine the condition of the workholding devices for wear of locating surfaces and loss of power of clamping mechanisms. This aspect is gaining increased importance with the availability and use of high speed, high production machine tools.[13]

In summary, there are important setup and workholding aspects for improved machining:

- Detection of setup improvement areas.
- Distinction between external and internal setups.
- Development of modular tooling.
- Practice of good workholding principles.

PROCESS PLANNING ASPECT

Process planning involves the development of the detailed sequence of tasks necessary for the manufacture of a product. An inaccurate process plan can result in a faulty product and high product cost. It is primarily the manufacturing engineer's responsibility to ensure that process plans are accurate.

The accuracy of machining speeds, feeds, and depths of cut should be examined with respect to the work material-cutting tool-machine tool combination. This necessitates the determination of the correct cutting tools to be used. The condition of the machine tool should also be examined with respect to the machining parameters. In manufacturing facilities with numerous machine tools of diverse age groups, the condition of each machine tool must always be borne in mind during its scheduling.

Machine loading imbalance should be avoided wherever possible by rerouting parts to other machines for equitable load distribution. In many industries, several different process plans may exist for the same part.[7] Therefore, some machines are loaded more often than others, resulting in load imbalance. Such loading increases queue lengths of parts waiting for processing, often starving downstream facilities.

The manufacturing engineer should be fully aware of the capacities and capabilities of the machine tools and the material flow patterns. Proper allocation of machines to jobs is then possible. Any redundant or bottleneck conditions should be eliminated. Continuous improvement of the process plan and material flow is a goal for the manufacturing engineer and operator: the following factors should be considered important for process quality improvement with respect to process planning:

- Accuracy of process plans.
- Use of standardized process plans.
- Equitable machine load distributions.
- Accurate estimate of machine capacities and capabilities.

WORK MATERIAL CONDITION

The responsibility for inspection of incoming work material quality primarily belongs to the receiving inspection department. However, the manufacturing manager should ensure that some incoming raw material characteristics are inspected wherever applicable:

- Variations in stock dimension.
- Variations in mechanical properties of stock, such as hardness.
- Surface quality of stock; this includes surface defects such as voids.
- Presence of inclusions.

The manufacturing manager should establish a good working relationship with the suppliers of raw material and inform them of incoming raw material quality problems. Additionally, a reduction in the number of suppliers should be attempted. This reduces the variability of raw material quality and establishes better control over raw material supply.

Establishing good quality control procedures for incoming raw material requires:

- Thorough examination of incoming raw material, mechanically and metallurgically.
- Establishment of harmonious working relationships with suppliers.
- Reduction in the number of suppliers.

VALUE-ADDED CONCEPT

The "value-added" concept emphasizes the establishment or knowledge of activities that lead to the addition of value to the work material. By clearly distinguishing value-added activities from the non-value added ones, elimination of process waste is possible. Value-added work can be defined as the work external customers would be willing to pay for.[14]

Thus, the manufacturing engineer/manager and the process operator should familiarize themselves with the various aspects of the philosophy of value-added work.

Charting Tools

Included in this classification are:

- Operation process charts.
- Process flow charts.
- Schedule charts.
- Operator process charts.
- Motion time analysis.

Operation process charts. The operation process chart is a graphic representation (see Fig. 16-6) of the points where mate-

CHAPTER 16
PROCESS IMPROVEMENT POINTERS

rials are introduced into the process, all sequences of inspection, and all operations with the exception of those related to material handling.[15] By such graphic representations the manufacturing engineer will be cognizant of all points of material entry and exit, and overall material flow. Such visual representation of the process assists in determining material flow time, the constituents (departments) of each process, and possible bottlenecks, and is thus helpful in identifying and eliminating process inefficiencies—thereby improving process quality.

An example of a detailed operation process chart is shown in Fig. 16-7 for a hypothetical mechanical assembly. The numbering starts with the first step on the part and continues until assembly (0-1 through 0-17). The time per operation is expressed in time measurement units (1 TMU = 0.00001 hours) and shown adjacent to each operation.

An operation process chart can be used by the manufacturing engineer to review the current process with respect to material, finishes, tolerance, and inspection. This may lead to alternative processing, tooling, and inspection avenues for process improvement. Operation analysis, an effective followup tool for such improvements, is discussed in detail later.

Process flow chart. While the operation process chart graphically depicts the overall flow, the process flow chart graphically represents the sequence of all operations, transportations, inspections, delays, and storage occurring during a process.[15] Different symbols are used to denote operations, transportation, inspection, delay, storage, and combined activities. Figures 16-8 and 16-9 are examples of process flow charts for a hypothetical manufacturing activity.

Table 16-1 lists six questions to be asked in sequence when analyzing process flow charts to improve a process.

A constructive approach in answering these questions, with corrective action if necessary, will lead to improvement in process efficiency and product quality.

Schedule charts. Schedule monitoring charts are also known as GANTT charts. The time-phased progression of activity by workcenters is displayed. Figure 16-10 provides an example.[16] Inspecting the horizontal axis for each type of work can reveal the lateness of the activity.

Operator process charts. This chart, also known as RH-LH chart, shows a graphic representation of the coordinated activities of the left and right hand. In Fig. 16-11, the vertical columns show simultaneous activities for each hand. The different symbols

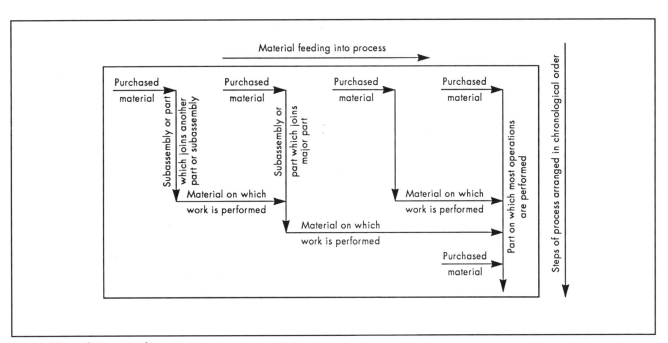

Fig. 16-6 Operation process chart.

TABLE 16-1
Questions for Analyzing Process Flow Charts to Improve a Process.

Question	Followed by	Action Expected[15]
1. What is the purpose?	Why?	1. Eliminate unnecessary activity.
2. Where should this be done?	Why?	2. Combine or change place.
3. When should this be done?	Why?	3. Combine or change time or sequence.
4. Who should do this?	Why?	4. Combine or change person.
5. How should this be done?	Why?	5. Simplify or improve method.

CHAPTER 16

PROCESS IMPROVEMENT POINTERS

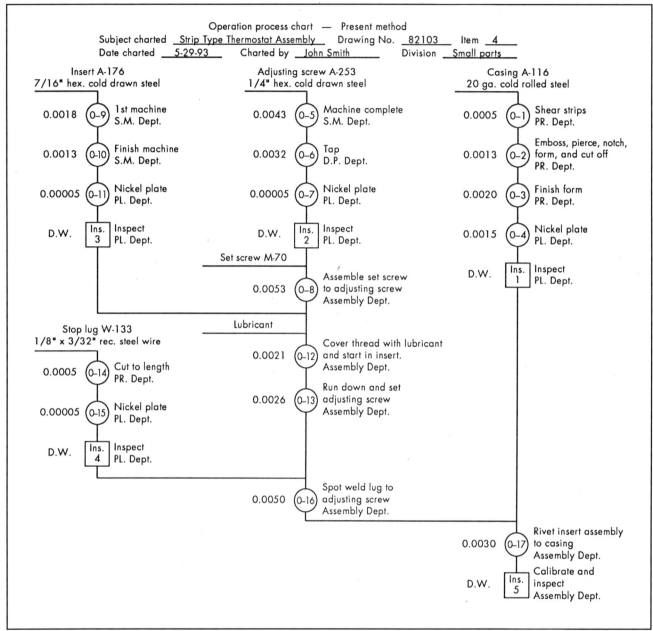

Fig. 16-7 Detailed operation process chart.

represent operation, transportation, hold, and delay. By comparing the total right and left hand work counts, process improvements may be achieved.

Motion time analysis. Process efficiency and cost competitiveness are influenced significantly by the correct sequence of operator and machine operations. Considering that 95% of the total flow time is consumed by nonmachining processes, there is a great incentive for reducing process costs by reducing unnecessary movements. One approach is through the Motion and Time Methods (MTM).

Work measurement techniques determine the time required by an operation or element of operation performed under given method and job conditions.[18] Standard operation times for each activity can be easily established by the worker. The RH-LH chart can be used for elemental motion activities, and motion times can be inserted in appropriate columns. Statistical analysis can be performed to determine the mean and the deviation of the time recordings. This data can then be used to establish standard times.

The utility of such standards is in establishing product costs, and in determining areas where processes may be improved by reducing or eliminating redundant motion and time.

Application of Statistical Quality Control Principles

The manufacturing engineer and the process operators should incorporate statistical quality control (SQC) principles plantwide.

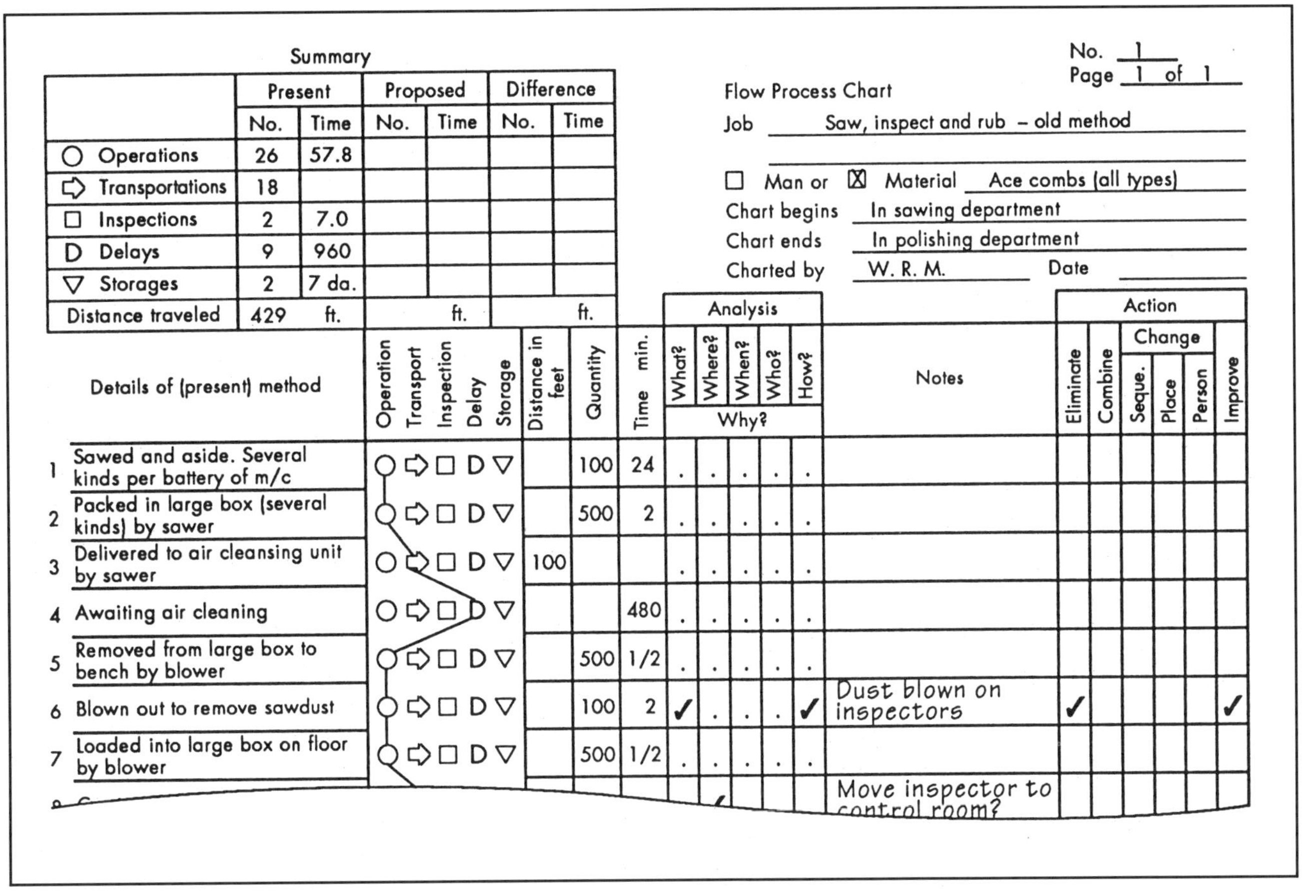

CHAPTER 16

PROCESS IMPROVEMENT POINTERS

	Summary																	
	Present		Proposed		Difference													
	No.	Time	No.	Time	No.	Time												
○ Operations	26	57.8																
⇒ Transportations	18																	
□ Inspections	2	7.0																
D Delays	9	960																
▽ Storages	2	7 da.																
Distance traveled	429	ft.		ft.		ft.												

Flow Process Chart No. 1 Page 1 of 1
Job Saw, inspect and rub – old method
☐ Man or ☒ Material Ace combs (all types)
Chart begins In sawing department
Chart ends In polishing department
Charted by W. R. M. Date

	Details of (present) method	Operation	Transport	Inspection	Delay	Storage	Distance in feet	Quantity	Time min.	What? Where? When? Who? How? Why?					Notes	Eliminate	Combine	Change Seque.	Place	Person	Improve
1	Sawed and aside. Several kinds per battery of m/c	○	⇒	□	D	▽		100	24	.	.	.	.	.							
2	Packed in large box (several kinds) by sawer	○	⇒	□	D	▽		500	2	.	.	.	.	.							
3	Delivered to air cleansing unit by sawer	○	⇒	□	D	▽	100			.	.	.	.	.							
4	Awaiting air cleaning	○	⇒	□	D	▽			480	.	.	.	.	.							
5	Removed from large box to bench by blower	○	⇒	□	D	▽		500	1/2	.	.	.	.	.							
6	Blown out to remove sawdust	○	⇒	□	D	▽		100	2	✓	.	.	.	✓	Dust blown on inspectors	✓					✓
7	Loaded into large box on floor by blower	○	⇒	□	D	▽		500	1/2	.	.	.	.	.							
															Move inspector to control room?						

Fig. 16-8 Material-type process flow chart.

SQC principles are extremely useful in detecting process shifts.

Control charting is the heart of SQC principles, and graphically displays quality characteristic measurements over time. Figure 16-12 shows a typical control chart[15] where the center line represents the average value of the quality characteristic when the process is in control—that is, when only chance causes of variation are present. The upper and lower control limits are established so that if the process is in control nearly all sampled points will fall between them. A point outside the control limits indicates that the process has deteriorated and corrective action is necessary. Two types of control charts are prevalent, the *average* chart and the *range* chart.

Some reasons for the use of control charts are:[17]

1. To improve productivity.
2. To prevent defects.
3. To prevent unnecessary process adjustments.
4. To provide diagnostic information.
5. To assist in determining process capability.

Therefore, if the manufacturing engineer, process operator, and all employees involved are knowledgeable about process control, significant process improvements may be expected. Additional knowledge of statistically designed experiments will also enable the manufacturing engineer and/or process operator to determine influencing factors for product/process quality. Management should seriously consider employee training in SQC principles.

Application of Cell Manufacturing (CM) Principles

Also known as group technology (GT), cell manufacturing is a manufacturing system where a family of parts is processed in a cell comprised of dissimilar machines.[19,20] This system is, therefore, clearly different from a functional manufacturing system where similar types of machines such as lathes, drills, etc., are grouped together. Functional systems exhibit typical performance characteristics, such as high work in process inventory, long travel times, poor job accountability and tracking, and poor job quality. These problems are mostly not observed in a CM system because of close proximity of workcenters and higher operator responsibility. Many successful CM implementations have been reported; it is possible to achieve significant process improvements by CM implementation. To consider CM, manufacturing engineers should consider the product types manufactured and analyze them for process similarity. An example of such analysis is shown in Fig. 16-13; this determines the cell composition. There are many other factors to be considered in CM implementation: product type and flow, product-mix stability, component manufacturing complexity, and most importantly, worker and managerial involvement.

Worker involvement is particularly important because worker fear of job loss may seriously jeopardize CM implementation. Apart from any psychological benefits, technical benefits such as means of setup reduction and other process improvement ideas may be generated through worker involvement.

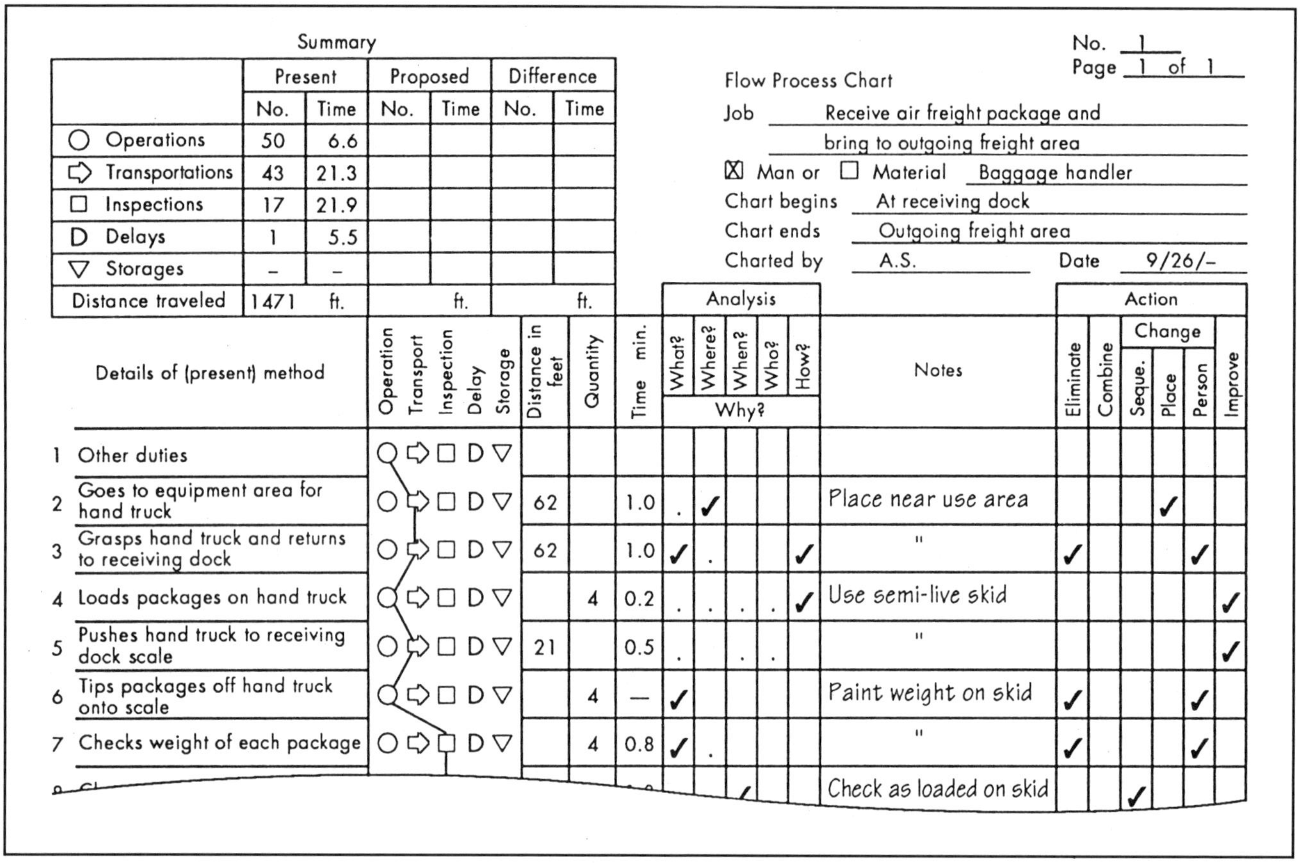

CHAPTER 16

WORKHOLDING

Summary						Flow Process Chart	No. 1 Page 1 of 1
	Present		Proposed		Difference		
	No.	Time	No.	Time	No.	Time	
○ Operations	50	6.6					Job Receive air freight package and bring to outgoing freight area
⇨ Transportations	43	21.3					☒ Man or ☐ Material Baggage handler
☐ Inspections	17	21.9					Chart begins At receiving dock
D Delays	1	5.5					Chart ends Outgoing freight area
▽ Storages	–	–					Charted by A.S. Date 9/26/–
Distance traveled	1471	ft.		ft.		ft.	

								Analysis							Action					
Details of (present) method	Operation	Transport	Inspection	Delay	Storage	Distance in feet	Quantity	Time min.	What?	Where?	When?	Who?	How?	Notes	Eliminate	Combine	Change			
											Why?						Seque.	Place	Person	Improve
1 Other duties	○	⇨	☐	D	▽															
2 Goes to equipment area for hand truck	○	⇨	☐	D	▽	62		1.0	.	✓				Place near use area				✓		
3 Grasps hand truck and returns to receiving dock	○	⇨	☐	D	▽	62		1.0	✓	.		✓		"	✓				✓	
4 Loads packages on hand truck	○	⇨	☐	D	▽		4	0.2	.	.	.	.	✓	Use semi-live skid						✓
5 Pushes hand truck to receiving dock scale	○	⇨	☐	D	▽	21		0.5	.		.	.		"						✓
6 Tips packages off hand truck onto scale	○	⇨	☐	D	▽		4	–	✓					Paint weight on skid	✓				✓	
7 Checks weight of each package	○	⇨	☐	D	▽		4	0.8	✓	.				"	✓				✓	
													✓	Check as loaded on skid		✓				

Fig. 16-9 Operator-type process flow chart.

WORKHOLDING

WORKHOLDING TECHNOLOGIES FOR CONTINUOUS IMPROVEMENT

In the effort to comply with the principles of continuous improvement, manufacturers are turning to sophisticated manufacturing systems. Workholding systems have undergone vast changes to meet the demands of these systems and the quality control standards they incur. They are designed to be utilized in manual, semiautomatic, and automatic manufacturing operations. To fully benefit from an investment in these workholding devices, early evaluation of part family size, the needed operations performed, run size, and changeover requirements aid in the proper workholding selection process.

Technological advancements in today's workholding devices make it easier to comply with the principles of continuous improvement. Current chuck designs offer a dynamic, real-time response to quickly adapt to changes in size, family, and operations performed.

To fully benefit from an investment in these workholding devices, evaluate some fundamental criteria up-front concerning the part family size, the needed operations performed, run size and changeover requirements. Once the parameters of a given manufacturing project are established, proper workholding selection will become more evident.

Part Considerations

When considering the proper workholding device for manufacturing, it is important to first evaluate the workpiece. The object is to increase quality while maintaining maximum flexibility. Certain parts have a high susceptibility to distortion when clamped—resulting in a high scrap rate. Other workpieces are part of a large family, where ease of production changeover is a major concern. Matching a workholding system's capability to a part's requirements will maximize overall productivity and quality.

Part rigidity. To minimize part scrap and downtime due to workholding, it is important to consider the composition and rigidity of the workpiece. Some workpieces distort under clamping pressure; there are chucks available that are designed specifically to suit fragile part applications.

One company's (ITW Woodworth) chuck design incorporates a durable urethane bladder, located within the chuck body. It is inflated via hydraulics or pneumatics in the machine spindle. When actuated, the bladder expands, flexing specially designed lightweight clamp fingers which contact and conform to the workpiece contour. A slight drawdown feature locates the part against a workstop.

The high-density placement of these fingers along a clamping surface allows the chuck to actually become homogeneous with

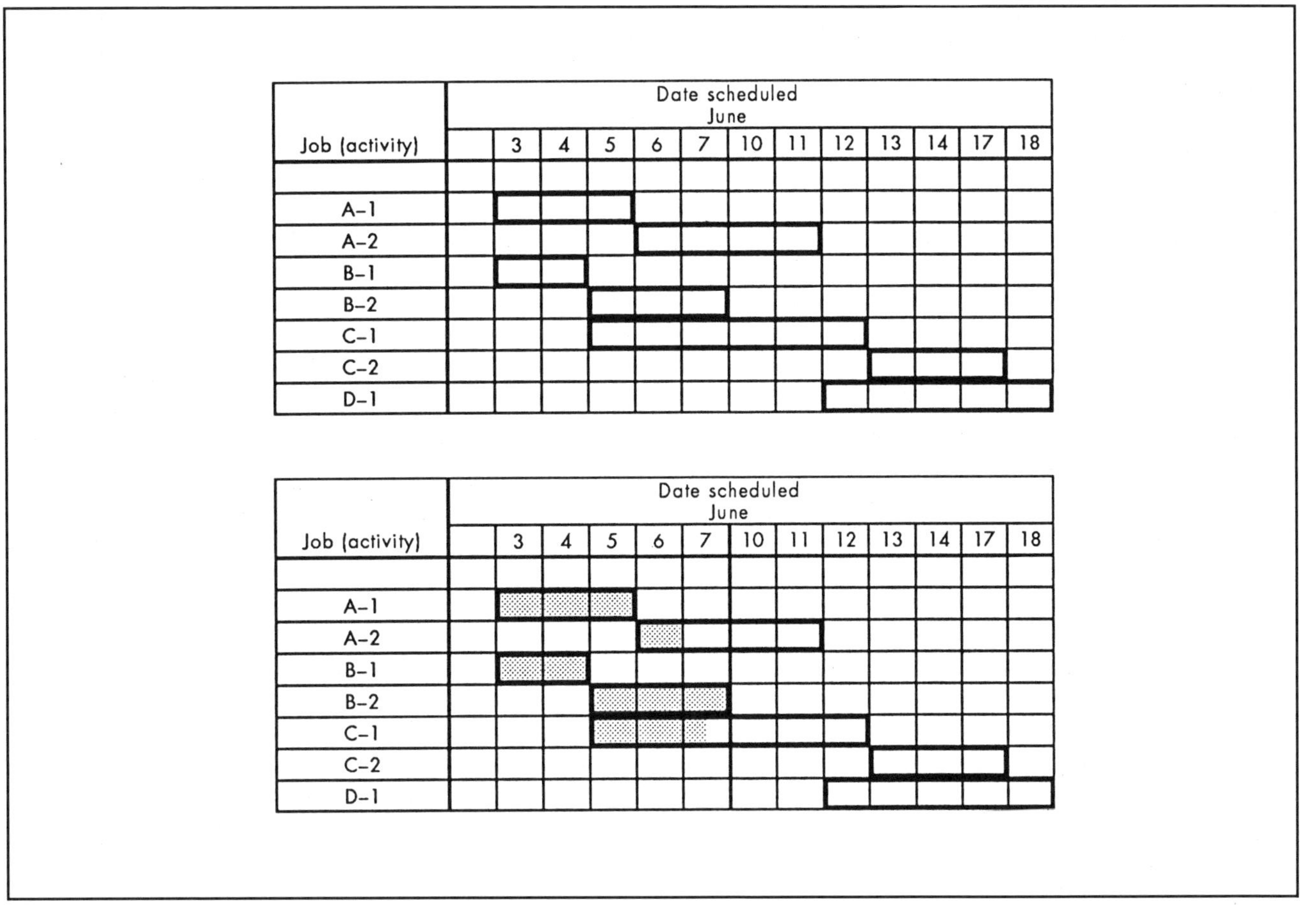

CHAPTER 16

WORKHOLDING

Fig. 16-10 GANTT chart showing actual versus planned schedules.

the workpiece. The result is increased part rigidity and resistance to deflection and distortion from force loads and driving torque requirements.

Single part variety. When there is no concern for multiple part capacity on a single workholding device, the chuck design emphasis can be focused on specific part machining operation performance. Working within the established parameters of a given part, many part-specific innovations can be incorporated into the overall chuck design to maximize performance and productivity.

One design on the market features short-stroke swivel clamps that move straight back, parallel to the machine spindle. There are many benefits to this design:

1. The lack of pivot points (found on standard chucks) diminishes the effects of extreme centrifugal forces created by high-speed machining operations. The result is greatly reduced jaw-force loss, and repeatable accuracy within 0.001″ (0.03 mm) TIR concentricity.
2. Workpiece change time is greatly reduced because the short-stroke design requires less jaw travel than conventional clamp designs.
3. The short-stroke clamp design limits the amount of clamp exposure by presenting less area for chip buildup that can inhibit clamp travel.
4. The straight line movement permits effective sealing of the chuck interior against contaminants.

In addition, this design has a smooth cylindrical profile to help reduce the air turbulence during high speed turning operations. The hazards of flying chips and coolant are minimized.

Multiple part family. There are workholding devices in place that offer flexibility of operation and/or changeover to accommodate families of parts. The most common of these is the diaphragm chuck. It utilizes the inherent strength and accuracy of spring steel to achieve chucking pressure. The principle of operation is a chuck-within-a-chuck system. A "master" chuck (or base chuck) is permanently mounted to a machine. The "workholding" chuck is inserted into the master and secured. Each part within the family can now have its own workholding device while being mated with a consistent master. The result is an increase in productivity and repeatable accuracy.

Air is introduced to the master chuck to move an internal piston forward about 0.030″ (0.76 mm). This movement is translated to a diaphragm which opens jaws through a flexing action. When the workholding chuck is inserted, the air is turned off to allow the jaws to move toward their relaxed position until they contact a qualified and standard locating diameter on the workholding chuck. Workholding chucks perform the same function on clamping diameters of parts through a second air supply.

Machining Operations Required

The number and type of operations performed on a given part and/or family of parts dictate the flexibility a given chuck needs to

CHAPTER 16

WORKHOLDING

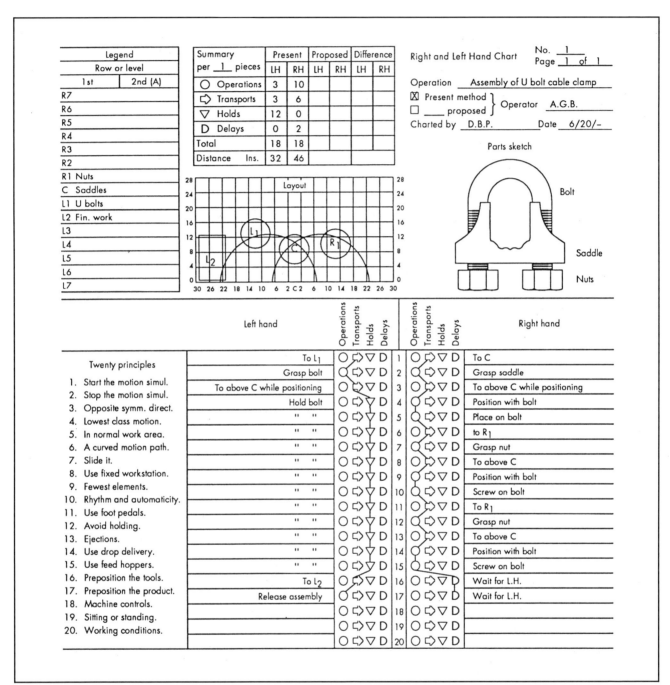

Fig. 16-11 Operator process chart.

optimize productivity. Excessive part handling increases cycle times and often leads to out-of-tolerance scrap due to re-chucking errors. Multiple machines increase production but this is costly and still does not address re-chucking errors. There are workholding systems that are capable of various levels of machining requirements.

Single operation. For single operations, there are several unique workholding devices that address the problems of excessive part handling by allowing increased machining capabilities within the same chuck.

One design (ITW Woodworth) has jaws which extend and retract. This retractable jaw feature allows one continuous turning operation of workpieces between centers over its entire length. There is no need to stop the spindle. Machining times are reduced to 50% of conventional chucking, which requires two operations and sometimes two machines for complete shaft turning.

These chucks are provided with a long, guided, spring-loaded drive center to accommodate locating center tolerance variation. Drive centers provide true between-centers machining for greater accuracy, eliminating part turnaround. A self-contained equaliz-

CHAPTER 16

WORKHOLDING

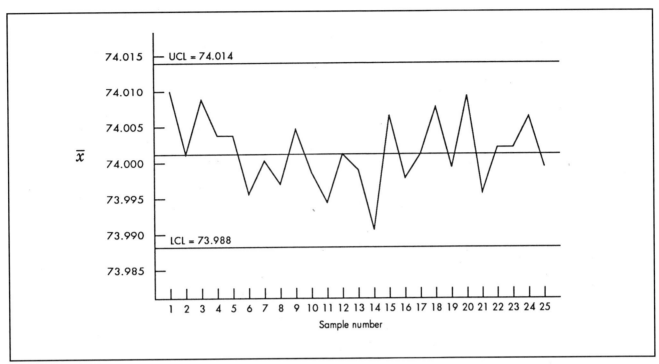

Fig. 16-12 A typical control chart.

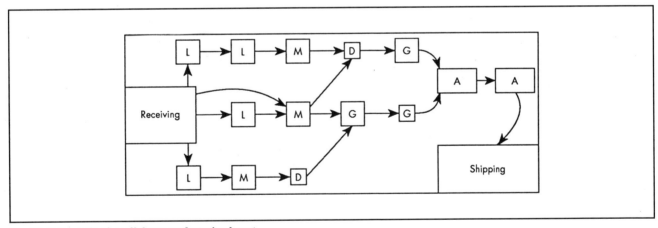

Fig. 16-13 Example of a cellular manufacturing layout.

ing drive pin mechanism adjusts to the end of the workpiece, even if there is a slight out-of-square from part to part, ensuring evenly distributed driving force. The chuck delivers 9000 pounds (40,000 N) of freely compensating chucking pressure, accuracy within 0.001″ (0.03 mm) TIR concentricity and speeds up to 4000 rpm. Centrifugal loss is minimized with two-piece spherical bearings which counterbalances the mass above and below the jaw arm pivot points.

Multiple operations. When a manufacturing process dictates multiple operations for proper parts manufacturing, workholding systems with built-in versatility may save the cost of new equipment. There are chucks available that are capable of both first operation (centering) and second operation (compensating) machining operations. This eliminates the need for chuck changeover from operation to operation—thereby increasing uptime.

Re-chucking errors can be minimized through the use of specially designed workholding devices that address the problem.

The chuck to use here would be one whose pressure can be varied from high to low, hydraulically or pneumatically, without releasing the workpiece. The result is differential chucking for both rough and finish turning in one operation. The design incorporates "positive" or "parallel" pull-back of clamp jaws to enhance accuracy by ensuring that every workpiece is squarely held against a workstop.

The chuck maintains accuracy even during re-chucking operations. A homing device allows the jaws to equalize and compensate for part variations, while the jaw actuator arms move uniformly toward center. Thus, a workpiece can be machined on one side, removed, rotated, and replaced on the same chuck for second side machining—with concentricities of 0.001″ (0.03 mm)

CHAPTER 16
WORKHOLDING

TIR. In many instances, one set of tooling can grip different diameters of a part (even the ID or OD of completely different parts) since this chuck can quickly and easily convert between external and internal chucking.

A chuck such as this can be changed from a centralizing to a compensating chuck in a few minutes. This feature provides first and second operation processing with the same chuck. The chuck design permits its jaws to float and engage independently of the chuck center through a virtually friction-free actuator.

Production Run Size

Ultimately, it is the production run size that dictates the return on investment (ROI) of workholding. The degree of workholding flexibility and ease of changeover required, all relate to the actual number of pieces to be made. A guideline on making the decision is this: minimal changeovers are best facilitated with a manual changeover chucking system. If there is a high number of changeovers, there is a need to facilitate them as quickly as possible. A pushbutton or pedal-activated system is recommended.

Long-run production. For long-run machining applications, it is easier to justify a customized workholding device for a given part. There is no need for highly sophisticated changeover automation. Although there are chucks available that offer a certain degree of flexibility and ease of changeover, the design emphasis is placed on specific part machining operation performance. Working within the established parameters of a given part, many part-specific innovations can be incorporated into the overall chuck design to maximize performance and productivity.

Short-run production. For short-run applications, chuck design emphasis can be placed on maximum versatility and ease of changeover. As mentioned earlier, there are workholding systems that are capable of first and second operations. This built-in versatility, coupled with unmanned or highly efficient manned workholding changeover mechanisms, provides the best productivity in most manufacturing operations.

Changeover Requirements

Manual changeover. A manual chuck jaw changeover system best serves medium-long and long production runs where the demands of automation and flexibility are not as high as those required in semiautomatic or fully automatic machining systems. It offers quick and simple changeover for increased productivity.

Quick-change jaws are available in what is called the "60-Second System." Changeover can be completed in about a minute. It consists of a master jaw, key locator, lock screw, and a jaw blank. The master jaw is integral with a chuck's actuator arm and a key locator is in the master jaw. The results are accurate positioning of the top jaws to the master. Individual jaws adapt for either ID or OD chucking and are mounted and locked or unlocked and removed by turning a single release screw one full turn. Another feature of this system is a more usable chucking area through the elimination of hold-down screws.

Semiautomatic changeover. For medium or intermittent runs, or where chucks may be too cumbersome to handle manually, a multiple system is required—even though the number changeovers may be few. For these situations, a semiautomatic changeover system may best facilitate productivity. One such system utilizes a multiple chuck—or chuck within a chuck system.

The master, or base chuck is usually of the diaphragm type that is permanently mounted to the machine. A mating chuck is the workholding device. The mating workholding type of chuck can vary with application. During changeover, the master chuck releases the workholding chuck for removal and replacement with a different workholding chuck. The result is an enhanced production capability for a wider variety of parts—all performed through the same master chuck already mounted to the machine. This function can be performed manually or robotically.

Automatic changeover. For maximum versatility of CNC or PC control of chuck changeover, a fully automated system is recommended. There are workholding systems available to facilitate highly accurate repeatability of chuck changeover. An added benefit is the off-line pre-qualification of a chuck's dimensional integrity. The resulting gains in machine uptime and manufacturing quality directly translate into increased productivity.

One type of chuck changeover is based on a precision coupling utilizing a spiroid gear. One half of the coupling is mounted to the spindle and the other half is attached to the back plate of the chuck. The two mate at the precision gear form at the spindle pilot. The gear teeth are concentric to the spindle pilot, and square and parallel to the back face of the rear coupling element.

The two halves of the coupling are mated and secured by one of two drawbars. The second drawbar performs the chuck actuation. There are no mounting screws, nor is it necessary to use a master part for centering. The position repeatability of the quick change elements is 0.0003" (0.076 mm) TIR—or less. Contamination from chips and cutting fluids is minimized through its sealed design at the spindle's coupling point.

ADVANTAGES OF PRESET WORKHOLDING, TOOLHOLDING, AND PART REGISTRATION TO REDUCE SETUP TIME

To maximize machine tool productivity, it is critical to reduce setup time by changing workholding and tooling between production lots of parts in as short a time as possible. Considering the sophistication of today's machines and the parts that they run, even part loading and positioning have become paramount to the reduction of nonproductive downtime. The accurate presetting of changeover components and prequalification of workpieces are necessary for efficient machine tool utilization.

A brief examination of changes in manufacturing philosophies and emerging production methods and technologies is required to fully appreciate the critical need for the presetting of workholding, toolholding, and part registration.

The overriding consideration for presetting technology must be placed at one point—an awakening of global competition over the last two decades has demanded of manufacturers that productivity become more efficient than in any previous period. This efficiency carries over to all levels of the manufacturing process, from higher on-the-floor machine tool productivity, to new process and quality controls for improved product performance, inventory management methods, and the management of manufacturing data provided by computer integration. Today, the numerous applications of NC equipment; the growth of flexible machines, cells and systems; automated parts handling; and SPC, CIM, and JIT methodologies are all assigned to diminish the waste of assets. High-speed cutting operations have helped reduce actual cycle time through tooling advances and machine control sophistication, but still machine downtime for setup and changeover is the major roadblock to improving productivity.

For years industry coped with the enormous hindrance that machine tool setup placed on productivity. In the past, the economic order quantity (EOQ) most often meant that manufacturers had to overrun and/or double run production lots to offset the effects of setup time. With a goal set for zero inventory levels

CHAPTER 16

WORKHOLDING

for both work-in-process and finished goods, these practices are no longer justifiable or economically feasible. And, as order quantities have decreased in an effort to reduce inventories, machine setups are now more frequent, increasing machine idle time and the complexity and conflicts of scheduling production. Even as machine operational technology has made dramatic improvements since the first NC machines of the 1960s, setup technology has lagged behind.

A recent survey by ITW Woodworth of its major metalworking customer base, to determine how available machine tool time is spent, indicates that as much as 50% is in a nonproductive idle mode (see Table 16-2). The majority of that time, as shown, is due to changeover, either tool changes, workholding setups, or parts handling.

Considering these figures, it is important to ask the question: How important is presetting to improving the productivity of present-day machine tools? Presetting should be considered one of the most critical factors for achieving reduced setup times and improving machine tool utilization and overall productivity. The benefits of presetting are perhaps most clear-cut but not limited to companies with large numbers of NC machines. With large tool and workholding inventories, the need for precise tool management becomes acute, and to maintain tooling costs within economically acceptable limits it is necessary to provide a controlled environment, to prepare tooling and workholding for upcoming jobs. Today the trend toward presetting is gathering momentum ln a wide range of shops for boring, turning, milling, and other metalcutting operations.

The second question to ask is: What is presetting? Presetting, simply stated, entails the off-line setup of tools, workholding, and workpieces to a known, accurate dimension that, when the tool and/or workholding device is moved to the machine, eliminates variation and reduces the need for extensive on-the-machine setting, trial cuts, offset compensation, and delays to production. The productive advantages then are broken down to three basic concepts:

1. Machine uptime is optimized as changeover time is reduced to a minimum. Considering part registry, even when loading successive pieces within the same lot or transferring pieces from machine to machine, parts handling time can be reduced and quality improved.
2. Quality is maximized while scrap-producing trial cuts are eliminated.
3. Presetting results in reduced tooling costs; for example, accurately preset tools can be set to extreme limits of part tolerance, resulting in more parts per tool index.

An analogy for what the process of presetting can mean to metalworking production can be found by examining another area of manufacturing, the assembly process. A clear perspective may be achieved by focusing on one aspect of assembly, the torquing of a nut onto a threaded fastener. Prior to the modern assembly line, when products were built virtually one at a time, workers spent a great deal of time selecting the proper bolt, nut, and tool, and transporting components. The actual time to tighten the nut, however, was predetermined by thread design and length of the exposed bolt, and typically accounted for only a small percentage of the assembly process. Today, the use of power tools has reduced the time needed to tighten the nut, but the majority of improvement for assembly operations has been realized by the development of standardized components, dimensions, tooling, and worker assignments. Present-day assembly is a merging of automation, computer, and human involvement, working with known and defined components to achieve optimum efficiency. The integration of modern technology and prearrangement has eliminated the major causes for delays between operations.

Applying this same concept to metalcutting machines, where cycle times are (for the most part) dictated by speeds and feeds (although technology such as new cutting tool materials is improving rates for metal removal), the area with the most room for improvement is changeover time. The recent advances made in presetting technology, theory, and reality have combined to make changeover more manageable, accurate, and rapid.

In general, the major concentration of presetting has focused on flexible, modern CNC machines and automated cells and systems. It should be noted, however, that most of the progressive logic that applies to these machines also pertains to manual

TABLE 16-2
How Available Machine Time is Spent

	Tool Manufacturers	Automotive Suppliers	General Metal Turning Operations
Tool change	10	5	17
Setup time	25	40	15
Parts handling	15	5	17
Total Changeover	50*	50*	49*
Run time	30	40	36
Scheduling	10	—	—
Maintenance	5	5	—
Other	5	5	15
	100%	100%	100%

*Results from an ITW Woodworth survey indicate that as much as 50% of available machine tool time is consumed by nonproductive operations such as tool changes, changeovers, and parts handling.

CHAPTER 16

WORKHOLDING

machining. To obtain maximum machine productivity, presetting should encompass three functions. They include *workholding* in which chucks, chuck jaws, fixtures, pallets, etc., must be changed each time a new lot of parts is to be run. The second area is *toolholding*, where part changeover necessitates the exchange of an entire tool holder. It also includes changes for tool wear and breakage during normal operation, and simple procedures such as the indexing of an insert, in which tolerances of milled pockets and tools may not lend themselves to precision tool setting.

Typically, after indexing an insert, the tool is backed off a predetermined amount, a trial cut is performed, the dimension is checked, an offset is entered into the machine (either manually or automatically), and a final cut is taken. Since cutting pressures can affect metal removal, the next part cycled, taking a cut at full depth, will also require inspection to verify that the compensation factor was correct. In the event that the offset was incorrect, at best the process has to be repeated. At worst, the workpiece has been scrapped. Taking into account that many machines are equipped with multiple tools and multi-spindles, and that tools frequently require dual offsets (radial and depth, for example), it is no wonder that production rates (from the survey in Table 16-2) scarcely reach 40% of machine uptime.

The third area to be addressed is *part registry*, which can be considered a natural extension of the capabilities of preset workholding devices. The concept of part registry—which is defined as the accurate positioning of workpieces within a workholding device to a specified location, precisely simulating machine mounting conditions yet prior to setting within the machine tool—has grown in importance as the complexity and quality requirements (spatial dimensions, true positioning) of machined parts has increased. This has impact upon the setup time of high-value, intricate, and fragile workpieces, where the loading, fixturing, and handling of a part are recognized as the critical elements of producing a component to specification. In addition, consideration should be given to applications involving heavy, unwieldy parts in which machine loading is time-consuming and adds operator fatigue (mental and physical) to the equation of machine productivity. Part registry also lends impetus to the trend toward chucking a part once for multiple machine processing and transferring between machines for improved quality and productivity.

In each of these areas, a variety of presetting methods and technologies are now being applied with varying results and levels of success in reducing machine setup time. Despite the differences in methods, the objective is the same: to increase uptime and quality output of metalworking machine tools.

Methods of Machine Tool Setup

The practice of shutting down a machine during setup comes from an earlier age when labor costs, capital investments, large inventories, and other financial considerations were less significant, and competition was less developed. Changing and machining chuck jaws on a lathe, installing a fixture, taking trial cuts with newly installed tools, or indexing a just-loaded part into position while the machine stood idle were accepted procedures.

Gradually, the importance of these elements were brought to the forefront and loomed larger in significance, recognized by management as detrimental to productivity, yet they were still tolerated. With much attention centered on rising labor costs, automation techniques and advanced CNC machining practices have greatly reduced the percentage of direct labor value-added, yet the percentage of value-added by indirect activities has increased substantially.

Today, many managers of metalcutting facilities will agree that machine tools, to reach acceptable earning levels, should be "producing chips" 85% to 95% of available time. To obtain this level, changeover should be at least 90% complete off-line before the actual machine changeover is to take place. Two possible methods on how best to arrive at these figures follow. The first is through the use of advanced machine systems that reduce setup via automation.

Automated Setups

Considering the sophistication of CNC controls and the use of computer-integrated manufacturing, many people believe that an efficient means to set tooling is still at the machine.

Tool storage systems and identification coding maintain data for every tool within the system. Tools are set only to approximate points; precision offsets are accomplished once the tools are loaded into the machine by means of automatic tool changers. Accurate probing functions can provide feedback to the control unit regarding insert and part location. The information is then used to make adjustments to programs. Sensors and other measuring systems can determine if workholding devices and workpieces are properly set. In this way, setups can be accomplished swiftly and accurately.

Four drawbacks to this method exist, however. First, because each machine tool oversees its own setups, the equipping of a multi-machine plant, system, or cell often results in extensive duplication and investment in probing technology. Second, to perform these tasks, the complexity of the machines and the controls requires highly trained technicians to operate and maintain them. Third, cycling through a probing function for part locations or touching a tool edge to a sensor to determine offsets is still an extra step that adds to a machine's nonproductive time. And fourth, since this method relies on sophisticated machine controls, it does not provide solutions to the needs of presetting and reducing the setup times for manual machine applications and existing NC equipment.

Presetting Off-line

A second proposed method to reduce setup time and, consequently, improve machine uptime, is to precisely preset the tools, tool holders, workholders, and workpieces before mounting them on the machine. Although variations exist as to actual equipment employed, the basic concept requires that the spindle coupling, or tool coupling, be duplicated at a staging area. This staging area can be located virtually anywhere within the plant—at machine sites, in the tool crib, or at designated tool stations. Depending upon application and precision required, each staging area will also require measurement devices, from simple dial indicators to optical comparators and laser systems.

Part registry requires one additional step—once the workholding device is preset, the workpiece must then be positioned and held in place. Locating surfaces of the part can then be inspected using familiar gaging techniques to determine and adjust location, concentricity, etc. With correct location achieved, the part remains in the workholder and the entire system is moved to the machine, or down a line of machines, ready to run.

The techniques of off-line presetting provide responses to the four problems previously discussed and encountered with CNC machine tool setups. First, duplication of setting equipment is unnecessary. Only the required number of staging areas are needed to meet application needs. Second, off-line presetting employs basic measuring systems that shop personnel are familiar with. Third, the complete process should not require additional

CHAPTER 16

IN-PROCESS GAGING SYSTEMS

machine time from production, other than the mechanics of removing and replacing the item being changed. In a modern quick-change system, a complete tool holder with preset insert, typically can be changed faster than indexing an insert. The fourth question answered by off-line presetting is the fact that it is not restricted to highly equipped and advance-controlled CNC machines. Manual machining, existing older NC equipment, and even the newer machines can make use of the presetting methods described.

To be efficient, once the components are preset the tooling and workholding must be of a quick-change design in an all-out effort to minimize changeover time. Typically, the retention systems must be simple to actuate so that operators can quickly engage and disengage the devices. Additionally, for simplicity of automated changing systems such as robotic handling, tool couplings may have to sacrifice some precision to accommodate larger positioning tolerances.

This points out the dilemma faced by proponents of off-line presetting—the loss of precision once the tooling leaves the setting station and is mounted on the machine. No matter how accurate the preset gaging, regardless of whether an insert is set to within one ten thousandth of an inch (0.003 mm), or that a chuck is running at 0.0005" (0.013 mm) TIR, the tool-to-machine mounting is critical to off-line preset technology.

The level of precision and, therefore, the presetting capabilities of toolholding and workholding systems are dependent upon the accuracy of the retention coupling. Theoretically the accuracy at the preset gage is transferred to the machine, but due to the discrepancies of mounting, this is not always ensured.

IN-PROCESS GAGING SYSTEMS

In-process gaging is a process control technique that improves quality and productivity, reduces operating costs, and helps maximize the inherent capabilities of production equipment.

In-process gaging systems are used most frequently on external grinders and ID grinders to locate workpieces, identify and control grinding wheel infeed rates, compensate for grinding wheel wear, and measure and control part size. Unlike other production gaging systems, in-process gaging continually monitors part size during metal removal. When the preset finished size of the workpiece is reached, the gage automatically stops the grinding machine. The advantage of this in-process monitoring is almost complete elimination of nonconforming parts.

Since practically all dimensions, conditions, and spatial relationships of both internal and external diameters can be measured while the part is being ground, in-process gaging offers an improved method of process control that extends the capabilities of the grinder. For example, a grinder capable of producing parts to 0.0002" (0.005 mm) can, with the addition of an in-process gaging system, generally produce parts to better than 0.00001" (0.0003 mm). It is possible to achieve repeatability of 0.0000004" (0.000001 mm) in production grinding operations using in-process gaging systems. To get the kind of repeatability and reproducibility required today, a gaging system must be capable of measuring tolerances at least 10 times better than the tolerances actually produced.

The economies of in-process gaging come when the need for high precision combines with large production quantities. A 100 piece parts run is usually the decision point, although if parts are complex, with many diameters to be ground, as few as 10 pieces could justify the installation of an in-process gaging system. Under 100 pieces, dimensional accuracy can be checked manually using a bench-type gaging system at little loss in productivity. If more than 100 parts are produced, in-process gaging is essential, since it is not reasonable to expect an operator to efficiently perform manual 100% inspection on 100+ parts. Also, the operating characteristics of the grinding machine can change significantly over the course of a 100-part run, causing the production of nonconforming parts. Wheel loss is one of the most critical factors, and without in-process gaging, wheel loss and other variables are difficult to compensate for.

RECOGNIZING PROCESS VARIABLES

Production grinding is subject to the effects of a number of variables. Coolant temperature, workpiece temperature, grinding wheel action (defined as metal removal rate, wheel breakdown, and wheelhead retraction speed), and the condition of the workpiece all affect dimensional accuracy. Those variables cannot be compensated for through the machine's own control system. Plus, any grinding machine has variances in accurately moving the wheelhead from one point to another time after time. In-process gaging helps eliminate the effects of these variables by working in a complementary fashion with machine controls.

KEY SYSTEM COMPONENTS

A typical in-process gaging system consists of a gage head, measuring arms, and tips that contact the part directly, as well as a controller/amplifier that serves as the communications link between the gage head and the grinding machine. Since in-process gaging is performed in a harsh production environment, the gaging system itself must be suitable in construction and performance to resist the effects of continuous operation in severe conditions.

Measuring arms should be as short as possible, and to the extent practical, in line with the center axis of the grinding machine. This configuration improves the thermal stability of the arms, and reduces elasticity and mass inertia—all three of which can affect measuring accuracy. Upper and lower arms should be parallel to each other. Parallel arms eliminate inaccuracies caused by the scatter inherent in gage head approach.

Tip condition is extremely important to accurate dimensional measurement. Carbide tips are the most frequently used because they are inexpensive and generally wear well. Since they are in contact with a moving surface, tips do wear. They can gradually shorten in the direction of measurement, and become concave after extended use. If excessive wear is a problem, carbide tips can be replaced with diamond or cubic boron nitride tips.

Tip measuring force, the force at which the tips press against the workpiece, is also critical. Measuring force should be just enough to cause the tips to ride easily on the workpiece. Less force increases the danger of tip flutter due to vibration caused by coolant between the tip and the workpiece. Excessive force accelerates tip wear and can mar the workpiece.

CHAPTER 16

IN-PROCESS GAGING SYSTEMS

The selection of a measuring and control amplifier is also important. The unit should have a resolution in the fine measuring range of 0.33 feet (0.1 m) or better. It must not be sensitive to temperature variations, voltage variations, or remote interference peaks commonly found near machine tools. It should have three switching circuits so that roughing, finishing, and fine finishing infeed rates can be controlled.

THE GRINDING CYCLE

To save time in a typical grinding cycle, the grinding wheel approaches the workpiece rapidly. At some point in the approach, the infeed rate changes to the coarse grind infeed rate. The grinding machine then removes a predetermined amount of stock at that rate.

At a pre-established point, the in-process gaging system switches the wheel to a fine grind infeed rate, usually between 0.67 and 2.62 feet/second (0.2 to 0.8 m/s), but determined in practice by process parameters. At the fine grind infeed rate, the infeed is slower and the wheel normally removes less stock. During coarse grinding, pressure builds between the wheel and the workpiece. The pressure often causes the workpiece to flex slightly. The slower fine grind infeed rate allows the part to straighten and cool while the wheel works more slowly in removing material to reach finished size. The slower rate also allows the grinding machine to stop more precisely when the gage communicates that it has reached the onsize condition.

Normally, when the grinding machine receives the signal from the gage that the on-size condition has been reached, the machine allows for spark out. Spark out is the dwell time of the grinding wheel on the part to allow it to continue to straighten and cool. Spark out also produces a finer surface finish.

TYPES OF IN-PROCESS GAGING SYSTEMS

There are two basic types of in-process gaging systems, contact systems and noncontact, or pneumatic, systems. Contact, or mechanical, gaging, where the gaging tips actually ride on the workpiece, is used for OD measurement, workpiece positioning, stock removal from faces, and for ID grinding. Contact gaging is also used extensively in match grinding where the ID is measured and information is communicated by the gaging system to the machine to set the zero for grinding the matching OD component. There are contact gaging systems designed to measure interrupted diameters, including odd and even numbers of gear teeth. This type of measurement is particularly valuable in the production of gear pumps where clearances are critical.

Pneumatic gaging is used in the same types of applications as contact gaging, but instead of contacting the part directly, pneumatic gages use precision air nozzles to measure differential pressure, comparing resistance to flow with a nominal air input. Pneumatic systems eliminate the tip wear inherent in contact systems, are self-cleaning, and contain fewer mechanical parts than contact gages. They also have the advantage of being able to measure much smaller IDs, down to 0.060" (1.52 mm), although they are not as precise as contact gages in the in-process function.

Pneumatic systems are frequently found on what are called in-process/post-process gages. Here, workpiece size is checked immediately after it is ground, and information is fed back to the machine control to compensate for any changes in nominal dimensions. The most common application for this type of system is in ID grinding, although pneumatic systems are being used more frequently in centerless grinding operations.

There is growing interest in in-process gaging for turning centers. Today, more and more turning centers have the capability to produce a finer finish part necessary for accurate in-process gaging. Also, the capability of turning machines to make tool adjustments has improved. Turning produces relatively large chips that present some difficulties for in-process gaging systems. However, these gaging systems can be installed on the machine in such a fashion that as soon as the part is finished, the gage measures it at that instant. Information for tooling changes can be quickly communicated to the machine controller.

In-process gaging systems are also available for flexible manufacturing operations. These systems can be quickly and easily reset to handle a range of external diameters.

In-process gaging systems can be used for pre-process inspection, although this practice is uncommon. When the workpiece is loaded into the grinding machine, the "pre-process" gage measures its diameter. If the workpiece is undersized, based upon a programmed size range, the gaging system tells the grinder to remove the part and load the next one.

GAP CONTROL SYSTEMS

Gap control is an enhancement to in-process gaging. Gap control eliminates "air grinding." Grinding operations are much more efficient when the grinding wheel can be brought up to the workpiece rapidly before grinding begins. Without a gap control system, the wheel will crash into an oversize part. Gap control stops the infeed before a severe collision occurs. It signals the machine to begin operations as soon as the wheel touches the part, irrespective of the beginning size.

A gap control system can also locate the workpiece and signal the grinder to take off material until the part gets to a nominal size for the in-process gage to be effective. An acoustic gap control system uses an advanced piezoelectric transducer to sense wheel contact with the workpiece and in milliseconds signals the machine to switch to the slower, coarse grind infeed rate.

The acoustic system also determines the condition of the wheel by analyzing the sound it produces during dressing. It can detect breakage or rounded corners, and can command the machine to continue wheel dressing until specifications are reached.

In some applications, an acoustic system is used to optimize the grinding cycle. Once the system is tuned correctly, it can command the grinding machine to either increase or decrease infeed rate to achieve maximum metal removal performance.

IN-PROCESS CONSIDERATIONS

In order to achieve the most value from an in-process gaging system, the variables affecting the process must be understood. Since gaging is being performed while the part is being ground, ambient conditions have an influence on machine control accuracy. Although in-process gaging systems compensate for these conditions, a certain amount of control over the production environment must be exercised by the operator.

Coolant temperature directly influences grinding accuracy. Coolant must not only maintain the workpiece and grinding wheel at a constant temperature, it must also maintain the elements in the gage head at a constant temperature. Different elements of the gaging head, such as the tips, fingers, and finger mechanisms, have different thermal constants. A major variation in coolant temperature can cause a temporary nonhomogeneous temperature distribution in these parts. These temperature swings can cause measurement inaccuracies which can affect part size control. In practice, good results can be achieved by keeping the coolant temperature stable and just above the ambient temperature.

CHAPTER 16

IN-PROCESS GAGING SYSTEMS

Workpiece temperature is also important. The grinding cycle is normally not long enough to ensure that the temperature of the workpiece will reach the temperature of the coolant. To help overcome this situation, workpieces awaiting grinding should be placed next to the machine for approximately two hours prior to the grinding operation. Masters used for setting and adjusting the measuring head, and reference bores used in match grinding, should also be placed near the machine to allow them to reach ambient temperature. Temperature equalization can be improved by spraying the workpieces with coolant for 10 minutes before grinding.

Coolant cannot prevent the surface of the workpiece from being heated to a certain depth during grinding, particularly during the roughing process. Inaccuracies caused by workpiece temperature variations can be overcome by extending the spark out or fine finishing time to dissipate the heat generated during roughing.

MEASURING FOR MACHINING

To control part size in any machining operation, it is necessary to measure the workpiece. How that measurement is performed, however, has changed significantly over the past decade and a half.

Before the advent of electronic or pneumatic gaging systems, specially designed and standard gages were used manually to measure workpiece dimensions. Many early snap, sight, and feeler gages were crude and could be easily influenced by the operator. That did not matter too much since the measurements were performed primarily to assure the part was of an acceptable size, without consideration to improving its quality. Under this type of system, the operator simply separated the good and bad parts. Turning machines and grinders were continually adjusted by the operator to keep the workpiece within the tolerance band prescribed by the gage.

While this type of manual gaging worked well for many years, the need for high-precision parts produced to close tolerances required improved gaging systems. In order to manufacture these more sophisticated components, machining houses looked for tighter control limits, improved part geometry, and methods to match components more precisely.

The Evolution of Size Control

Part size can be controlled in many ways, but the basic methods are:

1. The operator makes machine adjustments after measuring the part using a manual, offline bench fixture. The machine tool position is adjusted to change the part size, taking into consideration the measured part dimensions and the operator's experience with tool wear, machine capability, and part cycle.
2. An in-process gage contacts the part during the machining process and signals the machine to stop when the part dimensions reach a preset size. This method is used primarily in grinding operations.
3. An offline gage provides an offset size for the operator based on an SPC evaluation of measurement. A large number of parts are measured in order to establish a dimensional history, or database, which is analyzed by a computer to establish control limits.
4. A semiautomatic, post-process gage, where each part is hand-loaded by the operator, provides feedback to the machine for compensation. When a microprocessor readout is used, all aspects of the machining process are considered, including machine capacity, operating mode, time of day, and tool condition.
5. A fully automatic post-process gage accepts the part automatically from the machine. This method allows the machine to run without operator influence. The feedback algorithms vary, taking machine variables into consideration. Total part quality as machined is measured and complete historical data is retained for documentation.

Procedure Five is the progressive result of the preceding methods. It represents the most sophisticated and economical method of assuring part quality. The gage is the brain of all measurement systems, and when used in an automatic post-process system, it becomes a genius. The readout keeps constant vigil on gage accuracy and repeatability. Calibration (mastering) eliminates drift to maintain gage reproducibility.

The Correlation Challenge

All measuring systems are comparators since they are calibrated by a master whose dimensions were computed with those of a grand master. When a size measured on one gage is different from another, there is a correlation problem. Controversy over the size can only be solved by logical steps, with the final decision being based on which answer has the least chance of error.

Correlation is a challenge when parts must be interchangeable, especially when various methods of machining and measuring are used. The user must recognize that conflicts in measurement can occur. Parts must be toleranced and dimensioned with data consistent to machining and measuring, to keep conflicts to a minimum. Accurate locating points and gage position information are critical to accurate measurement.

Applications vary so much that a step-by-step example is not practical. However, there are some basic procedures to follow in calibrating gaging systems to eliminate correlation problems. First, use the same master to calibrate each gaging system. Second, contact size and location should be the same on each part, or so close that part geometry will not influence gaging results. Third, the same resolution should be used on each system. Fourth, part staging should be the same. Each part should be clamped in the same location, using the same clamping force. Fifth, the method of gaging must be considered, noncontact or contact, contact pressure and shape.

When machining and measuring are mated using today's technology, the result is the production of low cost, high qualityproducts. The post-process automatic gage, using statistical process control with a microprocessor readout, provides intelligent feedback to numerically controlled machines.

The machine with a gage becomes self-operating. The gaging system becomes a decision maker for machine control. It adjusts, trims, and shuts down the machine if necessary, and, in some applications, provides a diagnostic advisory to the operator. It also keeps track of itself to assure that it remains in calibration, and maintains its own repeatability and reproducibility.

With the need today to reduce costs and improve quality, manufacturers should consider and investigate the advantages of post-process automatic gaging.

CHAPTER 16

ADVANCED TOOLING

ADVANCED TOOLING

A recent survey by a leading U.S. cutting tool manufacturer found that of the tooling used by U.S. manufacturers, 45% to 55% are obsolete. This despite the availability and proven capability of much more advanced cutting tools.

The shift of consumers' tastes and expectations in the products bought, put demands on manufacturers unheard of several decades ago. It changed the production philosophy dramatically. Machine tool builders and cutting tool manufacturers scrambled for new innovative ways of machining. For a number of years, the capabilities of many chipmaking machine tools were limited by the prevalent cutting tools, with respect to both their design and material. With machine tools in the lead, tooling manufacturers were urged to develop new tools for faster metal removal, more accuracy, better surface finish, more consistency and predictability commensurate the power and overall capability of the machine tool, and extended life at the cutting edge. Then something interesting occurred, in that the relative position of machine tool and tooling manufacturers reversed considerably. The development of extremely hard cutting materials, such as a variety of solid and coated carbides, ceramics, polycrystalline diamonds, cubic boron nitrides, and cermet grades, has been accompanied by innovative designs of inserts, combined with newly applied cutter geometries and more accurate clamping systems. This technological leapfrog has enhanced precision machining and opened up the realization of meeting the most urgent manufacturing demands:

- Improving overall production quality.
- One pass machining.
- Fewer machining stations.
- Forgoing roughing operations.
- Faster machining.
- More flexibility of machining results.
- Better tool life.
- Cost savings.
- Better surface finishes and tighter tolerances.
- More productivity.

CUTTING TOOL MATERIALS

Ever since the turn of the century, there has been an increase in cutting speeds for the machining of workpiece materials of any kind. Over that time the development of new tool materials paralleled the development of new workpiece materials. The challenges posed by the chemical and physical compositions of new part materials had to be met by new cutting materials. It led to the development of high speed steel (HSS), then carbides, cermets, ceramics, artificial diamonds (PCD) and cubic boron nitrides (CBN).

The variable speeds and feeds could be better accommodated especially toward their highest spectrum (see Fig. 16-14). The need for higher cutting speeds as a decisive contributor to productivity became an important factor, and high speed machining over 5000 rpm is almost a given on modern manufacturing floors.

Of course, as new workpiece materials are developed and more complex alloys become dedicated raw materials to be shaped into complicated end products, the variety of cutting materials within the basic group increases to extend the tool life and simultaneously achieve the best possible machining results (see Fig. 16-15). Cutting tool manufacturers, striving for the "ideal" cutting tool material, can only move within the parameters of toughness and hardness, two opposing material characteristics.

There simply is no "ideal" cutting tool material (see Fig. 16-16). There is no single universal cutting tool material suited for all machining processes, machining all workpiece materials and achieving best results. One cutting material might come close to the "ideal" for one specific material and one specific process but fail when used with a different material and process, often even when there is only a slight deviation or minor change of one machining parameter.

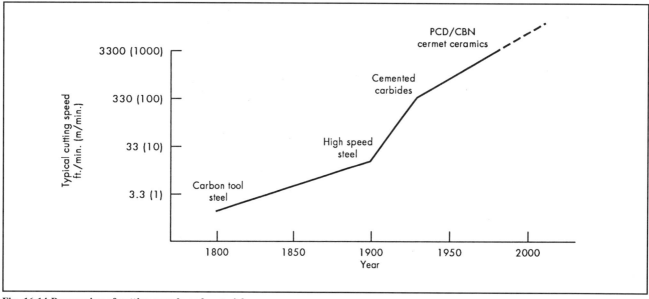

Fig. 16-14 Progression of cutting speeds and materials.

CHAPTER 16

ADVANCED TOOLING

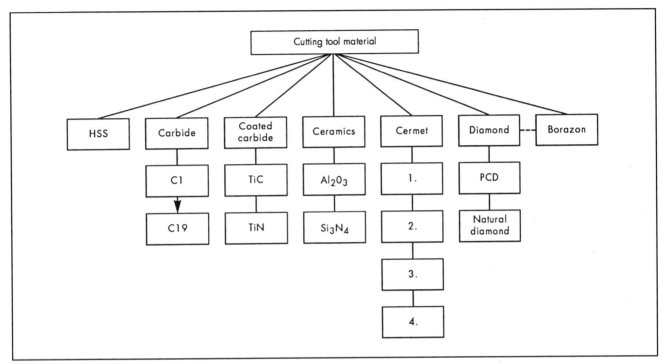

Fig. 16-15 Most prevalent cutting materials.

The most popular material to date still is cemented carbide, for it offers a useful and widespread compromise between toughness and hardness. While its uncoated versions are losing ground rapidly, specialized coatings further enhance carbides' use in manufacturing. However, more defined and specialized machining tasks, exerting extreme stress and abrasion and stringent requirements on finishes, are more and more tackled by "advanced" cutting materials (see Fig. 16-17), including ceramics, cermets, PCD, and CBN.

All of the present tool materials dramatically outperform their predecessors. Compared to high speed steel (HSS) in the early 1900s, today's super-HSS can cut up to 300% faster. Carbide tools can cut more than three times faster than super-HSS tools. Coated carbide's tool life is three to four fold that of uncoated carbide, and ceramic cutting tools can cut chips more than twice as fast as coated carbides.

The cutting zone temperature in most machining operations is directly proportionate to the cutting speed. That means the higher the cutting speed, the more the cutting material is exposed to heat and prone to chemical breakdown due to heat. The cutting material's resistance to heat is called "hot hardness." At 1200° F (650° C) for instance, cobalt-based HSS breaks down completely, while carbide hardness diminishes by 13% and ceramic by only 3%. That is why various cutting materials cover different speed and operating ranges.

The complexity and importance of cutting tool materials warrants individual scrutiny, for they do the actual cutting within the complex envelope of machining and manufacturing.

High Speed Steel

Tools of high speed steel, such as reamers, drills, taps and endmills, usually run on lower horsepower machines with limited spindle rigidity. Their advantage is their low price and ease of regrindability. In an advanced machining environment, they machine ferrous and nonferrous metal parts with less stringent part-print requirements.

Standard molybdenum HSS (M-1, M-2, M-7, M-10, M-15, M-42) are the most popular grades. To increase the hot hardness of HSS, cobalt (8% to 10%) is added—however, this at the expense of toughness.

So-called high premium HSS have even higher hot hardness; they can machine material with HRC 35 hardness. The higher hardness is achieved through a controlled and more uniform

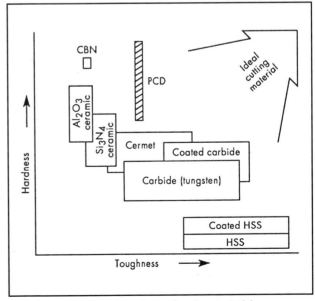

Fig. 16-16 Hardness and toughness of various materials.

16-23

CHAPTER 16

ADVANCED TOOLING

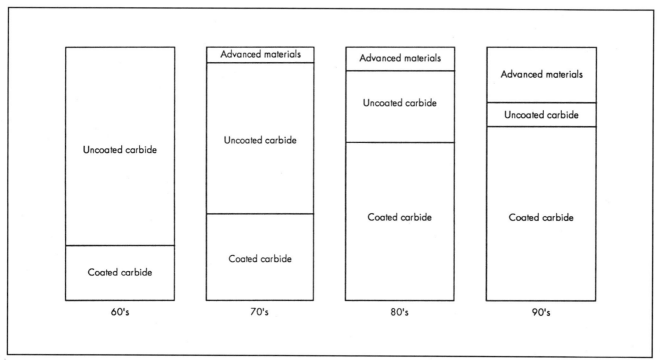

Fig. 16-17 Trends of usage of cutting tool materials.

microstructure when the steel alloy is molten, and by adding up to 10% vanadium and more sulphur to the alloy. The tool life of high speed steel can be improved significantly with titanium coatings. The drawback is that one of the original advantages, namely regrinding, removes the coating. To regain its effectiveness, the cutting tool must be recoated. The original tool size, however, cannot be maintained and eventual loss in size of the part machined will result.

The graph in Fig. 16-18 illustrates tool life at respective cutting speeds, comparing uncoated reground and TiN (titanium nitride) coated HSS tool material.

Carbides and Coatings

Carbides are superior to high speed steels, for they can run up to seven times faster and yet provide up to five times longer tool life. They come in different grades to accommodate different machining parameters and workpiece materials. Carbide tooling is versatile and can withstand high cutting speeds. Generally, it is the most widely used cutting material, also in advanced machining environments. What are carbides made of, in how many variations do they come, and what makes them universally usable?

Cemented carbides' raw materials are tungsten and carbon in powder form. Mixed at a temperature of approximately 3400° F

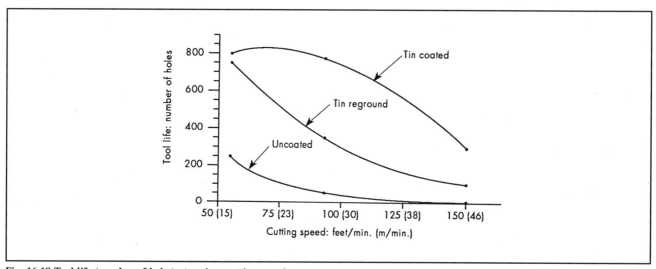

Fig. 16-18 Tool life (number of holes) at various cutting speeds.

CHAPTER 16

ADVANCED TOOLING

(1870° C), they carburize to tungsten carbide. Depending on what characteristic is desired, other additives are cobalt, niobium, titanium, and tantalum. The powder mix then is compacted and subsequently "baked" (sintered) at a temperature of about 3000° F. The end product is a precise, chemically defined carbide grade. There are a number of different grades for a wide range of operations. Per definition, grades C-1 through C-4, for example, contain carbide and cobalt binders, while grade C-5 through C-8 contain titanium and tantalum carbide. In general, higher grade numbers denote greater tool hardness, while lower grade numbers are indicative of more toughness. Lower numbered grades are more suitable for heavier, roughing cuts, while higher numbered grades are more wear resistant and for finishing cuts. Some cutting tool manufacturers have been experimenting with the size of the individual powder particles and have developed a "micrograin" carbide (smaller than 39 μin., 1 μm in diameter). Simultaneously increasing the cobalt content achieves a carbide grade that offers better than average toughness and good wear resistance. It is harder than any of the standard C-1 through C-8 carbides. Micrograin carbide tools are effective substitutes for regular carbide for extra high stock removal and high abrasive wear machining. Micrograin C-2 is the most popular carbide on the C-scale. This grade, in principle, retains the regular C-2 toughness and offers more wear resistance (low cobalt), resulting in longer tool life due to reduced chipping.

Coatings

The use of carbide coated tooling (inserts and tools), has increased machining parameters and tool life up to four times over uncoated cutting tool materials (see Fig. 16-19).

Popular coating materials include titanium nitride (TiN), titanium carbide (TiC) and aluminum oxide (Al_2O_3). Coatings are applied through chemical vaporized deposition (CVD) or physical vaporized deposition (PVD).[21]

A single coating layer has a thickness of 80 μin. to 200 μin. (2 μm to 5 μm). The coating provides an inert barrier that prevents diffusion of the cobalt from the carbide composition, thus prolonging tool life. Coatings increase the tool's versatility. Fewer grades are required to cover a broader range of machining applications, since available grades partially overlap the C-classifications for uncoated carbides.

Workpiece materials such as castings with heavy scale or sand, or most alloys harder than HRC 45, are usually machined with uncoated carbide; this also applies to nonferrous metals. In most other applications, they provide for good surface finishes and increases in productivity during turning, boring, drilling, reaming, and milling.

The success of single layer coatings has led to the development of multilayer or multiphase coatings, to combine the most favorable properties.

With multiple coatings, the first layer is usually TiC (titanium carbide), because of its good adhesion and similar thermal expansion to the substratum. The outer layer is selected in terms of maximum heat, wear, and cratering resistance as well as edge strength.

In a duplex layer, the outer coating is usually Al_2O_3 or TiN (titanium nitride). Triplex layers usually consist of a TiN layer to the carbide base, followed by an in-between TiN and an outer layer of TiN. This combination has excellent resistance to wear and cratering.

The thickness of the layers can vary depending on the material to be cut and the cutting method. Milling calls for a rougher outer layer than finish reaming. Another variation of applying coatings is partial layering. An example is shown in Fig. 16-20, depicting two cutting inserts with different partial coatings, "Z" for finish-machining steel and cast iron alloys, "X" for heavy cuts in steel.

Some manufacturers offer so-called glass- or diamond-coating. This technique is still in experimental stages and constitutes a "clean" carbon layer.

Coatings on carbide tooling have given this cutting material a different dimension and widened its application in all of chipmaking manufacturing.

Polycrystalline Diamonds (PCD)

Developed and introduced by General Electric in the early 1970s, it took more than a decade for polycrystalline diamonds to gain industry-wide acceptance. Manufacturing had to go through learning curves to find their optimum applications, while purchasing had to get used to their much higher pricetags, compared to conventional tooling material such as cemented carbides. Then, in the wake of global competitiveness calling for better quality and higher productivity, PCD cutting materials found their undisputed niche in machining.

A polycrystalline diamond cutter consists of a layer of diamond powder, sintered to one uniform mass of 0.02 to 0.03 inches (0.5 to 0.7 mm) thickness, which is pressed onto a substratum of cemented carbide. By using different grain sizes of diamonds, 177

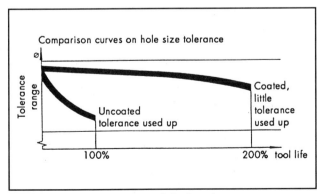

Fig. 16-19 Comparison curves on hole size tolerance.

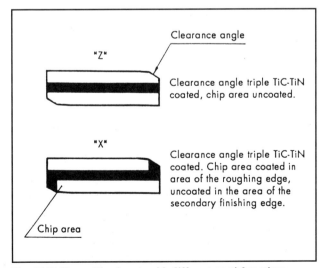

Fig. 16-20 Two cutting inserts with different partial coatings.

16-25

CHAPTER 16
ADVANCED TOOLING

μin. to 4000 μin. (4.5 μm to 100 μm) it is possible to optimize PCD for specific applications.

The microstructure of the PCD products influences their physical characteristics; the coarser the grain, the better the abrasion and impact resistance.

PCD tooling is used to machine nonferrous, nonmetallic materials. It is extremely popular in the manufacture of light metals, especially aluminum. Even abrasive high-silicon content aluminum (12% to 20% Si) can be machined with ease, and with excellent results in terms of surface finish and tool life at high speeds. Figure 16-21 shows two typical "Taylor" lines for cemented carbide (C-3 and C-4) and PCD 5 on AlSi alloy with 20% silicon. It illustrates the high cutting speed applicable and the minimal wear.

For the machining of ferrous materials, PCD is generally not used, due to its chemical reaction between tool and workpiece; one exception, though, is noteworthy. General Electric did extensive comparative machining of soft gray cast iron at lower surface speeds and found PCD's wear resistance better than any carbide grade.

Diamond tooling requires a rigid setup of machine tool and part fixturing, lighter feed rates, and smaller depths of cut. However, they should be run at very high speeds. This, combined with a good soluble cutting fluid, yields up to 300 to 400 times the tool life of carbide grades in finish-machining operations. PCD tooling can also be used for moderate roughing and semifinishing operations.

In summary, PCD-tipped cutting tools are applied in selective workpiece materials, due to their superior hardness, wear resistance, fracture strength, improved dimensional control, and excellent surface finish. This translates into high quality and reduced cost per part.

Cubic Boron Nitride (CBN)

Cubic boron nitride is produced by the high temperature, high pressure sintering (similar to PCD) of CBN particles and a binder material. The raw material is the cubic boron nitride grit that has been used for grinding abrasive ferrous metal. CBN cutting inserts are mostly dipped—that is, they are bonded to a carbide backing layer as opposed to solid, unbacked inserts. In hardness, CBN ranks second only to PCD. It has a low affinity to ferrous metals and remains chemically stable when machining them, even at high cutting speeds. Its mechanical and thermal properties make it a clearly superior cutting tool material for difficult to machine, hard ferrous alloys. Its hardness at 1300° F (700° C) is still better than that of carbide, or ceramic at room temperature, as can be seen in Fig. 16-22.

At high temperatures, carbide tools tend to soften at a rate similar to that of the workpiece material. This, combined with chemical reactions (oxidation) limits carbide's practical use to temperatures below 1300° F (700° C). CBN tools' temperature resistance and conductivity make them highly suited for running at high feed rates and cutting speeds where ceramics and carbide would break down.

CBN was developed to pre- and finish-machine hardened steel, hard cast iron (modular and malleable), and superferrous alloys with a high content of cobalt and nickel.

Parts previously finished through grinding, can now be more productively and economically machined through turning, milling and (fine) boring.

Depending on the microstructures of the workpiece materials (for instance, perlitic or ferritic or both), the machining parameters must be optimized. The same holds true for proper coolant. Depending on the application and the technology at hand, CBN tools can sometimes even run dry, if proper chip discharge is provided through pressurized air.

Interrupted cuts can occur for a number of reasons. The part design may cause air gaps between the tool and the material to be cut (for example, spool bores), or the workpiece material may contain soft or hard inclusions (such as casting sand particles). Tungsten carbide tools can handle interruptions only on materials in the HRC 50 to 60 range, and only at low cutting speeds. Ceramics are not an alternative due to their brittleness.

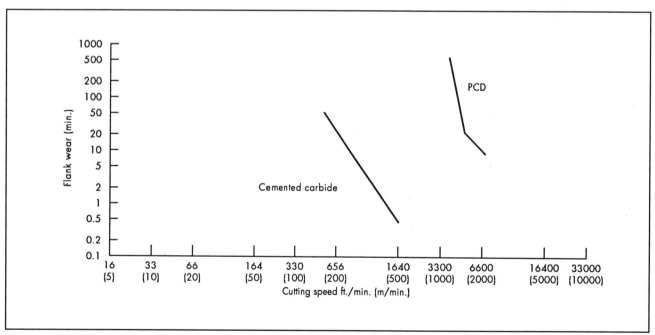

Fig. 16-21 Flank wear and cutting speed for cemented carbide and PCD.

CHAPTER 16
ADVANCED TOOLING

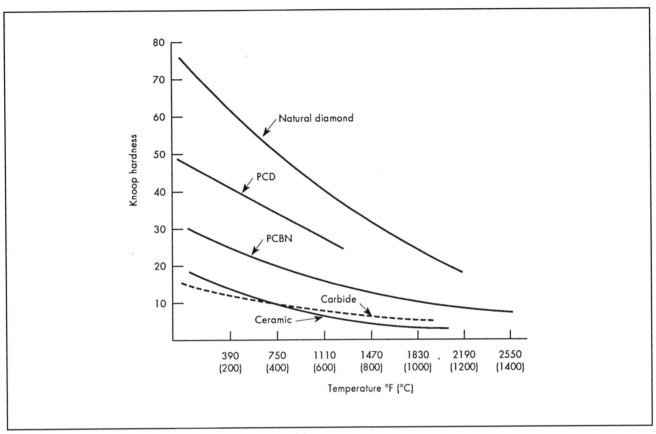

Fig. 16-22 Knoop hardness and temperatures for various materials.

Ceramics

Although proposed for machining in 1905 and patented some 10 years later, ceramics have only recently been accepted as a viable alternative to other cutting tool material, due to their nonuniformity and weakness (brittleness) and then they were used on older machine tools lacking the needed rigidity and accuracy.

Starting in the early 80s, many improvements have been made in the ceramics' microstructure (grain size, density), their processing, and variety of additives, resulting in cutting materials that can be applied at high cutting speeds and still yield extended tool life in milling, drilling, tapping, and turning operations.

The ceramic tools available consist of two base materials, aluminum oxide (Al_2O_3) and silicon nitride (Si_3N_4). In their purest form they are very limited for demanding and economic machining, primarily due to their brittleness. Another shortcoming, albeit of less consequence, is their low thermal shock factor, which somewhat limits the free use of coolants (see Fig. 16-23).

The right combination of additives, however, opens ceramics up for specialized, selective applications in which they are hard to beat, for they resist scaling and crater wear better than tungsten carbides. They can operate within a broader temperature range, which affects of cutting speed and stock removal, resulting in better surface finishes and defined chip control. The most prevalent ceramic cutting materials are:

- Aluminum oxide with zirconium oxide additives—Al_2O_3 + ZrO_2. Adding TiN to Al_2O_3 adds 30% more to its rupture strength, which is still below that of carbides. However, this ceramic can endure much higher speeds and is to be used for general machining of cast irons and steels up to HRC 60 (stainless, alloys, heat treated).

Interrupted cuts, high feed rates and high stock removal should be avoided.

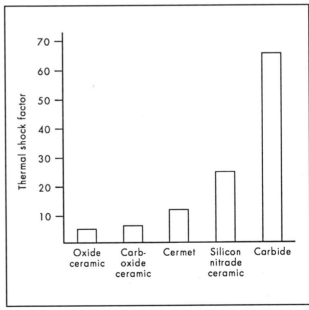

Fig. 16-23 Thermal shock factor for various materials.

CHAPTER 16
ADVANCED TOOLING

- Whisker-reinforced aluminum oxide—$Al_2O_3 + SiC$. The inclusion of silicon carbide (SiC) whiskers in Al_2O_3 based ceramics increases its toughness. SiC whiskers, which can be up to 50% of the content, increase the thermal stress coefficient of the ceramic. Furthermore, the different thermal expansion coefficients of Al_2O_3 and SiC, leading to inherent "push forces," counteract the "pull forces" generated through cutting.

  The higher toughness combined with high wear resistance make it a suitable cutting material for "tough" steel and even nickel and chromium-based alloys. Unfavorable machining conditions such as interruptions and high feed rates should be avoided, since they reduce ceramic tool life exponentially.

- Silicon Nitride—Si_3N_4. This material is excellent for machining gray cast iron. Its thermoshock factor is higher than that of any other ceramic. This, combined with high toughness, lets it successfully machine parts with interruptions, at high feed rates, and with regular coolant supply.

Silicon nitride does not cut steel well, since the temperatures generated are too high and it breaks down within relatively short machining times. The cutting edge of Si_3N_4 loses its sharpness shortly after the first cuts in any material before it stabilizes itself; with a small radius, fine-finishing is only advisable if there is an absolutely stable machine tool-fixture setup guaranteed.

Silicon nitride ceramics, also known as "sialons" are the superior cutting material for gray cast irons, increasing the cutting speeds up to 700% over carbides. Considering that silicon is abundantly available as raw material, it surely will be further developed to possibly enlarge its applications.

Cermets

Ceramic and metal, pressure bonded, result in cermet—ceramic particles dispersed in an oxide or carbide-based metal matrix. More specific, carbides containing either niobium, tantalum, or molybdenum, are added to a titanium nitride base to form cermet.

Generally, any optimum cutting condition is limited to a narrow temperature band of the specific cutting material. Below this band, built-up edges can occur. At temperatures exceeding the band, scaling (oxidation), diffusion (crater wear), or even plastic deformation become typical wear characteristics. Cermets resist this higher temperature condition much better than carbides. Since they also feature a relatively high fracture toughness, they perform extremely well at high cutting speeds in roughing and finishing (see Fig. 16-24).

Molybdenum, niobium, and tantalum give the cermet its hardness and toughness characteristics. Molybdenum is harder than tantalum and niobium is between the two. The three cermet grades are recommended for machining most steel and cast iron grades in the higher Rockwell hardness brackets (HRC 30-45).

The general trend on the modern production floor, toward higher speed machining, less stock removal, and better surface finishes, makes cermet a desirable cutting material. In addition, the low cost per cutting edge and the tool life expectancy have made cermet even more popular (see Fig. 16-25).

INSERT TECHNOLOGY

There was a time when the selection of the right cutting tool was made by trial and error. This time, of course, has long passed. The physics of metal cutting provide the theoretical framework by which to examine and design all aspects and elements of cutting tools. They are the workpiece, the machine tool, the cutting tool, the toolholder and the workpiece clamping. Their elements of machining interact with one another and are, therefore, dependent on one another: The intended shape of the workpiece influences the selection of the machining method. The geometry of the workpiece and the type of machine tool greatly determine work-

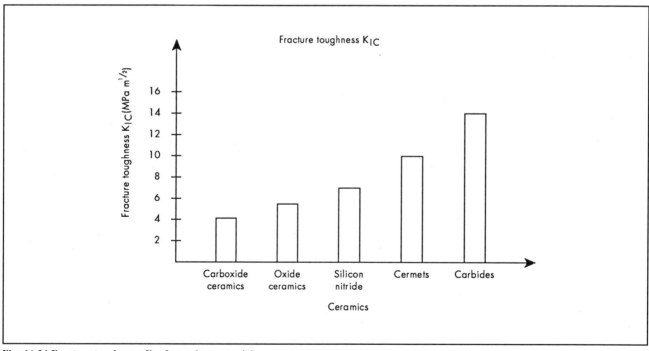

Fig. 16-24 Fracture toughness, K_{IC} for various materials.

CHAPTER 16

ADVANCED TOOLING

	Turning		Boring	
Material	4340 R_c 32		4340 R_c 32	
Cutting depth	0.010" (0.25 mm) per side (finish cut)		0.030" (0.76 mm) per side	
Cutting length	3.5" (89 mm)		4.0" (102 mm)	
Cutting diameter	1.937" (49.20 mm)		1.60" (40.6 mm)	
	Ceramic Al_2O_3	Cermet	Carbide C6 - coated	Cermet
Speed	912 SFM (287 m/minute)	912 SFM (287 m/minute)	500 SFM (152 m/minute)	1200 SFM (366 m/minute)
Feed	0.003 IPR (0.08 mm/rev)	0.003 IPR (0.08 mm/rev)	0.006 IPR (0.15 mm/rev)	0.006 IPR (0.15 mm/rev)
Tool life	15 parts	38 parts	2 pcs/index	5 pcs/index

Fig. 16-25 Comparative machining for cermet/ceramic and cermet/C6-coated carbide.

piece clamping. The workpiece material determines cutting tool material and geometry. Machine tool and cutting tool type often automatically specify each other. Their interaction and interdependence is obvious, however, their success is measured by what happens at the point of contact between workpiece and cutting blade.

The cutting blade, its physical composition as described above and its geometry thus, are the pivotal element in "chipmaking" manufacturing.

The cutting edge of traditional tooling used to be an incorporated part of the tool itself. One end of the toolbody was held in the machine, while the other, opposite end was ground to sharp, geometrically defined edges for metal removal. These "solid" tools, with intricate cutting geometries, still make up a sizable percentage on today's shop floors. Uncoated and coated solid HSS and carbide drills and endmills are used to rough-machine workpieces and are capable of holding only somewhat "wide" tolerances.

They are made at rather wide dimensional tolerances (American National Standards Institute (ANSI)-standards). Regrinding results in unpredictable dimensional and size losses. Once solid tools have lost their sharpness and size, they have to be discarded and scrapped.

Insert-tipped cutting tools have indeed revolutionized machining methods, in that they provide inexpensive cutter and tool standardization where it is needed and they fulfill a relevant prerequisite for custom-tailored tooling and tooling systems. Insert-tooling dominates all areas of chipmaking, that is, turning, drilling, milling, boring, reaming, etc.

Variety

There is a great variety of cutting inserts in terms of their shape, depending on the operation to be performed–roughing or finishing, the machining method to be used and whether or not they are indexable or of a special or standard type.

Indexable, standard inserts are classified by International Standards Organization (ISO) and ANSI and are denominated in a way that reflects shape, size, tolerance, type, thickness and chip breaking. A standardized, modern indexable insert is a well-balanced combination of angles, radii, flats and curvatures, reflecting the result of careful optimization of all fixed and variable machining conditions.

The area of custom-tailored tooling is wide open for creating the optimum cutting insert, particularly as to finish-machining. Here it means not just to make the proper selection from what is already available, but to design and create from scratch, to give true consideration to a specific, often brand new application. Given the specific machining parameters, the insert then has to be shaped accordingly.

Geometry

Figure 16-26 shows such a specialized cutting insert and its relevant, complex geometry for finish-machining a high-precision bore.

The "primary and secondary cutting angle" (for example, 30° and 3°) provide for quality surface finish. These angles are determined by the workpiece material and the machining data selected.

CHAPTER 16
ADVANCED TOOLING

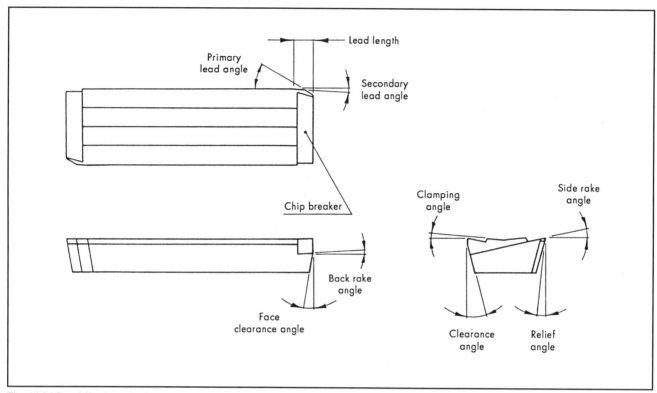

Fig. 16-26 Specialized cutting insert and its relevant, complex geometry for finish-machining a high-precision bore.

Variations in the "side rake angle" have a direct affect on the cutting force. As the angle is increased, forces are reduced (1% per degree). This angle also affects the direction of the chip flow.

The purpose of the "face clearance angle" and the "relief angle" is to prevent rubbing between the end face of the insert and the workpiece. Too wide an angle can weaken the insert, that is why it should not be greater than 5° to 7°.

The "clamping angle," as part of the locking system, prevents the insert from moving under normal cutting conditions.

The chipbreaker aids in breaking the chip to forego long, continuous chips which can cause damage to the insert (and tool) if not properly cut to size and discharged.

It would go beyond this section, to describe every specific and additional geometry of the respective cutting methods. Suffice it to stress again, that meticulously, geometrically designed cutting inserts have to accommodate the machining parameters at hand and secure optimum cutting.

Fatigue

Insert wear is a complex phenomenon influenced by many variables, which do not behave uniformly given the same parameters. Any slight change of any variable affects wear characteristics differently.

Causes of wear. The basic causes for wear are: abrasion, diffusion, plastic deformation, chipping, slow welding.

The most important factor leading to eventual wear and fatigue, is the temperature generated during cutting.

The rate and type of wear relative to the temperature generated is shown in Fig. 16-27.

Abrasion. The most common of all wear mechanisms caused by hard surface particles of the workpiece material, that is, sand inclusions, it causes continuous dislodging of microscopic pieces of the cutting material under any machining condition. The wear rate depends on the quantity, shape and distribution of the particles, as well as the materials' and the cutting tool's hardness.

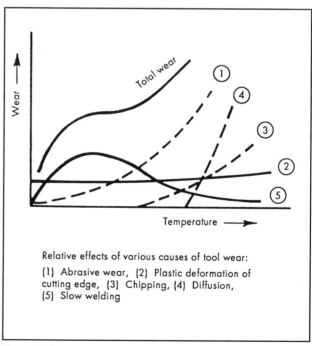

Relative effects of various causes of tool wear:
(1) Abrasive wear, (2) Plastic deformation of cutting edge, (3) Chipping, (4) Diffusion, (5) Slow welding

Fig. 16-27 Tool wear and temperature.

CHAPTER 16

ADVANCED TOOLING

Diffusion. The cause of wear mostly discussed among researchers. This complex phenomenon has so far only been fully explained as to machining metallic workpieces with carbides. It constitutes the chemical breakdown of the tool when reaching its critical temperature. At that point, there is an atomic transfer from the cutting insert to the chip, weakening the bond of the tool particles. The carbon diffuses to the chip followed by the cobalt, leaving the tungsten carbide unsupported, which is then pulled off by the velocity of the chip. The resulting crater wear of the cutting edge leads to the premature failure of the insert.

Plastic deformation. This form of wear mechanism occurs due to high pressure on the insert (high tangential and axial cutting forces) creating small depressions and bulging of the cutting edge. Greater tool pressure and subsequent higher cutting temperatures deform the insert edge area, practically leading to insert cracking and eventual failure.

Chipping. Brought about by subjecting the insert to intermittent loading–unloading machining cycles, for example interrupted cutting or eccentric, uneven stock removal. What happens is, that small parts of the inserts are detached from its surface and then chipped away, destroying the cutting edge.

Slow welding. It is associated with relatively low cutting speeds. As machining takes place at a temperature well below the recrystallization temperature for the material, the chip flow between insert and workpiece is irregular and laminar. In fact, there is a high resistance of the chips flowing along the insert face. This leads to a built-up edge in that area. When the built-up edge breaks away from the insert, it tears off small fractions progressively, albeit slowly, of the cutting material.

Main affected areas. Three areas of the insert are affected by wear caused during machining, namely: face, flank and edge.

Face. It is the area over which the chip passes during its formation. Away from the actual cutting edge, wear takes the form of a cavity or crater and as it gets deeper and wider, approaches the cutting edge, which will then abruptly break off and result in insert failure.

Flank. This is the clearance area of the tool along which the cutting edge is right at the actual chip separation point. Contrary to face wear, flank wear starts and moves away from it as it progresses into a "wear land."

Edge. Often, flank wear is the cause of edge wear. Edge wear by itself, however, occurs especially with finishing inserts, where the sharp edge of the tool provides a good surface finish. Edge wear is basically the rounding off of the sharp edge into bigger radii, eventually bringing the cut "out of spec," long before flank or face wear occurs.

All three, flank-, face- and edge wear, are determining factors of whether an insert is "worn out," has, in effect, reached its tool life limit. It is, therefore, imperative to continuously monitor the actual insert wear as it develops, so as to provide replacement before any damage caused by a worn insert can occur.

Lifespan And Cutting Chips

High-precision machining deals with many manufacturing variables. Cutting insert performance and its lifespan play a pivotal role along these lines. They generate the chips in metal-cutting machining (chipmaking manufacturing), which can be regarded as the interface between the tool (insert) and the workpiece. As such, they are an important variable of successful machining, which is often overlooked. While the type of chips and their formations are part of basic metalcutting literature, their true relevance within the realm of high-precision machining needs to be highlighted.

The process of chipmaking is basically material deformation or a plastic flow of material.

The criteria surrounding it, involves the properties of the workpiece material, the machining data applied, the cutting temperature, and of these, the thickness and length of the chips and their discharge out of the machining area are affected.

Now, one objective is to make room for the following, plastically deformed, material. This is done for two reasons:

1. To generate a smooth flow of the chip along and past the tool face, so as to not interrupt its continuous flow to carry with it a big part of the thermal stress.
2. To prevent a pile-up or build-up within the chip itself.

The other objective is to break the chip at a point where its size and form does not negatively impair the surface machined, leaves the immediate cutting area and is easily discharged from it.

With what criteria can chips be controlled?

Workpiece Material

In most cases a given variable to deal with. Brittle materials produce short, discontinuous chips. Breaking them at the right point and then properly disposing them is important. Ductile material, however, produce continuous chips, making chipbreaking insert geometries necessary. Besides the influences that geometry and machining data have on chip control (see below), it is prudent at the design stage, to specify material properties that can inherently assist in breaking up longer chips.

Additional amounts of phosphorus, lead and sulphur, to low carbon steels, for example, help along these lines.

Speed

The difficulty of obtaining the "right" chip is obvious when observing the effect speed has, for it is directly related to the temperature generated. If machining is done at lower speeds, the temperature at the tip of the insert is in a lower bracket and might be well below the recristallization point. This means, the hardness of chip and workpiece material is retained.

Increasing the speed and thus the temperature will soften the material and cutting is done more efficiently. If the temperature increases excessively, the wear characteristics of the cutting insert will rise exponentially. Generally, higher speeds make chips flow more smoothly, the steeper sheer angles, though, make them also thinner and more stringy. Chipbreakers can help solve the problem.

Feed

Of the machining parameters, the feed rate is mostly effective to control the flow of chips. At high feed rates, higher material distortion and deflection make the chip break easier. Although cutting tool manufacturers recommend certain machining data for specific workpiece material, quite often the best suited parameters as to chip formation and chip control can only be found empirically for a given machining task.

Coolant

The effects coolants have are three-fold:

1. They provide lubrication between the insert and the chip, lowering the coefficient of friction between the two and thus securing a smoother chip flow. Lubrication increases tool life and surface finish.

CHAPTER 16

ADVANCED TOOLING

3. They keep excessive temperature variations at bay, so as to forego their otherwise catastrophic effect on the cutting edge (see Speed).
4. They control chip flow and proper discharge. High-pressurized coolant (250 to 500 psi (1.7 to 3.4 MPa)) disposes of high chip volume on roughing stations and keeps the cutting area meticulously clean on finishing stations. Sometimes in combination with favorable machining data, it assists in breaking off chips due to shear pressure forces.

Geometry

Of any geometry, the defined chipbreaker, a ground-in corner (edge) of the insert, is most effective to manipulate the flow, direction and size of the cutting chips. It forces the chip to deflect at a narrower angle, causing it to break off after having caused a swarf which collides with the insert. The angle and position of the chipbreaker is dependent on the cutting method.

Negative rake angles of usually 6°, 8°, or 12° have a pronounced effect on chipbreaking, for they distort the chips so much, that they tend to "automatically" break off into the desired small pieces.

An often overlooked, albeit relevant aspect in terms of chip control, is the clamping of the insert in the toolbody. The further the clamp plate or clamp bolt protrude, the more obstruction to chipflow it signifies. Therefore, flat clamp plates and "sunken" bolts should be mandatory for any modern cutting tool. As is the case of the tool depicted in Fig. 16-28.

It furthermore and foremost, illustrates the arrangement of an insert that allows for the provision of a large chip area. It allows an unobstructed chip flow and ensures a complete discharge of same out of the machined bore.

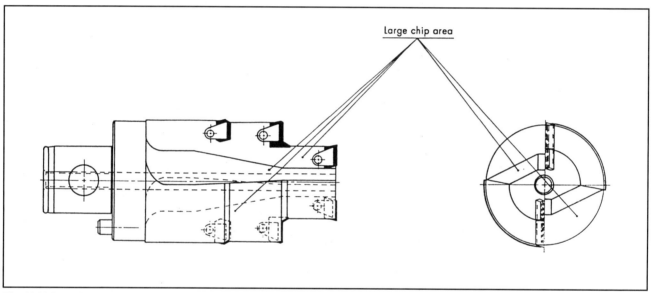

Fig. 16-28 Precision tool for heavy stock removal.

REFINED TOOLING MODULES

In a modern, numerically controlled manufacturing environment, productivity and quality are markedly determined by the cutting tools' performance, flexibility and reliability

Out of these pillar requirements, cutting tools need to be refined tooling concepts and designed into systems.

Performance can be described, for example, by the machining data applied and results achieved, such as cutting speed, feedrate, stock removal, chip volume or tool life or finish-machined geometries.

Reliability has geometrical and technological aspects. Geometrically, because the system has to secure positional repeatability and accuracy. Technologically it determines the predictability of desired machining results.

Flexibility–the variety of workpieces and frequent changeovers of parts within part-families, as well as the fast response time to design changes, require special tooling based on flexible, modular design.

MODULAR DESIGN

One of the objectives when machining complex workpieces, is to do as many different cutting operations as possible on one spindle station in one clamping mode. This necessitates a tool storage at that particular station, out of which the tools are phased in and out of the machine spindle as needed. In order to narrow down the number of tools needed to accommodate varying machine tool spindle designs, to cut idle, nonproductive machining time and facilitate the "management" of cutting tools,

CHAPTER 16

REFINED TOOLING MODULES

modular tooling systems were developed. They accommodate a wide variety of lengths, diameters and inserts and can be adapted to various basic holding attachments on greatly varying machine tools.

Figure 16-29 shows one such universal, modular tooling system. It covers a wide spectrum of turning, grooving, reaming, drilling, boring and milling operations and can be mounted in machine spindles of turning and machining centers. The system includes four basic types of components, taper-shanks (CAT, ANSI, ISO), rotating adapters extensions and reducers besides the designated cutting tool. Certainly, such systems satisfy the quest for flexibility standardization.

However, the selection of a precision tool system calls for more, it has to also satisfy stiffness and accuracy requirements. Leading cutting tool manufacturers generally guarantee a radial and axial tolerance of ±98 μin. (±2.5 μm) of one system module. The positioning of the cutting insert is specified to be within a tolerance of ±984 μin. (±25 μm). These individual tolerances from module to module, of course, add up and as a complete assembly can negatively affect accurate, precise cutting. So can the overall length of the tool system, for the more it protrudes from the machine spindle, the more it is prone to deflection during cutting (to deflect a tool bar by 0.001" with a diameter to length ratio of 1:1 takes an 11,000 lb (49 kN) force. In case of a 1:7 ratio it only takes 32 lb (142 N)).

Tool overhang, measured from the spindle face to the tip of the insert, has to be kept as short as possible. As a whole, today's modular cutting tool systems ensure precision machining in the H8/H9 ISO-tolerance.[23] To further develop the H6/H7 bracket other, additional, technical aspects have to be looked at.

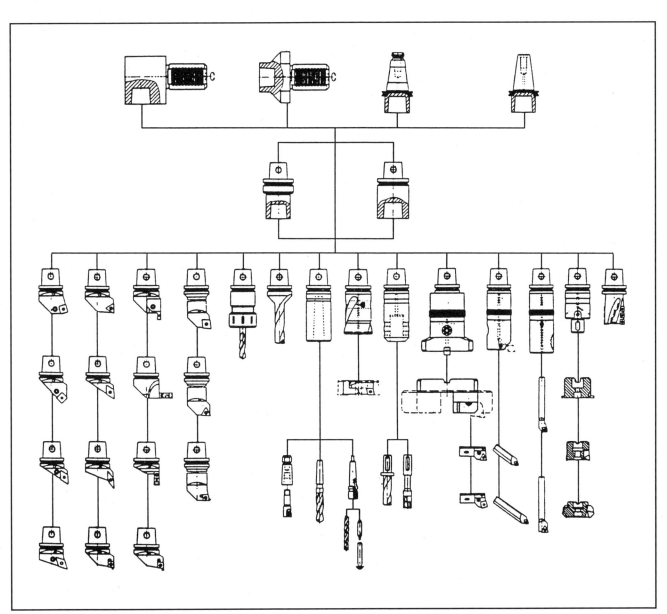

Fig. 16-29 Modular tooling system.

CHAPTER 16

REFINED TOOLING MODULES

MAP O3-PRINCIPLE

Again, cutting tools and tool systems in a free standing, unsupported cutting mode, are subject to deflection during machining. Deflection, as can be seen (see Fig. 16-30), increases exponentially through tool length (keeping the other factors of the equation constant).

Developed against the background of separating the guiding and cutting functions, known as the MAP O3-Principle, tools were developed featuring peripherally arranged guide pads (see Fig. 16-31).

Cutting forces are absorbed by the guide pads foregoing vibrations, which would otherwise impair the cutting operation. Vibrations or toolchatter can have negative, often disastrous effects, such as shortening the expected tool life, rough, uneven and unacceptable surface finishes, dimensional workpiece inaccuracies, or premature failure of spindle bearings or simply producing workpiece scrap.

Besides stabilizing the tool during cutting, the provision of guide pads opens up new avenues as to the insert design. Due to the tools rigidity, more intricate cutting blade (insert) geometries can be fitted, for instance, smaller radii at the tip and primary and secondary cutting edges, for even better finishes.

Tools can run at higher speeds and feed rates when entering the workpiece and during cutting, since the tool stabilizes itself right as it enters the bore.

Leapfrogging technology often stimulates other unrelated areas, offering solutions to other indirectly related problems. Here too, the separation of the geometrically defined guiding and cutting functions offers answers to otherwise unresolved ''interrupted cuts'' or bridging ''air gaps'' between workpiece materials

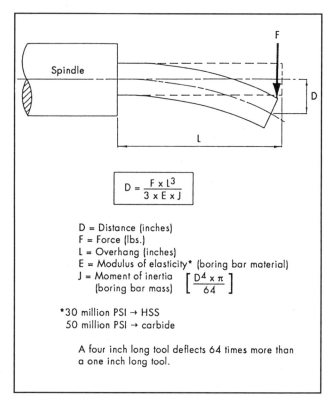

$$D = \frac{F \times L^3}{3 \times E \times J}$$

D = Distance (inches)
F = Force (lbs.)
L = Overhang (inches)
E = Modulus of elasticity* (boring bar material)
J = Moment of inertia $\left[\frac{D^4 \times \pi}{64}\right]$ (boring bar mass)

*30 million PSI → HSS
50 million PSI → carbide

A four inch long tool deflects 64 times more than a one inch long tool.

Fig. 16-30 Tool deflection.

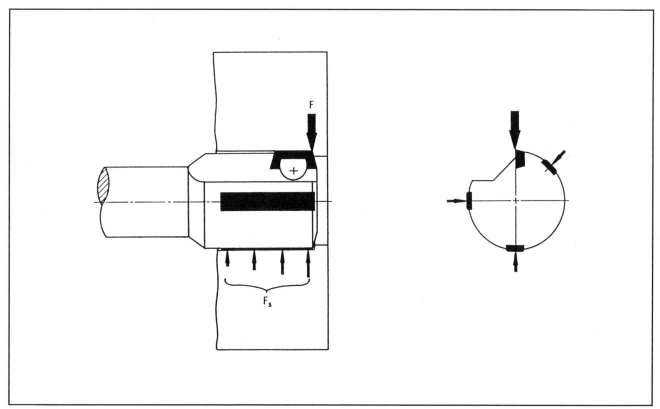

Fig. 16-31 Cutting force, F, at insert—support through guide pads, F_s.

CHAPTER 16

REFINED TOOLING MODULES

on the same centerline and the undesirable following of previously "out of line" machining.

Bridging "air gaps" through extended guide pads, overcoming interruptions in bores by providing additional, peripherally arranged guide pads, ensures high precision machining. Because of its rigidity, these tools do not follow the previous path of operation, but rather cut their own and thus automatically correct out of centerline conditions during the finishing cut. An important prerequisite for using "padded" cutting tools, is a clean 8% to 10% coolant emulsion, to prevent chip-pickup which can lead to scratch marks on smooth surface finishes. The guide pads are usually of C3/C4 carbide material. For use in modular iron and steel, cermet proves to be best suited, while in other alloys ceramic seems to be the better pad material. Especially when machining aluminum with a lower silicon content, a certain coolant lubricity has to be maintained for the (emulsion 6% to 8%). First initial tests with polycristalline diamond material are very promising, allowing coolant with hardly any lubricity.

TOOLBODY

Close attention has to be given to the meticulous design of the toolbody itself, since it is a vital part of any high-precision cutting tool system. It means to:

1. Optimize the chip flute and clamping areas.
2. Specify alternative toolbody materials.

Insert Clamping Area, Chip Flute Area

Any protrusion in the cutting and adjacent areas can destroy precision machining, for it lets cutting chips get caught, making them unmanageable. This invariably leads to temperature rise, which in turn promotes premature insert failure. Chips caught in the clamping area, swishing around the toolbody during cutting, leave unacceptable scratch marks on the surface machined.

High precision cutting tool system shown in Fig. 16-32 features toolbodies with well-opened up chip areas, as an almost natural continuation of a well-designed clamping system. There is no protrusion, since the insert clamping and adjusting mechanism are sunken "in the toolbody."

Fig. 16-32 Toolbodies with well opened up chip areas.

CHAPTER 16

REFINED TOOLING MODULES

Toolbody Material

The majority of toolbodies across the multiple cutting operations are of high speed steel or carbide in case of solid tooling. As to inserted tooling, it usually is regular tool steel (4140). Often intricate for finishing operations, it is imperative to specify a different body material, especially when vibrations or weight become specific issues.

Vibrations, inherently generated by the machining process, can be decreased through "heavy metal" body bars. Heavy metal is a high density material consisting of 90% to 93% tungsten with roughly equal rest-amounts of iron and nickel. Its density is about 17 grams/cm^3 with a hardness of HRC24 and a modules of elasticity of 40×10^6 psi (275 GPa). (Note: The equation for deflection, mathematically proves that a higher modulus of elasticity positively influences the rate of deflection.)

Padded tool bars of heavy metal can feature an amazing length to diameter ratio of 12:1 and still run "chatter-free" even at high speed (5000 to 6000 rpm).

Weight limitations are imposed on tooling systems, mostly in flexible manufacturing, when automatic system changes are required. The alternatives then are hollow bars and aluminum bars. Hollow bars can substantially reduce weight. However, in a precision-environment their use is sometimes limited, since geometrically defined coolant passages and provisions for "combination" operations cannot always be accommodated. Aluminum as a bar material then is a viable solution.

Figure 16-33 depicts an interesting design along these lines. An aluminum toolbody, surface-hardened, with incorporated steel-cassettes for the cutting inserts and guide pads. A light-weight and yet rigid tool bar.

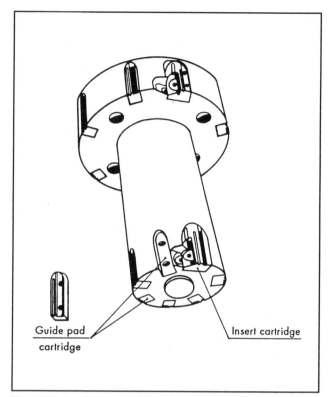

Fig. 16-33 Aluminum bar with steel cartridges.

INTERFACES: MACHINE TOOL/CUTTING TOOL

The making of machine and tool, or more precisely machine tool and cutting tool, creates (mechanical) interfaces which directly influence the machining results.

Interface 1: Toolholder in the machine spindle.
Interface 2: Tool shank connected to the toolholder (tool clamping).
Interface 3: Adapters (mounted onto machine spindle or between toolholder and cutting tool).

All three interfaces have to meet stringent requirements as to their criteria of stiffness (rigidity), repeatability, accuracy, reliability, concentricity, (centerline deviation), ease of handling.

ACCURATE TOOLHOLDING

Cutting tools are held in the machine spindle by toolholders which are either cylindrical or taper-shaped.

When machining with less precision and/or in a non-computerized numerically controlled CNC-environment, they are usually (collet) chucks mechanically holding and clamping the tools.

Precision manufacturing, however, calls for toolholders of a more defined and accurate design, albeit of the same shape as for the machine-end of it. The advent of flexible manufacturing and the enormous need for different tool variety, dedicated to individual machines, brought about the uniform design of the tapered toolholder. It automatically can be taken out of the machine spindle and replaced by another one through a tool changer.

The toolholder is fitted with a sleeve or ring on its opposite end, to accommodate the respective tool (see Fig. 16-34). This basic, "standard" interface meets the requirements for rigidity, accuracy and ease of handling. High-precision machining, for example one pass-machining of a critical spoolbore with geometric microfinishes, however, necessitates a more intricate tool clamping system as the second interface. Some toolholders have a well designed, built-in clamping system. The clamping is then done either hydraulically Fig. 16-35a or mechanically Fig. 16-35b.

Both clamping systems usually hold cylindrically shanked cutting tools. The high clamping force over a large clamping area is induced through hydraulic fluid as per Fig. 16-35a and an expansion sleeve as per Fig. 16-35b for the transmitting of high torques. The rigidity of the systems allows for high-speed machining and machining with unsupported cutting tools in a high-production environment.

While they are both designed for precision manufacturing, the mechanical system shown ensures absolutely consistent clamping pressure and high repeatability, for the axially induced force translates into a radial expansion of the clamping sleeve of its inner and outer diameters and thus to ID-OD-two surface clamping.

Both systems certainly are interesting alternatives, especially on high-production floors. However, mechanical systems, separating holding, clamping and adaptation through "modules" with tramming-in possibilities guarantee precision and flexibility at a favorable cost performance ratio.

CHAPTER 16

INTERFACES: MACHINE TOOL/CUTTING TOOL

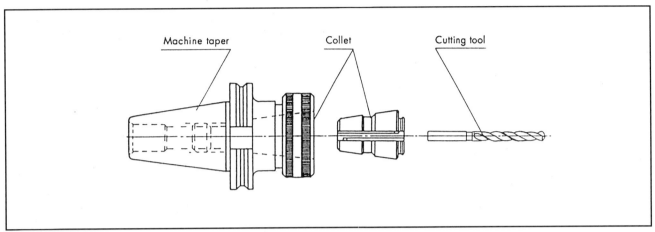

Fig. 16-34 Toolholder fitted with a sleeve or ring on its opposite end to accommodate the respective tool.

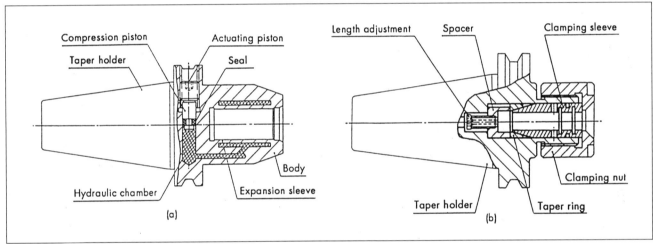

Fig. 16-35 Toolholding chuck: hydraulic clamping, *a*, and mechanical clamping, *b*.

PRECISION CLAMPING

Conventional shanks of cutting tools, where there is less emphasis on precision and/or non-numerically controlled (NC)/CNC-manufacturing, usually feature flats onto which a screw is driven to hold the tool in a holder. Or the tool features a threaded shank and a matching nut, horizontally screwed on, keeps the tool in a tight position to the toolholder. Another method is a self-locking taper as the tool shank drives into a mating taper of the toolholder. As outlined under "Accurate Toolholding," cylindrical tools usually run in toolholders with built-in clamping devices. Now, flexible and precision manufacturing demands more defined locking or clamping systems to:

- Guarantee a repeatability between tool changes of below 197 μin. (5 μm).
- Ensure an exact axial location and cutting insert orientation.
- Provide adequate stiffness and bending strength (large face contact area).

One such modular system is shown in Fig. 16-36.

When turning the thrust screw, the radial clamping force F1 drives the floating pin, which through its 90° taper, doubles the induced force F1 in the axial direction (forces F1 and F2 are equal) providing rigid clamping of the tool shank in the holder (the orientation pin secures the right assembly).

This interface thus meets the aforementioned three requirements.

IN-BETWEEN ADAPTATION

A composition of machine tool, toolholder, tool clamping and cutting tool as a system invariably and inherently calls out a certain runout, measured from the spindle face to the tip of the cutting insert.

The main contributing, individual components are:

- Machine Tool: Spindlebearing, spindle-diameter, manufactured spindle accuracy.
- Toolholder/tool clamping: Radial stability, repeatability, stiffness, clamping force, manufactured accuracy.
- Toolbody: Tool balance, material, length-to-diameter ratio, thrust face diameter, manufactured accuracy.

Modern, state-of-the-art machine tools, toolholders, tool clamping and cutting tools, offer precision to lesser or higher degree. In high-precision manufacturing the runout of the whole machining system has to be kept to a minimum striving for "zero" (0). The individual errors are a deviation of true radial and angular position relative to the theoretical centerline of the

16-37

CHAPTER 16

INTERFACES: MACHINE TOOL/CUTTING TOOL

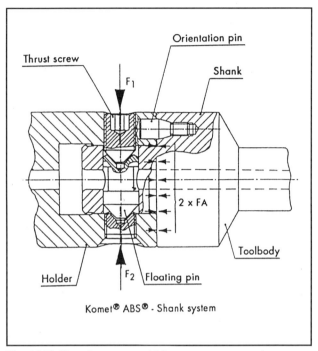

Fig. 16-36 Clamping system which guarantees repeated accuracy, ensures exact axial location and cutting insert orientation, and provides adequate stiffness and bending strength.

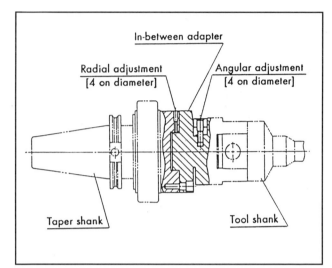

Fig. 16-37 In-between adapter.

system. To eliminate those individual errors and arrive at a runout-free condition, the tooling system has to be trammed in by means of an in-between adapter depicted in Fig. 16-37.

Mounted in-between toolholder and cutting tool, the adapter features two separate bolt circles, each for four (4) holes for bolts.

When manually rotating the cutting tool system, installed in the spindle, a dial indicator touching the tip of the insert indicates the deviation from the theoretical centerline. With four bolts on the periphery, the tool system can be trammed into "zero," eliminating both radial and angular runout. Provision for angular correction has to be made particularly as to longer tools exceeding length to diameter ratios of 8:1. Once the whole system is "clocked in" and showing "zero" runout, actual machining can begin.

OPTIMIZING PROCESSES AND PARAMETERS

The aforementioned tooling modules synthesized, offer formidable systems which can:

- Provide cost-cutting machining.
- Increase manufacturing productivity.
- Machine "world-class" quality.
- Eliminate excess machining stations.
- Lower the number of machining passes.
- Forego "secondary" operations.
- Accommodate high variety (flexible) manufacturing.
- Be instrumental to "lean" production.
- Ensure predictable machining results.

This, however, requires to optimize the machining processes and parameters of the tooling systems themselves and their peripheral determinants.

SINGLE STEP-MACHINING

Selecting the right part material depends on what design and manufacturing engineers expect from the material and the finished parts. The properties of specific materials and their cost to shape them into the desired form are the variables.

The most varying relevant physical material attributes are: Hardness (tensile strength), toughness, density, elasticity, thermal and electrical conductivity, heat resistance, durability, stiffness, corrosion resistance, specific weight, wear resistance, treatability (hardenability).

The machinability of the workpiece material to be chosen certainly plays a role in the decision-making process. Different alloy material can dramatically change the machinability in all directions. Percentages of nickel or chromium or molybdenum in steel, for instance, have specific effects on the way it can be applied. Just changing the carbon content of cast iron makes it more or less suitable for the respective workpieces.

Another example would be the effect of the silicon content in aluminum: The higher its percentage (up to 18%), the harder the materials surface becomes and the more abrasive it is during machining.

This leads to the "machinability" aspect of the workpiece material which is directly tied to the other variable: The cost of shaping the parts into its form. Contemporary literature rates "machinability" as good or bad and identifies the reasons why, usually citing the effect machining has on the part and cutting tool material at certain machining data. A typical case in point would be: Aluminum with 4% copper:

"Machinability as a rule is excellent when heat treated. The intermetallic particles are abrasive and call for wear resistant cutting edges."[24]

CHAPTER 16

OPTIMIZING PROCESSES AND PARAMETERS

Today, the aspect of "machinability" goes much further. The more important issue is not so much whether or not machining poses problems, but rather what the machining process yields. The trends that change the realm of machining with respect to the workpiece are:

New Metallurgical Techniques

Especially sinter and powder metals yield high density parts with superior tensile strength over conventional cast irons and steel. Sinter and powder processes make for better tolerance-controlled parts.

Accurate Casting And Forging

The demand for weight reduction and higher quality, notably in the automotive and aircraft industry, calls for more accurate molds with no excess material and no impurities. Investment casting is becoming ever more popular, for it allows the widest net part size range of any casting process, thin walls, undercuts and internal passages. Because the mold cavity is a refractoral material, besides ferrous and nonferrous metals, high melting point super alloys can be cast.

In forging, too, near net shape parts are formed through grain flow optimization. It increases impact strength and fatigue endurance limits, thus allows for thinner wall design, reducing excess weight and stock.

Light Weight, High Strength Material

Albeit seemingly diverging criteria–light weight and high strength–proper design and material selection can accommodate both. Aluminum in the automotive industry (engine and transmission) and titanium in aerospace (landing gear and wings) are two popular trends.

For automotive, less weight primarily translates into less fuel consumtion, for aerospace, too, but another important criteria is heat resistance. Thus, titanium aluminide or aluminum composites or other so-called superalloys and silicon or carbon-based composites offer just that: high strength, light weight and temperature resistance. Parts of all these materials are "cast" and "formed" with smooth surface finishes and tight-tolerance registration of cast-in holes and elaborate internal configurations.

Part Complexity

As design and manufacturing engineers try to squeeze more functions into small parts and subassemblies of sophisticated end products, they become more complex and difficult to machine. Examples are the antiskid feature of a car's brake system or the electronic part of automatic transmissions or the multifunction "communication" area of household products, etc. Satisfying varying consumers' tastes extends the workpiece complexities even further.

These main developments in the materials area have to be paid tribute to. Lightweight, accurately cast, complex workpieces of advanced materials cannot be "hogged out." They cannot undergo the conventional way of machining from roughing, semi-finishing, finishing and post-finishing. Parts of these configurations, ready for the "final touch" are costly. Any subsequent machining has to be done with precision, often fast and preferably in a single step.

Single step-machining is the finish-machining of a particular operation with one cutting tool. This means:

- Bringing surface finishes from 125 to 150 rms to 5 to 25 rms depending on the material.
- Removing uneven stock distribution.
- Accommodating highly differing workpiece material.
- Securing the most stringent requirements in straightness, roundness and perpendicularity.
- Overcoming interruptions, air gaps, irregular hard spots.

There is no room for error in a one step-machining process, therefore, the tooling system has to be reliable. It has to be rigid and yet feature built-in adjustability.

Cutting inserts need to be specifically designed for the respective machining task and particular attention be paid to primary and secondary cutting edges. The favorable developments in metallurgy and near net shape to net shape developments in material "forming" have opened up new avenues in the way quality workpieces can be productively and efficiently finish-machined in "one step."

ONE PASS-MACHINING

This method can take one step-machining further, in that it can provide multi-functional finishing in just "one pass." However, it also refers to semi-finishing operations. The emphasis here is on tooling systems that accommodate machining functions in one toolbody rather than have several tools do sequential machining involving several different cutting tools.

Two different applications depicted below illustrate the relevance of the one pass-machining principle in precision manufacturing.

Example 1: Alternator Cover (see Fig. 16-38)

Starting from the cast-contoured aluminum part, complete machining is done in one pass-semifinishing and one pass-finishing. Each doing several operations in "one pass," either consecutively or simultaneously, depending on the machine setup.

The tool and part pictures show the respective pass of machining diameters, chamfers, transitions and facing in one operation. It assures bore concentricity, per-pendicularity and closely held tolerances. It maintains these conditions part after part, thus assures geometrically identical part configurations even at high production rates.

Example 2: Brake Cylinder

Another example is complete one-pass machining of multiple bores, transitions and facing operations of a brake cylinder as shown in Fig. 16-39.

The basic, disc-type toolbody has mounted bodies in a planetary type design to forego any sequential machining passes.

Single step-machining is "invited" by the accuracy of the raw-machined workpiece. One pass-machining is a demand given by design, functionally and geometrically.

It is easy to see, that combination tooling systems capable of single step- and other one pass-machining, far surpass other conventional tooling methods for they ensure consistent quality, highly efficient, cost saving manufacturing.

Such high-precision systems, however, also call for a nontraditional way of tool management logistics, handling systems, knowledge and setting.

PROACTIVE FINETUNING

Figure 16-40 is an explosive view of a high-precision machining system. It is a composition of the elements described above: Heavy metal tool bar, guide pads, PCD cutting inserts, antilock breaking system ABS-shank configuration, adaptation for angular adjustment, adaptation for radial adjustment, coolant ring for coolant supply, V-50 machine taper.

It is built for complete finish-machining of high-precision bores in one step and one pass.

CHAPTER 16

OPTIMIZING PROCESSES AND PARAMETERS

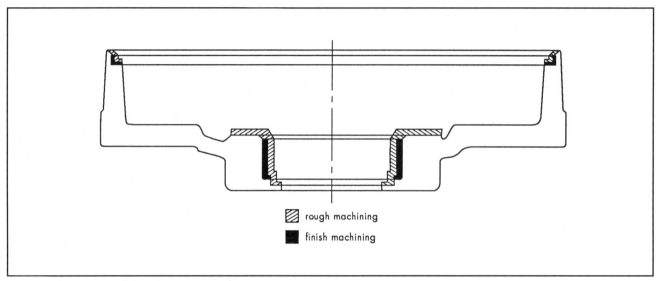

Fig. 16-38 One-pass machining of an alternator cover.

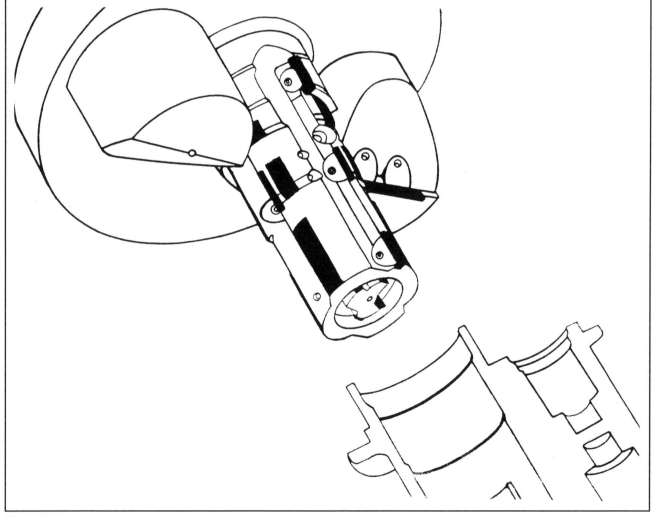

Fig. 16-39 One-pass machining of multiple bores, transitions, and facing operations.

CHAPTER 16

OPTIMIZING PROCESSES AND PARAMETERS

The system has to be prepared to meet stringent finish requirements in quality, medium to high volume production consistently and predictably. This goes deeper than the presetting of regular tool management and yet should be an integral and extended part of it.

"Tool management" as such is well described in contemporary engineering literature. Suffice it to say that it consists of the tools' identification, storage, handling, transportation, availability, (computerized) administration, cost control and setting.

Educated and dedicated setting or "proactive finetuning" starts with the absolute cleanliness of all individual components before their assembly.

The complete cutting tool system (see Fig. 16-40), assembled, then has to be placed into a setting fixture or "presetter." Again, there is ample literature describing them as part of "tool management." A few important notes as they relate to this section.

The setting device's or preset machine's function is to set the cutting edges dimensionally as predetermined (for example, per blueprint). In high-precision manufacturing it means setting within 40 μin. (1 μm). It should be done in a clean, dedicated area. To set tool systems with highest precision, setting devices should feature electronic, digital readouts (easy to read) and provide for setting in the vertical mode (weight). They should be equipped with ultrasensitive probes and/or an optical profile projector (accuracy). They should, preferably, also be equipped with an automatic compensation system for tool system runout which can then be compensated for during setting. Setting fixtures play a pivotal role in the finetuning process, for they proactively (presetting) and reactively (feedback) set the desired tool dimensions.

The setting procedure itself requires to set the primary and secondary cutting edges relative to a reference point of the toolbody. This includes step lengths and diameters. It is essential to proactively finetune them to one another, so that they stay within the close tolerance bands as per drawing and/or process sheet. Often, the tolerances to be held by one step-, one pass-machining systems vary. This has to be taken into account as well.

Another important aspect of "finetuning" is to only use the tolerances needed. Adjustable inserts can, within 40 μin. (1 μm), be raised or lowered from a mean diameter to stay at the upper or lower limit of the tolerance band. Lowering the insert to the low limit can add to the stability of the system. Raising it can increase tool life by allowing for more insert-wear. As described before, a complete cutting tool system incorporates a component to correct its runout. A system, checked and adjusted in a setting fixture to true zero runout will, invariably, show a certain runout when mounted into machine spindle.

If the runout is excessive, the system then can manually be "trammed" in through adjustable adapters to clock it as close to zero as possible.

As part of the finetuning process, it is usually done only once per system and spindle due to the expected repeatability accuracy of the cutting tool system itself. Proactive finetuning, in principle, works only if the production floor provides feedback from machining to tool setting and vice versa.

Accurate setting does not permit "missing" of even 40 or 80 microinches (one or two micrometers). Machining in turn has to apply the optimum machining data. Both stimulate the proper design of the cutting tool system (cutting geometry, system rigidity, etc.) and the machining peripherals such as coolant passages, or chip disposal.

Given the sensitivity of manufacturing within fractions of millimeters, proper handling and care of the cutting tool system as an assembly and the individual components must be secured within all areas of application.

Proactive finetuning essentially means to do it "right from the start," since any "work in process time," taken away from manufacturing for corrective action is costly. Human error and machine defaults, however, cannot be ruled out. It is, therefore, absolutely necessary to have two or three sets of cutting tool systems ready for machining: One currently in use, one as backup next to the machine and preferably one more in the setting area!

Proactive finetuning also answers the important call for predictability of machining results and tool life.

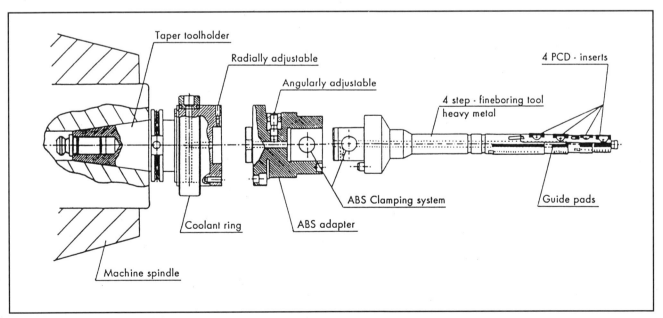

Fig. 16-40 High precision machining system.

CHAPTER 16

STRINGENT FINISH REQUIREMENTS

Manufacturing needs to know, that the workpiece at the end of the machining process, meets the geometries and finishes as expected.

Accurate setting of tooling systems with high repeatability also allows for predetermined tool life.

By preventing possible rework as scrap "at the end of the line" proactive finetuning, a necessity in high-precision manufacturing, thus also makes a contribution to cost savings.

STRINGENT FINISH REQUIREMENTS

TIGHT TOLERANCES, SMOOTH SURFACE FINISHES

In quality manufacturing there is an ever growing need for machining tighter toleranced parts. The average manufactured product of just 15 years ago, would fall short of today's requirements along these lines (see Fig. 16-41).

The quality of a finished, manufactured product is the reflection of its "workmanship," the way the individual parts are finished, fit together as the whole and its function.

On the production floor it is expressed as physical sizes, characteristics and features of parts through predescribed tolerances. Usually done in a structured and methodical engineering approach.

Tight tolerances and excellent surface finishes are pivotal manufacturing objectives, some aspects of which need to be looked into.

Triggering Tight Tolerances and Their Effect

In the wake of mass-produced products and assembly line setups, it became clear that the parts and components had to be dimensionally defined and toleranced to make them fit, also in random order. The aspect of allocating certain male and female parts and their interchangeability added to the concept of tolerancing. To make automatic assembly work, the parts then had to be produced with much closer tolerances. Following the logical quest for producing subassemblies, that do not have to be "run in" but rather be produced with predictable accuracy, they are produced with closer tolerances, yet. Controlled, closer tolerancing made products more reliable. Debugging them following production methods based on "trial and error" was no longer affordable. Too strong was the pressure on keeping manufacturing costs down and on being competitive. Do it "right the first time" had become the only sensible approach.

Today other, equally important demands have to be met:

- Improved safety.
- Energy saving.
- Weight and noise reduction.
- Emission control.
- Extended service life.

Design engineering, as a result, has to specify even tighter tolerances for practically all geometric part characteristics:

- Flatness.
- Roundness, straightness.
- Angularity, perpendicularity.
- Parallelism.
- Concentricity, symmetry.
- Surface- and line profile.
- Runout.

In addition, surface finish requirements increase. The tighter the tolerances of two mating parts, for example, the smoother the surface finish has to be. In fact, smooth, even surfaces are essential in the areas of:

- Material fatigue strength.
- Corrosion resistance.
- Sealing performance.
- Friction.
- Lubrication.
- Force distribution.

Tolerances and surface finishes usually complement each other. While the scope of (controlled) tolerancing is widely known, there seems to be less certainty as to some areas of surface technology.

Surface Texture and Integrity

Surface texture (surface finish) describes the quality of a workpiece surface through three different characteristics: Roughness, waviness and lay. Workpiece material, machine tool, cutting tool system, machining data, machining periphery all affect the surface texture (see Fig. 16-42). Roughness is universally used to indicate a certain finish requirement.

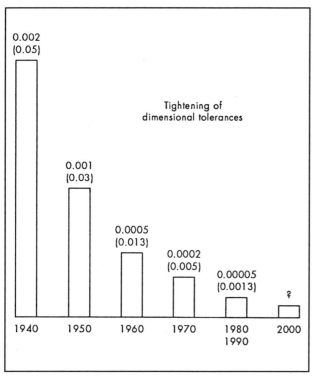

Fig. 16-41 The tightening of dimensional tolerances.

CHAPTER 16

STRINGENT FINISH REQUIREMENTS

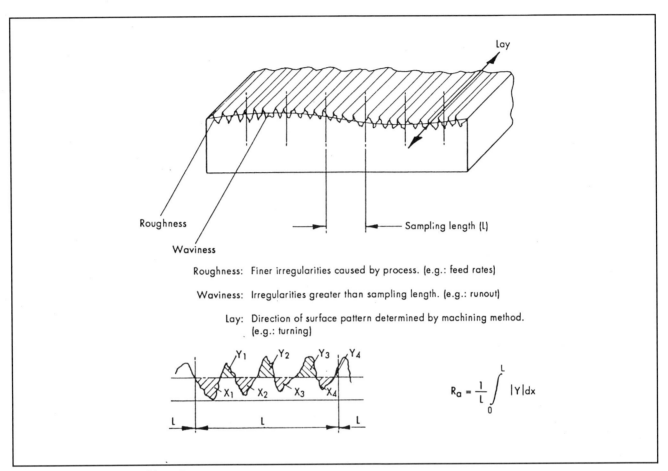

Fig. 16-42 Surface texture and integrity.

It is defined through different parameters and equations and given in numerical values:

R_a (R_q): Arithmetic average height from the mean (most commonly used).

RMS: Geometric average height from the mean.

R_t: The maximum peak to valley height.

R_z: Average distance between the five highest peaks and five lowest valleys.

Their approximate correlation is:

1 $R_a \cong$ 40 RMS; $R_t \cong$ (6 to 7) R_a;
$R_z \cong 0.85\ R_t$

Figure 16-43 shows the most common metalcutting methods and the roughness averages (R_a) they can cover (approximately).

For most applications and this includes high-precision manufacturing, the criterion of roughness as a measure of surface quality is sufficient. However, in cases where specific surface patterns are required, the surface texture needs to be defined, specified and controlled. An example would be to give a bearing surface a certain roughness profile of a certain peak and valley distribution for better oil retention, rather than low, average roughness. Different machining methods (for example, milling and turning) yield different surface textures. In specific cases surface textures have to be engineered and machined to provide functional surface characteristics.

Every material removal process has its own effect on the workpiece. While surface texture is the criterion that controls the finish of the workpiece surface, surface integrity checks the interior effects machining processes have on the workpiece material.

Critically stressed parts should have "tagged on" specifications that include surface integrity considerations. Surface integrity deals with the alteration of surface layers (plastic deformation) by the machining process and the effects this has below the geometrical surface of the workpiece. The machining process here can result in: Microcracks, recrystallization, residual stresses, hardness alterations, material inhomogeneities. The number of workpiece material and machining process combinations is sheer countless. However, the two surface integrity effects, with the most direct bearing on design and application are fatigue strength and stress corrosion cracking propensity. Both, in many cases need to be empirically determined for a particular application before specifying the right material and machining process and finishes.

Cost Consideration

"Design to cost," "manufacturing to cost" by simultaneously providing marketing with quality products might sound somewhat unrealistic at first glance and yet it is not. Over-specifying at the design stage and using the most intricate, expensive machining methods do not necessarily improve a product's quality. In fact, design has to be function oriented and manufacturing productivity oriented.

CHAPTER 16

STRINGENT FINISH REQUIREMENTS

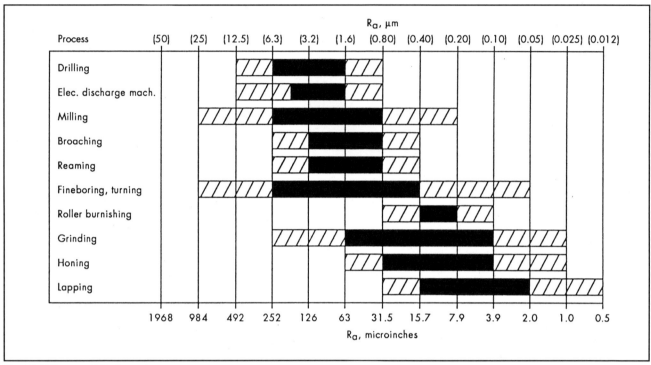

Fig. 16-43 Ranges of R_a for various processes.

Manufacturing cost increases with the number of machining passes, tighter tolerances and better part finishes. The other side of the spectrum is, that manufacturing without the proper processes, methods and finishes can be even costlier, due to premature part failure or shortened product life. "Overspecifying" and "underspecifying" are both unacceptable. Are there any cost guidelines engineers can go by? The answer is: Yes, there are guidelines but no, do not let these misguide them.

The truth is, many attempts have been made to empirically find cost guidelines. The upshot: Tolerance increments plotted over relative production cost, describe a reversed parabolic curve. That means production costs increase geometrically for uniform tightening of tolerances. However, relative cost is just that, relative. The same holds true for the relationship between machining cost and surface roughness. Both tight tolerances and good surface finishes can be achieved with advanced cutting tool systems at no "extra" production cost. The key is for the design and manufacturing engineering to avoid extremes. For example:

Design. Apply narrow specifications selectively and where they are needed rather than for the entire workpiece.

Manufacturing. A dark shadow on an otherwise smooth, per spec surface finish, does not necessarily call for rework only because of a nonfunctional, visual "flaw." As to surface texture, given the fact that machined surfaces are complex and parts with different functions sometimes require different texture profiles, the criteria for quality is not always the smoothest surface. Rather, especially when friction or oil retention or adhesion are required, the right distribution of peaks and valleys has to be specified. Coarser finishes are sometimes more desirable.

Surface integrity specifications should be specified only to the critical and highly stressed zones of the component part. Whenever high tensile strength and high stress resistance to extreme and alternating conditions (temperature) are relevant factors; the machining processes become more involved, since cost and productivity take a second seat to "safety." It is not the machining method per se that only causes cost increases. Raw (unmachined) material and post-machining processes as well as a too cautious machining approach (conservative machining data) add to the original production cost.

Achievable Parameters

Surface technology, dimensional configurations and tolerances are evidence of quality manufacturing. Industry-wide and the world over, precision-machined parts are dimensioned, produced and inspected in micrometers.

From design to manufacturing there needs to be a real understanding of what it means to manufacture within 40 μin. (1 μm).

Answers to the following questions need to be given before narrow tolerancing:

Do engineers have the right feeling for what a micrometer is? (see Fig. 16-44 gives an idea of its relative size).

Before specifying tight tolerances of any kind for any part, the following criteria are to be considered:

- Function of the part.
- Manufacturing capabilities.
- Industrial quality standard.
- Cost.
- Production volume.
- Measuring equipment.

Functional, not visual aspects dictate the narrowness of the finishing parameters.

The manufacturing capability is determined by the whole machining and manufacturing system. Often overlooked is the fact that in many cases the tolerance sum of the individual components

CHAPTER 16

STRINGENT FINISH REQUIREMENTS

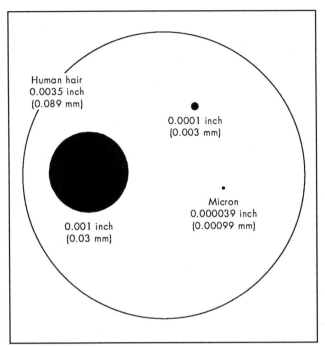

Fig. 16-44 Micron compared to other dimensions.

of the machining system is greater than the specified part tolerance. Capabilities have to be identified to specify feasible finishing parameters. The technological progress in the areas of metallurgy, machine tools and cutting tools have substantially narrowed down international industry tolerance standards for respective machining operations and component parts. IT6 and IT7 (ISO standards), for example, are current tolerance standards for quality finishes of finish-metalcutting processes.

Assuming specified close tolerances are a constant and production volumes are variables, the manufacturing approach certainly would be different. Holding close tolerances in a high volume production environment, calls for special machining processes and methods. It is here where the quality consistency is more difficult to achieve.

How about specifying finishes that cannot even be verified? Manufacturing of close-toleranced parts is an expensive proposition. It is, therefore, essential to have measuring equipment available that in itself "measures up" to the quality finishes produced. Verification at the end of the production process ensures the all important feedback of how manufacturing is doing and whether the part finishes demanded can indeed be met.

While manufacturing now has started to be quality-minded, people need to be cost conscious at the same time and they need to make sure that the quality finish standards can be maintained consistently.

SIX SIGMA AND C_{PK}-MANUFACTURING

Manufacturing has to be done with predictability and consistency to maintain quality at the same quality level.

A high quality level derives from well-designed and well-executed systems and processes. The actual quality performance has to be measured, compared to standards and the difference has to be acted on. This is called "quality control"—a collection of numerical values and statistical facts. Six sigma and process capability numerically describe how good a product and a process is.

Six Sigma

The mathematician Gauss proved the existence of natural distribution. He proved that every condition or event in life follows a certain pattern of variation, known as the bell curve (see Fig. 16-45). So do manufacturing processes.

When a product or process exhibits small variations, the sigma level is high and the likelihood of defect low. Six sigma is accepted as the highest level in manufacturing quality. It covers 99.73% of the area of normal distribution and when "a product is six sigma" it exhibits 3.4 defects per million, taking the typical variations of manufacturing into account. Six sigma, in other words, is a predetermined quality standard by which manufacturing has to produce its parts. As a control device it is the area into which sample parts, taken at random, have to fall at the end of the machining process.

Quality control, when sampling parts at the end of manufacturing, typically does so by forming a histogram, a vertical bar chart of frequency distribution. The histogram just like the distribution curve, highlighting the center and variations of data is popular, because of its simplicity in use and value of interpretation. The shape of the histogram, plotted within the specification limits, can "tell on the process."

Three histograms were plotted, the individual values of which form three differently shaped bell curves (see Fig. 16-46). The tolerances of all "three processes are well within the minimum/maximum-limit. Process A produces the greatest amount of variations while process C, however, produces the least. Process A produces parts that require rework. Although process B produces all pieces closer to the mean target, its higher variability over process C would result in increased assembly time.

Process Capability (C_{pk})

It measures the uniformity of machining and manufacturing processes by evaluating machine tools and cutting tools through the results achieved. Process capability analysis on the shop floor is a "natural" continuation of histograms and frequency distributions and takes it a lot further. Process capability studies identify variations of the process from the aim of it. To illustrate this, the following two example graphs are discussed (see Fig. 16-47).

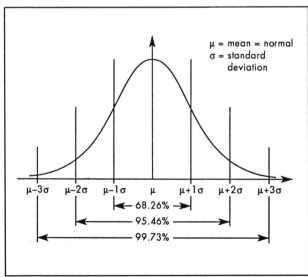

Fig. 16-45 Bell curve.

16-45

CHAPTER 16

STRINGENT FINISH REQUIREMENTS

The target diameter of a workpiece is 5″ (127 mm), with an allowable tolerance of ± 0.002, that is, a maximum diameter of 5.002″ (127.05 mm) and a minimum diameter of 4.998″ (126.95 mm). A sample of 50 parts taken off the machine, their diameter measured and then plotted down, yields a histogram, which reveals that all parts fall within a range of ±3 sigma (standard deviation). Thus 99.97% of all parts fall within six sigma in both cases.

C_{pk} by definition, is the measure of how well manufacturing and machining centers are on the target. As can be seen, Process I is just that, right on target. Process II, by contrast, is moved off center to the lower end of the specified tolerance. This offset from center is precisely what determines process capability and it is defined mathematically as follows:

$$C_{pk} = \frac{\text{Allowable tolerance range}}{2\text{x furthest away from mean}}$$

In the example shown, Process I yields a C_{pk} of 2.0, whereas process II a C_{pk} of 1.33. A C_{pk} of 1.33 has been accepted by most industrial manufacturers as a desirable and acceptable objective, while trying to work toward a higher quotient. If a machining process cannot meet the expected capability, machine tools might have to be overhauled or different ones specified or other cutting tool systems may have to be used.

True Position

Traditionally about 70% of all chipmaking involves holemaking of which about 70% are precision bores, usually for accommodating mating parts.

True positioning as a statistical tool in quality control is applied where tight tolerances, paired with function and interchangeability of mating parts are crucial.

The center of the bore by definition is its "true position," determined by distance specifications from another location of the part, for example, manufacturing bores.

A comparative measurement of several sample parts then finds the actual location of the bore plotting the deviations in the X and Y axes. The average total deviation stipulates the within or out of tolerance condition.

The example (see Fig. 16-48) shows high accuracy within tolerances.

Six sigma, process capability and true position are statistical measures to evaluate the outcome of a manufacturing process, compare it to a standard and/or a predetermined objective and reveal possible causes of nonconformance.

The purpose of inspection at large is product acceptance or the disposition of product based on its quality. This disposition involves several cardinal decisions, as shown in Table 16-3.

FIRST PART/GOOD PART, ZERO DEFECT-PRODUCTION

Ideal manufacturing produces 100%-defect free, right from its inception, good parts. Ideally, scheduled process performance is equal to actual process /performance. A tall order in an environment of many variables. Can manufacturing indeed meet the goals of the first part/good part and zero defect concept? The answer is "yes and no" or "yes, but." Clearly it is a desirable goal and if approached with realistic parameters, attainable.

A look at the influencing machining results, reveals the high complexity of manufacturing (see Fig. 16-49).

The concept of "doing it right the first time" is the right and

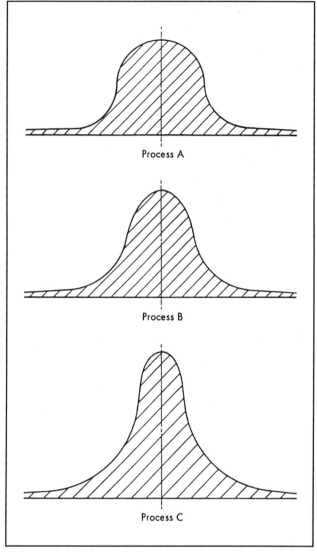

Fig. 16-46 The plotting and comparing of three processes.

only approach in modern, precision manufacturing. It means to strive for perfect results, looking for inconsistencies, correcting errors before they can happen, to eliminate rectifying, redoing, scrapping, apologizing. It calls for total quality management (TQM), the involvement of all levels and departments of the corporation and statistical process control (SPC) as part of TQM.

Right here is it where one needs to think for a moment. SPC-methods were developed as a response to the recognition that variations in the characteristics of manufactured goods exist. Inherent variabilities of every manufacturing process cause end products to differ from one another. If the variability is considerable, it is impossible to produce predictable results. Hence, the goal is to narrow down variability as much as possible and sensible. The demand of first part/good part when a new manufacturing method or process has been installed cannot be taken literally, it can be strived for to, over time, get closer to the "ideal." The same holds true for zero defect manufacturing. Setting it as a goal is imperative, to assure perpetual pursuit of coming close to achieving it. That is why people need to give both

CHAPTER 16
STRINGENT FINISH REQUIREMENTS

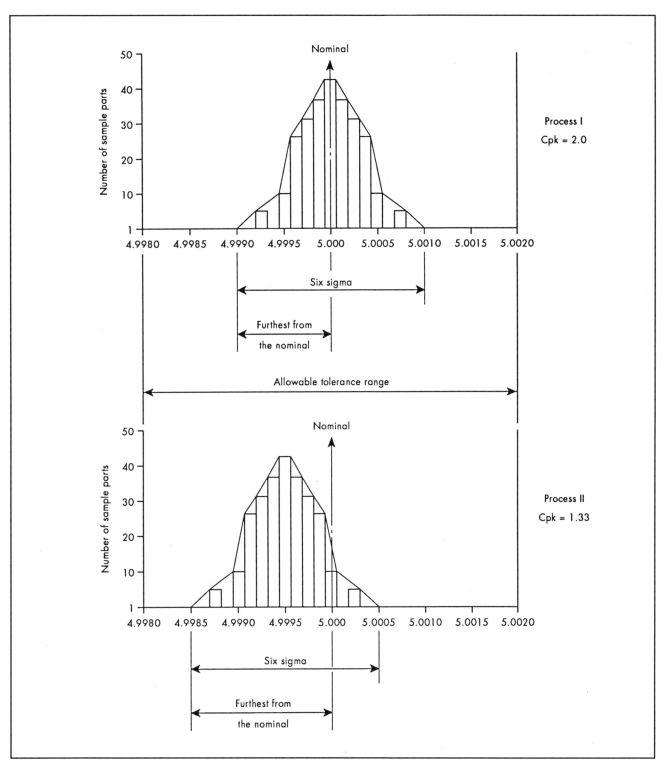

Fig. 16-47 Process capability, C_{pk}.

concepts a realistic range to work in, assign them a close and realistic "tolerance." The aspect of "design to producibility" and what one could call "realistic result targeting," play an important part in the realm of this concept.

Design to Producibility

It is design that specifies manufacturing. A number of manufacturing issues can be dealt with at the (early) design stage. Design is supposed to know the function of the product and it

16-47

CHAPTER 16

STRINGENT FINISH REQUIREMENTS

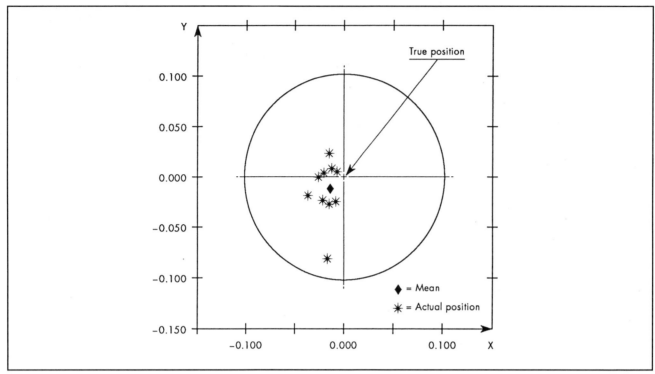

Fig. 16-48 Example of true position showing high accuracies within tolerances.

TABLE 16-3
Product Disposition Decisions

Name	Purpose
Conformance decision	To judge whether the product conforms to specifications.
Fitness-for-use decision	To decide whether nonconforming product is fit for use.
Communication decision	To decide what to communicate to outsiders and insiders.[6]

ought to lead manufacturing to producing it with producibility as the underlying principle. For this the following aspects are to be considered:

1. Explore simplicity. Design for maximum simplicity in functional and physical characteristics.
   Determine the best production methods. Seek the help of a manufacturing engineer to design for the most economical manufacturing methods available, with consideration for their inherent producibility limitations.
2. Analyze materials. Select materials that will lend themselves to low-cost production as well as to design requirements.
3. Minimize production steps. Design for the minimum number of separate operations in machining (such as rough turn, finish turn, grinding and polishing the same piece).
4. Eliminate fixturing and handling problems. Design for ease of locating, setting up and holding parts.
5. Employ the widest acceptable tolerances and roughest finishes. Specify surface roughness and accuracy commensurate with the type of part or mechanism being designed and the production method or methods contemplated.

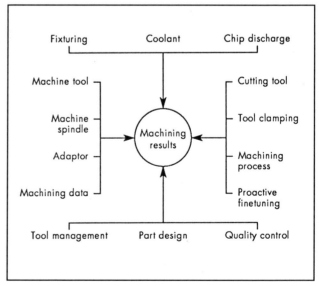

Fig. 16-49 Major influences on machining results.

CHAPTER 16

STRINGENT FINISH REQUIREMENTS

Realistic Result Targeting

It concerns the aspect of feasibility. What is the manufacturing floor's capability? Can it really and realistically hold the tolerance necessary?

It involves also the knowledge and consideration of process variation and tolerance accumulation.

Antiquated and outdated machine tools and tooling systems cannot produce parts and products of world-class standard. It is that simple. It is not enough to be quality minded, touting it and imposing it on manufacturing. Precision manufacturing equipment is the prerequisite of producing precision parts. It is the corporate management's responsibility to provide the proper "tools" commensurate the produced parts quality. Only then can productive manufacturing begin and end-results be cost-effectively achieved. A given, tight part tolerance can either be achieved with ease, predictably and repeatedly or with an extraordinary effort of processes and labor. The latter is doomed.

Statistical process control and process capability as part of it, recognize variations in the process of machining and manufacturing. Random variations such as temperature fluctuations and assignable variations such as workpiece material and variations in measuring are the contributors to varying product quality. Random variations usually are beyond control, assignable variations can be targeted and dealt with. The inherent variation of testing and measuring is an important contributor, most often overlooked.

The accumulation of tolerances within the manufacturing process, is another issue to be addressed. Again, every machine tool is built within a certain tolerance range. The same holds true for the tooling system and all other equipment contributing to the process (handling, fixturing, etc.). In addition, every machining step in itself is done within a certain tolerance limit.

The finished product incorporates the sum of all individual tolerances. Yes, it is possible to, say, consistently achieve a 98 μin. (2.5 μm) roundness of a cast iron connection rod bore in high-volume production. Provided, however, that all manufacturing modules have the inherent built-in capability for it. Only if tolerances are within process capability, can manufacturing results be predictable. It is prudent for corporate management to accept only the best of products. It is necessary for design to specify the best possible finishes. For manufacturing it is mandatory to have the capable "tools on hand" to produce the quality expected.

Six sigma and 1.33 C_{pk} are proven, realistic manufacturing goals. They need a high-degree of commitment and involvement of the entire manufacturing company, the most relevant prerequisite for first part/good part and zero defect manufacturing.

THE NEED FOR FURTHER ADVANCEMENT

Outlining and describing most modern high-precision machining systems and then demonstrating their practical, successful application in a world-class manufacturing environment, raises the question whether the optimal level has been reached the ultimate or whether there is room for improvement and if so where? The answer to this question is "no" and there is room for further, needed improvements surrounding high-precision machining.

Relevant areas where further technical advancements are visible and within grasp.

HIGH SPEED MACHINING (HSM)

Increasing productivity through faster production can be achieved by shortening nonmachining time optimizing machining processes and accelerating the actual machining time. The latter is done by increasing the rate of metal removal through increasing speed and feed rates. Where exactly high speed machining *per se* begins is difficult to specify and varies for the respective machining method (milling, turning, etc.) and the workpiece material.

Typically, milling and fineboring of aluminum is done with modern machine tools and cutting tool systems at around 500 m/min cutting speed and 1.5 m/min feed rate. Speeds of 5000 to 6000 rpm are common in finish-machining aluminum parts.

To make high-speed machining a worthwhile and economical proposition, cutting speeds of 5000 m/min at feed rates of up to 10 m/min would have to be applied. Or generally speaking, speeds of finish-machining operations would be between 12,000 and 20,000 rpm.

The manufacturing benefits are:

- Increase in material removal rates.
- Increase in machine uptime.
- Reduction in the number of machine spindles.
- Reduction in cycle time.

The trade-off: Machine tools and cutting tool systems have to be suited for HSM and, often overlooked, so does the workpiece material.

Machine Tools

Machine tool builders' main design considerations are dynamic stiffness, safety and main spindle bearing systems. The use of extreme high rapid traverse in the acceleration/deceleration phase, generates high massforces which can be easier controlled by using lighter weight material (fiber, composites) for the respective machine tool components.

Increased speeds and feeds set increased energy free, adding to safety concerns. Minimizing the impact of collisions and foregoing them altogether has to be a built-in and extensive safety feature of machine control.

Of the spindle bearing systems available–hydrostatic, hydrodynamics, pneumatic, electromagnetic and roller bearing–roller bearing is the most common in chipmaking HSM. According to Dr. Weck, University, Aachen,[25] they are the most promising, for they meet the requirements for high stiffness and low friction. As a "tandem O" arrangement, the angular ball bearings are preloaded axially and oil-air lubricated and cooled.

Currently there is extensive research and development in the areas of ballbearing materials (ceramics, coated carbide) and lubrication (grease) and scaling (contact versus contactless) to improve the reliability and ease of maintenance of the systems.

Workpiece Material

High speed machining is, in principle, the machining with extremes and thus any variable involved acts and reacts to its extreme and its limit.

CHAPTER 16

THE NEED FOR FURTHER ADVANCEMENT

Workpiece interruptions, material's microstructure variances, inclusions, hardspots, its hardness, its thermal conductivity when subjected to HSM, yield different machining results.

There are different high speed rates for different workpiece material. Aluminum and magnesium alloys as well as (fiber reinforced) composites can be machined at very high speeds. Ferritic cast irons, represent the other, lower end of the speed spectrum.

While other materials can be run in-between the data for aluminum and cast irons at high speeds, they pose challenges to the cutting tool material. This holds especially true for titanium alloys, the low thermal conductivity of which causes rapid diffusion at the cutting edge and premature tool failure. Applying speeds of 7000 rpm in aluminum is considered conventional, in case of hardened steel it would be the area of high speed machining. Again, the generation of extreme heat at the "tool tip" lowers its applicable speed rate.

Cutting Tool Systems

Whatever the machinability of a given workpiece material is, only the most sensitive considerations in the cutting tool system's design can secure successful, effective high speed machining.

During extensive tests in the 1920s and 30s, the engineer Carl Solomon found out that temperatures reach a certain maximum, specific to the material cut and then, after further increase of the cutting speed, the temperatures decrease. Indeed, his hypothesis that the temperature at the workpiece and the cutting material decreases to a point where it does not have an adverse effect on the machining process, proved to be true for many workpiece materials. At higher speeds, the chips' temperature rise over-proportionally, absorbing the heat generated and their thickness diminishes, facilitating chip control.

As cutting speeds increase, the cutting forces diminish and then hold steady at some point, depending on the workpiece material. This positively affects machining of thin-wall material. Lower stresses on cutting tools as a result of reduced cutting forces leads to longer edge life and accuracy.

And again, as previously stressed, the higher the applied revolutions per minute, the better the surface finishes. High speed machining inherently aids the cutting tools in their machining process, provided the system is stable and rigid: Stability at the cutting edge, rigidity of the system itself and especially at the interface to the machine tool spindle.

Their relatively low thermal and chemical stability disqualifies carbides from most high speed machining. Ceramics can cut up to 1000 m/min, however, their extreme brittleness and their poor resistance to thermal and mechanical shock leaves their use in HSM rather limited, too. CBN for ferrous metals and PCD for nonferrous metals are the choice cutting material for HSM, due to their superior wear resistance (high hardness, thermal and mechanical stability). Noteworthy is, that CBN's cutting speed in ferrous metals of up to 1000 m/min just about ends where PCDs in nonferrous metals begins.

Any cutting tool system, especially when running at 4000 rpm and higher, needs to be balanced to eliminate any excessive vibration. The more machining operations are designed into one cutting tool bar and the more clamping and adjusting features are built-in, the more asymmetrical the tooling system might be.

Charts plotting the maximum permissible, remaining unbalance in gm/kg over speed in rpm defines quality classes (Q1 to Q16), for the degree of sensitivity in balance-free machining.

To meet the demand for high repeatability and rigidity, particularly in flexible and high speed machining, the interface of cutting tool and machine tool (tool shank, adapter, toolholder) is of utmost importance. The technical aspects and developments along these lines are covered in the following subsection.

Most high speed machining has been done by aerospace companies (for example, wing milling) and the automotive industry (for example, aluminum engine and transmission components). There is a more widespread use of HSM in Japan and Western Europe. Lack of knowledge in the engineering arena, little technical background and routine case histories, so it seems, are preventing industry from using this alternative technology.

The benefits of HSM outweigh the cost for the initial investment:

- Three to five-fold increase in metal removal rates
- Up to 70% reduction in machining time.
- A 25% to 50% savings in cost for machining.

High speed machining can increase the productivity of some operations up to 100 times.[26]

Still, a lot more research has to be done to broaden the scope of HSM-applications, especially in steel cast iron and composite material.

The successful applications of pre- and finish-machining intricate aluminum parts should lead the way to gain more experience in other areas, so that this future oriented, valuable technique is optimized.

INTERFACES MACHINE TOOL/CUTTING TOOL

The precision tool shanks and American National Standard Institute (ANSI) and European-based International Standard Organization (ISO) taper toolholders, that accepted as industry-wide standards, have been under scrutiny in recent years, in search for alternatives better suitable for flexible manufacturing and, above all, machining at higher speed- and feed rates.

Today's standard interfaces have reached their limits relative to stiffness, accuracy and their capability to transmit higher bending moments and torque.

The high centrifugal forces generated during high-speed machining lead to positional inaccuracy. The taper in the spindle opens up more on the bigger than the smaller end of it. This change in geometries, results in an unequal taper seat and an axial movement of the tool shank into the machine spindle. Extraordinary axial forces then are necessary to pull the stuck tool out of the spindle. Furthermore, cutting forces and taper expansion can push the tool shank out of its concentric position, foregoing any machining at high speeds. In fact, this condition can become a safety hazard when the clamping mechanism fails, releasing the tool during machining.

Of course, predictable machining with a high degree of repeatability is, under these circumstances, impossible.

An interesting, new technology has evolved after several years of research and development to possibly replace the current shank and toolholder design. An international group comprising machine tool builders, tooling manufacturers, endusers and university professors just completed extensive tests and submitted to German Industrial Standard Institute DIN and ISO its newly developed machine tool/cutting tool interface proposal for standardization. It is said to meet the demands for:

- Transmitting forces and torque uniformly.
- Maintaining a constant and equal clamping force.
- Form and power clamping.
- Static and dynamic stiffness.
- Positional accuracy.

CHAPTER 16

THE NEED FOR FURTHER ADVANCEMENT

- Straightness and roundness.
- Safety at high speed.
- Securing a balanced condition.
- Repeatability.
- Ease of coupling and uncoupling.

Paying tribute to the aforementioned demands and the different machining methods from heavy milling to CNC-machining center and typical lathe work, the group suggests the design of a short taper and a double-cylinder shank (see Fig. 16-50a). Both configurations are hollow shank designs with a positive face contact for resting at the machine spindle face (see Fig. 16-50b).

Test-runs revealed that the short taper shank has slightly greater rigidity. It accepts a higher bending moment, can transmit higher torque and resists centrifugal forces and thermal stress rather well. However, a high (puls like) extraction force is necessary to take it out of the spindle. Also, producing the short taper shank calls for finetuning the contact face area to the taper to stay within closely-held manufacturing tolerances.

Advantageous for the double cylinder is its defined play to the spindle diameter. During machining at higher speeds the double cylinder expands slowly and thus ensures very accurate radial seating. The shank adjusts spindle expansions, caused by centrifugal forces, by following the expansion. In order to provide high

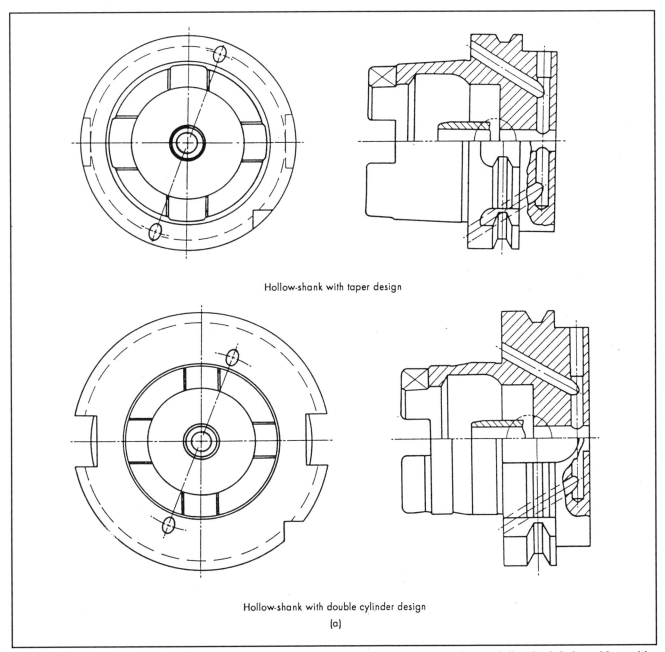

Hollow-shank with taper design

Hollow-shank with double cylinder design

(a)

Fig. 16-50 Hollow shank with taper design and hollow shank with double cylinder design, *a*. Both designs are hollow shank designs with a positive face contact for resting at the machine spindle, *b*.

CHAPTER 16

THE NEED FOR FURTHER ADVANCEMENT

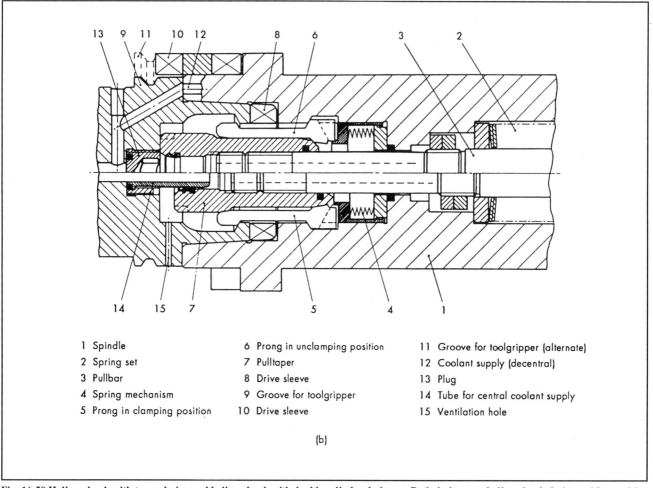

1 Spindle
2 Spring set
3 Pullbar
4 Spring mechanism
5 Prong in clamping position
6 Prong in unclamping position
7 Pulltaper
8 Drive sleeve
9 Groove for toolgripper
10 Drive sleeve
11 Groove for toolgripper (alternate)
12 Coolant supply (decentral)
13 Plug
14 Tube for central coolant supply
15 Ventilation hole

(b)

Fig. 16-50 Hollow-shank with taper design and hollow-shank with double cylinder design, *a*. Both designs are hollow shank designs with a positive face contact for resting at the machine spindle, *b*. (*cont.*)

radial accuracy and ensure a smooth, true fit, the production of the contours of the double cylinder is rather complicated and has to be done with highest precision.

Figure 16-50*b* shows the cross-section of a spindle with clamping GM engaged into the hollow shank. The four GM on the periphery pull the tool system straight into the machine spindle.

Practical machining applications will demonstrate which of the two, or even both systems, will eventually prevail. Fact is, that machine tool builders, tool systems manufacturers and endusers now can choose another, intriguing and promising machine tool/cutting tool interface.

The necessity dictated by process and proven practical applications, will decide on the scope of acceptance on the manufacturing floor. The systems suppliers' willingness to abandon their own, "homegrown," toolholding designs will play as much a role as will cost considerations.

CUTTING TOOL MATERIAL

Cutting material, being literally at the tip or forefront of technology of any machining system is the variable, manufacturing passes on to research and development for achieving higher productivity, better quality and cost savings.

While cutting material manufacturers will have to continue to strive for product improvements across the board, there are two main areas, they are concentrating on: Submicron-carbides and diamond-coating.

Submicron Carbides

Carbides of extreme small, fine and uniform microstructures. Compared to traditional carbides of the same compositions they are harder and have higher tensile strength. The higher the hardness, the less the rate of wear and tear. An increase in tensile strength means machining can take place at higher cutting speeds by simultaneously, increasing tool life.

Table 16-4 is indicative of developments under way, especially in Japan and Western Europe.

This grade of submicron, for example, is said to increase tool life four-fold and increase applicable cutting speeds up to 30%.

Submicron carbides seem to be well-suited for finish-machining hardened steel products, an area of much needed progress. In fact, they might represent a similar breakthrough technology for ferrous metals as carbide coatings are. More development is needed to possibly offer the whole carbide spectrum in submicron structures to cover more variety of workpiece material at higher speeds and also with a favorable performance/cost ratio.

CHAPTER 16

THE NEED FOR FURTHER ADVANCEMENT

TABLE 16-4
Characteristics of Traditional Carbides to Submicron Carbides

Typical Carbide (tungsten-cobalt)	Hardness, HV	Density, lbs/in^3 (g/cm^3)	Tensile Strength, psi (N/mm^2)	Microsize, μin (μm)
Traditional	1600	0.54 (14.9)	290,000 (2000)	39 to 118 (1.0 to 3.0)
Submicron	1800	0.54 (14.9)	435,000 (3000)	0.8 to 39 (0.2 to 1.0)

Courtesy Krupp-Widia

Diamond Coating

Increasing chipmaking of nonferrous metals has scientists and engineers the world over scrambling to upstage cutting materials currently in use. Even PCD, which have taken medium to high production volume machining by storm, are endangered species. The reason: Thin diamond coatings, also known as CVD-diamonds.

They are layers of diamond crystals of up to 1970 μin. (50 μm) thickness, usually grown on tungsten carbide by means of chemical vapor deposition (CVD). The process is this: A gaseous mixture of methane and hydrogen is injected into a reactor chamber of low pressure and high temperature. The gas is then "bombarded" with microwaves, which separate the various atoms to form a carbon plasma above the substratum (for example, tungsten carbide) onto which the carbon atoms are attracted. There they line up in a layer the size of one atom. Repeating the process layer after layer then forms a film of about a hundred microinches (several micrometers) in thickness on the substratum.

CVD-diamonds feature a dense polycrystalline structure with a hardness of 7000 to 12,000 kg/mm^2 and a coefficient of fraction 0.05 to 0.15. It is generally up to 15% harder than the normal polycristalline diamond (PCD). Its abrasion resistance is equal to that of PCD. Its thermal stability is said to be better than PCD, because of its purity in carbon and hydrogen. This means it can withstand very high temperatures without breakdown.

Generally, CVD-diamond's properties are said to be close to those of the natural diamond. For cutting tools it means limiting the formation of built-up edges, improving abrasion resistance, and being able to run even at higher speeds than with polycristalline diamonds. Figure 16-51 reflects comparative results of flank wear tests in low silicon (7.5%) aluminum at a cutting speed of 656 ft/min (200 m/min), a feed rate of 0.004 in/rev (0.1 mm/rev) and a stock removal of 0.02 in. (0.5 mm) on the diameter. The substratum for the diamond coating is silicon nitrite. Over a cutting length of 32,800' (10,000 m) the thin-diamond coated 20 μin. (0.5 μm) cutting insert showed by far the least flank wear and achieved the same surface finish as the PCD-insert.

Researchers at Hiroshima University, Japan, claimed that there was no coating separation during machining, however, at higher speeds a thicker diamond film on the same substratum performed better.

Diamond coatings have successfully been used in a variety of chipmaking methods such as drilling reaming, turning. However, they have hardly passed the development stage, despite possessing all the characteristics of PCD and more (possible provisions of multiple cutting edges per insert, a greater depth of cut and a favorable cost per edge ratio). The main problems researchers and engineers face are adhesion and peeling. Cobalt, an otherwise desirable ingredient of tungsten carbide to lend toughness to its composition, becomes a catalyst for graphitization during machining when coated with a diamond film. The result is poor adhesion to the substratum tungsten carbide which, depending on the machining data applied, causes the diamond film to peel off.

Clearly, the remaining research variables seem to be silicon nitrite, coating thickness and carbide's cobalt contents. Diamond coatings seem to adhere to silicon nitrite better than to tungsten carbide. However, silicon nitrite lacks the prized toughness of tungsten carbide. Thicker versus thinner or also less versus more multilayered coatings, have to be weighed and their merits be determined empirically through varying applications.

A low cobalt (less than 5%) tungsten carbide substratum with a multilayered film coating of up to 1,970 μin. (50 μm), used with rather conservative speeds of no more than 2500 to 3000 rpm, might be the cutting insert material seen by researchers as the most applicable in years to come.

The insert manufacturers, in turn, will need to apply economical and reliable grinding techniques to arrive at closely toleranced inserts and a super-sharp cutting edge. The latter is especially relevant for finish-machining operations.

If cutting tool manufacturers can offer reliable CVD-diamond coatings at reasonable cost well below that of polycristalline diamonds, then the enduser might find it attractive to have even existing carbide tools coated.

This also would translate into a sizeable productivity boost on the manufacturing floor.

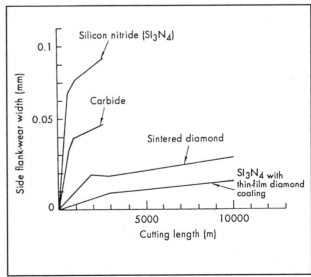

Fig. 16-51 Side flank-wear width to cutting length.

CHAPTER 16

LASER PROCESSING

LASER PROCESSING

The term "Continuous Improvement," when applied to laser processing, can be interpreted in two ways. First, laser processing in and of itself is part of the broad manufacturing trend toward continuous improvement. The second interpretation, is the continuous improvement of the laser machining process itself.

Continuous improvement in the laser industry is occurring across three dimensions. The first two are technology oriented; specifically ongoing developments in hardware and software. The third is simple operator ingenuity. Improvements in hardware and software can be seen in various recent state-of-the-art developments.

AUTOFOCUS

Because the efficiency and quality of laser processing depend largely upon maintaining proper focal length distance, the ongoing development of autofocus devices for laser systems has added significantly to process repeatability and quality assurance. There are two types of noncontact autofocus devices, one that is limited to conductive materials and one that can work on nonconductive materials.

One company (Laserdyne Division of the Lumonics Corporation) has developed a noncontact sensor that assures stability between the workpiece and the laser gas-delivery nozzle. The system is based on capacitance. The autofocus nozzle is positively charged, while the workpiece is negatively charged. A small current is passed between the nozzle and the workpiece. Any change in voltage is interpreted as a change in distance, and adjustments are made accordingly.

One feature of this system that sets it apart from conventional autofocus systems is that the sensor relays information directly to the CNC system that positions the axis. Previous methods had the autofocus as a separate entity outside the control loop of the CNC (see Fig. 16-52). This limited how quickly distance changes can be compensated for. Additional applications that are open when the autofocus device feeds information directly into the system's linear axis controller, including:

- Focus control, not only in beam direction, but also in any user-selectable direction (for example, cutting off normal to the surface).
- Determining the actual location of a feature.
- Detecting the presence of the workpiece.
- Determining which piece is ready for processing.
- Turning the system on or off from within the program (that is, when passing over a hole).

Lasers are now becoming the tool of choice for processing nonmetals such as plastics and composites. When processing such materials, focal distance is extremely critical. However, due to the nonconductive nature of these materials, a new method for assessing and adjusting focal distance "on-the-fly" had to be designed.

One company (H.G.G. Laser Fare) has developed an autofocus system for use specifically on Kevlar™, but it will work with other nonconductive materials as well. The device utilizes a low-powered HeNe laser, which is focused on the workpiece. The location of the beam is then used to triangulate the position of the gas-delivery nozzle. Changes in the distance from the nozzle assembly to the workpiece are interpreted by a function generator and are in turn translated into beam-axis motions by a Compumotor servo driver (see Fig. 16-53).

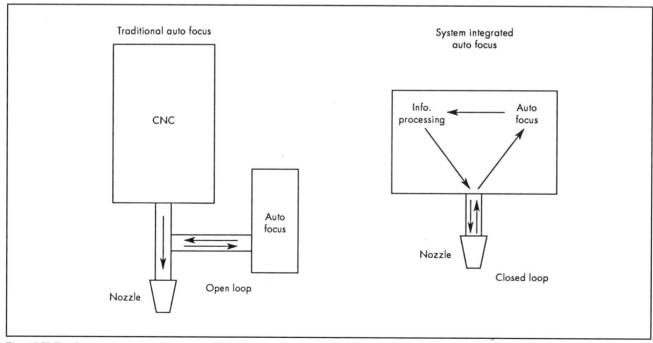

Fig. 16-52 Previous method of autofocusing.

CHAPTER 16
LASER PROCESSING

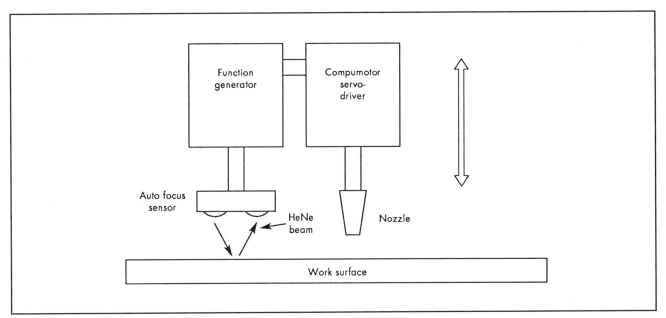

Fig. 16-53 New autofocus method utilizing a low-powered HeNe laser.

PART PROGRAMMING

One of the most important parts of the pre-production process is programming equipment to do the right thing. In laser processing, it is more aptly expressed as "going to the right place." Because the majority of laser systems are CNC driven, programming the drilling and cutting of complex hole patterns and shapes can be a rigorous and time-consuming task. This is typically done by manually positioning the equipment along the cut path and then storing the locations in a CNC program.

One company (Laserdyne) has developed a position-teaching system that can significantly reduce the part-programming process.

A camera is mounted through a quick change over (QCO) assembly right where the gas delivery nozzle typically is. The camera sends an image of the part to be programmed to a monitor on a hand-held remote control. A part of the remote control is shown on the monitor as well. The remote control has a three-axis joystick which the operator/programmer maneuvers to position the nozzle assembly. The operator simply jogs the assembly along the path where the beam is to travel. When used in conjunction with the system-feeding autofocus, the assembly can orient itself to be normal to the surface. All of the system positions can be stored with a single keystroke.

The camera has a resolution of 0.001" (0.03 mm) for aligning to the cut path. As a result, the teaching of 3D contours is three to five times more accurate than other methods.

There is also an ergonomic advantage to this part programming system. Because the positioning and teach-programming can be done remotely, there is no need for the operator to lean over the workpiece to view the part surface. The operator maintains a comfortable (and safe) position during programming.

COORDINATION OF BEAM CHARACTERISTICS AND POSITIONING SYSTEMS

Traditionally, the control of beam characteristic, such as power and pulse rate and positioning has been housed under different roofs. However, applications such as on-the-fly drilling have necessitated the development of systems that can coordinate beam on-time with beam positioning.

With this type of software in use, a laser system can be programmed to drill a specified hole pattern on-the-fly. By setting the beam parameters such that a hole can be produced with a certain number of pulses, the laser system can simply position itself in the proper location, open the beam shutter, burst-drill the first hole, and move the length of the drilling area at a rate of speed that is proportional to the pulse rate and the number of holes to be drilled. (Picture a sewing machine without thread.) For applications where it is possible to create a hole with one pulse, this development is especially beneficial (see Fig. 16-54).

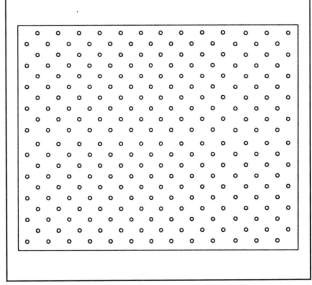

Fig. 16-54 Application where the use of one-pulse lasing is beneficial.

CHAPTER 16

LASER PROCESSING

Other applications of synchronization of beam characteristics and positioning systems include:

- Ability to produce sharp corners without melting them.
- Constant heat input (regardless of geometry) which results in minimum HAZ (heat affected zone).
- Constant cut width which provides increased accuracy.

ORBITAL NOZZLE ASSEMBLY

The way laser beam delivery typically works, either the lens assembly, the tooling configuration, or some combination of the two are manipulated to get the laser beam to the proper cutting location. While effective, there are limitations in speed and accuracy.

The orbital nozzle assembly (ONA) is an alternative to standard beam manipulation systems. Within the ONA is a servo-driven mirror. The servo rotates the mirror off-center causing the beam to move in a circle. The result is a rounder hole that can be drilled faster than with standard beam delivery systems. The device also serves as a gas-assist delivery system (see Fig. 16-55).

USE OF INERT GASES

Some applications, where oxide build-up is undesirable, require the use of an inert assist gas. In this section, the focus will be on inerting as it pertains to welding and cutting.

WELDING-LAMINAR BARRIER INERTING

Traditionally, laser welding with inert gases is done by flowing a stream of gas through a 0.15" (3.8 mm) diameter tube. While this method can be quite effective, with oxygen levels of twenty parts per billion obtainable, the method is not reliable. If the tube is misaligned even by one millimeter from the weld point, the oxygen levels rise substantially.

A new device, called an inerting ring (being developed by the Praxair division of Union Carbide and H.G.G. Laser Fare), can flood a small area with a virtually pure inert atmosphere when mounted on a beam delivery system. The ring, whose inner diameter is made of porous metal, is mounted so that the beam delivery assembly completely covers the top of the ring. Gas flowed into the ring flows downward out of the porous inner diameter. The resulting column umbrellas the work area with an atmosphere that has an oxygen level of ten parts per billion (see Fig. 16-56).

There are two distinct advantages which arise through use of the inerting ring. The most obvious is the reduction of unwanted oxide buildup. The second is that the inerting ring can flood hard to access areas that conventional "tube-feeding" cannot reach effectively. It may also be possible to accurately mix an atmosphere to have a specific oxygen content. This can both maximize weld depth while simultaneously controlling for oxide build-up.

CUTTING-HIGH PRESSURE INERT ASSIST

Inert gases can also be used to cut materials. While the laser provides the power for vaporizing the material, the gas pressure blows the material away. Because the inert gas does not contribute to the physics of the cutting process as does oxygen (which is typically used), a higher pressure is required than is usually the case when cutting with oxygen or an oxygen mix. Typically, the pressure required to cut with inert gas must be greater than 125 psi (862 kPa). The gases most often used are nitrogen, argon, helium, mixtures, and air.

Advantages to cutting with an inert assist gas include a minimization of the burr, heat affected zone (HAZ), and oxide buildup. This is especially beneficial with parts that are to be welded as cut.

The limitation to both of these methods is based in the amount of gas required to get each particular job done. Each application requires approximately three to ten times more gas than conventional methods.

SUMMARY

There are two fundamental aspects to continuous improvement in the laser industry. First is to be aware of new developments in the field. One must stay abreast of the latest in technology in order

Fig. 16-55 Orbital nozzle assembly.

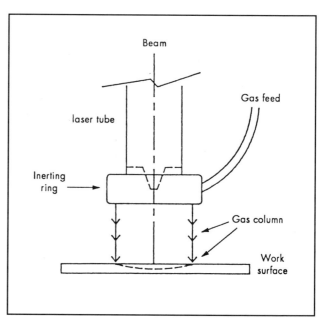

Fig. 16-56 Inerting ring.

CHAPTER 16

MACHINING COOLANTS

to remain viable and competitive.

Second, one must make use of plain old human ingenuity. Organizations that stress continuous improvement must keep this mindset instilled in all persons involved in the production process, from R&D to the person who loads the parts.

Continuous improvement does not imply a strict, regimented, stoic point of view. Continuous improvement is something that is not only good for production, it is good for morale. Those who are immersed in this mindset realize that it can be enjoyable and rewarding.

MACHINING COOLANTS

IMPROVING QUALITY AND PRODUCTIVITY THROUGH FILTER MEDIA SELECTION

Quality and productivity are closely associated with coolant clarity. This is an accepted fact in metalworking industries. Part quality characteristics such as surface finish and conformance to dimensional tolerances, and manufacturing productivity indicators such as tool wear and coolant system maintenance are correlated with the amount and size of contaminants in the coolant. This logic and the associated costs versus performance provide a basis for the economic justification of a coolant filtration system.

Filter systems using nonwoven filter media rolls have demonstrated the capability to handle large volumes of coolant flow, filter fine particulates without necessitating a pre-coat, and remove other contaminants from the coolant stream to achieve a functional level of clarity and stability. The component which does the actual filtering of the metalworking fluid is the nonwoven media, making it the single most important element in determining the efficiency of the filter system. Yet the selection of the filter media used is often an area that is not well understood nor given adequate attention, which can result in suboptimum operations and "hidden" costs.

The purpose of this section is to provide a foundation for the better understanding of filtration principles as they relate to the selection of filter media. A statistically sound, quantitative methodology is presented for conducting comparative evaluations of filter media and monitoring filtration system performance as part of a continual improvement process.

FILTER MEDIA SELECTION CRITERIA

In coolant filtration, the end user is generally interested in four primary criteria when selecting filter media: durability, efficiency, capacity, and cost:

Durability may be defined as sufficient strength to withstand the physical rigors of the filter system without tearing or rupturing in the filter bed, or causing difficulty in extracting from the filter system.

Efficiency may be defined as the retention of particles of a certain size and the rate of their removal.

Capacity may be defined as the amount of contaminants removed per unit area of filter media, before the media's permeability reaches its useful life.

Cost may be defined as unit price times usage, plus maintenance and disposal.

As these criteria relate to the filtration principles described in the preceding section, the following filtration process characteristics are important:

- Filter media structural integrity.
- Filter cake generation.
- Coolant flow rate through the filter media.

These process characteristics are significantly determined by the:

- Filter media manufacturing process.
- Filter media composition.

These process characteristics, and hence the selection criteria, can be evaluated in operation by measuring and monitoring the following process variables:

- Filtrate clarity.
- Index cycle times/filter media consumption.

In addition to:

- A review of media physical properties.
- Visual observation.

FILTER MEDIA PHYSICAL PROPERTIES

The physical properties generally considered important in the selection of filter media include the following:

- Basis weight.
- Thickness.
- Tensile.
- Tear.
- Mullen burst.
- Porosity.

Typical physical properties for each media grade originate with the media manufacturer. To properly compare these physical properties, it is necessary to check that the *same test methods* were used to generate the respective data.

Notably absent from this list of physical properties is the so-called micron efficiency of the filter media. This is because such efficiencies quoted are often misleading. Unfortunately there currently is no universally accepted standard method for testing and interpreting liquid filtration efficiency data, to determine what is often referred to as "nominal" micron ratings. When quoted a micron efficiency rating, it is important to request what level of efficiency this number was derived at and whether it was a single or multiple pass test (for example, that the media is 80% efficient at 20 microns in a single pass test). Reported nominal micron efficiencies are often arrived at through single pass testing, which does not reflect the realities of the filter system operation nor does it take into account the capability of the filter media to enhance the formation of a filter cake for greater efficiency. It is at best a relative number, useful for comparison only if the same test methods and interpretation of resulting data are applied.

FILTER MEDIA MANUFACTURING PROCESS CHARACTERISTICS

Figure 16-57 show the visual difference in similar basis weight filter media grades created by the four predominant nonwoven

CHAPTER 16

MACHINING COOLANTS

processes used to manufacture rolls for coolant filtration: wet laid, spunbonded, dry laid, and point-bonded. It is evident from visual examination that the wet laid process creates a more uniform web. This translates into a tighter, more consistent pore size distribution which is a major determining factor for dependable filtration efficiency. By comparison, spunbonded and dry laid filter media are typically less uniform due to the nature of the respective manufacturing processes, which can result in greater variation in pore size distribution and less consistent or efficient contamination removal.

The uniformity inherent in wet laid media also lends itself to multidirectional strength which is especially desirable in vacuum and pressure filters to prevent tears, splits, or ruptures in the filter bed. By comparison, the unidirectional formation of dry laid media results in good strength in the machine direction but weakness in the cross direction, which can invite splits in the media in the filter bed under a load. Filter media manufactured by the point-bonded process exhibits exceptional physical strength characteristics, however as much as 25% to 30% of the surface area may be impermeable due to thermal sealing, which can result in increased media usage and associated purchase and disposal costs.

A major determinant of filtration efficiency and capacity is the ability of the filter media to mechanically entrap contaminants while maintaining an operationally sufficient flow rate, and thereby promoting the formation of a filter cake. The differences in bulk or loft evident in the cross-sections of wet laid and spunbonded filter media of the same basis weight is shown in Fig. 16-58. The mechanical entrapment of contaminants within the loftier, fibrous structure of the wet laid filter media provides some depth filtration advantages. This characteristic enables the filter media to retain a larger quantity of contaminants without impeding flow as readily as if the particulates were captured only on the surface of the media. This advantage is graphically shown in Fig. 16-59. As a result, the loftiness (often referred to as caliper or thickness) of the filter media makes it less susceptible to blinding, enabling a superior filter cake to form, thereby improving filtration efficiency.

The "softness" of the finish of the filter media, which is evident to human touch, is also a result of both the manufacturing process and media composition. The softer the finish, in combination with loft, the greater the depth loading benefits. This characteristic enables the filter media surface area to be more fully utilized (including in the Z-direction), thereby maximizing filter media capacity and effective life.

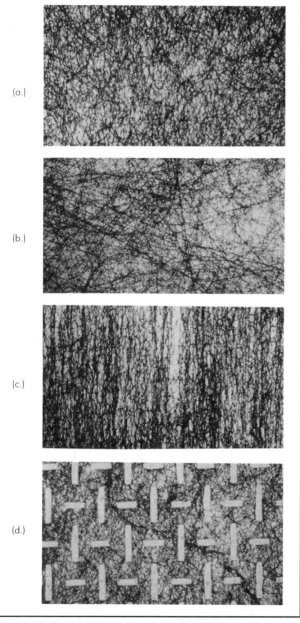

Fig. 16-57 Visual difference in similar basis weight filter media grades: (*a*) wet laid, (*b*) spunbonded, (*c*) dry laid, and (*d*) point-bonded.

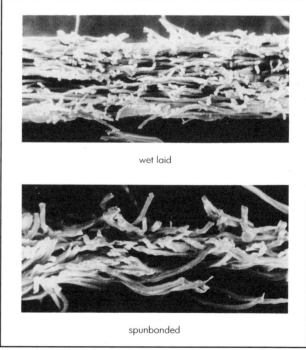

Fig. 16-58 Differences in bulk or loft evident in the cross-sections of wet laid and spunbonded filter media of the same basis weight.

CHAPTER 16
MACHINING COOLANTS

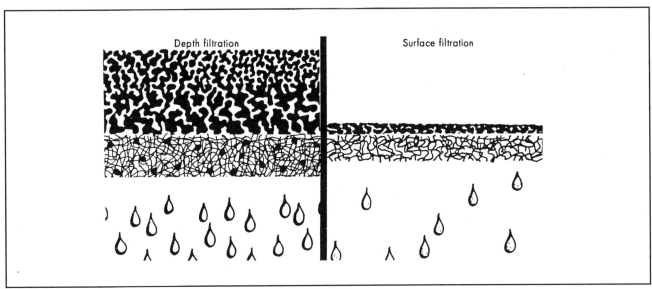

Fig. 16-59 Depth filtration versus surface filtration mechanical entrapment.

FILTER MEDIA COMPOSITION CHARACTERISTICS

Fiber type, fiber length, fiber diameter (or denier), fiber blend, and the binding system utilized in the manufacture of filter media determine its specific filtration characteristics and physical properties.

The type of fiber selected is determined by the intended end use. Polyester has been found to be most versatile and appropriate for metalworking fluid filtration applications. Polyester does not have an affinity to oil like polypropylene and nylon fibers, which can cause the filter media to blind over and prematurely index. Premature blinding results in suboptimum cake generation and filtration efficiency, and increased media usage and disposal costs. Polyester is less likely to swell and is better able to withstand the physical rigors of extended submersion, especially under pressure or vacuum, when compared to rayon or other cellulosic fibers. For metalworking coolant applications, a small percentage of cellulose in a polyester blend composition acts to absorb and remove tramp oil from the filter system without adversely impacting coolant flow rate. Polyester has also shown the ability to withstand the range of temperature and Ph operating conditions typical of metalworking coolants and washer solutions.

Fiber length and fiber diameter (denier), along with fiber type, play a significant role in determining the physical properties and strength characteristics of the filter media. *Sufficient* durability for the specific application, with optimal filtration efficiency and capacity characteristics, is of prime consideration in the manufacturer's selection of the fiber components.

Fiber diameter is a major determinant of the flow characteristics (porosity), pore size distribution, and formation uniformity of the filter media, which correlate with filtration efficiency. Table 16-5 shows the direct correlation between fiber diameter and porosity (cfm), with the smaller diameter fibers resulting in a tighter (more resistant to flow) filter media.

Table 16-6 shows the relationship between fiber diameter and pore size, with larger diameter fibers creating a more open sheet. Table 16-7 concludes with the direct correlation between fiber diameter and micron efficiency, with smaller diameter fibers resulting in higher filtration efficiencies in single pass testing.

The above data was compiled using wet laid polyester nonwoven media, however the correlations demonstrated are also applicable to other nonwoven manufacturing processes.

Figure 16-60 shows how small diameter (fine denier) fibers produce a visually tighter and more uniform media at the same basis weight. In the design of filter media, a performance tradeoff must be made between large diameter fibers for high flow rates and small diameter fibers for filtration efficiency. The wet laid manufacturing process is most versatile in its capability to manufacture nonwoven media with virtually any size or type of fiber or fiber blends that can be uniformly dispersed in a slurry. This enables wet laid filter media to be manufactured by blending various diameter fibers with the objective of providing sufficient durability for the intended application with optimal filtration efficiency and capacity.

The binding of fibers in the creation of nonwoven media for filtration may be a chemical, thermal, or mechanical process. The binding process is critical to ensuring the physical properties associated with filter media durability, and its chemical compatibility with the intended application. Adverse chemical reactions may not be immediately apparent but surface only after using a particular filter media for an extended period of time. Because of high costs often associated with chemical nonconformities, the consideration of only filter media specifically designed for and/or with a successful history in similar metalworking fluids applications can help to avoid such problems.

As with any product, quality (or consistency), assures dependable performance results. The required physical properties and chemical compatibility must be designed into the filter media, controlled during its manufacture, and assured through quality auditing. Quality improvement must be an ongoing process. A filter media supplier that can demonstrate such a quality system commitment is better able to serve as a partner in a company's efforts to continually improve the quality, productivity, and cost position of its operations.

In recent years there has been a proliferation of available roll media for coolant filtration in the marketplace. Frequently, these materials may have been designed for other primary uses or by companies not focused in filtration technology, and may be

CHAPTER 16

MACHINING COOLANTS

TABLE 16-5
Fiber Diameter versus Porosity (cfm)

Fiber Diameter (denier)	Fiber Length (mm)	Aspect Ratio	Basis Weight (g/m^2)	Porosity (cfm)
0.1	3.0	937	51.4	46
0.5	5.0	709	50.0	195
1.5	12.7	1025	52.0	493
3.0	19.0	1087	51.3	750
6.0	25.4	1025	51.7	943
15.0	38.1	972	51.0	1210

TABLE 16-6
Fiber Diameter versus Pore Size

Fiber Diameter (denier)	Fiber Length (mm)	Pore Size	
		Min. (μm)	MFP (μm)
0.1	3.0	10.74	17.63
0.5	5.0	21.56	35.06
1.5	12.7	44.34	64.47
3.0	19.0	105.30	189.20
6.0	25.4	139.10	241.40
15.0	38.1	n/a	n/a

TABLE 16-7
Fiber Diameter versus Micron Efficiency

Fiber Diameter (denier)	Fiber Length (mm)	Micron Efficiency		
		10 μm	20 μm	30 μm
0.1	3.0	99.76	99.96	99.94
0.5	5.0	37.91	76.5	99.13
1.5	12.7	26.11	52.40	81.04
3.0	19.0	20.53	23.07	39.91
6.0	25.4	21.27	9.5	23.46
15.0	38.1	5.84	16.41	21.41

underpriced as part of a short-term strategy to fill manufacturing capacity. This situation has invited filter performance problems and undependable supply channels. It therefore becomes important to the filter media purchaser to learn more about the *manufacturer* of these materials—the company's business history and stability, experience and long-term resource commitment to metalworking fluids filtration, and the quality system employed, prior to evaluating and specifying a particular filter media.

EVALUATING FILTER MEDIA

Given the important criteria of durability, efficiency, capacity, and cost in determining the optimal filter media for an operation, quantitative methods for comparative evaluation are desirable. While advanced technology is available, such as the use of particle counters and scanning electron microscopes for determining filtrate clarity and particle retention, this discussion will focus on an approach that uses common lab apparatus which enables a technician to perform an adequately discriminating evaluation in-house, on a timely and regular basis, and at minimal cost. The following methodology will also employ the use of statistical analysis to determine the significance of results and for ongoing process monitoring and control.

Durability is defined as sufficient physical strength to withstand the rigors of the filter system without tearing or rupture detrimental to the operation. In the manufacture of filter media, industry standard tests are routinely performed to evaluate the tensile, tear, and burst strength of the material. Typical values for these physical properties are available from the manufacturer of the filter media. Again, the use of the same test methods to compile this data must be verified for this information to have any value for comparative purposes. Standard test methods (for example, ASTM, TAPPI) are generally noted on physical data sheets supplied by the media manufacturer—or such comparative testing may be requested of the manufacturer or a test lab by providing a competitive sample (approx. 9″ × 11″ or larger). This comparison between the current and proposed media provides a relative measure of physical strength suitability prior to evaluation in operation. Should the actual physical strength characteristics required by the filter system be known, a direct comparison may be made with the typical properties provided.

In operation, visual observation can be used to evaluate filter media durability. If there are no holes or tears evident at the filter

CHAPTER 16
MACHINING COOLANTS

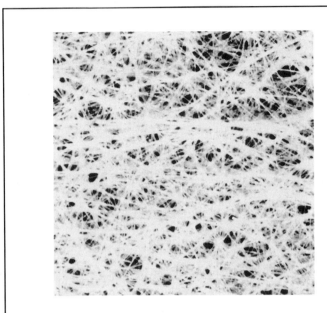 

Fig. 16-60 Fine denier, *a*, versus coarse denier, *b*.

discharge, the physical strength of the filter media is considered sufficient. Should holes or tears be evident here, it is then necessary to ascertain whether these voids occurred in the filter bed (damaging) or at the discharge (inconsequential unless hindering the discharge mechanism). This may be determined by observing the condition of the media after it emerges from the filter bed on the discharge ramp. Another indicator in vacuum or pressure filters is the vacuum/pressure gage, which may be providing atypical low or zero readings if the media is torn/ruptured in the filter bed. In gravity or vacuum filters that index on fluid levels, atypical low levels and long index cycles could also signal such a problem. Filtrate clarity provides a final indicator.

Where the media tearing or rupturing occurs can also help troubleshoot the system to improve filtration capability. Plotting the location and distance between voids in the media can assist in determining the root cause of this problem for corrective action. For example, edge tearing (see Fig. 16-61*a*) may be caused by improper media alignment or tracking in the system, media sticking to the seals, or the media roll being too wide for the filter. The location and distance between voids serve to determine if this problem occurs systematically or randomly. Systematic tearing, (see Fig. 16-61*b*) represented by a similar, equidistant pattern, may indicate the cause to be misalignment or damage in the flight conveyor system, perforated plate, or seals, or something caught in contact with the media. With long index cycles, the filter media may dry and stick to the discharge ramp or seals, also resulting in systematic tearing. Randomly occurring voids (see Fig. 16-61*c*) generally indicate insufficient media durability, given that the vacuum or pressure break system is operating correctly.

There are different types of media being marketed into filtration applications that have exceptional strength—often a requirement of the primary non-filtration application for which the media was designed. By using such materials to "solve" a tearing problem, filtration efficiency may be compromised. Identifying and eliminating the root cause(s) of the tearing, rather than just addressing the symptom, will enable filter media selection to optimize filtrate clarity.

Efficiency may be determined by measuring filtrate clarity, filter cake thickness, and index cycle times. The evaluation of these variables can also be used to determine filter media *capacity* and *cost*. It is understood that solids and tramp oil in a coolant system are detrimental to effective coolant life, and hence manufacturing quality and productivity. To determine the quantity of these harmful contaminants in the coolant after it has passed through the filter, simple tests can be conducted to measure suspended solids and tramp oil, as detailed in Tables 16-8 and 16-9, and supported by Figs. 16-62 and 16-63, respectively. The review of soiled filter pads from suspended solids tests under a microscope with a scale can provide evidence of the size of particulates passing through the filter media.

To provide an indicator of filter media capacity and usage (a major cost factor), index cycle times, or the time intervals between when new media is introduced into the filter bed, are observed and recorded. The longer the average index cycle time, the less filter media is consumed. Longer cycle times reflect greater filter media capacity and contaminant retention, provided filtrate clarity also shows improvement. If clarity deteriorates while cycle time increases, it is a signal that the filter media is more open which may have detrimental effects at the cutting tool/work interface.

Increased filter cake thickness also implies improved filter media efficiency and capacity, and should likewise correspond with improved filtrate clarity test results and longer index cycle times. Filter cake thickness can be measured, but is usually a qualitative judgement due to the difficulty in measurement precision. The relative dryness of the filter cake is another qualitative indication of a well operating filter system.

The thickness of the filter cake can be useful for filter media comparisons on the same filter system. However, when observing different systems it is important to understand that the type of filter cake produced is a direct reflection of the quantity and particle size distribution of contaminants coming from the man-

CHAPTER 16

MACHINING COOLANTS

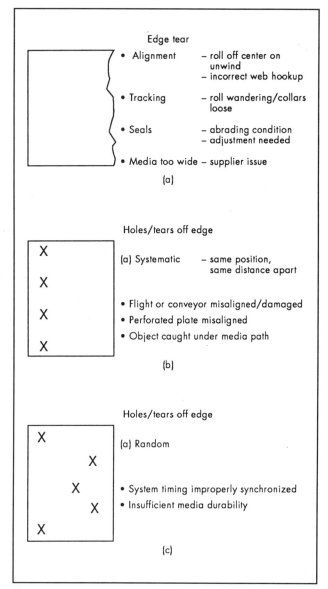

Fig. 16-61 Media tearing or rupturing: *a*, edge tearing; *b*, systematic tearing; and *c*, randomly occurring voids.

ufacturing process. Operations that create a fine, narrow distribution of particulates (for example, honing) or a relatively small quantity of contaminants (for example, parts washing), by their nature will not produce a thick filter cake, and therefore the filter media is the major determinant of filtrate clarity. In these operations, a heavier weight, tighter media such as 1.8 oz/yd^2 (61 g/m^2) or heavier is generally recommended. In operations creating a large quantity of contaminants with a wide particle size distribution (for example, rough grinding), a thick cake can be produced given that the initial filtration efficiency and flow characteristics of the filter media are adequate. In these operations, a lighter weight media such as 1.0 oz/yd^2 (34 g/m^2) to 1.5 oz/yd^2 (51 g/m^2) is generally more appropriate—a rule of economy being the selection of the lightest weight media with sufficient strength, acceptable initial efficiency, and good resistance to premature blinding.

COMPARATIVE STUDY SEQUENCE

When conducting a comparative study of filter media on an operating system(s), the following sequence is recommended:

1. Obtain samples and index cycle times with the existing filter media operating under a production load (equilibrium assumed) to establish baseline performance. A minimum of three to five filtrate samples should be drawn at different times, ideally at the same vacuum/pressure gage reading (record gage reading). It is important that the production equipment that the filter is supporting be operating steadily, and that the filter system is properly functioning (for example, appropriately indexing on vacuum rather than timer), prior to and during sampling. Filtrate samples should be taken at either the cutting tool/work interface, an appropriate plumbing juncture, or from the flow stream into the clean tank—assuring that all samples are taken from the same location. Index cycle times should be recorded using a stopwatch if observing, or by noting the difference in the index cycle counter (provided the filter is so equipped) or measuring the length of filter media used over set periods of time, preferably starting and ending with an observed system index.
2. Install the proposed alternative media roll. Index the splice through the filter bed. Allow sufficient time to pass to allow the system to approach or reach equilibrium—a minimum of one day. Do not record index cycle times or take samples during this time period, but record any noteworthy observations. It is not unusual, during this transition phase, for a more efficient filter media to result in shorter index cycles as it cleans the system.
3. Obtain samples and record index cycle times following the same methodology and under the same system conditions as in (1) above, with the proposed alternative media in operation.
4. Compare the results and determine the statistical significance of any differences.

STATISTICAL ANALYSIS OF EVALUATION RESULTS

Because time is often a constraint in evaluating filter media, assumptions are necessary which apply results from a short-term evaluation to long-term expectations for the filter system(s) studied and other similarly designed and applied systems in the manufacturing facility. It is therefore important to:

1. Determine if the results achieved in the short-term comparative evaluation are significant enough to justify a change in filter media.
2. Verify and sustain the expected improvements over the long term. Statistical methods can be appropriately applied for these purposes.

Using data collected from suspended solids testing of filtrate samples obtained when two different types of filter media were in operation, an operator may wish to determine with 95% confidence (or another acceptable level) whether the average clarity achieved by one filter media is significantly different from the average clarity achieved by another filter media. The use of the statistical *t*-test for this purpose is shown in Fig. 16-64. The use of the *t*-statistic for determining statistical significance can be applied to index cycle times or any other quantitative data collected while each respective filter media was in operation. This

CHAPTER 16

MACHINING COOLANTS

TABLE 16-8
Suspended Solids Test Procedure

Step	Procedure
1	Obtain a sample of the filtrate at the cutting tool/work interface or at another accessible location.
2	Pre-clean all lab apparatus that will contact the filtrate or lab filter pad.
3	Using forceps and a clean glass petri dish to avoid contamination, dry (15 minutes at 176° F, 80° C) and weigh a lab filter pad of the appropriate pore size (determined by clarity requirements) to the nearest 0.1 mg.
4	Mount the filter pad on the perforated base of the funnel unit and turn the vacuum on while assembling the reservoir. Setup is shown in Fig. 16-62.
5	Agitate the filtrate sample vigorously for at least one minute to ensure the suspension of all solids.
6	Pour the filtrate into a graduate. Use 25 ml increments. Empty the graduate slowly into the funnel reservoir. For comparative purposes, use the same sample volume, as large as will pass through the filter pad.
7	With the vacuum still on, use isopropanol to flush the filter assembly to assure the capture of all solids and remove any leftover oils from the filter pad.
8	Shut off the vacuum and very carefully remove the filter pad from the funnel assembly. Use forceps to prevent human touch. Place the filter pad on a clean petri dish.
9	Dry the filter pad for 15 minutes at 176° F (80° C).
10	Reweigh the filter pad to the nearest 0.1 mg.
11	Calculate the difference between the pre- and post-weighed filter pad. Use the appropriate multiplication factor to calculate the amount of sediment (mg/l) in the filtrate.

TABLE 16-9
Tramp Oil Test Procedure

Step	Procedure
1	Obtain a sample of the filtrate at the cutting tool/work interface or at another accessible location.
2	Agitate the sample vigorously for approximately one minute.
3	Fill the centrifuge tube with 100 ml of the filtrate.
4	Cushion and balance the tube in the centrifuge carrier.
5	Spin the sample for 30 minutes at 3000 rpm.
6	Allow the centrifuge to stop without braking. The resulting contents of the tube should be split into distinct layers (see Fig. 16-63).
7	Read the volume of free (tramp) oil on the surface of the sample and calculate the percentage by sample volume.

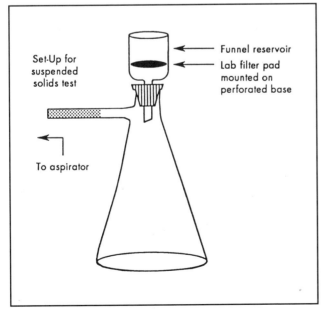

Fig. 16-62 Setup for suspended solids test.

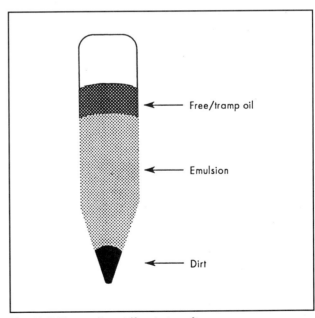

Fig. 16-63 Tramp oil centrifuge test results.

CHAPTER 16

MACHINING COOLANTS

use of statistics provides quantitative support for the selection decision, based on a finite time period for evaluation.

Whenever collecting numerical data, it is advisable to plot this information in time-ordered sequence on graph paper, as opposed to tabular listings, to more readily highlight any apparent trends or abnormalities. An example can be seen from the plotting of index cycle times shown in Fig. 16-65. Once a sufficient number of data points are collected (for example, 20 points), an initial assessment can be made as to whether the filter system is operating in a state of statistical control.

If a process is in statistical control, its variation is stable and its output is predictable. The resulting control chart then becomes a tool for ongoing monitoring, troubleshooting, and controlling the filtration process.

Statistical control is fundamental to continuous improvement. The control chart can be used to verify whether a change to the process, such as the use of an alternative filter media, has had an effect on the operation, and to characterize that effect. This is shown graphically in Fig. 16-66 (example shown is a typical pattern for index cycle time improvement). If the filtration process is not in statistical control, assignable causes need to be identified and eliminated. In the interim, while statistical control is not evident, the maintenance of a run chart for recording routinely accumulated data, such as filtrate clarity checks, will provide a better representation of changes in process operation than data compiled in a tabular format.

One noteworthy cause of misleading data or out-of-control signals can be the measurement system used. The measurement system must be able to provide adequate discrimination so that variation is properly reflected in the recorded data. As a general rule, any measurement device must be capable of measuring to one decimal point beyond any published specification (for example, if the desired measure is in mg, the measurement instrument should read to 0.1 mg). Obvious rounding of numerical data, such as is evident where all measures end in zero, is to be avoided. Inadequate measurement discrimination is also indicated when there are less than four possible outcomes within the control limits or a disproportionate frequency of zero readings in the Range chart. The usefulness of any data collected is dependent on the accuracy, repeatability, and reproducibility of the measurement system used.

An executive summary of filter media evaluation results is often desirable, such as represented in Tables 16-10 and 16-11. These results were from actual on-site evaluations. Table 16-11 also shows the cost benefit of one filter media over another, despite a higher unit price.

SUMMARY

The selection of filter media is a primary determinant of the operating efficiency and cost of the filtration system(s) and the resulting quality and productivity of the manufacturing operation(s) serviced by the filter(s). Metalworking coolant filtration, which by its nature is a chemical system in a mechanical world, may not be receiving the proper attention necessary to optimize results. Filter media may be selected based on subjective reasoning, such as confidence in or the salesmanship of the supplying source, rather than based on quantitative and statistically significant data. As a result of a lack of understanding and/or engineering time constraints, the "if it ain't broke don't fix it" syndrome may be endemic to an organization, which runs contrary to the philosophy of continual process improvement and world-class manufacturing competitiveness.

Filter media selection based on quantitative evaluation can be readily performed using simply understood concepts and procedures, as this section has described. In summary, because so much routinely depends on the filter media used in metalworking operations, it is recommended that the responsible engineer:

1. Understand the physical nature of the filter media: manufacturing process, fiber composition, typical physical properties data.
2. Know about the filter media manufacturer: company experience with this application and long-term commitment to supply, technical expertise and innovation, quality system.
3. Utilize the filter media supplier as an extension of engineering resources as appropriate to any evaluation or troubleshooting process.
4. Quantitatively evaluate to determine the optimal filter media for the operation: filtrate clarity (solids, tramp oil), media usage (index cycle times), maintenance (for example, tool wear), part quality (for example, dimensional conformance, surface finish).
5. Visually observe: media condition at the filter discharge, filter cake thickness/dryness, vacuum/pressure gage readings.
6. Use statistical methods to determine the significance of differences in the data collected as a basis for decision on filter media selection. Use control charts/SPC to confirm over time the improvements expected based on the short-term evaluation and subsequent selection of an alternative filter media. This data also becomes the baseline for future evaluations as part of the continual improvement process.

SLUDGE EVACUATION TO ELIMINATE COOLANT WASTES

Protecting distribution nozzles, hoses, and lines which feed coolant to cutting or grinding operations is obviously critical to the efficiency of overall system operation. That and proper filtration, it is recognized, will help prevent related fouling and help increase tool life. Keeping process solids from recirculating (with coolant) onto production parts will also reduce parts damage and rejection. Proper filtration is beneficial here, too. Nonetheless, even filtered coolant does commonly become contaminated and still requires disposal and replacement...and this is one of today's greatest problems/expenses in coolant management. System downtime—for cleaning and repairs—causes lost productivity and increases the labor burden. And coolant disposal—an EPA-regulated process—is an environmentally-sensitive and economically-expensive issue.

Filtration, when properly applied, can successfully "extend" coolant life by removing process solids from the recirculated fluid. Typically, however, most techniques do not address the total coolant environment; therefore, most filtration does not overcome all of the important elements and processes involved that directly create contamination and ultimately cause the need for coolant disposal/replacement. The key concern, it is now believed, is the overall accumulation of contaminants in any given coolant system. And this accumulation comes from a variety of sources and processes.

Process Solids

Created by the actual machining process, these grindings and fines are flushed from working machine tools and accumulate in the low-velocity areas of a system. In fact, many systems encourage such accumulation by utilizing weirs, baffles and

CHAPTER 16
MACHINING COOLANTS

Filter media 1	Filter media 2
50.8	35.1
68.7	30.4
37.2	37.5
48.2	33.6
57.1	26.3
$\bar{X}_1 = 52.40$	$\bar{X}_2 = 32.58$
$s_1 = 11.61$	$s_2 = 4.35$

$H_0 : \mu_1 \leq \mu_2$

$\alpha = 0.05$

$$t = \frac{\bar{X}_1 - \bar{X}_2}{S_p \sqrt{1/n_1 = 1/n_2}} \quad \text{where}$$

$$S_p = \sqrt{\frac{(n_1 - 1)s_1^2 + (n_2 - 1)s_2^2}{n_1 + n_2 - 2}}$$

Reject H_0 if $t > 1.86$

$t = 3.58$

H0 =	hypothesis
μ =	population average
α =	probability of rejecting if true
n =	sample size
$\bar{X}$ =	sample average
s =	sample standard deviation

We are 95% confident that the suspended solids in the filtrate through filter media 1 is not less than or equal to that through filter media 2 (translation: filter media 2 provides improved filtrate clarity).

Also using the t statistic:

$$CI_{0.95} = (\bar{X}_1 - \bar{X}_2) \pm t\,(s_{\bar{X}_1 - \bar{X}_2})$$

$$\text{where } (s_{\bar{X}_1 - \bar{X}_2}) = \sqrt{\frac{s_1^2}{n_1} + \frac{s_2^2}{n_2}}$$

$$= (7.03, 32.61)$$

CI = confidence interval

We are 95% confident that the true difference in average suspended solids between filtrate from filter media 1 and filter media 2 is between 7.03 mg/l and 32.61 mg/l.

Fig. 16-64 Statistical analysis of suspended solids in filtrate (in mg/l).

CHAPTER 16

MACHINING COOLANTS

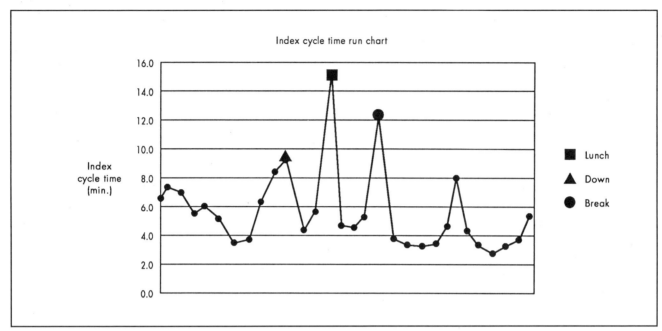

Fig. 16-65 Index cycle time run chart.

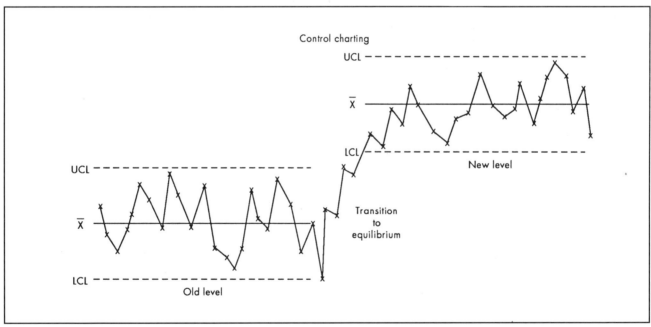

Fig. 16-66 Control charting.

TABLE 16-10
Coolant Filtration

	Test #1 Spunbonded	Test #2 Wet Laid
Filter System: Hydraulic vacuum		
Filtrate: Water soluble oil		
Operation: Cast iron grinding		
Duration:	48 hours	48 hours
Index Cycle Time:	3.8 minutes	21.4 minutes
Solids:	48.2 mg/l	33.6 mg/l

CHAPTER 16

MACHINING COOLANTS

TABLE 16-11
Coolant Filtration

Filter System: Hydraulic vacuum
Filtrate: Water soluble oil
Operation: Steel grinding

	Test #1 High Cellulose Blend	Test #2 Low Cellulose Blend
Duration:	10 days	7 days
Daily Usage:	100 yards	17.7 yards
Daily Cost:	$24.02	$7.16
Average Contamination Level	4.9 mg/l	0.4 mg/l

settling tank concepts to retard the recirculation of these solids. In addition, applied techniques for dealing with the accumulation of solids/sludge (that is, conveyor-equipped tanks and/or drag-outs) often achieve only limited success and therefore experience a residual accumulation of the finer solids.

Tramp Oils

Introduced to the system via the actual working parts or by way of leaks from system hydraulic/oil lines, this fluid can form a surface "blanket" which retains heat and actually retards cooling of the coolant solution.

Precipitated Cations

As the coolant solution experiences changes in temperature and velocity, the water in coolant precipitates calcium and mineral deposits. Also contributing to this process is the exposure to oxygen and the evaporation of water, thereby concentrating the precipitants, which ultimately become grit-like solids that also accumulate in the system.

In combination with the heat generated by a machine tool process, this accumulation of solids in a damp, dormant setting provides the environment for coolant's worst enemy: Bacteria/fungus. Even if techniques are chosen to periodically evacuate such accumulated solids, the retention time nonetheless breeds rapid bacterial growth. And it is this concern that reveals the most effective solution, identified simply as a concept of combined circulation, filtration and evacuation. These three processes, it is now believed, can be employed in concert, working continuously, simultaneously and without ever interrupting system operation or requiring downtime for routine maintenance.

Circulation

Unlike so many traditional systems, this process requires that contaminant accumulation be prevented, rather than encouraged and cleaned-up later. The process calls for keeping contaminants suspended in the recirculating coolant flow until it can be filtered. This is means evacuating coolant immediately after use from each individual machine to a central retention tank for all applicable machines, keeping coolant transfer flumes at a velocity to prevent sedimentation and applying directed turbulence to the coolant retention tank in order to influence all contaminants into the filters. This is directed turbulence is created by employing a manifold of eductors, fed by a side-stream of the pumps (which are also used to recirculate the coolant to each individual machine, as needed). The educators, operating with a "venturi-like" concept, feature the ability to generate four times the flow rate that is pumped to them. This is not only enables a smaller system pump to be utilized, but, more importantly, provides the increased flow necessary to continuously influence the otherwise-settling contaminant particles toward the pump intake. This sweeping action is further enhanced by employing the educators in such a pattern to prevent settling and accumulation of the solids. Each eductor system pattern is typically unique, adapting to the shape, size and configuration of the tank or pit chosen to accommodate the coolant from all operating machines. As a general rule, the tank and eductor pattern is optimally designed so that dirty coolant is fed into the tank nearest to the pump intake.

This coolant turbulence, it should be noted, continuously mixes the tramp oils with the coolant in the system to prevent any "blanketing" of the coolant reservoir...and actually encourages tramp oil evacuation from the system as the filtered/separated solid contaminants absorb these oils. Additionally, the overall filtration system includes, as a final stage, an oil-absorbing filter installed side-stream on the solids separator's outlet. This technique capably removes tramp oils to prevent any troublesome concentration. Circulation of the coolant through the filtration system for periods of time in excess of actual production time is highly recommended to enhance this feature.

Filtration

This is the fundamental building block of the entire system. The technique must effectively filter the solids and immediately evacuate the sludge from the recirculating system. Barrier-type filters are not appropriate because they capture the contaminant solids "in-stream," thereby accumulating the solids in such a manner that the system's circulating coolant is constantly—and undesirably—flowing through this accumulation. Unlike any other alternative, the unique performance of one particular liquid-solids separator has proven ideally suited for this purpose. Utilizing centrifugal action to remove the solids matter, one particular separator (LAKOS Separator) immediately removes solids from the process flow and bleeds those separated solids to an isolated collection vessel for convenient handling/disposal purposes. Once separated in this manner, the solids are no longer in contact with the recirculating coolant...and therefore, the solids are incapable of "re-contaminating" the coolant. No other technique affords the simple operation (no screens, cones or filter elements to clean or replace), the surety of non-mechanical failure (no moving parts to wear out) and such concentration of solids without excessive fluid loss.

Evacuation

A function of the overall system, this process of continuously evacuating the solids first from the individual machines, secondly from the coolant retention tank and then immediately from the separators—prevents the sludge from accumulating within the recirculating coolant and subsequently breeding harmful bacteria/fungi.

CHAPTER 16

MACHINING COOLANTS

Results of applying this concept to actual systems have been extremely successful. In a typical example, one machine tool operation with five centerless grinding machines experienced these specific benefits:

- Downtime, which averaged fifty hours per month for system cleaning, has been reduced to zero.
- Lost productivity, once overcome only by expensive overtime operation, has been eliminated.
- Disposal costs, averaging $350 per month for proper handling, have been completely eliminated. In fact, solids are now conveniently accumulated in a 55-gallon (208 l) drum and hauled away for salvage value (average: one drum per week).
- Coolant retention volume has been reduced from a total capacity of 800 gallons (3028 l) to only 320 gallons (1211 l)—and the coolant's operating temperature range of 85° to 95° F (29° to 35° C) performs with no loss of efficiency.
- Coolant make-up requirements have been reduced from 50 gallons (189 l) per month to only 20 to 30 gallons (76 to 114 l) per month.
- Wheel dressings, which once required routine downtime, have been unnecessary for periods of one year or more.

It is also noteworthy that this concept contributes dramatically to a much-improved work environment. Machines are cleaner. Routine downtime and messy cleaning procedures are unnecessary. Operators experience fewer skin rashes and health ailments as a result of the cleaner (bacteria-reduced) coolant. And operator morale is noticeably better, leading to increased productivity.

With no evidence of sludge accumulation or harmful bacteria/fungi, it is conceivable that a properly monitored system with this "circulation-filtration-evacuation" concept could promote virtually indefinite coolant life.

ON SITE CLEAN UP OF SMALL VOLUMES OF OILY WASTES

In the 1980s, technology was developed to treat small volumes of wastewater such as parts washer, machine coolant, vibratory and mop/floor scrubber water. The volumes involved range from as low as 20 gallons (76 l) a day in a small machine shop up to 2000 gallons (7600 l) a day in a larger shop.

With the development and implementation of pretreatment regulations, plant personnel have two choices, the waste liquid can be transported in bulk to an approved disposal site for treatment and disposal or the plant could decide to attempt treatment on site.

The cost to dispose of wastes by hauling it off in bulk is escalating. For a small waste generator, this cost can exceed $350.00 per 55 gallon (208 l) drum. In addition, the legal exposure of hauling and the declining options available as to where to take the waste, has made the treatment of waste at the point it is generated an attractive alternative.

Treatment Options

When evaluating on site, treatment options the following criteria should be used:

- What are the capital and operating costs?
- What maintenance is required?
- Can a permit be obtained for the process?
- Will the vendor of a proposed system guarantee it's performance and what is the performance guarantee?
- Can present plant personnel operate the proposed system?
- Can the residue from treatment be legally disposed of with no further treatment? If it can not, what additional treatment is required?
- What wastes can be treated?

Initially, for small generators, evaporation and ultra filtration were the systems of choice as those technologies offered solutions that were acceptable on the factory floor. Both required little or no operator involvement and in the case of evaporation, no sewer permits. As the environmental climate has become more stringent as to the landfilling of the sludge residue and air emissions from evaporation are questioned along with the maintenance and operating costs involved with these technologies, physical treatment methods have become much more common. This is not too surprising as larger flows of this type of wastewater are almost always handled on a physical/chemical basis.

The benefits of physical chemical treatment lie in the landfillability of the sludge residues, the lower capital and operating costs of the type of treatment, and the multiplicity of wastes that can be processed through it. The problem in the past was in the controls, equipment and chemicals required to do the job were greater than the skills available on the plant floor. The incorporation of prepackaged generic chemicals, programmable controllers, and simplified treatment procedures have made this option much more feasible and it is now within the existing skill level of the average shop.

An additional benefit is the limitation of liability that is incurred whenever the waste is reduced to a solid. Even if the waste is still hazardous, the reduced volume lessens the liability along with the cost of disposal.

Typical Wastes

Typical wastes in a factory that is forming, bending, machining, drilling, and finishing metal parts would consist of:

1. Machine coolant.
   - Soluble oils.
   - Semi-synthetic.
   - Synthetic.
2. Parts washer.
   - Pressure washer.
   - Caustic.
3. Vibratory.
4. Floor wash.
   - Mop water.
   - Floor scrubber.

This waste is usually collected in a central tank for convenient pick-up by the waste hauler. The ideal system would handle all of this combined waste and produce sewerable water and a dry landfillable "cake."

Waste Treatment

Physical/chemical treatment can be attained using simple or increasingly complex treatment schemes. The amount of cost savings or fail safe required will determine the level of automation, however, in all cases, a basic knowledge of waste treatment is required. This knowledge is best gained by learning how to "jar test." This test is performed on a sample of the wastewater to be treated and is executed as follows:

TO EXECUTE A JAR TEST:

1. Draw a small sample of the water to be treated and put 100 milliliters into a graduated cylinder.

CHAPTER 16
MACHINING COOLANTS

2. Add 20% coagulant one drop at a time, with mixing, shaking or stirring, and count the number of drops until coagulation begins. Twenty percent coagulant solution contains 0.7 oz (20 grams) of dry coagulant dissolved in 2.8 oz (80 grams) for water. One drop of 20% coagulant solution in 100 milliliters of wastewater is equivalent to 100 ppm or milligrams per liter or pounds per million pounds. Therefore, the number of drops required to initiate coagulation times 100 equals the parts per million required to treat the wastewater. The amount of coagulant required is directly proportional to the number of particles which must be removed from the water.
3. At this point, check the pH with pH paper to determine if the solution now lies in a favorable range which is 6.0 to 6.5.
4. If pH adjustment is necessary, it can be raised with hydrated lime or caustic soda and it can be lowered with the aluminum sulfate. Lime is best prepared at 5% by weight, four drops of this is equivalent to one drop of a 20% solution.
5. After pH correction (if step is required) and an embryo floc is present, add 0.2% solution of polymer (one lb. (0.45 kg) of anionic polymer in 60 gallons (227 l) of water), dropwise, with mixing, shaking and stirring, counting the drops until the embryo floc grows to a unit size which contains several embryo floc particles. One drop of 0.2% polymer solution in 3.38 oz (100 milliliters) of wastewater is equivalent to 1 ppm or 1 milligram per liter (mg 11) or one lb. per million lbs. of water.

The water phase of the wastewater sample should be clear after step 5 and the contaminants should all be in the floc. If this is not the case, repeat the sequence using more coagulant and polymer until a treatment schedule is found which is capable of clarifying the wastewater.

The test results from this procedure can then be incorporated into a "recipe" for treatment. For simple systems, an operator can be instructed to weigh out chemistry and put it into the water in the same sequence as the jar test. Then depending on the separation method selected, sedimentation or flotation, followed by filtration can be accomplished.

The sequence can be automated with a PLC or timers to eliminate human error. The degree or automation usually is dictated by economics. As the amount and cost of disposal of the raw waste increases, more money can be justified on capital expenditures.

After separation, the sludge residue needs to be processed to a landfillable solid. With sludges form a physical/chemical process, this can be easily accomplished by the following methods:

Less than 10 gallons (37.8 l) of sludge per day can be dried using an infrared dryer. Over 10 gallons (37.8 l) of sludge processing through a recessed plate filter press is recommended.

Process times are greatly accelerated and the energy cost to produce a "cake" in a press is much less than the cost of drying same. Recessed filter plate technology is recommended as it is the only technology proven to produce solids in excess of 40%. In many states, this is the minimum acceptable solids content. Figure 16-67 illustrates the amount of solids to be expected from the most popular filtering technologies.

Phenol Removal

If phenol levels in the wastewater are higher than allowable limits set by the sewer authority, the level can be reduced after primary treatment has been completed. The most common methods for phenol removal are:

Carbon adsorption; phenols are adsorbed on the surface of activated carbon granules. The advantage is the operator skill level required is low. The disadvantages are:

- The water must be clean otherwise the carbon will plug up.
- Difficult to predict break through (when the carbon is exhausted and must be replaced).
- The exhausted carbon may be a hazardous waste.

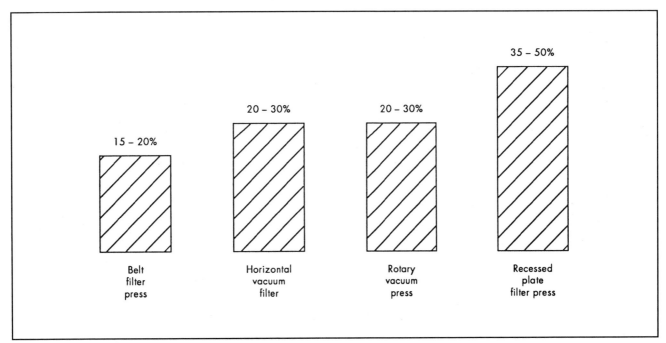

Fig. 16-67 Amount of solids to be expected from the most popular filtering technologies.

CHAPTER 16

MACHINING COOLANTS

- In any case, the exhausted carbon becomes another waste to be disposed of.

OXIDATION:

Phenols can be destroyed by oxidizing with potassium permanganate or hydrogen peroxide. Oxidation has the advantage of not producing a waste that has to be disposed of. However, a program of overtreatment must be initiated in order to assure consistently low phenol residuals. The disadvantages of chemical oxidation are:

- Chemicals used can be hazardous.
- Phenol residuals will be slightly higher after treatment vs. carbon adsorption.
- Requires higher operator skill.

Summary

The treatment of wastes at the point it is generated, on the factory floor, is becoming an attractive alternate to disposing off site. Plant management must have a working knowledge of what is possible to achieve at the plant level and approach treatment of wastes in the plant as another process right along machining, drilling, painting, etc., that must be planned on and executed.

References

1. A.D. Roberts, and S. C. Lapidge, *Manufacturing Processes* (New York: McGraw-Hill Book Company, 1977), pp. 58-104.
2. A. Bhattacharya, *Metal Cutting—Theory and Practice* (Calcutta, India: Central Book Publishers, 1984), p. 29.
3. V. Arshinov and G. Alekseev, *Metal Cutting Theory and Cutting Tool Design* (Moscow: Mir Publishers, 1976), pp. 119-121.
4. E.J.A. Armarego and R.H. Brown, *The Machining of Metals* (Englewood Cliffs, New Jersey: Prentice-Hall, 1969), p. 292.
5. R.J. Hocken, *Technology of Machine Tools. Machine Tool Accuracy*, Volume 5, 1980, UCRL-52960-5, Lawrence Livermore Laboratory, University of California, Livermore, CA.
6. Edward G. Hoffman, ed., *Fundamentals of Tool Design* (Dearborn, MI: Society of Manufacturing Engineers, 1984), p. 142.
7. T.C. Chang, "Manufacturing Process Planning," *Handbook of Industrial Engineering*, 2nd edition (New York: John Wiley and Sons, Inc., 1991), pp. 587-611.
8. R. Komanduri and R.H. Brown, "On the Mechanics of Chip Segmentation in Machining," *Journal of Engineering for Industry*, Volume 103 (February 1981), pp. 33-51.
9. Y.C. Shin, H. Chin, and M.J. Brink, "Characterization of CNC Machining Centers," *Journal of Manufacturing Systems*, Volume 10, Number 5, 1991, pp. 407-42-291.
10. J. Raja and D.J. Whitehouse, "Field Testing of Machine Tool Diagnostic Techniques Using Surface Metrology," *Annals of the CIRP*, Volume 32, Number 1, 1983, pp. 503-506.
11. J. Raja and D.J. Whitehouse, "An Investigation Into the Possibility of Using Surface Profiles for Machine Tool Surveillance," *International Journal of Production Research*, Volume 22, Number 3, 1984, pp. 453-466.
12. Jean V. Owen, "Moving to Modular," *Manufacturing Engineering* (February 1992), p. 51-55.
13. Jean V. Owen, "The Chuck Challenge," *Manufacturing Engineering* (October 1992), pp. 52-57.
14. W.E. Conway, "The Quality of Work can be Improved with Industrial Engineering," *Industrial Engineering* (April 1992), pp. 34-36.
15. George F. Raymond, "Charting Procedure," *Maynard's Industrial Engineering Handbook*, 4th Edition (New York: McGraw-Hill, Inc., 1992), pp. 3.3-3.22.
16. D.D. Bedworth and J.E. Bailey, *Integrated Production Control Systems* (New York: John Wiley and Sons, 1982), pp. 42-43.
17. D.C. Montgomery, *Introduction to Statistical Quality Control*, (New York: John Wiley and Sons, 1985), p. 102-118.
18. G. Nadler, *Work Design* (Richard D. Little, Inc., 1963), pp. 385-617.
19. J.L. Burbidge, *The Introduction of Group Technology* (New York: John Wiley and Sons, 1975).
20. M.P. Groover and E.W. Zimmers, *CAD/CAM: Computer-Aided Design and Manufacturing* (Englewood Cliffs, New Jersey: Prentice-Hall, 1984).
21. In CVD, the tools to be coated are placed in a gas chamber heated from 1200° to 1800° F (650° to 980° C). At high operating temperatures, the gases dissolve to form new compounds which deposit themselves on the tooling material.

    In CVD, the tools to be coated are placed in a vacuum chamber. The coating material is evaporated and ionized by several arcs to react with a gas such as nitrogen and thus creating a compound, TiN for instance. The reaction and deposition takes place at 400° to 800° F (200° to 427° C).
22. ISO standardizes tolerances as per capital alphabet letters for bores and small letters for mating shafts in accordance with respective diameters. The letters represent quality grades by numerical tolerance ranges. For example: Diameter 24 mm H7 signifies a tolerance of 827 μin. (21 μm) per definition. The lower the lower the numbered letter, the closer the tolerance band. H8 signifies a wider tolerance than H7, H6 a closer tolerance.
23. Sandvic Coromat
24. Fabrik 2000 Nr. 2, May 1991
25. John Agapiou, "Full Speed Ahead," *Cutting Tool Engineering* (April 1991), pp. 52-59.

Bibliography

Dixon, Wilfrid J. and Frank J. Massey, Jr. *Introduction to Statistical Analysis*. Fourth Edition. New York: McGraw-Hill Book Company, 1983.

Johnson, Peter R. "The Micron Rating of a Filter Medium: A Discussion of the Performance of Filter Media." *Fluid/Particle Separation Journal*. Vol. 2, No. 3 (September 1989), pp 157-161.

Joseph, James J. *Coolant Filtration*. E. Syracuse, NY: Joseph Marketing Inc., 1985.

Martin, John and Bryan Thomas. "Wet-Laid Technology in Nonwoven Filtration." *Filtration News* (September/October 1989), pp 32-33.

Traver, Robert W. "Measuring Equipment Repeatability—The Rubber Ruler." *Annual Convention Transactions*. Milwaukee: American Society for Quality Control Inc., 1962.

Wheeler, Donald J. and Richard W. Lyday. *Evaluating the Measurement Process*. Knoxville: Statistical Process Controls Inc., 1984.

CHAPTER 17

FORMING

CONTINUOUS IMPROVEMENT IN THE METAL STAMPING INDUSTRY

CHAPTER CONTENTS:

EQUIPMENT SELECTION 17-1

BLANKING OPERATIONS 17-1

LUBRICATION 17-2

QUICK DIE CHANGE 17-3

AUTOMATION 17-4

PRESS, DIE AND OPERATOR PROTECTION 17-7

MISCELLANEOUS 17-8

A flip answer to the question of how to improve stamping operations is to figure out how it should have been done originally, and then make it so. However, the initial cost of tooling and equipment for stamping operations is very high compared to other metalworking processes; mid-run remodeling must be able to be cost justified.

TQM and continuous improvement concepts certainly apply to the stamping industry. Quality of part and process go hand-in-hand, and quality business management is especially important when dealing with very expensive capital equipment. Training cannot be underestimated, and concern for safety must be the number one goal.

EQUIPMENT SELECTION

Selecting the proper type of press is just as important as choosing the correct tooling. Manufacturers have many choices: mechanical or hydraulic presses, gap- or straight-sided frame construction, a myriad of strokes and speeds, etc. In some instances, traditional stamping tooling and equipment is not the best option. Sometimes, part design, production rate, and production volume suggest nontraditional alternatives such as a press brake or turret punch and laser production cell, rotary or other slide machines, or such processes as hydroforming and Dualform.®

Despite new innovations, most stampings continue to be run on hydraulic and mechanical presses. Mechanical presses tend to be best at:

- Blanking parts at high speed (above 100 spm).
- Blanking parts automatically fed with a short feed length.
- Shallow drawing.
- Work requiring an easily controlled depth or stroke.

Hydraulic presses tend to be best at:

- Deep drawing.
- Short run work with frequent die changes.
- Blanking with form, coin, or other secondary operations in a single stroke.
- Lower speed, high tonnage blanking with long feed lengths.
- Work requiring a repeatable pressure rather than repeatable depth or stroke.[1]

When selecting a press for a job, whether existing or new, consider the following:

- Bed size.
- Tonnage—required versus available.
- Frame construction—gap (easy access from three sides) versus straight side. (more rigidity to support unbalanced loads).
- Action—stroke(s) and speeds required at various points along the stroke.
- Accessories—for example, quick die change, equipment, cushions, etc.

BLANKING OPERATIONS

Blanks can be cut from strip or coil stock in-house, or purchased from blanking companies or material suppliers. To be profitable cutting blanks in-house, several conditions must be met before starting:

- New equipment must be considered.
- Production orders must be sufficient to keep the equipment running at full capacity.
- Space must be available for a blanking press, a feed line, a cradle, a leveler, blank handling, and scrap handling.
- Space must be available for storage of coils ahead of the blanking line.
- Means must be provided for handling and disposing of generated scrap.
- Capital must be available to purchase and install the equipment required.[2]

Other factors, such as WIP inventory storage space, and overhead cranes and other material handling equipment may also have to be considered.

Purchasing blanks lends itself to a JIT operation. A good working relationship must be developed with the supplier, since late deliveries and unusable material cannot be tolerated. Proximity of the customer to the supplier is a key consideration.

It may be possible, after carefully considering all of the options, that a company presently cutting its own blanks would be better off purchasing them, and vice versa. A mix of purchased stock and in-house processed material may be the most cost-effective alternative.

Strict attention must be paid to the size of blanks. Blanks should be as small as possible while allowing

---

*The Contributor of this chapter is: **James A. Rumpf**, Assistant Professor, Manufacturing Engineering Technology, Ferris State University.*
*The Reviewer of this chapter is: **David Alkire Smith**, President, Smith and Associates, Monroe, MI.*

CHAPTER 17

CONTINUOUS IMPROVEMENT IN THE METAL STAMPING INDUSTRY

enough material to form the finished part. Unless it is a net size blank, some material, such as in the binder area and cutouts, generally gets trimmed away after drawing or forming operations. If segregated properly, the "trimmings" can be sold back as scrap, but it is much better not to pay for excess material at all. Another possibility is to use larger cutouts, or offal, as the stock for other, smaller parts.

The purpose of nesting blanks in a strip or coil layout is to utilize the highest possible percentage of the sheet area. Figure 17-1 shows some comparisons on a deep draw job, such as an oil filter canister, where some additional material from the edge of the blank is trimmed after forming.

The main goal is to use only as much binder or hold-down stock as needed, and only where it is needed. Different, changing combinations of blank layout, coil width, and scrap allowances make finding the one best way a continuous effort.

LUBRICATION

Drawing lubricants, sometimes called drawing compounds or "dope," are an important aspect of metal stamping, but they are not magic elixirs that solve problems created by second-rate die design or poor material. They are part of the indirect cost attributed to production, and properly applied, they reduce the friction between the tools and the stock, increasing productivity via better flow of material in the die and longer tool life.

Lubricants are actually combinations of ingredients. The major ingredient is usually oil, water, or a solvent-based carrier. To the carrier are added wetting agents, such as animal fats, fatty acids, emulsifiers, etc., and, optionally, extreme pressure agents consisting of chlorine, sulfur, and phosphorus.[3]

The proper use of drawing lubricants requires considering four main factors: tooling, stock material properties, lubricant properties, application methods, and secondary operations. For any particular type of material and die design combination, there is one best type compound to use.[4] Key points to consider for each factor:

- Tooling. Leading edges of the dies must not scrape compound off the incoming parts. Recirculating systems must be kept free of contaminants, and original mix and makeup water must not produce any unwanted chemical reactions such as the formation of acids. Auxiliary lubrication systems may be required, to adequately protect normally inaccessible areas of the die or to reapply lubricant for subsequent operations. Die materials such as carbide, chrome, urethane, Teflon,® etc., will each have individual reactions with various types of compounds.

- Stock material properties. Surface properties of the material can be divided into five categories: normal, active, inactive, and coated. *Normal* surfaces readily retain lubricant and require no special wetting agent. Cold rolled steel, hot rolled steel, and aluminum-killed steel all fall into this category. *Active* surfaces encourage chemical reactions which can lead to problems such as white rust, staining, etching, etc., under certain conditions. Brass, copper, terneplate, zincrometal, and tin plate fall into this category. Inactive surfaces such as nickel, aluminum, and stainless steel have low likelihood of chemical reaction, but require high film agents. *Coated* (nonmetallically) surfaces demand special care, as the lubricant must not cause any peeling, blistering, discoloration, damage, etc.

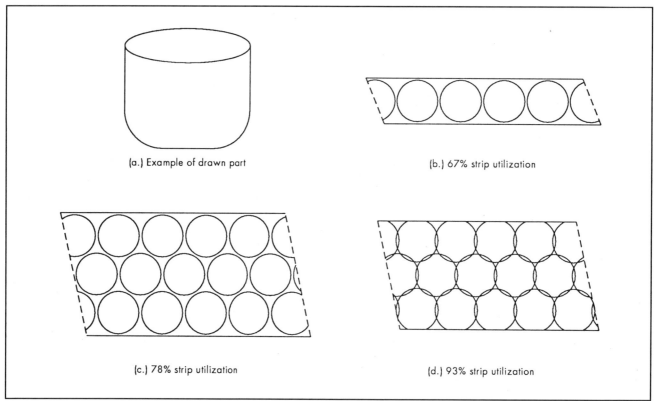

Fig. 17-1 Strip utilization.

CHAPTER 17
CONTINUOUS IMPROVEMENT IN THE METAL STAMPING INDUSTRY

- Lubricant properties. Lubricants fall into four basic categories: petroleum-based, water-soluble, solvent-based, and synthetic soluble. *Petroleum-based*, or oil-based, are best used where high load characteristics exist. *Water-soluble* lubricants can attain the overall performance of oil-based lubricants under certain conditions at a fraction of the cost, while conserving oil. *Solvent-based* lubricants can be used with different types of nonmetallic coatings and can eliminate the need for secondary cleaning operations. *Synthetic soluble* lubricants are used where extra high pressure conditions exist, and are also very clean.
- Application methods. Many methods are used to apply the lubricant, such as manually sponging it on, dripping it on the panels, rolling it on, and using an airless spray. The most important point is to ensure that just enough lubricant is applied and only where it is needed. Many times it is not necessary to coat the entire panel.
- Secondary operations. Very few stamped parts are single-piece final products; most often they must be assembled, welded, or otherwise attached to other parts. Lubricant residue can severely inhibit thermal joining processes, and certainly must be removed for painting and coating operations. Cleaning and/or degreasing may be the only, albeit expensive, alternative.
- Special considerations. Address disposal and safety issues before using any product. Different areas of the country have different regulations on the subjects of toxicity in the workplace, contaminated waste, etc.

The three main points to remember with lubricants are to (a) use just enough of (b) the proper compound in (c) just the areas that need it. This minimizes the expense of the lubricant itself, and the subsequent operations like cleaning, hauling, and disposing. With lubricants, the associated indirect costs, unchecked, can eat into profits.

QUICK DIE CHANGE

The cost of having expensive stamping equipment sit idle for any part of the workday can readily be equated to lost earnings and poor return on investment.[5] One key to continuous improvement in the stamping industry is increasing the portion of the workday the equipment is making parts by decreasing idle or downtime.

In many cases, changing over equipment from one part to another is the major component of pressroom downtime. Reducing the "hit-to-hit" time on a stamping press via quick die change has the obvious advantage of enabling JIT production, but also generates more press capacity, granting the opportunity to increase the production volume of existing parts or to take in new work.

Depending on accounting practices at individual facilities, different formulas for estimating payback can be used when justifying the expense of a quick die change program. If capacity is needed to increase volume of existing parts, use the formula:

$$\text{VIP} = \text{SP} \times \text{PR} \times \text{CPW} \times \text{PWPY} \times \text{RCOT} \quad (1)$$

where:

- VIP = value of increased production
- SP = selling price of the parts (average)
- PR = production rate in parts per minute
- CPW = changeovers per week
- PWPY = production weeks per year
- RCOT = reduction in changeover time (minutes)

Example: The average selling price of a part run over a line is $2.00; parts run at 8 parts per minute, there are 3 die changes per week, and production runs 50 weeks per year. Reducing the changeover time from 60 to 30 minutes is worth $2.00/part × 8 parts/minute × 3 changeovers/week × 50 production weeks/year × (60 minutes/changeover − 30 minutes/changeover) = $72,000 per year. Therefore, using a typical two-year simple payback, $144,000 can be expended to implement a quick die change program under these conditions.

If production volumes are fixed, excess capacity can be used to bring additional work into the line. The formula for to determining the value of the machine time made available is:

$$\text{VMT} = \frac{\text{MC} \times \text{CPW} \times \text{PWPY} \times \text{RCOT}}{60} \quad (2)$$

where:

- VMT = value of machine time
- MC = machine cost in dollars per hour
- CPW = changeovers per week
- PWPY = production weeks per year
- RCOT = reduction in changeover time in minutes
- 60 = the constant to convert from minutes to hours

Example: If the hourly machine cost for a line of four presses is $300.00/hour, there are 2 changeovers per week, 50 production weeks per year, and the changeover time can be reduced from 4 hours to 2 hours, then that machine time freed from changeover for production is worth ($300.00/hour × 2 changeovers/week × 50 production weeks/year × (240 minutes − 120 minutes)) / 60 = $60,000 per year. Therefore, using a typical two-year simple payback, $120,000 can be expended to implement a quick die change program under these conditions.

Increased productivity and production capacity are just two of the benefits of a quick die change program. The following benefits can also be used to help justify it:

- Improved Quality. Shorter runs allow earlier detection of defective parts. Standardization of procedures will improve quality by reducing mistakes.
- Reduced Setup Times. Reduces man-hour requirements during nonproductive time.
- Shorter Lead Times. Increased response to changes in production schedules due to short runs.
- Reduced Inventory Levels. Shorter lead times result in smaller finished goods inventories. Shorter production runs result in lower in-process inventories.
- Improved Safety. Consistent clamping forces, automated safety interlocks, automated material handling and standardized setups can increase safety in the workplace.[6] Every effort should be made to quantify these additional benefits to help the quick die change proposal through the approval process.

Traditionally, press line changeover has required coordinating the actions of a variety of tradesmen: die setters, material handlers (truck drivers and/or crane operators), die tryout, mechanical devices ("gadgets"), electricians, pipefitters, millwrights, inspectors, and production workers all may have a role to play. Increasingly, though, companies that approach or achieve the goal of single minute exchange of dies (SMED) employ workers with multidisciplinary responsibilities. These "production technicians" are trained and certified in some or all of the following skills: press operation, safety inspection, automation setup, quality check, metal finish, and die change.[7] Regardless of the actual trades involved, except for the smallest of dies, changeover is a job requiring teamwork by two or more people.

CHAPTER 17

CONTINUOUS IMPROVEMENT IN THE METAL STAMPING INDUSTRY

The first step in preparing for quick die change is taking inventory of all of the tasks required to changeover the press line, and categorizing those tasks as "internal" or "external" to the die set. Internal die set tasks are those elements of the procedure that must be performed while production is halted (for example, unclamping the die, adjusting shut height, etc.). External die set tasks can and should be performed before production stops (for example, positioning new dies and stock in the area).[8] To improve die set efficiency, external tasks must be performed externally, and as many internal tasks as possible should be converted to external tasks. Areas to be addressed include:[9]

1. Die storage, retrieval, and maintenance. If dies are not stored adjacent to the press line, a transportation plan must be devised. A coding scheme to match die to press is suggested. When dies are rotated out of production, a strict maintenance program must be followed, covering cleaning and inspection for broken, cracked, missing, or worn components. Repair or replacement at this time will ensure that the tool is ready for the next production run. A system tracking the number of hits each die has absorbed can alert the tool room to be ready for a more thorough inspection and overhaul.
2. Material handling. Ensuring that the right material, without edge or surface damage, arrives in time for the first-panel hit can make the difference between a smooth transition and a demoralizing delay.
3. Standardization (die and parts). Standardizing die lengths, widths, and locating methods makes it easier, and therefore quicker, to situate dies in the proper location and clamp them down. Standard die heights minimize the need to adjust the shut height of the press. If the dies cannot be standardized during the design stage, plates can be added to take up the difference. For related families of parts replacing die inserts rather than the entire die saves time.
4. Operator training. For quick die change methods to be followed quickly, each operator must be familiar with the controls of the particular machine, the steps of the die change procedure, and the proper sequence in which to perform them. Operator training also helps achieve "buy-in;" workers are more likely to support efforts that they understand.
5. Frequency of changeovers to a different part. Even with relatively few die changes, a quick die change program can bring excellent payback. The more frequent the changeovers, the more return on investment (see the previous formulas). However, even infrequently changed dies can benefit from quick die change concepts; many of them require no capital investment at all, just a degree of pre-planning.
6. Press condition. Precisely locating dies in a press with excess clearance in the gibs is a waste of time and effort. Regular inspection and maintenance of the mechanical portions of the press ensures that normal mechanical wear will not detract from performance. Likewise, the electrical, hydraulic, and pneumatic systems should be monitored, and updated with programmable logic controllers (PLCs) and automatic lubrication systems to ensure operator and tool safety. Frequently, retrofitting older presses to bring them up to the latest safety and performance standards is more economical than purchasing new equipment.
7. Automation. On manually operated presses, when the current run is completed the operator can just walk out of the way of the changeover team. In an automated system, however, equipment must be moved out of the way to access the die area. The tradeoff must be evaluated: the benefit of the automation during the production run (higher production rate, less direct labor, etc.) versus the drawbacks during changeover (increased time required). Whenever possible, the automation should be connected with quick-disconnect plugs and couplings. Both floor- and press-mounted equipment should swing or move out of the way during the changeover. Modular gripper arm inserts, vacuum cup booms, and reprogrammable controls on flexible devices require less time than changing entire unload-transfer-load cells. Proper care must be taken not to introduce pinch points and other hazards between production machines and die setting equipment.
8. Production planning. Production schedulers traditionally try to keep dies in the line running as long as possible between changeovers, to minimize the lost production due to die setting. With quick die change they regain some of that time, and smaller batches become economical to produce. If a variety of parts and handling techniques are used on a line, they can further aid the quick die change effort by rotating jobs through equipment with a minimum of equipment change between them, using group technology concepts in scheduling the sequence of dies through a press. Example: running two parts that use the same destacker back-to-back, then two other parts that use some other piece of equipment.
9. Quick die change equipment. The major items to consider for a quick die change program are:

   - Rolling or moving bolsters. Rolling bolsters move the die out from under the ram for removal by crane or cart, allowing safer die repair work. Some roll out in only one direction (generally front or back), while others have dual stations (side to side) allowing one half of the plate to hold a working die in the press while the other half is outside the column and available for staging the die for the next run (see Fig. 17-2).
   - Die carts. Die carts can cover more area than overhead cranes, and may serve more than one purpose.
   - Automatic die clamps. Automatic clamps eliminate the time-consuming task of individually unbolting the old die and bolting up the new one.
   - Outriggers. Usually used on presses too small for rolling bolsters, these attach to the edge of the stationary bolster plate and allow the die to roll out of the die space for removal.

AUTOMATION

Automation in the metal stamping industry refers to any mechanical handling device that loads, unloads, or performs other operations subordinate but necessary to the actual stamping of parts. Automation is typically cost justified based on the reduction of direct labor, but in some cases, there is no other safe or practical way to do some things manually (for example, load/unload a die with front and back aerial cams, handle extremely bulky parts, etc.). In such cases, automation is not just one option, it is the only option. Automation misapplied, whether alternatives exist or not, can be both counterproductive and costly.

To maximize productivity, engineers must find the correct balance between speed and flexibility, or between "hard" (dedicated) and "soft" (flexible) automation. Figure 17-3 shows the

CHAPTER 17

CONTINUOUS IMPROVEMENT IN THE METAL STAMPING INDUSTRY

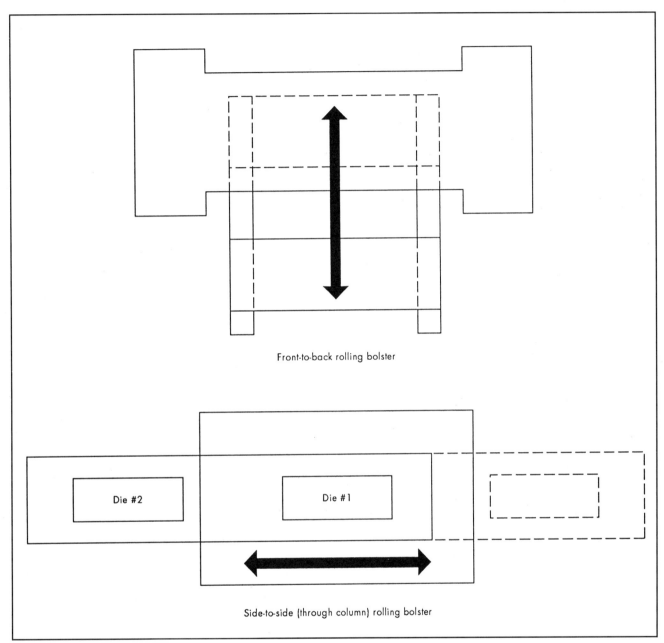

Fig. 17-2 Rolling bolster operations.

relationship between the three main factors guiding system selection: production rate, production volume, and product variance. A point representing evaluations of these factors is plotted, and the smaller the absolute distance between that point and the origin, the stronger flexible equipment is suggested. For each situation, the engineer responsible must determine what constitutes high and low, and how close is close enough to the origin. This figure shows that any one factor in the extreme can shade the decision toward dedicated, single-purpose equipment.

Primarily, automation is used to load and unload presses and transfer panels between presses. Besides special purpose machinery unique to a particular installation, major classifications of stamping automation include:

- Die assists—Simple devices such as lifters, kickers, ejectors, and air blasts, mounted in the die itself.
- Industrial robots—readily reprogrammable, multifunctional manipulators. Available in a variety of arm configurations, sizes, and carrying capacities, these devices are not limited to stamping plant applications and are the most flexible, hence reusable for other work.
- Robotic loaders and unloaders—special purpose units, usually press-mounted, that provide two-axis motion with a gripping function. These units have an electrical or hydraulic servo-drive, or adjustable, mechanical cams or linkages. They are reprogrammable in sequence and path within the limits of their component geometry.

CHAPTER 17

CONTINUOUS IMPROVEMENT IN THE METAL STAMPING INDUSTRY

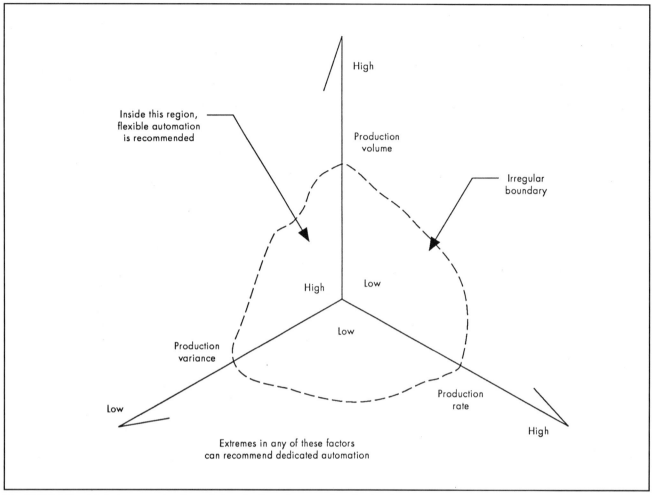

Fig. 17-3 Relationship between production rate, production volume, and product variance.

- Destackers—special purpose mechanisms designed to separate and transfer blanks from a stack, similar to dealing cards off the top of a deck.
- Shuttles—typically lift-and-carry or chain conveyor devices, primarily for the interpress transfer of parts, these sometimes can be used as loaders and unloaders as well. Lift-and-carry shuttles can be run through a die, enabling loading and unloading, if relief slots are provided for the shuttle rails. A subset of the shuttle, the walking beam, is the type of mechanism used in transfer presses and transfer dies for interdie transfer. These are different from progressive die mechanisms, in which parts are gradually stamped from an indexing strip that acts as a support web and carrier for the parts until they are separated at the last station.
- Extractors—sometimes called "iron hands," are simply press unloaders, usually pneumatic, that snatch a part off the die and drop it onto a table or conveyor.

Stamped parts must be consistently oriented on each die, making "finding" the part quite easy for the equipment. It is crucial that automation equipment not lose orientation of the part. For that reason, extractors are not compatible with other automation because of the extra expense of reorienting the part for the subsequent operation.

The following presents, approximately, a reasonable process of introducing automation in a metal stamping plant:[10]

First operations

1. Substituting calibrated coils for strips sheared from sheets.
2. Automatic or mechanized evacuation of product, slugs, and scrap.
3. Fully automatic feeding.
4. Elimination (at least drastic reduction) of secondary operations, by means of (a) progressive dies, (b) combination dies, (c) compound dies, (d) transfer dies, or (e) transfer presses.
5. Automatic counting, stacking, and storing of products.
6. Automatically checking the products, and avoiding rejects by stopping the press in case of trouble.

Secondary operations

1. Automatic evacuation of products and scrap (manual, individual, indirect feeding).
2. Manual indirect feeding.
3. Semiautomatic feeding.
4. Fully automatic feeding.
5. Automatic checking, stacking, and storing of products.

CHAPTER 17

CONTINUOUS IMPROVEMENT IN THE METAL STAMPING INDUSTRY

6. Automatically checking the products, and avoiding rejects by stopping the press in case of trouble.

Interconnected production line (additional operations)

1. Semiautomatic interoperational material handling.
2. Fully automatic interoperational material handling.

Some miscellaneous points of advice:

- Keep expensive robotic arms out from under press rams. Malfunctions do occur, and presses have been known to roll over while equipment is in the way. Only allow gripper or suction cup booms, the least expensive components to replace, to enter the die area.
- Purchase and use devices with as many modular components (grippers, cylinders, rotary actuators, etc.) as possible. This speeds replacement time and saves money by using "off the shelf" modules. Sometimes it is smarter to purchase an entire extra unit, which may be justified as a piece of training equipment, rather than to build up a spare parts inventory piece by piece.
- All equipment, whether press- or floor-mounted, should be easily transferred out of the way for die changes.
- Avoid magnetics if possible. Magnetized parts tend to pick up slivers that can enter dies and cause problems.
- Make good use of the area above the floor adjacent to presses. If not in the path of the overhead crane, controls, spare parts, workbenches, etc., can be put up on platforms, freeing valuable floorspace.
- Set speeds, and time all automation devices to just keep up with the slowest operation on the job—typically the initial draw operation. This press should be set to "hop," or run on continuous cycle. Slow, smooth operation will extend the life of automation, drive components, and the press clutch and brake assemblies.

PRESS, DIE AND OPERATOR PROTECTION

Protecting tooling and equipment from damage, and operators from injury, is paramount in the stamping business. Different types of sensors are used for different purposes.

Press Protection

Thermal devices are incorporated in electric motors to halt electrical flow under an electrical overload condition. They should not be relied upon to protect the presses they power. Tonnage monitors are electronic "watchdogs" that serve that function. Strain gages are mounted on critical components of a press (the inside of press columns has generally been found to be most effective) and wired to a controller that measures the voltage through the gages. The voltage readings from the gages are directly proportional to the strain they undergo, which in turn is directly related to the tonnage the press components absorb during the working part of the stroke. Tonnage monitoring can provide the following benefits:[11]

- Reduced setup time. Displaying the tonnage on the corners helps to balance the load and ensures uniform setup conditions time after time.
- Press and die overload protection. After load limits are set in the controller, it will immediately stop the press if those limits are exceeded. Slugs, misaligned parts, out-of-specification material, and many other reasons may trigger the halt command.
- Die underload protection. Underload limits can be set as well, and stop signals generated. Underloads may be caused by broken punches, misfeeds, improper shut height, etc.
- Increased tool life. Tonnage increase due to tool wear can alert maintenance personnel long before serious damage occurs.
- Preventive maintenance. Ram-to-bolster parallelism, angular deflection, and tie rod stresses can be checked while the press is running and may detect problems while they are still relatively minor.

Tonnage information is useful when timing shear operations for optimum performance. Once die timing is optimized to maximum tonnage required, presses can be run at maximum speed. By operating at maximum allowable cutting speed for the materials being worked, both quality and productivity is improved.[12]

Die Protection

Tonnage monitors can protect presses, but major damage can still occur to the press tooling before the ram comes to a complete stop. The use of a variety of types of sensors is the key to protecting dies from mishaps. The underlying philosophy is to use sensors to monitor everything that moves (such as material being fed in and extracted, slugs and scrap being shed, cams and cylinders extending and retracting, etc.) to make sure that it cycles and is positioned where and when it is supposed to. This involves tying in these discrete signals with the electronic equivalents of cam-driven switch boxes. The engineer or technician programs the controller to look for signals from individual sensors at certain angles of revolution or position of ram. If the signals do not occur when they should, the controller sends a stop signal (either normal or emergency stop), halting the press. These die protection devices typically handle a variety of types of signals and can provide several different kinds of outputs. The following six-step program is one recommended approach to setting up a die protection system:[13]

Step 1: Identify the problem areas. Using Pareto charts or other statistical tools, determine in which order die(s) are to be equipped.

Step 2: Select the sensors. The best controller does not make up for poor sensor selection. Types include:

1. Electro-mechanical sensors, which do not require external power, but do not work well in high speed situations:
   - Spring probes.
   - Limit switches.
   - Microswitches.
2. Electronic sensors, which have no moving parts to wear out, but require external power:
   - Inductive proximity sensors (include shielded and unshielded).
   - Photoelectric sensors (which include thru-beam, retro-reflective, diffuse reflective, and mini-light curtain).
   - Fiber optics.

Step 3: Install the sensors. Proper placement is necessary to actually sense the intended action.

Step 4: Determine the critical angle. The "point of no return" is the last point in the crankshaft rotation or slide travel at which the stop signal can be given to the press to prevent die closure. Very important!

Step 5: Set up the control. Properly time the inputs and outputs to provide the intended protection signals.

CHAPTER 17

CONTINUOUS IMPROVEMENT IN THE METAL STAMPING INDUSTRY

Step 6: Die protection is an ongoing process. Learning how to properly set up logic, place sensors, and tie in press controls is a continuous learning experience. Each press/die combination will have idiosyncracies; persistent attention is required.

Operator Safety

OSHA and various other government agencies have very specific regulations regarding operator safety and stamping presses. Lockout/tagout procedures for maintenance workers ensure they are safe from the accidental application of power while they work on a machine. Training and enforcement are the primary tools to best guarantee compliance.

Several methods exist to ensure production worker safety while maintaining peak operation efficiency. Operators must not enter the die area after the automatic cycle system has taken over, so actuation devices such as palm buttons must be placed a certain distance away based on the speed of the press, the position of the ram when the automatic control takes over, and a human speed constant. These may vary according to the area, so it is best to check with authorities.

Cumulative trauma disorders (CTDs) such as carpal tunnel syndrome can be avoided with light- or noncontact-actuation devices. Light curtains may also be usable for operator protection and/or cycle initiation. If solid guards are used instead of or in conjunction with control methods, they must properly protect the operator.

MISCELLANEOUS

Die designs should always incorporate quick die change and TPM features, such as easily accessible die handling devices and modular components. It may be helpful to address the designs with Design for Manufacturing, Design for Assembly, and Design for Automation concepts in mind; think of each die design and construction job as a production run of one unit.

Material specifications should be periodically reviewed for applicability to the job. Thickness variations from side to side and along the length of a coil can cause problems with shearing, draw beads, etc. Forming limit diagrams (see Fig. 17-4) should be available for all chemistries of material; design, construction, and maintenance efforts should be made to ensure operations stay in the "safe zone."

In many plants, compressed air losses due to leaking components such as cushions cost more than is perceived. Diaphragms, bladders, and other elements impervious to air seepage should be used if the pneumatic action cannot be eliminated altogether.

Stamping plant manufacturing engineers do not work themselves out of their jobs introducing continuous improvement into their operations. They do work themselves into higher profitability, possibly monetary bonuses, but definitely more job security. The definition of "world class" is ever changing, but not out of reach. Trying to achieve it requires constant attention to and refinement of product and process. The cost of this continuous improvement may be high, but it is never prohibitive. The alternative, in the long run, is always more costly.

References

1. Donald B. Dobbins, "Hydraulic Presses and Their Applications: Part VII of a Series," *MetalForming* (October 1991).
2. Robert K. Sterling, "Multiple Blanking" (SME Technical Paper MF76-402), (Dearborn, MI: Society of Manufacturing Engineers).
3. Joseph Ivaska, Jr., "Lubricants--A Productive Tool in the Metal Stamping Process" (SME Technical Paper TE77-499), (Dearborn, MI: Society of Manufacturing Engineers).
4. Federico Strasser, *Metal Stamping Plant Productivity Handbook*, Chapter 15, Troubleshooting, (New York, NY: Industrial Press, Inc., 1983).
5. James H. Woodard, "Designing the System to Match Your Needs" (SME Technical Paper MF90-120), Quick Die Change Clinic (December 1989), (Dearborn, MI: Society of Manufacturing Engineers).
6. "Things to Know About QDC," *MetalForming* (April 1993).
7. Dick Demara, "Quick Die Change Using the Team Concept (SME Technical Paper TE88-463), Quick Die Change Clinic (May 1988), (Dearborn, MI: Society of Manufacturing Engineers).
8. David Alkire Smith, "Continuous Improvement in the Stamping Plant Through Interactive Training" (SME Technical Paper MM90-05), (Dearborn, MI: Society of Manufacturing Engineers).
9. Ray Stover, "Preparing for Quick Die Change: What to consider before buying and installing Quick Die Change equipment," *STAMPING Quarterly*, Fall 1992, (Rockford, IL: The Croydon Group, Ltd.).
10. Federico Strasser, *Metal Stamping Plant Productivity Handbook,* Chapter 11, Ejecting Stampings and Scrap, (New York, NY: Industrial Press, Inc., 1983).
11. Ashok Bhide, "A Programmable Timing System for Transfer Applications" (SME Technical Paper TE90-273), Modern Technologies for Improved Transfer Press Systems Conference (June 1990), (Dearborn, MI: Society of Manufacturing Engineers).
12. David Alkire Smith, "Using Waveform Signature Analysis to Reduce Snap-Through Energy" (SME Technical Paper MF90-11), Die and Pressworking Tooling Clinic (August 1990), (Dearborn, MI: Society of Manufacturing Engineers).
13. Jim Finnerty, "Getting Started with Electronic Die Protection," *MetalForming* (January 1993).

Bibliography

Bakerjian, Ramon, ed. *Tool and Manufacturing Engineers Handbook, Volume 6, Design for Manufacturability*, Chapter 12. Dearborn, MI: Society of Manufacturing Engineers, 1991.

Bhide, Ashok. "Programmable System for Press Automation" (SME Technical Paper), 1988 Flexible Metal Stamping Operations conference proceedings. Dearborn, MI: Society of Manufacturing Engineers, 1988.

Boerger, Dennis. "New Approaches to High Speed Metal Stamping" (SME Technical Paper), FABTECH '83 Conference proceedings. Dearborn, MI: Society of Manufacturing Engineers, 1983.

Booth, Douglas E. "Robots--Their Use in Metal Stamping" (SME Technical Paper), FABTECH '83 conference proceedings. Dearborn, MI: Society of Manufacturing Engineers, 1983.

Booth, Douglas E. "The Use of Robots for Metal Stamping" (SME Technical Paper). Dearborn, MI: Society of Manufacturing Engineers, 1979.

Booth, Walter. "Press Room Guards and Barriers" (SME Technical Paper MM77-582). Dearborn, MI: Society of Manufacturing Engineers.

Carlson, Robert E. "Economics of Metal Stamping" (SME Technical Paper). Dearborn, MI: Society of Manufacturing Engineers, 1977.

Cubberly, William H., and Bakerjian, Ramon, eds. *Tool and Manufacturing Engineers Handbook, Desk Edition*. Dearborn, MI: Society of Manufacturing Engineers, 1989.

Davis, Richard L. "The Most for the Least" (SME Technical Paper). Dearborn, MI: Society of Manufacturing Engineers, 1974.

Farnum, Gregory T. "Rethinking the Stamping Process," *Manufacturing Engineering* (April 1987).

Fletcher, David S. "Proximity Sensors Specifically in the Metal Stamping Industry" (SME Technical Paper), FABTECH International '89 conference proceedings. Dearborn, MI: Society of Manufacturing Engineers, 1989.

Fredline, Jeffery R. "Troubleshooting Metal Stamping Presses" (SME Technical Paper), 1988 Predictive Maintenance for Presses conference proceedings. Dearborn, MI: Society of Manufacturing Engineers, 1988.

Fryjoff, Chet. "Computer-Integrated Manufacturing for Metal Stamping" (SME Technical Paper), 1985 Flexible Systems for Metal Forming Operations Conference proceedings. Dearborn, MI: Society of Manufacturing Engineers, 1977.

CHAPTER 17

CONTINUOUS IMPROVEMENT IN THE METAL STAMPING INDUSTRY

Gilchrist, Jack. "Present and Future Trends for the Stamping Industry" (SME Technical Paper). Dearborn, MI: Society of Manufacturing Engineers, 1977.

Harrison, Roger. "Safe Mounting Distance for Two-Hand Press Actuation, Part 1 of 2," *MetalForming* (December 1992).

Mallia, Robert P. "The Role of DC Feeding Machines in Flexible Metal Forming Systems" (SME Technical Paper), 1985 Flexible Systems for Metal Forming Operations conference proceedings. Dearborn, MI: Society of Manufacturing Engineers, 1985.

Mallick, Jr., G. T. "Acoustic Die Monitoring" (SME Technical Paper). Dearborn, MI: Society of Manufacturing Engineers, 1977.

Smith, David A. *Die Design Handbook*. Dearborn, MI: Society of Manufacturing Engineers, 1990.

Stauffer, Robert N. "Men and Machines—Komatsu's Approach to Pressworking." *Manufacturing Engineering* (May 1981).

Stephan, Peter H. "Automation in the Press Shop with Industrial Robots" (SME Technical Paper), 1992 PRESSTECH conference proceedings. Dearborn, MI: Society of Manufacturing Engineers, 1992.

Wallis, Richard A. "Solving Die Design Pressure Problems" (SME Technical Paper), 1985 Fabtech International Conference and Exposition conference proceedings. Dearborn, MI: Society of Manufacturing Engineers, 1985.

Weiland, John. "Producing Parts by the Fine-blanking Process to Improve Accuracy, Cost, and Repeatability" (SME Technical Paper). Dearborn, MI: Society of Manufacturing Engineers, 1988.

Wick, Charles. "Heavy Metal Stamping," article in August 1977 *Manufacturing Engineering*. Dearborn, MI: Society of Manufacturing Engineers, 1977.

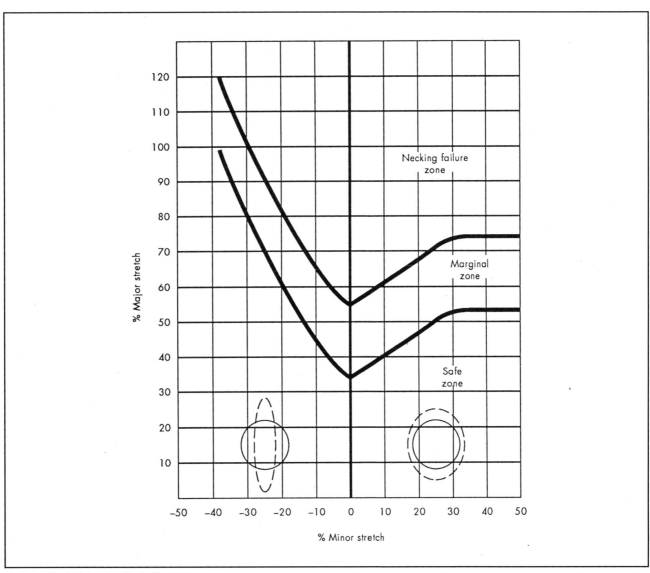

Fig. 17-4 Typical forming limit diagram.

CHAPTER 18

FINISHING

The term *finishing* can be defined as either a process for preparing or improving a material's surface (honing, grinding, buffing) or a process of material application (painting, anodizing, plating). Processes such as organic finishing, induction heating, soldering, brazing, power brushes, honing, deburring, autodeposition, painting, and waste minimization present opportunities for the implementation of continuous improvement techniques.

FINISHING/PAINTING PROCESSES AND LIQUID COATINGS

Operators of coating lines often misunderstand or fail to realize the need for continuous improvement. Many times people are afraid of taking any risk to make a change in their processes. The attitude is "if it isn't giving us a problem, then leave it alone."

This is a sure formula for eventual problems outside the basic area of production and getting products out the door. A firm not making improvements in a continuous manner is missing out on marketing opportunities due to competitive costing, improved quality, reaction time flexibility for emerging niche customers, and improved plant capacities.

Continuous improvement (CI) is a philosophy. It is a way of examining every part of an operation, not just isolated sequences or parts of the whole. CI should *not* be regarded as a justification for long-term major revisions or a major shift in products or markets. It is a discipline and a way of life where knowledge and quality of information are vital. Using continuous improvement techniques delivers solid performance and value.

Most of the continuous improvement procedures are not new to a finishing professional. These concepts have been performed for years by professionals who contribute to successful enterprises.

Today there are many new forms of substrates, processes, materials, and product needs, more than ever before in the history of finishing. With the main driving forces of world class quality; environmental, health, and safety regulations; and competitive pricing, finishing operations must react to survive. In finishing, there are many areas where improvements can be achieved.

Every business is composed of ingredients that make it an enterprise. They are such things as money, materials, manpower, methods, machinery, management, and marketing. All of these are areas for possible improvement.

MONEY AND MANPOWER

Money can be described in terms of money lost, money saved, and money available for reinvestment into the process. During the past four decades the finishing discipline has performed an approximate 180 degree turn where labor and materials are concerned. In the 1950s, about 65% of the costs of finishing came from labor, both direct and indirect. Materials made up the remaining 35%. Today it is much more normal to see material costs making up the 65% figure and labor the 35%. In many highly automated lines, materials and energy may take as much as 90% of the total costs. These costs of materials now include treatment and disposal due to environmental regulations.

The best practice is to prioritize *anything* that may improve material usage; the labor, quality, environment, safety, costs, and other considerations will begin to improve as a result. When dealing with materials, reducing waste must be a major effort. *The best gallon of paint or the best pound of chemical is the one you never use.* No one has to buy it, ship it, receive it, handle it, use it, perform maintenance on it, worry about it polluting or hurting someone, or

CHAPTER CONTENTS:

FINISHING/ PAINTING PROCESSES AND LIQUID COATINGS 18-1

PAINT PRETREATMENT AND PAINTING SYSTEMS 18-4

ENHANCED PAINTING EFFICIENCY THROUGH WASTE MINIMIZATION 18-7

DEVELOPMENTS IN ELECTRO-COATING 18-14

QC/QA FIELD APPLICATION FOR PAINTED PARTS 18-22

INDUCTION HEATING: HOW TO GET THE MOST OUT OF YOUR EQUIPMENT 18-24

PRIMER ON POWER BRUSHES 18-29

SURFACE IMPROVEMENT TECHNOLOGY 18-33

HOW TO FIND THE BEST HONING PROCEDURE 18-35

DEBURRING AND SURFACE CONDITIONING 18-41

AUTODEPOSITION: TOUGH COATINGS AND NO VOC 18-43

The Contributors of this chapter are: Kenneth A. Barton II, President, Sales and Engineering, QPAC-Nagel Precision Inc.; **Hans H. Fischer**, Product Engineer, Sunnen Products Company; **Robert D. Grear**, CMfgE, CM, Consultant; **Ronald G. Jacob**, Vice President and General Manager, Wright Metal Processors Division, Western Industries, Inc.; **Thomas C. Jones**, Technical Manager, Parker+Amchem; **John B. Kittredge**, Owner, John B. Kittredge Consultant; **Alan Monken**, Technical Services Chemist, Calgon Corporation; **George Pfaffman**, Vice President-Technology and Service Operations, Tocco, Inc.; **John C. Reseland**, National Market Manager, Calgon Vestol Laboratories; **Clifford K. Schoff**, Scientist, PPG Industries, Inc., Allison Park, PA; **William C. Wagner**, President, Wagner Consultants; **Frank J. Hettes**, BSME, Vice President, Research and Development, Weiler Brush Company.
The Reviewers of this chapter are: Rolf Bochsler, Vice President and CEO, Nagel Precision Inc.; **Joseph C. Byrnes**, Technical Communications Specialist, Sunnen Products Company; **George A. Fletcher**, Marketing and Product Development, Metal Koting-Continuous Colour Coat Limited; **Lee Gilbert**, Product Manager, Supfina Machine Co.; **Robert D. Grear**, CMfgE, CM, Consultant; **Brian T. Prylon**, Staff Project Engineer, North American Operations-Manufacturing Center, General Motors Corporation.

CHAPTER 18

FINISHING/PAINTING PROCESSES AND LIQUID COATINGS

dispose of its waste. This is the best improvement in the finishing shop provided it doesn't hurt product quality and put the company's reputation at risk.

MATERIALS

Material waste is reduced in various ways. Certainly, improved application methods are vital. What materials are the products made from? What lubricants are employed during their manufacture? What is the product configuration? How can we hang or transport product? What type of finish level is required? These are just some of the questions which will help to focus on reducing material waste.

Often suppliers do not support this policy; it can reduce their business. But, where waste is reduced, it is possible to move to a higher quality finishing product and still remain competitive. When applied with integrity, this policy will create a new and much stronger supplier relationship over the long run.

The Importance of Product Design

The design of a product is where finishing really begins. Many details which ultimately influence total finishing cost are found outside the paint department. If the finishing department does not become a part of the decision making areas long before the product reaches the paint line, then the opportunity to achieve improvements will be greatly minimized. A product must be designed so it can be manufactured, transported without damage, positioned to be cleaned and pretreated, drained properly without causing carryout, and hung in the most efficient patterns to apply coatings and cure properly. Products must fit the profiles of the facility as they pass through on conveyors, etc. Continuous improvement must begin in the design phase. In the end, consideration for these needs will permit production at the volumes and quality required and at the best value.

Substrate Selection

Selection of the substrate can materially affect tooling, lubricants, fixtures, racks, transportation, and the number and types of coatings required. It can impact on application methods, facilities, and energy required. Should the material be a metal or mixed metals, or some form of nonmetal? Will it be a single piece, or a variety of parts welded or bonded together? Must it be coated on one side only, or on all surfaces? What temperatures will it stand?

Finish Selection

The use of high-solids paints will dictate better control of application. If this is not done, any savings from using less paint will be offset with higher costs of overspray and maintenance. Combining higher solids paints and better transfer efficiencies may permit the coating of a product in one coat instead of two coats. This reduces labor, maintenance, and equipment costs, and improves plant capacity in the paint shop.

The use of higher solids—or any paint that reduces VOC—permits savings in the amount of energy and fresh air required in ovens. This saving is usually not gained simply by establishing air requirements. Many ovens are set at a very high rate of exhaust from previous high VOC products, and are not providing optimum energy values as environmental conditions improve. Lower VOC means less fresh air is required.

Failure to control film thickness will erode continuous improvement efforts. In some high-volume lines every 0.1 mil (0.0025 mm) additional film thickness can add over $100,000 annually to product costs. In most cases, this additional film thickness adds nothing to quality levels.

Moving to a waterborne finish to meet environmental regulations can have significant bearing on processes and costs. The definition of *clean* takes on a different meaning; waterborne products tend to demand a cleaner surface to coat than solventborne. The waterborne products are lower in solids, so more wet film is required to achieve coverage in many cases, compared to solventborne. Without an improvement in transfer efficiency, the paint bill will increase. By isolating the grounding of the material supply or moving to high-volume low-pressure (HVLP) equipment if applicable, waterborne products will probably yield an improvement in environmental considerations and about equal material costs. Stainless steel plumbing may be required for material supply systems.

Waterborne products also require more consideration for control of booth application environments such as humidity and temperature. The window of application tends to be tighter, and waterbornes may also require heated flash-off zones to remove the water. Failure to consider these areas will result in problems coming and going throughout the year. The savings here come from other areas such as pollution costs, and must be studied carefully against the *total* cost of finishing, not just a portion of it.

METHODS

Some techniques of continuous improvement can be performed on an existing line without requiring great sums of capital, and will yield valid improvements.

Plumbing

Extend the life of chemicals and rinses and save energy by adding some easy plumbing to older systems. Washers with more than one cleaner stage can be plumbed so the second cleaner serves as the makeup for the first cleaner when dropped. This will also enable the first stage replenishment to come from the second wash stage, if a cascading can be created. Rinse stages for cleaners can be used as replenishment water when cleaners are dumped. Depending upon the number of stages and chemicals used, various forms of conservation can take place to cut chemical costs. Careful control of the lubricants and soils make it possible to reduce the temperatures utilized in pretreatment.

Priming

Many products require the use of a primer for various reasons. The use of better generic resin primers may permit the elimination of finish paint on one side of a part, or the total elimination of top coat paint. Good cross-linking primers such as epoxies, polyesters, and polyurethanes can assist in this matter. Epoxy primers basically do not weather well when exposed to UV on exterior surfaces; if a part is on the interior of a product and not so exposed, an epoxy can be a good replacement for topcoat paint. If more durability is required for exposure, a polyester or polyurethane can be applied.

Primers are produced in colors other than the usual gray, black, or red oxide. This can reduce or eliminate the need for topcoat painting in some instances.

Loading

Improving the way parts are presented, in pretreatment and in the paint booth, can reduce waste and costs. It is not uncommon to find that only 25% of the space available for hanging parts is used.

Painting is basically planned upon how far a human being can reach. The average person can reach about four feet sideways and

CHAPTER 18

FINISHING/PAINTING PROCESSES AND LIQUID COATINGS

about four feet vertically while standing in one spot in a paint booth. This 16 square feet profile is a good way to gage productivity when hanging parts. How can parts be loaded to fill most of this area and still be able to coat all areas of the product? Looking at hanging patterns along these lines, it is fairly easy to determine what percentage of air is being painted. Simple loading techniques can increase capacity.

Better hanging of parts can reduce pretreatment costs and improve the quality of cleaning and phosphating. Often less energy is required. Paint can build on improving coverage and rejects. The extra time will often provide a better and more uniform cure for better quality. Most of this consideration applies to robots, reciprocator, and other automatic application methods, as well as manual lines.

MACHINERY

Improved transfer efficient application equipment is usually based upon either electrostatic principles or the high-volume low-pressure technology commonly referred to as HVLP. Each of the methods can be found in different forms designed to satisfy a wide variety of applications. They are normally 65% or more efficient. With added controls for fluid deliveries, atomization, distances, etc., they really do save materials and reduce waste. The HVLP systems especially are very inexpensive to convert to in many shops, and return their costs in a short time. They also can be used with waterborne finishes without the material system isolation that is required when converting electrostatic technology to waterbornes.

Other forms of high efficiency transfer come from such technologies as electrocoating and autodeposition finishing. These approach finishing by immersing products completely. In *electrodeposition*, the organic paint film is plated on the product electrically. In *autodeposition*, coating is plated chemically on the product. Both have excellent efficiency of coating and corrosion properties. Each can be used for either priming or finish painting. Each must be studied to determine if the level of finish painting will satisfy quality specifications of gloss, smoothness, and other properties.

One thing must be understood when comparing efficiencies and costs between such systems and conventional application systems. The electrodeposition and autodeposition systems are highly transfer efficient. However, there may be little actual savings in paint costs between them and older systems. These systems coat everything inside and out; more square footage may be coated than in other conventional systems that do not coat inside. This may realize a 50% increase in savings, and put it back on by a 50% increase in metal covered. Cost justification is often achieved by better ways of obtaining quality and improved corrosion values for hard-to-paint parts. These highly efficient systems are much more capital intensive initially, but do recover certain other costs since they do not require paint booths, etc. In any case a complete cost evaluation should be performed.

Manual or Automatic

In using HVLP and electrostatic systems there is a choice of either manual or automatic methods available. Energy costs can be reduced in the volume of air makeup required with such equipment. Automatic operation reduces certain labor costs but these must be offset by the nature of the controls necessary for automatic application.

Humans can react to changing conditions much better than an automatic applicator. The window of application must be controlled much more precisely with the automatic units. As with any improved transfer efficiency, there should be less material used and less waste created. Maintenance will go down in some areas such as paint sludge and filters, but could go up in such areas as guns, controls, and application devices. Capital costs are one-time, normally; labor costs are ongoing.

Defects and Rework

A finishing or paint line's function is to finish. It is not intended to repair or rebuild poorly manufactured or damaged product. Any attempt to allow defective parts to enter the paint shop will result in penalties to cost, quality, and throughput. Continuous improvement procedures provide paintable product.

Anything done to improve rework will have a tremendous impact on continuous improvement. Most older finishing operations dedicate a significant amount of their operations to rework capability. Up to 30% of the total cost of a finishing operation can come from doing something over again. The effort of continuous improvement must result in a much lower rate of defects.

Quite often, shops are operating within their budgets and people believe they are doing a good job. Unfortunately, there have been monies built into those budgets over the years to cover a high level of defects as "normal." These budgets have been subsidizing poor performance upstream and downstream by permitting product to come into their areas which is not proper. After efforts are made to analyze a finishing system from beginning to end for total costs, these performances can be improved. When efforts spread out into other operations, these financial and product quality improvements really reduce rework costs.

MANAGEMENT

Environmental regulations forced finishers into becoming better at their business; but improving processes has been good business for years. The waste minimization efforts of the EPA are not new concepts to a professional finishing person. It is just another way of saying "the best gallon of paint or pound of chemical is the one you do not use."

Start at the very beginning of the product production and use the method of looking at the total cost of painting. There will be many opportunities to find continuing improvement. These disciplines provide dozens of variables to consider and can be consolidated to maximize performance.

Continuous improvement is a way of doing business. It is a way of thinking and a way of life. It is winning in the long run, not the short term performance. Companies who consider this as unnecessary will eventually fail. They cannot continue to use the resources of this planet and be wasteful as in the past. Companies practicing continuous improvement concepts will gain opportunities not recognized by their less professional counterparts. In every problem lies the seed to the solution. Continuous improvement goes beyond problems—it can bring about improvements even when there are no problems to solve.

Leadership must be capable of embracing this philosophy. It has to be willing to empower people to recognize, explore, input, and implement change on a continuing basis. It must train personnel in this method of operation and provide the atmosphere for interaction. Companies should trust the people to carry out their duties.

While finishing begins with design, continuous improvement begins with management. Management has to show by precept and deed that it believes in, and wants, continuous improvement.

CHAPTER 18

PAINT PRETREATMENT AND PAINTING SYSTEMS

PAINT PRETREATMENT AND PAINTING SYSTEMS

As with any other operational system, the finishing line can benefit from application of continuous improvement principles. The use of proper maintenance procedures in paint pretreatment and painting systems will eliminate many of the typical problems encountered in day-to-day operations. Critical examination of the finishing operation will often reveal obvious system flaws. Once corrected, these changes can greatly influence efficiency. These actions, combined with organized programs for opportunistic improvements, will help to meet continuous improvement objectives without the need for total system replacement or overhaul.

The primary action to take in making improvements is to evaluate each aspect of the finishing system, from raw material handling to waste handling, in terms of its original operation and design specifications, its current performance, and its expected performance. All systems originally had clearly defined performance parameters, although with the passage of time these specifications may be lost or forgotten. If these can be established (or re-established), the current performance of the particular component can be compared to what was expected and/or achieved originally. Bringing a system or component back to its intended level of operation can often result in immediate improvements and increases in efficiency. It is necessary, however, to evaluate the expected performance of the system in terms of the original design; many times frustration results from trying to "optimize" a particular component or system to achieve better or different results than it was designed for. In this way, system limitations can be recognized and acknowledged, pointing out areas where improvements may be required.

Once areas for system or component improvement are determined, these improvements must be organized in a controlled and logical fashion. Since the availability of both capital and manpower is typically at a minimum for both maintenance and system improvements, plans for implementation should be well thought out with regard to each component. It is typically much more cost efficient to perform ongoing maintenance to keep systems optimized than to rebuild them. If major overhauls are necessary, it is important to coordinate changes to optimize performance while minimizing the capital and labor costs. Since many of the components of the finishing line are interrelated, in terms of both performance and operation, it is important to coordinate an improvement in one area so as not to degrade the efficiency of another. It is also necessary to review the *expected* optimum performance of a component or system versus the *desired* performance. Setting realistic objectives for improvement will eliminate the cost and frustration resulting from trying to achieve results greater than the optimum performance of a given system. Although the typical metal finishing line could include virtually any process between the front door of the plant and the shipping dock, the areas and systems considered here will primarily involve the paint pretreatment system and the paint system itself. Other processes and aspects of metal finishing will be mentioned only as they impact these primary systems.

PAINT PRETREATMENT

The first discussion area for application of the principles of continuous improvement is paint pretreatment (or prepaint treatment). This area could be extremely wide if all systems and situations were taken into consideration, but the discussion will be limited to the following areas:

- System performance.
- Raw material.
- The cleaning process.
- Conversion coatings.
- System maintenance and controls.
- Chemical maintenance and control.
- Waste handling.

Each of these areas is examined in terms of how to determine the need for improvements and suggested means for setting the improvement objectives. Before looking for "improvements" to make for various systems, however, the current system and its performance must be examined.

System Performance

One of the most common problems encountered in pretreatment systems is a deterioration in performance. This may become apparent with poor results in the cleaning of parts, nonuniform conversion coatings, paint performance problems, or higher-than-normal rates of rejected or returned parts. The first response to this deterioration is typically to look for problems with the cleaning or conversion coating chemicals or with the paint; however, often the system itself is not operating correctly. Spray washers, particularly, are susceptible to hidden maladies, such as scaled recirculation lines and plugged or misdirected nozzles. This can result in diminishing levels of performance. The gradual deterioration may not be readily noticeable until lines become fully clogged and result in some catastrophic event; in the case of a washer, the event in question would be the blockage of spray nozzles. Ongoing maintenance programs can reduce or eliminate this problem long before operation is affected. The same is true for the routine checking of nozzle alignment; misaligned nozzles are capable of producing uneven cleaning or conversion coatings which can lead, in turn, to paint problems. The key point is to maintain correct operation of the system or process at all times, through normal routine maintenance programs.

Raw Material

Once it is established that the finishing system is operating "normally" and maintained with an ongoing program, there are other areas to explore for improvement. The first area of investigation is the raw metal stock used. Various types and grades of metal are used in manufacturing, including hot- and cold-rolled steel, galvanized steel, aluminum, and other less common alloys and cast metals. The discussion here will stay focused on steel, the most common metal encountered in general manufacturing. Steel for light manufacturing is most typically purchased as either coil stock or flat stock. Cold-rolled steel is most common in light manufacturing operations, as it can be readily drawn, bent, drilled, and otherwise manipulated into finished forms. Most typically, cold-rolled steel will be coated with some variety of mill oil from a "pickle-and-oil" process. Steel vendors use different types and grades of oils in their pickling-and-oiling processes, some of which become very difficult to remove under otherwise normal conditions; foreign-milled steels often have especially tenacious oils, due to the requirement to retard rusting during

CHAPTER 18

PAINT PRETREATMENT AND PAINTING SYSTEMS

shipping. Knowing what oils are present on a particular steel stock allows preparation for the removal of these oils in the cleaning process. This reduces problems resulting from the use of different vendor materials. There are also numerous differences in the impurities found in some steels; for example, many offshore mills provide steel containing a higher carbon content. Higher than normal carbon levels will result in more carbon smut on the metal, increasing the difficulty of cleaning and possibly decreasing the end quality of the manufactured product. One way to assure continuous quality in the end product is continuous monitoring of the metal stock entering the plant, and purchasing material meeting established specifications. Savings from buying off-specification steel can quickly be erased when rejection rates and/or processing costs increase.

Cleaning Process

The first aspect of metal cleaning that should be evaluated is whether or not the cleaning processes and products are the correct ones for the desired end results. Any given system, even when optimized, has some limitations of performance. Quality of cleaning usually is directly proportional to the cost applied to achieve it. Parts to be powder-painted or plated cannot be effectively cleaned in a one-tank, one-step process. On the other hand, steel I-beams for construction use do not require the cleaning achieved by a seven-stage spray washer. It is therefore important to realize just what level of cleaning can be expected from a given system type or configuration, and to compare this to the desired end result.

Pick the proper cleaning product. As with the equipment, selecting the correct cleaning product is critical to the end result. Many times, when problems in cleaning are traced back to the chemical program used, the problem lies in the selection of the cleaner rather than its quality. Metal cleaners vary widely in strength and ingredients, with the latter being responsible for the functionality of the product itself. Heavy levels of "natural" fats and greases (those from vegetable or animal sources) normally require alkaline cleaners with high concentrations of sodium or potassium hydroxide (caustic). Removal of other oils and greases may require a product with a high concentration of surfactants to disperse the oils. Particulate soils, like carbon smut, may require products containing chelates. With the wide variety of cleaning chemicals available, it is possible to choose a cleaning compound specifically for the soil types present in a particular process (which include mill oils, drawing compounds, lubricants, etc.). The same is true for choosing the correct compound to match the cleaning system, with individual products being designed for immersion or spray systems and high- or low- temperature operation. The important point here is to find a cleaner designed to work effectively in the type of equipment currently in use, formulated to remove the types of soils actually found in the manufacturing process, and effective in terms of cost and use.

Conversion Coatings

The next area to be examined for possible improvement normally represents the most critical area of prepaint treatment—the conversion coating process. Conversion coatings are a way of chemically changing the metal substrate to improve paint adhesion and corrosion inhibition. The conversion coating processes most commonly encountered when working with steel parts are iron phosphating and zinc phosphatizing (or phosphating). The basic principle behind both is the application of an inorganic metal phosphate crystalline coating to the clean surface of the metal. In the case of iron phosphatizers, products may be specifically designed for combination cleaning and phosphatizing, for immersion or spray phosphatizing, or for both iron and soft metal (aluminum, zinc, etc.) phosphatizing. Zinc phosphatizers are more restrictive in their equipment requirements (not allowing combination cleaner/phosphatizer products) but are more varied in the types of coatings that can be achieved with regard to crystal structure and amount present on the surface.

System Maintenance and Controls

In terms of continuous improvement objectives, it is important to maintain the normal quality control parameters for the performance of conversion coatings and paints. The most common measures of paint performance in this regard are salt spray performance, coating weight, and paint adhesion. Salt spray testing represents a laboratory measurement of accelerated aging or weathering of painted surfaces. Established performance specifications normally exist for a particular paint as applied over a particular conversion coating process. Performance should be measured and recorded periodically to assure that the system is producing the desired results. This analysis should be part of an established quality control/quality assurance program that includes paint performance measurements and evaluation of the cleaning processes previously discussed. This type of program can normally be established in conjunction with a chemical vendor, but is ultimately the responsibility of the manufacturer; maintaining product quality must always be an internal function, not one left to outside vendors.

Chemical Maintenance and Control

Proper ongoing maintenance and control of the pretreatment equipment is important to continuous quality improvement. Equally important is the maintenance and control of the chemical programs employed in the pretreatment processes. When the correct chemical products have been instituted in the system, it is imperative that the operating concentrations be monitored and controlled to assure continuing quality. Included in these measurements would be the pH and conductivity of each solution and rinse stage (and the makeup water) and the temperature and spray pressure (if applicable) of each separate stage. Statistical process control (SPC) is one means currently in use for gathering and recording such information in a systematic method, but even the continued use of simple log sheets for recording product concentrations and additions can help to keep proper control of most processes. Automatic control and feed systems are available for many chemical programs. These systems relieve some of the problems associated with operator inconsistencies, but still need a degree of monitoring by plant personnel. The tighter the controls on chemical and system parameters, the more uniform the quality of the pretreatment that the system will produce. This results in less downtime, less repairs, and fewer product rejects.

Waste Handling

It is important to look at the types of waste handling required by the processes in use. Many metal finishers are involved with (or are contemplating addition of) in-house waste treatment systems. Although doing internal waste treatment adds another level of complexity to chemical programs, it also provides better insurance against fines and citations for waste materials which exceed local limits. Handling waste materials in-house requires strict monitoring and recording of results on a daily basis to maintain both quality and compliance. One way to assure the best treatment results is to limit or eliminate troublesome materials from the

CHAPTER 18

PAINT PRETREATMENT AND PAINTING SYSTEMS

processes producing the waste to be treated. Some of the bigger problems (such as the chlorinated solvents used in vapor degreasing) have already been virtually eliminated through more stringent regulations. Other formulations, such as those heavily chelated, may cause difficulties in separating metals from the processed waste water. If there are particular restrictions on effluent, one way to cope is to prescreen the pretreatment products before their use in the operation. The other option is to work closely with a waste treatment vendor to establish a waste handling program dealing with the present materials. As chemical vendors become more attuned to dealing with environmental concerns, it may be possible to find coordinated systems containing "treatable" pretreatment products and companion waste treatment chemicals.

PAINTING SYSTEMS

As with the pretreatment system, areas of improvement in the painting system can be broken down into the equipment and the chemicals. Equipment issues surrounding painting systems include the paint application equipment chosen, the paint booth type, and the mechanism(s) in place for sludge/paint waste removal. The area of paint chemistry is broad and application-specific; for the sake of brevity, differences in paints are touched on only briefly in terms of improvement possibilities. Chemical treatments, with regard specifically to detackification of paints, are germane to this discussion.

One of the driving forces behind improvements in painting systems has been the issue of VOC reduction. Since most volatile organic compounds (VOC) are attributable to the solvents found in paints, there has been a strong movement to reduce this solvent level over the past several years. This means developing a way to meet the desired reduction of solvents, including conversion to powder painting, conversion to water based coatings, conversion to high-solids "compliance" coatings, and conversion to high-efficiency application equipment. To convert to powder coating requires the installation of paint booths specifically for powder use, along with handling systems for the powder paint and its reclamation. Conversion to water based coatings is a less drastic alternative, at least with regard to capital investment, but water-based paints have generally been slow to meet all of the performance requirements of solvent bornes like baking enamels and urethanes. Switching to high-solids compliance coatings has done much to reduce VOC emissions from paint booths, with few problems for the users except in terms of detackification; fortunately, detackification technology has managed to keep pace with the development of higher solids coatings. Higher efficiency spray guns have also been developed to reduce paint overspray and, therefore, paint usage and VOC levels.

IMPLEMENTING THE IMPROVEMENTS FOR PAINTING

Major improvements in painting systems are costly and rather radical. Some day-to-day operational improvements may be implemented in:

- Paint application.
- Coating quality.
- Paint booth operation.
- Sludge handling and removal.

Each of these areas can be examined for potential improvements and methods for implementing changes.

Paint Application

The use of high solids paints and high efficiency spray guns has effectively reduced VOC levels by reducing the amount of volatile solvent present and the amount of overspray produced. However, painting of the manufactured parts needs to be monitored. The newer spray guns are rated for improved application efficiency based on set parameters of distance from work and duration of spray. If these parameters are violated by sloppy or inconsistent painting, better efficiencies will not be achieved and greater levels of overspray may result (due primarily to the perceived lesser coverage of the tighter-patterned guns and nozzles). This situation can be improved by monitoring both wet and dry film thicknesses on a routine basis and recording daily paint usage as a function of parts sprayed. By taking this step, overall paint usage (typically a major expense) can be controlled and waste reduced. A greater consistency of paint coating may also result in better control over finished quality.

Coating Quality

Hand in hand with monitoring the paint usage should be the monitoring of paint coating quality. A number of tests can be routinely run to evaluate the quality of paint coatings, including salt spray, paint adhesion, and impact testing. Specific tests giving meaningful information should be determined and run on a scheduled, routine basis. The specifications for the paint performance tests may be based on recommendations from the paint vendor, but it is important to establish the realistic performance of the particular system rather than accepting theoretical specifications. These tests can give both a measure of current performance (and possible indications of where problems lie) and a way to determine where improvements could be implemented to produce specific results.

Paint Booths

One area providing many opportunities for improvement is the booth where the painting takes place. Paint spray booths are often considered a "necessary evil" rather than an important piece of equipment in the painting process. Although the booth's operation may not directly affect the coating quality, the efficiency in which it operates can affect both the downtime of the entire system (which will affect the level of throughput of finished goods) and the quality of the work done by human painters. The areas of concern and potential improvement are in the physical booth design and operation and in the chemical treatment program used for paint detackification. As with cleaning systems, it is necessary to perform scheduled routine maintenance on paint booths to keep them operating as designed. Since many of the high maintenance areas are often hidden inside the booth (such as internal baffles and recirculation lines), the necessary servicing of these areas often gets overlooked or is done superficially. This typically results in a deterioration in performance of the booth, opening the way to environmental violations, fines, and potential health and safety hazards. Also as with cleaning systems, paint spray booths are designed to serve a specific purpose in a particular situation; as situations change it may become necessary to modify the booth or treatment program. The change to high-solids compliance coatings and water based paints has resulted in a need for different chemical programs. The most successful chemical treatments developed for both of these areas are the organic polymer systems; these systems often require some minor modifications to booth circulation or agitation.

Sludge Handling and Removal

Integral to paint booth operation is the area of paint waste handling and removal. As regulations regarding disposal have tightened in this area, the cost of disposing of paint waste has dramatically increased. Where once it was possible to dump water and sludge down the drain, separation of water and sludge is now required, for economic reasons as well as practical ones; with the current regulations in effect, virtually nothing but uncontaminated water can be sent to the sewer. The polymer detackification programs meet many of the improvement objectives, since they normally offer greater ease of removal than older conventional technology. Innovations in skimming equipment and continuous removal systems allow reductions of both the waste generated and the labor needed to maintain paint booth systems, even the older existing ones. One thing implemented in every case, however, should be an ongoing program of waste volume measurement and tracking. With this in place, it is possible to isolate the costs associated with the paint booth wastes and to allow rational comparisons of different sludge reduction methods and programs.

The keys to quality operation of the finishing line are continual monitoring of system performance, comparison to an established standard, and correct maintenance of this system at expected efficiency levels. Once the system is stabilized and maintained to controlled standards, it is possible to examine the operation for other improvement opportunities. Continuous improvement objectives can be met without the need for high capital expenditures; a smoothly run system is all that is needed, with a clear idea of desired performance. In simplest terms, the first step in continuous improvement of the finishing line is to operate the system as it was designed.

ENHANCED PAINTING EFFICIENCY THROUGH WASTE MINIMIZATION

INTRODUCTION

Inefficiency within the industrial product finishing system produces waste, and waste of any hazardous or nonhazardous material is not acceptable. Waste reduction is the sensible alternative, because it usually offers a permanent and economically advantageous solution to waste management problems.

No matter how elaborate or expensive a finishing system may be, its potential can be totally negated if it is improperly used. A solid working knowledge of every variable within the finishing system is of paramount importance if all waste streams are to be minimized and overall painting efficiencies maximized.

ENVIRONMENTAL CONCERNS

As environmental mandates become more and more restrictive, industrial finishers are being forced to make changes (sometimes drastic) in their finishing processes to be in compliance with environmental regulations.

The Federal Environmental Protection Agency (EPA) defines *waste minimization* as ''the reduction (to the extent feasible) of any solid or hazardous waste generated or subsequently treated, stored, or disposed of.'' One major source of waste during paint application is solvent emissions. The major factor affecting these emissions is the amount of volatile organic compounds (VOC) contained in each gallon of paint and the transfer efficiency of the paint from the application device to the target. In the U.S., the new Clean Air Act continues to impose much tougher restrictions on VOC emissions.

It has been reported that hazardous waste disposal costs increased by 1,940% during the past four years. The Resource Conservation and Recovery Act (RCRA) is imposing ever tighter regulations regarding hazardous waste which will increase operating costs. Many of the new rules and regulations have been designed to discourage treatment and disposal by increasing treatment standards and compliance requirements while limiting disposal options. Many previously feasible disposal options have become cost-prohibitive, making waste *reduction* economically significant.

Early EPA regulations focused on treatment and disposal rather than waste minimization. Today the focus is on source reduction/waste minimization at every level of the finishing process. A sound waste minimization program is one effective method of lowering costs, improving production, and increasing profits while coming into environmental compliance. Therefore, painting efficiency and waste minimization become synonymous.

In any discussion of paint line waste, paint transfer efficiency tends to dominate. While transfer efficiency is of paramount importance, it must be coupled with a multitude of other important variables which constitute finishing efficiency.

Painting efficiency should not be confused with *transfer efficiency* (TE). TE is only one part of painting efficiency. It is defined as the percentage of paint sprayed by a given device which ends up on the part/product being coated.

Clearly, TE by itself is not an adequate measurement because it does not take into consideration the variables which most directly affect the viability and overall efficiency of any painting operation. Unless these variables are thoroughly understood and adhered to, it will be impossible to maximize painting efficiencies and the generation of waste will continue to increase. Painting variables include, but are not limited to:

- Quality of the materials of fabrication.
- Fabrication processes.
- Conveyor line speed.
- Conveyor loading efficiency.
- Conveyor elevations.
- Pretreatment.
- Water entrapment.
- Air blow-off.
- Dry-off oven (time, temperature, mass).
- Cool-down zone.
- Temperature of ware entering spray booth.
- Spray booth size (air movement, CFM, and velocity).
- Climatic conditions.
- Coating temperature.
- Viscosity.
- Rheology and solvent system.
- Operator knowledge.
- Application equipment.
- Paint transfer efficiency.
- Wet and dry paint film thickness.
- Color change efficiency.
- Flash-zone.
- Bake oven (time, temperature, mass).

CHAPTER 18

ENHANCED PAINTING EFFICIENCY THROUGH WASTE MINIMIZATION

- Cool-down zone.
- Rejects due to a variety of defects.
- System maintenance and personnel training (see Table 18-1).

Companies that consider environmentalism as a problem will increasingly be paralyzed and pay dearly because of it. Conversely, those companies viewing finishing problems as opportunities to improve and strengthen the system will continue to grow and prosper.

Discarding finishing equipment, without an attempt to maximize its overall efficiency, is in most cases extremely expensive and totally unnecessary. There are ways the finishing industry can use to minimize and/or eliminate waste without investing in new equipment. One effective method of uncovering the existing opportunities for enhancing painting efficiency is to conduct a systematic finishing system analysis or audit. To illustrate this point, an actual finishing system analysis was selected. It contains a wide variety of concerns that have a direct bearing on waste and overall painting efficiency.

This audit highlights the deleterious results of operating an automatic electrostatic system (E.S.) in excess of its design parameters. The integrity of any finishing system is only as strong as its weakest component; when the weakest component fails, the entire system suffers.

PAINT FINISHING SYSTEM ANALYSIS— A CASE STUDY

This case study is intended to illustrate some of the factors affecting the paint finishing process. Important areas include:

- Conveyor loading efficiency.
- Metal pretreatment.
- Water entrapment.
- Dry-off oven.
- Cool-down zone.
- Prime booths.
- Flash zone.
- Automatic booths.
- Automatic electrostatic spray equipment.
- Reciprocators.
- Viscosity control.
- Fluid flow/atomizing air.
- Paint/solvent loss.
- Automatic spray stations.
- Results and benefits of waste reduction.

Conveyor Loading Efficiency

Extremely poor line loading efficiency results from several factors: placing parts in a single-hung manner when they should be double or triple-hung; excessive gaps between parts; and as many as 40 color changes per shift with each change requiring 32' (10 m) of an empty conveyor (See Table 18-2.) As a direct result of this deficiency, the conveyor speed was increased from a range of 12 to 15 fpm (4 to 6 mpm) to 32 fpm (10 mpm). While the intention was to increase production, new problems were created.

Metal Pretreatment

Four-stage power washer: With a line speed of 32 fpm (10 mpm), the parts put through the washer are exposed to the spray for only a fraction of the intended time, producing partially cleaned and inadequately phosphated parts. The effectiveness of each stage is greatly diminished if there is any delay. A delay allows contamination of stages two, three, and four while wasting vast quantities of cleaning and iron-phosphatizing chemicals.

Water Entrapment

Many of the fabrications do not have weep-holes, and water is entrapped because there are no air-knives to dislodge the water. As a result, the operating temperature of the dry-off oven has been increased. This increased temperature, while partially aiding the drying process, presents the equally serious problem of hot parts entering both the prime and topcoat booths. Instead of relying on

TABLE 18-1
Finishing System Variables

- Quality of steel, aluminum or other substrate(s).
- Fabrication processes.
- Conveyor line speed.
- Conveyor elevations.
- Conveyor loading efficiency.
- Racking.
- Pretreatment:
  washer design.
  washer maintenance.
  mechanical problems.
  chemical problems.
  quality problems.
- Water entrapment.
- Air blow-off station.
- Dry-off oven.
- Cool-down zone.
- Temperature of ware entering spray booth(s).
- Spray booth type.
- Spray booth size, air movement, volume (cfm), velocity.
- Climatic conditions.
- Temperature of coating.
- Viscosity of coating.
- Coating rheology.
- Solvent system.
- Operator knowledge and techniques.
- Application equipment.
- Fluid flow rate per gun, fluid pressure.
- Atomizing air pressure.
- Turbine rpms Disc.
- Disc Flow rate–oz. per min.
- Reciprocator speed.
- Vertical triggering efficiency.
- Horizontal triggering efficiency.
- Manual touch-up operator efficiency.
- Fan pattern size.
- Distance — application device to ware.
- Paint transfer efficiency, basic.
- Wet paint film thickness.
- Dry paint film thickness.
- Color change efficiency.
- Fluid handling system.
- Flash zone.
- Bake oven — time, temperature, mass, efficiency.
- Cool-down zone.
- Make-up air distribution.
- Reject level, dirt, and other defects.
- System maintenance.
- Personnel training.
- Coatings — value analysis.
- Paint film rejects — causes, and cures.

CHAPTER 18

ENHANCED PAINTING EFFICIENCY THROUGH WASTE MINIMIZATION

TABLE 18-2
Conveyor Loading Efficiency and Hourly Productivity
Conveyor "Line Speed F.P.M." × 60 Min. ÷ Parts (Single Hung)
"Hanging Centers" = Hourly Production

Line Speed (feet per minute)	Hanging Centers																	
	6"	8"	12"	16"	18"	24"	30"	32"	36"	40"	42"	48"	54"	56"	60"	64"	66"	72"
3	360	270	180	135	120	90	72	68	60	54	51	45	40	39	36	34	33	30
4	480	360	240	180	160	120	96	90	80	72	69	60	53	51	48	45	44	40
5	600	450	300	225	200	150	120	113	100	90	86	75	67	64	60	56	55	50
6	720	540	360	270	240	180	144	135	120	108	103	90	80	77	72	68	65	60
7	840	630	420	315	280	210	168	158	140	126	120	105	93	90	84	79	76	70
8	960	720	480	360	320	240	192	180	160	144	137	120	107	103	96	90	87	80
9	1080	810	540	405	360	270	216	203	180	162	154	135	120	116	108	101	98	90
10	1200	900	600	450	400	300	240	225	200	180	171	150	133	129	120	113	109	100
11	1320	990	660	495	440	330	264	248	220	198	189	165	147	141	132	124	120	110
12	1440	1080	720	540	480	360	288	270	240	216	206	180	160	154	144	135	131	120
13	1560	1170	780	585	520	390	312	293	260	234	223	195	173	167	156	146	142	130
14	1680	1260	840	630	560	420	336	315	280	252	240	210	187	180	168	158	153	140
15	1800	1350	900	675	600	450	360	338	300	270	257	225	200	193	180	169	164	150
16	1920	1440	960	720	640	480	384	360	320	288	274	240	213	206	192	180	175	160
17	2040	1530	1020	765	680	510	408	283	340	306	291	255	227	219	204	191	185	170
18	2160	1620	1080	810	720	540	432	405	360	324	309	270	240	231	216	203	196	180
19	2280	1710	1140	855	760	570	456	428	380	342	326	285	253	244	228	214	207	190
20	2400	1800	1200	900	800	600	480	450	400	360	343	300	267	257	240	225	218	200
21	2520	1890	1260	945	840	630	504	473	420	378	360	315	280	270	252	236	229	210
22	2640	1980	1320	990	880	660	528	495	440	396	377	330	293	283	264	248	240	220
23	2760	2070	1380	1035	920	690	552	517	480	414	394	345	306	296	276	259	251	234
24	2880	2160	1440	1080	960	720	576	539	500	432	411	360	319	309	288	270	262	240
25	3000	2250	1500	1125	1000	750	600	561	520	450	428	375	332	322	300	281	273	250

elevated temperatures to drive off excess moisture and/or entrapped water, strategically placed centrifugal-type (knife edge) blowers would be much more efficient and cost-effective.

Dry-off Oven

The oven is operated at elevated temperatures which would not be required if an air blow-off station(s) were installed. Millions of BTUs are wasted, as parts exit the oven at approximately 240°F (116°C) and enter topcoat booths at 120°F to 130°F (49°C to 54°C). These temperatures are extremely deleterious to applying coatings in an efficient manner. Spraying paint onto hot parts causes the solvents to evaporate out of the coating before it can flow and level on the part, causing dry spray and excessive solvent consumption. In most situations, wet paint particles provide excellent coating flowability and excellent E.S. attraction, but dry paint particles give poor flowability and diminished E.S. attraction.

Cool-down Zone

Dry-off to prime booths. There is not adequate time. All parts should enter the spray booths as close to ambient temperature as possible. Consider using fans for cooling.

Prime Booths One and Two

The distance from the center line of the conveyor to the filter media is 3' (0.9 m). This distance should be 5' (1.5 m) so that paint rebound, offspray, and overspray would be somewhat drier as they strike the filter media. Excessive filter loading is also due to the following:

- Excessive fluid flow rate per gun.
- Fan pattern per gun (too wide).
- Excessive atomizing air pressure.
- Extremely hot parts; solvent is driven out of coating too rapidly, causing excessive paint waste and solvent loss to atmosphere.

Flash Zone

The distance from the prime booth to the topcoat booth is only 29' (9 m), and at a line speed of 32 fpm (9 mpm) the flash is only 55 seconds. This is not adequate flash-off. The distance should be at least equal to three to five minutes of travel. Consider the use of fans for cooling parts.

Automatic Booths One and Two

The company incorporated both automatic and manual E.S. spray equipment into its existing finishing line to increase painting efficiency, but the booths' dimensions created a difficult situation. The measurements were as follows:

- Top of the conveyor to the floor is 7' 0" (2 m).
- Top of the booth to the bottom of the conveyor shroud, 13" (33 cm).
- Bottom of the shroud to the top of ware (part), 22" (56 cm).

CHAPTER 18

ENHANCED PAINTING EFFICIENCY THROUGH WASTE MINIMIZATION

- Top of ware to floor, 49″ (124 cm).
- Maximum hanging height for ware is 36″ (91 cm) See Fig. 18-1.

When sizing a spray booth for automatic E.S. spraying consider the following:

- The hanging height of the largest fabrication.
- The clearance from top of the part to the top of the conveyor should be 3′ (0.9 m).
- The clearance from the bottom of the part to the floor should be 3′ (0.9 m).

As shown in Fig. 18-2, a part with a hanging height of 4′ (1.2 m) would require a spray booth with an interior height of 10.0′ (3.1 m). The width of the booth is determined by adding both the length of the E.S. spray gun reciprocator mounting bar and the width of the fabrication.

A maximum distance of 5′ 0″ (1.5 m) is required from the center line of the conveyor to the exhaust chamber (filter media or water curtain); this distance should never be compromised.

Automatic E.S. Spray Equipment/Reciprocator One and Two

This equipment has been placed in a very uncomfortable atmosphere. The area is too small, with a lot of close ground potential: the conveyor shroud is 22″ (56 cm) above the top of the fabrication to be painted, with a concrete floor 13″ (33 cm) below the bottom of the hanging fabrication and only 24″ (61 cm) from the conveyor center line to the booth water curtain. Negatively charged particles will always seek out and travel to the nearest ground. Due to the extremely close proximity of the conveyor shroud to the hung parts, a large amount of paint is attracted to the shroud and drips onto passing parts, causing rejects. Parts subassemblies are entering the spray area at temperatures of 120°F to 130°F (49 °C to 54 °C). Adding a slow-evaporating tail solvent may help to balance viscosity and flowability, but spraying hot parts is difficult as the heat tends to evaporate much of the coating solvent before it has a chance to properly cover the surface.

It should also be understood that like-charged (negative) particles oppose each other; as a result, solvent evaporation is accelerated at a much faster rate than in nonelectrostatic application equipment.

One of the factors for successful E.S. spraying is proper solvent use. It is the solvent which reduces the material's viscosity and provides the conductivity and flowability in a coating material. Without viscosity control and flowability, E.S. attraction will be minimized, and waste maximized (or increased) through offspray, rebound, and overspray. Other factors to include are:

Grounding. Consider the ground integrity of all part hangers and hooks. Check if hangers have too much paint buildup. A poor ground is a safety hazard as well as a detriment to paint transfer efficiency via E.S. attraction. Hangers should be checked and changed daily. Eliminate factors such as: poor grounding, excessive spraying, hot parts too close to potential ground, and accelerated solvent evaporation due to negative (E.S.) charge imposed on atomized paint particles.

Electric eyes. The parts-sensing electric eyes are attached to the spray booth, which vibrates excessively due to two large exhaust fans. This vibration is causing misalignment of the eyes, resulting in a faulty signal which makes the automatic E.S. guns misfire. The eye-mounting poles should be rigidly mounted to eliminate any vibration. This, coupled with fine-tuning horizontal and vertical triggering efficiency, will markedly enhance TE.

Viscosity Control

Viscosity is the measure of a fluid's resistance to flow. Water is low in viscosity and flows readily. Molasses is high in viscosity

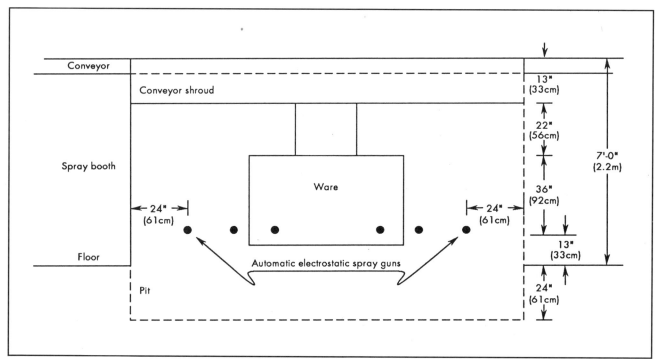

Fig. 18-1 Improper installation of automatic electrostatic reciprocator. (*Courtesy, Wagner Consultants*)

CHAPTER 18
ENHANCED PAINTING EFFICIENCY THROUGH WASTE MINIMIZATION

and flows slowly. Low-viscosity paints require less atomizing air, which reduces the forward velocity of the atomized paint. The lower the forward velocity, the greater the electrostatic attraction. The greater the attraction, the greater the reduction of offspray, rebound, and overspray—resulting in increased coverage per gallon (3.8 L) of paint.

Successful electrostatic spraying starts with viscosity control and paint flowability. High-viscosity paints have little or no flowability and low viscosity paints normally have good flowability.

If each atomized paint particle mechanically impinged or electrostatically attracted to the product is small, light, and wet enough, the result will be excellent attraction, flow out, and leveling. This will provide a good contiguous paint film and greatly enhanced paint area coverage.

Paints which tend to hide (cover) poorly at 1.2 to 1.5 mils (0.03 to 0.038 mm) dry film are usually high in viscosity, have poor flowability, and therefore have not flowed out properly. The same paints at lower viscosity may have very good hiding at 0.8 to 1.0 mil (0.020 to 0.025 mm) thickness when well atomized and retain good flowability. Many of the high solids, high-viscosity coatings flow-out and level very well when compared to the standard low viscosity, low-volume solids/high solvent coatings. Generally, electrostatic "wrap around" tends to be slightly drier than paint which has been deposited on the front surface.

Control of coating temperature is also an important tool in controlling viscosity and flowability. Accurate control of a coating's temperature will ensure a constant viscosity at the point of application regardless of climatic conditions. Climatic conditions affect viscosity and a host of variables controlling the quality of the finish. When the proper controls are not incorporated, the tendency is to spray at a much higher viscosity than is recommended. In summary:

High viscosity produces:

- Increased resistance to flow.
- Higher atomizing air pressure.
- Increased forward velocity.
- Faster evaporation of solvent from coating.
- Excessive dry spray.
- Increased offspray, rebound, and overspray.

This results in very poor E.S. attraction with waste of coating material.

Low viscosity produces:

- Less resistance to flow.
- Lower atomizing air pressure.
- Reduced forward velocity.
- Enhanced coating flowability.
- Reduced offspray, rebound, and overspray.

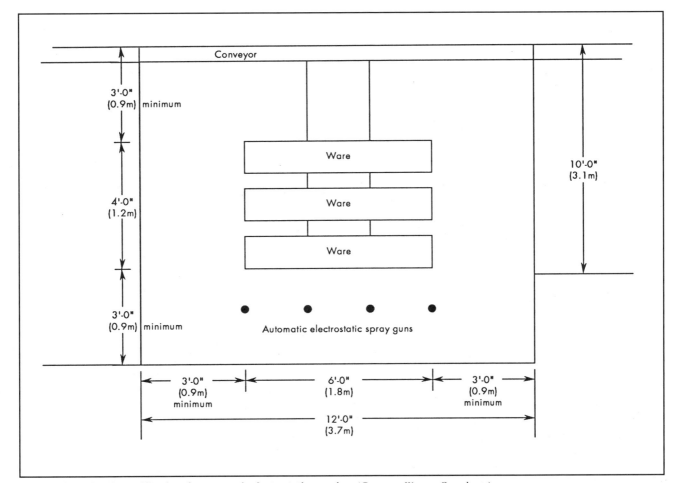

Fig. 18-2 Spray booth specifications for automatic electrostatic spraying. (*Courtesy, Wagner Consultants*)

CHAPTER 18

ENHANCED PAINTING EFFICIENCY THROUGH WASTE MINIMIZATION

The final results are an excellent E.S. attraction with maximized utilization of coating material. Thus, a proper balance is needed between temperature, viscosity, and coating flowability, coupled with the operator's knowledge of the proper setting for a proper spray pattern, proper atomizing and fluid pressures, proper distance (spray gun-to-target), solvent evaporation rate (gun-to-target), and evaporation from applied film.

The atomizing air pressures used have a direct bearing on the evaporation rate of a coating's solvent. The higher the atomizing air pressure, the faster the rate of solvent evaporation. A coating containing fast-evaporating solvents is counterproductive when used with the high-atomizing pressure spray method on hot parts. It increases cost per unit through increased reject rates, decreased production, increased paint costs due to reduced coverage of coating applied, increased emissions of VOCs, and increased generation of hazardous waste coupled with the high cost of hazardous waste disposal.

Fluid Flow/Atomizing Air

Efficient paint spraying begins with controlling fluid flow; always flow the minimum amount of paint required to meet a given production requirement.

Example.
Line speed 32 fpm (10 mpm); processing 640 parts/hr × 28 ft^2 (2.6 m^2)/part = 17,920 ft^2 (1,665 m^2)/hr to be painted.

Coating material.
One gallon (3.8 L) of a 35% volume solids (VS) coating applied at a 0.9 mil (0.02 mm) thickness and 65% TE will cover 456 ft^2 (6.6 m^2) of surface.

Electrostatic equipment.
Two reciprocators opposed and staggered. Each reciprocator has six automatic E.S. air atomizing spray guns: 2×6 = 12 guns.

Metal Coverage:
17,920 ft^2 (1665 m^2) at 456 ft^2 (6.6 m^2 coverage) = 39.29 gal (148.7 L) ÷ 60,456 ft^2 (6.6 m^2) min. = 0.6548 gal/min (2.4 L).
128 oz.(7.6 L) × 0.6548 = 83.81 oz (4.9 L)/min. ÷ 12 guns = 6.98 oz (1.6 L)/min/gun. This calculation shows the required flow rate (see Figs. 18-1 and 18-2).

When computed on the basis of the number of square feet to be processed, dry film thickness, volume solids of coating, TE, and number of spray guns involved, the flow rate (oz/min/gun) is very small.

Once the minimum flow rate is determined, the proper atomizing air pressure is the minimum amount of air pressure required to atomize the fluid properly with the least amount of forward velocity.

A proper spray pattern should be uniform in shape, size, dispersion of fluid, and particle size, with sharply defined boundaries. A minimum amount of partially atomized random particles should be present outside the spray pattern boundary. Keeping atomizing pressures as low as possible will aid in the formation of a correct and uniform spray pattern.

Paint/Solvent Loss

Manual Spray Stations. Outlet manifolds (25 colors) and two spray stations—one at each manual E.S. spray station. Because each fluid hose (a total of 50 between the two spray stations) is dead-ended (noncirculating), they are flushed each morning to purge the settled paint.

Volume: Approximately 19 (1.25 L) oz/hose/flush × 50 hoses = 950 oz. ÷ 128 = 7.42 gals/day.

7.42 gals. × 240 days = 1781.25 gals. wasted/year
Paint mixed: two parts paint, one part solvent

Paint: 1187.50 gals × $16.08 cost/gal = $19,095.00
Solvent: 593.75 gals ×   1.85 cost/gal = $ 1,098.43
           1781.25 gals   Annual Loss $20,193.43

This loss could be dramatically reduced by eliminating all 50 hoses and installing a single 0.25" (6.4 mm) manifold pipe, and ball valves connecting fluid pressure outlets to said manifold, with a solvent flush at one end of the manifold and a dual block fluid manifold with ball valves at the outlet end (see Table 18-3).

Automatic Spray Stations. The automatic E.S. lines are flushed each morning. This operation is wasting the following amount of paint:

Circulating loop paint loss:
25 lines (approx. 18′ × 3/8″ [5.5 m × 9.5 mm] tube and/or hose) × two stations = 50 lines.
Note: these lines are dead-ended off the circulating loop.

Volume: Approx. 684 oz/daily flush
684 oz. (18 L) ÷ 128 oz (3.8 L) = 5.34 gal/day wasted
5.34 gals (20 L) × 240 days = 1281.60 gals (4851.3 L)/year wasted.

Paint Mixed 2:1
Paint: 854.40 gals × $16.08 avg. cost/gal = $13,738.75
Solvent: 427.50 gals × $ 1.85 avg. cost/gal =   790.97
           1281.60 gals   Annual Loss $14,529.62

TABLE 18-3
Volumetric Content of Various Hose Sizes

Actual I.D. (in inches)	Length (in feet)	Content Gallons	Content Fluid Ounces
3/16	6	0.06	0.8
3/16	10	0.1	13
3/16	25	0.4	51
3/16	50	0.7	90
1/4	15	0.5	64
1/4	25	0.6	77
1/4	100	0.25	32
5/16	15	0.6	77
5/16	25	0.1	13
5/16	100	0.4	51
3/8	15	0.8	102
3/8	25	0.14	18
3/8	100	0.6	77
1/2	2.5	0.2	26
1/2	15	0.1	13
1/2	25	0.2	26
1/2	100	1.0	128
3/4	15	0.4	51
3/4	25	0.6	77
3/4	100	2.3	294
7/8	15	0.4	51
7/8	25	0.7	90
7/8	100	3.0	384
1	15	0.6	77
1	25	1.0	128
1	100	4.0	512

CHAPTER 18

ENHANCED PAINTING EFFICIENCY THROUGH WASTE MINIMIZATION

Color Change Paint Loss:
Paint loss from color change valves to E.S. spray guns on reciprocators:
Approx. 30.4 oz/color change
30.4 oz./color change × 25 changes/day, average = 760 oz.
760 oz.×240 days = 182,400 oz. ÷ 128 oz. = 1,425 gals/yr.

Paint Mixed 2:1
Paint: 950 gals × $16.08 avg. cost/gal. = $15,276.00
Solvent: 475 gals ×   1.85 avg. cost/gal. =   878.75
1425 gals   Annual Loss $16,154.75

Paint Loss Summary
Outlet Manifolds, Manual Spray Stations $20,193.43
Circulating Lines to Reciprocals   +14,529.62
Color Change to E.S. Guns (automatic)   +16,154.75
Approximate Annual Loss $50,877.80

Summation

This analysis reveals a TE efficiency of approximately 22% with an annual cost of approximately $467,855 for paint and solvent.

$ 467,855.42 = 12 month paint and solvent cost
$–102,928.19 = 22% paint transfer efficiency
$ 364,927.23 = 78% 12 month paint and solvent *waste*.

This annual waste figure does not include indirect labor required to clean and maintain spray booths, chemicals for water wash booths, static guard paper, paint filter pads and roll media, or disposal of paint waste, paint sludge, dirty filter pads, and roll filter media. The total cost for these items could reach $100,000 or a grand total of approximately $464,905 lost each year.

While the figures are staggering, they can be dramatically reduced by working on the items covered in this analysis. The cost of system fine-tuning, system modifications, personnel training, a partial new system or a total new system (excluding the E.S. equipment and conveyor system) could be paid for out of the vast savings to be realized by undertaking such a program.

The data compiled from this audit became the foundation of "Enhanced Painting Efficiency Through Waste Minimization, A Continuous Improvement Program." As a result of the audit, the problem areas within the system were identified, quantified, and prioritized. Thus each variable could be adjusted.

Results and Benefits

Line loading efficiency was improved to the point that:

- Painted steel per year was increased by 3,889,443 ft^2 (368231.29 m^2) while line speed was reduced from 32 fpm to 20 fpm (9.8 to 6.1 mpm).
- Chemical costs were reduced by over $500.00 per week, due to reduced chemical drag-out in a four-stage washer.
- Paint consumption increased by only 756 gallons (2862 L).
- Solvent consumption decreased by 11,094 gallons (41995 L).
- VOC emissions decreased by 98,450 lbs (44656 kg).
- Paint transfer efficiency increased from 22% to 65%, using the same application equipment.
- Dry-off oven temperature was reduced approximately 50%.
- Parts now enter spray booths at close to ambient temperature.
- Rejects were reduced by approximately 73%.

Fine Tuned

Annual $ Loss	Old System	New System	Savings	Annual Savings
Manual System	$20,193.43	$263.58	$19,929.85	
Auto. System	14,539.27	–0–	14,539.27	
Color Changer	16,154.75	1,397.40	14,757.35	
	$50,887.45	$1,660.98		$49,226.47
• Hazardous waste disposal costs reduced by				$30,415.00
• Solvent reduction 11,094 gals (41,995 L) =				$20,523.90
• Paint costs reduced from 0.0386¢/ft^2 to 0.0298¢/ft^2 (0.01187¢/m^2 to 0.0091¢/m)				$157,788.45
			Total Annual Savings	$257,953.82

Bottom-line
- Quality of finished product enhanced.
- Rejects drastically reduced.
- Hazardous waste minimized.
- Operating costs reduced markedly.
- Profits greatly enhanced.

Conveyors

Perhaps the biggest cause of paint waste and lack of productivity in an automatic electrostatic (E.S.) system is poor conveyor loading. Conveyor grades, hanger lengths, spray booth dimensions, trolley centers, conveyor speeds, hanging patterns, and conveyor loading have a direct effect on the efficiency of an E.S. system. The conveyor should always present the maximum amount of metal to the E.S. guns. Open spaces between parts should be exactly enough to achieve good flange coverage, and no more.

To maximize conveyor loading efficiencies and flexibility, the design of part hangers, hooks, fixtures, and load bars is of paramount importance. With built-in flexibility, hanging centers for various-size parts can be readily adjusted to maximize production efficiencies.

To minimize and/or eliminate existing conveyor loading problems, production scheduling, fabrication, painting, and assembly personnel should meet at least once a week to review product mix and paint (color) requirements and develop as many one-color production runs as possible. The department heads must have a clear understanding of any problems, and work together as a team to develop as smooth a flow as possible.

Case Study Summary

While the sample audit did not analyze the quality of the incoming materials of fabrication or the actual fabrication processes, these two variables must be tightly controlled via rigid quality control standards. It makes no sense to fabricate a product from substandard material; it will end up as a very costly reject at the final inspection station.

Rejection costs include, but are not limited to: raw material, fabrication, material handling, chemical cleaning, water and its disposal, energy (dry-off and bake ovens), paint, makeup air, hazardous waste disposal, and greater short- and long-term liability. As a rule, a rejected subassembly or unit is either reworked or scrapped. Reworking generally more than doubles the cost, as the reworked unit takes the place of a new subassembly or unit that could have been processed—thereby reducing production. Scrapping a rejected unit has a myriad of associated costs which increase unit costs and lower profits.

Small, light, wet paint particles sprayed on well grounded

CHAPTER 18

DEVELOPMENTS IN ELECTROCOATING

ambient temperature parts or subassemblies conveyed through a booth with proper clearances and exhaust cfm/velocities will enable the operator to fine-tune each variable and maximize E.S. attraction. Each deviation from the above lessens the overall painting efficiency. Electrostatic equipment, like any equipment, requires periodic maintenance and fine-tuning to operate at peak efficiency.

Waste minimization and overall painting efficiency will be in direct proportion to the degree of continuous attention given to each variable within a finishing operation. A program which continuously focuses on fine-tuning will truly offer cost-effective opportunities to enhance overall painting efficiency while eliminating and/or minimizing each waste stream, in many cases using existing application equipment.

DEVELOPMENTS IN ELECTROCOATING

Developments in the organic finishing industry have resulted in cost reductions, quality improvements, and reduced emissions of volatile organic compounds (VOCs).

One significant improvement in organic coatings has been the development of electrocoating. Compared to mandrel spraying, flow-coating, or even powder coating, it can deliver more consistent quality, better coverage, and lower pollution.

Electrocoating, often referred to as E-coating, is a process for painting metal parts by charging them with a negative electrical charge and submerging them in a tank containing a formulated mixture of paint, water, and other materials. The bath usually contains a relatively dilute (10 to 25% weight solids) waterborne dispersion of film-forming resin(s), cross-linking agent(s), pigments, and additives. The paint particles carry a positive charge. This process is called *cathodic;* however, it can also be an *anodic* process where the polarities are reversed. These opposite charges cause the paint to be attracted to the metal in a uniform coating over the entire part, even in places difficult to reach with other methods. The process may be part of a continuous conveyor line or may be done on an individual or batch basis via a hoist system.

Organic electrocoatings are waterborne paints deposited by an electrodeposition process.[1,2,3] They are used as primers or one-coats, mainly on steel and zinc coated steel, but also on aluminum, magnesium, copper, and other metals. They also can be used on nonmetallic conductive surfaces such as plastics primed with carbon black pigmented paint. In their use as metal primers, organic electrocoats have helped improve the corrosion resistance and lengthen the lifetimes of automobiles, appliances, agricultural equipment, and many other objects.

Organic electrocoating is an electrochemical process, but it is not identical to electroplating. The similarity lies in the fact that both involve an anode and cathode immersed in an aqueous medium through which current is passed. Organic electrocoating is not a plating process, however. There is slight dissolution of the metal accompanied by electrolysis of water at the paint-metal interface. The electrolysis reaction produces H^+ in anodic electrodeposition and OH^- in cathodic electrodeposition. In either case, the paint particles attracted to the object are destabilized, and precipitate to form an irregular, but adherent, coating. Subsequent baking allows the coating to flow out, cures the film, and further improves adhesion.

Advantages of organic electrocoating are:

- Rapid process—coating can be deposited in two minutes or less.
- Can be highly automated.
- High degree of paint utilization.
- Uniform film thickness.
- Film thickness independent of bath solids, viscosity, or shape of coated article.
- Low fire hazard.
- Low solvent emission.
- Flash time not necessary.
- Deposit free from sags and runs.
- Ability to reach and coat normally inaccessible areas.
- High film tenacity even when wet.
- Good corrosion resistance (epoxy primers).

Disadvantages of organic electrocoating are:

- High installation cost.
- Restricted to coating conductive surfaces.
- Low pigment/binder ratios necessary.
- Air drying difficult to achieve (but anodic air dry systems are available).
- Color changes not practical.

A key aspect of continuous improvement in finishing is recognition that the total system must be taken into account, not just one part of it. Therefore, one cannot develop or choose an electrocoat system without considering the other components of the finishing process such as cleaning, pretreating, and topcoating. Achieving an optimum finish depends on cooperation between component suppliers and finishers to control and optimize the components and their interactions. Certainly, continuous improvement in finishing depends on making use of improved substrates, pretreatments, etc. as well as improved electrocoats.

Another key factor is the design of the object being painted. Design has great influence on paintability and subsequent durability. Paints and other materials cannot be used to make up for poor design. On the other hand, good design plus an optimized finish will give corrosion resistance and long life.

With regard to the electrodeposition process itself, the object being painted must allow paint and the electric field to reach all recessed areas. Additional holes may have to be drilled or stamped in the part or structure to achieve this. Auxiliary anodes and increases in paint throw power may be needed as well.

Objects subjected to weathering and corrosive agents must be designed so as not to trap dirt, water, salts, and other contaminants. A well designed vehicle protects interior areas from intrusion of contaminants, allows rapid drainage of electrolytes that do get in, and has a minimum of recesses and other dirt and electrolyte traps; it will suffer much less from corrosion than will a poorly designed vehicle, regardless of the materials used.

TYPES OF ORGANIC ELECTROCOATS

There are two types of electrocoats: anodic and cathodic, depending on whether the article being painted is an anode or cathode (has a positive or negative charge). Anodic electrodeposition was the first to be commercialized, but has been largely superseded by cationic electrodeposition for automotive and appliance applications.

CHAPTER 18

DEVELOPMENTS IN ELECTROCOATING

Anodic Electrodeposition

This technology has been used commercially in the U.S. since 1962 and is useful for many applications, particularly where little or no corrosion resistance is required. The part to be painted is the anode, with a positive electrical charge which attracts the negatively charged paint particles (actually micelles containing the polymer(s), cross-linker, pigments, and additives).

The advantages of anodic electrodeposition are:

- Lower cost.
- No hydrogen evolution at the paint-metal interface.
- Lower temperature cure.

The disadvantages of anodic electrodeposition are:

- Phosphate pretreatment dissolution/spalling.
- Poor hydrolysis/saponification resistance.
- Metal dissolution (may discolor paint, affects film and bath conductivity).
- Corrosion resistance not as good as cationic electrodeposition; due to excessive metal dissolution, to pretreatment disruption, and lack of coating corrosion resistance.[1,2]

Cathodic Electrodeposition

This newer technology has largely taken over from anodic electrodeposition, particularly for high volume production and where good corrosion resistance is required. A wide range of products is available: grey or black pigmented epoxy-urethane and epoxy-polyesters for use as primers, and brightly colored or clear glossy acrylics for one coat or topcoat applications.

In cathodic electrodeposition, the article being painted is the cathode which has a negative electrical charge to attract the positively charged paint micelles. Hydrogen gas is given off at the cathode, but except at extremely high voltages or over zinc-iron alloy-coated steel, the amount of hydrogen is so small and the gas is so volatile that it does not become trapped in the paint film. Questions have been raised as to whether cathodic electrodeposition can cause hydrogen embrittlement of steel parts, but no evidence for embrittlement ever has been found.

The advantages of cathodic electrodeposition are:

- No dissolution of phosphate pretreatments on cold-rolled steel. Only slight dissolution over zinc coated steels.
- Excellent hydrolysis/saponification resistance.
- Low level of metal dissolution.
- Better corrosion resistance than anodic electrodeposition[3,4].

APPLICATIONS

Organic electrocoats are applied in different ways. They are either a primer, one-coat, or two-coat (dual electrocoat) systems.

Primers

Organic electrocoats are useful as primers particularly for appliances, motor vehicles, and agricultural equipment, because they tend to reach and coat deep recesses and other normally inaccessible areas that are difficult to paint by other means.

With cathodic primers, the paint first is deposited on the areas nearest the anode. The deposited film is an insulator and no more paint deposits there. The driving force of the electric field created by the potential difference between the electrodes becomes concentrated on the more remote (and unpainted) areas. This preferential deposition on the successively decreasing bare metal areas produces a remarkably uniform primer film thickness in all areas. This uniform thickness is maintained during the bake because the very high solids (98% or more by weight) content of the film guarantees there is not excessive flow and little tendency for the film to be attacked by refluxing water or solvent in enclosed areas.

Primers usually are applied as thin films of 0.4 to 1.4 mils (10 to 35 μm) depending on the application. A thickness of 0.5 to 0.6 mils (12.5 to 15 μm) usually is sufficient for corrosion protection, but thicker films may be needed for appearance, chip resistance, or other properties. Film thickness is controlled via the residence time, applied voltage, bath temperature, and solvent content of the bath. Raising any of these parameters increases film thickness.

Electrocoat primers give excellent adhesion, including wet adhesion, and therefore resist corrosion. They also resist the spread of corrosion if damage to the finish allows corrosion to begin. Low coverage on sharp edges, due to flow away from these edges in the oven, has occasionally led to corrosion problems, but new high edge coverage primers prevent this.

One-coat

In a one-coat electrocoat system, the coating acts as both primer and topcoat. This places considerable demands on the coating. Along with color, gloss, and possibly outdoor durability, there is the usual need for good adhesion and corrosion resistance. However, often some corrosion resistance has to be given up to get the colorful high gloss appearance.

Exterior applications. These include agriculture and lawn equipment, transformers and other electrical equipment, some automotive parts such as wheels and windshield wiper assemblies, lawn and patio furniture, fasteners, etc. Outdoor durability, i.e., resistance to UV exposure from the sun and other weathering effects, is needed along with at least some corrosion resistance. Acrylics and durable epoxies are used for these applications. Anodic and cathodic acrylics have exterior/interior versatility and the widest possible color choice.

Interior applications. These include indoor articles such as computer parts and accessories, metal office and laboratory furniture, bathroom cabinets, microwave ovens, circuit boards and other electronic and electrical equipment, tool boxes, toys and also interior or underhood automotive parts. Interior applications do not require paints with good UV and weather resistance, but may need corrosion resistance if exposed to a humid atmosphere (as in a bathroom, kitchen, laundry room, or some laboratory applications) or subject to splashing by electrolyte (underhood car parts). Detergent and stain resistance also are required for some applications. Cationic epoxies are commonly used along with anionic and cationic acrylics.

Two-coat or Dual Electrocoat

Two-coat electrocoat systems are relatively new and uncommon, but are quite practical and have considerable potential for special effects and special applications. The first coat is a primer, usually a cationic epoxy, with much greater baked film conductivity than a normal primer. However, the wet film conductivity is not overly high, so the deposition characteristics, throw power, etc. are not unusual. The second coat is a topcoat which deposits on the conductive primer substrate. The topcoat usually is an acrylic, but also can be an epoxy. It can be pigmented or clear and could even act as the base coat or clear coat for a color plus clear application that gives exceptionally high gloss and distinctiveness of image.

PAINT SPECIFICATIONS

Suppliers of electrodeposition coatings are continually synthesizing new resins and developing new paint formulas. This leads

CHAPTER 18

DEVELOPMENTS IN ELECTROCOATING

to new products, many of which have improved properties. Advances in recent years include formulations free from lead and other heavy metals, lower cure temperatures, better sharp edge coverage, greater throw power (better coverage in recessed areas), lower solvent levels (VOCs as low as 0.5 lb/gal or 1 kg/3.8 L and dropping toward zero), better corrosion resistance, tougher films, lower and higher practical film builds (0.2 to 4 mils or 5 to 100 μm), and better color/higher gloss for one-coats. Electrodeposition coatings continually are being applied to new substrates and products previously painted by other technologies.

THE ELECTROCOAT BATH

Organic electrodeposition differs from other paint application methods in a number of ways. One major one is the electrodeposition bath is both a storage tank and an application instrument (see Fig. 18-3). In addition, the electrodeposition paint is a low viscosity colloidal dispersion prone to settling and sensitive to a variety of contaminants and effects. Monitoring and control of the electrodeposition bath are critical to the ultimate quality of the paint film as well as the efficiency and stability of the bath. This is true whether dealing with a small lab bath or a large automotive or appliance tank. Continuous improvement depends on continuous attention to the electrodeposition process and related processes such as cleaning and pretreatment.

Some general guidelines include cleanliness and good housekeeping in and around the bath area and of bath equipment (including filters and ultrafiltration units), maintenance of a high level of agitation in the bath, monitoring and control of parameters such as solids, pH, conductivity, solvent level, coulombic yield, pigment/binder ratio, acid and amine levels (milliequivalents per gram), and temperature. These measurements and the data from them should be a large part of the statistical process control (SPC) program for the electrodeposition bath.

Temperature control is very important. High bath temperatures (85 to 100°F or 30 to 40°C) give greater film thicknesses, better flow during and after deposition, and for high-build electrocoats, less tendency to pinhole over zinc-coated steels. However, high bath temperature also can cause pigment and resin flocculation, molecular weight increases, and other changes that lead to settling and kick-out in the bath and on coated parts, clogging of filters, film build problems (low or variable), mottling, streaking, and other appearance problems. It is very important to stick to the paint manufacturer's recommendation regarding bath temperature.

One device to monitor electrodeposition baths, rarely used but highly recommended, is a light microscope. Periodic examination of pigment and resin dispersion quality by microscopy of droplets of paint at 200-400x with transmitted light can be used to identify trends—either in the direction of flocculation and destabilization or toward improvements in dispersion quality (which can happen over time). Reflected light microscopy at lower powers (5-100x) can be used to identify and characterize defects if they occur in the paint film. Some of the defects caused by bath stability problems are quite distinctive, none more so than "horizontal settling" in which fine flocs or "seeds" give a sandy appearance.

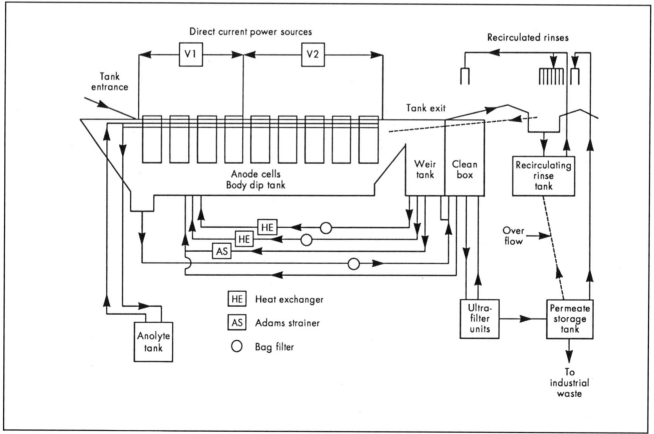

Fig. 18-3 Schematic of cathodic automotive electrocoat tank.[5]

CHAPTER 18

DEVELOPMENTS IN ELECTROCOATING

A recent advance in electrodeposition bath testing is the introduction of a device for in-tank measurement of voltage, current density, and temperature.[6] This instrument is self-contained and submersible. It shows great promise for the measurement of the effective voltage, localized current density, and temperature a few millimeters from the paint surface.

ULTRAFILTRATION

Another unique aspect of organic electrocoating is ultrafiltration. This process involves pressure-induced filtration through membranes of very fine porosity. The ultrafilter acts as a "kidney" for the electrodeposition process, removing soluble contaminants and low-molecular-weight resin. Ultrafiltration has been a key factor in the success of this painting technique and its continuous improvement.[7] By returning paint solids to the bath, ultrafiltration improves the efficiency of the electrodeposition process by 15 – 20%. By removing soluble salts and controlling conductivity, ultrafiltration decontaminates the bath and controls deposition. Permeate removed in the concentration process is used for rinsing, and rinse water itself can be concentrated. This critical feature of the electrocoat process makes very high paint utilization possible and minimizes waste.

Improvements in ultrafiltration membranes and equipment in recent years have contributed to increased process efficiency and improved bath control and film quality.[7] However, such advances can be negated by poor monitoring and maintenance of ultrafiltration units.

ANODES AND ANODE CELLS

In cationic electrodeposition, there are banks of cells down each side of the bath (see Fig. 18-3). These cells contain the anodes, the opposing electrodes for the process. They also are electrodialysis cells that remove excess acid solubilizer to maintain bath composition and pH. For many years anode cell technology changed very little, but rapid advances have taken place recently which are an important part of the continuous improvement of the electrocoat process[8]. Developments include new anode cell and anode designs, new dialysis membrane and anode materials, changes in the sizing and positioning of anodes, and the use of auxiliary anodes. These advances have helped improve film thickness uniformity, smoothness, and color control and have led to a realization of the impact of the anode cell system on the quality of the finish. More improvements in anode cell technology are expected in the future.

COATING CURE

Following electrodeposition and rinsing, parts go to an oven for curing of the coating. In the oven, the coating quickly flows out to give a smooth coating and more intimate contact with the substrate. Later, cross-linking reactions begin which change the coating from a semisolid mixture of pigment particles, linear polymer chains, and smaller molecules to a hard, tough, chemically resistant film with a three-dimensional structure.

The term *cure* is ambiguous and the determination of cure is subjective. To cure simply means to dry and/or cross-link a paint film enough to produce the properties required by the customer. Adequate cure of a given coating for one application may not be adequate for another. If the paint meets the requirements, then it is cured. If the oven temperature is too low, the time in the oven too short, or certain parts take too long to heat up, then the properties will not be achieved.

It is possible to overbake or overcure a coating. A high bake temperature or longer bake time than specified may cross-link the coating to the point of brittleness or may lead to thermal degradation. However, the cure window (acceptable cure temperature/time range) for most electrodeposition coatings is wide. Also, high bake temperatures may improve corrosion and detergent resistance due to better adhesion of the more rigid structure and lower permeability of the more tightly cross-linked film. However, it never should be assumed an overbake situation gives superior or even specified properties. Film properties should be tested on parts baked over a range of times and temperatures.

Typical cure schedules for cationic electrodeposition coating are 340 to 350°F (170°C) for 30 minutes, but there is considerable interest in lower cure temperatures to save on energy costs and to protect heat-sensitive plastics and other nonmetallic parts of electrocoated objects or assemblies.

Paints curing at 300 to 325°F (150 to 165°C) are not unusual and lower cure temperatures are practical in some systems. Anionic electrodeposition paints, mainly acrylic-melamines, tend to cure at lower temperature, 250 to 300°F (120 to 150°C). Products curing at 180 to 200°F (80 to 90°C) exist and even air dry paints are available.

Before changing to a low cure product, the finisher should discuss this with the metal supplier. Many metal products are hardened or strengthened by heat and depend on a high-temperature electrodeposition bake for the development of ultimate properties. It may be that the savings in energy or other advantages of a low cure temperature may be outweighed by losing substrate properties or having to change substrates.

REPLACING SOLVENT-BASED FINISHES WITH ELECTROCOAT

E-coat is most widely used as a primer, although it is commonly used as a finish coat in high volume applications such as fluorescent fixtures, wheelbarrows, and toolboxes. It provides excellent resistance to chipping, cracking, and abrasion. The finish also provides superior corrosion protection; it can withstand over 500 hours in a scribed salt spray test and a 1000-hour humidity test.

In most cases, where E-coating can be used in place of another finishing process, the switch will bring a substantial cost reduction. However, there will be high capital costs in the beginning. The biggest cost reduction in switching to electrocoating comes from its greater transfer efficiency (TE). This typically ranges from 95 to 98%, depending on the type of tooling and part configuration. The small percentage of coating not applied to the parts may be dragged out of the bath and end up on a drain pan. Some of it will coat the tooling (racks) used to hold the parts, but one coat insulates the rack from subsequent coating. In this area, savings can be realized in cleanup. Racks used in spray or flow-coating processes receive one coating per pass and require more frequent cleaning.

Improvements in transfer efficiency can be made by modifying the part design to eliminate areas that collect coating, create puddles, or result in air pockets. Tooling design also comes into play in part orientation and establishing proper grounding. Orientation can be done with spring fingers for contact points; hooks fabricated from square stock will provide two corners for grounding.

These are the main areas where improvements in transfer efficiency can be made. However, electrocoating's 95–98% efficiency compares to 60–70% for spray coating on large, flat parts. This can drop as low as 30% on small parts, and often it becomes necessary to apply 1.5–2 mils of other coatings to

CHAPTER 18

DEVELOPMENTS IN ELECTROCOATING

achieve a minimum of 0.7–0.8 mils on all surfaces. Since electrocoating overcomes most of these inefficiencies, it is more effective to concentrate on other areas for further improvement. Electrocoating is good for light sheet metal or low-mass metal materials.

The energy costs of makeup air in paint booths are another factor to consider. Drying of solvent-based coatings requires high exhaust rates to prevent the accumulation of solvent in flash-off zones and ovens. Especially in northern climates, the cost of filtering and heating makeup air and the BTUs that are lost through exhaust fans can be quite high. Ovens (even some flash-offs) for electrocoating need air too. Electrocoating ovens are usually longer and hotter than for solvent systems, hence, more air is needed. The big savings is not in the oven but in eliminating the need for paint booths.

Another gain provided by electrocoating, especially important in meeting environmental regulations, is the substantial reduction in volatile organic compounds (VOC). Spray primers generally approach 3–3.5 lbs (1–1.6 kg) of VOC per gallon (3.8 L), with an upper limit of three and one half for those classified as "extreme performance." These are usually the newest high-solids coatings; traditional spray coatings tend to run between 4–6 lbs (1.8–2.7 kg) of VOC per gallon (3.8 L) as they come from the can and often must be "reduced" with additional solvents. By contrast, the bath used in E-coating contains less than two lbs (7.6 kg) of VOC per gallon (3.8 L).

If the water is excluded from the VOC calculation, even water-base spray or dip coatings may not be environmentally friendly. These coatings still contain a small amount of solvent, contributing a surprisingly high percentage of the solids. Powder coatings, as applied, would appear to produce less VOC than electrocoating, but the process of polymerization generates volatile compounds. The subject of VOC reduction is more complex than it appears, and decisions must be made carefully to achieve the best possible results without sacrificing corrosion protection and other attributes.

Flow-coating and dip tanks, using water-base coatings with some solvent to obtain better flow, offer some quality improvements over conventional spraying. However, both are subject to an effect termed *wedging*, which results in a thicker coating toward the bottom of a part than at the top. This is caused by the flow of the coating as it drips and dries. Because E-coating delivers uniform and controlled deposition, it provides greater uniformity and higher quality than other organic processes. Its greater "throwing power" results in better part coverage.

E-coat also provides excellent corrosion protection. While anodic acrylics or even cathodic acrylics offer improved durability for exterior finishes, the greatest corrosion protection is afforded by cathodic epoxies used as a primer. The two biggest users of E-coat primer, the automotive and appliance industries, are also two of the most concerned with corrosion. In these cases, E-coat is used as a primer to protect against extreme corrosion, while additional finish coatings provide the wide variety of colors desired by the consumer on exposed surfaces. Paint costs can depend on if the part is painted on only one side or on both sides.

Disadvantages and Constraints

Since E-coating requires a dedicated line or tank and a post-rinse for each color, it is only practical as a topcoat where long runs of the same color can be made. Even if the expense of setting up separate lines is not a factor, shorter runs are discouraged by the difficulties of maintaining stability in a line not used full-time. Switching colors in an E-coat bath is not a practical alternative in most cases. While it does not require a large amount of paint (10 to 20% paint solids), disposal of the previous material can be complicated. With 25- and 40-micron filters used in the line to filter out dirt, the solids cannot be filtered out. The solids in E-coat materials are only about 10 to 20%, but these must be precipitated out for proper disposal. With environmental regulations, the objective is to have no waste that needs disposal. Selecting the right process or combination of processes for the product will avoid creating excessive waste. Higher transfer efficiencies and better equipment design will help keep it at a minimum.

USING ELECTROCOAT AS A PRIMER FOR POWDER COATING

Because of its excellent adherence and corrosion protection, E-coat is ideal as a primer for most other coatings, including exotic powders that would normally require special nylon or vinyl primers. In fact, when used under powder, E-coat provides a combination of improved corrosion resistance, reduced powder usage, and better powder coverage. The epoxy resins can be used without a topcoat for areas not exposed to direct sunlight. Under direct sunlight, epoxy E-coat primer will chalk, losing its gloss and eventually turning from black to gray, although its corrosion protection is unaffected. Surfaces out of sight or not exposed to sunlight can be protected against corrosion with an overall E-coat, while exposed surfaces are over-coated with powder or liquid. In many cases, such parts can be hung back-to-back for the final spray, improving efficiency by spraying only the visible areas and coating two parts for the price of one (if the equipment permits two-sided painting).

An unexpected benefit realized by one equipment manufacturer who was painting the product's exterior a light color. Originally, the firm used a conventional primer in matching color. When converting to E-coating for various parts, the company found that the coating's standard black color allowed the quality staff to monitor topcoat coverage more effectively. Normal powder or liquid coatings of approximately one mil (0.0254 mm) thickness were adequate to hide the base coat.

Many contract coating firms use a zinc phosphate pretreatment to meet the rigid quality demands of the automotive and appliance industries, where the E-coat process is already widely used. This provides a bonus for equipment manufacturers, who can often upgrade their coating results without paying a premium by using an existing line that meets the standards.

Conventional powder coatings will generally cost from two to three times the cost of E-coating, when applied to typical equipment parts. While cost figures given for powder coating sometimes look impressive, they are usually based on an optimal minimum film surface that is difficult or impossible to achieve in normal production. Real parts, with recesses and corners, take more coating to achieve coverage by spraying. Initial transfer efficiency in production may be lower for powder coating than with E-coating. With powder recovery and reuse, final transfer efficiencies vary due to the loss of fines, contamination, losses during color changes (done by some industries), and multiple coats on the hangers. Using an electrocoat primer to provide corrosion protection will make the role of the powder topcoating less critical—primarily for appearance—and eliminate the need to apply excess powder (amount of powder use is often related to appearance requirements). This waste reduction translates into lower costs.

Some economies of scale can be achieved by ordering primed parts in larger quantities and stocking them after E-coat has been

CHAPTER 18

DEVELOPMENTS IN ELECTROCOATING

applied. Then shorter runs of various color topcoats can be applied as needed. If they are stored out of direct sunlight, either indoors or outdoors, the primed parts will be protected against corrosion until needed with no chalking of the coating. Open outside storage and subsequent finish coating over chalked E-coat should be avoided, however.

USING ELECTROCOAT AS A PRIMER FOR VINYL OR NYLON POWDERS

Some powder coatings, particularly vinyl and nylon powders, require a primer for adequate adhesion and optimum performance. Traditionally, this has been accomplished with vinyl-based liquid primers that are either sprayed or dipped and cured or partially cured prior to applying the powder topcoat. Experimentation has shown that epoxy E-coat works well in many cases as a primer for nylon and vinyl powder coatings. It provides the benefits of improved corrosion protection for inaccessible areas and also reduces VOC emissions.

An E-coat primer functions as a supplementary coating, protecting inaccessible areas impossible for powder to reach, or areas that do not require a finish coat for appearance. It is common to apply vinyl or nylon powders over a primer by preheating the parts and dipping them into a bed of powder fluidized with compressed air. While this helps assure more even coverage and provides a desired heavy film build ranging from 5 to 20 mils (0.127 to 0.5 mm), the powder still has difficulty reaching into air pockets, nooks, and crannies of complex parts.

To improve transfer efficiency, an electrostatic fluidized bed may be used. In this situation, a portion of the bed is electrostatically charged, and excess powder then falls from the preheated part back into the bed. This approach generally will not yield as heavy a film build as a plain fluidized bed. With either type of powder application, however, a primer still is necessary for nylon or vinyl resins. The types normally used have a high solvent content and often must be reduced with still more solvent to achieve a thin, uniform coating for low-volume production. Where epoxy E-coat primer is used, VOC emissions can be reduced to under 2 lbs (0.9 kg) per gallon (3.8 L) from as high as 3 to 6 lbs (1.4 to 2.7 kg) per gallon (3.8 L) with vinyl spray or dip primers, and coverage is much more uniform.

APPLYING ORGANIC-BASED DRY FILM LUBRICANT ON WIRE

A definite quality improvement can be made by preapplying an organic-based dry film lubricant to wire used to make continuous hinges. These "piano hinges" are widely used on cabinet or compartment doors of various types. During manufacture, the hinge parts are formed around the wire pin in a continuous process, and the hinges are cut to length after the forming operation. Applying a dry film lubricant to the wire before it is built into the hinge prevents squeaking and avoids the need to apply messy lubricants later.

Either steel or aluminum wire can be used. Preparation consists of coating the steel wire with zinc phosphate, or cleaning and etching the aluminum. While the zinc phosphate protects the steel wire against corrosion, it is extremely prone to squeaking if not lubricated. This is true of the aluminum to a lesser degree. For this reason, a dry film lubricant is applied to the wire after pretreatment and then post-cured at a high temperature. Because the hinge leaves are folded over the wire, rather than pushing the wire through as with other larger hinge pins, the lubricant does not get scraped off. The pretreatment also helps retain the lubricant.

While the nature of the dry film lubricant application makes it difficult to achieve an even coating, normal inconsistencies actually can be an advantage. As the hinge is formed, any excess coating gets trapped in the barrel and serves as a reservoir of permanent lubricant to prevent squeaking and eliminate future need for oily lubricants.

REPLACING HIGH-VOC LIQUID ARCHITECTURAL COATINGS

Kynar®-based resins are widely used for architectural finishes because they provide extremely good corrosion protection and weatherability on aluminum window and door frames, large panels, trim, and related building materials. Because of their ability to provide excellent color control, they often replace anodizing as the preferred architectural finish. A complete color selection has been developed, including exotic colors and metallics. Unfortunately, they also are high in VOCs, and suitable alternatives will be needed as environmental restrictions increase. In addition, many recent color treatments incorporate metal flakes and clear topcoats, often requiring a three-step finishing process with a VOC output of 4 to 6 lbs (1.8 to 2.7 kg) per gallon (3.8 L). This may go even higher if the material must be reduced with solvents for application. EPA exemptions have been temporary and short-term, and experiments have been tried to produce powder, water-base, and high-solids versions without much success. Afterburners and incinerators needed to clean up high VOC levels are costly. A new recirculating air paint booth with capture equipment has a high initial cost.

One answer is a series of high-performance polyester powder coatings used for architectural applications. Color choices are similar to the Kynar® coatings and performance appears to be equivalent as well. Application techniques are similar to other powder coatings, but superior pretreatment is essential. The chromate conversion process, while similar to other pretreatments, usually requires several added steps.

The architectural field already has accepted some polyester and silicone-based acrylic resin liquid coatings with performance substantially less than Kynar. The polyester powders exceed these requirements. Manufacturers and custom coaters are working to develop and test the polyester powders to the point where they will soon match Kynar performance in the field. Environmental considerations also are favorable, since powder coatings are not considered to have VOC problems and are a one-step process that cures at less than 400°F (204°C), rather than a two-step or three-step process, each curing at over 450°F (232°C).

In the field, another advantage of the polyester powders becomes apparent, as they are easier to touch up if scratched. The Kynar coatings, because of their fluorocarbon base, present a nonstick surface difficult to repair. The process often requires three separate coats, including primer, topcoat, and clear coat, to come close to matching the finish. It is much easier for manufacturers to develop a two-component liquid that will match and adhere to the polyester powders and simplify on-site touch-up.

USING ELECTROCOAT AS A PRIMER FOR MILITARY CARC COATING

Chemical agent resistant coating (CARC) was developed for the special needs of the military and incorporates two unique features. As its name implies, it can be cleaned up after exposure to chemical, biological, or nuclear contamination; a urethane-based topcoat makes cleaning and re-use possible. It also incorporates some special antiradar properties that help hide the coated equipment from enemy radar. In fact, during the Persian Gulf

CHAPTER 18

DEVELOPMENTS IN ELECTROCOATING

War, special coatings were developed and supplied on an emergency basis for this purpose.

CARC coatings are typically four-part applications, consisting of a wash primer, a white epoxy primer, either a moisture-cured or two-component polyurethane topcoat, and possibly a camouflage coat. The epoxy primer generally has a seven-day window for finish coat application. After that time, it cures so hard the topcoat will not adhere.

Until recently, military specifications called for any approved coatings to be field-applicable to accommodate repairs. Now, E-coating is acceptable to replace the first two steps, although the original process is still used in the field. For new components, however, this reduces costs substantially, first because the transfer efficiency is greater, and second because one step can be eliminated. A phosphate pretreatment is part of the electrocoating process, and the E-coat primer provides excellent corrosion protection. The military specification is written for a gray nonlead and nonchrome, high-build primer. However, exceptions are being written regularly, particularly for color, since most E-coat primer systems in operation are black. More and more applications are converting to high-build coating, which allows the buildup of thicker depositions (1.2 to 1.5 mils – 0.030 to 0.038 mm) simply by adjusting the voltage in the system. Earlier E-coat compounds could not approach the capability of these high-build materials. Lower film-build epoxy primers (0.6 to 0.8 mil – 0.015 to 0.020 mm) also have been approved with test results. Cost and environmental considerations are important reasons for replacing the catalyzed primer used in the original process. When a batch of this material is manually mixed and not used completely before its pot life expires, the excess must be discarded. It becomes a waste, with all its attendant disposal problems.

Elimination of the vinyl acid wash primer step also helps cut VOC emissions dramatically. Vinyl acid wash primer is the only coating now allowed by the EPA and most states to exceed 3.5 lbs (1.6 kg) per gallon (3.8 L). While now permitted up to 6.5 lbs (2.9 kg) per gallon (3.8 L) because there has been no alternative for it, this is on an annual review basis only.

Quality improvements achieved by using E-coat with CARC finishes are primarily in increased corrosion protection, certainly a big factor for the severe service conditions found in the military. While film thicknesses up to 1.8 mils (0.02 mm) are specified for the original two-step primer process for CARC, exceptions to 0.8 mils or less are granted for electrocoating because of its superior corrosion resistance.

REPLACING CADMIUM AND ZINC PLATING WITH E-COAT ON TUBULAR PRODUCTS

While nonorganic cadmium or zinc coatings provide excellent durability and corrosion protection, they also are expensive. Further, on some types of parts, the nature of the plating process makes it hard to achieve the coverage needed to protect all surfaces. Tubular products represent one such example.

In many cases, E-coat can provide the right combination of physical durability, corrosion protection, and cost reduction with an increase in quality. Compared to most organic coatings, E-coat provides a hard finish, although not as hard as a plated finish. For this reason, it is not suitable for permanent protection of screw threads, for example. For corrosion protection, however, E-coat applied over zinc phosphate pretreatment generally matches or exceeds the corrosion protection of plating racked parts with zinc or cadmium, at the same or lower cost.

The advantage of using E-coat for tubular parts is its superior throwing power. Inorganic coatings, being conductive, continue to deposit on the most accessible outside areas of a tubular part rather than depositing on the inside. E-coat is nonconductive and self-insulates as it is deposited, so it seeks out the noncoated conductive areas such as inside tubular sections. This gives it the ability to coat into a tube a distance of 10 times or more the diameter of the opening, whereas plating will barely cover one diameter. This same tendency applies to other inaccessible areas, of course, but tubular parts are the most dramatic example of improvement. Parts should be designed to eliminate air pockets or puddles whenever possible.

While it is possible to achieve additional throwing power with plating by using internal electrodes, this is a costly alternative requiring capital investment in dedicated tooling. Further cost reductions can be achieved because the high costs of environmental measures that affect plating do not apply to E-coat. The costs of installing and maintaining water treatment and hazardous waste disposal operations, for the cyanide-based baths and similar processes used in metal plating, make it economically advisable to find suitable alternatives whenever possible.

An additional consideration for welded zinc-plated products is the dilemma of either plating over a weld, which may cause quality problems, or welding after plating and burning off the coating. Where corrosion protection is the primary objective, E-coat can be applied at either stage of manufacture. A partially finished part may be coated prior to welding to provide corrosion protection during handling. Burning off the E-coat, when necessary, is much less hazardous than burning off zinc or cadmium plating. Later, the weld surface or the entire part can be recoated easily. In either case, proper cleaning of nonconductive weld slag is essential to achieve a quality coating.

PAINT FILM TESTING AND OTHER MEASUREMENTS

A very important factor in the improvement of a process is testing of the process results or products. Paint testing methods are not the most accurate or precise of tests, but if used properly, they can be an important and effective part of statistical process control. Test methods must be standardized. It does no good to test if every person has a different version of the method. It is necessary to use ASTM tests or to develop in-house standard test methods and stick to them.

Other than for film thickness, parts themselves rarely are tested because of the cost of sacrificing them and the fact that such tests often are difficult to do. The usual method is to run panels through the line and do the tests on the panels. Common tests include film thickness, solvent rubs for cure (minimum of 100 double rubs with acetone or methyl ethyl ketone), crosshatch adhesion (ASTM D 3359), direct and reverse impact, and measurement of color. Appliance coatings need to be tested for stain and detergent resistance; auto primers for smoothness (which can be done on the car) and stonechip resistance (on panels). Many coatings must pass corrosion testing: salt spray (ASTM B 117) and/or cyclic scab corrosion.

Most testing is done during coatings development, for product submissions, or when problems occur, but periodic coating and testing of panels is a good way to monitor the system and provide SPC data. At least one paint supplier removes a gallon of paint from each bath once a month (once a week when a problem exists) and takes it back to the lab to do coat-outs. Film thickness and color are noted every time and the panels kept for reference. Other tests are run on an as-needed basis depending on problems or questions that arise.

CHAPTER 18

DEVELOPMENTS IN ELECTROCOATING

Two paint system measurements that test properties critical to paint performance are throw power and pump stability. These are not standard bath monitoring tests, but are important steps in paint development. Throw power is the ability of a paint to deposit films in recessed areas. A number of methods for measurement are described in the literature[2-4, 9-12], the two most widely used in the U.S. being the *Ford Cell Test*[9] and the *GM Cell Test*.[10] Most tests involve measuring the thickness or height of the paint deposited on the inside of a hollow metal tube or box open only at the bottom which has been immersed in a laboratory electrodeposition bath.

Pump stability testing usually involves 24 hours of continuous circulation of a paint specimen through a high shear impeller pump at 90°F (32°C). After pumping, the paint is filtered and the amount of sludge determined. This test measures the shear stability of the colloid composition of the paint.

Coatings Defects

All organic coatings show surface defects from time to time.[13] Electrodeposition coatings are no exception, but they tend to have fewer defect problems than many coatings. Also, a considerable number of electrodeposition coats are primers where appearance is not as critical and many minor problems are handled by sanding. Solving and preventing defect problems is important to any continuous improvement program, however.

Since electrodeposition coatings give fairly uniform films of moderate thickness, sagging is rare. Orange peel is common, although smoother substrates and a better understanding of the causes of coating roughness and bumpiness are leading to smoother coatings.

Historically, edge coverage has been a problem with electrodeposition coatings. Many formulations flow easily because they have low viscosity early in the baking, and they cannot resist the surface tension forces at the edge. The result is flow away from the edge and a thin (or no) coating at the edge. The simple remedy always has been to load the coating with clay or some other extender or pigment, but this gives poor gloss and/or severe orange peel—one problem is exchanged for another. Recently, new formulations giving high edge coverage and fairly smooth surfaces have appeared.

Solvent or water popping is very rare in electrodeposition coatings and only occurs with excessively high film builds and/or with very hot ovens. The usual remedy is to ramp the bake temperature and allow volatiles to escape before the coating skins over. There is a pop-like defect called *galvanized gassing* that has caused considerable trouble in finishing over zinc coated steels, particularly electrogalvanized, in the last few years.[14] This defect is caused by gassing from voids, blisters, flaps, or other defects in or under the zinc layer. When topcoated, galvanized gassing often gives defects identical to solvent pops. It usually is necessary to cross-section the defect and examine it with a microscope before a definite diagnosis can be made.

Cratering can occur in electrodeposition coatings due to oil on the substrate, in the bath, or oozing out of hem flanges and other closed areas in parts being painted. The cause sometimes is very difficult to identify, but better metal cleaning and bath filtration with special oil filters usually improve the situation.

A defect that often looks like cratering is rupture or pinhole gassing/cratering which occurs at high voltages and/or over zinc coated steels, particularly, zinc-iron alloy coated steel.[5,15] Although the mechanisms with anionic and cationic coatings are different, the defects respond to lower voltage in both cases. The mechanism for formation of the defect in anionic coatings is joule heating in which the deposition process at high voltage heats water in the precipitating film to the boiling point. The vapor blows its way out, leaving craters, pinholes, and bubbles. In cationic coatings, the defects are caused by electrical discharge—sparks—in the bath accompanied by evolution of hydrogen and water vapor. The combination of sparking and gassing produces a variety of defects; some of these flow out in the oven, but many are frozen in as the coating cures.

One surface defect unique to electrodeposition coatings is called *hashmarking*. This defect is a series of horizontal lines or layers in the paint looking somewhat like geological strata. It appears that the paint is deposited in jerky steps as the part is immersed and this is exactly what happens. The part is so "hot" electrochemically, and the deposition so rapid and complete, that very tight stripes (which do not flow out much on baking) are coated out as the part drops down into the bath. Remedies include removing the first anode or two to give a "cold" or low voltage entry, or changing the cleaning process to make the metal less electrochemically active.

Problem Solving

Problem solving with electrodeposition coatings is similar to any coating system: make careful observations and experiments to define the problem and determine its cause, *then* attack and solve it. Too often an observer jumps to conclusions and applies a remedy for the wrong problem. Skillful problem solving is an important part of continuous improvement, particularly when the solutions are well documented and are communicated to paint and equipment development people.

Tools for defect/problem identification include:

- Eye and experience (can fool the observer, however).
- Hand lens or pocket microscope.
- Laboratory light microscope.
- Stereo microscope (5-100x).
- Metallurgical microscope (reflected light, 100-500x).
- Cross-section of defect areas.
- Biological microscope (transmitted light, 100-500x) to look at paste or bath specimens.

Determination of causes is more complicated. Success depends a great deal on careful observation and inspired detective work. Depending on the problem, observations may include walking and examining the paint line from the entry of the substrate to the exit of the finished goods, close examination of the substrate for cleanliness and to determine whether it is what it is supposed to be, and careful monitoring of the paint manufacturing process. Initially, such observations should be done by eye by an experienced observer. They should be augmented by a hand lens, microscope, or other tools where helpful. Photography of the defects and related effects can be useful, but usually must wait until examples of the problem are brought back to the lab. There the full range of tools can be applied: light and scanning electron microscopy (SEM) for documentation of defects, x-ray fluorescence (usually in conjunction with the SEM) and Fourier transform infrared spectroscopy for identification of chemical species in and around defects, ion analysis for contaminants and, occasionally, sophisticated surface analysis techniques such as ESCA, Auger, and scanning ion mass spectroscopy (SIMS) to identify low levels of contaminants and distinguish between different chemical structures (such as whether silicon, *Si*, is present in a silicate or in a silicone fluid or surfactant). Other chemical analysis techniques can be applied to bath samples.

A very important part of problem solving is the recreation of defects. If you can produce defects in the lab, you usually can

CHAPTER 18

QC/QA FIELD APPLICATION FOR PAINTED PARTS

solve the problem. This normally is done in a small (one to five gallons) lab bath, but some testers have had good results with a one liter bath using material directly from the production tank. The small volume makes it easy to introduce contaminants, try additives, change temperature or other conditions, and do other experiments without taking a lot of time or wasting (and having to dispose of) a lot of paint.

Finally, when solving problems, the coatings literature should not be forgotten. A few hours in the library may save days or weeks in the lab. Computer information and data banks can be searched using key words to locate and print out relevant references. These can be looked up and read, or copies can be obtained from other libraries. In-house reports and memos related to past problems can be very useful since many problems tend to reappear every few years. When a problem is solved, the problem and its solution should be documented carefully and a retrievable report (one referred to in a computer information bank or card catalog) should be filed. Many successful problem solvers keep detailed files of problems attacked and solved with copies of reports, photographs, spectra, etc. Good detectives use their eyes, their minds, and their files.

FUTURE TRENDS AND IMPROVEMENTS

The trends already apparent with organic electrocoats will continue into the 21st century: smoother coatings, better edge coverage, better throw power and uniformity, better corrosion resistance without heavy metals, and lower cure temperatures. Two-coat electrocoats will become more common, including color plus clear systems. A wider range of chemistries, colors, and film thicknesses will be available. New high throw power coatings will give better value, because coverage of inside or deeply recessed areas can be accomplished without depositing excess coating on the outside surface. Electrodeposition coatings will be used more in the electronics industry, and will be applied to a variety of new substrates and products in a number of other industries. Continuous product and process improvement will be accompanied by continuous widening of the application of these versatile materials.

QC/QA FIELD APPLICATION FOR PAINTED PARTS

Quality control (QC) and quality assurance (QA) programs are important for measuring process improvements for painted parts. Performance characteristics, field-test techniques, standards, and understanding the benefits of quality assurance are also needed.

QUALITY CONTROL

Quality control in today's manufacturing environment ranges from simple visual checks to the implementation of a series of laboratory evaluations, based on ASTM methods and aimed at meeting standards. The decision on which program to use depends on the degree of quality desired by the customer. The producer's job is to make sure that the product meets the customer's standards. The science of quality control is only a tool for the measurement of success or failure in achieving objectives. Quality control, however, is a necessary step to quality assurance.

QUALITY ASSURANCE

In the case of painted parts, quality assurance (QA) is the knowledge that the "marriage" of pretreatment and topcoat is strong and will withstand the scrutiny of customers and prospects.

The strength of this marriage can best be assured in two ways:

1. Commitment of the manufacturer to producing the highest quality possible. This could mean regular periodic control tests in the plant and dedicated personnel to perform them.
2. Support from all vendors involved in the process. This includes pretreatment vendors, paint suppliers, applications/equipment people, and possibly outside consultation.

PERFORMANCE CHARACTERISTICS

Two fundamental performance characteristics of painted parts should be tested and measured:

- Corrosion resistance.
- Coating adhesion.

These are not mutually exclusive characteristics, and each affects the other's performance. It is mandatory to conduct specific tests for each one to completely understand the relationship of the substrate pretreatment, the topcoat, and the ensuing quality of the finished product.

Corrosion Resistance

Salt spray (fog) testing. Salt spray is the most common of the corrosion tests performed in the industry. Salt spray testing is conducted on painted or coated specimens which are subjected to a corrosive environment. The tests are conducted according to American National Standard ASTM B117 and ASTM Designation D1654. Typically, the test is performed in an on-site laboratory, by chemical and/or paint suppliers or independent testing companies. Salt spray is an accelerated destructive test method which mimics severe outdoor exposure on painted parts. The controlled atmosphere used in this test requires special equipment. The basics of the method subject a painted part or panel to a concentrated salt spray, or more correctly, salt fog, at a precise temperature for a period of time agreed to by the parties involved. ASTM B117 spells out detailed methods for correctly handling the test specimens and constructing the salt spray cabinet. The evaluation of the test specimens falls under ASTM D1654. This standard requires that the topcoat of the part be scribed or cut through to metal before exposure to the corrosive salt fog environment. The procedures used to evaluate the results of the exposure measure how well the paint film and pretreatment work together to keep corrosion from expanding from the scored line, and also measure the moisture permeability of the paint film.

Blister creepage. The first evaluation, called *blister creepage*, is measured as a rating of 0 to 10. Each number in the rating relates to the distance corrosion has expanded away from the scribe. A rating of 10 means there was no creepage and a rating of 0 means there was more than 5/8" (16.0 mm) expansion of corrosion from the scribe area. Incremental ratings are measured in 1/32" (0.79 mm).

Body blister. The second procedure, *body blister*, rates both the paint's ability to prevent moisture penetration and the conversion coating's capacity to prevent corrosion. An unscribed area of

CHAPTER 18

QC/QA FIELD APPLICATION FOR PAINTED PARTS

the panel is reviewed for corrosion spots, blisters, and other damage, and given a rating of 0 to 10. The rating is based on the percentage of a prescribed area of the panel experiencing failure. No failure gets a 10 rating, and a 0 rating means that greater than 75% of the area observed showed body blisters.

As a general rule, every hour in salt spray represents one week of outdoor exposure. The number of hours of exposure, and the desired ratings for both blister creepage and percent of body blisters, are specified through the agreement of all parties (vendors, suppliers, manufacturer, customer). It is important to include the blister creepage and body blister ratings as part of the specification, as it is these values that qualify the salt spray test.

Disadvantages of using salt spray. Many companies use salt spray results as the only measurement in their efforts towards quality control. This has some drawbacks, such as:

- Equipment expense and maintenance.
- Time lag to get results. This is dependent on standards. A 1000-hour test must be in the cabinet for over 41 calendar days.
- Causal factors have not been tested to determine the reasons for any failure.

Other corrosion resistance performance evaluations include *humidity testing* and *water immersion*. Though these tests are not the norm in industry as a whole, they sometimes replace salt spray as the definitive test method.

Coating Adhesion

The second fundamental performance area tested is *adhesion*. Adhesion performance can affect corrosion resistance, and vice versa; but certainly, if the paint does not stick, the marriage has failed. There are three basic adhesion tests adequately describing the interrelationship of the substrate, pretreatment, and topcoat matrix.

The three tests, *cross-hatch*, *conical mandrel*, and *impact*, will quite adequately show how well the pretreatment and paint film are bonded and able to work together under stress. Typically, the better the adhesion, the better the corrosion resistance.

Cross-hatch test. This method measures adhesion by using tape on the coating (ASTM Designation D3359-83). This test is designed to measure the bond between coating films and metallic substrates. For parts pretreated chemically (phosphatized), this test also measures the bonding performance between the conversion coating and the paint film. The test should be done on actual production parts if possible. Panels representative of production may be used also. The procedure involves very simple equipment: a razor knife, a straightedge, and tape that is 1" (25.4 mm) wide and rated to approximately 40 ounces per inch (1.1 kg per 25.4 mm) width adhesion strength (masking tape will work for field applications.) The test consists of cutting through the paint film cleanly eleven times vertically and eleven times horizontally. The scribe lines should be one mm apart, and must intersect at a 90° angle. This provides 100 squares in a grid. The tape is then applied over the grid and rubbed firmly with the eraser end of a pencil. (One end of the tape should remain free of the surface in order to pull the tape off after about a minute.) Pull the tape rapidly but smoothly at 180° to the surface and evaluate. Results range from a 5B rating, which means no loss, to 0B which means greater than 65% loss. *Loss* is defined as "removal of one of the squares." To help ensure that the edges of each square are not damaged before the test is finished, a clean, sharp, unchipped blade is recommended.

Conical Mandrel or T-bend Test. Also known as "Elongation of attached organic coatings with conical mandrel apparatus, ASTM Designation D522-60 (reapproved 1979)," this test method is as the name implies. A painted panel is bent around a metal cone, and the distance of the longest crack produced is measured. By using graphs in the ASTM Standard, a percent elongation is obtained. This percentage requires a correction for film thickness. Bending of the panels shows the degree of flexibility of the coating and the surface it is applied to, as well as the level of adhesion of the film to be expected from the product in the field.

Bump or Impact Test. This test is also called "standard evaluation of organic coating from results of rapid deformation, ASTM Designation D2794." The name of the test is descriptive of the method and the representation it is trying to achieve. By physically striking a painted part or panel and observing the effects, the test indicates how a product would withstand a rock hitting it, a tool dropping on it, or another incident of impact. This test measures both adhesion and coating brittleness. The test is performed by placing a panel over a 5/8" (15.9 mm) hole in the apparatus base plate. A 2 lb (0.9 kg) weight is dropped through a graduated tube onto a 5/8" (15.9 mm) rod, which is rounded at the point of contact with the panel and resting on the panel surface. The impacted area is measured and rated on the concave side for direct impact and on the convex side for reverse impact. The measurement is in inch-pounds pressure, taken from the graduation marks on the slide. Failure is recorded when paint loss is evident from applying and removing pressure-sensitive tape over the contact point.

Two more tests can help account for or give indications that failures have occurred or are imminent. These are:

Film Thickness or DFT. *DFT* is the measurement of dry film thickness of nonmagnetic organic coatings applied on a magnetic base (ASTM Designation D1186). This test of the amount of paint applied, usually measured in thousandths of an inch (mils), can give an indication of how finished parts will perform in salt spray and adhesion testing. For example, if a lower coating thickness is applied, the ability to keep moisture away from the substrate is impaired, and therefore the part is more readily corroded. If the film thickness is greater than standard, the normal cure times for the paint may be inadequate and may change the characteristics of the finished part. Dry film thickness can be measured easily with the use of a magnetic tester, or for quick reference a 0.0 to 5.0 mil caliper may be used. Because all paints are moisture-permeable to some extent, properly cured paint films at different thicknesses will lend different corrosion resistance performance values; usually, the thicker the film the better the resistance.

Pencil Hardness. Film hardness determined by a pencil is done according to ASTM Designation D3363-74 (reapproved 1980). This method allows for the use of drawing leads or pencil leads of varying hardness to determine film hardness. Procedures for this test are specific yet easy. Leads are sharpened flat and then pushed over the panel at uniform pressure at a 45° angle, using the hardest lead first (6H) and different hardnesses sequentially until no defacement of the film is noticed. The point of film rupture reveals the *gage hardness*. The pencil hardness that does not mark the surface would be considered the *scratch hardness*.

Most paints have a prescribed hardness based on film thickness and bake-off times and temperatures. This is where pencil hardness is a very good investigative test. If the dry film thickness is within specification, yet has a higher film hardness than expected, the bake-off or curing method conditions may need to be changed. Undercured paints also cause problems with adhesion;

CHAPTER 18

QC/QA FIELD APPLICATION FOR PAINTED PARTS

however, this usually results in fewer salt spray hours and/or a lower rating. The pencil hardness test can help to determine if parts are over- or undercured. Together, corrosion resistance and adhesion are the basis of quality assurance. The six specific tests outlined can be looked at as the brush strokes that blend the color and complete the picture.

ASTM COMPARED TO GMP

Field application of QC/QA for painted parts takes root in ASTM testing. Because of plant needs for production, manpower shortage, or other factors, it is not always possible to perform the prescribed tests in strict accordance with ASTM standards. Also, many manufacturers do not have facilities to conduct all these tests. However, suppliers of high-quality finished goods must not rely only on marginal quality controls for the assurance that production is consistent. A quality assurance program using good manufacturing practices (GMPs) replicating ASTM controls, is usually sufficient to give fair warning of pending problems. Table 18-4 describes the tools needed and the tests to perform in the field to take the first step toward quality assurance through quality control.

OTHER QUALITY ASSURANCE CHARACTERISTICS

A good quality assurance program has other ingredients, including:

- Vendor support for training personnel.
- Proper documentation.
- Objective setting and review.

Users of pretreatment chemicals and paints expect vendors to help train personnel in the proper and safe use and handling of materials, the techniques of testing, and the integral role of the quality assurance program in overall performance. Vendors should also be prepared to perform salt spray tests periodically. Monthly or quarterly tests are the best for good quality assurance. Regular service calls should be part of the program. Service reports (Fig. 18-4) should be part of the documentation package for evaluating such things as solution analysis, quality tests, visual inspections, equipment operations, etc. Periodic reporting by vendors allows for continuous feedback and better control of the process. It is the manufacturer's responsibility to support and document the results of the test program. To assist in these efforts, suppliers should be able to provide forms and/or protocols. In documenting the tests, panels should be processed daily (minimum). The field QC tests conducted should be logged in and kept for review with vendors. With this commitment to comprehensive documentation, trends affecting the quality of finished goods will become evident. Setting objectives must be a team effort. Manufacturer and vendor must clearly define the goals needed to satisfy the customer. The goals must be meaningful, measurable, realistic, and achievable to work.

TABLE 18-4
Quality Assurance Field Tests

Test	Tools Needed	
	In-plant	Preferred
1. Cross hatch (Adhesion)	Razor knife Ruler Masking tape (30 lb/sq in)	Laboratory x-hatch tool ASTM tape
2. Bend Test (Tests paint adhesion and flexibility)	Shop vise	Conical mandrel
3. Impact (Tests paint adhesion and flexibility)	Small ball-peen hammer	Laboratory impact test
4. Pencil Hardness (Measures paint hardness)	6H–6B drawing pencils	Same
5. Paint Thickness	0.00-5.00 mil Calipers	Magnetic tester

BENEFITS OF QUALITY ASSURANCE PROGRAMS

Many tangible benefits may be expected from the implementation of a field-applied quality assurance program:

- Used proactively, it is a great defense against costly reject rate increases and downtime.
- It can help increase production, through smoother operation and better understanding of the process by operators.
- It keeps customers satisfied and feeling confident that the products mean quality.
- It provides justification for capital expenditures. For example, the results may show that below-standard performance can be traced to the equipment. With this program and all the supporting documentation, justifying new equipment expense is much easier.
- This type of program adds a great deal of credibility to a product when prospective customers see the active pursuit of quality.

A common action word is *value-added*. It is becoming ingrained in businesses throughout the world as the only way to perform into the future. A field operational QC/QA Program for the manufacture of painted parts will add value to products and make an operation better than it was before.

INDUCTION HEATING: HOW TO GET THE MOST OUT OF YOUR EQUIPMENT

Induction heating is entering its sixth decade as a proven production tool for process and mass heating operations. The technology is used worldwide by durable goods manufacturers in a wide range of applications, from hardening automotive and aircraft components to curing adhesives, continuous seam annealing of welded pipe, and high speed sealing of oxygen-free food packages

CHAPTER 18

INDUCTION HEATING: HOW TO GET THE MOST OUT OF YOUR EQUIPMENT

Chemical tests:	Stage 1	Stage 3	Stage 5	Paint booths
Product(s)				
Concentration				
Temperature				
pH				
Pressure				
Last purge				
Feed systems				

Quality tests:	Desired rating	Observation
Salt spray		
X-hatch		
Impact		
90° bend		
Pencil hardness		
Paint thickness		

Visual tests:	Observation
Conversion color	
Metal type	
Passivation	
Water break	
Wash jets	
Carry over	
Tank floats	
Test reagents	
Test equipment	
Oven temperature	

Inventory:		
Application	Product name	On-hand
Cleaner(s)		
Conversion		
Sealer		

Comments:

_____
Calgon Vestal Chemical Specialist   Account Representatives   Test date

Fig. 18-4 Sample service report. (*Courtesy, Calgon Vestal Laboratories*)

CHAPTER 18

INDUCTION HEATING: HOW TO GET THE MOST OUT OF YOUR EQUIPMENT

and tamperproof medicine packets.

Properly designed and maintained induction heating equipment has demonstrated extremely long service life; many systems placed in operation 25 to 30 years ago remain in service today. However, the efficiency of older equipment is well below what can be expected from state-of-the-art systems now available from leading builders of induction heating equipment. Induction heating makes possible the incorporation of appropriate thermal processes without skilled operators. Often, all that is required is the push of a button.

The major components of an induction heating installation include power supplies, induction workstations or heat stations, process controls, inductor coils, work handling equipment, and ancillary systems.

POWER SUPPLIES

Induction heating requires an alternating current (AC) power supply to induce heat energy into a workpiece. The rate of alternating current can vary from 60 (or below) up to 450 kilohertz (cycles per second). Some applications use frequencies in the megahertz range; these are quite specialized and normally are not found in durable goods manufacturing operations.

Higher frequencies are used for shallower depth heating or hardening. Lower frequency systems achieve deeper depths of heating or hardening, or are used for heating in forging and melting operations.

The majority of induction heating power supplies currently in service are 7.5 to 500 kW motor generator sets. Motor generators typically are operated in an audio frequency range of one to 10 kHz. For applications requiring frequencies in the 200 to 450 kHz range, electronic vacuum tube oscillators rated from two kW to 800 kW are widely used.

Motor generator sets and oscillator tubes are old technology. Today, both of these high frequency power sources are being made obsolete by solid state low and high frequency power supplies which offer substantial benefits in all operating areas. Improved power electronic devices, such as SCRs and transistors, make solid state units very responsive and extremely reliable. Solid state units also are more energy efficient in both the operating and standby mode, and require 30 to 50% less floorspace than conventional equipment.

From an operating standpoint, solid state power supplies are user friendly for improved utilization and uptime. Recent systems incorporate shop floor-programmable microprocessor control to simplify and speed setup and provide real time monitoring of equipment operation. Solid state systems currently are available in sizes from a few kWs up to a megawatt (or greater if required) and frequencies up to 400 kHz. Available output capabilities now match the power requirements of virtually all production induction heating operations.

Power Supply Operating Tips

To obtain the maximum performance from induction heating power supplies, equipment manufacturers and users offer the following advice:

1. Seek the assistance of an induction heating power supply manufacturer and the local electric utility, for an energy analysis of existing or proposed equipment. Energy savings and reduced maintenance costs often can justify upgrading from motor generators and oscillators to solid state units.

2. Before purchasing a used motor generator set, investigate its prior repair history. It often is uneconomical to repair used equipment to acceptable operating standards.

3. Consult with equipment suppliers when changing applications. Power supplies may require additional capacitors or transformer upgrades; upgrade or replace protective devices on motor generator sets. Replace mercury arc rectifiers on radio frequency oscillators to solid state units. On radio frequency units, convert to more efficient and durable ceramic oscillator tubes to ensure required performance and safety and extend service life. The upgrade or retrofit of audio frequency units, and particularly radio frequency units, should be performed by a competent equipment manufacturer. This will ensure the use of properly specified components and avoid compromising the safe operation of equipment.

Coming Trends

The major trend in induction heating power supplies is the expanding application of solid state technology. As improved and lower cost electronic devices become available, and device count drops, the reliability and responsiveness of equipment will increase. The user benefits from the lower purchase price and increased performance. Power supply controls technology also will have a major impact on future operations. Microprocessor controls with built-in diagnostic capabilities are available today and fast becoming standard equipment for both captive and commercial induction heating shops. The move to pushbutton access will replace specialized operator skills.

INDUCTION HEATING WORKSTATIONS

The induction heating workstation houses system components such as transformers and capacitors. These devices are required to match the output of the power supply to the induction heating coil. This allows full power to be delivered to the workpiece.

It is recommended that workstation components be located as close as possible to the work/inductor. This allows the power supply to be located remotely, if necessary. This frees floorspace for automation and materials handling equipment, and permits induction heating workstation tooling to be integrated into a manufacturing line. Transmission between the power supply and workstation components is at high voltage, so transmission lines can be relatively long (several hundred feet with audio frequency units) without compromising induction heating performance. It also should be noted that as frequency increases, the structural size and mass of load-matching components and line transformers decrease and, therefore, are more compact.

Because of increasing attention to specialized installation requirements for production equipment, the induction heating industry is responding to this concern with modular packaging of components. Today, workstations can be incorporated into a power supply. These unitized systems reduce floorspace requirements from 30 to 50% and allow equipment to be easily integrated into new or existing manufacturing systems for in-line induction heat processing.

PROCESS CONTROLS

A major advantage of induction heating over gas-fired systems is its use of electrical energy and its available precise, quantifiable controls. The equipment can incorporate on-line, highly responsive process controls and monitoring capability. Present process control technology offers the end users a great deal of flexibility in:

CHAPTER 18

INDUCTION HEATING: HOW TO GET THE MOST OUT OF YOUR EQUIPMENT

Power Regulation. Most solid state power supplies incorporate output power regulation, comparable to the output voltage regulation available with motor generator sets. Power supply output is held constant throughout the heating cycle. Users also can specify controls which allow switching between regulating power, volts, and current.

Energy Control. As companies become more sophisticated in the application of induction heating equipment, they frequently add output energy monitoring and control to systems. The energy monitor multiplies the in-phase component of voltage and current to produce kilowatts. The energy monitor incrementally summarizes this value to provide the user with the total kilowatt seconds being absorbed by the load. This allows real time monitoring or controlling of the process and is a valuable tool for fine-tuning operating parameters to improve quality.

Time Control. Time is a critical factor in induction heating. It is recommended that individual electronic timers or an electronic time-monitoring function be incorporated into programmable logic controllers.

Progressive Heating Control. In process heating operations where the part is incrementally heated (or progressively heated, as in scan hardening), electronic drive units which accurately control scanning speed are recommended. These units typically incorporate feedback correctional control capability. The most sophisticated units use the latest design servo drive systems.

What The Future Holds

Major strides currently are being made by the induction heating industry to provide systems which support statistical process control (SPC) and statistical quality control (SQC) requirements.

Leading manufacturers also are utilizing techniques of cause and effect analysis, failure mode analysis (FEMA), design of experiments (DOE) and associated Taguchi procedures to establish optimum process parameters. Also new are "smart" sensors, standard-type sensors supported by a microprocessor. Smart sensors incorporate into algorithms such factors as load matching, signature analysis, and nondestructive test data for trend analysis.

Artificial Intelligence

Users of induction heating equipment should begin to investigate systems incorporating a defined logging system. This allows collecting process data for individual components on an ongoing basis. Data logging systems include input of all relevant process and workpiece parameters, which are then evaluated by a suitable computer algorithm. Specific parameters for part processing are reviewed, depending on input variables such as material chemistry and prior metallurgical structure changes.

Artificial intelligence will have a significant impact on induction heating in coming years. Such systems may allow an operator to "dial up" required process parameters (heat level, time, etc.) for even a single part, eliminating the time-consuming and costly process of cutting and polishing samples for laboratory analysis, and speeding all other operations involved in the induction heating of a workpiece.

INDUCTION HEATING COILS

The inductor coil is the most critical component of an induction heating installation. It is specifically designed and formed to generate the thermal profile required to produce specific results or hardening. In hardening applications, it is the vehicle for quench applications and performs the critical function of proper cooling.

Designing an induction heating coil requires skills in mechanical, electrical, and metallurgical engineering, as well as a significant amount of hands-on experience. Simple coils are usually fabricated from copper tubing. More complicated coils are precision machined and fabricated to incorporate intricate water cooling and quench chambers. A number of factors influence coil performance, whether designed in-house or outsourced:

- Virtually all coils are water cooled. This is required because of the high current densities used in generating the electromagnetic field. These currents can run as high as 20,000 amps; lack of water cooling can cause an almost instant coil failure.
- The current flow in the part is essentially a mirror image of the current flow through the coil. The challenge is to ensure that the electrical requirements of the coil, and its electric current flow and density, are compatible with acceptable mechanical strength, rigidity, and durability.
- Coil performance can be enhanced by using flux field concentrators, which focus the electromagnetic field densities to produce the proper current flow in the part. Concentrators can be steel laminations, or a composite, iron-based material which can be machined to the configuration of the coil.

In most applications, a flux concentrator will improve workpiece quality and reduce energy requirements.

Coil Design Tips

1. Do not undertake designing other than a simple induction heating coil without the assistance of a competent coil designer.
2. Do not attempt to repair or rebuild an induction heating coil without the assistance of a competent coil manufacturer.
3. When possible, obtain coils from a supplier with the in-house capability to test coils on production equipment and make a complete metallurgical analysis of sample parts.
4. Consider at what point in the manufacture of a component induction heating is performed. Frequently, induction heat treating occurs toward the end of the manufacturing process. Therefore, most value-adding operations have been performed—so an improperly heat treated part results in expensive scrap.
5. Do not automatically purchase coils from the low bidder. Weigh the durability and life of a coil against the cost of replacing, rebuilding, or downtime required for changeover.
6. Proper design, maintenance, and repair of inductor tooling becomes increasingly important as induction heating systems become more automated or are integrated into manufacturing systems.

Another issue to review in seeking higher productivity from induction heating equipment is inductor coil tool change time. To meet the requirements of reduced downtime and additional flexibility for multi-application operations (both manual and automatic), leading induction heating manufacturers have developed a technique for "Quick Change" inductors. This addresses the situation in unionized facilities where as many as three different skilled trade classifications may be required to make an inductor change. A properly designed quick change arrangement incorporates completing the electrical, cooling water, quench fluid, and

CHAPTER 18

INDUCTION HEATING: HOW TO GET THE MOST OUT OF YOUR EQUIPMENT

mechanical support functions into a single operational device. The optimum arrangement is to make this connection using no special tools or wrenches. This type of mechanism has been available for low power, low current radio frequency applications. However, leading induction heating equipment manufacturers have developed improved designs that are now available for high power applications as well.

New Techniques

Leading suppliers of induction heating coils are using improved techniques for mathematical analysis and modeling of internal current flow and thermal energy transfer. The objective is to optimize inductor performance and durability for improved coil life. It is important for users to continually evaluate the capabilities of their coil suppliers. For example, with available computer-aided design (CAD) and computer-aided manufacturing (CAM) techniques, fully machined coils now are available. Such coils exhibit improved durability and mechanical rigidity, and consistent dimensional accuracy for quality assurance.

There have been major improvements in the materials for flux field concentrators. In the past, metal laminations were the only alternative; today, machinable composite materials are available. There also have been improvements in the methods of applying flux concentrators to a coil, and some available flux concentrators today can be removed from coils for reuse on rebuilt or similar coils.

Following the concept of equipping CNC machine tools with automatic tool changers, manufacturers of induction heating equipment are developing automatic inductor tool changing systems. At this time, such equipment is expensive, but as the use of these flexible manufacturing systems expands, they will become more cost-effective and find wider application.

WORK HANDLING EQUIPMENT

Induction heating systems are reliable and adaptable to a wide range of manufacturing arrangements. Equipment can be manual, semiautomatic, or totally automated. In the laboratory, all process parameters (power/time/quench) can be programmed to develop innovative arrangements to optimize the manufacturing system for low cost processing.

A major advantage of induction heating is that equipment can be integrated into existing or new manufacturing systems. Systems can be manual, in-line automation, flexible automation (robots), or any combination of these work handling techniques. The function of part-handling equipment is the proper, consistent, and accurate placement of the part in the induction heating coil. Proper part location is normally critical to the performance of the process. In a scanning application, the handling equipment also must properly move the part through the coil. Specialized innovations may be required to control size and shape during heat processing. These include lateral restraints and clamping chucks to mechanically restrain parts during heat treating processes.

ANCILLARY EQUIPMENT

To complete an induction heating installation requires an adequate supply of good quality cooling water and, in hardening applications, the addition of a quench fluid recirculating system.

Cooling Water

An adequate supply of good quality, properly cooled water is required in most induction heating operations. An inadequate water supply, or improperly cooled water, is the major source of equipment downtime. Experienced users of induction heating equipment report that cooling water problems account for up to 70% of equipment downtime.

The characteristics of good water for cooling induction heating equipment include:

- Acceptable carbonate hardness limits to prevent precipitation on internal cooling surfaces, which reduces heat transfer and restricts water flow. The generally accepted carbonate hardness limit is 10 to 12 grains per gallon (3.8 L) (maximum).
- Acceptable levels of electrical conductivity. The cooling circuits in the power supply, and particularly on oscillator-type units, are subject to a relatively high voltage differential in the water-cooled circuit designs. If the electrical resistivity exceeds the specified levels, electrolysis can occur. This causes corrosion of copper components and fittings and results in internal circulating currents which reduce output.
- Adequate and acceptable incoming cooling water temperature levels. These requirements must be met to ensure that the internal heat generated by components is properly removed and dissipated.

Quench Recirculating Systems

The quench recirculating system provides the coolant for proper quenching of the part, after heating to obtain the required metallurgical characteristics. Adequate flow is important. Maintaining the proper quench temperature also is critical if proper hardness is to be achieved without cracking.

Today, most induction heating systems use water-based polymer solutions as a quench medium. Mixture concentration levels can be critical and must be maintained to ensure the quality of the process. The concentration should be monitored regularly.

There are two issues to address in selecting a quench system: filtration and foaming. Improper filtration and foaming can cause havoc with even the best designed induction heating installation. In the quenching operation, some scale is generated along with other debris which may exist on the surface of a part. The recirculating system must be equipped with suitable filters or strainers to ensure that this debris does not get into the integral quenching orifices that are part of the inductor or inductor assembly. When considerable amounts of scale are present (for instance, in hardening a forged surface), additional care must be taken to slow the recirculating velocity in the quench tank to allow some of the debris to settle out, reducing the load on the filtering devices.

Foaming, on the other hand, is caused by agitation and air retention resulting from specifying too small a recirculating tank. Foaming will impact the quenching action and contribute to "soft" parts.

Based on field experience, the recommended tank size is five times the gallon per minute capability (a 200-GPM – 757-LPM pump should have a 1,000-gallon – 3785 liter tank). This ratio can be reduced, but then other special precautions must be taken to ensure control of the problems of scale and foam-generating agitation.

SUMMARY

Manufacturing engineers are beginning to address more function-driven manufacturing process arrangements to optimize cost reductions, along with higher quality and more reliable manufacturing systems. These innovative concepts are supported

CHAPTER 18

PRIMER ON POWER BRUSHES

and enhanced by concurrent engineering programs with product engineers, to improve the manufacturability of parts. The manufacturing community is beginning to recognize that innovative integration of the heat treat into the manufacturing process can result in synergistic improvements, including elimination of operations and replacement of expensive operations with less costly types. This has resulted in the induction heat treat operation being moved further along in the manufacturing process. There now is a substantial number of cases where it is the last operation.

Applications incorporating induction as a "post" process add an additional requirement to the induction heat treat operation: it must be responsible for controlling dimensional changes to significantly tighter tolerances, since there are no subsequent corrective machining operations. Removal of surface material by grinding after heat treat can remove some of the beneficial residual stresses, and grinding must be tightly controlled to prevent surface damage. Therefore, elimination of grinding improves ultimate workpiece quality.

Induction hardening and heat processing applications are at the leading edge of emerging innovative manufacturing technologies. As manufacturing engineers and product designers become more familiar and aware of the capabilities of induction heating, there will be substantially more opportunities for component cost reduction, durability improvements, and improved quality.

PRIMER ON POWER BRUSHES

In the 1990s, power brushes will continue to be a major method for surface finishing, surface roughening, deburring, edge blending, and cleaning castings, machined parts, and other manufactured parts (see Fig. 18-5).

The use of power brushes is expected to continue to increase in the 1990s and the first decades of the 21st century, because brushes can comply with the many complex shapes produced for modern technology, provide fine surface refinements, enhance other advanced machining operations, and help meet increased quality requirements. Robotic applications, using brushes to finish surfaces, deburr, and blend edges, are a particularly fast-growing field because of the brush's ability to produce consistent results with typical robot operating tolerances.

GENERAL ADVANTAGES OF POWER BRUSHES

At present, power brushes offer many advantages, which the manufacturing engineer should carefully consider, in both automated and nonautomated applications:

- They remove surface contaminants from metal without disturbing base material.
- They are durable.
- In edge blending, they produce a smooth edge instead of two secondary sharp edges.
- They don't load up when working soft material.

A power brush can remove burrs and tool marks while maintaining the high surface finish and tight tolerances required in automatic machining operations.

On the other hand, brushes are not always the best tool for every application. For instance, heavy burrs, caused by dull cutting tools displacing large amounts of metal, are better removed by abrasive tools or other cutting tools such as rotary burrs. Deburring and finishing large numbers of small parts are better suited to mass finishing techniques such as vibratory finishing.

TYPES OF BRUSH

Most industrial wire power brushes use either medium carbon heat-treated steel wire or 302 stainless steel wire for the bristle material. Nonferrous fill materials are available for nonsparking and other special needs. The two most widely used wire forms are crimped and twisted knots.

Crimped wire, a waveform shape, produces a brush in which the wires interact with each other to reduce vibration damage to the individual wires as they impact on the workpiece. This configuration also allows a larger number of wire tips to contact the workpiece. The knot-style brush uses bundles of straight wire twisted into individual knots or short cables. This produces extremely high impact forces for removing heavy scale and surface contamination.

A recent development in brush fill materials is abrasive-filled nylon filament. This filament produces a truly flexible abrasive product with the flexibility of a brush and the cutting ability of an abrasive tool.

Elastomer-bonded or encapsulated wire-filled brushes, available in various configurations, use wire bristles bonded together with an elastomer material. These products have characteristics similar to dense-filled crimped wire brushes; the bristles can move and are still somewhat flexible. However, these products are more aggressive than standard brushes.

Encapsulated brushes work well in applications where precise control of the brush is necessary, such as stripping insulation from copper wire and deburring key slots and other restricted areas.

DESIGNS

Wheel brushes are circular in shape and generally no wider than 2" (51 mm). Standard wheel brushes range from 0.625" to 15" (16 mm to 381 mm) in diameter and may be filled with materials such as steel, stainless steel, brass, bronze, and nickel silver wires; natural fibers such as tampico and animal hair; or synthetics such as nylon (abrasive-filled nylon) and polypropylene. Both crimped and twisted knot wire forms are used in wheel brushes.

Cup brushes are available in three basic types: miniature, 0.375"–1" (9.5 mm–25 mm) diameter (stem mounted); crimped wire, 1.75"–6" (44 mm–152 mm) diameter, for lighter jobs; and knot, 2.5"–6" (64 mm–152 mm) diameter, for heavy-duty jobs requiring aggressive brushing action. These brushes are generally used on portable power tools.

End brushes are generally used in portable air and electric tools for light-to-heavy-duty work where space is limited. Diameters range from 0.156" to 1.125" (4.0 mm to 28.6 mm).

Tube brushes are generally twisted-in-wire (or bottle brush) types for use on portable tools and drill presses to clean and finish holes and internal threads. Diameters range from 0.188" to 1.25" (4.8 mm to 32 mm) although other sizes are available.

Wide-face brushes are identified as brushes where the face is greater than the diameter. This type of brush may be created by stacking a number of wheel brushes on a common arbor, or may

CHAPTER 18

PRIMER ON POWER BRUSHES

be a factory-assembled integral unit. Wide-face unit brushes are generally specially designed for specific customer applications.

Miniature brushes are small versions of end, wheel, and cup types used for brushing miniaturized components. Sizes range from 0.156" to 1.5" (4.0 mm to 38.0 mm) and fill materials may be animal hair, synthetics, or wire. These brushes are generally mounted on a 0.094" or 0.125" (2.4 mm or 3.2 mm) shank (see Fig. 18-6).

WHEN TO USE WIRE BRUSHES

Wire brushes should be considered for heavy-duty applications such as removing large burrs. The ends of the wires act like tiny tool bits, impacting the work surface, knocking off burrs and work-hardening it slightly. Wire brushes are always used in weld cleaning operations because they remove slag without removing the weld metal and assure a clean surface for painting or a subsequent weld pass. There are two wire configurations for power brushes. *Crimped* wire is used for less aggressive applications. *Straight* wire, which is always twisted into a knot, is for more aggressive applications.

Crimped wire is often best for light-to-medium-duty applications (see Fig. 18-7). Tempered high-tensile carbon steel is the best material for most applications, lasting as much as 10 times longer than other nonheat-treated materials. Stainless steel should be used on stainless steel and all nonferrous metals.

Knot types are best for heavy-duty applications, such as gear deburring, and produce uniform results on jobs requiring removal of scale and surface contamination (see Fig. 18-8). These types, including a special "stringer bead" style, are also used for cleaning welds.

A manufacturer of steel pinion gears increased productivity 800% by replacing a hand-held power-deburring tool with an automatic skiving and brush deburring system using two knot-type steel wheel brushes on a rotary turret.

Manufacturers use crimped-type wheel brushes for such applications as:

- Cleaning and removing rust from cast iron cooking utensil covers: 15" (381 mm) narrow face.
- Removing burrs from grooves of castellated automotive nuts: 4" (102 mm) narrow face wheel used on automatic machine.
- Cleaning contact connections before lead dipping of resistors: 6" (152 mm) narrow face).
- Removing rubber baked on large truck and bus tire molds during curing of the tire 12" (305 mm) narrow face.

Tips on Using Wire Brushes

As cutting tools, power brushes have efficient speed and pressure for specific operations. In most operations utilizing wire brushes, the highest speed and lightest pressure will assure the fastest cutting action and longest brush life.

Using as fine a wire as possible usually results in longer brush life. Increasing the speed of a brush increases the face hardness and cutting action; therefore, a fine wire brush rotating at a higher speed will often produce the same results as a coarser wire brush rotating at a slower speed.

When using a wire-filled brush, periodically reverse the direction of rotation to take advantage of the self-sharpening action which will result. This is done by removing the brush from the spindle, turning it side for side, and remounting securely. Figure 18-9 shows some recommended surface speeds for brushing applications.

When brushing a recessed area or the inside diameter of a tube, use an end brush as follows:

- Insert the brush.
- Start the brush rotating.
- Turn off the power before removing the brush.

This will prevent the brush from flaring out and permit it to fit into the recessed area again.

Stainless steel (302 series) brushes can attract carbon steel particles from the air or workbenches and become contaminated.

- Deburring
- Edge blending
- Surface finishing
- Cleaning
- Roughening for adhesion
- Weld preparation and cleaning

Fig. 18-5 Primary power brush applications. (*Courtesy, Weiler Brush Company*)

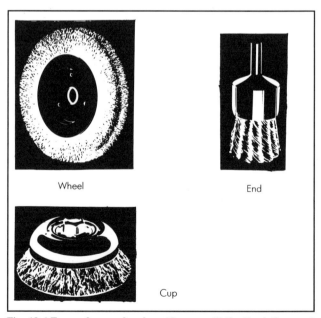

Fig. 18-6 Types of power brushes. (*Courtesy, Weiler Brush Company*)

CHAPTER 18

PRIMER ON POWER BRUSHES

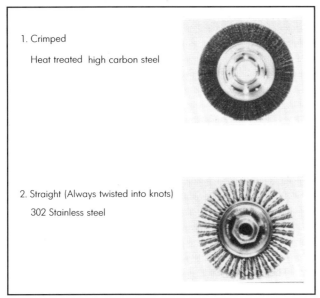

Fig. 18-7 Crimped and knot-type wire power brushes. (*Courtesy, Weiler Brush Company*)

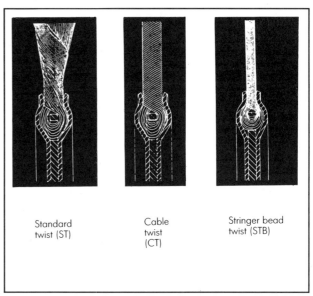

Fig. 18-8 Close-up view of knot-type wire power brush. (*Courtesy, Weiler Brush Company*)

To prevent such contamination, wash the brushes in a degreaser and keep them wrapped in plastic. Never use a carbon steel brush on stainless steel parts.

If a brush application requires high pressure (downforce) between the brush and workpiece, brush life may be shorter than expected. A shorter-trim brush will experience less wire deflection under pressure and will last longer. Figure 18-10 lists performance variables affecting the results.

For both wire and abrasive nylon filament brushes, avoid excessive pressure, which causes over-bending of the filaments and heat buildup resulting in reduced brush life. Instead of greater pressure on the brush, try:

1. A brush with a more aggressive cutting action.
2. Higher speed (increased rpm, increased brush diameter).

WHEN TO USE ABRASIVE-FILLED NYLON BRUSHES

In abrasive-filled nylon brushes, the filaments act as flexible files or a flexible abrasive and allow the sides to perform the majority of the work. These brushes are usually made with crimped filaments. The crimp keeps the filaments separated, and gives the brush a full even face.

Abrasive-filled nylon brushes, made from heat-stabilized material with abrasive grit distributed throughout the filament, are usually the best brushes for following complex or deep machined and drilled parts, and smoothing their rough surfaces—including holes and indentations. Grit sizes normally used range from 46 grit to as fine as 600 grit.

Abrasive-filled nylon brushes are often the economical approach to a manufacturer's finishing problem, deburring and finishing in one step instead of three or four, furnishing a microfinish of nine or ten RMA, and lasting longer because they "run cooler" without conducting heat into the brush. Nylon is resistant to most chemicals; hydrocarbons, oils, and most organic solvents have no lasting effects.

This type of brush is entering wide use in industries, like aerospace, which demand fine deburring, finishing, and edge blending. Abrasive-filled nylon brushes are also excellent for deburring aluminum and powdered metals, for decorative finishing of aluminum, stainless steel, brass, and hard plastics, and for edge breaking on carbide cutting tool inserts.

The use of abrasive-filled nylon brushes has increased markedly during the last five years and is expected to continue this dramatic increase during the decades ahead.

Use the Right Abrasive

Silicon carbide grit provides the best combination of hardness, sharpness, and toughness economically available. Silicon carbide is a harder and sharper grain than aluminum oxide, is more friable, and does a better job of finishing metal. It is recommended for most abrasive-filled nylon applications. Aluminum oxide grit is tougher and less likely to fracture, and is generally used for finishing softer metals.

Shapes of abrasive-filled nylon brushes include wheel, cup, end, and tube brushes. Recently a new composite hub construction has been introduced in wheel brushes to take the place of conventional "metal hub" construction which is best suited for wire filled power brushes. Advantages of this new composite hub are:

1. All the filament material used in the one piece molded hub is exposed and can be used during the life of the product. Typical prior construction wasted 20% of the fill material in the metal mounting hub.
2. The user gets 20% more brush for the same cost.
3. The molded one-piece hub construction does not damage the filaments (which can happen with metal hub components).

Typical cases where abrasive filled nylon brushes are usually preferred:

- Deburring aluminum extrusions for heat sinks when the extrusions contain numerous machined holes.
- Deburring the ends of aluminum extrusions sawed to length.
- Removing flashing from aluminum and powder metal die castings.

18-31

CHAPTER 18

PRIMER ON POWER BRUSHES

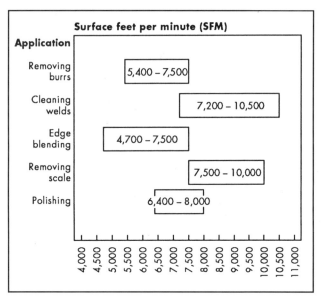

Fig. 18-9 Recommended surface speeds for brushing applications. (*Courtesy, Weiler Brush Company*)

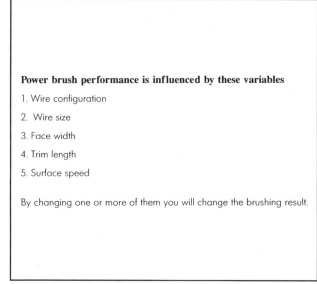

Fig. 18-10 Power brush performance variables. (*Courtesy, Weiler Brush Company*)

- One-step finishing between coarse grinding and a final, extremely fine finish.

Applications of these brushes include everything from deburring flutes and piccolos to blending and finishing trailing edges of jet engine blades or deburring and rounding off corners on carbide inserts used in high performance cutting tools.

More Tips

1. Abrasive filled nylon brushes should be run at a lower (under 3500 sfpm) speed range, while exerting enough pressure on the brush to cause the sides of the filament to cut as well as the tips. Optimum cutting action results when the workpiece penetrates, or "mushes" into the face of the brush.
2. Use rectangular filaments for more contact and impact with the workpiece, in such applications as carbide edge honing, deburring injector nozzle cut-outs, and finishing decorative hinges.
3. Use round uncrimped filaments for such applications as deburring the slots in cast magnesium computer cage backs, and putting a uniform scratch finish on aluminum chair bases.
4. Use silicon carbide abrasive grit for most applications, because its harder and sharper grain finishes metal more effectively. Power brushes are used as a cleaning tool, not as a grinding tool. They don't remove any base material (see Fig. 18-11).
5. Choose aluminum oxide abrasives to prevent carbon contamination of soft metals. This type of abrasive brush is usually preferred for jet engine blades or other aerospace products.
6. A small amount of a special lubricant will prevent smearing at higher surface speeds, allowing the brush to be operated up to 10,000 sfpm at twice the work load operated before. Temperature of the nylon filament should be kept below 150°F (66°C) if possible; this will give the maximum brush life.

THE BOTTOM LINE

Advances in power brush materials, design, and construction are continuing as leading edge manufacturers enhance brush life and increase operator productivity.

Top-performing wire brushes now incorporate a solid steel ring design which locks the wires in place. Knot-style brushes have exact strand counts for perfect balance.

Manufacturers utilize computers throughout their operations to aid such design advances, control equipment, and provide immediate information on business operations. In brushes as in everything else, knowledge saves money and increases productivity.

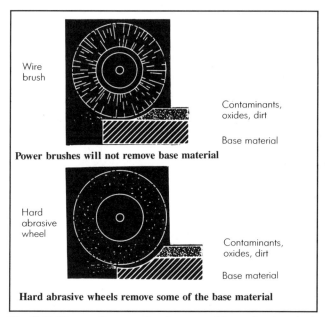

Fig. 18-11 The use of a power brush compared to an abrasive wheel. (*Courtesy, Weiler Brush Company*)

CHAPTER 18

SURFACE IMPROVEMENT TECHNOLOGY

Surface improvement technology, as it is applied to metal removal, has grown substantially in the past few years. Tighter tolerances, new workpiece materials, and new and improved abrasives are the main reasons for this growth. This type of surface finishing (metal removal) has several names describing very different processes. However, they all have one common function; all are low-velocity abrading processes using lower cutting speeds (compared to grinding) and lower pressure to minimize heat, resulting in excellent surface, size, and geometry control.

- *Polishing* is the use of abrasives to create a smoother surface with little, if any, geometry improvements (seal areas, low load-bearing areas).
- *Lapping* is the process of folding the workpiece material. It incorporates a loose abrasive for very fine surface finish and small amount of stock removal. (Generally used on flat applications.)
- *Honing* is generally associated with inside diameter surface finishing. It uses abrasives to improve surface finish and geometry characteristics such as roundness, taper, and sizing. (High wear areas and applications needing close tolerances).
- *Superfinishing* generally incorporates vitrified abrasive products such as diamond and CBN. It applies a light pressure with a high reciprocating action to achieve improved surface finish and geometry characteristics (outside diameter applications with high load areas and tight tolerances).
- *Micro-finishing* has grown significantly in the past few years. Generally associated with abrasive tape as the media, it has replaced most polishing applications where high loads and tight tolerances are required.

Other surface finishing methods such as brushing and buffing improve the workpiece appearance and sometimes are incorporated with other processes to enhance results. These processes have deviations within their category which allow varying process results.

CHOOSING THE PROPER PROCESS

In most cases, the process can be selected by an experienced person. However, it is becoming more common in areas of reliability, efficiency, and performance that a certain amount of development is incorporated into the discussion. To properly define process and machine type, first establish the workpiece condition coming to the process and the desired process results.

In an example, a decision had to be made regarding the use of superfinishing or microfinishing processes. The workpiece was designed with a load-bearing outside diameter, close surface finish requirements, and additional geometry requirements such as roundness and flatness. The incoming workpiece's outside diameter (O.D.) varied in surface finish and geometry due to improperly maintained grinders.

Reviewing the factors affecting the selection, superfinishing was eliminated due to the incoming part condition (an oil hole or groove which could damage stone tooling and cause irregularities in the surface finish), unmanned machine, and surface finish requirements. Abrasive tape microfinishing, using a hard back-up shoe, was chosen.

The incoming workpiece condition varied in surface roughness. Abrasive stone superfinishing requires a rough incoming surface to dress the stone; without this condition the stone loads up and becomes dull. Abrasive tape, on the other hand, indexes new abrasive each cycle which allows a more forgiving incoming surface requirement. If damaged by oil holes or other part characteristics, the abrasive again is indexed and new abrasive is in place for the next incoming part. Workpiece geometry such as concave or convex surfaces can wear a stone's contact surface improperly; abrasive tape, with a hard backup and new abrasive media indexed in place each cycle, eliminates this problem (see Fig. 18-12).

This machine will be placed in a unmanned transfer line. Abrasive tape can be monitored for breakage and nonindexing, while stone cannot. The superfinishing stone process could have been chosen in different circumstances: manned machine, simple design, high production rate, lower tool cost, and less tool wear.

CHOOSING THE PROPER METHOD

The following outline will serve as a check list for various aspects of the microfinishing/superfinishing selection process.

Incoming workpiece condition
Surface finish
Form size
Flatness

Outgoing workpiece condition
Surface finish
Roundness
Form
Size
Flatness

Production rate per hour

Estimated lifetime production

Surface finish process
Abrasive media
Machine type
Coolant
Tool cost

Operator involvement

Workpiece handling method

Workpiece orientation during finish operation

Gaging and measurement
Choose proper surface finish measurements that best apply to this application (RMS, RA, RZ, etc.).

Process monitoring

Machine uptime

Machine maintenance

Various abrasive materials
Carborundum
Emery
Garnet
CBN
Diamond

18-33

CHAPTER 18

SURFACE IMPROVEMENT TECHNOLOGY

Criteria	Finishing stone	+/−	Finishing tape	+/−
Roundness	Before / After 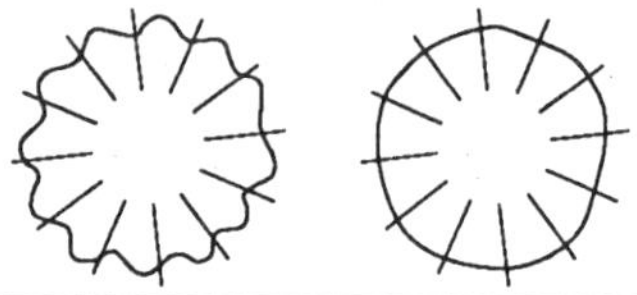	+	Before / After 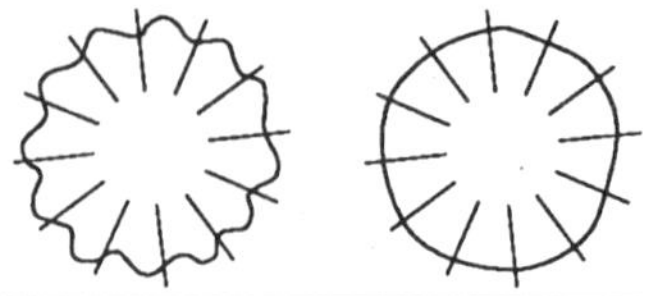	+
Straightness	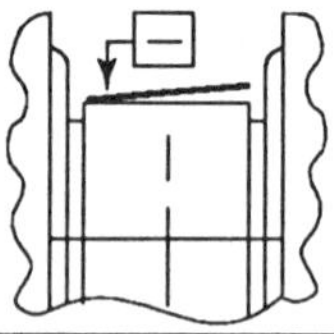	+	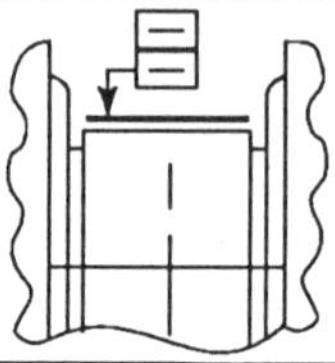 Straightness plus constant convexity	+ +
Roughness	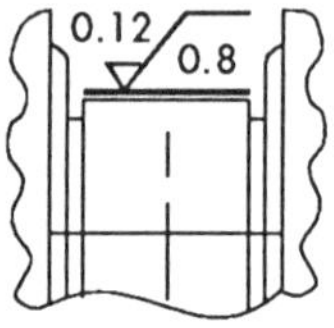 0.12 / 0.8 — Dispersed	+−	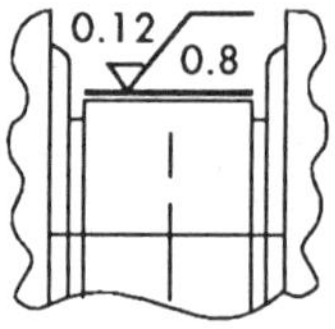 0.12 / 0.8 — Constant	+ +
Oil bore	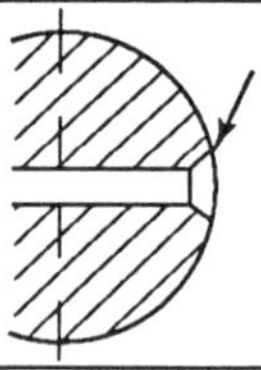 Sharp edges	−	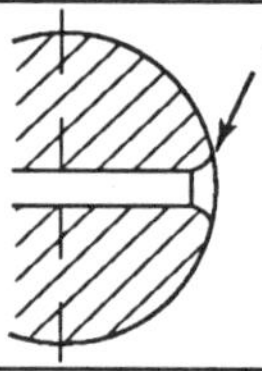 Round edges	+
Costs of perishable tools	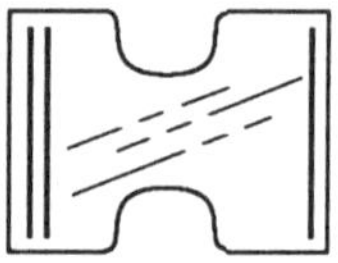 Special formed	−	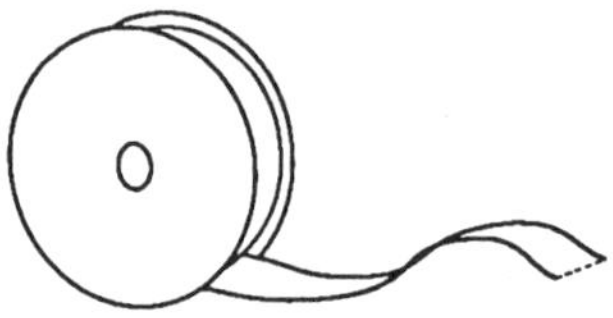	+
Tool detection	Nonpracticable stone weave indicator only	−	1. Tape feed control. 2. Broken tape detection. 3. Tape end control.	+
Micro-structure	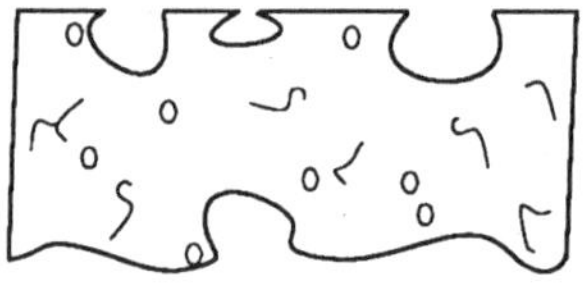	+	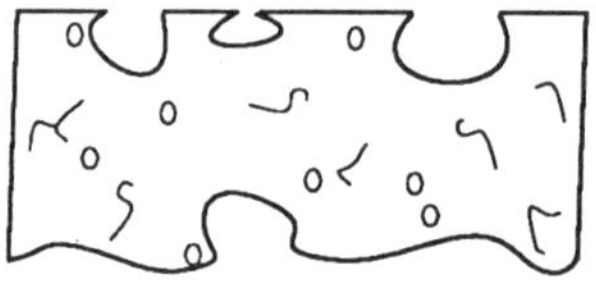	+

Fig. 18-12 The use of a finishing stone compared to finishing tape. (*Courtesy, QPAC-Nagel Precision Inc.*)

CHAPTER 18

HOW TO FIND THE BEST HONING PROCEDURE

PCD
Aluminum oxide
Silicon carbide

Coated tools
Aluminum oxide
Titanium oxide
Titanium carbide
Hafnium

Coolants
Soluble oil
Straight oil
Synthetic

Type of measurement
RMS, RA, RMAX, etc.

Process parameters
Abrasive media
Abrasive backing
Surface speed
Pressure
Oscillation (rate and distance)
Duration
Tool index
Coolant
Dwell
Coolant type

Temperature in-process control
Gaging for size, taper, etc.
Off-line gaging for surface finish

HOW TO FIND THE BEST HONING PROCEDURE

Finding the best honing procedure means the user must first understand the following points:

- What honing is.
- How the honing process works.
- How to select the proper abrasive and honing oils.
- How to select the spindle speed and stock removal rate.
- What to look for when looking at the cross-hatch pattern.
- How to use statistical process control.
- How to pick the best machine.

HONING

Honing is an abrasive machining process which is designed to improve the accuracy of cylindrical, usually internal, surfaces. It is characterized by large areas of abrasive contact, low cutting pressure, low velocity, floating part or tool, and automatic centering of the tool by expansion inside the bore.

Boring is a machining process which is designed to enlarge holes, usually to improve the accuracy of a drilled hole. It is characterized by small area of contact, high pressure, fixed part and tool, and the ability to move or change the center line. Both processes have valid applications depending on requirements. In some cases both processes may be used—boring to remove large amounts of stock and to establish or move the center line, followed by honing to improve bore geometry or provide a particular surface finish and crosshatch pattern.

Ten common bore errors which are normally associated with machining, heat treating, or chucking are illustrated in Fig. 18-13. Various honing processes can correct these errors with the least possible amount of material removal compared to other machining processes. In addition, honing is also beneficial for an eleventh error—*Surface Integrity*.

Surface integrity refers to the quality of the surface of the metal after machining. As illustrated in Fig. 18-14, most machining processes are abusive to the machined material and may fracture the crystals of metal to a depth of about 0.002" (0.05 mm).

The honing process is relatively gentle to the material being worked, providing a sound base metal surface finish as illustrated in Fig. 18-15.

HONING PROCESS

Honing requires no chucking or alignment. The process allows the part to float on the tool, aligning itself with and being supported by the tool. In this way, honing can remove the bore errors caused by other less accurate machining processes.

ABRASIVE SELECTION

In the past, selecting the proper abrasive for a honing job required choosing from a long list of possible types of abrasive, bond strength (usually called hardness), and grit size.

These lists described the conventional abrasives, aluminum oxide and silicon carbide. The superabrasives, namely cubic boron nitride (CBN) and diamond, greatly simplify the choice of the best honing stone. Diamond is required for honing glass, ceramic, and tungsten carbide. Practically all other materials hone best with CBN, whether the material is hard or soft. In hard steel, especially high speed tool steel, the CBN stone cuts much faster than aluminum oxide stones. CBN saves so much labor cost that its use is almost mandatory. In addition, the CBN stone will create better accuracy; for straightness, because it cuts so freely, and for exact size, because it wears so slowly. Part after part hones to the same diameter, with only very infrequent feed-up for tool wear. Mandrels with diamond plated shoes are available, making this almost-automatic size repetition even longer lasting between feed-ups. Speed of stock removal in hard steel is a multiple of that achieved with aluminum oxide, the saving of time varying with the type of alloy being honed. The greatest improvement is seen when honing high speed tool steel. In soft steel, there is not a great gain in cutting speed; the main advantage in using CBN in soft materials is the repetition of size. Further savings are realized by the much longer stone life of the superabrasive stone and the need to buy only one stone rather than a large number of different stones of conventional abrasive.

SPINDLE SPEED

The spindle speed for honing is slow, compared to grinding, but slow spindle speed does not translate into slow stock removal. The large abrasive contact area of honing as opposed to the point contact of grinding, together with the fast stroke of honing makes up for the lack of spindle speed. Refer to Fig. 18-16 for the recommended spindle speeds for a typical honing machine.

The two advantages of these slow spindle speeds are cleanliness and safety. The typical honing machine does not have a hood to keep honing oil inside the oil pan. High spindle speeds would fling

18-35

CHAPTER 18

HOW TO FIND THE BEST HONING PROCEDURE

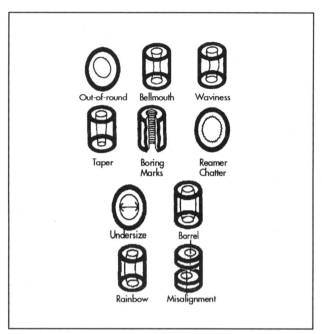

Fig. 18-13 Ten common bore errors. (*Courtesty, Sunnen Products Company*)

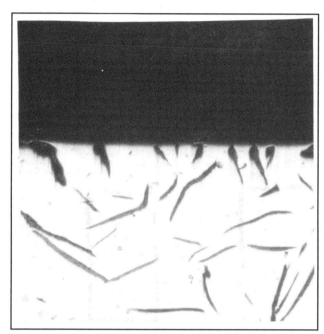

Fig. 18-15 Sectional view of honed cylinder. (*Courtesy, Sunnen Products Company*)

oil out of the machine. The safety consideration deals with the fact that the honing tool supports the workpiece. If the workpiece is heavy in relation to a small diameter honing tool, the tool could bend and throw the workpiece off, which could be dangerous at high spindle speeds. There seems to be no other reason to limit spindle speed for honing. Faster spindle speed equals faster stock removal, and this holds true for any type of abrasive.

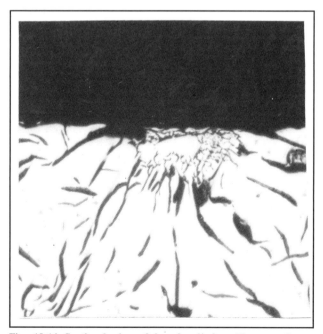

Fig. 18-14 Sectional view of bored cylinder. (*Courtesy, Sunnen Products Company*)

CROSSHATCH PATTERN

A characteristic feature of a honed surface is the crosshatch pattern which is seen in Fig. 18-17. The pattern is generated on the bore surface as the part is stroked back and forth over the rotating honing tool. The faster the stroking rate (spm) in relation to the tool rotation (rpm), the larger the crosshatch angle. Limitations in the stroke speed capability of the honing machine will make it highly unlikely that the crosshatch angle will be too large. Although exact crosshatch angles can be calculated by controlling the ratio of spindle speed to stroke rate, using the formula shown in Fig. 18-18, most blueprints have ignored this angle, as it has never been proven to make a difference in the functioning of a honed part. Regardless of whether or not crosshatch is called out on the blueprint, it should be pointed out that the pattern created does make an excellent oil retention and bearing surface.

Stock Removal Rate

The formulas shown in Table 18-5 can be used to estimate the amount of time in minutes required to remove a given amount of stock, using a honing machine which provides three horsepower at the spindle.

Tables 18-6 and 18-7 show how to estimate honing time in seconds for small parts. Honing machines for small parts, 0.0063 to 2.5" (1.5 to 64 mm) internal diameter, usually have only one horsepower at the spindle, and the limit for stock removal speed is the strength of the honing tool.

It is very important to find the correct cutting pressure and rate of feed for the stone being used. Use the lowest cutting pressure that gives good cutting action. To determine which cutting pressure will result in the lowest cost when doing production honing, try different pressure settings and tabulate the results, as shown in the example in Table 18-8. A cutting pressure of three produced the best total cost per part in this example.

Table 18-9 shows how to get the desired surface finish by selecting the proper grit size honing stone. The honing machine

CHAPTER 18

HOW TO FIND THE BEST HONING PROCEDURE

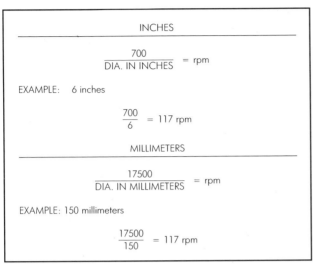

Fig. 18-16 Spindle speed. (*Courtesy, Sunnen Products Company*)

```
A – Average Crosshatch Angle
S – SPM (Strokes per Minute)
R – rpm (Revolutions per Minute)
L – Stroke Length
D – Part Diameter

TANGENT A/2 = (0.6366) (L)(S)/(D)(R)
```

Fig. 18-18 Cross hatch angle formula. (*Courtesy, Sunnen Products Company*)

operator reads the desired finish off the blueprint and acts accordingly. But if the workpiece is still in the design stage, the designer has an important decision to make. Selecting a finish finer than necessary would add unnecessary cost to the manufacture of the part. A finish done too finely may cause malfunctioning of the component. A very smooth surface is unable to hold an oil film, and moving parts run dry, with very undesirable results. Of course, too rough a finish has other undesirable results.

Most often, the surface finish is listed as microinches Ra, or micrometers Ra. Ra stands for roughness arithmetic average. Many other units of surface finish definition are in use, such as Rt (total roughness, peak to valley). There exists an interesting relationship between Ra and Rt. An Rt of about ten times Ra indicates a good honing job. If the ratio is much more than ten to one, the machining marks preceding the honing operation have not quite cleaned up, or there has been welding (pick-up, galling) between the honing tool and the workpiece (something usually caused by unsuitable honing fluid, as discussed under Honing Fluids). Lately there has been more interest in a finish which is rough and fine at the same time, called *plateau finish*. It is created by first rough honing with a very coarse stone, right up to finish size. This is followed by a second honing operation with a very fine stone, such as 600 grit. This second operation takes only a few seconds, depending on the percentage of plateau desired; 50% is often called for. The plateau operation removes so little metal that it does not measurably change the final diameter.

The Ra definition does not measure a plateau surface. A newer definition, called roughness of the core (Rk) measures the main portion of the surface and lists the small peaks in the surface as Rpk and the oil-retaining valleys as Rvk (see Fig. 18-19).

SURFACE FINISH FORMULA

The finer the grit of the honing stone, the finer the honed finish will be. It should also be pointed out that the rougher the grit, the lower the cost of honing. Therefore, if there is a large amount of stock to be removed and a fine finish is required, it is usually better to hone in two operations. The first operation uses a coarse grit stone for fast stock removal, while the second operation is done with a finer stone to obtain the specified surface finish. This

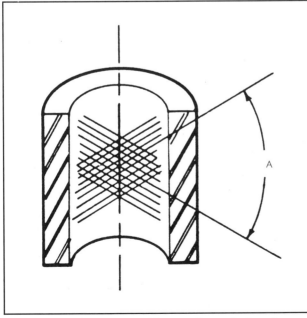

Fig. 18-17 Cross hatch pattern. (*Courtesy, Sunnen Products Company*)

TABLE 18-5
Stock Removal Rate

Length = L
Diameter = D
Require Stock Removal = RSR
Honing Time in Minutes = Min.

MILLIMETERS

$$\frac{L \times D \times RSR}{3400} = \text{Min.}$$

EXAMPLE: $\frac{100\text{mm} \times 1200\text{mm} \times 0{,}28\text{ mm}}{3400} = 10$ Min.

INCHES

$$\frac{L \times D \times RSR}{.21} = \text{Min.}$$

EXAMPLE: $\frac{48\text{in} \times 4.0\text{in} \times .011\text{in}}{.21} = 10$ Min.

CHAPTER 18

HOW TO FIND THE BEST HONING PROCEDURE

TABLE 18-6
Time Required for Honing

Workpiece Diameter	Honing Time	Workpiece Diameter	Honing Time
In.	Factor	mm	Factor
0.0625	100,000	1.5	5.88
0.1250	40,000	3.0	2.44
0.1875	25,000	5.0	1.61
0.2500	16,700	6.0	1.18
0.3750	10,000	7.0	1.00
0.4375	8,300	8.0	0.83
0.5000	7,400	10.0	0.65
0.6250	5,500	12.0	0.50
0.7500	4,750	15.0	0.38
0.8750	3,750	18.0	0.31
1.0000	3,450	20.0	0.28
1.1250	2,950	25.0	0.21
1.2500	2,750	30.0	0.18
1.3750	2,500	35.0	0.15
1.5000	2,200	40.0	0.13
1.7500	1,950	50.0	0.10
2.0000	1,640	60.0	0.09
2.2500	1,450		
2.5000	1,230		

TABLE 18-7
Formula

Finish Diameter = FD
Length = L
Stock Removal = SR
Honing Time Factor = HTF
 (see Table)
Seconds Required To
 Hone = SEC
 FD × L × SR × HTF = SEC

Example: 1″ Diameter,
2″ Length, .005″
Stock Removal:
1 × 2 × .005 × 3450
 = 35 seconds

is another advantage of CBN honing stones, because they show less difference in cutting speed between coarse and fine grits than conventional abrasives; it might be economically acceptable to dispense with the rough honing operation.

The formulas shown in Table 18-10 are an aid in determining how much material to leave in the bore for finishing. As the example shows, to go from a rough-honed 50 to 10 microinches (1.25 to 0.25 micrometers) *Ra* finish, it will be necessary to remove 0.0004″ (0.01 mm) of stock, as measured on the diameter. Removing less material will not produce the desired finish, and removing more is a waste of time and money.

HONING FLUIDS

Commonly, the fluids used for honing are referred to as "coolants." This is a misnomer, because cooling is not one of the strong points of any honing oil or water-based product. The most important reason for using a honing fluid is its chemical activity. The honing fluid must be inactive at normal temperatures, so it does not corrode anything. But it must instantly become active when the temperature comes close to the melting point of the metal being honed. This high temperature occurs in microscopic spots at the points of cutting action, and results in welding of the metal guide shoe to the metal being honed. These tiny weld spots would be torn apart by the force of the honing machine, and the results would be rough surface finish and rapid wear of the honing stone and guide shoe. This welding problem is especially likely to happen with high-alloy materials, such as stainless steel. Capable "coolant" prevents welding by chemically changing the hot spots from metal to a nonmetallic compound, which cannot be welded.

Honing Oils

Tests have shown that honing action is actually faster when the oil is hot. The idea that cold oil could guarantee the exact size of the finished part, without having to consider the shrinking of the bore diameter when the part cools off, proved to be unsuccessful, because the input of heat during rapid honing is vastly greater than the very limited cooling ability of oil.

Another common myth is that honing tools without guide shoes are immune to welding problems. This is a honing fluid problem, for if the honing fluid does not have enough chemical activity it will permit stone loading. Metal chips created by the cutting action of the honing stones stick to the stone surface. When there is stone loading, there will be metal-to-metal contact and welding, with the same undesirable results as when using a honing tool with metal guide shoes.

Water-based Honing Fluids

With tighter government regulations and greater concern for safe, biodegradable products, water-based honing fluid technology has developed new products that may offer an alternative to conventional oil-based honing fluids. Tests on specially developed water-based honing fluids in new and conventional honing machines look very promising. Unlike traditional water-based products used for grinding or cutting, these new products were developed for use in honing machines. But water-based fluids need replacing very frequently, and the cost of legally disposing of used fluids is rising rapidly. Honing oils, on the other hand, don't need changing. Honing swarf can be separated from the oil by using filters, magnetic separators, or simply gravity.

TABLE 18-8
Cutting Pressure versus Cost

Cutting Pressure	Seconds for 0.004″ Stock Removal	Stone Wear per Part	Stone Cost per Part	Labor Cost per Part at 1ᶜ/Sec	Total Cost per Part
2	30	0.0001	1 cent	30 cents	31 cents
2-1/2	20	0.0003	3 cents	20 cents	23 cents
3	10	0.0005	5 cents	10 cents	15 cents
3-1/2	5	0.0015	15 cents	5 cents	20 cents

CHAPTER 18

HOW TO FIND THE BEST HONING PROCEDURE

TABLE 18-9
Approximate Surface Finish

Material	Abrasive Type	Approximate Surface Finish in Microinches (μ") Ra								
		Grit Size								
		80	100	150	220	280	320	400	500	600
Hard steel	Aluminum oxide/silicon carbide	25	--	20	18	12	10	5	3	1
	CBN (Borazon®)	--	55*80	45	30	28	--	20	--	7
Soft steel	Aluminum oxide/silicon carbide	80	--	35*55	25	20*35	16	7*10	4*8	2
	CBN (Borazon®)	--	65*100	--	70*80	--	--	25	--	--
Cast iron	Silicon carbide	100	--	30*40	20	12	10	6	5	3
	Diamond	--	--	--	80	--	--	50	--	20
Aluminum, brass, bronze	Silicon carbide	170	--	80	55	33	27	15	12	2
Carbide	Diamond	--	--	30	20	--	--	7	--	3
Ceramic	Diamond	--	--	50	40	--	--	20	--	15
Glass	Diamond	--	--	95	70	--	--	30	--	15

Material	Abrasive Type	Approximate Surface Finish in Micrometers (μm) Ra								
		Grit Size								
		80	100	150	220	280	320	400	500	600
Hard steel	Aluminum oxide/silicon carbide	0,65	--	0,50	0,45	0,30	0,25	0,12	0,08	0,03
	CBN (Borazon®)	--	1,40*2,00	1,15	0,75	0,70	--	0,50	--	0,18
Soft steel	Aluminum oxide/silicon carbide	2,00	--	0,90*1,40	0,65	0,50*0,90	0,40	0,18*0,25	0,10*0,20	0,05
	CBN (Borazon®)	--	1,60*2,50	--	1,75*2,00	--	--	0,65	--	--
Cast iron	Silicon carbide	2,50	--	0,75*1,00	0,50	0,30	0,25	0,15	0,12	0,08
	Diamond	--	--	--	2,00	--	--	1,27	--	0,50
Aluminum, brass, bronze	Silicon carbide	4,30	--	2,00	1,40	0,85	0,70	0,40	0,30	0,05
Carbide	Diamond	--	--	0,75	0,50	--	--	0,18	--	0,08
Ceramic	Diamond	--	--	1,27	1,00	--	--	0,50	--	0,40
Glass	Diamond	--	--	2,40	1,80	--	--	0,75	--	0,40

*If two values are shown: the first number is for small parts, honed on machines with one horsepower or less; the second number is for large parts, honed on machines with two or more horsepower.

Always use a honing fluid which was properly blended for the process being used. Years of research, blending, testing, reblending, and evaluating have gone into the development of quality honing fluids.

STATISTICAL PROCESS CONTROL

The ever-increasing demand for better quality of all manufactured products is forcing more and more manufacturers into applying statistical process control (SPC) to their manufacturing methods. Subcontractors must use SPC because their customers demand it, and original equipment manufacturers (OEM) demand it so their products can have the quality they need to compete in the world market. In the past, the machine operator produced workpieces inspected later and found to be within blueprint tolerance—except for some parts that were not within tolerance. 100% inspection is much too expensive for large quantity production; the solution is for the operator to measure each part immediately after making it, correcting the machine if the measurements indicate a need for correction. Of course, the operator has to use a gage to measure the accuracy. Today's gages will automatically feed each measurement into a computer capable of statistical quality control (also known as statistical process control). This computer might be a large mainframe; but a small, portable computer makes it possible for the machine operator to see immediately when a pattern of results develops. This pattern may require a machine adjustment before workpieces are made outside of tolerance (see Fig. 18-20). The chart, produced by the computer printer, is called the *X-Bar and R chart*, and it enables the machine operator to prevent making scrap parts. The printer can also make another chart, called a *histogram*. It will go with the finished parts to the customer, who needs to do no incoming inspection because the chart will show such values as standard deviation, Cr, Cp, and Cpk values, which tell a statistician everything about the quality of the workpieces.

HONING MACHINE SELECTION

Both horizontal and vertical honing machines are available. In general honing, no proof is available that one type of machine gets better results than the other, either in speed or in accuracy attainable. There are, however, some obvious limitations. For instance, to hone a 10' (about 3 m) long tube, a vertical machine would have to be at least 26' (about 8 m) tall, and it would be difficult to find a building to fit it in. On the other hand, short heavy parts, for example with a 6 by 10" (150 by 250 mm) bore,

CHAPTER 18

HOW TO FIND THE BEST HONING PROCEDURE

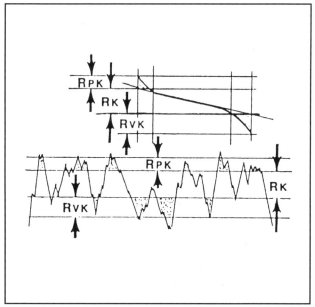

Fig. 18-19 Definition of *Rk*.

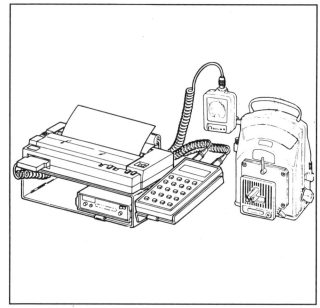

Fig. 18-20 Gage, processor, and printer.

are better suited to vertical machines. This is because the operator can easily lift the tool in and out of the bore and support the weight of the tool.

There are four common types of honing machines, ranging from the simplest, basic honing machine to automated, robotized versions, selling for a hundred times as much.

With the basic honing machine (see Fig. 18-21) the operator is required to hold the part by hand while stroking the part back and forth over the honing tool.

With a power stroked honing machine, the operator puts the part into a fixture and starts the spindle. The machine hones the part; then the operator stops the machine and unloads the finished part. Another improvement is the automatic honing machine which gages the bore diameter while honing and stops the honing operation when the desired size has been reached. The single spindle machine offers a fast, accurate, and inexpensive alternative to conventional multistroke honing in limited applications. The multispindle high-production honing machine (see Fig. 18-22) incorporates computer numerical control (CNC), computer touch screen, and single stroke honing tools. All these machines require the operator to make decisions about things like stroke length, spindle speed, cutting pressure, and tool selection.

More sophisticated honing machines have a computer screen which ask questions about part diameter, part length, material, hardness, stock removal and so on, all information which can be

TABLE 18-10
Stock Removal for Surface Finish

Surface Finish in Micrometers (μm) Ra

$$\frac{\text{Existing Finish} - \text{Desired Finish}}{100} = \frac{\text{Required}}{\text{Stock Removal}}$$

EXAMPLE: Existing finish = 1.25 μm;
Desired finish = 0.25 μm

$$\frac{1.25 - 0.25}{100} = 0,01 \text{ mm}$$

Surface Finish in Microinches (μ″) Ra

$$\frac{\text{Existing finish} - \text{Desired finish}}{100,000} = \frac{\text{Required}}{\text{Stock Removal}}$$

EXAMPLE: Existing finish = 50 μ″; Desired Finish = 10 μ″

$$\frac{50 - 10}{100,000} = 0.0004 \text{ in.}$$

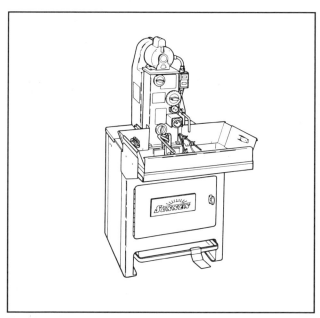

Fig. 18-21 Basic honing machine.

CHAPTER 18

HOW TO FIND THE BEST HONING PROCEDURE

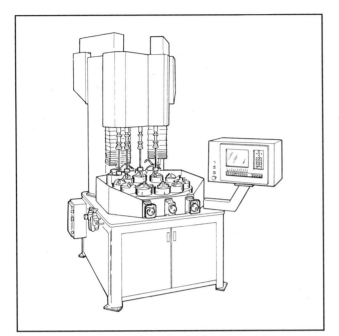

Fig. 18-22 Multispindle high-production honing machine.

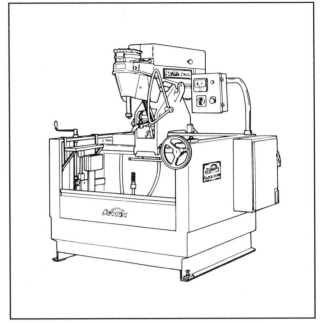

Fig. 18-23 Vertical honing machine.

read off the blueprint. When these questions are answered via the computer touch screen the machine automatically sets itself up to the proper spindle speed, stroke length, stroke rate, crosshatch angle, sparkout time, and tool retraction, and will recommend the proper tool to use. From then on the process is completely automatic, and the machine will remember up to 100 setups.

There are machines which can be incorporated into a complete system which includes a workpiece magazine and a robot. The next category is the custom-built honing machine which will do everything, including loading, gaging, and unloading; some machines even segregate the parts according to differences in size, label them, and present management with a printout of statistical quality control.

A custom-built machine can cost five to ten times as much as a stock machine. Furthermore, since it is custom-built, it is not available from stock, and replacement parts may be hard to obtain. The machines mentioned were designed for small parts, things that could be held in one hand. Large-diameter workpieces are best honed on a vertical machine (see Fig. 18-23), if they are not over 16" (400mm) in length. Longer parts may require a horizontal machine such as the one shown in Fig. 18-24. These machines also come with computer numerical control (CNC), and can handle workpieces up to 394" (10,000 mm) in length with bore diameters from 0.990 to 23.5" (25.1 to 600 mm).

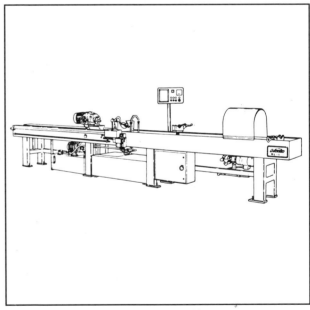

Fig. 18-24 Horizontal or tube honing machine.

DEBURRING AND SURFACE CONDITIONING

Deburring is, simply, the removal of burrs. But what is a *burr*? This is a somewhat elusive question because everyone knows what a burr is, but can't explain it to anyone else, and no two people really agree on the definition.

The burr, edge, and surface conditioning technology (BEST) division of SME has defined a *burr* as follows:

"A burr is an undesirable projection of material that results from a cutting, forming, blanking, or shearing process."[16]

Deburring, then, becomes the removal of undesirable projections of material. In other words, if a burr is defined as "all of the sharpness along edges" (to prevent cutting anyone during assembly or use), the burrs are gone when the edges have been dulled so

18-41

CHAPTER 18

DEBURRING AND SURFACE CONDITIONING

they will no longer cut anyone's finger. Much of the projecting material may still be there.

Surface Conditioning is the adjustment of the surface contours to meet requirements. Often no adjustment is required. Sometimes a major change is required as, for example, developing a high quality, preplate finish on a part with a fairly rough surface. In this instance the customer may specify a maximum surface roughness requirement of six microinches (0.152 micrometers) or specify that all scale or oils are to be removed. The conditioning of the surface is easily accomplished by several of the deburring and surface conditioning processes.

DEBURRING AND SURFACE CONDITIONING PROCESSES

The following are the more popular methods of deburring and surface conditioning:

Abrasive flow machining uses a putty-like, abrasive-laden material pumped through the critical areas of a fixtured part or parts. By controlling abrasive type, size, and quantity, and the volume and rheology of the material passing through the part, exceptionally high-quality surfaces with exacting radii can be developed on critical parts.

Blast finishing uses a variety of abrasives, such as aluminum oxide, glass beads, plastics, or steel shot, propelled by air, water, or a rotating wheel, to abrade localized areas of parts.

Very fine micro-sized blasting can be done on intricate, small components—or enormous parts can be descaled by these means. Some blast methods markedly improve the compressive stresses in surfaces, extending working life.

Brush deburring uses rotary brushes made of natural fibers, steel or other wire, or abrasive-filled monofilaments to scrub localized internal or external areas of parts. The high speed brushes can develop radii, smooth surfaces, and remove surface contamination.

Buffing, like brushing, uses rotary tools to work localized areas of parts. Buffs are made of natural materials like cotton and sisal, or they can be synthetic. Compounds are rubbed into or sprayed onto the buffs, to add fine abrasives or to help smooth and brighten surfaces. Buffing is often used as a preplate finish on large, relatively regular surfaces.

Coated abrasive belting can remove quantities of a part's surface sufficient to alter its contour. Normally this is used when very rough surfaces are to be made quite smooth. For example, forgings can be made significantly smoother. This method helps remove die mismatch and can materially upgrade the quality of rough-cast surfaces.

Electrochemical deburring is used on fixtured parts to remove burrs in intersecting holes. The fixture locates the electrode near the burr and flowing electrolyte completes the circuit, "deplating" the unwanted metal.

Mass finishing processes are several. In general, a number of parts are distributed in a mass of abrasive or nonabrasive media in a machine which imparts a differential action between media and parts. A compound and water solution is used to keep the mass clean, prevent corrosion, etc. Several commonly used machines are:

- *Vibratory tub* and *round bowl* machines are the most numerous in the field because they perform deburring easily, rapidly, and economically while developing very fine surfaces when desired.
- *Centrifugal disc* is faster than the vibrators and can develop beautiful surfaces. Media and part size may be a bit limiting.
- *Centrifugal barrel* is very fast, but is a closed machine requiring more operator skill. It is best used when vibratory cycles exceed eight to 10 hours.
- *Thermal deburring* employs natural gas and air literally to burn the burrs off a part or group of parts. A heavy-duty machine seals a pressure chamber containing a basket of parts. Air and natural gas are metered in and ignited. The temperature of the burrs rises above that required for auto-ignition, and the burrs actually catch fire. They burn until the burr approaches the mass of parent metal, dropping the temperature enough to put out the fire. It is exceptionally fast and, for small intricate parts, can be very economical.

WHY UPGRADE OR IMPROVE THE PROCESS?

The best reason for upgrading is to be sure that the company is in fact, using state-of-the-art, lowest cost, and/or best quality methods. If the process is an old one, chances are some improvements have been made. Improvements can be in materials, materials handling equipment, deburring process equipment, and even in the manufactured parts.

If the manufacturing process has been making changes—even subtle changes—in the parts made, these might have an important effect on the deburring and surface conditioning processes. For example, a change in the cutting, tapping, or other lubricant used can contaminate the solution in vibratory systems if the fluids are no longer compatible. If the manufacturing arm is getting better control on its drills and other cutting tools, the burrs are smaller.

The deburring process can be shortened to take advantage of these improvements. The same holds true for surface roughness; better control of die casting dies means less heat check and smoother parts. Smoothing these surfaces takes less time and therefore costs less.

These are ways to keep product quality up as well as to improve it gradually and consistently. It often takes only a simple change to improve quality, and it can often improve costs as well. This kind of double benefit is possible only to those who are looking for improvement.

If managers are looking for it, reading the appropriate trade journals, visiting appropriate trade shows, talking with the better sales people, they keep current in their technology. When changes can be made that have an effect on operations, they should be ready to upgrade the process.

Upgrading and Improvement Problems

Naturally, there are problems with upgrading these processes. The most common problem in deburring is a change in personnel.

Changes in personnel mean the loss of continuity in the control of the process. Those "lost people" knew all of the background information about the processes. Reinventing the wheel for every change is expensive. The learning curve is steep, but improvement is essential.

To upgrade a deburring area, take a good look at the processes running today. Reach into the mass and grab a handful of wet media. Wiggle it around in your hand a bit and drop it back into the machine. Is your hand dirty? If so, it's very probable that your parts are as well. Look to the solution system to fix this problem.

Now look at the parts. Are they clean all over? Is a gray smut collecting in recesses? Is the media glassy and not cutting properly? Again, in all cases if so, fix the solution system.

Are the parts being deburred correctly? Not too much or too little? Is the media properly sized? Is it the right type and shape?

CHAPTER 18

AUTODEPOSITION: TOUGH COATINGS AND NO VOCS

Are there lodging problems you didn't have before? Is the surface roughness of the parts adequate and as you desire? Are there separation problems? Most of these problems are due to improper media. Do some research and find out what media was used in the process originally. If different from what you are now using, why did you change? Should you have changed? Did Purchasing substitute a similar, but cheaper, product?

If the media is not the problem, then look at your parts. Are they, in fact, the same as they were when the process was originally set up? Are the burrs bigger? Or smaller? Or different in shape? Perhaps the same solution system no longer seems to be cleaning your parts. If these are possible answers, then your parts are different. Changes in machining practices or the lubricants used could make these differences.

AUTODEPOSITION: TOUGH COATINGS AND NO VOCS

The metals finishing industry is undergoing major changes. The primary reasons for these changes are increasingly stringent environmental regulations; rising costs for material, labor, and waste handling; increasing quality requirements; and the need to paint preassembled workpieces, which are often composites of metal and heat-sensitive plastics.

To meet these challenges, more and more finishers are turning to autodeposition, a waterborne process that depends on chemical reactions to achieve deposition.

Autodeposition features no heavy metals or solvents, low-temperature cure, and high corrosion resistance. It uses a mildly acidic paint bath that solubilizes positively charged metal ions from the workpiece's surface. The ions react with negatively charged paint particles dispersed in the coating bath to form a deposit on the workpiece surface.

Unlike conventional systems, the autodeposition process does not include phosphate conversion coating stages and associated rinse cycles. The equipment used for other stages (cleaning, rinsing, painting) is the same as used for other immersion finishes. In fact, several companies have built autodeposition lines by converting stages previously used in other systems like phosphating and plating.

The physical and corrosion-resistance properties of autodeposited coatings depend on the chemical nature of the organic resin used to make the paint particles. One type of coating is formed from products based on a polyvinylidene chloride latex resin. The low-cured (220°F–104°C) films are flexible and hard, and provide excellent corrosion resistance. The process is totally free of solvents or heavy metals. Another type, based on acrylic chemistry, offers resistance to a variety of solvents at high temperatures. This process uses a small amount of coalescing solvent (1.6 lb/gal–0.19 kg/L VOC) and chromium in the final sealing rinse. It also cures at a higher temperature (350°F –177°C) than the polyvinylidene chloride latex resin. The deposition mechanism and chemical control of the two systems are identical.

BENEFITS

Waste treatment reduction. Since the polyvinylidene chloride latex resin does not contain VOCs (volatile organic compounds), it is not necessary to scrub or otherwise control emissions. Elimination of phosphate pretreatment eliminates the need to precipitate (and landfill) potentially hazardous metals like zinc and nickel. Discarded paint solids can be flocculated, filtered, and treated as nonhazardous waste. Further, treatment of the overflow from alkaline cleaners and associated rinses uses standard methods. The minimization of waste also reduces treatment and disposal costs.

Cost savings. Comparative analyses with conventional paint systems show that autodeposition systems occupy 30 – 35% less floor space, while providing a finish capable of withstanding rugged conditions. The savings result from the elimination of phosphate pretreatment and rinses, as well as reduced oven length because of lower cure temperature requirements. This means a lower capital investment. Moreover, because of the process's simplicity, existing systems can be converted at relatively low cost. A new system, requiring less space and equipment, will cost less. The low-temperature cure also saves energy.

Further cost-saving areas include rack stripping and maintenance. Once a rack is coated, the metal is protected from further reaction and coating deposition, preventing additional coating formation. Fewer stages mean reduced maintenance of both equipment and chemical process baths. Autodeposition requires no electrodes, rectifiers, or other potentially troublesome equipment.

Versatility. Many automotive components combine organic materials with steel. The higher cure temperatures required for conventional solvent paints can destroy these materials. Solvent-free, low-cure autodeposited films eliminate this problem and easily coat plastic/metal components. Examples of successfully coated parts include fully assembled leaf springs and seat components with plastic handle covers and nylon connections.

Autodeposition also enables painting of severely recessed or blocked areas that are difficult to coat by other methods. Wherever liquid wets the surface, a protective coating is produced. This deposition mechanism avoids the problem of electrical shielding, which limits the application of electrophoretic coatings. In electrocoating, the electric field weakens as it enters a recessed area (because of the Faraday cage effect), resulting in gradually diminishing film builds. Electrostatic spray and powder paint systems suffer from the same electrical shielding effect.

FUTURE DEVELOPMENTS

Autodeposition can produce high-performance coatings on zinc without compromising quality. This requires modifying the paint bath chemistry and paying close attention to the cleaner formulation, to prevent excessive substrate dissolution. Efforts are under way to develop the process further. Also under investigation are barrel coating techniques for bulk coating small workpieces such as fasteners.

Pilot trials have uncovered many prospective topcoats for use over autodeposition-primed parts. Examples include water-reducible alkyds and acrylics and low-cure urethanes. Since any primer-topcoat system has critical capability requirements, it is necessary to thoroughly evaluate topcoat candidates for application and performance prior to production trial.

CHAPTER 18

AUTODEPOSITION: TOUGH COATINGS AND NO VOCS

DESCRIPTION OF AUTODEPOSITION

Autodeposition is a waterborne process which depends on chemical reactions to achieve deposition. This process has been in commercial usage since 1973. Presently, there are numerous installations in operation, coating a variety of fabricated steel parts for the automotive and general industrial markets.

The autodeposition bath consists of a mildly acidic latex emulsion polymer, de-ionized water and other proprietary ingredients. The bath solids may vary between 3% and 5% by volume. The bath viscosity is close to water, with no organic solvents in the coating bath. Another characteristic of the autodeposition bath is its acidic nature; the pH range is typically between 2.5 and 3.0. The chemical phenomenon of the autodeposition process can be stated very simply. The mildly acid bath attacks the immersed steel parts causing an immediate surface reaction that releases iron ions. These ions react with the latex in solution causing a deposition on the surface of the steel parts. The newly deposited organic film is adherent yet quite porous: the chemical activators can rapidly diffuse to reach the surface of the metal allowing continued formation of coating to provide a high degree of protection and corrosion resistance.

The coating thickness of the autodeposition film is time- and temperature-related. The film or coating continues to grow as long as ionic species are being produced at the coating/metal interface. Initially, the deposition process is quite rapid, but slows down as the film begins to build or mature. So long as the part being coated is in the bath, the process will continue, however, the rate of deposition will decline. Different patterns of growth result depending on the particular latex used in the coating matrix and the resin solids of the bath. Typically, film thicknesses are controlled at 0.3 to 1.0 mils (0.0076 to 0.0254 mm).

A unique feature of the autodeposition process is the formation of a uniform film over all of a surface even in difficult to reach areas. In an autodeposition bath, the local anode sites are coated first since this is where ionic species are generated most rapidly. Once these sites are coated, the local cell polarity is reversed, and they become cathodic in relation to the surrounding uncoated areas. The polarity continues to reverse during coating and the result is further deposition where the coating is thinnest. This is where activator diffusion occurs most readily. In essence, the autodeposition film is continuously balanced providing uniform thickness throughout.

The film formed by the autodeposition process is unique for a latex vehicle. Latex films form by coalescence. However, the degree to which the film forms in autodeposition, via the coalescing action, is the basis for its uniqueness. Coated parts can be water rinsed immediately after being removed from the bath, with very little material loss. On leaving the bath, the coating consists of two layers: one is a very cohesive reacted layer, while the other is composed of undeposited excess polymer and activator from the bath. The chemical reaction continues in the second layer, resulting in film deposition and increased film thickness instead of solids loss to dragout. This unusual feature strongly limits carryover of contamination into the rinse tanks and greatly reduces the demand for waste treatment.

A final characteristic of autodeposition is its "wetting power" or ability to coat wherever the liquid touches the metal parts. Unlike cathodic electrodeposition (CED) where electrical energy is required to "throw" the coating into recessed areas, autodeposition will coat those areas if the coating bath can penetrate. Very simply, parts tubular in shape, assembled parts, or parts that have intricate designs are easily coated and protected by the autodeposition bath. It should also be mentioned that the autodeposition process does not require a phosphate stage or even a chromic rinse section. The elimination of these stages leads to considerable savings in required floor space, energy, and operating costs. Considerably lower temperatures are required to cure the coated parts. These are just some of the many advantages of using the autodeposition process.

AUTODEPOSITION COATING CHEMICALS

Autodeposition is a waterborne process which depends on chemical reactions with metal surfaces to achieve deposition of a paint coating. Autodeposition does not require a pretreatment process (for example, using phosphate to prepare the surface).

Process Sequence

A typical process sequence for an autodeposition finishing line is as follows:

Stage 1—Alkaline spray clean
Stage 2—Alkaline immersion clean
Stage 3—Plain water rinse
Stage 4—De-ionized water rinse
Stage 5—Autodeposition chemical coating (pH 2–3)
Stage 6—Plain water rinse
Stage 7—Reaction rinse
Stage 8—Curing oven

The process is adaptable to conveyorized or indexing hoist equipment. Cleaning, rinsing and oven curing are accomplished in accordance with good industry practice. All of the process equipment with the exception of stage five is identical to that employed in other finishing operations. A description of stage five follows.

Coating Time

The workpieces to be coated are immersed in the coating solution for a specified time. Normally 60 to 120 seconds applies coatings between 0.5 and 1.0 mil (0.027 and 0.0254 mm). The longer the immersion time, the thicker the coating deposited.

Temperature Requirements

The temperature of the coating bath should be controlled within a range of 68 to 72°F (20 to 22°C). Heat transfer surface temperatures should not exceed 90°F (32°C), and cooling transfer surfaces should not be lower than 40°F (5°C).

Heat transfer surfaces should be adequately protected from damage by the work being processed. Coil materials should be 316 stainless steel coated with Kynar® or Teflon®, or can be totally constructed of Teflon or polypropylene.

Heating/cooling coils in the autodeposition bath are required as a safeguard against accidental heat/cold carry-in, and/or if ambient plant conditions warrant (heating or cooling).

Agitation Requirements

Agitation of the coating bath is obtained by means of side-mounted, variable-speed mixers. Agitation is important to provide good films on any recessed areas (tubular shapes, preassembled parts). Film thickness variations can also be accomplished (without any change in bath chemistry) by changing agitation velocities; higher film thicknesses are normally achieved by higher agitation.

The type and quantity of AC variable speed mixers, and the power requirements for the controller, can vary from installation to installation. The final design is recommended by the chemical supplier and is based on tank capacity, tank geometry, and work package density.

CHAPTER 18

AUTODEPOSITION: TOUGH COATINGS AND NO VOCS

In general, a gentle agitation of the coating bath is provided by properly spaced mixers with AC variable frequency drive 1/2 hp motors. The agitation should maintain bath flow past the surface of the parts being coated at 40 to 50 fpm (0.20 to 0.25 m/s). High shear of the bath must be avoided. Examples are shear produced by centrifugal pumps, bearing surface below the bath surface, and high-speed impellers.

Pumping Requirements for the Coating Bath

A diaphragm pump— a 2" (5.1 cm) air diaphragm pump with polypropylene wetted parts, Teflon® diaphragm, valve balls, and valve seat O-rings, for example—should be used to add resin replenisher to the bath for paint solids make-up/adjustment. (The additions can be made directly from drums or from a bulk storage tank through a day tank.) The pump is also used to transfer the autodeposition bath to a hold tank for yearly tank maintenance.

The pump should have a PVC screen protection, to prevent any residues with particle size in excess of 1/2" (12.7 mm) diameter from entering. The pump should be mounted near the outside of the tank below the operating level of the coating bath, and the suction and discharge lines should enter the top of the coating bath (not through the tank wall). Provision should be made for flushing of the pump with de-ionized water after use.

Piping Recommendations

All piping should be rigid (nonplasticized) Schedule 80 PVC. Pipe flow velocities should not exceed 8.0 fps (2.4 m/s). Steel pipe with a suitable liner (Kynar® or Hypalon®) may be used if more strength is required. A 3" (7.62 cm) flange connection should be used on the drain of the coating tank.

Part Transfer Zone

Ambient relative humidity conditions (<40% in winter months) in most installations require a humidified vestibule between stages five and six. Moisture is produced by the use of misting nozzles supplied with de-ionized water at 40 to 50 psig nozzle pressure. In addition, excessive drain times (usually encountered when line speeds are low or when the vertical dimension of the work is extremely long) mandate the use of de-ionized water misting nozzles. The nozzles should be only the low-volume atomizing type which will yield 1.5 gallons (5.6 L) per hour. The de-ionized water supply line should be provided with a five-micron cartridge-type filter.

The line design has freshly coated parts emerging from the coating bath remain in air (60% relative humidity minimum) for a period of 45 to 180 seconds prior to entering the subsequent plant water rinse bath. This short dwell allows supernatant paint to continue to react with the metal surface, and greatly reduces the loss of paint solids to the subsequent water rinse.

Make-up and Replenishment

An adequate de-ionized water supply to the coating should be provided. Two additional fill lines and feed pumps are required to maintain bath chemistry.

The tank is made up with autodeposition coating chemicals at 10 to 15% by volume plus de-ionized water at 85 to 90% by volume. Any bath volume loss during production must be restored with the proper chemicals plus de-ionized water.

Tank Volume

The tank volume is based on tank geometry, rack geometry, part density, and production rate. The tank height should be based on the height of the work, plus at least 18" (46 cm) from the bottom of the work or rack to the tank bottom, plus 6" (15 cm) of freeboard. The tank width should allow for work passage plus clearance to allow for work movement and accessories located on the side of the tank (heating coils, mixers, skimming trough, piping, etc.).

Tank Materials

A mild steel tank lined with a durable, acid-resistant material, approved by the chemical supplier, is required. Typical liner materials are three-ply (soft, hard, soft) natural rubber or fiber-reinforced organic composite coatings.

Waste minimization and high quality finishes begin with a well trained workforce that understands the value of controlling and constantly fine-tuning each variable within the system. Attention to and control of these variables are the key to improved quality, greater productivity, lower unit cost, enhanced profits, minimization of all waste streams, and environmental compliance.

References

1. G.E.F. Brewer, J. Paint Technology., 45 (587) 37, 1973.
2. W. Machu, *Handbook of Electropainting Technology*, Electrochemical Publishing Ltd., London, 1978.
3. M. Wismer et al, *Cathodic Electrodeposition*, J. Coatings Technology., Vol. 54 (688), pp 35–44, 1982.
4. P.E. Pierce, J. Coatings Technology., 53 (672), 52, 1981.
5. C.L. Coon and J.J. Vincent, *Processing Parameters Influencing Paintability of Precoated Metals with Cathodic Electrocoat Primer*, J. Coatings Technology., Vol. 58 (742), pp 53–60, 1986.
6. M. Podany, Proceedings, ELECTROCOAT '92, Gardner Publications, Cincinnati, OH, 1992, p. 11-1.
7. J. Allshouse, Proceedings, ELECTROCOAT '92, Gardner Publications, Cincinnati, OH, 1992, p 7-1.
8. K. Legatski, Proceedings ELECTROCOAT '90, Gardner Publications, Cincinnati, OH, 1990, p. 15-1.
9. G.E.F. Brewer, M.E. Horsch, and M.F. Madarasz, J. Paint Technology., 38 (499), 453, 1966.
10. D.R. Hays and C.S. White, *Electrodeposition of Paint. Deposition Parameters*, J. Paint Technology., Vol. 41 (535), pp 461–471, 1969.
11. G.E.F. Brewer and L.W. Wiedmayer, *Thickness of Electrocoats Inside of Cavities Versus Size and Shape of Openings*, J. Paint Technology., Vol. 42, (550), pp 588–591, 1970.
12. N. Furano and Y. Ohyabu, Progr. Org. Coatings, 5, 201, 1977.
13. P.E. Pierce and C.K. Schoff, *Coating Film Defects*, Federation of Societies for Coatings Technology, Philadelphia, PA 1988.
14. C.K. Schoff, Proceedings, ELECTROCOAT '92, Gardner Publications, Cincinnati, OH, 1992, p. 10-1.
15. C.K. Schoff, J. Coatings Technology, *Electropaint-Substrate Interactions*, Vol. 62, (789), pp 115–123 (1990).
16. S. J. Slawinski, *The Development of a Burr Definition*, SME Technical Paper MR79-743.

CHAPTER 19

ASSEMBLY

INTRODUCTION

In the past, quality was a measure of defective units transferred to the customer. Today, quality is defined simply as "exceeding customer expectations." It is becoming common to use the capital Q to designate the expanded definition of quality. Small q quality refers to the ability to produce a product without performance defects, while *Quality* refers to the ability of a product line to meet customer expectations in all areas. This definition encompasses all aspects of the customer's interaction with the product, including both the objective and subjective evaluation of product and company performance. Quality is now a much broader field of study and application. The techniques used to improve customer satisfaction with complex assembly operations are in their infancy. Many companies find implementation of Quality programs difficult since they require widespread behavioral change. This section discusses:

- Quality prerequisites; foundations are laid before making any major efforts for improving Quality.
- Successful techniques and programs for helping assembly operations achieve high levels of customer satisfaction.
- Methods useful in obtaining the behavior modifications necessary for the successful implementation of these techniques and programs.
- A "road map" for determining which improvements will give the greatest positive impact for assembly operations, and how to proceed to reach Quality objectives.

Assembly is defined as a series of operations performed on a given set of equipment (the line) to fabricate a set of components into a finished product, ready for shipment to a customer (internal or external). Assembly is a process of subprocesses. It is characterized by a flow chart describing a set of repetitive steps (the subprocesses) that transform the input into the output.

Total quality is defined as the integration of all departments, functions, employees, and activities toward the satisfaction of a company's customers and the continuous improvement of its processes.

PROCESS DEVELOPMENT AND SPECIFICATIONS

A quality assembly process starts at product development. Managers and designers of the assembly process must do an excellent job of translating customer expectations into a measurable set of product specifications. This is done with a disciplined and repetitive product development process; otherwise, mediocre product and process design could result.

Techniques such as quality function deployment (QFD) or failure modes and effects analysis (FMEA) are successful in guiding development teams through the translation of a customer's subjective and objective desires into consistent product specifications. If product specifications do not meet customer expectations, the assembly process is doomed before it starts. An example of a modern product development process is shown in Fig. 19-1.[1]

Customers want more than just a product that works. Surveys of customers, from many different markets and industries, almost always include the following requirements on their lists of what it takes to give them satisfaction:

- High-quality product.
- Reliable product.
- On-time delivery.
- Short delivery lead-times and the ability to change orders as required.
- Fast response when a problem is identified.
- Fair price.
- Other services that are the responsibility of areas outside of assembly, such as sales support and technical literature.

Although these new dimensions of product performance may not be covered in official specification

CHAPTER CONTENTS:

INTRODUCTION 19-1

IMPROVING ASSEMBLY SYSTEMS 19-17

CONTINUOUS IMPROVEMENT FOR FEEDERS 19-19

IMPROVING THE QUALITY OF MACHINE-MADE BRAZED OR SOLDERED JOINTS 19-27

AUTOMATE TO UPGRADE SOLDERING AND BRAZING OPERATIONS 19-29

IMPROVING PRESS FORGING CAPABILITIES 19-36

GUIDELINE FOR OPTIMIZING Nd-YAG LASER WELDING 19-45

THE INVENTION OF HOOK AND LOOP FASTENERS 19-51

The Contributors of this chapter are: **Harald N. Bransch**, Senior Applications Engineer, Applications Laboratory-Industrial Products, Lumonics Corporation; **Stephen J. Casalou**, President, R.A. Casalou Inc.; **R. Michael Clark**, President, Black and Webster Assembly Equipment Division-Air Hydraulics, Inc.; **Marian Estreicher**, Chief Engineer-Mechanical Presses, Siempelkamp Pressen Systeme (Germany); **Klaus Hilgers**, Chief Department Engineer-Mechanical Presses, Siempelkamp Pressen Systeme (Germany); **Derek Lidow**, Executive Vice President, International Rectifier Corp.; **Charles M. Reyle**, President, C.M. Reyle and Company; **Ronald J. Ruhl**, Sales and Marketing, Vibratory Feeder Division-Automation Devices; **Bruce R. Williams**, Vice President, Fusion Inc.
The Reviewers of all or part of this chapter are: **Patrick J. Billarant**, President, Aplix Inc.; **Donald E. Hegland**, Editor Publisher, Assembly magazine.

CHAPTER 19

INTRODUCTION

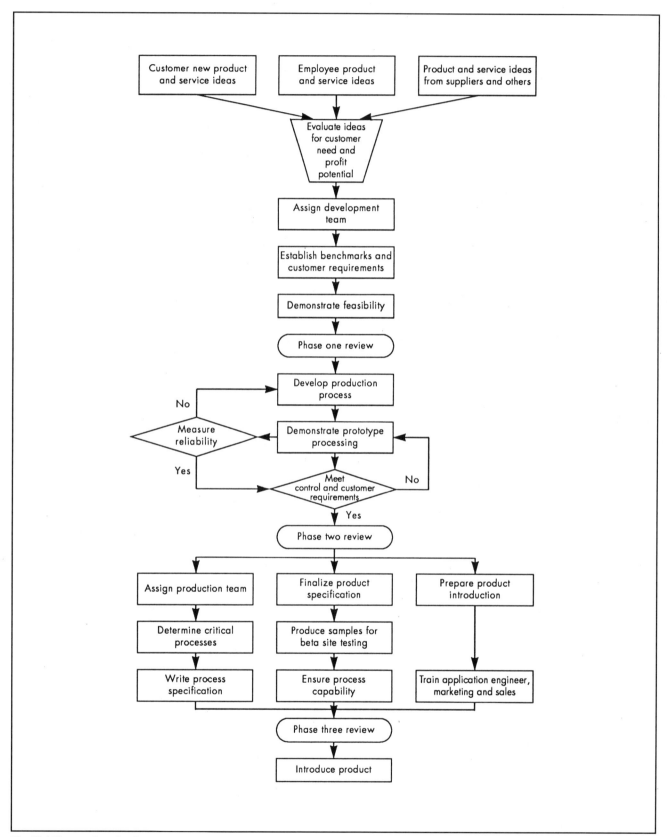

Fig. 19-1 New product development process.

CHAPTER 19

INTRODUCTION

documents, the product's design and assembly process must meet the customer's expectations in each of these performance areas. An assembly process can no longer be considered high quality just because it produces a rugged product with few defects. Today, to meet customer expectations and claim a high-quality reputation, new products must be assembled in a cost-effective, responsive, and reliable manner.

PROCESS DESCRIPTION AND DOCUMENTATION

Quality in an assembly area is achieved when the assembly process is properly described and documented. The ability to describe and document the assembly process is a critical requirement:

- To ensure repeatability.
- To train personnel.
- To re-start the process in the event of a move, expansion, or other disruption.
- As a baseline from which to make improvements or solve problems.

In other words, a good description of the assembly process is critical to achieving consistency regardless of conditions. A process description or specification must therefore serve everyone who operates the process: engineers, technicians, and operators. A process specification only understood by an assembly engineer is not effective as a training or reference tool for everyone involved in the process.

A complete description of the overall assembly process must include the following sections:

Materials. The materials section should include descriptions of the gases, liquids, and solids used in the assembly process.

Manpower. The manpower section describes the personnel and the competencies they must demonstrate in order to perform the assembly process.

Methods. The methods section must include descriptions of all steps taken to prepare, perform, check, and improve the assembly process. Methods must be described so that all personnel can understand the requirements and perform the process repeatably, and know whether the operation was performed correctly. Finally, the methods section must describe what to do when the operations no longer yield the desired results.

Management. The management section should discuss the over-arching processes into which the given assembly process must interface. Examples of management systems are scheduling and accounting. Although often described in documents different from the process specification, these management systems may dictate severe constraints on how the assembly process can and cannot be performed.

Machinery. The machinery section lists and describes all the equipment used by the process. Document and define what is done in each of these areas in terms simple enough for an average operator to understand. Experienced operators often do the best job of writing assembly specifications, since they have the most hands-on experience with how the process really works.

OBJECTIVES OF AN ASSEMBLY OPERATION

Even if the assembly process is specified so that the product will meet or exceed customer expectations, and the process is described and documented for repetitive performance, this does not mean that the process is effective. An effective assembly process must meet a wide range of economic, performance, and quality criteria. The common performance measures for an assembly line are:

Throughput. Throughput is a measure of how many units are produced within a certain period of time (in units/hour). Throughput measures the capacity of an assembly process. It is a critical measurement because it relates directly to the economic potential of the process (how much can be made?) as well as the ability to meet a given level of customer demand.

Quality. Typically, quality is measured in terms of how many parts per million (PPM) of the finished product do not pass outgoing inspection or do not meet specification. Of course, this measurement is only as good as the specification, defect definition, and inspection procedure used on the assembly line. A low PPM number to a loose specification or worse, to a set of specifications the customer does not care about, does not indicate a high-quality product. A good specification must:

- Correlate directly to a customer requirement.
- Be measurable to a defined tolerance or accuracy.
- Be compared with the output product to yield a definitive answer "yes" or "no" whether a defect exists.

The trend is growing to measure quality with C_p or C_{pk}. C_{pk} is known as the *process capability index*; it derives from standard statistical process control (SPC) measures. A C_p of two indicates a process whose average output is a distribution entirely within the three-sigma limits of the process mean. Such a process is also referred to as a *six sigma* process and it will yield, on average, only 6.4 PPM defective parts. Figure 19-2[2] shows the relationship between C_p, C_{pk}, sigmas, and PPM.

Reliability. Reliability is a measure of how many parts fail per X hours of standardized testing (for example, FITS = failures in ten to the nine hours). Such measures are only useful if the tests correspond to actual usage and abuse. Although use tests are commonly performed to gain reliability data, it can be impractical to wait for the results to become statistically meaningful. The development of accelerated tests therefore often becomes a priority, but the results of the accelerated reliability tests must correlate with "real life" usage and abuse. Such correlation is confirmed using established scientific techniques and models.

Performance to schedule (PTS). PTS measures what percentage of the scheduled output was shipped on time, by product type. The ability to produce on time, as scheduled, relates directly to customer satisfaction. PTS is a measure of how well yields, schedules, and cycle times are controlled through the assembly process. PTS is measured over various "time fences." (A time fence is that period over which the production schedule remains fixed to plan production.) Time fences vary according to types of business and methods of manufacture, sometimes remaining fixed over the entire period of the longest material lead-times, and sometimes over just a few hours. The time fence and PTS definition are chosen to best reflect how the assembly process should perform and achieve customer satisfaction. Made-to-order product lines will need to factor in material ordering cycles, while make-to-stock products may include only the time required to release the order to the assembly line.

Cost. How much did it cost to produce a unit ready for shipment? Costing is the subject of many books and an entire specialty of the accounting profession. Costing gets complicated because not all costs are directly associated with a particular unit or batch (called direct costs). When costs are general in nature and defy easy association with a particular unit or batch, cost accountants attempt to construct a cost model of how to allocate these indirect costs to individual units of production.

CHAPTER 19

INTRODUCTION

Chart #1 — Overall yield vs. sigma (Distribution shifted +/- 1.5σ)

Number of parts (steps)	± 3σ	± 4σ	± 5σ	± 6σ
1	93.32%	99.379%	99.9767%	99.99966%
7	61.63	95.733	99.839	99.9976
10	50.08	93.96	99.768	99.9966
20	25.08	88.29	99.536	99.9932
40	6.29	77.94	99.074	99.9864
60	1.58	68.81	98.614	99.9796
80	0.40	60.75	98.156	99.9728
100	0.10	53.64	97.70	99.966
150	----	39.38	96.61	99.949
200	----	28.77	95.45	99.932
300	----	15.43	93.26	99.898
400	----	8.28	91.11	99.864
500	----	4.44	89.02	99.830
600	----	2.38	86.97	99.796
700	----	1.28	84.97	99.762
800	----	0.69	83.02	99.729
900	----	0.37	81.11	99.695
1000	----	0.20	79.24	99.661
1200	----	0.06	75.88	99.593
3000	----	----	50.15	98.985
17000	----	----	1.91	94.384
38000	----	----	0.01	87.880
70000	----	(Use for benchmarking)	----	78.820
150000	----		----	60.000

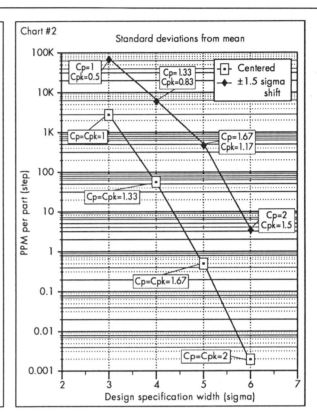

Chart #3

Total defects per unit	Rolled throughput yield (%)
5.3	0.5
4.6	1.0
3.9	2.0
3.5	3.0
3.2	4.0
3.0	5.0
2.3	10
1.9	15
1.6	20
1.4	25
1.2	30
1.0	37
0.9	40
0.8	45
0.7	50
0.6	55
0.51	60
0.43	65
0.36	70
0.29	75
0.22	80
0.16	85
0.10	90
0.05	95
0.00	100

Rolled throughput yield (%) = $100\, e^{-d/u}$

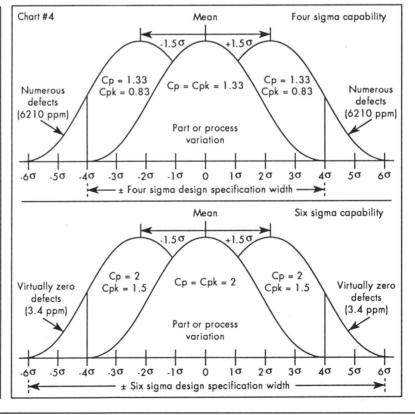

Fig. 19-2 Relationships between C_p, C_{pk}, sigmas, and PPM.

CHAPTER 19

INTRODUCTION

Cost relates directly to overall quality in two ways:

1. A customer will be unhappy if the cost is higher than expected.
2. A high-quality line usually is a low-cost line because it runs more efficiently.

Flexibility. Flexibility relates to how responsive an assembly line can be to changes in customer demand. Since forecasting orders is not a perfect science, some flexibility is a requirement for any assembly line. Indeed, many marketplaces demand a high degree of flexibility from suppliers. *Flexibility* can be measured in several ways:

- How many different bills of materials (BOM) and routings can be produced on a set of equipment? This measurement method determines the flexibility to make different products utilizing similar processes and equipment.
- How long must the equipment sit idle between BOM/routing changes? This measure determines the throughput loss in changing from one product to another.
- How big are lots compared to minimum order size? This relates to the flexibility to change among product styles, when a product style change requires virtually no line stoppages.

The ability to supply a product the way customers want it requires a high flexibility measurement, since many different variations and product types are assembled on the same line. It is common today to see assembly lines that can produce thousands of different product variations.

The economic viability of an assembly line that makes many different variations relates to the second method of measurement, since a shutdown of an assembly line to change from one product type to another is costly. Techniques were developed to make even the most complicated changes in equipment setup in a few minutes or less. Manufacturing managers now realize that setup changes must be managed like pit stops at car races. Single minute exchange of dies (SMED) teaches managers and technicians how to minimize setup times.

Getting the product to the customer fast, without prior notice, can be accomplished with inventories, or, less expensively, with minimum cycle times. Small lot size, the essence of the third method of measurement, relates directly to the ability to achieve short cycle times. Often, however, it is more direct simply to measure cycle time.

Assembly cycle time. Assembly cycle time is defined as the time required to assemble a product for shipment, starting with a complete set of unassembled components. Assembly cycle time is often measured as a multiple of the "theoretical" cycle time, which is defined as the processing time, assuming no queue time. Cycle time is now recognized as a critical measure of the efficiency and the design integrity of a given assembly process. Everything must work correctly for a process cycle time that approaches theoretical goals.

Total cycle time. Total cycle time is defined as the time required to process a customer order, purchase and receive raw materials, assemble, and ship the product. The total cycle time is not usually measured as a multiple of a theoretical standard, because it involves the entire product supply chain and is much longer than the assembly cycle time. Total cycle time is an excellent indicator of how well a specific assembly process is integrated with the company's scheduling and control systems, suppliers, and customers.

Management must set clear targets to meet all these critical assembly line parameters. They relate not just to customer satisfaction, but also to profitability and economic return. Without clear performance targets, assembly line personnel cannot make the day-to-day decisions that will lead to the best overall results for the customer or company.

ASSEMBLY VARIABLES

Given a defined process, the input of the following items can be controlled to reach targeted performance levels.

Equipment. Selection of equipment is critical to the assembly operation. All variables can be affected by equipment capability, uptime, productivity, flexibility, automation, and, of course, cost. Understanding the needs of the process is important to selecting the proper equipment.

Maintenance. Equipment will not operate without adequate maintenance. Maintenance includes the methods, techniques, and investments made to keep equipment tuned to peak performance.

Headcount. The number of people performing the assembly processes affects throughput and cycle time. It may have a detrimental or positive effect upon quality and reliability.

Training. The ability of a workforce to perform and control processes relates directly to the amount of training received. Training may be received through training classes or past experience. Training is the investment a company makes in improving worker productivity. A high level of training allows for greater empowerment, which, in turn, results in greater line flexibility, improved quality, and reduced costs. Training is not simply a number of class hours or years of experience; rather it is a set of competencies that a given group of personnel can perform reliably.

Scheduling and tracking systems. Everybody needs to know what is happening, particularly when something deviates from the plan or operation's control. The overall scheduling and tracking system may involve many subsidiary systems for finance, production control, purchasing, inventory, and order entry. A "system" is not defined by a computer program or set of forms, but rather a level of management attention that results in overall system accuracy, response time, and ability to spot exceptions before they cause problems.

Management control structure. The type of "command and control" structure used to control the assembly line should be closely related to the type of customer served. When the overall system is cross-functional and worker controlled, more responsive and lower-cost products are assembled.

Redundancy. Having idle capacity is the most costly and least productive method of achieving throughput, reliable service, and flexibility. Table 19-1 shows how these variables relate to the performance of a given assembly process. Changing any one of these input variables will create changes in the performance of the assembly line. Optimization of the input variables is very different for different industries. Assuming that an assembly line needs to improve its performance in one or all of the general performance categories, maintenance and training give much greater improvement for a given level of investment—provided the improvement efforts are managed properly.

QUALITY PREREQUISITES IN ASSEMBLY

For a process to produce with consistency (the cornerstone of quality) certain fundamental conditions must exist. Certain conditions should surround the assembly operation as a matter of course:

1. The process must be capable and in control. *In control* means the critical process parameters are in statistical

CHAPTER 19

INTRODUCTION

TABLE 19-1
Effect of Assembly Variables on Process Performance

	Equipment	Maintenance*	Headcount	Training*	Systems	Control	Redundancy
Throughput	++	++	++	+	+	—	++
Quality	++	++	—	++	+	++	—
Reliability	+	++	—	++	—	—	—
Cost	++	+	+	+	+	++	—
PTS	+	++	—	++	++	+	++
Flexibility	++	+	+	++	+	++	+
Assembly cycle time	+	++	—	++	++	++	+
Total cycle time	+	+	—	+	++	++	+

Where: ++ = major beneficial effect, + = beneficial effect, — = little or no beneficial effect.
*Columns represent variables giving highest return on investment.

control. *Capable* means that the process has a C_{pk} that meets customer expectations. A process that is not in control must be brought into control. Once in control, its C_{pk} is measured. If the C_{pk} is too low, the process needs improvement to the point where it can produce the product consistently.
2. The equipment must work properly. Quality is not achieved with malfunctioning equipment.
3. Staffing must be available. Just like raw materials, trained staff must be available and ready to perform and/or monitor the assembly process. With an incomplete staff, the assembly operation cannot be completed as specified.
4. All components must be available when required. Many assembly processes fail to achieve their performance targets because of an inconsistent supply of materials to the line.
5. Schedules cannot continually change. A process works when it can be finished. Scheduling that changes within the assembly cycle time, resulting in the starting and stopping of work, causes massive inconsistency and poor quality.
6. There must be adequate visibility. Preparation is vital. A line without adequate visibility to prepare material, manpower, supervision, etc., for production is compromised in its ability to perform as specified.

These are the basic conditions. They must be in place before the assembly process can continuously improve. These are critical management issues. They involve creating the proper organization and systems with adequate attention and support from all the company's employees and suppliers.

TECHNIQUES USED TO IMPROVE ASSEMBLY PROCESSES

Design of experiments (DOE). DOE is a set of statistical techniques used to determine factors contributing to the variability of any process. DOE narrows down what is needed to improve a process to reach performance goals. Design of experiments is typically used to determine the process variable(s) with the biggest impact upon final product performance. DOE is the adaptation of techniques of classical statistics to reduce the number of trial-and-error experiments performed to determine the effect of several variables. DOE is essential for effectively analyzing complex assembly processes.

Failure mode effects analysis (FMEA). Knowing what can go wrong, before it goes wrong, helps quality, reliability, product design, and equipment uptime. Developed by NASA, FMEA[3] is the complete mapping of process variables' effects on product quality, reliability, and performance. FMEA pays particular attention to the potential effect of variables, to spot defects that could occur under special conditions. A sample FMEA is shown in Fig. 19-3.[4]

The variables affecting the critical aspects of product performance are the ones that must be controlled the most. Investing heavily in controlling variables with minimal effects upon product performance is not in the customer's best interest.

Poka-yoke. Poka-yoke is the design of piece parts and operations to make it impossible to assemble a product improperly. The term means "mistake-proof" in Japanese and was developed in 1961 by Shigeo Shingo.[5] The technique teaches how processes and products are designed so they cannot be assembled improperly, using operational checklists, sensors, and piece part design to achieve zero defects. An example of an operational checklist is a work aid or tool that automatically presents to the operator the correct part type and number oriented the proper way. As Fig. 19-4 shows, sensors are also extensively used in poka-yoke systems to ensure a correctly performed operation. Another example of poka-yoke is the design of parts that fit properly only when assembled the correct way. This can be achieved by varying the diameter and shapes of holes, ridges, etc. In complex assembly operations, simple and cost-effective tools, checklists, and sensors can dramatically reduce hard-to-find defects.

Problem solving. A standardized methodology for solving problems was developed by Walter Shewhart at Bell Laboratories before World War II. Often referred to as the plan-do-check-act (PDCA) cycle, it specifies a series of steps to take to ensure any problem is properly solved. Many versions of this technique now exist. Figure 19-5 shows the flow chart that describes the Ford 8-D team oriented problem solving process.[6]

Quality function deployment (QFD). QFD[7] is a technique used to design an assembly process to perfectly match the needs of the customer. QFD, developed at the Mitsubishi shipyards in the early 60s, is a technique of capturing customer expectations and optimizing process variables to achieve these expectations. Competitive product offerings are also taken into consideration in developing these optimizations. Although originally developed for external customers, the process can be used repeatedly through

CHAPTER 19

INTRODUCTION

Design failure mode and effects analysis	Part/assembly name: Glass envelope Part/assembly number: E10935 Engineer: Brady Customer/application/other: General use			Supplier/vendor name: Scheduled production release date: 9/30 File name: Envelope Printed: Dates: 4/1				FMEA number: 75 Page 1 of 1	
Function(s) and specification	Potential failure mode(s)	Potential effects of failure mode on end product and end user	S E V	Potential causes of failure	L I K	In place and planned cause preventions or detections	E F F	R P N	Recommended corrective actions	Response accepted by	A P R
Provides: leakproof, pressurized container for inert gas atmosphere translucent envelope for maximum light transmission shockproof means of installing bulb in socket safe enclosure for end of life burnout smooth surface for identification marking	Crack	Allows oxygen to penetrate envelope; destroys inert gas atmosphere; filament burns up: lose source of light	7	Tension stress at glass surface with flaw present	8	Product specification to include correct cooling rate per specification #107	2	112	Reexamine allowable stress levels Reduce specified stress levels in thin glass area		

Fig. 19-3 Sample FMEA.

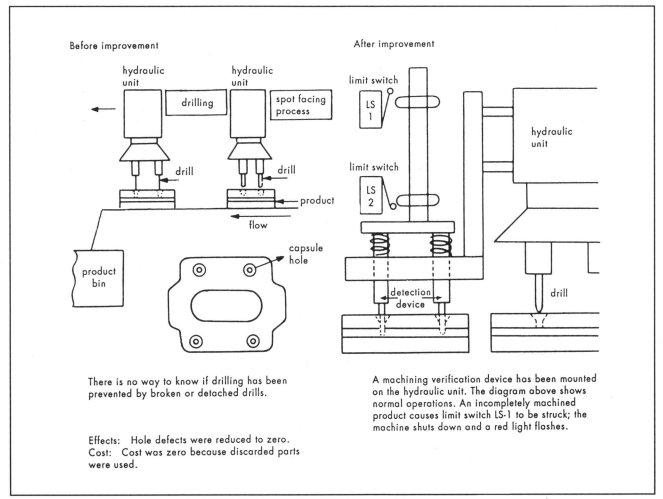

Fig. 19-4 Preventing missed capsule holes.

CHAPTER 19

INTRODUCTION

any manufacturing chain by applying it to internal customers as well. A sample QFD output is shown in Fig. 19-6.[8]

Single minute exchange of dies (SMED). Developed by Shigeo Shingo in Japan in the 60s, SMED is set of techniques used to minimize equipment setup. The utilization of SMED greatly increases line flexibility by minimizing the time needed to change between products, thereby reducing economic lot size. Toyota was so successful with SMED that they reduced lot size to one, achieving the ultimately flexible and responsive assembly line.

The reduction of setup time to near zero is accomplished by preparation, practice, and standardization. Preparation reduces distraction, practice reduces confusion and mistakes, and standardization reduces complexity of the operations. Figure 19-7[9] shows how bolts and attachment techniques may be standardized to speed setup.

Statistical process control (SPC). Developed at Bell Laboratories over 50 years ago, SPC is a set of statistical analyses that determine how a process is performing and whether intervention is required to allow the process to produce the desired output. SPC is very well known today and forms the basis of most quality programs. It is the most efficient and cost-effective method of controlling any process.

As with any tool, the effectiveness of these techniques depends upon how well they are understood and utilized, and how the organization responds to the results. Many organizations fail to realize all the benefits of the advanced techniques they use because of poor follow-through.

A typical example is the use of SPC. When SPC charts are maintained by technicians or engineers rather than line personnel, the process is controlled as tightly as possible and will not perform

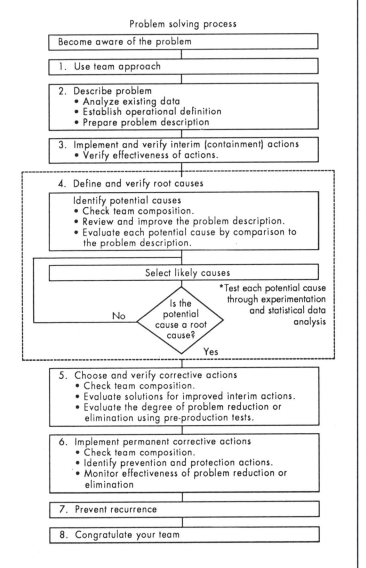

Fig. 19-5 Team oriented problem solving.

CHAPTER 19
INTRODUCTION

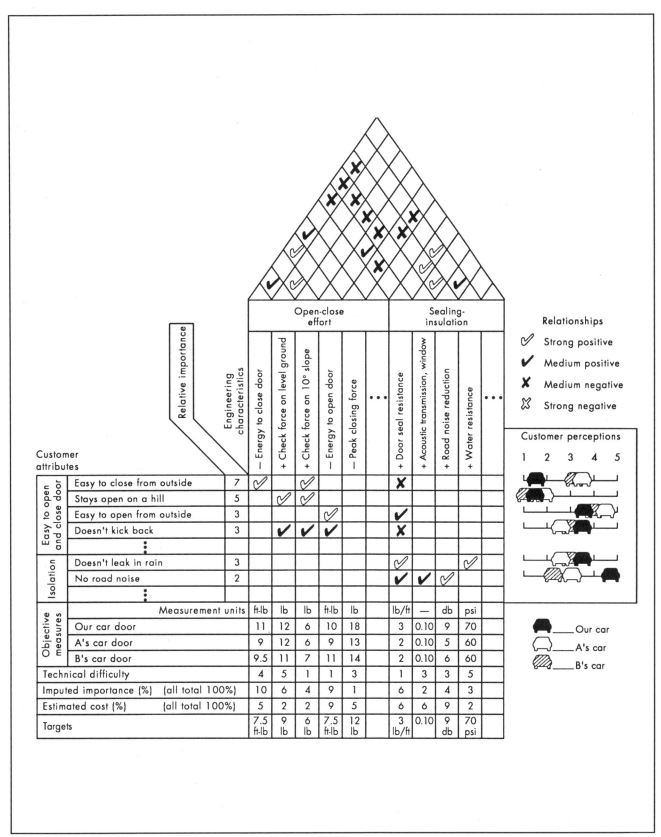

Fig. 19-6 Sample QFD output.

CHAPTER 19

INTRODUCTION

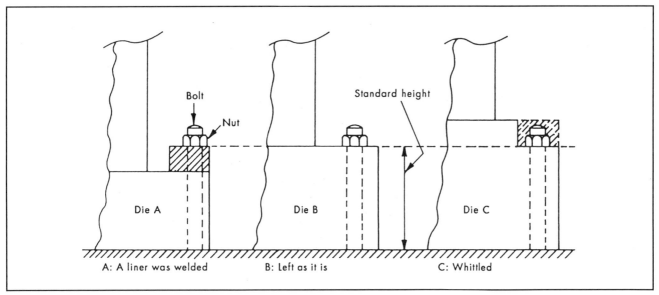

Fig. 19-7 Standardizing die-holder height reduced the need to exchange fastening tools.

to its potential. SPC allows line personnel to monitor their own processes and to know at the earliest possible moment, with statistical certainty, when intervention is required. Operators can easily be trained to perform many of the necessary interventions themselves, and know whom to call in cases where the process has a major technical or mechanical problem. Teaching SPC, and then not training operators to intervene as necessary, results in disappointingly small improvements in overall performance.

PROGRAMS FOR CONTINUOUS IMPROVEMENT OF ASSEMBLY QUALITY

Over the past twenty years, several company-wide improvement programs have proven successful in yielding positive quality results in a wide variety of companies. Each of these programs has a different focus and will optimize a particular aspect of an assembly operation's performance. Implementation of these programs requires patience and a great deal of organizational focus and energy, but results in commensurate gains in overall performance.

Total Productive Maintenance (TPM)

TPM is an operating philosophy, with a developed set of supporting techniques, that focuses organizational resources upon ensuring that equipment operates with 100% effectiveness. These techniques were developed in Japan at Nippon Denso in the 60s.[10] TPM teaches that equipment can cause loss in six different ways:

1. Equipment failure.
2. Setup and adjustment.
3. Idling and minor stoppages.
4. Reduced speed.
5. Process defects.
6. Reduced yield and scrap.

Figure 19-8 shows how all these forms of equipment loss relate to one another and total equipment effectiveness. Reducing the six forms of equipment loss requires an ongoing effort to initiate programs for preventive maintenance, maintenance prevention, and maintainability improvement, supported by total participation at all levels of the organization. In essence, equipment effectiveness becomes the shared goal of all employees, from equipment design, maintenance, and selection, through to parts ordering and scheduling. Employees typically dedicate 3% of their time to measuring and improving equipment effectiveness. Techniques such as SPC and FMEA are widely used to perform these improvement tasks.

The result of TPM is consistent throughput, more uniform product made to tighter tolerances, and reduced scrap.

Work Cells

Cross-functional layouts help worker productivity through team empowerment. An outgrowth of the Toyota Production System[11] pioneered in the 60s and 70s, work cell equipment is controlled by a team, rather than an individual. Cells require operators to become proficient in using a multitude of equipment with a high degree of cross-training. The teamwork and improved communications among operators within a cell result in simplified systems and dramatically reduced cycle times. The higher degree of operator skill, understanding, and responsibility results in assembly problems detected and fixed quicker, without the intervention of technical personnel.

Cells require, but also inspire, worker empowerment. When properly supported with systems, tools, and training, cells can become a major source of product and process innovation.

Manufacturing Resource Planning II

Manufacturing resource planning (MRP) II is the integration of management planning, scheduling, forecasting, capacity planning, and buying to achieve the defined objectives of a manufacturing operation; it is not a computer program, but an operating philosophy.[12]

Figure 19-9 shows the flow chart of an MRP II system. Although widely associated with ordering materials for an assembly operation, the program affects assembly quality in the following ways:

- Improves on-time delivery.
- Reduces total cycle time.
- Improves visibility and reduces scheduling variations.
- Reduces direct and indirect product costs.

CHAPTER 19
INTRODUCTION

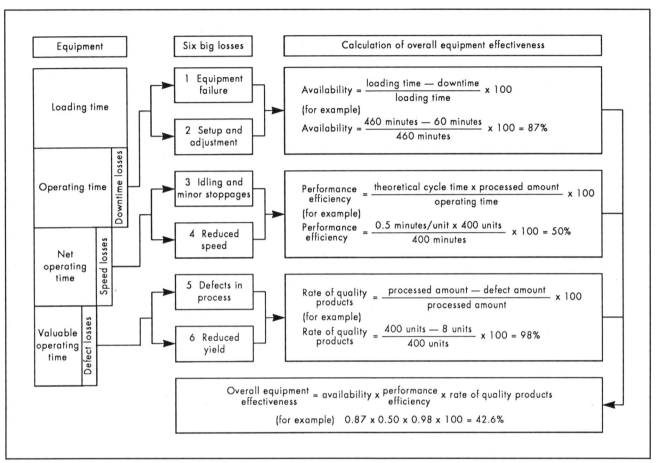

Fig. 19-8 Overall equipment effectiveness and goals.

MRP II requires the complete integration of production related activities to get a high degree of predictability and control. The integration of diverse activities such as sales, design, and purchasing requires total management focus and support. Maintaining a single company-wide data base with a high degree of integrity poses the biggest challenge. This data base includes financial, planning, productivity, design, product flows, bills of material, and scrap assumptions, as well as forecasts of customer needs.

Competitive Cycle Time

Competitive cycle time (CCT) focuses on cycle time as the major cost and performance driver of all processes. Popularized by the book *Competitiveness Through Total Cycle Time*,[13] the program prioritizes problem solving efforts by the magnitude of their effect on the process cycle time. This single-minded focus makes management decisions easier to make (but not any easier to implement). Any problem takes time to detect and resolve, and the bigger the problem, the more it lengthens cycle time; CCT forces an organization to prioritize its problems in a logical manner.

Figure 19-10 shows the three interdependent cycles that operate within a manufacturing organization: design/development, strategic thrust, and make/market cycles. Improvements in these cycle-time loops are accomplished by tracking cycle time and then reducing steps, waiting times, and process problems using standard process improvement and problem solving methodologies. Reductions in the cycle times of the three overall cycles result in significant improvements in flexibility, cost, scrap, and reliable short lead-time delivery.

Theory of Constraints

Theory of constraints (TOC) teaches how to reduce the expenditure of production and assembly resources by optimizing the throughput of any bottleneck processes. This technique, made popular by *The Goal*,[14] makes efficient use of lean management because the methodology focuses on localized areas, rather than an entire organization. Achieving throughput and the resulting revenue, with minimal effort, resource, and cost, is the goal referred to. Optimized throughput is achieved by identifying the single biggest bottleneck, most often a production bottleneck. The organization then mobilizes and focuses to achieve maximum throughput with scheduling, thoughtful maintenance, utilization of backup equipment, expanded hours of operation, and other methods. Application of the theory of constraints results in a more predictable, more flexible, and less costly assembly operation.

The prescribed five-step process for TOC improvement is:

1. Identify the constraint.
2. Decide how to exploit the system constraint.
3. Subordinate everything else to the above decision.
4. Elevate the system constraints.
5. If, in a previous step, a bottleneck was broken, go back to step one.

CHAPTER 19

INTRODUCTION

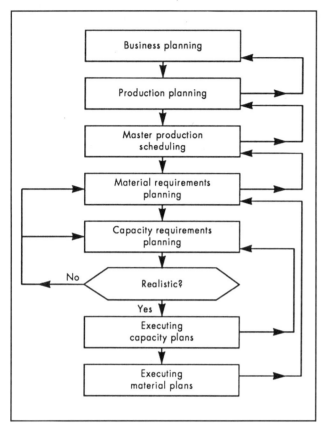

Fig. 19-9 MRP II flow chart.

The Goal also warns that although breaking through bottlenecks improves the wellbeing of any company, traditional accounting techniques (designed to give credit for inefficiency) may obscure the true extent of the performance improvement.

Just-in-Time or Kanban

Just-in-time (JIT) is the dynamic linking of all assembly operations to maintain constant line balance. This process was developed as part of the Toyota Production System. JIT accomplishes this linking of manufacturing processes via a simple "pull" scheduling system. A pull scheduling system is one that allows product to be manufactured only when it is required by the operation downstream of it (as opposed to being outputted by the operation preceding it). A JIT system defines that the inventory staged before a given operation must be below a set level (near zero) before pulling more parts from the previous operation. Product moves so swiftly in a JIT environment that systems for tracking and dispatching product can be manual and are often based upon simple cards passed from one station or cell to another (called Kanban in Japan).

Companies utilizing JIT have the lowest cycle times and inventories, while maintaining the greatest flexibility. The keys to operating with JIT are a workforce that is empowered to respond to problems quickly, and equipment that operates reliably.

ACHIEVING TOTAL QUALITY IN ASSEMBLY

Total quality management (TQM) is a philosophy of management for sustaining continuous improvement throughout an organization. It is not a specific prescription for assembly success; it is the integration of all actions within a company to achieve customer satisfaction. As such, all the programs listed above complement TQM. For TQM's success, each employee within an organization must understand the methods that the company will use to define and exceed customer expectations.

Employees at all levels are generally reluctant to accept empowerment for the achievement of ill defined or rapidly changing objectives. TQM will fail without a clear improvement path that is understood by all employees. A successful program clearly defines program goals that directly relate to improved customer satisfaction and are easy for every employee to understand. These programs keep the overall goals simple. A general classification scheme helps to indicate a clear path for all employees to channel their continuous improvement efforts.

Table 19-2 shows the relationship between input variables and assembly process performance. Each of these major TQM programs optimizes a particular row-and-column set. The programs make certain assumptions about which aspects of performance precede others, determining their singular point of focus. All these programs demonstrate impressive results when properly implemented by assembly operations.

These techniques share the following characteristics:

1. Require the total support of top management.
2. Closely coordinate actions throughout large areas of a company beyond the given assembly operation, typically requiring teams with members from a wide range of departments.
3. Require specialized training.
4. Require changes in behavior throughout an organization.

More than anything else, they all involve a single-minded focus by the entire organization on the achievement of a well defined result. High Quality is achieved with a single-minded dedication to adhere to a well defined and understood process, to achieve products with unique performance/cost trade-offs. Furthermore, an assembly process will constantly improve if the latest information received from any and all sources is processed using the techniques described.

Who authorizes all these changes? Contrary to popular belief, TQM does not need to start "at the top." Some empowerment exists at any level of an organization. The key is to operate within the existing empowerment.[15] Any empowerment is enough to make some improvement; it may not be enough to begin one of the company-wide programs listed. An improvement effort starts by following these steps:

1. Write down the goals to be accomplished. Start small; choose projects with a high probability of success.
2. Write down the support required to accomplish improvement goals.
3. Ask all managers, who must furnish either direct or indirect support, for their comments on the plans. Modify the plans according to any constraints they place upon the support they pledge to furnish.
4. Let operators or employees know of the goals of the project and the support they will need to furnish. Solicit their inputs and determine if any modifications are required.
5. Update managers and employees on a regular basis about the progress of the project. Discuss problems as well as progress.

Using this process, empowerment will grow slowly as the project progresses toward its goals. This process will not yield unlimited empowerment; the project will be limited to the level of support it has received.

CHAPTER 19
INTRODUCTION

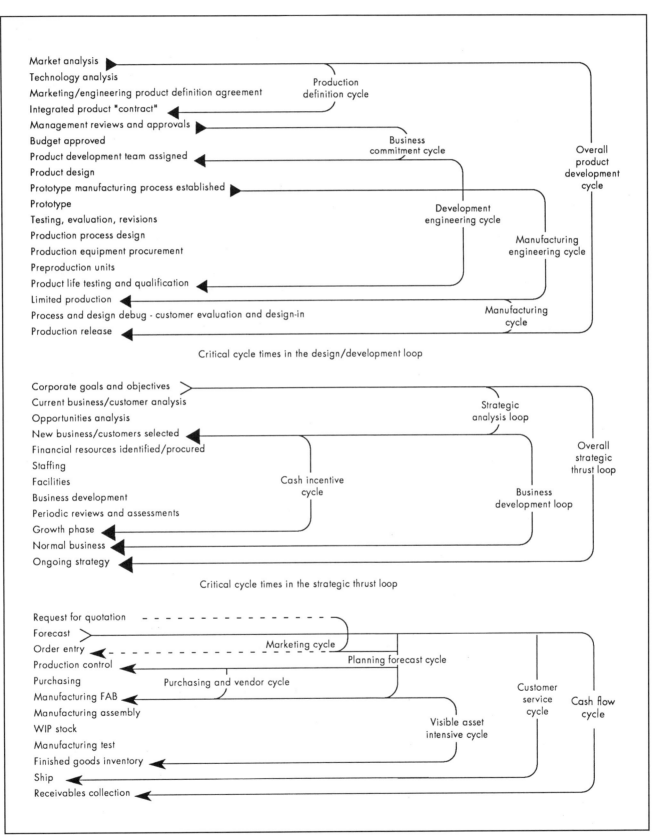

Fig. 19-10 Critical cycle times.

19-13

CHAPTER 19

INTRODUCTION

TABLE 19-2
Input Variables and Assembly Process Performance

	Equipment	Maintenance	Headcount	Training	Systems	Control	Redundancy	Program focus
Throughput	++	++	++	+	+	—	++	TPM, TOC
Quality	++	++	—	++	+	++	—	
Reliability	+	++	—	++	—	—	—	
Cost	++	+	+	+	+	++	—	
PTS	+	++	—	++	++	+	++	MRP
Flexibility	++	+	+	++	+	++	+	JIT
Assembly cycle time	+	++	—	++	++	++	+	Cells
Total cycle time	+	+	—	+	++	++	+	CCT
Program focus		TPM		Cells	MRP, JIT	CCT TOC		

BEHAVIOR MODIFICATION

How does an organization get single-minded focus? The word "change," particularly when others request it, usually generates fear. How to achieve change is the subject of research, books, and training classes.[16]

The following simple three steps work well in achieving behavior modifications at any level of an organization:

1. **Ask questions that reinforce the desired activity.** Dictating behavior modification does not work; do not try. Asking open-ended questions, with the genuine intent of learning another person's perspective as to why, how, or when an activity will take place, naturally starts a learning process that minimizes defensive behavior. Once the position of another person is understood, ask what actions they would recommend to establish the actions and behaviors required to implement the targeted improvements. This generates involvement and a shared sense of responsibility. Feel free to give opinions, state objectives, and ask for the other person's reaction. The key is to keep discussions looking forward and free of blame or judgments in the mind of the person who needs to change.

2. **Give recognition and rewards.** Recognize and reward the desired behaviors. Again, the rewards and recognition are those things that will help the people making the changes feel good about what was accomplished. The optimum form of recognition and reward is dependent upon local culture and environment. In general, recognition from a highly respected individual or the peer group will have a long-lasting impact upon performance; monetary reward is the least personal form of recognition and has the shortest lasting impact.

3. **Make it easier.** Making it easier to perform processes in the prescribed manner sends the important message to an employee that the company cares about the changes requested, and that the company will change itself to support the desired behavior modifications. Training and modifications in support systems and procedures are classic methods used to make the new behaviors easy to sustain.

All three steps may be pursued in parallel, or in series. The application of these three steps will have a linear impact upon the magnitude of the changes achieved.

GENERAL CLASSES OF ASSEMBLY OPERATIONS

Table 19-3 divides assembly operations into four classes. Improvements in systems and visibility (planning) go together with manpower and process. These classes are based upon experience, not scientific study.

In general, focus the improvement activities over all aspects that make a process—manpower, methods, machine, and materials. This chart could have more columns for such activities as training and vendors. However, once the line workers gain empowerment, they take care of these areas very well. It is in the areas of systems, visibility, and process design where they need help.

An assembly operation can easily determine its class. By using the criteria that define the next highest class, an improvement path is defined that requires an organization to assimilate all the quality improvement techniques in a step-by-step fashion. Each "box" can serve as a specific set of objectives for management or improvement teams.

IMPROVEMENT ROAD MAP

A series of questions will point out actions to take to improve an assembly process:

Who's responsible? Determine who wants continuous improvement to take place within the organization and who will take responsibility for implementing the necessary changes. Determine what level of empowerment is available to the group for making improvements in the assembly area. Work with the empowering person, team, or organization to write down improvement objectives.

Are the assembly prerequisites present? Without these conditions, a solid foundation for continuous improvement cannot exist. The six assembly requirements are:

- Maintain statistical control or a statistically capable process.
- Operate with well maintained equipment.

CHAPTER 19
INTRODUCTION

TABLE 19-3
Classes of Assembly Operations

	Implementation				Results		
Class	Control	Systems	Visibility	Equipment	Quality	Predictability	Cost
A	Workers control process, layout, schedule	Daily control, small lot sizes	Schedules tie directly customer	Equipment operators define maintenance	Low PPM $Cpk>1.5$	Precise on-time delivery to tight intervals	Benchmark
B	Workers control process	Integrated inventory, WIP, scheduling with weekly refresh	Forecasts accurate over total cycle time	Equipment maintenance works closely with operators and supervisors	<100 PPM $1<Cpk<1.5$	Good delivery to weekly intervals	Competitive
C	Workers participate in improvement activities	Reliable forecast of output available	Must order material to inaccurate forecasts	PM defined and performed	>100 PPM	Good delivery to monthly intervals	Variable
D	Manager decides	Disjoint	Forecasts unavailable or not useful	No PM, downtime not measured	>1000 PPM	Inconsistent delivery	High

- Assign adequate staff.
- Maintain adequate part supplies.
- Establish stable schedules.
- Use techniques that anticipate demand.

Failure to meet these prerequisites will generate customer complaints, even if the product meets their technical specifications, because of an unpredictable supply. From the point of view of the manufacturer, inefficient (or unstable) assembly lines will produce at high cost and low quality levels and will continually distract upper management attention from more productive activities.

Getting the six prerequisites in place can resemble wrestling alligators, as the issues interrelate to a great extent. Focusing on making improvements to problems in order will sort out issues in the quickest and least confusing fashion (see Fig. 19-11).

As the process comes under control and improves its capability to meet specifications, it becomes easier to determine true staffing and material needs. Similarly, with staffing and materials in place, it becomes easier to schedule and to get the support of other organizations to improve visibility. The financial investments needed to accomplish these tasks are modest and pay back quickly. Often large cost savings are mentioned as the key benefit of implementing TQM. People or organizations making such claims focus their estimates on the returns manufacturing operations see from putting in place these six prerequisites. Most investment is in the form of training and time for organized problem solving.

The Theory of Constraints, MRP II, and Competitive Cycle Time help companies raise their assembly operations to a level where the six prerequisites are solidly in place. The other programs mentioned will best help organizations after these basics are established.

Many assembly operations never get beyond this point as it takes a great deal of time, attention, and energy to put a solid foundation in place. Management can become distracted or complacent and feel that these problems are a normal part of running an assembly operation.

Do the appropriate people in the organization understand the techniques so they may use them effectively? Tools help accomplish objectives more efficiently. People dig holes without shovels, but shovels make digging easier. Similarly, techniques are tools that allow any continuous improvement effort to proceed with a minimum of energy and investment.

It is not necessary for all employees at all levels to understand all these techniques. Top management needs a general understanding of the tools to help support line management and engineering in their implementation. Line management needs to understand the benefits of using these techniques to decide which techniques bring about the highest priority improvement opportunities. Line management should be tasked with the responsibility of choosing which tools the assembly line will begin to use, as well as getting the targeted improvements.

The point of understanding techniques such as SPC or QFD is only for the purpose of enabling an organization to use these

CHAPTER 19

INTRODUCTION

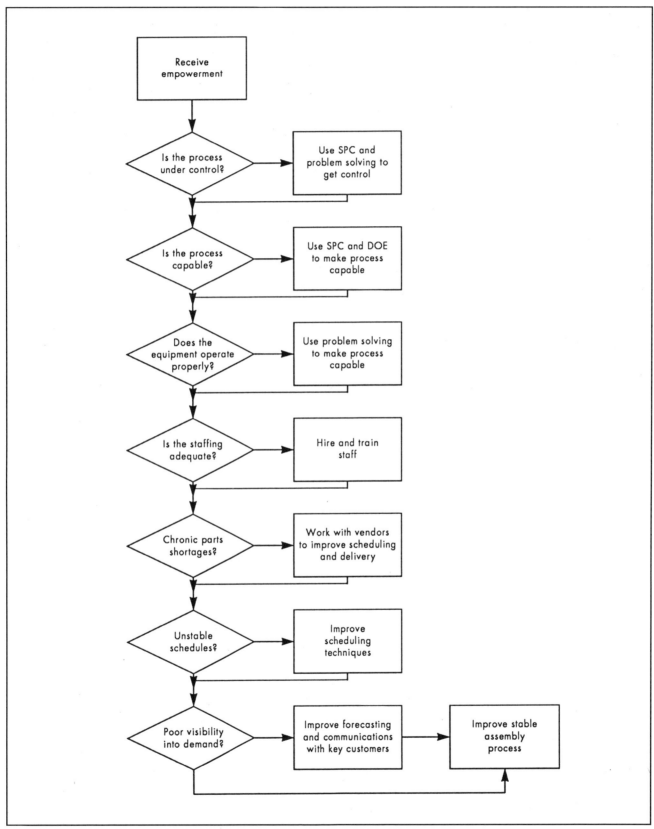

Fig. 19-11 Flow chart—making a stable assembly line.

CHAPTER 19

INTRODUCTION

techniques effectively for achieving well defined objectives. The challenge becomes getting line operators to change their behavior to make the use of these techniques effective. Handing a shovel to a person doesn't result in digging the right holes. Behavior modification addresses the issue of how to get an employee to use a new tool effectively.

Will customers always be satisfied with the product when it meets its specifications? Operations that continue to assemble the same product family for several years can easily lose sight of additional desires that customers develop; what satisfied a customer a few years ago may not do so today.

Take the time to meet with customers and expand product specifications to include all their expectations, even nontechnical ones. QFD does a good job of leading a group through such issues.

Is the assembly process described completely, and in terms that any operator can understand? Ask the operators performing the most sensitive or critical operations to describe what they do, how they do it, and why they do it the way they do. Unless all operators performing sensitive operations can answer these questions in a coherent fashion, process documentation and training is needed. A small team of experienced assembly operators can take the responsibility for such a project.

Are the objectives of the assembly line completely set and known? Do the supervisors and engineers responsible for the assembly process know and understand the expectations for performance in terms of throughput, quality, reliability, PTS, cost, flexibility, and cycle times? Do the supervisors and engineers measure these key parameters and plan to implement improvements to reach those objectives? If the objectives are not clear, then all responsible parties must sit down and agree upon what can and should be expected in terms of performance. The general classes of assembly operations can form a foundation for crafting the improvement objectives.

Does a plan exist to improve the use of equipment, maintenance, headcount, training, scheduling and tracking systems, management control structure, and minimize redundancy?

Supervisors and engineers responsible for a given assembly area must be able to answer the question, "How will the improvement objectives be met?" Line management can follow one of the continuous improvement programs or craft a new improvement plan. Table 19-3 describes the general classes of assembly operations. This can form a foundation for a set of custom improvement plans.

Programs such as MRP II, competitive cycle time, and theory of constraints will offer the greatest improvement for Class C or D operations. TPM, work cells, and JIT will give the most benefit to Class C and B operations. Class A operations must use their experience and vision to experiment with new techniques and programs to continuously improve beyond their benchmark levels.

Do all employees understand the plan and what they must do to make it happen? All employees within an operation should answer these two questions: "What improvements are you planning to accomplish for supporting the improvement activities of your area?" and, "Why?" Universal understanding of the improvement plans are often overlooked as a critical success factor in improving operations; typically only a few people fully understand the improvement plans. Employees must understand the improvement plans to the point where they know what they need to do at any moment, under most common circumstances, to move their area of responsibility toward the overall goal. Without a general level of understanding, it is very difficult to get the behavior modifications and the focus required to achieve a high level of Quality. Direct understanding makes people feel they are part of the improvement activity; second-hand understanding makes a person feel like an outsider. An outsider will either become a nonparticipant, or a person trying to show why it was wrong to exclude him. In either case, these people cannot be counted on to improve the Quality of the processes they perform or see performed.

The key is to brief all employees on the improvement plans. This takes time and some ability to explain concepts in terms consistent with the knowledge and terminology used by any given group of employees. Responsibility for briefing a group can be delegated to persons with demonstrated ability to give simple explanations. Asking the two questions will give the necessary feedback regarding how widely and how well the plans are understood.

SUMMARY

The techniques and programs described form a basis from which any assembly operation can begin to improve. No generic improvement plan exists, but the programs and techniques described should allow any person or group to determine a starting point from which to launch continuous improvement activities. The key to the successful implementation of any TQM or continuous improvement effort is to sustain the new behavior patterns required by the implementation of the techniques and programs discussed. Ask questions, give recognition and rewards, and make it easy to engage in the desired behaviors, and the program will yield results beyond your expectations.

WHAT'S NEXT?

Management control techniques will become greatly simplified as companies begin to become proficient in their use. An organization is most limited by its management's ability to react, so this becomes a good focus for future efforts. Management's ability to react is in turn limited by its ability to sort through all the data it receives, make a decision, plan for its implementation, and then deploy it throughout the organization. Few techniques exist in these areas; this is a new frontier.

IMPROVING ASSEMBLY SYSTEMS

In all kinds of economic times, it is the duty of manufacturing to keep improving the method of making its product. Machining processes are improved by adapting new technology machine tools to do the work; in most cases, there is a standard machine tool to accomplish that task. Assembly operations are not as adaptable. There are not necessarily machines that work faster than those made 10, 20, or more years ago.

To improve productivity in assembly systems, there are many areas to look for weaknesses. These areas can include machine operator/assembler movements, material staging, quality of parts

CHAPTER 19

IMPROVING ASSEMBLY SYSTEMS

coming into assembly, the operator, the machine, or the product design (see Fig. 19-12). In most cases, answers do not float to the surface without the manufacturing engineer doing a thorough job of analyzing the assembly process.

All assembler movements must be broken down into quantitative measures. How long does it take to move the component parts to the tooling station? What happens to the assembly when complete? Is there dead space in any function of the assembler? By analyzing each movement, the manufacturing engineer can begin to see which movements might be improved and assign priorities for dealing with changes.

Proper material staging for assemblers can mean more efficient use of time for making a product. Detailed planning and layout of the work area should take the place of using machines wherever they sit. Operator involvement here will give a unique (and first-hand) viewpoint on how the process could improve.

Operators can lend insight to the project that is not gained from charts and graphs. Encourage a team approach toward the area where the operators spend a good portion of their working hours. Operators want to be productive and will respond if given the chance. Have them discuss the good and bad about the process. If there is initial resistance, slowly coach them into wanting to help make their jobs more satisfying. Tremendous gains in productivity and quality are realized by operators who "own" their jobs.

Where do rejects most often occur? This is a very important question that needs a specific answer. Trace problems back to the source. A lack of operator skill or interest may inhibit full efficiency or productivity. If component parts are bad, a prior machine may be the culprit—the wrong type, not properly maintained, or just worn out. An informal (or formal) SPC of the machined components can pinpoint the exact problem if parts are not held within tolerance.

Some component parts are over-engineered, under-engineered, or designed for failure in manufacturing. The manufacturing engineer must take a practical look at how each component fits into the completed assembly framework, and work with design engineering to change specifications where possible. It is imperative to have a positive communication link between manufacturing engineering and product engineering. Both viewpoints, melded into a common objective, can find solutions to the problems that surface. Each department must play to the strengths of the other and work together.

In most cases, the greatest time consuming step in the assembly process can realize the largest productivity gain. If the total cycle time is eight seconds (the press takes 1.5 seconds and operator load/unload uses 6.5 seconds) then the first look should concentrate on how the components get into and then out of the press. A 20% decrease in load/unload time will give a better return than a 50% decrease in press time (see Fig. 19-13).

Machines are designed to do repetitive work quickly and efficiently, with consistent results from assembly to assembly. Manual movements of the assembler are candidates for some level of automation. Unfortunately, there are budgets to follow that may make some automation attempts unfeasible. Improvement doesn't mean going from a 1950 assembly process to a current technology in one step. There may be insufficient money, and a lack of technical and maintenance expertise. Improvements can begin by concentrating on "first step" moves.

A few hundred dollars invested in a fixture or tooling nest redesign can make the load/unload function faster. When the component parts changed for a product improvement, were the assembly fixtures and/or tooling temporarily "rigged" to keep production flowing? Has that temporary fixture been in use for months or years?

On a high-production assembly of 2,000,000 units per year, saving 1/2 second in assembler load/unload time will save 277 3/4 hours per year. Considering manufacturing costs of $30.00 per hour (labor plus burden), that 1/2 second savings translates into $8,332.50 cost savings. Most engineers would willingly invest $2,000 to redesign/rework a fixture to save over $8,000.

SEMIAUTOMATIC ASSEMBLY SYSTEMS

Automation of an assembly process can produce additional parts with better and more consistent quality than a strictly manual process. However, this optimum approach is not always attainable because of capital investment constraints or component design (or both). Semiautomation is a positive step toward improving a manual assembly process. A machine can be designed to do *anything* a human can do. The problem is that some manufacturing assembly processes need "touch" to put components together. Machines that can simulate the human "touch factor" are usually on the high end of the capital investment graph. By concentrating on the simple movements of the assembler, automation can be added in steps and still realize a cost savings impact on the overall cost of the end product.

A semiautomatic assembly requires more operator interface. This is justified by comparing the combined cost of semiautomatic machines plus labor with the large capital investment required for a fully automatic machine.

A sometimes hidden expense is the startup cost for a new fully automatic or semiautomatic assembly process. From the date of machine order to the date of complete system production capability, the semiautomatic system is weeks, months, or in some cases even years quicker than a completely automated system. Another hidden expense is the amount of maintenance and level of technical expertise needed to support a fully automated assembly system. Flexibility is an advantage for the semiautomatic process.

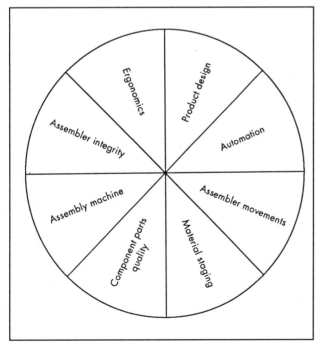

Fig. 19-12 Areas for improvements in assembly systems.

CHAPTER 19

IMPROVING ASSEMBLY SYSTEMS

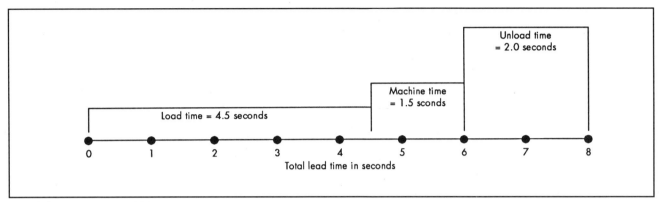

Fig. 19-13 Timing the assembly process.

A family of parts is accommodated with a simple tooling change, taking minutes instead of hours. Future changes in the assembly or component parts can fit into a semiautomatic system with lower re-investment costs in capital and/or labor. If a new assembly design obsolete products in manufacturing, the write-off (if any) is easier on the bottom line with the dedicated semiautomatic process.

FULLY AUTOMATIC ASSEMBLY SYSTEMS

The greatest reduction of unit cost and the most consistent quality are the major benefits of a fully automated assembly system. There is much higher potential for maximum product output. But a fully automatic process is a more rigid system, and less friendly to changes in components or future designs.

When the annual output justifies a fully automated system, the payback is easy to calculate. Total labor savings, final assembly consistency, and high-volume output can offset the high initial investment.

ASSEMBLY CELLS

Many product lines have a variety of sizes within the same design framework. Combining these similar assemblies into a dedicated assembly cell can provide manufacturing efficiencies and benefits. Instead of running batches of final assemblies through the shop, a carefully designed assembly cell can achieve the desired assembly within the family, almost on demand.

Tooling is made for easier interchanging or retooling with a simple changeover. These simple changeovers do not need the expertise of a setup person; they usually can be done by the assembler/operator. With the proper designed tooling, a new assembly is in production in minutes.

Benefits of an assembly cell include reduced inventory investment, reduced lead times, less warehouse space required, lower final assembled product cost, and a greater familiarity with component parts and how they affect final assembly. In a typical cell, one assembler may be responsible for two or more machines. This assembler becomes an expert on the manufacturability of the entire family of parts and is able to make further improvements based on experience. Quality improves because of a more streamlined effort and centralized information flow. The manufacturing engineer can become more effective in improving the assembly systems by concentrating efforts toward similar objectives, and working with fewer production people. Answers and solutions come quicker and with greater accuracy.

ERGONOMIC ISSUES

Weary assemblers at the end of a shift are not as productive. To reduce or eliminate the fatigue factor, the breakdown of operator movements must include ergonomic issues:

1. What makes the assembler tired or sore?
2. Is there unnecessary lifting and/or walking?
3. Does the machine "fit" the operator?
4. Is there proper lighting?

The proper machine "fit" should consider the optimum operating height, and the physical size and ability of the operator. In some cases, a machine is picked to do a job because it was available from the corporate warehouse, and not because it was the ideal machine for the job.

Specially designed ergonomic activators or hand-held tools represent a high initial investment, but the company is repaid over and over again by productivity gains and decreased workman's compensation costs.

SUMMARY

In a manufacturing and assembly environment, it is imperative to focus on particular problems or opportunities, break down each assembly operation into individual functions, and go about the job of improving one step at a time. First, document a benchmark as the assembly operation now exists; then, accurately identify the problems or areas most in need of improvement, and develop a plan for implementation of new and better assembly processes. Improvements in a single process have a positive effect on the overall assembly operation. Grouping assemblies into families can gain inventory efficiencies and lower final product cost.

CONTINUOUS IMPROVEMENT FOR FEEDERS

Continuous improvement efforts for automated assembly should consider:

- How to select the right feeder.
- How to tool a feeder bowl.

CHAPTER 19

CONTINUOUS IMPROVEMENT FOR FEEDERS

- Out-of-bowl orienting and conveying for in-line feeders.
- Centrifugal feeders.
- Feeder bowl linings.
- Muffle module.
- Feeder stands.
- Bulk storage hoppers and automatic level control switches.
- Placement of units.

HOW TO SELECT THE RIGHT FEEDER

Selecting the right feeder includes several factors:

- Bowl diameter.
- Bowl capacity and loading.
- Feed rate.
- Orientation.
- Base unit.

Bowl Diameter

Generally, the diameter of a vibratory feeder bowl should be at least ten times the length of the part it is to feed. In some instances, a smaller ratio is used depending on the part. For example, a 6″ (15 cm) diameter bowl will accommodate and feed a 3″ (8 cm) long bar-type part; however, the bowl would not offer sufficient storage of parts. In addition, the part would "chord" the track in an ineffective feeding attitude (see Fig. 19-14).

Good feeding characteristics come primarily from part contact with the helical in-bowl track. It is better practice to employ a 24″ (61 cm) or 30″ (76 cm) diameter bowl for the 3″ (8 cm) bar, so that the part may ride in a V-form track and stay away from the bowl wall. The larger bowl would also provide storage for an adequate supply of parts, eliminating the need to constantly replenish the bowl.

Bowl Capacity and Loading

Feed rates may be affected by part weight. As the bowl empties, the feeder's amplitude of vibration will sometimes increase. In addition to the weight factor, overloading can change the ability of parts to separate and feed in the proper attitude. Optimum part levels are regulated by installing an automatic storage hopper with a level control switch. The level control switch senses the need for parts in the bowl. The bowl does not get overloaded, but is never without parts (see Fig. 19-15).

Feed Rate

Feed rate is simply the number of parts discharged from the bowl in a given period of time. Feed rate is generally expressed in parts per minute (PPM) and depends upon:

- The configuration and the mass of the parts.
- The amount of orientation required.
- The number of in-bowl tracks.
- In-bowl and out-of-bowl power assists (such as air jets and miniature vibrators).
- The type and size of the base unit used.
- The setting of a variable-speed solid-state controller.
- The level of parts maintained in the bowl.

One easy way to increase total output from the feeder bowl is to specify a bowl having two or more in-bowl tracks. The parts output of the bowl is significantly increased by the number of tracks it contains. Standard bowls are available with as many as 10 tracks. Figure 19-16 illustrates several bowl configurations.

For best performance, equip the feeder with an automatic auxiliary supply hopper and a level control switch, to keep a predetermined level of parts in the feeder bowl.

Orientation

Orientation is defined as controlling the attitude of parts so that they are in the correct position when they arrive at the designated workstation. This is normally done by passing the parts through a series of engineered obstructions or guides in the bowl track. However, out-of-bowl aids such as orienting rolls and "warped" tracks may also be invaluable in orienting headed or tapered parts, or to invert parts 180° prior to reaching the workstation.

The feeding characteristics—and therefore, the orientation—of certain parts can be seriously affected by overloading the feeder bowl. For instance, washers normally feed radius-to-radius, but will feed face-to-face in an undesirable pattern if too many parts are in the bowl.

Pneumatic air jets, used in an efficient manner, can assist in orienting parts and increasing the number of oriented parts which reach the discharge of the bowl. They are frequently used to reject nonoriented parts, to separate parts, or to tip parts as they enter a discharge chute.

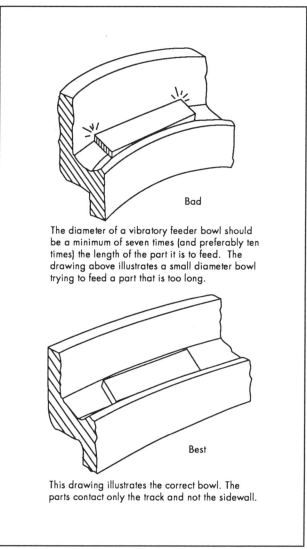

Bad

The diameter of a vibratory feeder bowl should be a minimum of seven times (and preferably ten times) the length of the part it is to feed. The drawing above illustrates a small diameter bowl trying to feed a part that is too long.

Best

This drawing illustrates the correct bowl. The parts contact only the track and not the sidewall.

Fig. 19-14 Bowl diameter.

CHAPTER 19

CONTINUOUS IMPROVEMENT FOR FEEDERS

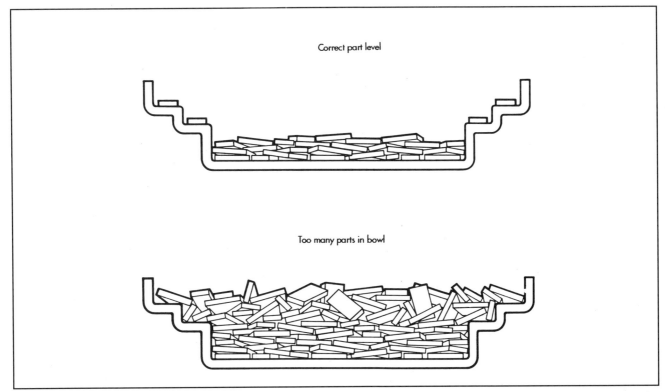

Fig. 19-15 Bowl capacity.

Base Unit

The electromagnetic base unit is selected according to the size and mass of the feeder bowl required for a given application. Consideration should be given to the average weight of the parts in the bowl at any one time.

Special lightweight bowls are available for some applications. Things to bear in mind when considering such bowls are:

- A lower total weight may permit the use of a smaller base unit for a given size bowl, with a corresponding reduction of mounting space and initial investment.
- A larger bowl may be mounted on a given base unit, providing more in-bowl parts storage and a greater opportunity to orient parts as they travel a longer track.

For some parts which require a greater linear movement, consider the use of a VFC "FA" base unit. Fast-angle feeders have a reduced spring angle which provides more linear (or horizontal) motion and less vertical motion than standard design feeders. Consult the manufacturer for specific recommendations.

HOW TO TOOL A FEEDER BOWL

Tooling a feeder bowl must consider:

- Basic tooling principles.
- Part size and weight.
- How to analyze a part.
- Design for automation.
- Precautions.

Basic Tooling Principles

Tooled feeder bowls provide a means for systematically placing parts into a single line of oriented parts as they leave the discharge point of the bowl. As the drive unit is energized, the parts will separate and work toward the outer wall of the bowl, then up the inclined track(s) inside the bowl.

When the parts arrive at an obstruction or guide, they either pass or fall back into the bowl, depending upon their orientation at that point. After the parts pass all the tooling, they should be properly and consistently oriented for delivery from the bowl.

The type of tooling used depends on the configuration of the part and its "natural" feeding characteristics. Analyze the part thoroughly. When there are too many parts in the bowl, natural orientation is limited.

Part Size and Weight

The way a part feeds is affected by its material, proportions, and dimensions. A part that is nominally 1" (2.5 cm) wide and 2" (5 cm) long and of reasonable thickness will normally feed well. The same part, if only 0.001" (0.0025 mm) thick, is too light to maintain consistent contact with the track, and will not follow an optimum feed pattern. Well proportioned parts as thin as 0.0015" (0.0038 mm) can be fed at a reasonable rate.

Reasonably heavy parts of good proportions will feed well, regardless of their overall size. Feeder bowls are available for handling cylindrical parts up to 7" (18 cm) long and 2" (5 cm) in diameter. Larger diameters are accommodated if their length is less than 7" (18 cm).

How To Analyze A Part

Consider the length, width, thickness, material, weight, and shape of the part. Look for any outstanding feature which may serve as a guide for orientation. Examples include angles, grooves, flanges, bosses, projecting pins, and concave or convex surfaces. These features should also be considered in relation to

CHAPTER 19

CONTINUOUS IMPROVEMENT FOR FEEDERS

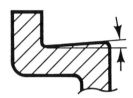

Positive Tracks have a less than 90° included angle between the track and the side wall of the bowl. Track angle is expressed in degrees of slope above the horizontal plane - usually 7°, 8°, 15°, or 60°.

V-Form Tracks contain grooves of 60°, 90°, 120°, or 150° as the included angle. Some run the entire track length, others only the latter part of the track. These tracks are designed to feed cylindrical parts.

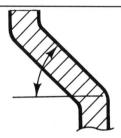

Hi-Negative Tracks have a 60° negative angle in the last quadrant of the bowl. Some follow a transition from a positive angle to the negative angle. Typical parts oriented in these bowls are flat, irregular parts or parts with a projection on one side.

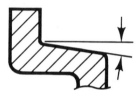

Negative Tracks have a greater than 90° included angle between the track and the side wall of the bowl, expressed as degrees of slope below the horizontal plane - usually 15°, 30°, or 60°. Typical parts to orient include thin rectangular stampings and shallow pan or cap type parts.

Radius-Form Tracks generally have a groove for the entire length of the track, but some just for the last quadrant. They are designed to feed a cylindrical part whose length is equal to or greater than its diameter. The track's radius should match the profile of the part being fed.

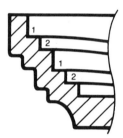

Multiple Track bowls have two or more tracks. Each track carries parts from the bottom of the bowl to its discharge. The discharge of stock bowls can have up to ten tracks. Advantages include higher production rates and the simultaneous delivery of fixed numbers of parts to the bowl's discharge.

Fig. 19-16 Bowl configurations.

the center of gravity of the part; they determine the "natural" feeding pattern of the part. The design of the part should work effectively with the feeder tooling.

Design for Automation

Consider making a small change in the part. Figure 19-17 illustrates a case where this principle was applied. There was an initial cost to modify the part, but that cost was more than offset by the savings. Inexpensive parts feeders replaced the slow and costly orientors originally quoted, parts fed almost twice as fast, and the outlook was favorable on the feeder's long-term reliability.

Precautions

Some parts are inherently difficult to feed. Lead wires on electronic components, for instance, tend to tangle. Unfired ceramics and unsintered pressed metal parts are very fragile and may be damaged. Parts coated with heavy oils tend to adhere to conventional bowl tracks and must be supported on thin rails.

Fortunately, many parts can be classified into one of eight categories, according to their configuration and dimensions. Figures 19-18 through 19-25 show typical tooling techniques when feeding these parts.

OUT-OF-BOWL ORIENTING AND CONVEYING FOR IN-LINE FEEDERS

To obtain the proper out-of-bowl orientation and in-line feeder conveyance the following must be considered:

- Eliminating gravity tracks.
- Precise track alignment.
- Track mounting options.
- Quiet feeding.
- Varying feed rate.
- Orienting rolls.
- The right parts.

Eliminating Gravity Tracks

The transfer or conveyance of oriented parts is no longer dependent on a gravity track, as the in-line feeder provides the motivating force assuring a continuous flow of parts. A substantial reduction in cost is realized through the elimination of the complex engineering and manufacturing of gravity tracks. System development is further enhanced by installing feeder bowls at lower levels more convenient to the operator for parts replenishment.

CHAPTER 19

CONTINUOUS IMPROVEMENT FOR FEEDERS

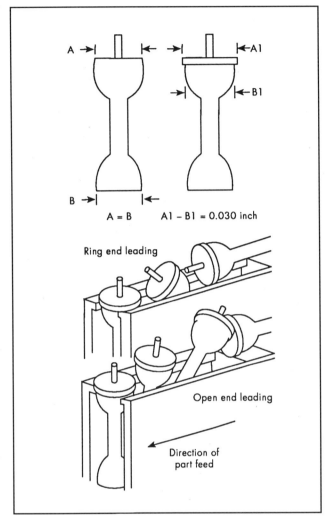

Fig. 19-17 Adding a ring to the closed end of a part resulted in less costly and more efficient feeding.

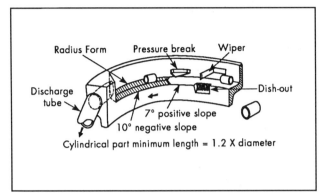

Fig. 19-18 Cylindrical parts.

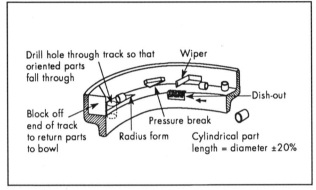

Fig. 19-19 Cylindrical parts.

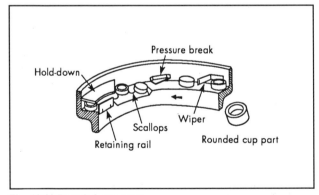

Fig. 19-20 Rounded cup parts.

Precise Track Alignment

The design of an in-line feeder all but eliminates any transfer of vibration to the machine's base. The in-line feeder becomes an integral part of the automated system and forms a stable mounting media for the attachment of a track.

In-line feeders do not require rubber feet. They are fastened directly to the machine surface. A firmly fastened unit maintains and assures precise track alignment for effective parts transfer from a feeder bowl to a machine, from one machine to another, or from a machine to an operator.

Track Mounting Options

Track rigidity is important. The portion of the track overhanging either end of the feeder should not flex and defeat the feeding action of the driving mechanism. Several in-line feeders can be placed in series for longer tracks, or side by side for wider tracks. Tracks can be offset from the center of the top plate.

Quiet Feeding

Because the in-line feeder mounting base is a nonvibrating member of the system, vibration transmission is minimal when it is firmly fastened to the machine base. The mounting media does not become a noise radiator as it does when using other types of vibrating feeders.

The track load rating is the weight of the finished track less the parts. Select an in-line feeder with a load rating equal to or slightly greater than the track weight. This way you can add weight to the in-line feeder to compensate for a lighter track.

The center of gravity of the track should be as close as possible to the midpoint of the in-line feeder top for best results.

Varying Feed Rate

The parts feed rate is varied either mechanically or through the use of an optional controller. When applications require a frequent

CHAPTER 19

CONTINUOUS IMPROVEMENT FOR FEEDERS

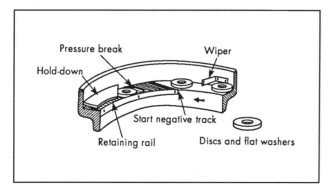

Fig. 19-21 Discs and flat washers.

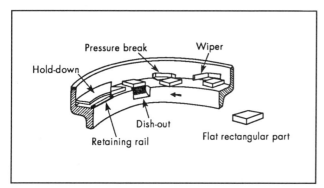

Fig. 19-22 Flat rectangular parts.

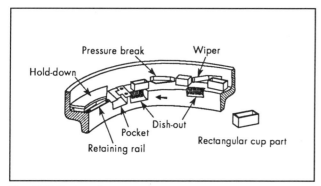

Fig. 19-23 Rectangular cup parts.

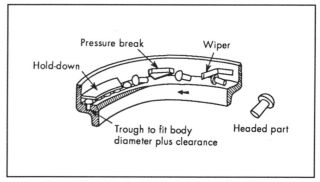

Fig. 19-24 Headed parts.

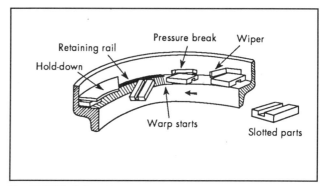

Fig. 19-25 Slotted parts.

change of feed rates or fine adjustment, the use of the optional controller is recommended. If a lower but fixed feed rate is required, the in-line feeder coil is adjusted.

Orienting Rolls

Orienting rolls are an effective way to orient parts outside of a vibratory feeder bowl. They also convey parts in a quiet and gentle way. Spacing between parts may be adjusted and this lends itself to other uses where parts are inspected or gaged.

Each roller is precision ground and hard chrome plated. Optional coatings can be applied for special applications like the handling of food or pharmaceutical items. Abrasive resistant coating may also be applied to the rollers. Total indicator runout should not exceed 0.002" (0.005 cm) when a measurement is made between the rollers at any point along their axes.

Stock-orienting rolls are driven from a gear-reduced motor by an O-ring type belt. The belt rides over pulleys so that one roller revolves clockwise while the other turns counterclockwise. This action breaks the friction between the parts and the rollers. The rollers are set at a slight angle so the parts flow downward and away from the point where they are dropped.

The Right Parts

As in most uses of orienting devices, employing the right part is essential to success. Figures 19-26 through 19-28 illustrate the symmetry, dimensional qualities, and proportions that let parts fall between the rollers and hang reliably.

CENTRIFUGAL FEEDERS

Centrifugal parts feeders are appropriate for high speed precise feeding of parts, with quiet operation and gentle handling. They provide operating advantages not available in other feeders. Feed rates are adjustable from a few parts per minute up to one thousand per minute, depending on the part.

The attributes of centrifugal feeders include low maintenance and variable speed control. They generally outperform vibratory-type feeders. Centrifugal feeders are ideal for manufacturing, assembly, packaging, capping, inspecting, and counting operations. Optional air drives are available for hazardous locations.

FEEDER BOWL LININGS

New lining options offer noise reduction and some degree of bowl surface protection. Often part feed rate improves. The rapidly increasing broad application of vibratory parts feeders in

CHAPTER 19

CONTINUOUS IMPROVEMENT FOR FEEDERS

diversified industries require specific bowl lining and coating properties. Detailed studies of many feeder bowl materials, coatings, and linings for both specific and general purpose applications found polyurethane is the most cost-effective general purpose coating.

Polyurethane linings have exceptional dampening properties. They provide a cushion between the parts in the bowl and the bowl itself (the primary source of feeder noise). Also, the lining reduces noise radiation from the bowl interior and exterior surfaces (the secondary major noise emission source). In addition to these features, polyurethane is highly resistant to abrasion.

See Fig. 19-29 for general recommendations on various linings, along with their characteristics. Linings are divided into two main classifications—soft and hard. The soft linings reduce the noise emission of vibratory feeder bowls by providing a cushion between the part and the bowl, and by lessening the surface noise radiation characteristics of the bowl. In all cases, the linings are mechanical bonds as opposed to chemical bonds. Surface preparation prior to coating is important for good bonding. For some linings, cleaning and roughening is adequate; on others, chemical preparation or priming is required. (Bowl linings are *not* the solution when the wrong bowl is selected or when the tooling is not correct.)

MUFFLE MODULE

The muffle module has proven to reduce the sound emanating from vibratory feeders by as much as 25 DbA.

Mounting rings prevent the movement of sections of the muffle to avoid interference with the feeding and conveyance of parts.

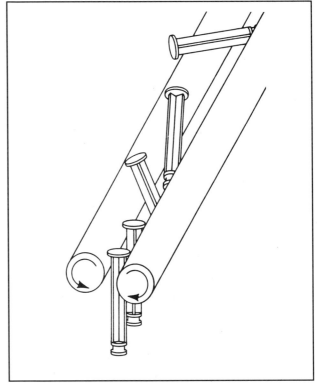

Fig. 19-26 Orienting part for correct delivery.

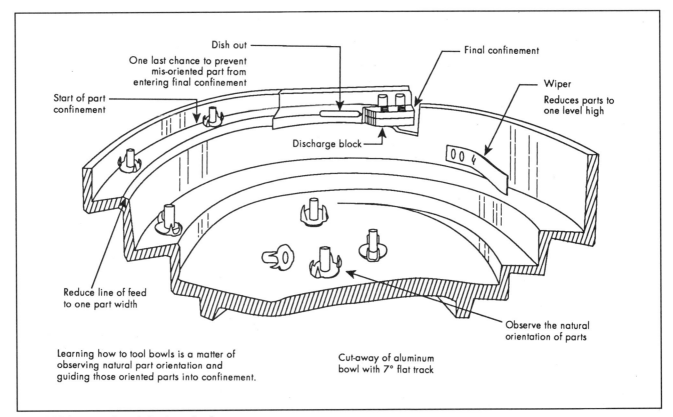

Fig. 19-27 Cutaway view of aluminum bowl.

CHAPTER 19

CONTINUOUS IMPROVEMENT FOR FEEDERS

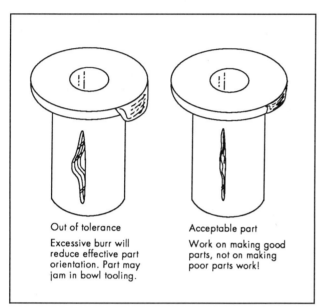

Fig. 19-28 Excessive burr can reduce effective part orientation.

FEEDER STANDS

Vibratory feeders need a rigid stand. Feeder stands of rigid, heavy welded steel construction provide maximum stability for supporting vibratory feeders. They permit positive feeder placement and assure the full application of the base unit vibration to the bowl. Feeder stands should occupy minimum floor space in close proximity to operating equipment. They may be supplied without the reinforced base for welding directly onto a machine base.

BULK STORAGE HOPPERS AND AUTOMATIC LEVEL CONTROL SWITCHES

Bulk storage hoppers are either delivered to the side or pan fed. They also have a switch for controlling the number of parts in the hopper.

Side Delivery Hopper

Side delivery hoppers are available in several sizes. Some models use a level control switch to signal the hopper that parts are needed by the feeder. A vibrator, built into the hopper floor, operates on command to provide a regulated flow of parts to replenish the feeder bowl and eliminate overloading.

One hopper has a fixed-height pedestal for mounting on the same platform as the feeder. Drop discharge height and chute adjust for proper gravity feed.

The hopper bin is isolated on elastomeric vibration mounts to prevent noise transmission through the pedestal. Both the hopper door opening and chute angle should be adjustable. A controller can provide a means of adjusting the amplitude of vibration.

Pan Feeder Hopper

These hoppers have two distinct attributes: they are *not* sensitive to the weight of parts in the bin section of the hopper, and they transmit very little vibration into the parts in the bin. With 60 pounds of parts in the bin, less than five pounds of loading is felt by the pan feeder portion of the hopper.

Precise parts delivery is possible with a variable output controller as a power source for the pan feeder hopper. This action is further improved when the adjustable door is lowered to limit the number of parts discharging from the pan feeder's tray.

Automatic Level Control Switch

An automatic level control switch can maintain a predetermined optimum level of parts in the feeder bowl and may also provide a control signal to operate remote conveyors, hoppers, bins, and other electrical or mechanical devices.

The complete assembly consists of a rigid support column, an adjustable-height support arm, and a free-moving sensor arm attached to the switch body. The end of the sensor arm, shaped like a paddle, rides on the parts in the feeder bowl. A low level of parts causes the sensor arm to rotate down, activating the level control switch contacts.

Proximity Detectors

The detector can be a photoelectric switch designed as a medium-speed proximity detector. The probe emits an invisible beam of light in the infrared spectrum. The emissions are sent out as a series of pulses at a fixed rate of speed. The receiver probe detects these emissions. Its circuit is timed to be synchronous with the pulse rate of the source probe. This feature significantly reduces false signals from external light sources within the plant.

When the detected part remains long enough in the sensing area, the relay will drop out. The normally closed relay contacts are now connected and can control an external device such as a parts feeder.

Application

Proximity detectors are used for any medium speed detecting application, where interrupted movement of an object or material

Properties	Soft Linings			Hard Linings		
	Polyurethane	Nylon 11	White Epoxy	Irathane	Metalized Stainless	Electroless Nickel
Abrasion resistance	good	excellent	fair	excellent	good	good
Noise reduction	excellent	good	good	excellent	poor	poor
Resistant to:						
Mineral Oil	good	good	good	excellent	poor	excellent
Weak Acid	fair	fair	fair	excellent	poor	excellent
Alkali	fair	good	fair	excellent	poor	excellent
Improvement of part feed rate	excellent	fair	good	good	excellent	good
F.D.A. approved	no	no	yes	no	no	yes

Fig. 19-29 Lining characteristics.

CHAPTER 19

MACHINE-MADE BRAZED OR SOLDERED JOINTS

is to be detected. Many machine builders use the probes to detect high and/or low levels of parts in feeder tracks, bins, and hoppers. The relay's contacts can sound an alarm or send a signal to controllers or microprocessors. Other applications include parts counting, liquid levels in sight tubes, and broken tool detection.

PLACEMENT UNITS

Placement units are designed to repetitively lift and transfer small parts by lifting, swinging in a 90° arc, and lowering. In the second half of each cycle, the action reverses. Positive control over parts movement is maintained for all of the 90° arc, with a high degree of accuracy and without the need for programming.

Application

One of the primary uses of the placement unit is for the assembly of parts. However, users also report the following applications:

- Using as a pick and place unit.
- Reaching in or out of a press.
- Orienting parts via the 90° transfer.
- Supplying parts to multiple lines.

IMPROVING THE QUALITY OF MACHINE-MADE BRAZED OR SOLDERED JOINTS

Improving the quality of machine-made brazed or soldered joints starts by:

1. Reviewing the basics.
2. Determining the cause of the problem.
3. Considering other important factors.

REVIEWING THE BASICS

When planning to improve the quality of brazed or soldered assemblies joined on an automatic or semiautomatic brazing or soldering machine, basic factors must be considered. What are the joint requirements—strength, seal, or cosmetic? How about total joint cost—can it be improved? These questions must be answered before making any changes or adjustments that might affect the desired characteristics of the joint.

DETERMINE THE CAUSE OF ANY PROBLEMS

Once the joint requirements are understood, what is causing the most problems in the machine brazing or soldering operation? Is the percentage of unacceptable parts produced by the machine too high, or is there room for improvement in this area? Can the total cost of the brazed or soldered joints be improved? How about the machine downtime? Is too much time being spent in machine maintenance or tooling setup? To obtain maximum efficiency of the automatic brazing or soldering machine, all of these points should be checked.

OTHER IMPORTANT FACTORS

For continuous improvement of any automatic brazing or soldering process, there are several areas for review. This guarantees that each area is utilized as efficiently as possible.

- Filler metal used, and how the filler metal and flux are deposited on the joined parts: are these the best possible materials and methods?
- Part-holding fixtures: are they in good repair? Do they hold the parts in proper relationship? Can they be made easier to load or unload?
- Heating and cooling system: is it in good repair? Can additional controls be added to improve the quality of joints?
- Mechanical and electrical machine components: are these in good repair? Do any need replacing?
- Operator training: bring the quality of the worker up to the quality of the machine.

These five areas should be reviewed on a continual basis, to keep an automatic brazing or soldering operation running as efficiently as possible.

Filler Metals

First, take a look at the filler metal used to join the component parts. Brazing or soldering filler metals are made up of either pure metals or a combination of metals alloyed together to provide the desired characteristics of strength and operating temperature. Many metals and metal alloys are available for brazing and soldering, and from these a selection is made based on joint requirements. The cost of filler metals depends upon the cost of the base metals used to make up each filler metal. Often there are several different filler metals—each made up of various base metals—that will meet a joint requirement. When materials used in these filler metals are priced as low as lead at $0.35/lb. ($0.77/kg) and as high as gold at $350/oz. ($27.00/gm), it's easy to see that filler metal cost can vary greatly. Therefore, the makeup of the filler metal and thus its cost must be carefully evaluated when selecting a filler metal to meet specific joint requirements.

With these costs in mind, the filler metal used is evaluated from the standpoint of joint strength; ease of capillary flow; brazing or soldering temperature required; temperature requirement of the part itself when in service; and material cost. Each of these points should be reviewed to see whether an alternative brazing or soldering alloy might provide better characteristics in any of these areas. Determine that the braze or solder filler metal used is the best available to meet all the joint requirements. The next question is, how is it applied to the joint? How is the flux applied? If the filler metal is applied to the joint by manually positioning preforms onto the part and then manually brushing a brazing or soldering flux onto the part, this is labor intensive. Automatic operation of these functions would allow the possibility of decreasing the machine cycle time and increasing production, assuming the machine is capable of operating at a higher production rate.

Application Methods. Automatic application of brazing or soldering filler metal is easily accomplished in a single step using paste brazing or paste solder alloys. These are a scientific combination of powdered filler metal, a suitable fluxing agent, and a neutral binder to keep the mixture in homogeneous paste form. Paste filler metals can be automatically dispensed in predeter-

CHAPTER 19

MACHINE-MADE BRAZED OR SOLDERED JOINTS

mined amounts as dots, stripes, or special shapes depending upon the joint configuration.

Another method of applying brazing or soldering filler metal is with automatic wire feeding mechanisms, using the filler metal in wire form. In this case, flux is applied to the parts to protect them from oxidation during heating. At the final heating station, the filler metal wire is fed automatically onto the part.

Both the paste and wire feeding methods are used extensively on automatic brazing or soldering machines. The selection of the best method for a particular brazing or soldering operation depends on many factors, including the joint design and the accessibility of the joint. Each of these points should be considered before deciding on any method.

Part-holding Fixtures

A second area to look at is the part-holding fixtures on the machine. Before considering any changes or modifications that might improve the operation, the total fixture concept must be understood. Basically, the fixtures are on the machine to hold the parts together in their correct relationship while they are bonded together. They must convey the parts through the various stations on the machine where the work is done. This includes application of flux and filler metal; heating to brazing or soldering temperature; cooling the parts back to room temperature; and any additional operations (machining, leak testing, etc.) performed on the automatic brazing or soldering machine.

If the component parts are self-locating, the fixture method becomes relatively simple. However, if the component parts must be properly positioned together by the fixture, and dimensional tolerances are very tight, the process can become fairly complicated. Once the requirements of the fixtures are understood, it is possible to evaluate them. Each point is checked to see if an improvement can be made in that particular area.

The fixture itself must also not interfere with the heating, and therefore, not be too cumbersome. If the fixture acts as a heat sink and draws heat away from the joined component parts, this will slow down the heat cycle and reduce the production rate of the machine. Fixtures should be designed so that a minimum amount of the heat used for brazing or soldering gets into the fixture itself. The fixtures should be checked to spot any heat sink points that could be relieved without impairing its function.

The materials used for the fixtures should withstand repeated heating and cooling cycles with a minimum of deterioration. Is there excessive wear on specific areas of the fixture, or is the total fixture worn and in need of replacement? If excessive wear occurs, stainless steel, or some of the more exotic high-temperature resistant tooling steels such as Inconel, may be used for fixtures on brazing or soldering machines.

When the component parts are brazed or soldered together and must emerge from the automatic machine with close dimensional relationships maintained on the final assembly, the fixture design becomes very critical. If the fixtures do not hold the component parts together correctly, there are several points that must be considered.

First, as the parts are heated they will expand due to the heat. The amount of this expansion depends upon how hot the parts become. Depending upon the composition of the base metals, brazing temperatures of 1200°F (649°C) and higher will cause considerable expansion in the component parts. The part fixtures must take this into account and compensate for it. When springs are used, care must be taken to ensure that they are not made ineffective by the heat. Recent developments in the ceramic industry produced ceramic springs that will withstand most brazing temperatures without losing elasticity. Another possible solution is to use a counterweight system to replace springs, and thus avoid constant fixture maintenance.

Is there room for improvement in the fixture design itself? Can it be redesigned to minimize the heat buildup in the fixture, without jeopardizing the ability of the fixture to hold the component parts in proper alignment? Can a better material be used for the fixture or for parts of the fixture, to prevent wear or heat erosion? Can the fixture design be improved for ease of loading and unloading the parts that are brazed or soldered? All of these considerations are necessary, to provide long fixture life and a minimum of required maintenance.

Heating and Cooling

Several different types of heating systems are used on automatic brazing and soldering machines. Most commonly used is a burner system, using different types of combustible fuel gas. For low-temperature work, natural gas and air are usually used. When higher temperatures are required, oxygen is used with natural gas or propane. Induction heating is another heat source often found on automatic brazing and soldering machines, and both high and low-frequency induction generators can be tailored to specific operations. Resistance heat, radiant heat, and even forced hot air are sometimes used.

The heat source on an automatic brazing or soldering machine brings the parts for joining to the liquidus temperature of the brazing or soldering filler metal. It is essential that the base metals in the joint area reach this temperature, but not necessary for any other portion of the assembly to do so. Therefore, heat localization is important, from a standpoint of machine efficiency and operating cost. There is no need to heat any other area of the parts being joined. This is an important point to check when evaluating the operating efficiency of a brazing or soldering machine. Is the heat source providing the necessary heat only where needed?

Monitoring instruments. Instruments, that show readings for the critical elements used to create the heat output, should be on the machine. For gas combustion systems, gas flow meters and manometers show the pressure of the combustible gas. For electrical heat systems, meters show the power output. With these instruments available, the heat output of the machine can be continually monitored. An important area to check and evaluate is how much adjustment or "tweaking" is done with the heat system to keep the machine operating properly. Wherever possible, controls of the heat output should be locked in and changed only by a setup person well trained in the operation of the machine. Another system, now used on many automatic brazing and soldering machines, automatically monitors the temperature of the parts in the final heat station. When the proper temperature is reached, this information is fed back to the heat control system, and the heat is discontinued at that point. This is generally done through optical pyrometers that focus on the part in the final heating station. Then, temperature limit settings are made so that (in a combustible gas system) when the maximum temperature is reached the burners either retract or go to a low flame. For electrical heating systems, the power is turned off at this point. Controlling the heat input into the parts joined can be a major factor in upgrading an older brazing or soldering machine.

Evaluating the best methods of heat delivery. Is the best method of heat input used on the machine? This should be evaluated, from the standpoint of getting the heat into the part as rapidly as possible and restricting it mostly to the joint area, and considering the cost of the heat generating source. This cost includes the initial cost, the operating cost, and the maintenance

CHAPTER 19

MACHINE-MADE BRAZED OR SOLDERED JOINTS

cost. The advantages and disadvantages of the system should be reviewed to see if improvements can be made in this area.

Proper maintenance is a must for any heating system on an automatic brazing or soldering machine. Regardless of the type of heat used, if the system gets out of adjustment due to poor maintenance, excessive scrap rates occur. Although this point seems quite obvious, often maintenance of the heating system is not a high-priority item. Burners must be kept clean and free from carbon buildup when combustible gas heating systems are used; otherwise the BTU output of the burners will vary. Flux condensing on induction coils will cause shorting of the coil, unless continual maintenance is provided. A general review of the total heating system is an opportunity to look for areas where the continual operation of the machine creates conditions that change the heat output. Maintenance schedules should be set up to make certain that these areas are corrected on a scheduled basis.

The cooling system on an automatic brazing or soldering machine is used to cool the parts back to room temperature after joining. It must also keep the fixtures cool and prevent heat build-up in the machine itself. This sounds like a simple task, but improper cooling can cause problems with the brazed or soldered joints. Too rapid cooling can cause stress in the joints as the component parts shrink back to normal from their heat-expanded size. This will weaken the joint. If water comes in contact with the molten filler metal, it will cause an uneven surface and perhaps make the joint unacceptable from a cosmetic standpoint. Therefore, programmed cooling should be used, taking as much time as possible to bring the part back to room temperature, without jeopardizing the machine production rate. Ambient air or low-pressure forced air should be used until the filler metal is solidified, and for as long as is practical after solidification. After this, an air/water mist will provide more rapid cooling, without shocking the assembly. Finally, a flow of water over the part and the fixture will bring the part back to room temperature.

Depending upon the stations available to accomplish the cooling, the three separate operations—ambient air, water/spray mist, and water flow—might have to be sequence-timed over one or two stations on the machine. If only one or two stations are available for this sequential cooling, another new development—refrigerated air provided by a vortex system—can be used in place of the ambient air to cool the brazed assemblies more rapidly. This system operates on the inverse pressure/temperature relationship; it utilizes compressed air and emits a lower pressure refrigerated air directed at the brazed assemblies. This provides a faster BTU transfer from the brazed assemblies, and allows a faster cooling rate, without the danger of shocking the brazed assemblies or creating possible stresses in the brazed joints.

Machine Components

Proper maintenance of the basic machine should be scheduled at preset intervals, to minimize downtime on the machine. When an automatic brazing or soldering machine is not operating because of required machine repairs, this reduces the efficiency of the total operation. Something as simple as a well planned and scheduled maintenance program for the basic brazing and soldering machine could go a long way toward increasing the efficiency of the operation.

Operator Training

After reviewing the various work areas of an automatic brazing or soldering machine to see if continual improvement can be made, there is another very important factor to consider—the human factor. How well trained is the machine operator (or operators) in the operation of the machine? How well do the setup people understand the various areas on the machine? If there is a lack of full understanding of the equipment by either the machine operator or the setup people, the best possible machine will not produce finished assemblies at the quality standard the machine was designed for.

Too often, after spending considerable money to purchase a highly technical automatic brazing or soldering machine, insufficient money or time is spent to educate the people who will be involved with the machine. Training of machine operators and setup personnel is most important, and is a major point to consider when improving the quality of machine-made brazed or soldered joints. This training is important initially, when the machine is put into operation, and very important on a continual basis. All automatic brazing and soldering machines will perform differently to some extent after they are in operation for a period of time, as the machine is "run in." The machine operator and setup personnel are the ones closest to the machine, and are the best qualified to give feedback and make recommendations regarding improving the quality of the machine operation. Unless these people are trained, they will not understand the reasons behind each operation of the machine. They must know what is and what is not an acceptable joint and what factors affect the ability of the machine to produce acceptable joints. Without this full understanding by the operating personnel, a valuable asset is lost.

Taking a little time to go over the key areas that make up an automatic brazing or soldering machine, first to fully understand the machine and then to make improvements in the areas where they are warranted, will result in a higher yield of quality parts, and operating personnel who are well trained in the operation of the machine.

AUTOMATE TO UPGRADE SOLDERING AND BRAZING OPERATIONS

Automation for improving soldering and brazing operations includes a review of:

- Preliminary process data.
- Basic considerations for automating soldering/brazing operations.
- Automation approaches.
- Automatic brazing/soldering machine types and design.
- Fixture design.
- Furnace operations.

INTRODUCTION

Demands for increased production, higher quality, and lower costs are well known in today's world of manufacturing. Domestic and global competition introduced greater challenges for improved

CHAPTER 19

AUTOMATE TO UPGRADE SOLDERING AND BRAZING OPERATIONS

product design, production planning, process methods, and more effective labor and management utilization.

In production soldering and brazing operations, various approaches are available for increased productivity, improved quality, higher yield, and reduced costs. Many of these operations are still performed by expensive manual methods that were used thousands of years ago. Soldering and brazing are among the earliest methods of joining metal.

Some metals formerly considered difficult are now production brazed or soldered, such as titanium, zirconium, and nickel. Ceramics are also production brazed. Various aluminum alloys, once considered somewhat exotic for soldering and brazing, are now automatically brazed using open flame and furnace procedures. New production soldering and brazing filler alloys, fluxes, and techniques were developed to meet these requirements.

Various factors require consideration in a brazing or soldering automation review, focusing on mechanical joints involving ferrous and nonferrous base metals.

The factors outlined provide a "stair-step" outline for a review of brazing or soldering operations under serious consideration for automation. These steps can also be used for in-depth examinations of present operations for general process improvement. As an example, an existing brazing or soldering operation may be operating by a process not designed for—or conducive to—the results desired. This situation could be caused by various changes in the process throughout the years.

A correctly designed automated soldering and brazing process should:

- Remove operator skill from the operation.
- Reduce part handling.
- Provide a smooth production flow of high-quality parts.
- Reduce rejects and rework.
- Reduce costs.
- Increase productivity.

PRELIMINARY PROCESS DATA

Prior to undertaking a brazing/soldering automation and/or improvement project, thoroughly research records for answers to the following questions:

1. When was the entire soldering/brazing process line last reviewed (including all individual upstream and downstream operations)?
2. Since that review, how many process changes were made?
3. Did each process change take into consideration all possible effects upon the soldering/brazing operations (changes in base material, fitup, filler alloy, heat system, cleaning process and chemicals, etc.)?
4. Were tests actually conducted to ascertain if these changes affected the brazing/soldering process?
5. How did the changes effect the brazing/soldering operations?
6. If detrimental, what corrective action was taken?
7. Did the corrective action solve the problem?
8. Has the original part been changed since the soldering/brazing operation was initially established? How?
9. Have new parts been added to the operation?
10. How many operators were required when the brazing/soldering operation was initially established?
11. How many operators are now required?
12. What was the initial hourly brazing/soldering production requirement?
13. What is the present hourly production rate?
14. What was the original yearly brazing/soldering production requirement?
15. What are the present yearly production requirements?
16. What is the present labor rate and overhead?
17. What major changes have occurred in machine, fixtures, heat system, components, etc.?
18. Have brazing/soldering operations been moved since they were originally installed?

BASIC CONSIDERATIONS FOR AUTOMATING SOLDERING/BRAZING OPERATIONS

Basic theories of soldering and brazing are important aspects of an operation under consideration for automation. A comprehensive understanding can be gained from various American Welding Society (AWS) soldering and brazing manuals, and other technical publications.

One major feature of the brazing and soldering processes is the unique characteristic of the molten filler alloy to flow around and into the joint area by capillary attraction. Often this important feature is overlooked when the assembly joint is designed. Capillary attraction and gravity play important roles in soldering and brazing, especially in automated machine operations.

Major factors requiring serious analysis for automating a soldering or brazing application are:

Joint Base Metals. The joint base metals (steel, copper, aluminum, etc.) each require review for their strength characteristics, brazing compatibility with each other, compatibility with various filler alloys, creep, specific heat, and material heat coefficients. These factors all interact within the brazing/soldering process, as well as later during the assembly's end service use.

Size and Mass. The size and mass of the assembly base metals have direct influence on the machine concept, the type of heat system utilized, output, parts handling approach, total machine cost, and number of operators required.

Joint Requirements. Joint requirements, such as strength, fatigue, cosmetics (appearance), shock resistance, service temperatures, corrosion resistance, and other special requirements, deserve examination. Incomplete specifications require resolution. The plant must have quality control personnel, procedures, and correct equipment to test upstream and downstream production parts for all requirements involved in the project. The assembly joint should have proper design, fit matching, and bonding surface area to provide the joint strength requirements. A well designed "self-locating" joint provides numerous benefits to the overall soldering/brazing operation. This is especially true if the joint is positioned during the heat cycle to take advantage of both gravity and capillary attraction. Furthermore, self-locating assemblies simplify fixture design and reduce machine cost, and help the operator accurately and repeatedly load each fixture correctly, for reduced rejects and increased machine output.

Filler Alloy. Brazing and soldering filler alloys are available in many standard and nonstandard compositions. Filler alloys that melt and flow below 800°F (427°C) are considered *solders*, and filler alloys that melt above 800°F (427°C) are known as *brazing filler alloys*. American Welding Society (AWS) and vendor specifications describe specific compositions, solidus/liquidus ranges, and other characteristics. A thorough examination should be made to ensure that the correct filler alloy is used for each application. Items to review include joint design, strength characteristics with base metals and filler alloy, service requirements, corrosion resistance, and ability to wet and flow into the joint

CHAPTER 19
AUTOMATE TO UPGRADE SOLDERING AND BRAZING OPERATIONS

formed by the base metals within the specified temperature. Filler alloys are available in several forms: wire, shims, preforms, rings, balls, slugs, powder, or paste. Pastes usually contain both a powdered filler alloy and flux. They can be "custom-blended" for special applications. Paste alloys are readily adaptable for automatic machine operation. They can be automatically dispensed in various deposit sizes and configurations. Furthermore, since they contain both filler alloy and flux, separate fluxing operations (sometimes requiring additional operators) are eliminated. Wire feed systems are also available for machine operation.

Metallurgy. Metallurgical considerations include a review of the various intermetallic reactions during and after the brazing/soldering process, such as liquation, phosphorus and sulfur embrittlement, carbide precipitation, and stress cracking.

Flux Activity. Flux activity plays a critical role within the soldering and brazing processes. The flux must remove light oxides from the base metals, and provide a path for the filler metals to wet and form the joint. Also, the flux must prevent reformation of oxides on the base metals during brazing and soldering.

Some fluxes provide activity and oxide protection in very high-temperature ranges involving long time cycles. Others are used for mild fluxing applications, such as dilute atmosphere furnace operations.

Flux residue removal is an important consideration for many soldering or brazing operations. When clean parts enter the soldering/brazing process, less residue will remain in the joint area for postcleaning. Furthermore, precleaned parts provide a basis for consistently acceptable soldered or brazed assemblies, especially when the application requires sealed joints.

The flux must be compatible with the base metals, filler alloys, and machine heat system, as well as EPA and OSHA requirements.

Clean Parts. Parts presented to the brazing/soldering operation should be clean and dry. They must be free of grease, oils, metal chips, special coatings, dirt, ingrained drawing compounds, and base metal surface oxides. Many of these contaminants are difficult to detect and remove.

Rust inhibitors and/or other protective coatings may or may not affect soldering/brazing operations. However, the residues may etch and dull the base metals for downstream processes such as plating or painting. Thorough tests should be conducted to determine whether these products offer problems with brazing/soldering operations or other downstream processes.

Many believe fluxes remove all of the various contaminants mentioned; this is not necessarily true. The major function of flux is to remove light surface oxides on the joint base metals and promote "wetting" within the joint area. The flux must also prevent reoxidation of base metal surfaces during the heat cycle.

Surface oxides are a major problem for soldering and brazing operations. The problem is aggravated when components supplied for the soldering/brazing process sit in warehouses or undergo long transit distances, in varying atmospheric conditions that promote metal oxidation, prior to reaching the user. Aluminum and stainless steel surface oxides are difficult to detect, and terne plated steel's mottled surface also presents surface oxidation detection problems. Aluminum and other reactive metals tend to reoxidize immediately after cleaning. Therefore, precleaned parts should be presented to the soldering or brazing process without delay.

Soldered and brazed assemblies sometimes require removal of flux residue due to continued flux activity or the revival of activity caused by the hygroscopic characteristics of the residue. It is advantageous to remove flux residues while parts are still warm from the brazing/soldering process and have not hardened. Therefore, including the postcleaning process within the automated brazing/soldering machine project often merits serious consideration.

Regulations. Recent EPA and OSHA regulations have placed restrictions and time limitations on the use of CFC cleaning processes. Many companies are now aggressively seeking other cleaning methods, such as ultrasonic systems requiring less toxic aqueous cleaning chemicals. One system utilizes ultrasonic cavitation to remove contaminants, including surface oxides, ingrained drawing compounds, and entrapped metal chips, within very short time cycles. The process cleans assembly surface areas internally as well as externally, and is especially effective when used on complex assemblies containing multiple soldered or brazed joints.

Costs. A cost justification analysis should include present and future total project production requirements for year, month, week, day, and shift. Note the number of shifts the process will be operating each day.

A review of the present system should include a "part mix list" reflecting production requirements for each part, size of production lot runs, and number of part changes per shift, or per day. Job shop operations require special machine and fixture consideration for the changeovers required. Figure 19-30 shows a simple cost comparison for brazing/soldering automation justification.

It is common to find that an automated brazing machine approach provides capacity for an assembly, with additional machine time remaining. Rather than having an idle machine, determine the feasibility of incorporating another assembly for soldering or brazing on the same machine.

This chart is an analysis of the manufacturing costs for joining a specific assembly. Data relative to the present operation is placed in the "Present Process" column.

The "Automated Process" column contains projected costs of automating the process.

A comparison of the two processes will reveal cost savings.

	Present Process	Automated Process
1. Total yearly production		
2. Number of operators		
3. Operation production rate		
4. Operator wage and overhead rate		
5. Material cost		
A. Total hourly production rate		
B. Total man hours		
C. Labor cost		
D. Material cost		
E. Total yearly cost		
Estimated Yearly Savings		
• Capital investment.		
• Estimated first year savings.		
• Estimated succeeding years savings.		
• Equipment amortization period.		
• Other features (improved quality, improved output, cleaner worker/plant environment, etc.).		

Fig. 19-30 Cost comparison analysis.

CHAPTER 19

AUTOMATE TO UPGRADE SOLDERING AND BRAZING OPERATIONS

AUTOMATION APPROACHES

Basic brazing/soldering machine functions provide the assembly nesting fixtures (if required), the heat system, a method of conveying the assembly through the heat system, and other secondary operations such as assembly or testing.

Part transport methods through the machine operations include shuttles, rotary index tables, belt conveyors, racetracks, and overhead conveyors. Pick-and-place units, or robots, may supplement the conveyor for loading and unloading, especially when the brazing/soldering machine is designed for a combined on-line system operation.

Large, bulky, and/or heavy assemblies may require a heat system that shuttles, or transverses, the heat source throughout the joint area.

The machine's heat system provides the necessary BTUs to increase the temperature of base metals and allow filler alloy joint formation by capillary attraction. A well designed and established heat system removes the cost of operator skill from the brazing/soldering operation. The heat system is the heart of the brazing machine, since it provides a critical function within the soldering or brazing process. A poorly designed heat system may be extremely difficult to overcome.

Several types of heating are available, such as gas fuel systems, electric elements, infrared, induction, and resistance. Combinations of electric and gas fuel systems may provide the required heat, as well as a reducing atmosphere within the furnace.

Gas Fuel Systems

Gas fuel systems include natural gas/air, natural gas/oxygen, propane, acetylene, and other gas fuels. These fuel types offer various BTU output. Higher BTU gas fuel systems are attractive for some operations; they reduce the heat cycle time required for the brazing/solder cycle (which may increase machine output).

Gas fuel systems direct gas mixtures into burners arranged in flame patterns, sequenced to bring the assembly joint up to the required brazing or soldering temperature. Installation and maintenance costs are usually moderate compared to other heating methods. Also, gas fuel systems are flexible: burners can be adjusted, or controlled, for variations in production part configuration and size.

Electric Heating Systems

Electric heating systems utilize element, induction, and resistance approaches to provide the required soldering/brazing kilowatts (BTUs) into the assembly joint area. Electric systems use conduction, convection, infrared, or combinations of these methods to attain the required joint soldering/brazing temperatures.

Induction systems do not contact the assembly. Special coils induce high frequency (or low frequency) fields into the joint, causing the base metals to increase in temperature. A typical layout is shown in Fig. 19-31. The heat can be localized within the joint area at a fast cycle. Coil design is critical to accomplish the required heating. The position of the part within the coil area must be maintained for each part cycle, so part configuration, number of joints, and joint location are limiting factors. Also, requirements for large or long hairpin coils can lead to expensive perishable upkeep maintenance. Induction machines can be high cost and dedicated heating systems.

Resistance heating systems utilize electrodes that contact and exert pressure on the joint area while a low-voltage, high-amperage current is passed through the base metals. The resistance of the base metals to the electrical current causes the joint temperature to rise. The heat cycle is localized and fast. However, the heat cycle time and penetration are determined by base metal thickness, pressure, and the resistance equipment used in the application.

Both induction and resistance heat systems require a thorough examination of the part design, base metal composition, and production requirements (present and future). This is most important when a machine is under consideration for several different assembly configurations.

Incorporation of Other Operations on Brazing Machine

Additional operations may be incorporated in the soldering/brazing machine sequence, such as automatic placement of filler alloy, fluxing stations, and part wash stations (when required). Automatic part loading and unloading deserve consideration.

The use of PCs for machine automation has increased significantly. They are also applicable for brazing machine approaches. Upstream and downstream production operations that can be incorporated on the brazing/soldering machine, such as assembly, staking, and testing, may eliminate cost centers and parts transfer operations.

Incorporation of parts cleaning as an in-line sequence of the brazing/soldering machine operation offers advantages worthy of serious attention.

AUTOMATIC BRAZING/SOLDERING MACHINE TYPES AND DESIGN

Automatic brazing/soldering machines are either the shuttle or rotary-index type.

Shuttle Machines

Shuttle machines are usually one-fixture or two-fixture machines that offer limited production capacity. They have a fixtured part and fixed pattern heat station to eliminate the need for operator brazing skill. The part is shuttled into and out of the heat system in a timed sequence. Or, the heat system moves into and away from the fixtured part joint area during the brazing/soldering cycle.

A gas fuel or induction heat system is normally utilized with this machine approach. Depending on part base metals, mass, and configuration, production of 40 to 100 parts per hour is possible with a shuttle machine.

The concept is a suitable solution for low-volume production or preproduction runs. These simple machines can be used to confirm fixture designs for production soldering and brazing machines.

Manufacturers of bulky assemblies with part handling problems (such as casket lids or fuel tanks) may find the shuttle machine approach appropriate for their production soldering or brazing operations.

Rotary Index Machines

Rotary index machines offer an economical and flexible approach to automating a soldering or brazing operation.

The rotary index brazing/soldering machine consists of a rotary indexing table mechanism with part-nesting fixtures mounted on the tooling plate. An indexing circular motion transports fixtured parts throughout the sequenced brazing function, including part loading, filler alloy placement, fluxing, heating, cool down, and unload. Other secondary operations may be included (see Figs. 19-32 and 19-33).

CHAPTER 19

AUTOMATE TO UPGRADE SOLDERING AND BRAZING OPERATIONS

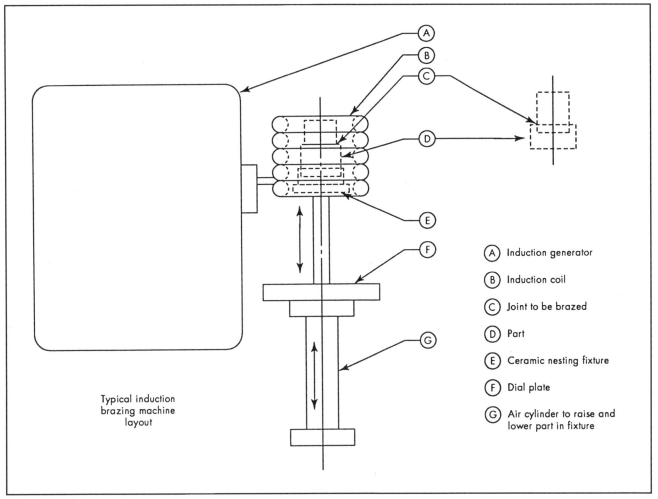

Fig. 19-31 Typical induction brazing machine layout.

The machine heat system can be gas fuel, induction, or resistance. Gas fuel systems offer flexibility and are less costly to install and operate on the rotary index machine. Multiple burners, installed on the manifold, direct flame heat patterns to the part at several stations. The part joint is heated as required, and then cooled while fixtured.

Induction and resistance systems may be considered when localized heat is very important. However, they are somewhat less flexible when the machine is used for several assembly configurations, or parts with varying joint locations.

Gas fuel system machine output is usually limited by operator part loading time. Heat stations, and the number of burners at each station, are incorporated on the system manifold as required by the part base metal, size, mass, and required machine output.

Rotary index machines can be set up for multiple parts by using dual fixtures, or using special index drive arrangements which operate with several part fixtures mounted on a common tool plate. Usually, a machine designed for one part can be converted to another assembly by fixture alterations, or fixture replacement. Rotary index drive units offer an economical approach for transporting parts through the brazing/soldering sequence. A correctly loaded quality index table will provide many years of accurate and reliable service.

Other secondary production operations may be incorporated on a rotary index brazing/soldering machine, such as riveting, staking, spinning, testing, cleaning, and sequenced automatic parts feeding for an assembly prior to, or following, the brazing/soldering sequence.

FIXTURE DESIGN

The requirements for good fixture design vary by application, with the range of special possibilities almost unlimited. Certain rules for durability and performance, however, apply in nearly every case. A few of these basics form the standard for a fusion brazing or soldering fixture (see Fig. 19-34).

FURNACE OPERATIONS

Furnace approaches consist of three basic types: nonatmosphere, controlled atmosphere, and vacuum.

Nonatmosphere

Nonatmosphere furnaces utilize electric element, infrared, or gas fuel heat systems to provide required soldering/brazing temperatures. Gas fuel systems consist of burner arrangements, or gas infrared burners.

19-33

CHAPTER 19

AUTOMATE TO UPGRADE SOLDERING AND BRAZING OPERATIONS

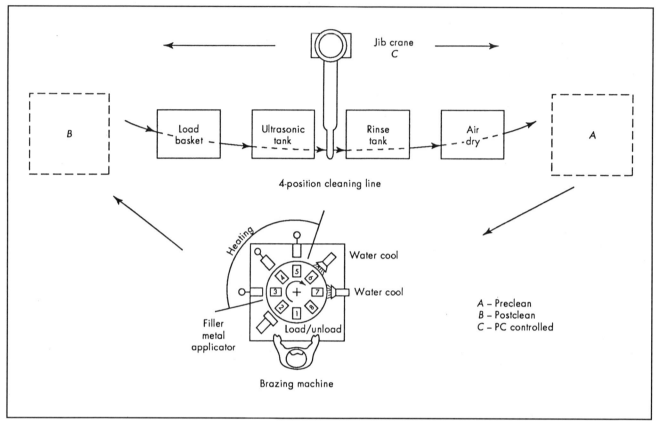

Fig. 19-32 Rotary indexed brazing/soldering machine.

Nonatmosphere furnaces are box or retort type. They are generally used in batch operations, or laboratory activities. Nonatmospheric furnaces or ovens with conveyors are used for high-volume production. However, if possible, parts should be self-locating to preclude the use of fixtures.

Controlled Atmosphere Furnaces

Controlled atmosphere furnaces provide required gas atmospheres within the heat chamber to reduce part base metal surface oxides. Also, they provide a protective atmosphere to prevent further surface oxidation, or other intermetallic reactions between the base metals and filler alloys at high-brazing temperature exposure. A controlled atmosphere furnace in correct operation normally provides high-purity brazed joints, and the assemblies brazed in a controlled atmosphere furnace usually do not require postcleaning.

Furnace gas atmosphere depends upon the part base metals, filler alloy, brazing temperature cycle, and joint quality required. Atmospheres can be provided by gas generation systems (combusted fuel gas, dissociated ammonia). Pure gases are also used for atmospheres such as hydrogen, nitrogen, or argon.

Controlled atmosphere furnaces are available as box (retort), or high-volume systems with conveyors. Special attention should be given to the use of self-locating parts.

Controlled atmosphere furnaces require precise control of the furnace heat system temperature cycle, dew point, atmosphere gas mixture, gas generator operation (if included with furnace), and conveyor speed for each of the various assemblies brazed.

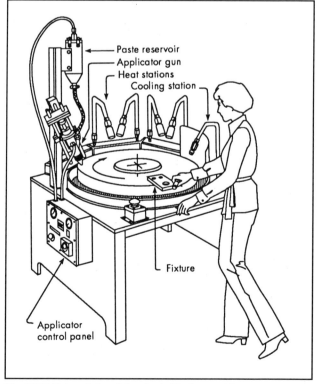

Fig. 19-33 Rotary indexed brazing/soldering machine.

CHAPTER 19
AUTOMATE TO UPGRADE SOLDERING AND BRAZING OPERATIONS

As noted, the requirements for good fixture design vary by application, with the range of special possibilities almost unlimited. Certain rules for durability and performance, however, apply in nearly every case. Shown here are a few of these "basics" which form the standard for a brazing or soldering fixture.

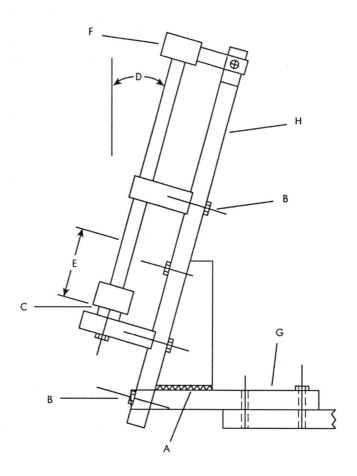

A. Points of high stress are gussetted and welded for maximum strength.
B. Part supports are notched, doweled, and screwed into position as required for precise alignment.
C. Contact with the assembly is minimized, and support points are located as far from the joint area as possible, to minimize the heat sink effect.
D. The part is inclined from vertical, so that gravity will assist in distributing flux and filler metal via capillary attraction from a single application point.
E. Clear access is provided to the joint area, to facilitate paste application, heating, and cooling.
F. When a hold-down device is required, a counterweight or mechanical clamp is used instead of springs which could lose resiliency. Ease of operation speeds loading and unloading of parts. Mechanical or air-operated assists may also be used.
G. The fixture baseplate is doweled and screwed to the brazing or soldering system, to maintain alignment through each production step.
H. All components are machined from Series 300 stainless steel, to resist corrosion during repeated heating and cooling cycles. Titanium, Inconel®, ceramic, or other materials may be used at specific points for added durability.

Fig. 19-34 Fixture design.

CHAPTER 19

AUTOMATE TO UPGRADE SOLDERING AND BRAZING OPERATIONS

Vacuum Furnaces

Vacuum furnace brazing is utilized in the electronics industry, especially in high-purity multiple joint applications using pure platinum, gold, silver, palladium, and other exotic filler alloys.

Vacuum furnaces are used in applications involving difficult-to-braze base metals and filler alloys, such as various types of stainless steel, titanium, and aluminum alloys. Multiple joint configuration assemblies composed of these materials are candidates for vacuum furnace brazing.

Vacuum brazing produces high-quality joints, especially in a retort setup with controlled atmosphere.

Other Furnace Considerations

Automating furnace operations requires the consideration of production assemblies that are self-locating and do not require fixtures. This is especially true for furnaces with conveyors. Lengthy conveyors require numerous fixtures, especially when the throughput and return lengths are considered; even simple fixtures are costly, and their upkeep could be expensive. Fixtures also add to the furnace loading and heat requirements.

When fixtures are required, some important points must be thoroughly reviewed. The fixtures should be constructed of stainless steel, Inconel, or other materials that provide long service life within the temperature ranges specified. Also, the fixture materials should have minimum coefficients of expansion throughout the brazing cycle, to prevent fixture distortion during the brazing and cool-down cycles. Simple universal-type fixtures should be used for multiple-part job-shop brazing operations.

Variations in furnace loads (part base metal, mass, configuration, etc.) create swings in furnace operations. Furnace swings are a nuisance for production schedules. A furnace operation provides maximum efficiency when used loaded in a continuous (24 hour daily) schedule. Furnace startups, testing, trial runs, and balancing are very time-consuming, and furnace operations require competent maintenance and operating personnel.

Therefore, consideration of a furnace operation deserves a serious review of overall production requirements, product mix, and other points. Furnace systems are high-cost capital equipment.

SUMMARY

Reviewing a soldering or brazing operation requires a thorough examination of various aspects of the process. Many variables enter the process, both upstream operations and downstream postbrazing operations. A working knowledge of these upstream/downstream operations is essential.

Various upstream activities, such as precleaning, joint design, and fit, become important since they directly affect downstream soldering and brazing operations. Soldering/brazing operations become a "collection station" for upstream problems, such as drawing, bending, endforming, cleaning, worn dies, poorly designed joints, erratic gas/air pressure, out of specification parts, poorly designed fixtures, etc.

Therefore, the brazing/soldering review team should carefully examine all upstream operations involved with the brazing/soldering process to preclude later "surprises."

It is important to work with established and reputable vendors within the field, especially companies with multi-interests who work with their customers throughout the long term. They furnish engineering for the process equipment design, construction, and operation, as well as field service. Also, they possess experience with the latest filler alloys and techniques, working with many applications.

IMPROVING PRESS FORGING CAPABILITIES

Improvement in equipment productivity is measured by the number of dollars in revenue produced for each dollar of capital invested in equipment. This is the primary reason for improving the forging capabilities of presses. One possible improvement results when quick change tool systems are used for *clutch-controlled screw presses*.

QUICK CHANGE TOOL SYSTEMS

The objectives of a quick change tool system are:

- Increased uptime of press.
- Setup tools in the tool room, *not* at the press.
- Elimination of the need for adjustments by positive tool location, using matched edges, keys and keyways, etc.
- Repeatability of setup for uniform product dimensions.
- Die resinking capabilities.
- Suitable for the forger's products, product cost, product complexity, and run (or batch) size per setup.
- Cost efficient.
- Easy to manipulate.

Typically, a quick change tool system is composed of a combination of these components:

- Tool holder of special design.
- Mini-bolsters, cassettes, or similar die containers.
- Dies/die inserts.
- Clamps, lifters, positioning devices, and other miscellaneous items.

SYSTEM COMPONENT GROUPS

The quick change system contains the following component groups:

1. Upper tool holder.
2. Lower tool holder.
3. Two pairs of upper and lower cassettes.
4. Railbound, battery powered tool changer cart with two load/unload stations.

Press Operations

Phase I: Press is in operation.

The upper tool holder is mounted in the press ram and the lower tool holder is mounted on the press bed. Both cassettes of the first cassette pair are located in the tool holders with forging dies installed. The cassettes are clamped in place by spring elements with hydraulic release function (see Fig. 19-35).

Phase II: Changing of tools.

1. Stop press. Switch controls to "Tool Changing" mode. The preheated second cassette pair with new dies is on the tool changer cart in its "Park" position, away from the press.
2. Lower the press ram onto lower tooling.

CHAPTER 19

IMPROVING PRESS FORGING CAPABILITIES

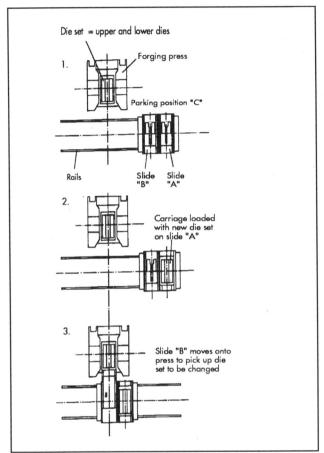

Fig. 19-35 Changing of tools.

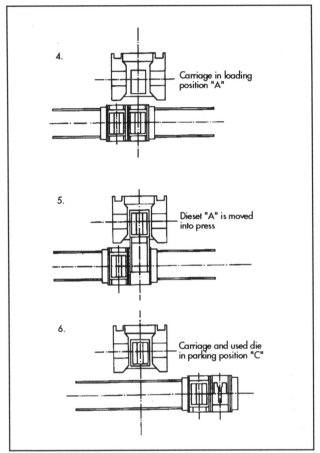

Fig. 19-36 Changing of tools.

3. Hydraulics loosen the cassette clamps; the upper cassette now rests on the lower one.
4. Move ram to top position.
5. Lift cassette pair up 2-2.5″ (50-60 mm) using either bed kick-outs or additional lifting cylinders. Leave in this position.
6. Move tool cart into the "Unload" position in front of the press. A lock assures positive positioning of the cart relative to the press. Extend charge/discharge arms mounted on cart into the press under the cassette pair.
7. Lower the bed kick-outs or lifting cylinder. The tool package now rests on top of the charge/discharge arms.
8. Retract charge/discharge arms, with tool package, from the press back onto the tool cart.

Figure 19-36 illustrates moving the tool cart to the "Load" position, so that a fresh set of tools is in front of the press.

1. With charge/discharge arms, move the cassette with new tooling into the press.
2. Lift the cassette off the arms with the bed kick-out or lift cylinders.
3. Retract the charge/discharge arms.
4. Move the tool cart to the "Park" position.
5. Lower the bed kick-out or lifting cylinders. Lower the ram. Two keys and keyways each, in the upper and lower tool holders, assure correct and positive location of the cassettes in the tool holders.

6. Clamp the cassettes in both tool holders.
7. Raise the ram to top position.
8. Switch controls to "Run" mode. Recall the program with the forging parameters for the new part from the CPU of the press controls. The press is ready to forge again.
9. Load the next cassette/tool set onto tool cart in its "Park" position away from the press.

Press Operation Requirements

One person is needed for tool change. Real time is about five minutes total. The five-minute time period does not include cleaning the press prior to changeover, or the time needed for peripheral equipment, infeed systems, die spraying/lubrication system, trim press, etc.

HYDRAULIC TOOL CLAMPING

Quick change tool systems demand great attention to the method of clamping or locking the tooling solidly in place in the forging press. The tool clamping method must be designed:

- To produce sufficient clamping force to positively lock the tools in place.
- With precisely repeatable absolute and relative (top to bottom) tool positions.
- To allow continuous monitoring of the clamping status and quick clamping and unclamping of tools.

CHAPTER 19

IMPROVING PRESS FORGING CAPABILITIES

- For easy service and maintainance. Hydraulic quick tool clamping systems require additional attention to operating safety and reliability.

The system described here has been developed for a cassette-based quick tool changing system for a 2,756 short ton (2,500 metric ton) clutch-controlled screw press. The potential combined mass of top and bottom cassettes and dies in such a press is as much as six to 10 short tons (five to nine metric tons). Because of this, the tool changing system uses hydraulics extensively.

Figure 19-37 (*a through d*) shows a front to back section through the upper and lower tool holder and cassette arrangement that constitutes the quick tool change system.

Cassettes can be designed to accept one or more forging dies. Items (*a*) are hydraulic lifting cylinders. Items (*b*) are hydraulic cylinders for lateral positioning of the cassettes.

Items (*c*) are hydraulic swing clamps that secure the cassette to the tool holder, and items (*d*) are locator or centering pins for cassette alignment. They also protect the extended swing clamps from contact with the cassettes during cassette installation. Figure 19-38 is the left to right section through the same arrangement. In Figure 19-39, the hydraulic lifting cylinders feature rollers on top. The cylinders lift the lower cassette out of the tool holder, or lower the cassette into it. When fully raised, the top rollers of these cylinders match up with recessed rollers in the front of the tool holder. The cassette is then pulled out onto a roller table.

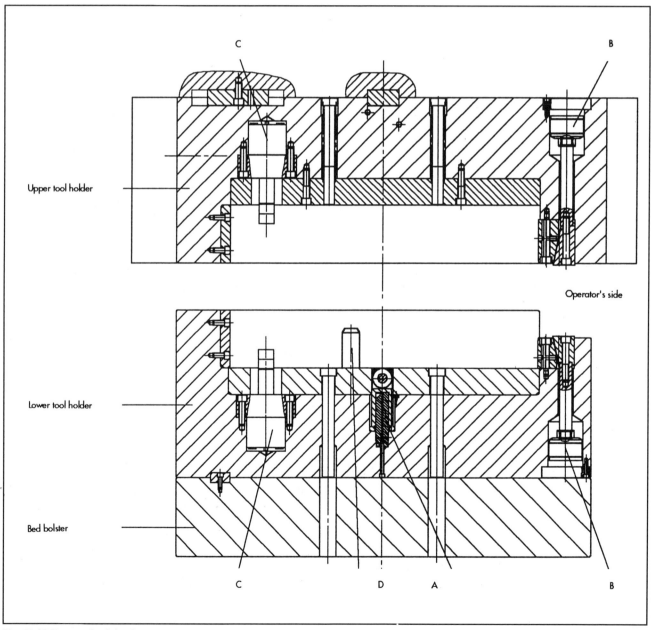

Fig. 19-37 Front-to-back section.

CHAPTER 19

IMPROVING PRESS FORGING CAPABILITIES

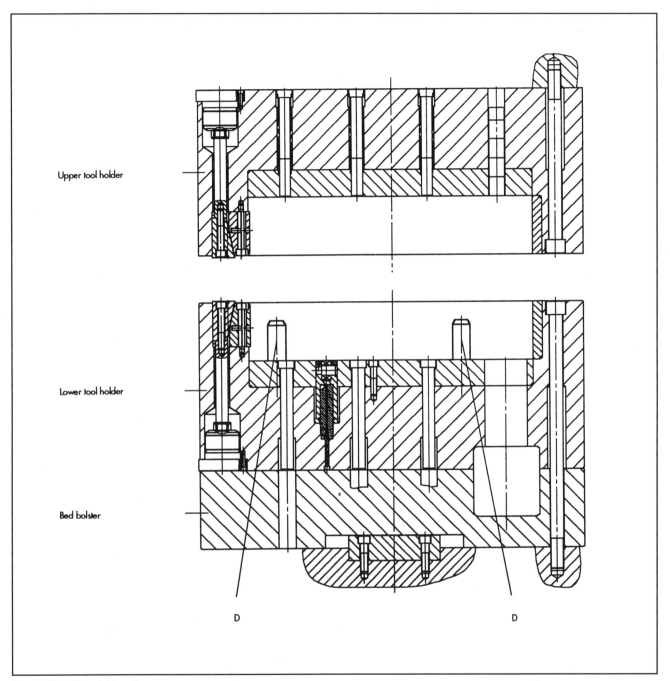

Fig. 19-38 Left to right section.

Figure 19-40 shows the hydraulic cylinders for lateral positioning. They locate the cassettes against matched edges in the tool holder, and take up side forces resulting from the forging operation. Pressure switches control end positions and positioning force.

Figure 19-41 is a close-up of the hydraulic swing clamps that hold the cassettes in place. The clamping force depends upon the maximum possible separation force, which has to be taken up with absolute certainty. Pressure switches control terminal positions of the clamps and clamping force. Controllable check valves act as additional safeties for the hydraulic hoses of the upper tool holder.

QUICK TOOL CHANGE SEQUENCE USING CASSETTES

1. Blow or wipe away scale, dirt, metal scrap, flash, etc., from die area.
2. When changing both upper and lower cassettes, lower ram onto bottom cassette. When removing lower cassette only, leave ram in top position.
3. Release lateral tool positioning and clamping cylinder pressure. Upper cassette can only be released if it is in solid contact with the bottom cassette.

CHAPTER 19

IMPROVING PRESS FORGING CAPABILITIES

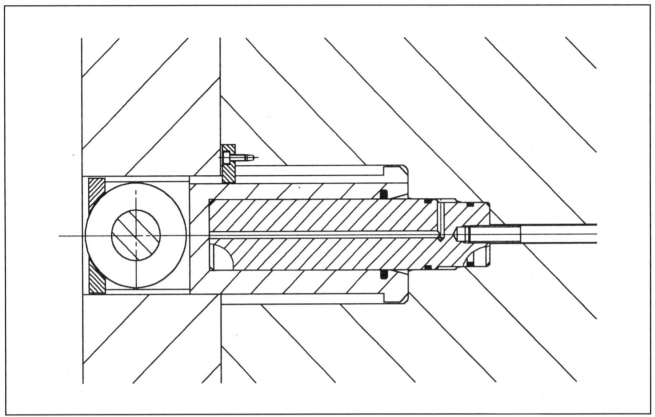

Fig. 19-39 Hydraulic lifting cylinder.

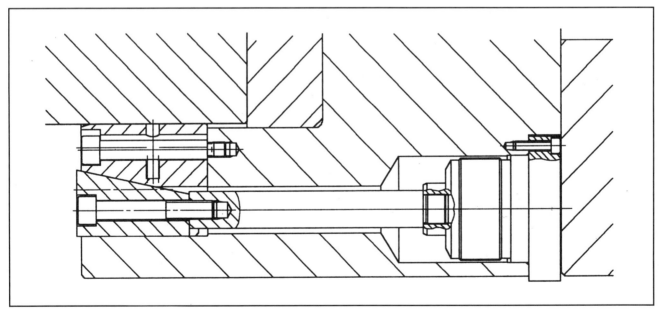

Fig. 19-40 Hydraulic cylinders for lateral positioning.

4. Move ram back to top position.
5. Upper cassette rests on lower cassette. Lift cassettes with hydraulic lift cylinders.
6. Lower front centering wedges in lower tool holder.
7. Remove both cassettes together from the press. This can be done by using an eye bolt in a bolt hole and a rope to pull the assembly out onto a roller table either by manual methods, a fork lift, or other means. Robot or other die set

CHAPTER 19

IMPROVING PRESS FORGING CAPABILITIES

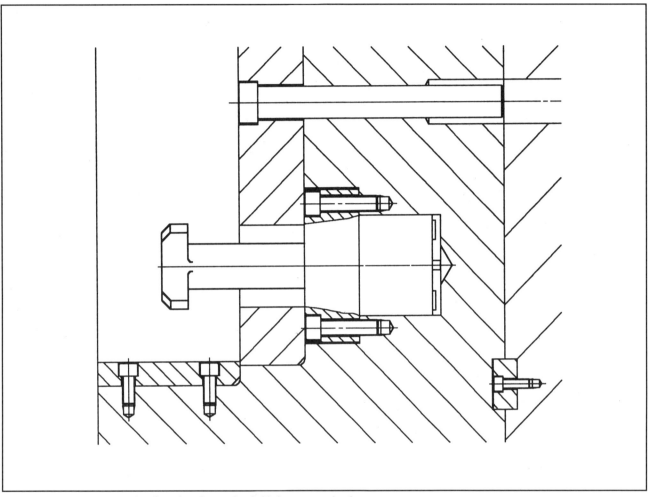

Fig. 19-41 Close-up of the hydraulic swing clamps that hold the cassettes in place.

handling equipment may be used. A roller table is most handy to move the cassette sets around the shop, but there are size and load limitations on these tables.

8. Thoroughly clean the inside of both tool holders of dirt and other foreign matter. Pay special attention to cleaning the matched edges, or cassette positioning and locating could be impaired and cassettes and/or tool holders may be damaged.

The new cassette pair can now be installed, reversing the sequence of the procedure.

This particular quick tool change system, if correctly employed, will make it feasible to cut the downtime (from last good part made with preceding tools to first good part made with the fresh tools) to about 15 to 20 minutes for presses up to 3472 short tons (3,150 metric tons) capacity. This is of particular advantage when changing multiple-die tool sets.

Depending upon part configuration and process parameters, a trim station can be integrated into the forging press for in-process part trimming. When trim tooling is designed as a die set, trim tool changeover can be done at the same time as the changeover of the forging tools. The elimination of a separate trim press simplifies material flow, manufacturing sequence, and the layout of the forging cell.

A few details of hydraulic quick tool change systems have been proven in operation for years:

- A hydraulically actuated clamp that can be quickly removed. Tool changeover is done manually (see Fig. 19-42).
- A roll-out, roll-in bolster arrangement for quick tool changing, used primarily for large press sizes (see Fig. 19-43).
- A quick tool change system with hydraulic lifting cylinders only. A forklift or similar equipment is used to exchange the tool cassettes (see Fig. 19-44).

It is important to understand that no single quick tool change systems design is suitable for all applications. The systems must be job-tailored for success and economic viability.

IMPROVING PRECISION IN FORGING BY INTEGRATING TOOL HOLDERS INTO A PRESS RAM GUIDING SYSTEM

To a large extent, the precision of the guiding systems, for the press ram and press tools or tool holders, significantly influences the quality of a forging.

In the case of separate guiding systems for ram and tooling or tool holders, under load both systems cannot be of the exact same precision during the critical moments of the forming process.

CHAPTER 19

IMPROVING PRESS FORGING CAPABILITIES

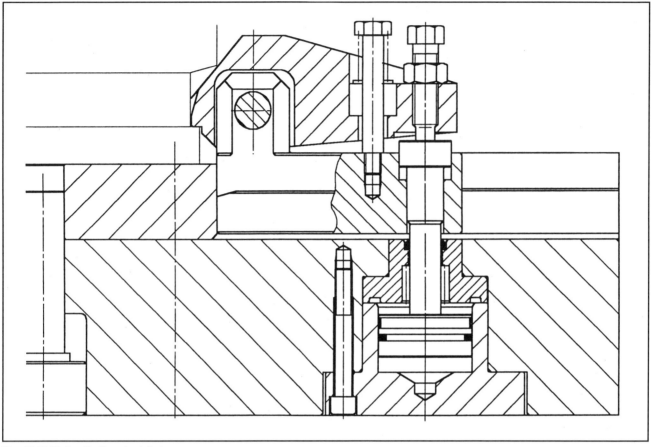

Fig. 19-42 Hydraulically actuated clamp.

Actually, the more precise of the two guiding systems will severely diminish the effectiveness of the other. This may lead to excessive die mismatch, underfill, or other quality problems.

By integrating the tool holders into the ram guiding system, these problems can be avoided.

In Fig. 19-45, one single system securely guides the ram and upper tool holder, as well as the lower tool holder, almost to the die parting line. Each mating pair of upper and lower guides is located on common reference ridges cast into the press frame. They are positioned in an *X*-arrangement that is insensitive to heat influences.

Such a guiding system increases the total guided length of the ram by 20 percent (dimension $a2$ in the right figure) compared to previous designs (dimension $a1$ in the left figure).

During off-center load forging, the increased guided length of the ram reduces resultant horizontal forces ($F2$ in the right figure) correspondingly by about 20%, compared to resultant horizontal forces ($F1$) in the left figure.

The integrated ram and tool holder guiding system shown in Fig. 19-46 greatly reduces resultant horizontal forces during off-center forging; it also relocates the point of attack of the remainder of these forces to a lower, more advantageous position. This further reduces horizontal loads acting upon the press frame and diminishes horizontal frame deformation.

As a result, ram tilt and die mismatch are minimized.

Using load data in a finite element analysis program clearly demonstrates the great improvements possible with tool holders integrated into the ram guiding system. When the guiding system of the tool holder is integrated into the press ram guiding system, the horizontal press frame deformation is reduced by 70% under identical load conditions as compared to the original press frame design (see Fig. 19-47).

Figure 19-48 is the plan view and side elevation of an upper and lower tool holder with guides integrated into the ram guiding system. In this specific case, mechanical clamps are employed to fasten the tools in the holders. Matched edges are used on the right and rear inside face of the tool holders; wear plates are used in the rear position.

In the lower tool holder, there are four hydraulic lifters with rollers on top to lift tool assemblies to the tool holder level front. Flush-mounted rollers in the front permit easy removal of upper and lower tooling in one package onto a roller table.

Figure 19-49 demonstrates the reduction in free length (from dimension $b1$ to dimension $b2$) of press frame uprights, thus increasing frame stiffness in an area very critical for making precision forging possible.

Tool holders are integrated into the ram guiding system, and heat insensitive *X*-arrangement for the guide gibs allows ram clearance of only 0.002" (0.05 mm) in production. The result is reduction of horizontal press frame deformation by 70%.

The stated improvements in tool holder guidance greatly reduce forging tolerances and dramatically improve the repeatability of forging dimensions and preset forging parameters—both very important for economic employment of quick change tool systems.

CHAPTER 19
IMPROVING PRESS FORGING CAPABILITIES

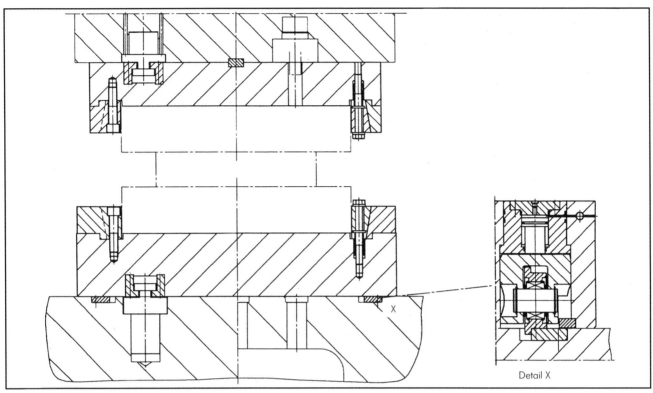

Fig. 19-43 Roll-out roll-in bolster arrangement.

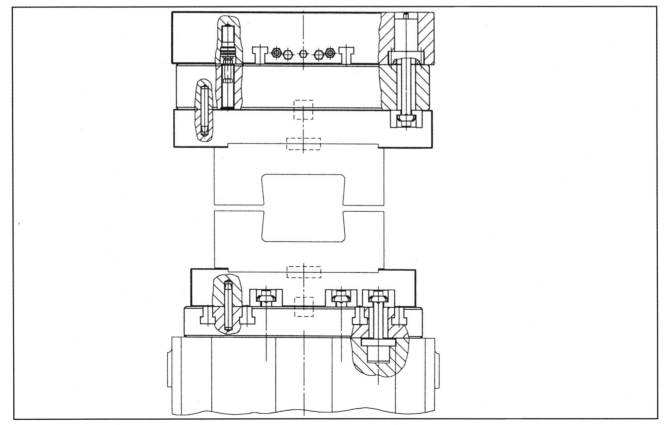

Fig. 19-44 Quick tool change system.

CHAPTER 19

IMPROVING PRESS FORGING CAPABILITIES

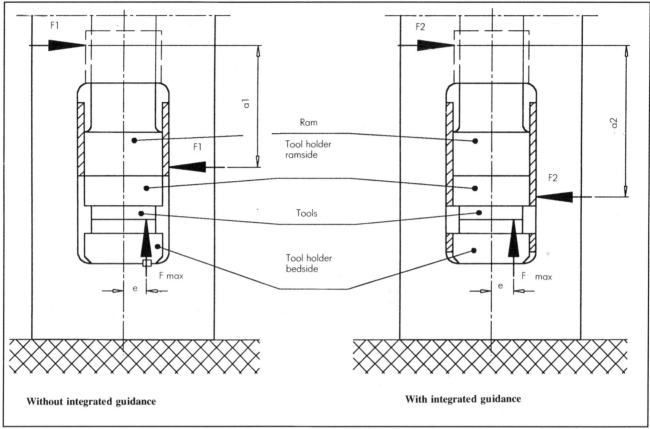

Fig. 19-45 Integrating tool holders into the ram guiding system.

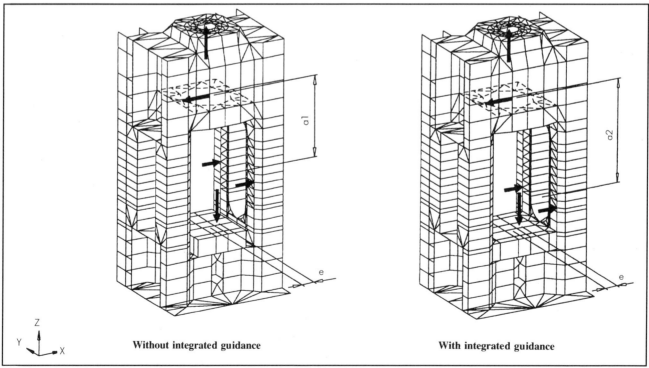

Fig. 19-46 Integrated guiding system.

CHAPTER 19

IMPROVING PRESS FORGING CAPABILITIES

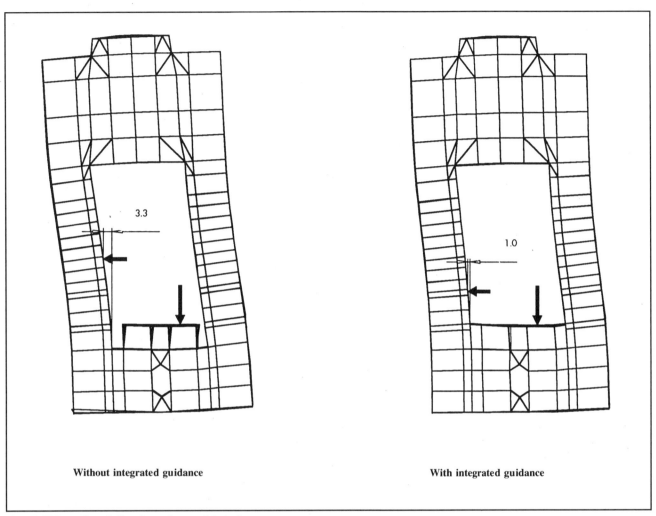

Fig. 19-47 Ram tilt and die mismatch are minimized.

A GUIDELINE FOR OPTIMIZING Nd-YAG LASER WELDING

To make the best use of Nd-YAG lasers for welding, the following factors are important:

- Laser parameters.
- Material considerations.
- Pulsed Nd-YAG laser welding.
- Optimizing laser welding parameters.

INTRODUCTION

Welding with a pulsed Nd-YAG laser is an efficient and economical joining process for a wide range of parts found in the electronic, automotive, aerospace, nuclear, and consumer product industries. Because of the small diameter of the focused beam and the precise control of the laser energy, the best candidates for laser welding are typically small parts required for high-reliability applications.

There are several variables used to control and optimize any laser-based process. However, pulsed Nd-YAG laser processing has additional variables in the nature of pulsed operation that add flexibility and complexity to the operation. The objective of this article is to review the major welding-process parameters, and describe an optimization scheme that maximizes process yield.

LASER PARAMETERS

Laser processes are best optimized by considering the physical laser quantities such as pulse energy, peak power, pulse shape, average output power, and power intensity at the surface of the workpiece. Other parameters such as pulse width, repetition rate, and spot size are useful in controlling the previous parameters.

To illustrate laser pulse parameters, Fig. 19-50(a) shows a rectangular laser pulse with a typical initial overshoot. The time required to deliver the energy is the *pulse duration* or *pulse width*,

CHAPTER 19

A GUIDELINE FOR OPTIMIZING Nd-YAG LASER WELDING

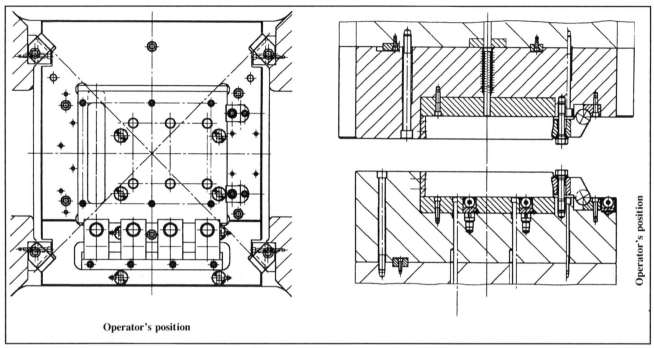

Fig. 19-48 Plan view and side elevation of an upper and lower tool holder with guides integrated into the ram guiding system.

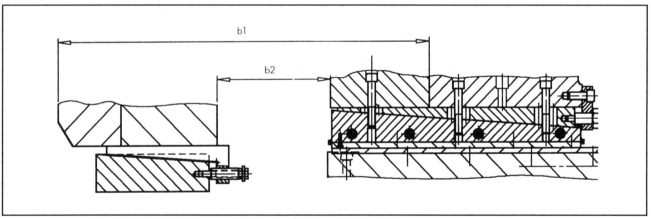

Fig. 19-49 Reduction in free length increases frame stiffness.

and it is controlled by the power supply. The *energy* is controlled by adjusting the height or peak power of the pulse until the desired energy output is achieved. Energy monitors fitted on industrial laser units will display the pulse-energy output of the system. The final peak power is calculated by dividing the pulse energy by the pulse width. On some laser systems, depending on the power supply, the pulse shape is controlled by defining changes in the instantaneous peak power during the pulse, as shown in the output trace in Fig. 19-50(*b*). This capability is known as *pulse shaping*. The number of pulses per second is the *pulse frequency*, and it is also controlled by the power supply. The average output power of the system is the pulse energy multiplied by the pulse-repetition rate.

Average and peak power are independent, limited only by the maximum output of the laser system. For example, a typical Nd-YAG laser specification is 550 W average power, 7 kW peak power, 55 J maximum pulse energy. Such a laser could be run at 28 J pulse energy, 10 ms pulse width (therefore 28 J/10 ms = 2.8 kW peak power) and a repetition rate of 14 Hz (therefore 28 J × 14 Hz = 390 W average power). Similarly, for a 28 J pulse, the minimum pulse width could be 4.0 ms (28 J/7 kW = 4 ms) and the maximum repetition rate could be 19.5 Hz (550 W/28 J = 19.5 Hz).

The optical system of a Nd-YAG laser also has several parameters, such as beam quality, conventional or fiber-optic beam delivery, lens type, *f*-number, focal length, and final focused-spot size. Full discussion of fiber-optic and conventional laser system optics is found in other references.[17,18] Of all optical parameters, the focused-spot size is the most important. The simplest control of spot size is changing the focal length of the process lens, because the diameter of the focused-spot size is directly proportional to the focal length of the lens.

CHAPTER 19

A GUIDELINE FOR OPTIMIZING Nd-YAG LASER WELDING

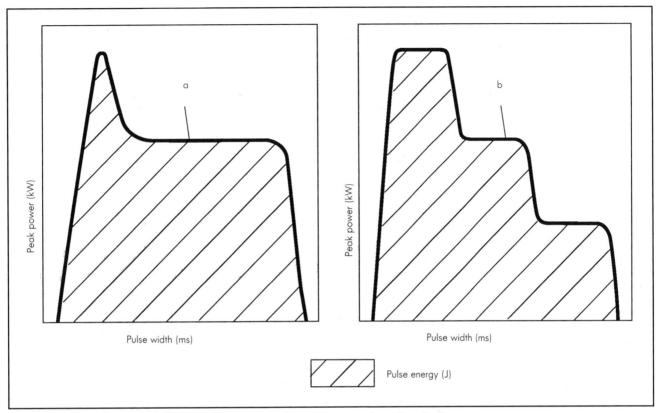

Fig. 19-50 Nd-YAG laser pulse for (*a*) rectangular and (*b*) a three-sector, ramp-down shaped pulse.

The spot size and the peak power are related in an important process parameter referred to as the *power intensity*, defined as peak power divided by the spot area. The power intensity generally defines the materials processing regime, and the specific range depends on the material processed. For metalworking applications, low power intensity causes surface heating without melting—useful for heat treating. Higher power intensity causes surface melting and some vaporization—useful for welding. Still higher power intensity causes significant vaporization and ejected material, and is useful for cutting and drilling.

The flexibility in configuring the system's operating characteristics offers several options for optimizing the output conditions of a pulsed Nd-YAG laser operation. Understanding the main controls available is essential to quick and effective selection of the best laser parameters.

MATERIAL CONSIDERATIONS

In addition to specific laser parameters, it is useful to review general characteristics of metals commonly subject to laser welding. Component materials are generally not a variable parameter when optimizing a laser welding process. There are occasions, however, where a change to laser welding provides an opportunity to alter material specifications to improve the process. In some cases, a material change is required for successful laser welding. The main differences between laser welding and other welding techniques are the laser's high-power intensity and higher weld-cooling rate; and laser welding is generally autogenous (no filler metal is added to the joint).

Stainless steel is the material most commonly encountered for laser welding applications. Laser welding is well suited for stainless-steel joining for several reasons. The Nd-YAG laser wavelength is absorbed readily by stainless steel, yet the high viscosity of the weld pool resists splattering when using a high-intensity beam; this enables penetrations up to 0.26" (6.5 mm) with a 1000W pulsed laser. In general, components made from stainless steel are more precisely manufactured; consequently, joint tolerances are typically within the range required for laser welding. If weld cosmetics are important for stainless steel components, argon cover gas is used to prevent oxidation.

Austenitic stainless steels (AISI-300 series alloys) have high nickel contents which are beneficial for laser welding. Types 301, 304, 304L, 316, and 318 are readily welded with Nd-YAG lasers, with 304L and 316 the leading candidates. Welds in free-machining stainless steel, such as AISI-303, tend to result in hot-cracking when sulfur and phosphorus segregate to the weld center line. Ferritic stainless steels (AISI-400 series alloys) are generally not laser welded because high cooling rates promote martensite formation, and consequently, brittle welds and cracking. Preheating and pulse shaping prevent weld cracking but they cannot cure all problematic incidences.

A wide range of components are produced from plain carbon steels (PCS), and there are special considerations when welding these alloys with pulsed lasers. Similar to ferritic stainless steel, PCS with carbon levels above 0.40% tend to form martensitic regions because of the high cooling rates, and this will result in brittle welds. The viscosity of PCS is lower than stainless steel, and this results in a greater tendency for splatter. Consequently, lower peak-power pulses are favored. Cover gas is rarely used when laser welding steel components, because weld cosmetics are not critical and slight oxidation enhances energy absorption.

CHAPTER 19

A GUIDELINE FOR OPTIMIZING Nd-YAG LASER WELDING

Aluminum alloys are reflective to laser wavelengths and have high thermal conductivity. Nd-YAG lasers can weld aluminum alloys readily, because high peak power (2 to 10 kW) can overcome the high thermal conductivity. Absorption of Nd-YAG laser wave length is higher, compared to that of the CO_2 laser. High cooling rates of pulsed Nd-YAG laser welding cause hot-cracking when joined with a filler of A14047 alloy; add sufficient silicon to the joint to prevent hot cracking. Depending on part requirements, the use of a cover gas such as argon or nitrogen may be optional. Using no cover allows the weld pool to oxidize, enhancing energy absorption and increasing penetration. However, cover gas allows better flow of molten filler metals and reduces weld defects such as pinholes.

Titanium and zirconium components can be welded advantageously with Nd-YAG lasers.

Some of the reasons are:

- It is possible to repeatedly produce porosity-free crack-free welds.
- The wavelength is transmitted through glass windows into inert-atmosphere welding chambers.
- Occasionally, the cooling rates are fast enough to prevent oxidation with only local cover gas. The greatest challenge when welding these highly reactive alloys is ensuring adequate cover to prevent oxidation. Oxygen levels above 100 PPM will increase penetration drastically, or result in burning. Therefore, repeatable welds cannot be produced without consistent cover gas shielding.

Kovar™ is commonly used for applications involving glass-to-metal seals. This alloy is welded in a similar manner to ferritic stainless steels, where 0.08" (2.0 mm) deep welds are produced with a 400-W laser. Kovar parts are often plated with gold or nickel to improve corrosion resistance. Gold-plated Kovar can be welded without cracking as long as there is an underplating of electrolytically plated nickel. Electroless nickel contains phosphorus that produces porosity, and direct gold plating results in grain-boundary segregation and weld cracking.

Copper, silver, gold, and platinum based alloys have very high thermal conductivity and reflectivity. Although these alloys are difficult to weld in general, a 400-W Nd-YAG laser can deliver high peak power pulses, up to 20 kW, with 0.01 to 0.02" (0.3 to 0.5 mm) spot sizes to achieve up to 0.02" (0.5 mm) penetration. Nickel-plated copper is laser welded more efficiently because the plating acts as a better absorber than the base metal.

In general, the metallurgy of an autogenous laser weld must be considered carefully for new designs and when changing from current joining processes. A more complete description of metallurgical and plating considerations is found in other references.[19]

PULSED Nd-YAG LASER WELDING

Pulsed Nd-YAG lasers are used for both spot-welding and seam-welding applications. Spot welding is the simplest form of pulsed laser welding; several laser-system parameters contribute to the weld characteristics.

There are two types of spot-welding modes, conduction and penetration. Conduction mode involves power intensities sufficiently high to cause melting, but not high enough to vaporize metal. As shown schematically in Fig. 19-51(a), the top surfaces of conduction-mode welds are generally very smooth, without a great deal of surface ripple. The final weld shape resembles a half-sphere which is modified by fluid flow of the weld pool. Consequently, penetration is limited to approximately one half of the top-surface weld diameter. Because of the limited penetration, conduction-mode laser welding is well suited to welding materials under 0.020" (0.5 mm) in thickness.

At higher power intensities, vaporization occurs, and the vapor pressure and surface tension flow promote formation of a vapor cavity within the weld pool. This condition produces a penetration-mode weld, which is shown schematically in Fig. 19-51(b). The shape of the cavity captures beam energy more efficiently, and this enhances penetration. As the power intensity increases, material is ejected from the weld pool during formation of the vapor cavity. The greatest penetration with acceptable surface cosmetics occurs at the point where a slight amount of material is ejected from the weld pool.

At very high power intensities, there is a transition from welding to drilling, as most of the metal is ejected from the weld pool. As shown in Fig. 19-51(c), this results in top-surface undercut or drilled welds. For a given optical system, the specific power intensity or peak power required to produce the various weld modes is highly dependent on the material type, and less dependent on the surface condition of the part.

OPTIMIZING LASER WELDING PARAMETERS

Ideally, the laser welding process is optimized around some specific measurable characteristic for guiding the direction of parameter development. Objective characteristics such as penetration, weld strength (tensile or shear), or leak rate (for hermeticity), are desirable quantities for optimization, because the mean and standard deviation for a sample can be determined easily. Subjective characteristics, such as weld shape or cosmetic quality, are difficult to optimize in conjunction with measurable characteristics. In some cases, subjective characteristics are coded according to ranges (good = 1, limit = 0, bad = -1) for comparison with other objective characteristics.

When optimizing spot welding operations, the first option is to determine the required laser spot size that will achieve the required weld size. A very small spot diameter would dictate use of a short focal-length lens, while a larger spot size would favor a longer focal-length lens. Once an optical system is selected, the peak power is varied to produce test welds, to determine the range that produces the desired weld mode. It is not uncommon for first attempts to produce shallow surface melting (indicating higher peak power is required), or a hole (indicating that lower peak power is required). In some instances, the optical system may produce spot sizes that require peak powers outside the normal operating range of the laser. This typically occurs when a long focal-length lens is used, and insufficient melting occurs even at the highest-rated peak power. In such a case, the optical system is modified to reduce the spot size and increase the power intensity. An adequate optical-system and peak-power combination must be determined before other factors can be optimized.

The next step is to optimize the pulse energy required to produce the required penetration. As shown in Fig. 19-52, pulse energy is the basic quantity that determines the volume of melted metal. Therefore, increasing pulse energy increases weld penetration. For the peak power determined earlier, the pulse energy is controlled with pulse width. Ideally, the pulse energy is high enough to produce the required weld penetration, and no more. The energy output of a laser is limited by the maximum mean power. Excess energy for a process means that the pulse repetition rate is reduced so that the mean power limit is not exceeded. If a welding operation is specified with too much energy, then the repetition and the process rate must be reduced.

CHAPTER 19
A GUIDELINE FOR OPTIMIZING Nd-YAG LASER WELDING

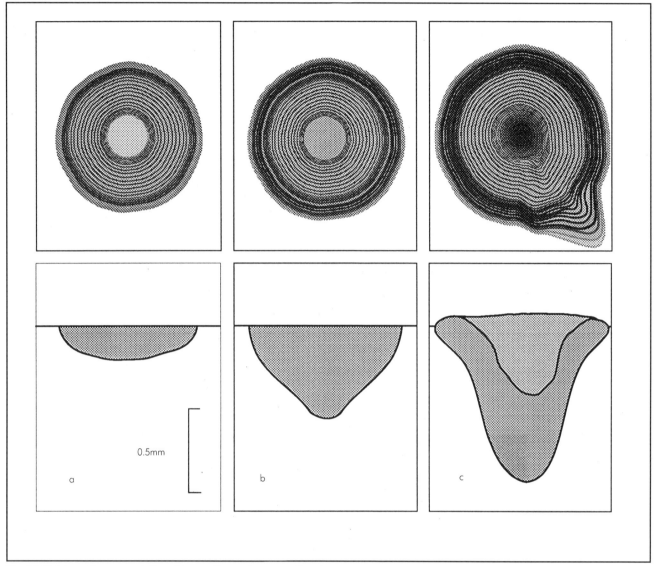

Fig. 19-51 Schematic representation of top surface and cross-section views.

Optimizing pulse energy can generally be done with a basic, rectangular pulse shape. Once the pulse energy is established, further process improvements are achieved with pulse shaping. Pulse shaping is the ability to control the instantaneous peak power within a single weld pulse. This is done by dividing a single pulse into sectors, each with a height and width. In general, pulse shaping is used to reduce pulse energy, improve weld cosmetics, improve consistency, and reduce or eliminate porosity, pinholes, and cracking. There is an unlimited range of pulse shapes, and this makes optimizing this parameter difficult.

Some guidelines assist in converging on the "best" pulse shape. For general welding operations, start by comparing the effects of rectangular pulses with three- or four-sector ramp-up and ramp-down pulse shapes, all at the same energy and total pulse width. Ideally, one of the shapes will produce better characteristics and can then be used as a basis for further optimization over a narrower range. For example, if a four-sector ramp-up shape was most effective, then a second iteration, with two-, three-, and five-sector ramp-up pulse shapes, would determine the next best shape. Other pulse shapes may be derived from other standard practices, such as adding a prepulse to clean the part surface, and adding preheat or post-heat sectors to control cracking. Once an alternative pulse shape is selected, the pulse energy may require modification to re-establish optimum conditions.

With laser spot-welding operations, the initial test repetition rate of the laser can exceed the production rate. If this occurs, further optimization will minimize the repetition rate of the laser, thus minimizing maintenance requirements. This is a straightforward step with fiber-optically delivered laser systems. However, systems employing conventional optics experience thermal lensing with changes in mean power. Thermal lensing decreases the final spot size with decreases in mean power, and vice versa. For example, if the process was optimized at 400 W, and only 200 W was required for production, the welding mode shifts from penetration welding to drilling as the spot size is reduced. If this happens, another iteration of the optical system selection is

CHAPTER 19

A GUIDELINE FOR OPTIMIZING Nd-YAG LASER WELDING

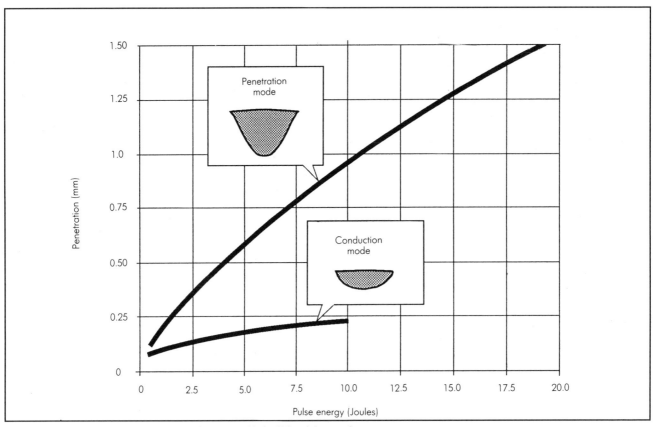

Fig. 19-52 Penetration versus pulse energy for spot welds in 304 stainless steel.

required. The main objective is to generate the original spot-size, and other parameters can still be used.

Multiple-pulse spot welds are slightly more complex than single-spot welds, because the total number of pulses becomes an additional process parameter. Generally the main objective of optimization is to decrease the process time for each weld. In this case, the optimization procedure for single-spot welding is to modify the process so that the product of the number of pulses and single-pulse energy is minimized. In this case, total weld time (number of pulses per rep rate) becomes another weld characteristic (in addition to tensile strength, for example), when varying the combinations of number of pulses and pulse energy. Statistical design of experiments becomes a useful tool when trying to produce the best weld in the least time by controlling the number of pulses, pulse energy, pulse shape, and mean power.

Seam welds are optimized in a similar manner to spot welds, because seam welds can be made in the conduction mode and the penetration mode. In general, for laser powers below 500 W, seam welding is considered as a series of overlapping spot welds where the weld pool solidifies between pulses. Seam welds with various percent overlaps are shown schematically in Fig. 19-53. The minimum weld overlap for consistent seam welding is approximately 50%, although this increases to 70% overlap for hermetic seam-welding applications. Extremely cosmetic welds may have 80 to 90% overlap.

Weld speed and percent weld overlap are important parameters to consider when optimizing seam welds. Weld speed, V, and percent weld overlap, L, can be related to repetition rate, R, and weld diameter, D, in the equation: $V = D \cdot (1 - L/100) \cdot R$.

This equation implies that as weld speed increases, the percent overlap decreases for a constant repetition rate, or the repetition rate must be increased for a constant weld overlap.

Considering these guidelines, the maximum possible weld speed is limited by the maximum repetition rate of the laser. This is a function of the maximum laser mean power and the pulse energy. Therefore, the first step to optimize seam welds is to determine the approximate peak power and pulse energy for the given penetration and cosmetic requirements. Although individual welds solidify between pulses, weld speed also has an effect on penetration because of residual heating. Decreasing weld speed increases percent overlap and slightly increases penetration. Therefore, once an approximate pulse energy range is established and the repetition rate is set for the maximum laser power, then the speed can be modified to achieve the desired penetration or weld strength.

The maximum seam-welding speed occurs at approximately 50 to 60% weld overlap and this is a guide when optimizing weld speed. If the speed is increased to the point that overlap is reduced below 50% and still achieves required penetration, then the pulse energy can be reduced and the repetition rate increased. The weld is repeated at an increased speed until the required penetration is obtained. If the pulse energy is too low, the weld speed is decreased to obtain the required penetration, and the weld overlap will increase toward the 90% range. If this occurs, the pulse energy is increased and the repetition rate decreased. The final optimum weld speed may depend on several other factors that shift the weld overlap percentage, but it should be close to the 50 to 90% range for most applications.

CHAPTER 19

A GUIDELINE FOR OPTIMIZING Nd-YAG LASER WELDING

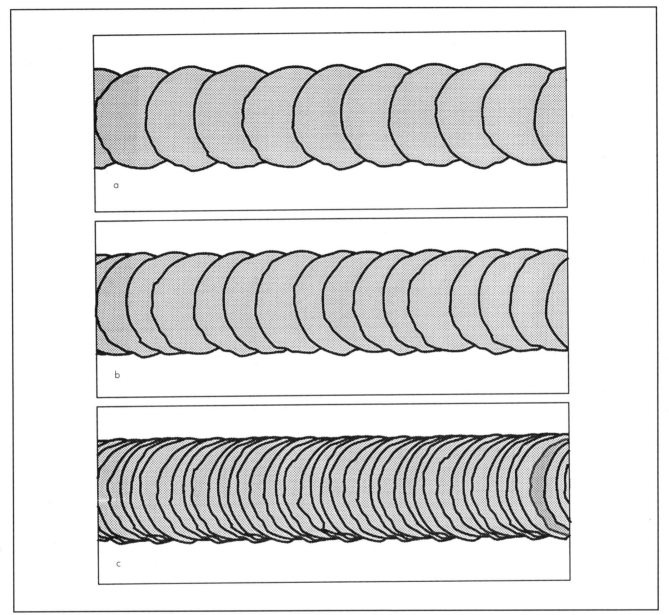

Fig. 19-53 Schematics of seam welds produced with weld overlaps of (a) 50%; (b) 70%; (c) 90%.

CONCLUSION

Pulsed Nd-YAG laser welding is a flexible joining process that involves many parameters and requires careful investigation to optimize. Understanding the aspects of pulsed laser parameters, weld metallurgy, and parameter control are prerequisites for optimizing a laser welding system. In addition to in-house experimentation and testing, samples can be produced by a laser manufacturer or a contract shop, so that the other laser parameters and cycle times can be determined. Ultimately, the set of parameters that consistently produces the best parts will become clear.

THE INVENTION OF HOOK AND LOOP FASTENERS

Fasteners have undergone many changes and improvements throughout history. One modern fastener is the hook and loop method. This fastener has evolved because new ways were needed to solve an old problem: how to attach one object to another. The hook and loop fastener is a good example of the results of the continuous improvement process.

CHAPTER 19

THE INVENTION OF HOOK AND LOOP FASTENERS

INTRODUCTION

By now, most people have heard the story of Georges de Mestral, the Swiss engineer, who while hunting one day in 1949, noted that his trousers were covered with cockleburs. The tenacious manner in which they clung to the fibers of his pants caused him to be curious enough to examine them under magnification. To his surprise, he found that the cocklebur was comprised of many miniature hook-type elements and his trouser fibers formed small loops in which the cocklebur's hook elements became entangled. This gave him the basis for his invention which he named Velcro® (*VEL* for the loop elements as found in *VEL*our and *CRO* for the hook elements as found in a *CRO*chet hook). Today, the term Velcro designates a particular brand name. This has given rise to an array of other generic terms for these materials such as: hook-and-loop tape fasteners, self-gripping fasteners, and touch fasteners. Little could Mr. de Mestral have imagined that his invention would become one of the most useful inventions of the 20th Century.

THE EVOLUTION OF HOOKS

The first hooks were formed from monofilaments and woven into a base tape. Their shape replicated the cocklebur's, in that the hook elements were all facing in the same direction. It was not realized until the product debuted (in a raincoat closure) that this design yielded a performance characteristic of good latitudinal shear strength in one direction, but diminished performance in longitudinal shear, tension, and peel strength. The problem with this design was further complicated by the necessity for sewing the hook tape facing in the correct direction. The principles of continuous improvement dictated that the hook elements should be oriented to provide strength in all directions and modes of performance. Thus, the second generation of woven hook elements were bi-directional, alternating every other row (see Fig. 19-54). Over the years, the performance characteristics of the hook have been continuously improved by altering the diameter, raw material, heat setting, and cutting of the monofilaments. Today, there are many variations of the original product; though appearing similar to the untrained eye, they perform at widely different levels, as dictated by the application.

SPECIAL FEATURES OF HOOK AND LOOP

One of the attributes of woven hook and loop tapes is their ability to be cycled thousands of times while still maintaining a relatively high percentage of their initial engagement strength. The primary reason for this is the flexible nature of the hook monofilament. The hook monofilament and the loop fiber filaments are round, and therefore, able to slip free from each other without severe damage. Unfortunately, higher cycle life tends to yield lower tension, shear, and peel values. Engineers and product designers, looking at new and different applications, often found little need for high cycle life but desired greater strength. The demand for stronger products, with lower cycling requirements, spurred the development of the "mushroom" type fastener (see Fig. 19-55). Changing the monofilament from nylon to polypropylene and searing the monofilaments with heat caused the polypropylene stems to melt into a mushroom shape. When the mushroom was mated with a low pile loop, the gain in shear strength was 70% over that of regular woven hook and loop. The higher strength, which is caused by the mushroom shape, tends to tear more of the loop filaments. Depending upon the loop used, a relatively low number of cycles can damage enough loop filaments to reduce overall strength significantly. One added advantage that was realized with the mushroom was its ability to form a much thinner side profile closure.

CONTINUOUS IMPROVEMENT IN HOOK STRENGTH

Efforts towards continuous improvement in hook strength did not stop with the advent of the mushroom. In the late 1970s, hook products were molded or extruded from 100% plastic resin. These types of hook tapes utilized what is referred to as an "arrowhead" shape (see Fig. 19-56) and the strength far exceeded that of both traditional hook and mushroom. These stronger hooks reduced the number of usable cycles to only a few (again, depending on the type of loop used) due to their highly destructive effect on the loops. The strength of the plastic hook strips was of major interest to automobile producers. The first automotive application of these high-strength fastening materials was for headliners. Prior to hook and loop, headliners were held in place with various types of clips. The clip system was a blind location and attachment system; consequently, it was not unusual to ship cars from the assembly plant with less than all of the clips properly engaged. The hook and loop system offered positive engagement in a blind attachment system, and there were other benefits as well. It was found that the front and rear trim moldings could be eliminated, which reduced material and installation labor costs and also improved the appearance. This also offered the interior designers an opportunity to move the interior courtesy light (dome light) away from the center of the headliner, since the primary purpose for a center location was to prevent the headliner from sagging.

Another area where hook and loop contributed to continuous improvement was in automotive seating. In the early 1980s, auto manufacturers began using hook and loop fastening systems for cover-to-molded-seat-pad retention. Seat design engineers saw a

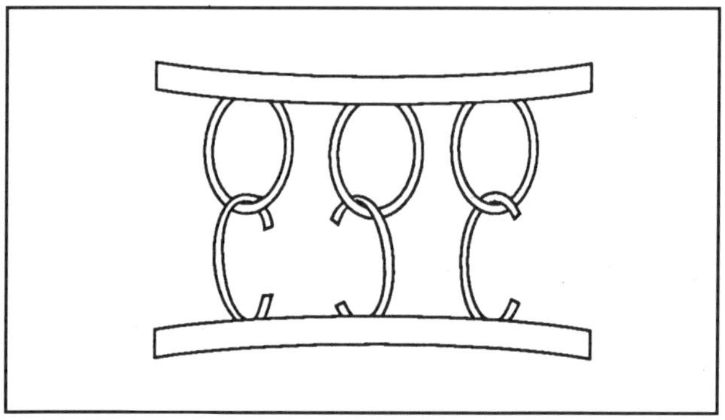

Fig. 19-54 Bi-directional hook elements.

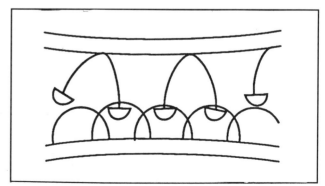

Fig. 19-55 Mushroom-type fastener.

CHAPTER 19

THE INVENTION OF HOOK AND LOOP FASTENERS

number of advantages over the traditional wirepocket and "hog ring" attachment system, such as:

- Reduced assembly labor.
- Potential for automation.
- Design flexibility to retain the cover without the need for unwanted sewn lines.
- More comfort in the "insert" areas of the seat when the design calls for a "tied-down" appearance.
- Ergonomically easier to assemble.
- Easy repositioning of the seat cover to eliminate wrinkles and correct improper alignment.

The first seats produced for cars built in North America, that utilized hook and loop, employed hook strips that were glued to the urethane foam pad. Within two years of the implementation of this application, hook strips were developed that could be molded into the seat pad. This was an important development, since gluing was labor intensive and presented environmental and worker-related concerns due to the solvents present in the adhesives.

Mold-in Strip Fasteners

The first mold-in strip fasteners were of a composite structure which used a 1″ (25 mm) wide plastic hook backed with a 2″ (50 mm) wide nonwoven material. The purpose of the nonwoven was to provide a porous material that the foam seat pad could bond to. The wider nonwoven backing provided a 0.4″ (10 mm) wide selvedge area on each side of the hook, where a protective shield could be affixed. The protective shield formed a barrier to prevent foam intrusion onto the hooks during the molding process. In 1985, a new generation of mold-in hooks was introduced which eliminated the need for laminated backing materials. The concept of a double-sided hook tape made from a single extrusion (see Fig. 19-57) proved to be cost efficient and offered other advantages:

- Improved bonding characteristics to foam.
- Fewer subsurface voids caused by the hook strips.
- Better composite flammability.
- Softer, more flexible parts to provide greater seating comfort.

Currently, mold-in hook products are offered in narrower versions 1″ (25 mm) wide which are molded into deep trenches in the seat pad to create sharp tie-down appearance features. Additionally, some models employ mold-in strips which are die-cut into curved shapes. Today hook and loop systems are found in the majority of U.S. produced vehicles. Japanese auto producers have incorporated this concept into nine different models produced in Japan, and European car manufactures have also begun using hook and loop for their seating programs. The shape of the arrowhead hook was refined over the past decade, and the variety of plastic resins used has grown significantly. Depending on factors such as temperature, flexibility, strength, and cost, design engineers can choose products made from such resins as: polypropylene, polyester, polyolefin, and nylon.

LOOP MATERIAL DEVELOPMENT

During the same period of time that the various hook configurations were emerging, the loop was going through its own development cycle. This is not to suggest that hooks and loops underwent developmental changes independent of each other; as mating parts, they evolved together as a system and individually as components within a system. For some unexplainable reason, loop quite often does not get the same recognition, in terms of its contribution to the strength and performance of the total system. The fact is that one of the principal means of changing a hook's performance is to change the loop it is mated to. Loop materials are usually of either woven or knit construction; however, certain types of nonwovens and carpet were successfully used as loop, especially in automotive applications. Woven loop normally has a selvedge edge and is manufactured in standard widths up to 4″ (100 mm). Knit loop has no selvedge and is slit from roll widths up to 79″ (2000 mm).

Napping Process for Loop Material

The first loop material was subjected to a process called *napping*. Unnapped loop was characteristically very uniform, and this actually was a detriment to peel strength. By subjecting the loop to a severe brushing process, the filaments are reoriented and raised so that they become more random in height. This has the effect of increasing initial engagement and peel strength with only a slight degradation in tensile and shear values. In 1967, a new concept in loop was introduced which incorporated the best elements of napped and unnapped loop. By using yarns which were texturized prior to weaving, loop was made to perform as well as napped or unnapped without additional processing or risk of damage from overnapping (see Fig. 19-58).

Texturized Loop

The texturized loop is especially desirable for applications where appearance is a design consideration. This configuration of loop gives the appearance of a *plush pile* carpet and has even been called "hook and pile" in some engineering circles. Though pleasing to the eye, this loop, when mated to arrowhead-type hooks, is among the strongest of all known hook and loop combinations.

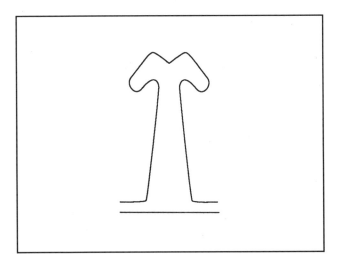

Fig. 19-56 Arrowhead shape.

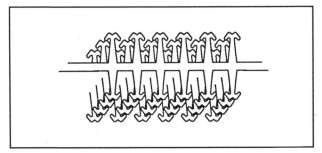

Fig. 19-57 Double-sided hook tape.

CHAPTER 19

THE INVENTION OF HOOK AND LOOP FASTENERS

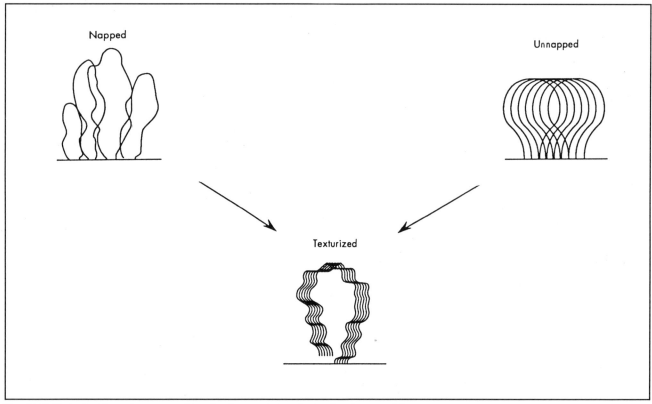

Fig. 19-58 Texturized loop elements combine the characteristics of napped and unapped loops.

Scrim

Scrim (or wide loop sheeting) was developed to work with the hooks that were molded into the foam seat pads. Design engineers wanted the trim cover to follow the contour of the seat pad, without unwanted sew lines in the trim cover, which were inevitable if narrow woven or knit loop was to mate with the hook in the insert area of the cover. Though it is possible to glue narrow loop to the inside of a seat cover, this is labor intensive and always presents a problem of location. Scrim loop sheeting solved this problem by covering part or all of the inside of the seat cover. The scrim loop sheeting is adhered, by adhesive or flame laminating, to the "plus pad" or foam portion of bilaminate trim cover material, and thus forms a trilaminate. The advantage to this approach is that the cover will mate with the hook in the seat pad regardless of its location. The down-side is that it requires that loop be used over the entire seat cover while only mating with a small amount of hook in the seat pad. Also, scrim loop sheeting will not work in those areas of the trim cover where a "join-seam" occurs over the top of a hook strip. Narrow loop, wrapped and sewn around a "join-seam" or to a "design-sew," is still the preferred method of attachment in these areas.

CONCLUSION

In the automotive industry there are numerous other applications where hook and loop is used as a part of the continuous improvement effort. Beyond seats and headliners, there are iterations and variations of de Mestral's invention on interior floor carpets, trunk carpets, floormats, storage compartments, pillar covers, rear package shelves, speaker grills, consoles, map pockets, seat cover closures, armrest and headrest cover closures, visors, door panels, electronic components, and more. Many of these applications required a development engineering activity. Perhaps it was an improved plastic resin, a special pressure sensitive adhesive, a higher UV resistant dyestuff, or a total change in the shape of the hook or the loop filament.

Outside the automotive arena, other industries such as medical, aerospace, military, atomic energy, home furnishings, and athletic equipment have spurred new hook and loop product development. There are hook and loop materials that conduct low-voltage electricity, and high-temperature materials with stainless hooks and Nomex™ loop. Hook and loop has gone into space with astronauts and to the deepest depths of the oceans. This is what continuous improvement forces producers to do, to survive in a market that is constantly changing while at the same time demanding more sophisticated products, manufactured at higher quality levels and produced on shorter lead times. The hook and loop industry is proving every day that continuous improvement does work.

References

1. International Rectifier, *The International Rectifier Quality Plan*, (El Segundo, CA, 1993).
2. Mikel Harry, *The Nature of Six Sigma Quality*, Motorola Government Electronics Group, (Phoenix, AZ).
3. Department of Defense, *Procedures for Performing a Failure Mode, Effects, and Criticality Analysis*, US MIL-STD-1629A.
4. *Failure Mode and Effect Analysis*, (Ohio: Technicomp Publishing, 1990).
5. Shigeo Shingo, *Zero Quality Control: Source Inspection and the Poka-Yoke System*, (MA: Productivity Press, 1986).
6. Ford Motor Company, *Team Oriented Problem Solving*.

CHAPTER 19
THE INVENTION OF HOOK AND LOOP FASTENERS

7. Yoji Akao, *Quality Function Deployment*, (MA: Productivity Press, 1990).
8. John R. Hauser & Don Clausing, "The House of Quality," *The Harvard Business Review*, (May-June, 1988).
9. Yasuhiro Monden, *Toyota Production System* (GA: Industrial Engineering and Management Press; IEEE, 1983).
10. Seiichi Nakajima, *Introduction to TPM: Total Productive Maintenance*, (MA: Productivity Press, 1988).
11. Yasuhiro Monden, *Toyota Production System* (GA: Industrial Engineering and Management Press; IEEE, 1983).
12. Oliver Wright, *Manufacturing Resource Planning: MRP II; Revised Edition*, (VT: Oliver Wright Limited Publications, 1984).
13. Philip Thomas, *Competitiveness Through Total Cycle Time: An Overview for CEOs*, (New York: McGraw Hill, 1990).
14. Eliyahu Goldratt, *The Goal: Excellence in Manufacturing*, (New York: North River Press, 1984).
15. Michael Beer, R.A. Eisenstat, and Bert Spector, "Why Change Programs Don't Produce Change," *Harvard Business Review*, (November-December 1990).
16. The following references deal with behavior modification in more detail or from a different perspective:
    Bradford, David and Cohen, Allan, *Managing For Excellence*; (Wiley; New York, 1984);
    Byham, William, *Zapp! The Lightening of Empowerment*, Ballantine Books, New York, 1992;
    Hunsaker, Philip, and Alessandra, Anthony, *The Art of Managing People*, (New York: Simon and Schuster, 1986).
17. S. L. Gorscak, H. N. Bransch, T. R. Kugler: "Designer's Guide for Laser Hermetic Sealing," *Hybrid Circuit Technology*, (August 1991).
18. T. J. Culkin, T. R. Kugler: "Designer's Handbook, Industrial Lasers, Parts one and two," *Photonics Spectra*, (October and November 1987).
19. W. Koechner, *Solid State Laser Engineering*, (New York: Springer-Verlag, 1988).

Bibliography

American Welding Society, Inc., *AWS Brazing Manual*, (New York: Reinhold Publishing Corporation).

American Welding Society, Inc., *AWS Soldering Manual* (New York: Reinhold Publishing Corporation).

Andrew D. Althouse, Carl H. Turnquist, & Bowditch, *Modern Welding* (South Holland, IL: The Goodheart-Wilcox Publishing Company, 1992).

Bruce R. Williams, *Need A Brazing Machine* (Willoughby, OH: Fusion, Inc,).

Robert E. Vargo, *The Ultrasound Cleaner-The Prime Mover*, Davenport, IA, Swen Sonic Corporation.

The Aluminum Association, *Aluminum Brazing Handbook*, Washington, DC.

The Aluminum Association, *Aluminum Soldering Handbook*, Washington, DC.

CHAPTER 20

ACCIDENT PREVENTION AND CONTINUOUS IMPROVEMENT

This chapter will discuss the importance of establishing an ongoing Accident Prevention Program and how it is an essential part of any Continuous Improvement Program. Accident prevention is recognized as one positive method of cost control and how to improve safety, productivity and efficiency. Other benefits of a safety program include fewer employee injuries, less equipment damage, reduced scrap and rework, etc.

The National Safety Council estimated that in the United States during 1991 there were:

- $63.3 billion dollars in work-related accident costs (direct and indirect costs).
- 1,700,000 disabling work injuries.
- 75,000,000 days lost due to work-related injuries.
- 9900 work fatalities.

These figures do not include minor injuries that have occurred. Practical suggestions will be given regarding making continuous improvements in the areas of:

- The cost of industrial accidents.
- Accident prevention programs.
- Ergonomics (material handling / repetitive motion tasks).
- Layout safety planning.
- Pre-planning of industrial facilities.
- Machine safeguarding.
- Preventive maintenance.
- Job safety analysis.
- Lockout / tagout.
- Product safety.
- Slip and fall prevention.
- Personal protective equipment.

Instituting effective safety programs makes sense from a business, moral, professional and legal viewpoint (there are a number of Federal, State and local regulations that require their existence). Designing, manufacturing, and marketing safe products and services is also necessary to stay competitive in international markets.

THE COST OF INDUSTRIAL ACCIDENTS

Before discussing various methods of improving manufacturing methods and equipment, it is essential to understand the costs of injuries. As soon as someone is injured, costs begin to accumulate. It is important to account for direct and indirect accident costs to understand how they affect the overall company and why it is important to prevent injuries from occurring. Investment in safe work methods and equipment can prevent needless injuries and save large amounts of capital that can be used for other purposes.

DIRECT ACCIDENT COSTS

The following describes some typical direct accident costs and additional ones may be thought of:

- If a minor injury occurs, it may require only prompt first aid. However, it is expensive to set up and maintain a well equipped first aid room with either a properly trained individual, an industrial nurse, or a physician in charge.
- If the injury is serious, the individual may be at home or in a hospital for days, weeks, or months. There may be continuing medical treatment and surgery.
- The injured person also receives compensation benefits according to the rates established in the particular state. Payments may continue until the individual is well enough to return to work.
- The employer is responsible for the cost of first aid, medical treatments, and compensation benefits. The company may purchase an insurance policy to cover some of these expenses or set-up a self-insurance plan. Either way, these direct costs affect the entire organization. Establishing an effective accident prevention program can reduce the number of injuries, their severity, and their effect on the company.

It is a mistake to visualize the profit or loss of accident prevention solely in terms of the narrow area of direct or insured costs. As substantial as they may appear, they are only a part of the overall accident costs. In many situations, the indirect costs may far surpass the direct costs.

INDIRECT ACCIDENT COSTS

Following is a list of indirect costs that are normally uninsured, can be difficult to estimate in dollars, and sometimes can be very expensive:

CHAPTER CONTENTS:

ACCIDENT PREVENTION PROGRAMS 20-2

ERGONOMICS 20-4

CUMULATIVE TRAUMA DISORDERS OF THE HAND AND WRIST 20-8

LAYOUT SAFETY PLANNING 20-13

PRE-OPERATIONS SAFETY PLANNING CHECKLISTS FOR INDUSTRIAL FACILITIES 20-25

MAKING MACHINES SAFER 20-30

JOB SAFETY ANALYSIS 20-34

PRODUCTS LOSS CONTROL 20-39

OTHER SAFETY CONCEPTS 20-43

The Contributors of this chapter are: John W. Russell, CSP, ARM, PCMH, Director of Industrial Service, Loss Prevention Department, Liberty Mutual Insurance Group.
The Reviewers of this chapter are: Neal Freedman, Senior Loss Prevention Consultant—Industrial, Loss Prevention Department, Liberty Mutual Insurance Company; *John J. Lavallee*, President, J. Lavallee Associates, Inc., Worcester, MA.

CHAPTER 20

ACCIDENT PREVENTION PROGRAMS

1. Production time lost by an injured employee; wages paid for:
   - Time spent in getting first aid.
   - Time not worked on that day, if the person does not return to work.
   - Poor quality or less productivity when the person returns to work.
2. Time lost by executives and supervisors in:
   - Seeing that the injured person is taken care of.
   - Investigating the circumstances of the accident.
   - Rearranging production schedules.
   - Making a report of the accident.
3. Time lost in:
   - Repairing, replacing, clearing away, rearranging or cleaning up equipment / material that was damaged or disarranged by the accident.
4. Production time and product lost if a:
   - Machine is shut down.
   - Continuous process is halted.
5. Production time and product lost by:
   - Other employees near or at the scene of the accident.
6. Cost of:
   - Overtime wages to make up for lost production.
   - Material spoiled or damaged by the accident or a substitute employee.
   - Reprocessing of product.
   - Transferring, hiring, or retraining of a substitute person.
   - Temporarily accepting lower quality and/or quantity of production from a substitute employee.
7. Other costs such as:
   - Public liability claims.
   - Cost of renting replacement equipment or facilities.
   - Loss of profit on contracts canceled by customers.

These costs can obviously be serious and substantial. The majority of them interfere with smooth, continuous production operations. Some safety individuals and organizations have shown that the indirect or "hidden" costs may be four to ten times as much as the direct costs. Calculating the cost of several recent accidents that have occurred will substantiate the fact that accidents can be very expensive and they should be prevented whenever possible.

ACCIDENT PREVENTION PROGRAMS

There is nothing mysterious about effective accident prevention. It is not merely a matter of posting bulletins, issuing rules, and holding meetings, but a definite responsibility of every supervisor and employee. It must be approached in the same logical manner that is used to increase production and efficiency, reduce costs, and eliminate waste.

Efficient production methods and safety cannot be separated. Satisfactory personnel relations cannot be maintained without a sincere interest in accident prevention on the part of both management and supervision.

Company management must establish and support an ongoing safety program. Without top management support, the success of the safety program is greatly reduced. It is important that each department and employee is given safety responsibilities and that they are held accountable for them. No one person or small group of individuals can control hazards, reduce injuries, and improve productivity; it requires a team effort that involves everyone.

The following provides a few examples of how each department can support the overall safety program:

- The Engineering Department can help design equipment and facilities that minimize safety hazards and enhance efficiency.
- The Manufacturing Department can reduce hazards through changing manufacturing methods, scheduling production, and identifying hazards.
- The Purchasing Department can make sure that purchased materials and equipment meet Federal, State, and local codes and regulations.
- The Quality Control Department can make sure that the products that are being produced meet safety requirements and are made using safe work methods.
- The Maintenance Department can ensure that equipment and facilities are kept in good condition and that preventive maintenance methods help prevent production interruptions and potential injuries.

Figure 20-1 illustrates a partial checklist that can be modified to be used for inspecting departments (additional hazards may exist). It is recommended that supervisors supplement company safety programs with their own plan for control of accidents. This will include some of the activities shown below:

1. Engineering controls.
   - Thorough investigation of all accidents and near misses to determine hazards, causes, and corrective measures.
   - Analysis of accident history to identify trends and needed changes.
   - Inspection and maintenance of machinery and equipment.
   - Job Safety Analysis (JSA) to identify safer, improved procedures.
   - Review of plans and specifications for new installations, to identify and eliminate safety hazards.
2. Stimulation.
   - Personnel contacts with all employees to develop safer attitudes.
   - Special personal supervisory attention to careless employees.
   - Constructive supervision that includes helpful criticism.
   - Delegation of specific safety responsibilities.
   - Ask employees for their safety suggestions.
   - Set a good example by following safety rules and wearing personal protective equipment.
3. Education.
   - Proper placement and training of new and transferred employees.
   - Instruction in approved methods of operation, rules, and regulations.
   - Periodic retraining of employees.
   - Use of posters, bulletins, displays, and warning signs.
   - Short meetings to discuss special problems and hazards.

CHAPTER 20

ACCIDENT PREVENTION PROGRAMS

	yes	no
Material handling—are heavy/awkward objects moved mechanically?	___	___
Material handling equipment—is it being used and in good condition?	___	___
Machines—are hazardous points on machines guarded?	___	___
Lockout—is equipment locked out during specific repairs?	___	___
Lubrication—is lubrication done while equipment is shut off?	___	___
Hand tools—are they in good condition and used for the proper tasks?	___	___
Materials—are materials stored or piled properly?	___	___
Aisles—are aisles marked and kept clear of items?	___	___
Floors—are floors free of tripping and slipping hazards?	___	___
Housekeeping—is the facility neat, clean, and free of scrap?	___	___
Electrical cords—are they well insulated and in good condition?	___	___
Platforms—are railings and toeboards properly designed and fastened?	___	___
Stairways—are treads and handrails secure and in good condition?	___	___
Ladders—are ladders in good condition and used properly?	___	___
Personal protective equipment—is it being used and maintained properly?	___	___
Illumination—is adequate lighting provided in the department?	___	___
Ventilation—are hazardous dusts, vapors, etc. removed properly?	___	___
First aid—do employees secure first aid for minor injuries?	___	___
Accidents—are all accidents investigated to prevent recurrence?	___	___
Enforcement—are unsafe practices corrected and instructions given?	___	___
Cooperation—Do employees offer suggestions and comply with rules?	___	___

Fig. 20-1 Partial checklist which can be modified for use for inspecting departments.

- Provide opportunity for all employees to serve on the inspection committee on a rotating basis.

To be successful, every program must be carefully directed and strictly enforced. The supervisor must be prepared to take disciplinary action, if necessary.

SPECIFIC PROBLEMS

The supervisor should also develop special plans for the control of specific problems that may arise. The same technique should be used that is employed on production problems. From knowledge of the accidents that have occurred, and familiarity with seasonal or infrequent operations, the supervisor should anticipate those problems that require special consideration.

An example of this planning is shown below. It was developed by the supervisor of a machining department who recognized the need for controlling an increase in accidents caused by lifting.

1. Engineering controls.
   - Request monorail and hoist for milling machine area.
   - Install abrasive tread on the floor around the oil separator.
   - Install scrap conveyor and collection hopper to eliminate need for manual handling.
   - Review layout of finished stock inspection workstations.
   - Standardize height of material handling trucks and storage racks to eliminate lifting.
   - Provide a conveyor for moving tote pans, instead of manually pulling them.
2. Stimulation.
   - Prepare a chart showing the recent increase in the number of lifting injuries.
   - Obtain a prize award from management for the best employee suggestion on eliminating strain injuries.
   - Discuss every lifting accident with injured workers.
   - Secure written reports from individuals who have been involved in a lost time injury involving manual material handling during the past six months.
3. Education.
   - Hold meetings with employees to discuss proper methods of manual material handling.
   - Prepare special job instructions for individuals handling heavy or awkward items.
   - Personally talk with employees seen performing unsafe lifting tasks.
   - Ensure all employees are trained on newly installed hoists/monorails, and periodically review training received. Training records must be kept on file by the responsible department.

DEVELOPING THE SAFE ATTITUDE

Employees will reflect the behavior and attitude of their supervisor. It has been established that a careless employee can get hurt on the best of equipment and that a careful employee can work safely under adverse conditions. The supervisor must develop a proper attitude among employees in addition to maintaining a comprehensive safety program. A safe employee attitude can be encouraged in the following manner:

- Show a personal interest in the safety of each person.
- Stress safety when interviewing and orienting new employees.
- Indicate hazards in all new methods and jobs.
- Delegate people to perform specific accident prevention duties.
- Appoint a safety committee and rotate members.
- Require the wearing of personal protective equipment and assist in its selection and purchase.
- Issue written safe work procedures.
- Require a written or verbal report from injured employees and emphasize that the sole purpose is to prevent recurrence, not to place blame.
- Post safety messages that apply to local problems.
- Reinstruct employees who have developed unsafe working habits.
- Warn violators of safety requirements.
- Invite and implement suggestions from all workers.

CHAPTER 20

ERGONOMICS

It is equally important that the supervisor avoid those things that discourage employee cooperation with the safety program. The following conditions can result in an unsafe attitude:

- Failure of the supervisor to set a good example.
- Improper or vague instructions.
- Poor discipline.
- Delays caused by poor planning.
- Inattention to complaints and grievances.
- Lack of physical safeguards.
- Poor arrangements and lax housekeeping.
- Too many rules.
- Overemphasis on speed.
- Failure of supervisor to adopt employee suggestions.
- Poor, inadequate, or no training provided.

CORRECTING EMPLOYEE INDIFFERENCE

Any person who refuses to support the supervisor's safety program, or assumes an attitude of not being concerned about other people's welfare, requires the serious attention of the supervisor. The careless employee is probably the most dangerous type of individual in an organization, and the supervisor should use several or all of the following methods for improvement:

- Discuss the effect of an injury on the employee's family.
- Demonstrate the loss of income that accompanies disabling injuries.
- Point out the unnecessary pain and suffering that may be involved.
- Appeal to pride; mention that good workers are seldom hurt.
- Show the possibility of others being injured when someone is careless.
- Assign a specific safety duty to develop a sense of responsibility.
- Secure cooperation by explaining the supervisor's responsibility for uninterrupted production and safety in the department, and ask for assistance in accomplishing the desired results.
- Issue warnings and take disciplinary action.

OVERCOMING THE NEGLECT OF MINOR INJURIES

The importance of prompt first aid cannot be overemphasized. Every supervisor has observed the results of neglecting minor injuries. Immediate first aid treatment for all minor injuries should be a strict rule of every supervisor. Employees who fail to cooperate with this requirement should receive the immediate personal attention of their supervisor. The following points can be demonstrated to impress offenders with the importance of this precaution:

- Infection and other serious complications frequently result from failure to secure prompt first aid treatment.
- There is more discredit attached to failure to obtain first aid than to being injured.
- Complying with the rules is part of the job.
- The supervisor complies; there are no exceptions.

ERGONOMICS

Ergonomics is the process of fitting the job to the person. Safety professionals agree that the problem of strains and other disorders of the back, arms, hands, and wrists is dramatically increasing. The proportion of these types of injuries is expected to rise from the current one-third of all worker's injuries to one-half by the year 2000. It is essential to design jobs following ergonomic guidelines to improve safety, efficiency, and productivity.

People vary greatly in their abilities. For example, they differ in how they are able to:

- Perform stressful physical work on a daily basis.
- Operate controls and read displays.
- Climb ladders.
- Walk over or work on different floor surfaces.

When jobs are built upon the capabilities and limitations of individuals, productivity can be enhanced and job satisfaction can be improved. Occupational ergonomics seeks to create a match between the worker and the demands of the job.

The Americans with Disabilities Act (ADA) also addresses making accommodations for persons with disabilities, who may not have all of the capabilities ordinarily required by a job. The workstation could be made accessible to the individual or the job could be redesigned to modify or eliminate problem tasks, as long as the essential job functions are not changed. Additional information on this topic can be obtained by contacting the U.S. Equal Employment Opportunity Commission and U.S. Department of Justice, Civil Rights Division, in Washington, DC.

This section will discuss how to establish an effective Ergonomics Program and practical engineering and administrative controls for improving manual material handling and repetitive motion tasks.

ERGONOMICS PROGRAM GUIDELINES

Many companies are finding the need to develop ergonomic programs to control the rising incidence of cumulative trauma disorders. These programs may vary from one organization to the next depending on resources available, degree of exposure, and top management's commitment to controlling the problem. However, there are five basic program elements which should help establish a program.

1. Appoint an Ergonomics Team.
    - Set up an ergonomics team. Most companies do not have full time ergonomists, doctors, occupational health nurses, physical therapists, and other staff to implement and manage an ergonomics program. An ergonomics team, utilizing the skills and knowledge of those persons directly involved in the operations, has proven to be a successful alternative.
    - Membership on the ergonomics team may vary, depending on the organization's structure, the talents required to address the hazards, and operational limitations. Productive team members may come from a range of both managerial and nonmanagerial personnel. Management and staff functions may include safety, personnel, medical, executive or operations management, engineering, maintenance, purchasing, and department

CHAPTER 20
ERGONOMICS

supervision. The involvement of affected employees should be encouraged because they can provide a great deal of insight.
- Develop a written statement which states the team's purpose, goals, and activities. Also, emphasize top management's commitment to the program and the degree of support that will be provided. Outline the basic tasks to be performed by the team including as a minimum: monitoring progress, proposing procedures, analyzing accidents and complaints, setting priorities, reviewing results of task evaluations, proposing budgets, and reviewing suggestions and recommendations.
- Meetings should be held on a planned, periodic basis. Document the minutes of each meeting including the agenda and attendees.

2. Appoint a Coordinator.
   - The coordinator could be chosen by the ergonomics team. The general responsibilities of the coordinator would be to:
   - Act as a liaison between management and the ergonomics team.
   - Chair the ergonomics team.
   - Provide reports to management.
   - Maintain a library of pertinent information on ergonomics.
   - Maintain records as appropriate.
   - Be responsible for documents proposing or authorizing changes, such as purchase orders.

3. Expand Safety Education and Training.
   - Provide education and training on ergonomics for management and employees. In determining the training needs, keep in mind that the type of training will vary, depending on the group involved.
   - Train the ergonomics team and management on the overall program organization, program management, principles of ergonomics, worksite evaluation, and hazard abatement.
   - Train supervisors and employees to recognize signs and symptoms of cumulative trauma disorders, identify ergonomic deficiencies in their area, and be familiar with the company's reporting procedures.
   - People have legitimate concerns that training for hourly employees can lead to a rash of cumulative trauma disorder claims. But, the information received through outside sources may be sensationalized, adding problems that could have been avoided through a proactive stance.
   - Outside consultants and insurance company loss prevention specialists can provide resources such as scripted presentations, videos, and pamphlets to assist in the training efforts. Key management personnel should also be invited to seminars and institutes for ergonomics training.

4. Evaluate Jobs.
   - Evaluate those jobs which have produced, or have a potential for producing, cumulative trauma disorders. Expand the scope to include jobs in the planning stage. The purpose of a job evaluation is to:
     a. Identify significant risk factors associated with cumulative trauma disorders.
     b. Recommend solutions, both physical and administrative, for the elimination or reduction of those risk factors.
   - These evaluations should be one of the activities of the ergonomics team. Selection of which jobs are to be evaluated can be based on several factors such as workers' compensation claims, turnover, leaves of absence, complaints, and group medical records.
   - An ergonomic job evaluation includes observational analysis. Since most workers perform repetitive tasks very quickly, videotaping the task can be helpful. This allows the job to be viewed in slow motion as many times as needed to accurately identify stressful postures and awkward movements.

5. Establish Medical Controls.
   - This program element differs from the other elements in that it must be supervised by a physician or occupational health nurse familiar with recognition and treatment of cumulative trauma disorders.
   - Early case detection is the primary medical control for the prevention of cumulative trauma disorders. An overall program addresses other areas also, such as treatment, return to work, and monitoring.
   - Educate employees to recognize the symptoms of cumulative trauma disorders and encourage them to report any symptoms to their supervisor as soon as possible, while the condition is still reversible.
   - Baseline examinations are recommended for all new and transferred employees to jobs known to involve repeated biomechanical (muscle, tendon, joint) stress. Subsequently, if symptoms develop, baseline data will be available against which changes can be evaluated. Re-evaluations should be conducted annually for all employees in jobs known to involve repeated biomechanical stress. This would include jobs for which symptoms, complaints, or claims have been made.
   - Medical surveillance based on the use of questionnaires, supplemented by appropriate examinations, may also prove to be of benefit in detecting early cases before significant impairment and disability develop.

MANUAL MATERIAL HANDLING TASKS

More than one fourth of all compensable work injuries are associated with manual handling tasks. The majority of these injuries involve low back pain. Compensable low back pain is the most prevalent type of Workers Compensation (WC) claim, accounting for 16% of WC claims and 33% of WC costs.

Results from research studies have shown that the ergonomic approach of good job design, one that matches the job to the capabilities of the worker, is the most effective of any current approach in reducing industrial low back pain. Good job design can reduce up to one third of low back claims, and up to two thirds of claims among workers performing heavy manual handling tasks.

Ergonomics research has identified the following risk factors associated with compensable low back pain:

- Manual handling tasks involving excessive weights, forces, and frequencies.
- Body motions that require bending, twisting, and reaching.

Each of these risk factors is considered in the overall evaluation of manual handling tasks. Experimental results reveal that there is no single maximum object weight that applies to everyone, because strength varies greatly among individuals. Simply stated, some workers are stronger than others, and what is maximum for one is not maximum for another.

CHAPTER 20

ERGONOMICS

Consequently, the best way to evaluate object weight is in terms of what percentage of the working population can be expected to perform the task without overexertion. The higher the population percentage is for a given weight, the lower is the risk of injury—conversely, the lower the percentage, the higher the risk.

A low-percentage (high-risk) task means that while many workers may still be able to handle the weight, only a few can perform it without overexertion. The ideal task will fit 90% or more of the population. Population percentages of less than 75% are considered risk factors, and are not consistent with good safety and health practices.

Lifting objects from the floor causes a worker to be particularly vulnerable to low back pain; therefore, population percentages less than 90% are considered risk factors for lifting tasks from the floor. Details of Liberty Mutual experiments have been published in technical journals, and reprints are available upon request.

Task Evaluation

Manual handling tasks are evaluated by determining task weights, forces, distances, frequencies, and duration, as well as the body motions involved in performing the tasks. These data are then compared with ergonomic data developed from studies conducted in the Research Center. Low back pain risk factors are identified if the population percentages are less than recommended, if significant body motions are present, and/or if task duration exceeds recommended levels.

Ergonomic evaluation not only assists in determining whether redesign is needed for existing jobs, but is also useful as a criterion in the design of new operations. The redesign of existing tasks can usually be shown to be cost effective, with a relatively short payback period based upon low back pain cost savings alone.

Task Redesign

The primary emphasis is on ergonomics—the fitting of the job to the worker. Research has shown that the application of ergonomic principles to job design is the most effective approach in reducing the incidence of compensable low back pain in industry.

Designing the job to fit the worker reduces the worker's exposure to the risk factors of low back pain, and consequently reduces the medical and legal problems of selecting the worker for the job. Good job design also places less reliance upon the worker's willingness to follow established training procedures, such as lifting properly. Good job design is effective because it reduces the probability of initial and recurring episodes of low back pain, allows the worker with moderate symptoms to stay on the job longer, and permits the disabled worker to return to the job sooner. Finally, good job design usually increases the overall efficiency and productivity of a task or operation.

Other Control Approaches

Although job design is the most effective approach for controlling low back pain, it does not provide complete control. The job design approach is applicable to most operations; however, there are jobs that are more difficult to design and control, such as police work and firefighting. These jobs require greater dependence upon pre-placement testing and selection of workers. Selective techniques include medical examination, strength and fitness testing, and job-rating programs.

Education and training represents another control approach for compensable low back pain. Traditionally, training programs have been directed toward the workers. Equally important is the education of management in responding to low back pain when it does occur. This type of training emphasizes concern for the worker, avoidance of adversary relationships, appropriate follow-up procedures, and establishment of early return-to-work programs.

There is no simple solution to the complex problem of low back pain in industry. An effective control program must be a combination of all approaches. The organizations that have been most successful in controlling compensable low back pain have all used multiple approaches. Although low back pain cannot be completely prevented at the present time, the use of multiple approaches can reduce the problem to controllable levels.

MANUAL HANDLING TASK REDESIGN

These general principles, as outlined, constitute the ergonomic recommendations that should be used as a guide for the redesign, and for the initial design of manual material handling tasks. The significant body motions, and the weights and forces involved in manual handling, have been identified as low back injury risk factors. Both require consideration when task evaluation and design is done.

1. Minimize significant body motions.
   - Reduce bending motions.
      a. Eliminate the need to bend by:
         - Using lift tables, work dispensers and similar mechanical aids.
         - Raising the work level to an appropriate height.
         - Lowering the worker.
         - Providing all material at work level.
         - Keeping materials at work level (for example, do not lower anything to the floor that must be lifted later).
   - Reduce twisting motions.
      a. Eliminate the need to twist by:
         - Providing all materials and tools in front of the worker.
         - Using conveyors, chutes, slides, or turntables to change direction of material flow.
         - Providing adjustable swivel chairs for seated workers.
         - Providing sufficient work space for the whole body to turn.
         - Improving layout of work area.
   - Reduce reaching out motions.
      a. Eliminate the need to reach by:
         - Providing tools and machine controls close to the worker, to eliminate horizontal reaches over 16 inches.
         - Placing materials, workpieces, and other heavy objects as near the worker as possible.
         - Reducing the size of cartons or pallets being loaded, or allowing the worker to walk around them, or rotate them.
         - Reducing the size of the object being handled.
         - Allowing the object to be kept close to the body.
2. Reduce object weights and forces.
   - Reduce lifting and lowering forces.
      a. Eliminate the need to lift or lower manually by:

CHAPTER 20
ERGONOMICS

- Using lift tables, lift trucks, cranes, hoists, balancers, drum and barrel dumpers, work dispensers, elevating conveyors, and similar mechanical aids.
- Raising the work level.
- Lowering the operator.
- Using gravity dumps and chutes.

b. Reduce the weight of the object by:
- Reducing the size of the object (specify size to suppliers).
- Reducing the capacity of containers.
- Reducing the weight of the container itself.
- Reducing the load in the container (administrative control).
- Reducing the number of objects lifted or lowered at one time (administrative control).

c. Increase the weight of the object so that it must be handled mechanically:
- Use the unit load concept (such as bins or containers, preferably with fold-down sides, rather than smaller totes and boxes).
- Use palletized loads.

d. Reduce the hand distance by:
- Changing the shape of the object.
- Providing grips or handles.
- Providing better access to object.
- Improving layout of work area.

- Reduce pushing and pulling forces.
  a. Eliminate the need to push or pull by:
  - Using powered conveyors.
  - Using powered trucks.
  - Using slides and chutes.

  b. Reduce the required force by:
  - Reducing the weight of the load.
  - Using nonpowered conveyors, air bearings, ball caster tables, monorails, and similar aids.
  - Using four-wheel hand trucks and dollies with large diameter casters and good bearings.
  - Providing good maintenance of floor surfaces, hand trucks, etc.
  - Treating surfaces to reduce friction.
  - Using air cylinder pushers or pullers.

  c. Reduce the distance of push or pull by:
  - Improving layout of work area.
  - Relocating production or storage area.

- Reduce carrying forces.
  a. Eliminate the need to carry by converting to pushing or pulling:
  - Use conveyors, air bearings, ball caster tables, monorails, slides, chutes, and similar aids.
  - Use lift trucks, two-wheel hand trucks, four-wheel hand trucks, dollies, and similar aids.

  b. Reduce the weight of the object by:
  - Reducing the size of the object (specify size to suppliers).
  - Reducing the capacity of containers.
  - Reducing the weight of the container itself.
  - Reducing the load in the container (administrative control).
  - Reducing the number of objects lifted or lowered at one time (administrative control).

  c. Reduce the distance by:
  - Improving layout of work area.
  - Relocating production or storage areas.

WORKPLACE DESIGN

Good workplace or workstation design will help reduce worker fatigue, exertion, and musculoskeletal disorders that can result from awkward postures and static loads. Good ergonomic design can also increase the efficiency and productivity of an operation.

The general ergonomic principles listed below should be considered in the design of a workstation.

1. Keep elbows down—shoulder abduction angle no larger than 30° (the angle between the upper arm and the torso when the arm is raised sideways away from the body).
2. Keep hands down—below shoulder level.
3. Avoid long reaches—not over 16″ (406 mm).
4. Provide elbow supports for long reaches. Padded supports are preferred.
5. Avoid using first 3 in. (76 mm) of work surface.
6. Avoid tilting head forward greater than 15° from torso. Use head supports if greater angle is required.
7. Avoid tilting trunk forward.
8. Avoid sharp edges on work surfaces.
9. Visual scanning should require eye movements only. Head movements will contribute to fatigue.
10. Allow for change in posture—at worker's discretion (for example, sit/stand work station).
11. Avoid muscle overloading. The maximum force that can be exerted drops rapidly with time as shown (see Fig. 20-2):

    For occasional activities, less than 50% maximum voluntary exertion.

    For repeated activities, less than 30% maximum voluntary exertion.

    For continuous application, less than 15% maximum voluntary exertion.

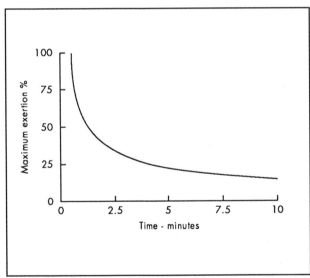

Fig. 20-2 Maximum voluntary muscle exertion versus time.

12. Avoid foot controls for standing workers. If necessary, electric and pneumatic foot switches are preferred. Design for operation with either foot. Avoid elevated or mechanical foot actuating devices if possible, but if used provide a foot rest at the same height as the foot control.
13. Avoid hard floors for standing workers—supply floor mats, and provide a footrest of up to 8 in. (200 mm) high.

CHAPTER 20

CUMULATIVE TRAUMA DISORDERS OF THE HAND AND WRIST

14. Provide suitable seating:
    - Chair base: 5 point base. Seat height: Seated work station—adjustable between 16 and 20.5 in. (406 and 521 mm). Sit/stand work station—adjustable between 27 and 32 in. (686 and 813 mm), including adjustable footrest between 11.5 and 13.5 in. (292 to 343 mm).
    - Seat size: 15 to 17 in. depth (381 to 432 mm), 17.7 in. width (450 mm); "waterfall" front edge.
    - Seat slope: adjustable 0° to 10° backward slope.
    - Back rest size: at least 12.5 in. wide (318 mm).
    - Back rest height (lower edge): adjustable between 3 and 6 in. (76 and 152 mm) above seat.
    - Back rest depth: adjustable between 14 and 17 in. (356 and 432 mm) from front edge of seat.
    - Back rest tilt: adjustable ±15° about vertical axis.
15. Provide suitable work surface height (adjustable, if possible) as shown in Table 20-1.

TABLE 20-1
Suitable Work Surface Heights

Standing Operator	Male	Female
Precision work (elbows supported)	43 to 47 in. (1090 to 1190 mm)	41 to 44 in. (1040 to 1120 mm)
Light assembly work	39 to 43 in. (990 to 1090 mm)	34 to 39 in. (860 to 990 mm)
Heavy work	33 to 40 in. (840 to 1020 mm)	31 to 37 in. (780 to 940 mm)
Sitting Operator		
Fine work (for example, fine assembly)	39 to 41 in. (990 to 1040 mm)	35 to 37 in. (890 to 940 mm)
Precision work (for example, mechanical assembly)	35 to 37 in. (890 to 940 mm)	32 to 34 in. (810 to 860 mm)
Writing or light assembly	29 to 31 in. (740 to 780 mm)	28 to 30 in. (710 to 760 mm)
Coarse or medium work	27 to 28 in. (690 to 710 mm)	26 to 28 in. (660 to 710 mm)

These recommended dimensions also apply to machine feeding heights.

CUMULATIVE TRAUMA DISORDERS OF THE HAND AND WRIST

The hand, in combination with the wrist, is a very complex, multipurpose tool which is required to perform powerful and repetitive motions, as well as intricate acts. Consequently, the hand is subject to a variety of injuries which accompany these tasks, such as abrasions, punctures, and scrapes. Another type of hand and wrist injury is caused by repetitive motion. Certain types of industrial jobs require the hand to perform the same motion over and over again. Some of these motions may damage the hand or wrist if repeated often enough. There are also many off-the-job activities which can cause the same disability. Current evidence has shown that the combination of high force and high repetitiveness of the task significantly increases the risk of Cumulative Trauma Disorders more than either factor alone.

This section will discuss the common types of cumulative trauma disorders, and techniques that can be used to reduce the number and severity of these injuries in the industrial environment. There are a number of practical modifications that can be made to reduce injuries and improve efficiency and production.

WHY IS THERE A GREATER INCIDENCE NOW?

During the past ten years there has been a dramatic increase in the incidence of cumulative trauma disorders reported through the State Worker's Compensation Systems. The U.S. Bureau of Labor Statistics estimates that the incidence of CTDs more than doubled between 1979 and 1986. In some industries, CTDs annually affect as much as 5% of all full-time workers.

It is often difficult to pinpoint all the factors which have led to this dramatic increase. Certainly, a combination of different factors is involved. Some of the more probable reasons include:

- Higher production rates. The pace of many job activities has increased, often as a result of new workplace mechanization and automation.
- Shift towards service and high-technology jobs. Many tasks in these expanding industries tend to be more repetitive, prolonged, and labor intensive.
- Increased awareness by medical practitioners. What in past years may have been considered merely a "sore hand," is now recognized under a specific diagnosis.
- Greater reporting. People suffering from these disorders are more likely to seek medical treatment, due to increased public concern about the problem.
- Expanded WC laws. All State Worker's Compensation Laws have now been expanded to recognize disorders which develop gradually and in a cumulative fashion.
- Knowledge of occupational causes. Expanded research has led to a greater understanding of the work-related factors which may contribute to the development of CTDs.

TYPES OF CUMULATIVE TRAUMA DISORDERS

There are 27 bones in the hand and wrist. Ligaments connect the bones to each other and tendons connect muscles to the bones.

CHAPTER 20

CUMULATIVE TRAUMA DISORDERS OF THE HAND AND WRIST

When a muscle contracts it pulls the tendon, which moves the bone to which it is attached.

Most of the muscles which operate the hand are found in the forearm. Consequently, there are many long tendons running from the muscles in the forearm through the wrist to the bones in the hand. Portions of these tendons are enclosed in sheaths, known as the synovial membranes, which protect and lubricate the tendons and permit them to slide back and forth freely.

Excessive back and forth movement of the tendon can lead to inflammation of the tendon which is known as tendinitis. The inflammation of the tendon sheath is termed tenosynovitis, which is one of the most common repetitive motion injuries to the wrist. Either tendinitis or tenosynovitis may also be caused by a direct blow to the tendon and its sheath. A specific type of tenosynovitis involves the sheath of the thumb tendon and is known as DeQuervain's disease.

Tenosynovitis is the most frequent cause of carpal tunnel syndrome, a leading type of cumulative trauma disorder. Carpal tunnel syndrome is characterized by numbness, tingling, or a burning sensation in the first three fingers. More advanced cases can result in atrophy (wasting) of the thumb muscles and loss of sweat function. The carpal tunnel is a narrow wrist passageway formed by the carpal bones, their ligaments, and the transverse carpal ligament. Through this tunnel pass nine tendons, the median nerve, and blood vessels (see Fig. 20-3). Swollen tendon sheaths in the carpal tunnel, from inflammation or injury, can compress the median nerve and cause abnormal sensations in the fingers. Additionally, anything that compresses the median nerve can cause carpal tunnel syndrome, including tool handles that press into the center of the palm or other changes within the tunnel itself such as bony growths, cysts, fractures, or hemorrhages.

Another type of repetitive task injury is known as trigger finger, wherein the finger becomes locked in the flexed or bent position. Straightening the finger is accomplished only with difficulty and pain, and occasionally with a "popping" sound. This situation is caused by a nodule formed on the tendon, and is believed to result from frequent flexing of the finger against substantial resistance.

TYPICAL EXPOSURES

There are certain industrial operations where repetitive hand and wrist injuries may be noted. However, do not assume that a problem exists in a particular operation until investigation reveals actual injuries or worker complaints. Tasks which have frequently produced injuries include but are not limited to:

- Bench assembly of electrical and electronic equipment such as computers, typewriters, communication equipment, and radio and television transmitting and receiving equipment.

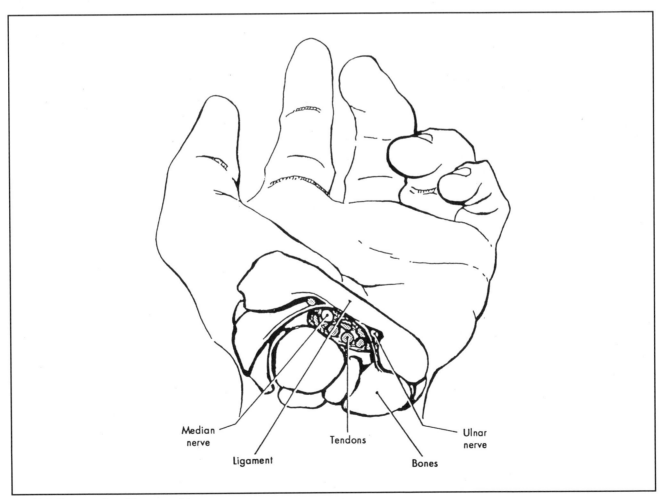

Fig. 20-3 Cutaway view of human wrist.

CHAPTER 20
CUMULATIVE TRAUMA DISORDERS OF THE HAND AND WRIST

- Repetitive manual pushing of machine controls.
- Buffing or grinding hand-held objects.
- Etching or engraving of glassware such as bowls, vases, and goblets.
- Cutting, sewing, or folding goods.
- Wrapping and packaging various small goods.

CONTRIBUTING FACTORS

Recent studies performed at the University of Michigan and Liberty Mutual have identified six risk factors as the most important for developing repetitive motion disorders:

1. High frequency of repetition of the task.
2. High forcefulness of exertion.
3. Awkward postures. These involve any deviation of the wrist away from a straight line, including ulnar (towards the little finger) or radial (towards the thumb) deviation, and palmar flexing (towards the palm) or extension (away from the palm). Other awkward postures are finger pinching, extreme reaching, and forearm rotation.
4. Mechanical pressure. This involves sharp edges of tools or work surfaces jutting into the palms and/or wrists.
5. Vibration. This may result from impact tools, power tools, bench mounted buffers, grinders, etc.
6. Exposure to cold, from working in cold environments or handling cold products.

The most significant single risk factor is high repetition. But the risk of injury is increased dramatically when two or more risk factors are present concurrently, particularly when a job involves both high repetition and high forcefulness. In that case, the odds of sustaining a cumulative trauma injury are over 30 times what they would be in the absence of those risk factors.

TECHNIQUES FOR REDUCING CUMULATIVE TRAUMA INJURIES

The basic technique for reducing hand and wrist injuries is to reduce the exposure of the hand to repetitive motion. This can be accomplished through good work practices, training procedures, medical evaluation, work station design, and tool selection.

Work Practices

Exposure to repetitive tasks can be reduced by:

1. Reducing task frequency. This is the most obvious approach, but also the one that meets the most resistance since it interferes with production.
2. Rotating workers. Workers can be rotated among different types of jobs, so that the hands do not perform the same repetitive motions all day long.
3. Alternating hands. If the job is performed primarily by one hand, it should be designed so that either hand can perform it.
4. Reducing wrist motion. Jobs should be designed so that they can be performed with straight wrists. Hands are stronger and less vulnerable to injury when the wrists are kept straight. Highly repetitive tasks involving wrist action should be mechanized where possible. Some other engineering controls include:
   - Position the work and the worker to eliminate awkward postures:
     a. To reduce wrist flexion, either raise the work or lower the worker.
     b. To reduce wrist extension, either lower the work or raise the worker.
- Make workstations and seating adjustable to allow for changes in posture.
- Angling or tilting the work toward the worker may eliminate wrist deviations.
- Use fixtures and jigs to support workpieces. Make them adjustable, so that the fixture can be angled to reposition the part, instead of bending the wrists.
- Locate tools and parts within easy reach.
- Round surface edges to avoid sharp protrusions.
- Keep parts bins below elbow height to avoid bent wrists.
- Design jobs to reduce hand force and frequency of repetition.
- Reduce forces needed to turn knobs and valves. Those requiring power to turn should be designed for a palmar grip.
- Mechanize or automate the job, when possible, to eliminate hand movements, particularly for high frequency manual tasks.
- Whenever possible, heavy tools should be mounted on automatic retractors or overhead balancers.
- Jobs should be designed to use either hand. One hand should not be used quite a bit more than the other.
- Minimize hand movements by eliminating rehandling, and/or by combining operations at a single location.

Training Procedures

Tenosynovitis has been called the training disease because it is quite prevalent among new workers who are still in training. The basic preventive approach is to let new workers gradually break into the job. Apparently, a gradual break-in period will allow the hand to adjust to the increased stress of repetitive tasks. Do not require new workers (or newly placed workers) to maintain the same pace as experienced workers. It is important that the workers are shown the correct methods for performing their jobs and that management is trained to understand the problems and their controls.

Medical Controls

Medical control issues should be discussed with health professionals, to further identify cumulative trauma disorder loss sources and to recommend solutions. Employees should report any persistent pain in their arms and hands. Early treatment of disorders can lead to quick recovery. Waiting too long may result in a need for surgery or possibly permanent damage. The guidance of a health professional is needed before using splints or medications.

STRESSFUL POSTURES ASSOCIATED WITH CUMULATIVE TRAUMA DISORDERS

Stressful posture is one of the primary occupational risk factors for the development of cumulative trauma disorders (CTD) of the upper extremities. Postures are considered stressful when the body's position requires the use of force or causes stretching or compression of body tissues. Jobs should be analyzed to see whether the shoulder, elbow, wrist, and hand positions shown in Figs. 20-4 to 20-7, respectively, occur frequently.

CHAPTER 20

CUMULATIVE TRAUMA DISORDERS OF THE HAND AND WRIST

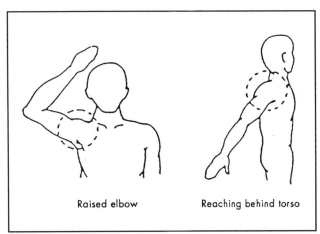

Fig. 20-4 Stressful postures for prolonged or repeated exertions of the shoulder.

STRESSFUL POSTURES FOR PROLONGED OR REPEATED EXERTIONS

Jobs that require the worker to repeatedly reach and work with the arms above shoulder level or with elbows away from the body can influence the development of shoulder problems. Shoulder ailments can also be caused by repeatedly reaching behind the body or throwing materials over the shoulder into a receptacle.

Motions that are particularly stressful to the elbow include inward or outward rotation of the forearm, especially with the wrist bent. Rotation that turns the palm downward is called pronation. Rotation that turns the palm upward is called supination.

Four wrist postures can be stressful, particularly when combined with forceful exertion such as a tight grip on a tool or other object.

There are three principle types of grips: power, hook and pinch (precision) grip. Using a pinch grip repeatedly should be avoided because it takes more strength to hold an object between the fingers and thumb than it does with a power or hook grip. Exerting this force puts more tension on the tendons.

The (preferred) power grip may be harmful if combined with frequent motion and excessive force. The hook grip (such as holding a tray with one's fingertips) is less stressful than the pinch grip but not as good as the power grip.

HAND TOOLS

The use of hand and power tools is often a factor in repetitive tasks which result in carpal tunnel syndrome and other hand and wrist motion disorders. While most tools are satisfactory for general purpose or mechanic's use, they may be entirely inappro-

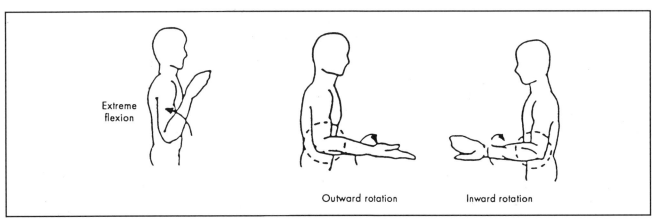

Fig. 20-5 Stressful postures for prolonged or repeated exertions of the elbow.

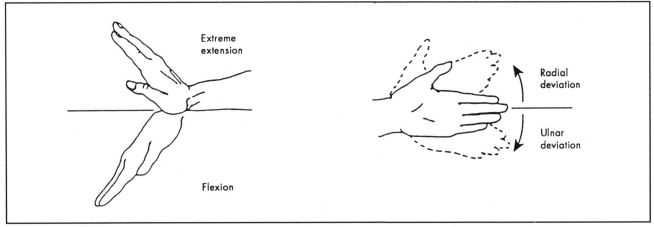

Fig. 20-6 Stressful postures for prolonged or repeated exertions of the wrist.

CHAPTER 20

CUMULATIVE TRAUMA DISORDERS OF THE HAND AND WRIST

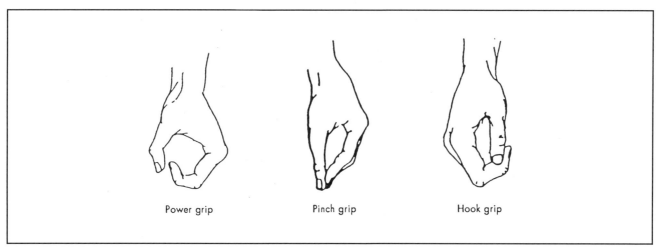

Fig. 20-7 Stressful postures for prolonged or repeated exertions of the hand.

priate for continual or repetitive use in a production situation. Hand tools vary considerably in shape, size, and weight, from tiny tweezers to heavy grinders. The design of these hand tools influences the positions and motions of the hand.

A large variety of tools of the same type or purpose is generally available from which to select a proper tool for a specific repetitive task. However, before selecting a tool, the first consideration should be good ergonomic design of the workstation (see information on this in this chapter.) The design should be one that keeps the wrists straight or in a natural position, and which minimizes wrist bending and flexing as much as practical during the operation. This could involve fixtures to hold, tilt, or rotate the work and will help simplify tool selection.

When selecting a hand or power tool, the following ergonomic principles should be considered:

1. General.
   - Select a tool that will allow for straight wrist operation. It is much better to bend the tool than to bend the wrist. For example, power tools may have pistol grip or in-line handles. In-line handle tools could have straight, offset, or right-angled heads. Hand tools such as pliers can be obtained with bent handles (the so-called ergonomic pliers) or in curved-nose models.
   - Select a tool which can be used with either hand. This helps accommodate left-handed workers, or where the type of task permits, allows the worker to change hands to reduce the repetitive stress on either hand. Some types of hand tools normally designed for right-handed use are also available for left-handed use.
   - Select tools with handles of slip-resistant materials or design.
   - Handles (or tool parts used to hold the tool) should have a diameter between 1.0 in. (2.5 cm) and 1.75 in. (4.4 cm). The recommended diameter for a power grip is 1.6 in. (4.0 cm); hook grip (for example, holding a suitcase) is 0.80 in. (2.0 cm), and for a precision grip (for example, holding a pencil) at least 0.25 in. (0.6 cm).
   - The handles of pistol-grip tools should be at an angle of approximately 78° from the horizontal, to reflect the axis of the grip.
   - Powered tools should be used whenever possible. Repeated manual exertion can cause discomfort and injury.

The following principles apply primarily to either power or hand tools, but could apply to both in a few instances.

1. Hand tools.
   - Avoid all sharp edges and corners on tool handles.
   - Avoid narrow tool handles that concentrate large forces on small areas of the hand.
   - Many tools are obtainable with rubber or plastic foam-covered handles or grips. This cushioning helps to cover sharp edges, increases the diameter of narrow handles, and provides a slip-resistant and in general, more comfortable grip. On some power tools, it is also used to dampen effects of tool vibration.
   - Select different sized tools for workers with different sized hands.
   - Avoid form-fitting handles that fit only one size of hand. Select conventional scissors and shears carefully, as finger loops can cause high pressure on the thumb and fingers; or use power cutters instead.
   - Select tools with a power-grip handle whenever possible. Avoid the pinch or precision grip.
   - Avoid short tool handles that press into the palm of the hand; the handles should be long enough to support the entire set of fingers, 3.75 in. (9.5 cm) or greater, and a hand grip of 4.5 in. (11.5 cm) or greater.
   - Clearance around the handle should be from 1.2 in. (3 cm) to 2 in. (5.1 cm) with an additional 1 in. (2.5 cm) allowed for a gloved hand.
   - Plier-type tools:
     a. Closing space between the plier handles should be 1 in. (2.5 cm) or greater, 2 in. (5.1 cm) or greater for two-handed tools. Alternatively, a guard should be placed around the danger area.
     b. Plier-type handles which must be closed or squeezed during tool operation should have a handle opening (open grip span) less than 4.5 in. (11.5 cm). If high grip forces are required, the handle opening should be in the range of 2.5 in. (6.3 cm) to 3.5 in. (8.9 cm).

     Grip forces should not exceed 28 lbs (12.7 kg) for tasks requiring sustained effort (one minute or less) without discomfort, 22 lbs (10.2 kg) with gloves. For tasks of long duration with frequent, intermittent use, 14 lbs (6.3 kg), 11.1 lbs (5 kg) with gloves.

CHAPTER 20

LAYOUT SAFETY PLANNING

c. Spring action should return the handles to the open position.
3. Power tools.
   - Select tools where the index finger will not be overused for triggering operation. Consider tools with a blade or wand-type trigger where all four fingers together operate the trigger, or with a thumb trigger (the thumb is much stronger).
   - Tools that are used or held intermittently for short periods of time should not exceed 29 lbs (13 kg) with 10 to 12 lbs (4.5 to 5.5 kg) recommended maximum weight. Tools that are used or held continuously should not exceed 4.4 lbs (2 kg). Heavier tools should be counterbalanced.
   - Select balanced tools. The center of gravity of the tool should be aligned with the center of the grasping hand so that the hand will not have to overcome rotational moments or torques of the tools.
   - Select tools designed to minimize vibration, rotational torque, or impact forces on the hand and wrist.

LAYOUT SAFETY PLANNING

The layout of a facility can affect efficiency, productivity, and the ability to accomplish continuous improvement. This section provides practical guidelines for evaluating layouts of areas that range from individual workstations and departments to larger plant areas, with emphasis on smaller work areas.

Since efficient layouts help minimize accidents and injuries, it is important to be able to suggest changes for present and proposed designs. To analyze layouts, it is essential to understand the process and material flows as well as the relationships between activities of different work areas.

When new equipment is purchased, machinery is moved or a new addition/facility is planned, it is important to consider safety and ensure that current and future safety hazards are controlled or eliminated. It is much easier (and less expensive) to design a work area properly than to redesign it after changes have been made.

Efficient layouts:

1. Minimize material handling tasks, thereby causing fewer back injuries.
2. Provide for planned flow patterns and less rehandling.
3. Reduce accidents such as slips/trips/falls.
4. Minimize noise, vapors, etc.
5. Allow for:
   - Maximum use of space.
   - Improved machine accessibility.
   - Higher productivity.
   - Improved efficiency.
   - Planned life safety.
   - Improved quality control.

TYPES OF LAYOUT PROJECTS

There are many times that a layout project is appropriate as well as beneficial. Projects of this nature require imagination and creativity. Identifying present and future safety and manufacturing hazards is essential for providing recommendations that will control or eliminate the exposures. Sometimes individuals think that layout projects are only for large or new facilities, but this is not true. Frequently, the situation involves changing a particular process by moving or installing equipment, tables and/or materials.

For example, a company may decide to install a lift table at the end of a conveyor line so that the employee does not have to bend over to load boxes onto a pallet. Analyzing accident data might indicate that there have been a number of seemingly unrelated injuries that have occurred in this area. After evaluating the task and layout, it may become apparent that the injuries could be related. The employees may have tripped or slipped on items that were sitting on the floor while carrying boxes from the conveyor. They may have bumped against machines or sharp objects. Changing the layout and installing the lift table may help reduce or eliminate more than just back injuries (see Figs. 20-8 and 20-9).

Several types of layout projects may be encountered:

- Production method change/automation.
- Replacing outdated equipment.
- Installation of new equipment.
- Product design change.
- Relocating a department.
- Minimizing crowded conditions.

It is important to become involved in this type of project early, so that an evaluation can be done at the right time. If the project has been in progress for some time, it may be difficult to make changes or submit appropriate recommendations. Normally, it is not necessary to wait until blueprints, diagrams, or specifications have been developed.

WHO IS RESPONSIBLE FOR WHAT?

The size of the company normally determines who is involved in the layout planning. Generally, the layout function is a staff service connected with the manufacturing or production department. In a small facility, there is usually no formal Facility Layout Department. Most of the work will be done by the combined efforts of engineers, general manager, supervisors, and sometimes the company president. Occasionally, an outside architectural firm will be hired to perform the work. In a large facility, there may be a staff of individuals who spend all of their time working on plant layout projects.

Whether the company is large or small, the same information and decisions are necessary. If the finished industrial layout is going to be safe and productive, many individuals have to work together as a team. Even if an architectural firm or outside consultant is hired, it is important to understand that the consultants may not be entirely familiar with their operations and particular safety needs. In this situation, a combined safety planning effort—with the input of the company, outside consultants, and insurance company safety representatives—can be very effective.

PROJECT PLANNING

The layout project should be organized carefully. It is important to do initial fact-finding, identify current and potential

CHAPTER 20

LAYOUT SAFETY PLANNING

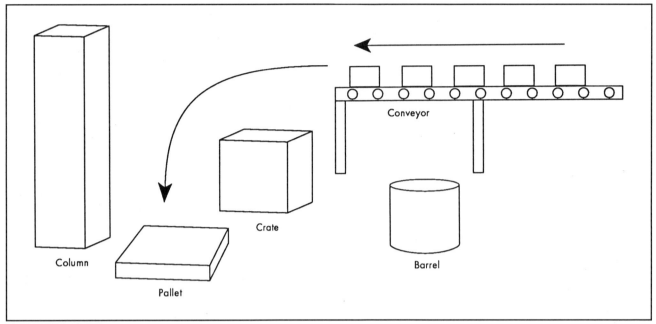

Fig. 20-8 "Before" view of palletizing task.

exposures, set priorities, develop a plan of action, submit recommendations, and follow up. After determining the size of the layout budget, it is important to gather as much pertinent information as possible. This includes products being made, services being provided, flow of materials, equipment, project timetable, safety considerations, etc. This will help define the scope of the project and what aspects should be reviewed.

Sometimes a blueprint of a current or proposed layout is available. At first glance, the thought of reviewing the plans may be overwhelming. Without enough information, someone unfamiliar with the facility should not be required to instantly develop observations and recommendations. Before discussing different types of layout concepts, it is important to understand the basic steps that are used to coordinate layout/handling plans.

MATERIAL FLOW DESIGN

Companies achieve higher productivity when the items moving through them do so smoothly and efficiently. It is important that a flow pattern be planned, rather than allowing it to develop on its own over a period of time. Before a flow pattern can be designed properly, there are a number of factors to be considered, including:

- Materials or products (number of different parts and operations).

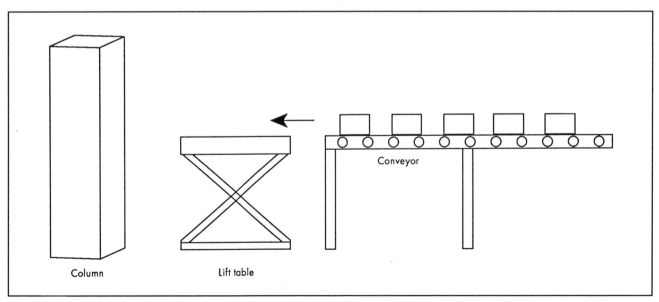

Fig. 20-9 "After" view of palletizing task.

CHAPTER 20

LAYOUT SAFETY PLANNING

- Production frequency, speed, and volume.
- Material handling methods (equipment, items being moved, distances).
- Processes (type, sequence, and requirements).
- Space requirements (size, shape, constraints).
- Other factors (number of employees, flexibility, expansion capability).

Each layout project must be analyzed carefully so that a flow pattern can be designed to accommodate as many of these factors as possible (see Fig. 20-10).

To make the review easier, an overall process flow plan may sometimes be available. Depending upon the project, it is very important to know the flow in the area being studied, as well as how it affects the flow of the entire facility (see Fig. 20-11).

Layouts can be based on product flow, process flow, or group flow.

Product Flow Layout

Production equipment is located in the same sequence as it is used to produce the parts (see Fig. 20-12). This is advantageous when a single item or a small number of products are manufactured in high volume. This type of layout decreases material handling and the overall manufacturing cycle time. Assembly line operations are characterized by product flow layouts.

The problem of expediting the completion of specific parts is lessened. Fewer difficulties in reporting production quantities, clerical work, and interdepartmental problems.

Disadvantages include requiring similar machine tools or operations in several locations. This increases investment in manufacturing equipment and tooling, and decreases flexibility in assigning employees.

Process Flow Layout

Equipment is arranged in groups of similar machine tools (see Fig. 20-13). This is advantageous when a company produces a large number of dissimilar products in low volume, such as a job shop. It decreases the investment in manufacturing equipment and tooling and results in maximum utilization of equipment. There is more flexibility if a machine breaks down or if there is an overload in the production schedule.

Disadvantages include increased material handling, increased manufacturing cycle time, and difficulty in expediting parts that are behind schedule. It may also be more difficult to report production quantities and handle interdepartmental questions.

Group Flow Layout

Equipment is arranged in a group, but it is not always used in the same sequence. One team of workers completes the product.

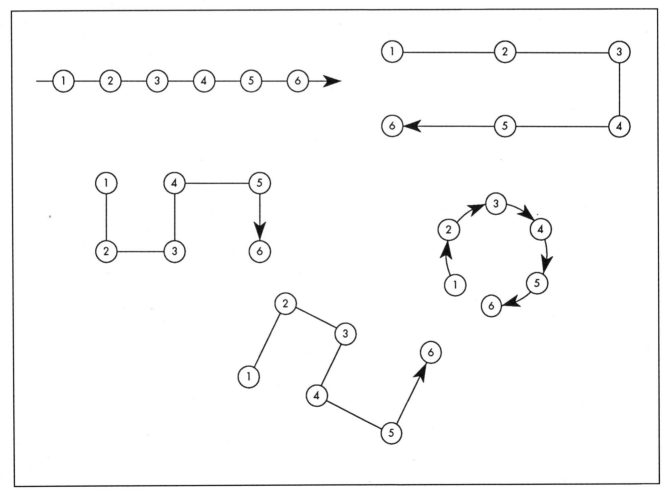

Fig. 20-10 Flow patterns.

CHAPTER 20

LAYOUT SAFETY PLANNING

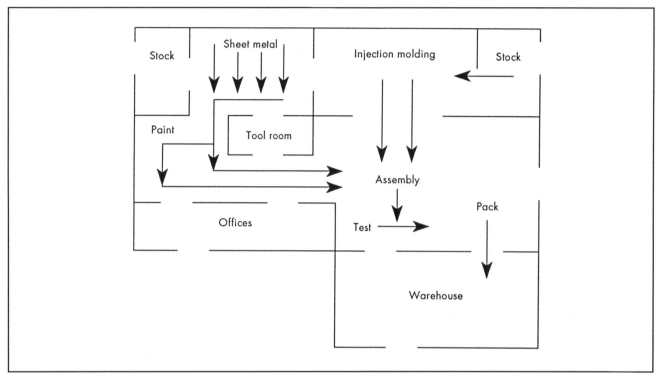

Fig. 20-11 Flow through a facility.

This may promote more of a team approach, better efficiency, and product quality (see Fig. 20-14).

One disadvantage is that parts behind schedule may disrupt the production of the group. Another is that similar machine tools may be needed in each group.

CHOOSING THE FLOW METHOD

It is important that the individuals planning the layout work together to decide which method or combination of methods is the best for the project.

When analyzing the material flow within one workstation or between different areas, it is important to see if there is excess handling or backtracking involved (see Fig. 20-15).

When reviewing a blueprint, there may be a tendency to think of a two-dimensional facility. Layout planning should be done with all the different levels of the building in mind. Figure 20-16 shows that equipment, parts, building structures, etc., must be taken into consideration when designing a process flow layout.

ANALYZING MATERIAL FLOW

A number of techniques are commonly used for evaluating and planning flow processes and layouts. They can be used for the analysis of existing layouts or for planning a new facility. Not all of the techniques discussed will be useful in every type of layout project.

The following techniques will be reviewed:

- Assembly chart.
- Operation process chart.
- From-To chart.

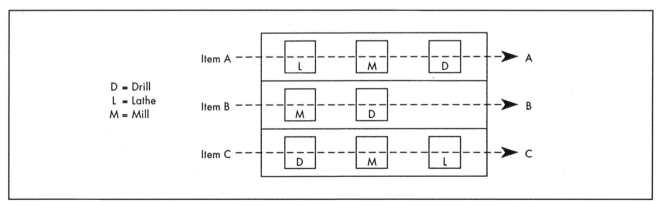

Fig. 20-12 Product flow layout.

CHAPTER 20
LAYOUT SAFETY PLANNING

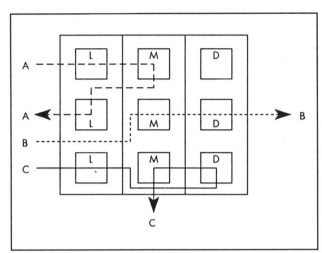

Fig. 20-13 Process flow layout.

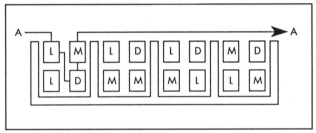

Fig. 20-14 Group flow layout.

- Flow process sheet.
- Activity relationship chart.
- Relationship diagram.

Assembly Chart

The Assembly Chart represents the sequence in which parts flow into the assembly of a product. It shows an overall picture of the assembly process, the order in which parts are used, and an impression of the overall material flow pattern (see Fig. 20-17).

Comparing the assembly chart with the actual layout of equipment and processes can indicate if there is unnecessary rehandling of materials that can be corrected.

An assembly chart is constructed using the following steps:

1. Determine the operation or assembly steps for the product. This information can be obtained from the Parts List or Bill of Materials and by observing the tasks.
2. Draw these steps as circles and connect by vertical lines (Top Circle = First Step, Bottom Circle = Last Step).
3. If any components are assembled at a particular step, they should be drawn as a circle that is placed to the left of the step and connected with a horizontal line.
4. Check to make sure that all steps and components have been entered. The finished chart should show the actual tasks from start (top) to finish (bottom).
5. Study the chart to see if any steps can be combined or eliminated to improve safety and lessen manual material handling.

Operation Process Chart

This illustrates the individual operations that are performed on each part or sub-assembly (see Fig. 20-18). Besides building upon the assembly chart, it shows each operation in the production sequence. It indicates which parts are related to each other and which parts can possibly be manufactured in adjacent areas. This is useful when a product does not have a large number of components. Comparing this chart to the actual layout can show how unnecessary rehandling of the materials can be eliminated.

The construction of an operation process chart includes the following steps:

1. Determine the sequence of operations that are performed on the parts. This can be obtained from the Production Routing or Production Sequence Sheets and by observing the tasks.
2. Draw these steps so that materials are represented by horizontal lines and actual manufacturing operations are shown as circles connected by vertical lines.
3. Subassemblies are handled as in the assembly chart.
4. Check to make sure that all materials and operations have been entered.

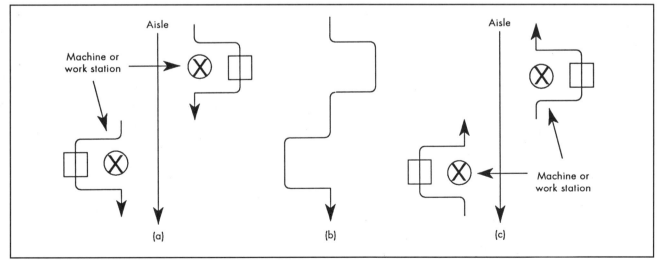

Fig. 20-15 Checking work flow for excess handling or backtracking.

20-17

CHAPTER 20

LAYOUT SAFETY PLANNING

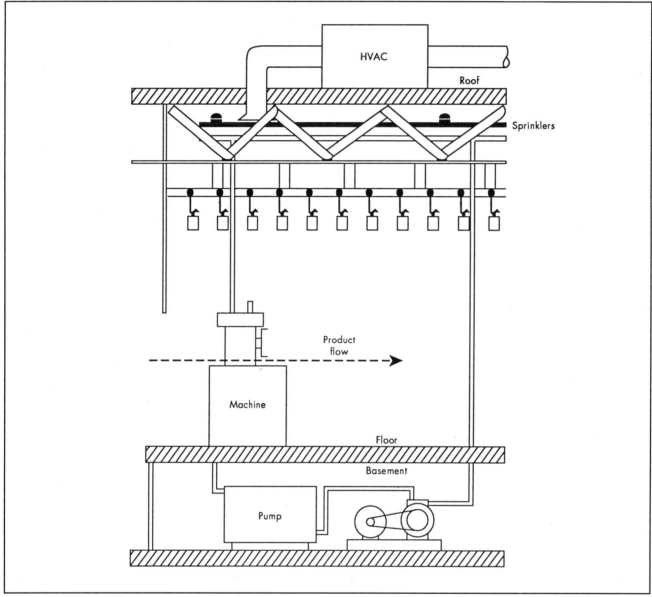

Fig. 20-16 Consideration of equipment, parts, building structures, etc., when designing a process flow layout design.

5. Study the chart to identify steps that could be eliminated or combined to lessen rehandling of materials.

Figure 20-19 is a graphic representation of an operation process chart. It identifies material feeding into the process, and the steps are arranged in chronological order. Reviewing this chart may indicate tasks that could be combined or modified to minimize material handling hazards and injuries.

From-To Chart

This is helpful when a number of items flow through one area. It shows the volume of movement between activities, processes, or departments; dependency of one activity on another; and backtracking of parts. This chart is similar to the mileage charts that are commonly found in road atlases. The numbers usually represent factors such as unit loads, distances, weights, volume, or some other type of information. Backtracking of materials or extra rehandling of items over long distances can be identified and corrected.

A From-To chart is constructed according to the following steps:

1. Determine the activities, tasks, machines, or departments that should be analyzed. Obtain the necessary information (number of loads/transports, distances, etc.) and also by observing the tasks.
2. Enter the activities, tasks, etc., across the top and down the left side, in the same order using a sequence that represents the actual tasks, flow of material, etc.
3. Write numbers in the squares to represent the number of moves from one task to another, such as five trips from Machine #1 to Machine #2.

CHAPTER 20

LAYOUT SAFETY PLANNING

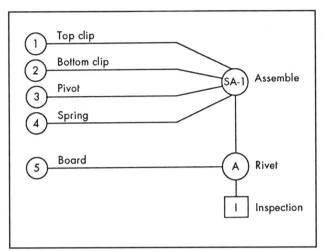

Fig. 20-17 Clipboard Assembly Chart.

4. Total each column and row checking that corresponding columns and rows equal each other and that grand totals agree.
5. Note that entries in the squares below the diagonal line indicate backtracking of parts. Entries directly above the diagonal represent movement between adjacent areas. Empty boxes directly above the diagonal indicate areas being skipped and possible extra handling of parts.

Figure 20-20 shows an example of the present layout of a printed circuit board department that has various machines and workstations. In order to analyze the material flow, a From-To Chart is prepared (see Fig. 20-21).

Reviewing the chart shows that the boards backtrack and there is rehandling as follows: seven trips between Machine C and B, nine trips between F and E, five trips between G and A per shift. Recommendations for combining tasks and/or relocating machines/tables could be submitted. This could reduce unnecessary material handling and potential injuries. This same method can be used to analyze material flow between departments.

Figure 20-22 shows one possible layout that could be used to reduce material handling distances and improve efficiency.

Flow Process Sheet

This sheet is used for tracking the flow of one item at a time. It shows the number of steps that are performed in a given process. It also indicates other information such as if the step is an operation, transport, inspection, delay or storage. It helps identify loss sources and hazards and provides the opportunity for determining suitable control actions.

The steps to make a flow process sheet include:

1. Determine which task will be broken down and obtain the necessary information from the operation process chart and by observing the task.
2. Enter the information onto the sheet and select the appropriate activity symbol for each step. Connect the symbols with a continuous line.
3. Enter the other data such as handling equipment and method, distances, etc.
4. Study the sheet to identify steps that could be combined or eliminated.

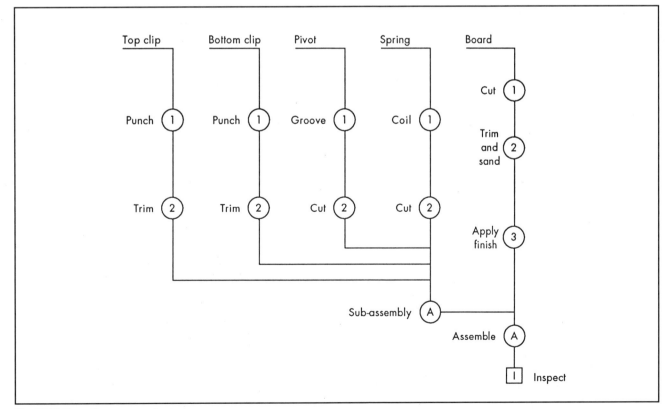

Fig. 20-18 Operation process chart.

20-19

CHAPTER 20

LAYOUT SAFETY PLANNING

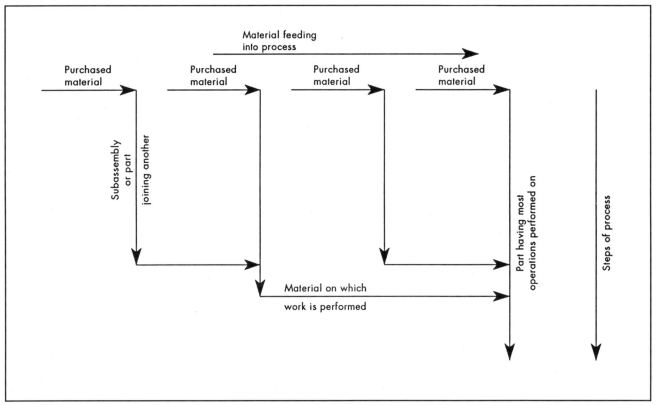

Fig. 20-19 Graphic representation of operation process chart.

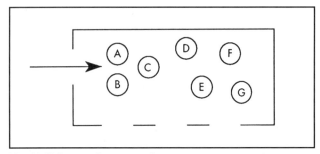

Fig. 20-20 Present department layout.

Figure 20-23 shows an example that analyzes the tasks that take place in one portion of the printed circuit board department that was studied in the From-To chart example. The steps are observed and recorded onto the flow process sheet.

Reviewing the sheet shows a number of unnecessary material handling transports and storage motions. Carts could be recommended to move the boards from one area to another, instead of manually lifting/lowering and carrying the totes. Possibly the inspection task could be combined with another step to minimize material handling and improve efficiency and productivity.

Activity Relationship Chart

This is used to evaluate relationships between departments and activities. It shows whether certain areas should or should not be located next to each other. It also indicates the closeness of relationships and the reasons why. This may provide an opportunity to combine tasks to reduce material handling, or move dependent operations closer together.

An activity relationship chart is made following these steps:

1. Determine the tasks that will be broken down by obtaining the necessary information from the From-To Chart and by observing the tasks.
2. Enter the activities onto the sheet and write in the importance of the relationships and reasons why.
3. Enter the definitions for the closeness values onto the sheet.
4. Study the sheet and see if there is a correlation between the information and the actual or planned layout. There may be opportunities to combine activities or move them closer together to eliminate excess material handling.

Figure 20-24 shows the present layout of equipment in a department. The activities are determined along with the importance of their relationships and entered onto the chart (see Fig. 20-25).

Reviewing the sheet shows the relationships and their importance. Comparing this to the present layout can indicate long travel distances between operations and opportunities for combining activities. This same technique can be used to analyze small work areas, activities within one department, or between different departments.

Relationship Diagram

This is a block diagram of the areas that are included on the activity relationship chart. It visually indicates the strengths of the area relationships and allows the opportunity to note any work areas, departments, or tasks that could be combined or moved closer together. This could help reduce rehandling of materials and manual handling of items over long distances.

20-20

CHAPTER 20
LAYOUT SAFETY PLANNING

Location Superior Electronics Corp.

Items charted Computer Circuit Boards - Boards Moved Per Shift

Date _____ By Tom Adams

From \ To		A	B	C	D	E	F	G									
		1	2	3	4	5	6	7	8	9	10	11	12	13	14	15
A	1		30	5		1											36
B	2			30		1											31
C	3		7		30		1										38
D	4					30		1									31
E	5						30										30
F	6					9		30									39
G	7	5															5
	8																
	9																
	10																
	11																
	12																
	13																
	14																
	15																
Totals		5	37	35	30	41	31	31									210

Comments _____

Fig. 20-21 From-To chart.

A higher number of lines between areas indicates a stronger need to place these tasks closer together. For example, four joining lines indicate a need to have two tasks or areas close together; whereas one line shows a lower priority.

A Relationship Diagram is constructed using the following steps:

1. Determine the tasks that will be diagrammed, and obtain or prepare an Activity Relationship Chart showing the necessary information.

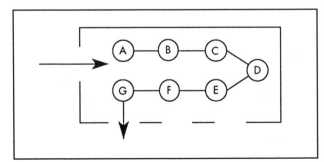

Fig. 20-22 Proposed layout.

20-21

CHAPTER 20

LAYOUT SAFETY PLANNING

Company: Superior Electronics Corp.					Department: Circuit Boards					Date:		
Location: Boston, MA					Material: Boards					By: T. Adams		
Process description:												

Step	Operation	Transport	Inspect	Delay	Storage	Description of step	Handling equipment	Distance	Operators	Handling method	Population % M/F	Significant body motions	Improvement opportunity
1	O	⬆	☐	D	▽	Carry raw material	Totes	50'	1	Manual	40/20		Use hand truck
2	O	⇧	■	D	▽	Inspect	None		1	Manual			
3	O	⬆	☐	D	▽	Carry to machine A	Totes	12'	1	Manual			Use hand truck
4	O	⇧	☐	D	▽	Lower to floor	Totes	24"	1	Manual	25/10	Bend	
5	O	⬆	☐	D	▽	Store on floor			1	Manual			Store on table
6	O	⬆	☐	D	▽	Lift 1 board to machine	None	32"	1	Manual	90/25	Bend	Use part slide
7	●	⇧	☐	D	▽	Cut board	None		1	Manual		Reach	
8	O	⬆	☐	D	▽	Lower board to tote	None	32"	1	Manual		Twist	
9	O	⇧	☐	D	▽	Store on floor	Totes		1	Manual			
10	O	⬆	☐	D	▽	Carry to machine B	Totes	20'	1	Manual			Use conveyor
	O	⇧	☐	D	▽								
	O	⇧	☐	D	▽								
	O	⇧	☐	D	▽								
	O	⇧	☐	D	▽								
	O	⇧	☐	D	▽								
	O	⇧	☐	D	▽								

Fig. 20-23 Material flow evaluation chart.

CHAPTER 20

LAYOUT SAFETY PLANNING

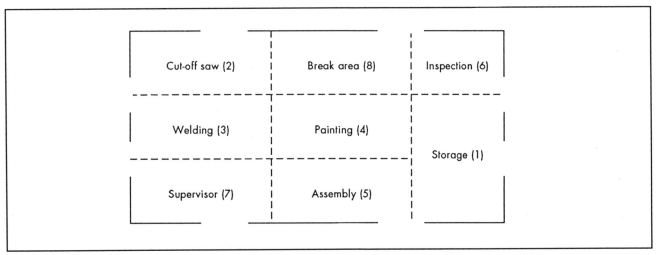

Fig. 20-24 Present layout.

2. Draw the steps shown on the chart as blocks. Draw lines between them to represent the strength of the relationship (4 Lines = Absolutely Necessary, 3 Lines = Very Important, 2 Lines = Important, 1 Line = Ordinary Closeness, No Lines = Unimportant, 1 Wavy Line = Not Desirable).
3. Study the diagram to identify any opportunities to combine activities or move them closer to eliminate excess material handling.

Figure 20-26 is a graphic example showing the information contained on the previous chart.

After drawing the diagram, the next step is to combine this information with space requirements for each area. The blocks are changed in size to reflect the space needs, and are positioned to reflect important relationships, as seen in Fig. 20-27.

If the diagram shows areas far away from each other that have absolute or very important close relationships, there may be excessive material handling required. If the diagram indicates that departments with undesirable closeness are next to each other, the layout should be revised if possible.

Figure 20-28 is based upon the previous modified relationship diagram and illustrates one proposed layout.

SPACE REQUIREMENTS

Sometimes it is not feasible to build a brand new facility when additional space is needed. There are a number of factors that should be considered when determining space requirements. A decision should be made to work within the present boundaries or add additional space. Although it is important to use space wisely and limit distances between machines/operations, it is also important to provide enough room for individuals and materials to move safely and efficiently.

If space is to be added to the facility, it is important to determine where it should be placed. Maintaining a smooth and efficient process flow helps in maintaining a safe work environment. Figure 20-29 shows various flow patterns.

AISLE DESIGN FACTORS

One important factor in determining space requirements is the careful planning of aisle widths and locations. In many cases, the space required for aisles may exceed the room needed for machinery and storage areas. Safety depends upon providing adequate aisles for movement of employees, equipment, materials, products, vehicles, scrap, etc. In case of emergencies, it is necessary to have proper aisles for evacuation and access for firefighters and equipment.

Aisles can be identified as main, cross, personnel, departmental, service, maintenance, or access for elevators/electrical panels/sprinkler valves, etc. Here are some general guidelines for aisle planning:

- Width depends upon product size, handling equipment, machinery size, storage practices, one- or two-way traffic, pedestrians, volume of traffic, etc.
- Large plants may have main aisles 12' to 20' (3.6 m to 6.1 m) wide for anticipated heavy traffic.
- A 10' (3.0 m) aisle can handle two loaded forklift trucks passing each other and allow clearance for a pedestrian.
- For one-way traffic, the aisle should be at least 3' (0.9 m) wider than the widest vehicle.
- For two-way traffic, the aisle should be at least 3' (0.9 m) wider than twice the width of the widest vehicle.
- A minimum of a 6' (1.8 m) turning radius should be adequate for small industrial trucks.
- Personnel and interior aisle widths should be a minimum of 2.5' to 3' (0.8 m to 0.9 m) and larger depending upon the number of occupants, travel distances, building codes, etc.
- Aisle layout should prevent blind corners.

Location depends upon building size/shape, distance to doors, equipment locations, fire walls, column spacing, access to elevators, floor load capacity, use of aisle, etc.

Spacing depends upon equipment size, storage unit size, type of building, number of aisles, type of columns, flexibility, layout, expansion, building codes, etc. Make sure that all applicable Federal, State and local codes and regulations are met.

COMPUTER SOFTWARE

Although computer software layout programs can be very helpful, a large amount of time and information is needed to use them. Even if someone else uses the programs, it is beneficial to understand the programs available, and their benefits and limitations.

CHAPTER 20

LAYOUT SAFETY PLANNING

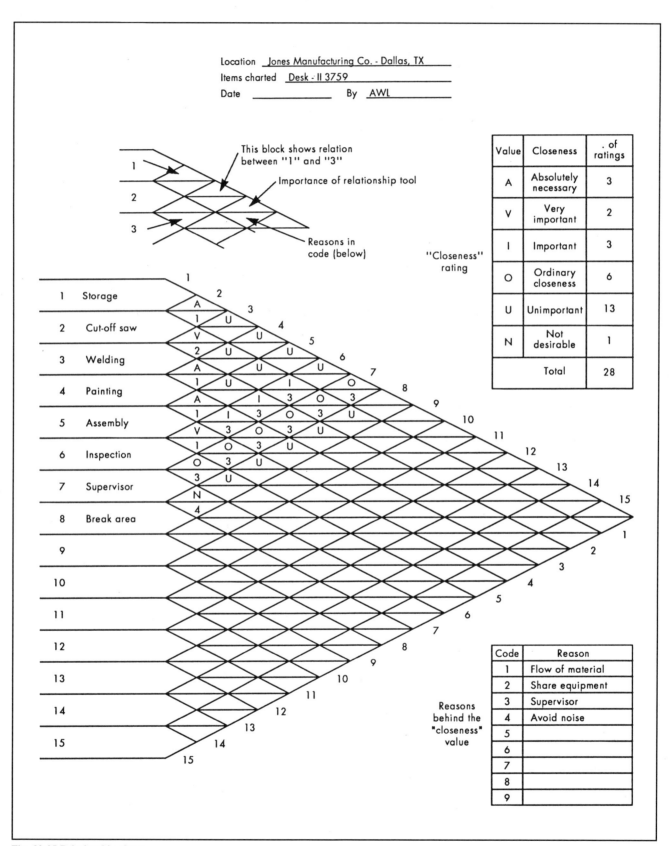

Fig. 20-25 Relationship chart.

CHAPTER 20

PRE-OPERATIONS SAFETY PLANNING CHECKLISTS

One group of programs analyzes flow and solves handling problems by applying simulation, waiting line theory, and other analytical procedures. A second group of programs prepare new or improved layouts by comparing relationships between departments and placing important ones close together. A third group automates drafting by using interactive computer graphics (CAD/CAM).

Additional information on these types of software can be obtained by contacting various computer suppliers and referring to trade publications and magazines.

PRE-OPERATIONS SAFETY PLANNING CHECKLISTS FOR INDUSTRIAL FACILITIES

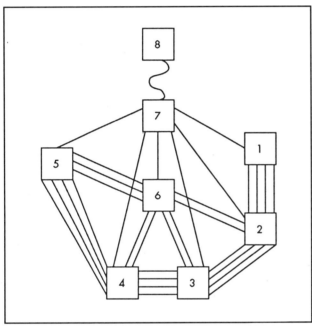

Fig. 20-26 Initial relationship diagram.

The possible scope of safety pre-planning may be illustrated by the following case histories.

1. A metal stamping company planned to purchase 200 new die sets for its power press operations. The new dies were being acquired primarily to achieve better quality tolerance and improved production efficiency. From a safety standpoint, it was suggested that the outside die builder investigate the feasibility of attaching barrier guards directly on the new dies, appropriately designed to protect the point of operation while allowing unhindered feeding of parts. The die designer was also asked to explore the possibility of semiautomatic slide feed arrangements for parts feeding. Both suggestions were implemented, with the result that press injuries decreased by 80% and productivity increased over the following two-year period. It was cost effective and practical to institute these safeguarding features at the time of die design. Trying to modify the dies in this way after they were already in use would have been nearly impossible.
2. An electronics company was planning a new production line for assembling a computer memory unit. Each unit weighed approximately 72 lbs (33 kg). By doing a material flow analysis as part of pre-operations safety planning, it was determined that each unit would have to be manually carried to separate test and inspection stations at the end of

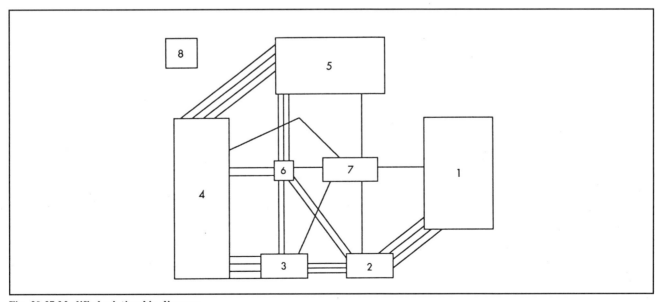

Fig. 20-27 Modified relationship diagram.

CHAPTER 20
PRE-OPERATIONS SAFETY PLANNING CHECKLISTS

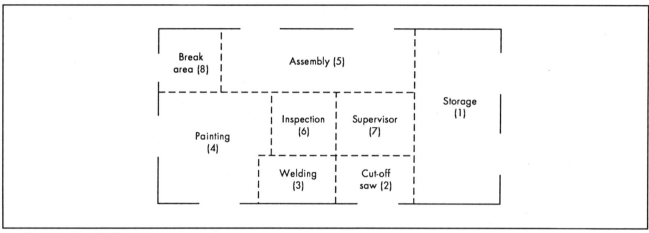

Fig. 20-28 Relationship block diagram.

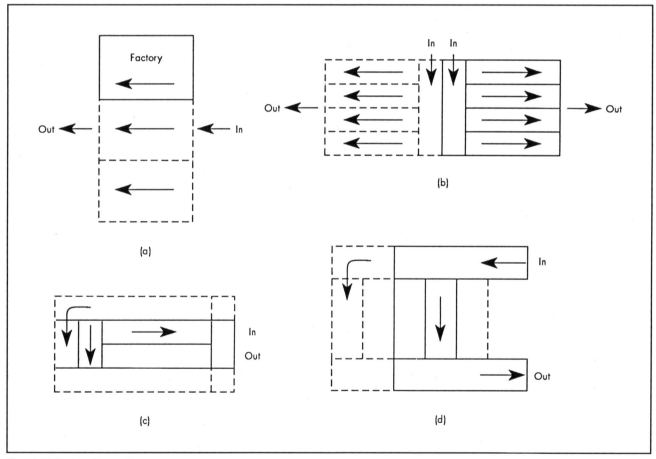

Fig. 20-29 Various flow patterns.

the assembly line. Before the plans were finalized, it was suggested that the test and inspection procedures both be performed at a single workstation, with the units transferred to that station by conveyors and ball transfer tables. This idea was incorporated into the final plans for the assembly line, eliminating a strenuous manual handling task which may have resulted in serious workers' compensation losses. The chance of accidentally dropping and damaging the unit was minimized and the operation was more efficient.

3. A new plant was being built for a manufacturer of bathroom medicine cabinets. A technical review of the layout plans for the new plant revealed that a railroad siding would be located at the front of the plant, directly

CHAPTER 20

PRE-OPERATIONS SAFETY PLANNING CHECKLISTS

adjacent to the main employee walkway and entrance. No provisions had been planned for isolating the track from the walkway, or for protecting employees from railcar switching activities which would occur only a few feet away. This potential hazard was pointed out and the construction plans were modified to include a concrete wall separating the siding from the employee areas.

4. A clothing manufacturer was considering expanding warehouse floorspace by adding a mezzanine to its existing facility. In a pre-planning study, it was discovered that the number of stairways planned for access to the mezzanine did not meet the life safety code requirements. Also, no prior consideration had been given to guardrail protection around the perimeter of the mezzanine to protect against falls. The company's plant engineer greatly appreciated being notified of these deficiencies in time to avoid a very costly retrofit which would have otherwise been necessary.

Two safety pre-planning checklists are shown in Figs. 20-30 and 20-31. They summarize some of the important concerns encountered during pre-operations planning for manufacturing plants and for warehouse/storage facilities, and can be used as an outline for recognizing potential hazards and controls. Not every category contained in the checklists will apply to each project. For this reason, it is important to custom-tailor the pre-operations study to the unique hazards and conditions anticipated for the new project. Use the checklists only as a first step, to indicate areas where more thorough inquiry is needed.

---

1. Layout and material flow.
   - Are adequate space provisions made for people and equipment?
   - Can areas of congestion be eliminated?
   - Is there an established plan for efficient flow of materials?
   - Has a material flow analysis been done?
   - Can any planned material transports be shortened or eliminated?
   - Have provisions been made for temporary and in-process storage?
   - Are aisleways and pedestrian walkways adequate?

2. Workplace design.
   - Have workplaces been designed with employee comfort in mind?
   - Will any excessive reaching, twisting, or bending be required by workers?
   - Is there a good ergonomic match between heights of worker and working surfaces?
   - Can adjustability be provided for heights of worker or work surfaces?
   - Are seated employees provided with comfortable, adjustable chairs with backrests?
   - Are storage compartments provided to avoid cluttered workstations?

3. Manual materials handling.
   - Have excessive manual handling tasks been identified and targeted for control?
   - Can any proposed manual tasks be eliminated or automated?
   - Have weights and forces been kept to a minimum?
   - Are tasks designed so that all movements are between knuckle and shoulder height?
   - Are tasks designed to keep all movements close to the body?
   - Is there any unnecessary rehandling of items which can be eliminated?
   - Are carrying distances minimized as much as possible?
   - Are materials kept off the floor to reduce low bending?
   - Is there too much reliance on two-person handling of heavy loads?
   - Have provisions been made for loading/unloading stock from machinery?
   - Have provisions been made for use of mechanical aids to reduce hand labor?
     - Lift Tables — Handtrucks — Load Positioners
     - Conveyors — Forklift Trucks — Stackers
     - Hoists, Cranes — Carts — Robots

4. Stores and warehousing.
   - Have efficient methods been established for loading and retrieving stock from storage racks?
   - Can stock be kept on pallets or otherwise unitized to allow mechanized handling?
   - Are manual stock-picking locations easily accessible without excessive reaching or bending?
   - Do storage racks conform to ANSI and RMI standards?
   - Are the racks stable, with sufficient load capacity, and protected against vehicle collision or earthquake?
   - Is there adequate protection against stock falling off racks onto persons below?
   - Is there any potential need for automated storage and retrieval systems, or for carousels?
   - Can liquids, powders, etc., be handled in bulk or semi-bulk systems?

5. Shipping and receiving.
   - Have efficient methods been established for loading/unloading trucks and railcars?
   - Is there sufficient space provided near the loading docks for staging of loads?
   - Is each bay provided with (acceptable) dock plates or (preferable) built-in dock levelers?
   - Have provisions been made for chocking of vehicles or restraining them with ICC-bar locks?
   - Has a restricted area for outside drivers been provided?
   - Is there a mechanical system for opening bay doors, to avoid manual strain?
   - Has an efficient traffic pattern been developed for safe movement of incoming/outbound vehicles?

6. In-plant vehicle traffic.
   - Is the layout and width of aisleways sufficient to permit uncongested operation of vehicles?
   - Have special pedestrian walkways been established to prevent conflict with vehicles?
   - Are areas of restricted driver visibility provided with convex mirrors or traffic signals?
   - Are the floor conditions adequate for safe operation of vehicles?
   - Have protective measures been established for employees in vehicles, to prevent falls?
   - Do forklift truck operations conform to the requirements stated in ANSI Standard B56.1?
   - Has a preventive maintenance program been established for vehicles?
   - Is the battery charging area protected against explosion, electrical, and handling exposures?

7. Falls/elevated surfaces.
   - Is adequate drainage provided, both inside and outside the plant?
   - Have measures been taken to control the leakage of water, oil, and other liquids at their source?
   - Will floors have anti-skid surfaces?
   - Can raised, anti-slip grated platforms be provided for especially slippery areas?
   - Can ramps be substituted for short stairways?
   - Are conveyor crossovers or underpasses provided where necessary?
   - Have guardrails and perimeter protection been provided for the edges of platforms, balconies, floor openings, etc.?
   - Are ladders or steps provided where necessary for employee access to heights?

Fig. 20-30 Pre-operations safety planning checklist for manufacturing plants.

CHAPTER 20

PRE-OPERATIONS SAFETY PLANNING CHECKLISTS

- Will all ladders be secured against slipping?
- Have provisions been made for ice and snow removal, especially at employee entrances/exits?

8. Housekeeping.
   - Have procedures been established for regular clean-up of wastes and scrap?
   - Have provisions been made for emergency clean-up of spills or overflow?
   - Have methods been designed to confine dust, oil, and other contaminants at their source?
   - Will there be a regular preventive maintenance program for equipment?
   - Can the use of extension cords, temporary wiring, portable heaters, and fans be eliminated?
   - Has a plant self-inspection program been established?
   - Are all housekeeping responsibilities assigned?
   - Have enough waste containers and bins been provided?

9. Environmental factors.
   - Will the illumination be adequate?
   - Is proper heating and air conditioning provided to achieve worker comfort?
   - Will noise exposures be controlled by engineering and/or administrative means?
   - Have all potential sources of vapors, dusts, fumes, and mists been identified?
   - Have sufficient exhaust ventilation capacities been provided for elimination of airborne contaminants?
   - Are there radiation hazards (or lasers, microwaves, etc.) which require special controls?
   - Can measures be taken to minimize skin contact with chemicals to prevent dermatitis?
   - Have provisions been made for supply and use of any needed personal protective equipment?
   - Will operations generate hazardous wastes? How are they disposed of?

10. Fire prevention.
    - Will there be a well designed automatic sprinkler system?
    - Do plans include an automatic fire detection and alarm system?
    - Have portable extinguishers been provided?
    - Do controls seem adequate for common hazards (electrical power, heating, air conditioning)?
    - Will any special precautions be needed for the use and handling of flammable liquids, gases, or dust?
    - Have protected noncombustible storage areas been provided for flammables?
    - Have plans been established for the isolation of ignition sources, and smoking restrictions?
    - Is there any major catastrophic potential from explosion, high pressure, etc?
    - Do the proposed building plans ensure:
      a. Adequate fire walls and partitions?
      b. Control over the spread of smoke and heat?
      c. Protection for vertical and horizontal openings?
      d. Minimization of concealed space?
      e. Reduction of highly flammable building or interior finish materials?
      f. Building security against arson, vandalism, etc.?

11. Life safety.
    - Do all building design features conform to the life safety code (NFPA 101)?
    - Is there an emergency alarm and evacuation plan?
    - Are at least two exits easily accessible from every location in the building?
    - Have plans been established for emergency lighting and power?
    - Are the floor area and aisle widths sufficient for the proposed occupant load?

12. Electrical safety.
    - Will all electrical installations be performed by a qualified electrician in conformity to the National Electrical Code (NFPA 70)?
    - Will all electrically powered machinery and power tools have protective grounding?
    - Can circuits be equipped with ground fault circuit interrupters?
    - Is electrical bonding provided for areas of expected static buildup (filling/discharging operations, moving belts, etc.)?
    - Have live overhead electrical cables been protected by nonconductive insulation?
    - Will each machine have its own main power disconnect switch that can be locked out?
    - Is all electrical equipment approved by Underwriters Laboratory?
    - Are any special precautions needed for electrical usage in wet locations?

13. Machine safeguarding.
    - Have all machines been designed for "hands-out-of-die" operation?
    - Is there any further potential for automatic feeding or removal of parts?
    - Will the point-of-operation of each machine be provided with a protective guard or device?
    - Can hinged or removable covers be electrically interlocked?
    - Do the planned safety features on every machine fully meet all ANSI and OSHA requirements?
    - Can any required safeguards be supplied directly by the machine manufacturer?
    - Are all machine controls convenient, readily identified, and protected against accidental activation?
    - Have provisions been made for complete machine de-energization during maintenance, repair, adjustment, and setups?
    - Have procedures been established for regular maintenance of all machines and safety equipment?
    - Will personal protective equipment be furnished (eye protection, hairnets, hand tools, etc.)?

14. Acquisition of new equipment.
    - Have all needed safety features been included in the purchase order?
    - Does the equipment fully meet all industry and government safety standards?
    - Is the equipment of appropriate capacity, speed, and design for the intended use?
    - Do operating controls for the new equipment conform to the requirements listed in NFPA 79?
    - Have ergonomic factors been considered in equipment selection and specification?
    - Will complete operating and maintenance instructions for the new equipment be obtained?

15. Administrative controls.
    - Will special employee or supervisory training be required for the new operations?
    - Has a list of applicable safety rules and safe work methods been established?
    - Will any special emergency facilities or first aid equipment be needed?
    - Have plans been made for regular safety inspections of the new operations?
    - Will special work schedules or job rotation plans be needed?

Fig. 20-30 Pre-operations safety planning checklist for manufacturing plants. *(continued)*

CHAPTER 20

PRE-OPERATIONS SAFETY PLANNING CHECKLISTS

1. Layout and site selection.
    - Has adequate floor space been allocated for:
        a. Temporary storage and staging of stock?
        b. Aisleways and pedestrian walkways?
        c. Pallet storage and repair?
        d. Damaged or returned merchandise?
        e. Checking, receipt validation, and marking of stock?
        f. Packaging or repacking of items?
        g. Equipment storage and maintenance?
        h. Accumulating waste and scrap?
        i. Employee areas and sanitation facilities?
    - Can areas of congestion be eliminated?
    - Is there an established plan for efficient flow of materials?
    - Can any planned material transport distances be shortened?
    - Will storage of chemicals and toxic substances pose any threat to neighboring areas?
    - Are there any outside hazards involving road traffic, rail operations, etc.?

2. Rack storage.
    - Do storage racks conform to ANSI and RMI standards (MH 16.1)?
    - Do the racks have sufficient load capacity for the proposed storage?
    - Are the racks protected against collapse, vehicle collisions, and earthquakes?
    - Is there adequate protection against stock falling off onto persons below?
    - Have efficient methods been established for loading and unloading stock from storage racks?
    - Has a plan been developed for periodic inspection of rack condition?
    - Can stock be kept on pallets or otherwise unitized to allow mechanized handling?
    - Is there any potential need for automated storage and retrieval systems, or for carousels?
    - Can liquids, powders, etc. be handled in bulk or semibulk systems?
    - Will hard hats be required around rack storage areas?

3. Order-picking.
    - Are manual order-picking slots arranged so that:
        a. Heavier items are in the most easily accessible positions?
        b. Faster-moving items are in the most easily accessible positions?
        c. "Pick slots" are more easily accessible than "reserve" slots?
        d. There is a "unidirectional" pattern by the order-picker with no backtracking?
    - For manual picking from elevated storage positions, have provisions been made:
        a. To provide ladders or elevating picking vehicles?
        b. To use only approved picking platforms with guardrail protection?
        c. To provide lifelines for protection against falls?
        d. To strictly prohibit riding on pallets or the forks of industrial trucks?
    - Are order-picking positions designed to minimize bending, reaching, twisting, and crouching in awkward positions by the order-picker?
    - Have order-picking locations been designed to minimize the travel distance between the pick slot and the "consolidation" vehicle (cart, pallet, truck, etc.)?
    - Will stock numbers and serial numbers be readable without manually readjusting the stocked items?
    - Can replenishment operations be designed that do not require manual unloading or consolidation of stock?
    - Has a system been developed for identifying orders that can be held for full-pallet shipment, to eliminate split-case picking?

4. Manual materials handling.
    - Have excessive manual handling tasks been identified and targeted for control?
    - Can any proposed manual tasks be eliminated or automated?
    - Have weights and forces been kept to a minimum?
    - Are tasks designed so that all movements are between knuckle and shoulder height? Are tasks designed to keep all movements close to the body?
    - Is there any unnecessary rehandling of items which can be eliminated?
    - Are like items kept together to avoid unnecessary mixing of products, and subsequent sorting?
    - Are carrying distances minimized as much as possible?
    - Are materials kept off the floor to reduce low bending?
    - Is there too much reliance on two-person handling of heavy loads?
    - Have provisions been made for use of mechanical aids to reduce hand labor?
        —Lift Tables —Handtrucks —Load Positioners
        —Conveyors —Forklift Trucks —Stackers
        —Hoists, Cranes —Carts —Robots

5. Conveyor systems.
    - Do all conveyor systems conform to the requirements stated in ANSI standard B20.1?
    - Are protective safeguards provided for all nip points, inrunning rolls, and power transmission components?
    - Are pop-out rolls or gap plates provided for the transfer points (live roller, belt to roll) of all powered conveyors?
    - Are stop buttons or shutdown cables conveniently located at every employee work location along the conveyor?
    - Have provisions been made for positively de-energizing powered conveyors during cleaning, jam-clearing, and maintenance?
    - Are conveyor crossovers or underpasses provided where necessary?
    - Are conveyor heights designed for optimum handling (29''-31'' for straight-arm carries, 40''-44'' for bent-elbow handling)?

6. Powered vehicles.
    - Is the layout and width of aisleways sufficient to permit uncongested operation of vehicles?
    - Have special pedestrian walkways been established to prevent conflict with vehicles?
    - Are areas of restricted driver visibility provided with convex mirrors or traffic signals?
    - Are the floor conditions adequate for safe operation of vehicles?
    - Have protective measures been established for employees in vehicles, to prevent falls?
    - Do forklift truck operations conform to the requirements stated in ANSI Standard B56.1?
    - Has a preventive maintenance program been established for vehicles?
    - Is the battery charging area protected against explosion, electrical, and handling exposures?

7. Shipping and receiving.
    - Have efficient methods been established for loading/unloading trucks and railcars?
    - Is there sufficient space provided near the loading docks for staging of loads?
    - Is each bay provided with (acceptable) dock plates or (preferable) built-in dock levelers?
    - Have provisions been made for chocking of vehicles or restraining them with ICC-bar locks?
    - Has a restricted area for outside drivers been provided?
    - Is there a mechanical system for opening bay doors, to avoid manual strain?
    - Has an efficient traffic pattern been developed for safe movement of incoming/outbound vehicles?

8. Outside storage.
    - Have specific areas been designated for any needed outdoor storage?
    - Will racks or other provisions be made to ensure an orderly outside storage arrangement?
    - Are outside storage areas protected from vehicle traffic?
    - Are flammable materials protected from direct sunlight, and ignition sources?
    - Are sensitive materials protected from wind, rain, freezing?

Fig. 20-31 Pre-operations safety planning checklist for warehouse and distribution centers.

CHAPTER 20

PRE-OPERATIONS SAFETY PLANNING CHECKLISTS

- Are outside storage areas protected from vandalism?
- Do yard conditions permit safe employee movement, free from potholes, ruts, or other major hindrances?

9. Falls/elevated surfaces.
   - Is adequate drainage provided both inside and outside the warehouse?
   - Have measures been taken to control the leakage of water, oil, and other liquids at their source?
   - Will floors have anti-skid surfaces?
   - Can raised, anti-slip grated platforms be provided for especially slippery areas?
   - Have guardrails and perimeter protection been provided for the edges of platforms, balconies, floor openings, etc.?
   - Are stairs provided with handrails, non-slip treads, and sufficient width and clearance?
   - Are ladders or steps provided where necessary for employee access to heights?
   - Will all ladders be secured against slipping?
   - Have provisions been made for ice and snow removal, especially at employee entrances/exits?

10. Housekeeping and maintenance.
    - Have procedures been developed for regular clean-up of waste and refuse?
    - Have provisions been made for emergency mop-up of spills and leaks?
    - Have methods been devised for preventing leakage of oil, water, or grease at its source?
    - Will aisleways be marked off and kept clear?
    - Can the use of extension cords, temporary wiring, portable heaters, and fans be eliminated?
    - Will there be a regular maintenance program for identifying and repairing floor defects?
    - Is there any potential for the use of mechanical cleaning or sweeping equipment?
    - Are storage compartments provided for tools and equipment, to avoid cluttered work areas?
    - Will there be regularly scheduled preventive maintenance for powered equipment?
    - Will the illumination be adequate?
    - Is proper heating and air conditioning provided to achieve worker comfort?
    - Have sufficient ventilation capacities been provided for removal of forklift vapors or other airborne contaminants?
    - Has a warehouse self-inspection program been established?

11. Fire prevention.
    - Will there be a well designed automatic sprinkler system?
    - Is there adequate clearance (at least 18") between stacked material and overhead sprinkler heads? Is rack storage protected?
    - Do plans include an automatic fire detection and alarm system?
    - Have portable extinguishers been provided?
    - Are controls adequate for common hazards (electrical power, heating, air conditioning)?
    - Will any special precautions be needed for the handling of flammable liquids, gases, or dusts?
    - Have protected noncombustible storage areas been provided for flammables?
    - Have plans been established for isolation of ignition sources, and smoking restrictions?
    - Is there any major catastrophic potential from explosion, high pressure, etc?
    - Do the proposed building plans ensure:
      a. Adequate fire walls and partitions?
      b. Control over the spread of smoke and heat?
      c. Protection against internal and external exposures?
      d. Protection for vertical and horizontal openings?
      e. Minimization of concealed space?
      f. Reduction of highly flammable building or interior finish materials?
      g. Building security against arson, vandalism, etc.?

12. Life safety.
    - Do all building design features conform to the life safety code (NFPA 101)?
    - Is there an emergency alarm and evacuation plan?
    - Are at least two exits easily accessible from every location in the building?
    - Have plans been established for emergency lighting and power?
    - Are the floor area and aisle widths sufficient for the proposed occupant load?

13. Administrative controls.
    - Will special employee or supervisory training be required for the new operations?
    - Has a list of applicable safety rules and safe work procedures been established?
    - Will any special emergency facilities or first aid equipment be needed?
    - Have provisions been made for the supply and use of any needed personal protective equipment (hard hats, safety shoes, etc.)?
    - Will medical screening be needed for employees assigned to the new operations?
    - Have plans been made for regular safety inspections of the new operations?
    - Will special work schedules or job rotation plans be needed?

Fig. 20-31 Pre-operations safety planning checklist for warehouse and distribution centers. *(continued)*

MAKING MACHINES SAFER

Machine related injuries produce the highest number of permanent partial disabilities and rank third among all industrial accidents. The majority of these injuries are caused by individuals performing unsafe acts, using incorrect work procedures, or bypassing safety guards and devices. Employees cannot perform a job safely and efficiently if they are assigned to a dangerous, unguarded machine.

Although there are many different types of machinery that may be used in the manufacturing process, there are some common principles that should be followed to improve the safeguarding and use of the equipment. It is important to make sure that:

- The correct machine is used for each job.
- Machine hazards are recognized and eliminated or controlled.
- Adequate guards and safety devices are installed and maintained.
- Supplementary methods are used to improve conditions in and around the equipment.
- Only authorized and trained individuals operate, adjust, and repair equipment.
- Safe work methods are used by the employees and are monitored by the supervisor.

CHAPTER 20

MAKING MACHINES SAFER

- New or modified equipment is inspected and meets applicable safety standards before use.
- Horseplay is not tolerated.

GET THE RIGHT GUARD OR DEVICE

Before installing guards or devices on equipment, it is essential to identify all hazards associated with using the machinery. If insufficient time and thought are given to analyzing the equipment and jobs, serious accident exposures may be overlooked. A Job Safety Analysis or other hazard identification method can be used to systematically determine hazards, controls, and safe work procedures. The hazard analysis should also look at other factors such as improving ergonomics and minimizing occupational health exposures.

It is important to determine what safety controls are feasible and appropriate before implementing them. If the guards and devices are not practical, the employees may remove them to perform their jobs. Some safeguards and retrofit kits may be available from the original machine manufacturer. After the safeguards have been installed, it is important to monitor them for effectiveness to decide if they need further changes.

The use of guards, devices, and other supplementary protective measures may not result in 100% protection. It is better to improve the safety of the machine 40% or 50% than to do nothing until 100% protection can be obtained at one time. Take advantage of every margin of protection available as soon as possible.

MACHINE SAFEGUARDING

The basic principle of machine safeguarding is to prevent a part of the operator's body (and other employees and visitors) from contacting:

- Point of operation hazards (machining area).
- Inrunning nip points (belts, pulleys, etc.).
- Rotating parts (gears, shafts, etc.).
- Shear points.
- Flying chips.
- Other hazards (electrical, hydraulic, pneumatic, etc.).

Several of these are illustrated in Fig. 20-32.

GUARDS AND DEVICES FOR MOVING PARTS

Moving machine parts and mechanical equipment such as the point of operation, gears, belts, pulleys, sprockets, shaft ends, etc., should be guarded by fixed, movable, or interlocking shields, barriers, or enclosures. Guarding can sometimes also be obtained by changing the location of a moving part such as a belt, pulley, or shaft.

In some situations, presence-sensing devices can be used to protect the point of operation. Other safety devices include two-hand controls, pullbacks, restraints, and hand feeding tools. The guards and devices should prevent the operator(s) from having to reach into the point of operation. Figure 20-33 shows a number of these restraints and guards.

GUARDS VERSUS PRODUCTION

Some individuals oppose the use of a guard, based on the mistaken idea that all guards slow down work and interfere with production. It has been proven repeatedly that modern safeguards do not have to interfere with regular work. In many cases, they actually speed up production.

SUPPLEMENTARY PROTECTIVE METHODS

There are cases where for one reason or another it is not practical to completely guard a machine, or where the installation or construction of a suitable guard is delayed. In such situations, develop supplementary methods to improve conditions in and around the machine. Instruct, train, and supervise the workers in safe work procedures so that the safest possible working conditions can be obtained.

For example, hand tools may be used for feeding operations, operator seating may be installed, slippery flooring eliminated, improved lighting provided, or shields installed to stop flying particles and liquids at their source.

COST OF MACHINE ACCIDENTS

Machine accidents can involve serious personal injuries that result in direct and indirect costs. Preventing machine injuries can help eliminate the following costs:

- Direct Costs. Medical treatments, compensation benefits, and insurance premiums.
- Indirect Costs. Loss of production, training of substitute employees, machine repairs, material spoilage, time investigating the accident, etc.

The total of direct and indirect costs can be very substantial, which is in addition to the pain of the injured worker and family members.

MACHINE MAINTENANCE

Proper maintenance helps assure maximum operating time of equipment at a minimum cost with the safest working conditions for operating and maintenance personnel.

Effective maintenance involves:

1. Routine, repetitive maintenance necessary on a day-to-day basis to keep equipment clean, lubricated, and operating.
2. Preventive maintenance which involves inspection and overhaul of equipment on a predetermined schedule before breakdown occurs.

Routine Maintenance

The mechanical failure of a machine, such as a power press, may be the cause of a serious accident. Routine maintenance is primarily to keep equipment and the workplace in the same operating condition without a major change. If conditions are good, maintenance keeps them that way. If conditions are not good, routine maintenance will not improve them. The fact that maintenance is routine in nature does not detract from its importance, either in maintaining production or preventing injuries.

It is an important activity, and a good maintenance program will go a long way in creating an efficient and safe working environment. But a good maintenance program should be planned and organized in the same way as any other important activity.

Regular Systematic Inspections of Machines and Related Equipment

Preventive maintenance inspections should be made periodically at intervals depending on the equipment and the extent of use. They may be made yearly, monthly, weekly, or perhaps daily by personnel of the maintenance crew. Machines and equipment selected for inspection should include those that are subjected to exceptionally hard wear, or are of a kind that, should failure occur, would result in serious trouble either in loss of production or in accidents.

CHAPTER 20
MAKING MACHINES SAFER

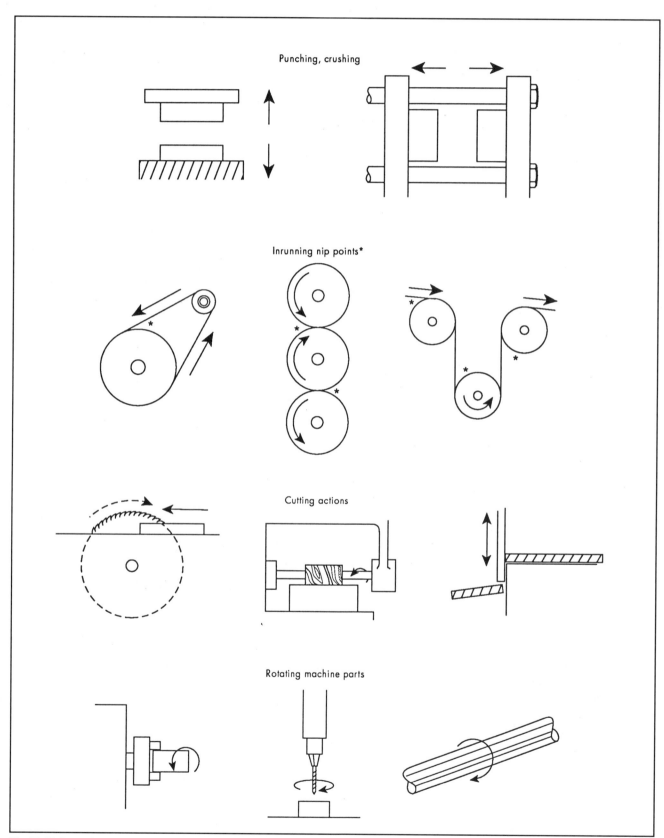

Fig. 20-32 Operations to consider in machine safeguarding.

CHAPTER 20

MAKING MACHINES SAFER

Checklists should be developed and used as a means of determining how inspections should be made. Work with the maintenance department and equipment manufacturer in setting up a complete inspection procedure with written records. These inspections should include a check of guards, devices, other safety controls, and operating conditions.

A Program of Periodic Replacement of Essential or Critical Parts

Successful preventive maintenance is based on an established regular system for replacement of worn or defective parts and equipment disclosed during routine or scheduled inspections. For many machine parts, forecasts of replacements are possible.

Guards and devices require the same inspection and maintenance that all mechanical equipment should receive. The upkeep of the guards and replacement of missing or broken parts are essential for the safe operation of the machine. Guards that are not operating properly should be repaired or replaced promptly. Makeshift arrangements should not be used since they require constant repair and do not provide real protection. Guards should be strong, rigid, and built to withstand wear and tear.

Keeping Records of Inspections, Replacements and Findings

Inspection reports provide permanent records which are an important guide in maintenance. An inventory of all machinery, containing important data and specifications, can be very helpful in determining minimum service requirements and in establishing maintenance schedules.

Repairs and Replacements as Indicated From Inspections, or in Accordance With Replacement Schedules.

It is important that prompt corrective action is taken, following inspections and according to predetermined schedules, to avoid untimely breakdowns or employee injuries.

Follow Through to Evaluate Progress

It is important to determine that inspections are completed as scheduled and reports are completed to show the findings. Periodic adjustments may be required in scheduling work to maintain an even and constant workload.

GOOD HOUSEKEEPING

Good housekeeping is essential to orderliness, efficiency, and accident prevention. Machines should be positioned so that operators do not have to stand or sit in aisles where traffic is heavy. There should be sufficient storage space at the machine for raw stock and finished parts. Arrangements for handling scrap should not interfere with operations or clutter surrounding floor areas. Containers, carts, conveyors, etc., for parts can be used to keep material off the floor and also reduce material handling. Machines and floor surfaces should be kept clean and free of grease and oil.

PERSONAL PROTECTIVE EQUIPMENT

Study each machine operation to determine the proper protective devices for the operator's eyes, face, hands, feet, etc. This equipment should be provided to and worn by the individuals. Supervision should make sure that the equipment is used and maintained properly.

SELLING MACHINE SAFETY TO THE EMPLOYEE

Employees sometimes mistakenly feel that a guard is a reflection on their skill and experience. It is important that they understand that the guard is placed on the machine to give protection to employees when they accidentally put their hands where they should not be. Operators may not like a new machine guard at first because it changes the usual habit of operation.

The employees should be involved in the selection of safeguarding methods to learn from their experience of operating the equipment. They are more willing to make the safeguards work if they are part of the decision making process. The supervisor should demonstrate the guards for the operators and encourage them to become accustomed to them.

SUPERVISION

The most expensive and best designed guards provide no protection if they are not used. Once they have been installed, explain to the employees that the safeguards provide protection and that they should be used in accordance with instructions.

There are guards and devices on some machines which have to be adjusted for different types of operations and users. Other guards must be taken off for repair or adjustment of the machine. Make sure that the safeguards are replaced after the repairs have been completed, are left on during machine use, and are not bypassed.

MACHINE SAFETY TIPS

- Use the correct machine for the job, do not overload it, and operate within designed speeds.
- Eliminate the need for the worker to reach into danger areas by providing automatic or semi-automatic part feeding and part/scrap removal methods whenever feasible. Hand feeding tools should be used when automated methods are not practical.
- Provide machine controls mounted in proper positions, protected from accidental actuation, and properly colored and labeled. Emergency stop buttons should also be provided (more than one, if necessary).
- Provide properly adjusted guards and devices that meet or exceed OSHA and ANSI standards.
- Make sure that there are written lockout procedures, all power sources have been identified and can be locked out, and the procedures are enforced. Procedures must be established for working on machines that have the same power sources or are controlled by other pieces of equipment (automated cell operations).
- Keep the areas around the machines free of trip-and-fall hazards, and provide adequate room for operating and repairing the equipment.
- Establish a good ergonomic match between the worker and the machinery, and provide material handling equipment where necessary.
- Enforce the use of personal protective equipment, and do not allow machine operators to wear long, unrestrained hair, jewelry, or loose clothing around moving machinery.
- Establish written safe work guidelines, train employees regarding these requirements, and update training on a

CHAPTER 20

JOB SAFETY ANALYSIS

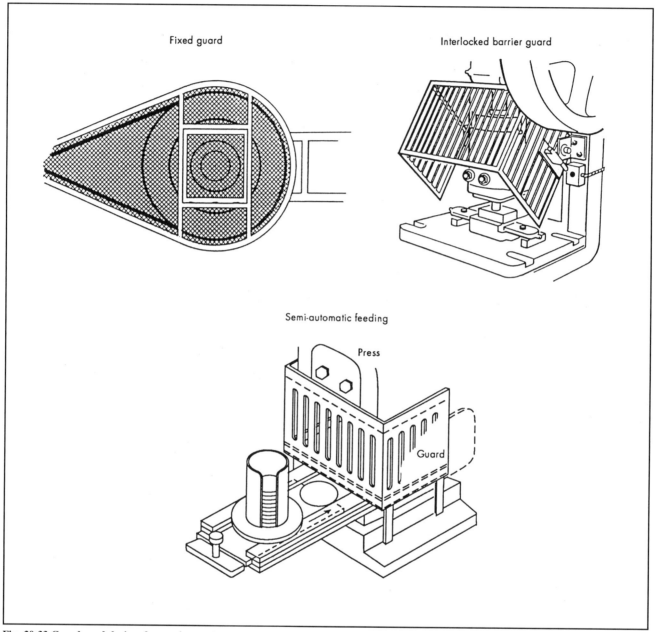

Fig. 20-33 Guards and devices for moving parts.

periodic basis. Individuals who do not feel well or have unusual personal worries should be supervised and cautioned about day-dreaming. Supervise workers to ensure that safeguards are used and not removed or bypassed, problems are reported and corrected, unsafe work habits are corrected, and safety requirements are enforced.

JOB SAFETY ANALYSIS

Job safety analysis (JSA) is a valuable tool that can help reduce accidents and at the same time improve the efficiency of operations. Supervisors wear many hats and have many responsibilities such as maintaining production, training employees, investigating accidents, and making safety surveys. JSAs can provide help in carrying out each of these activities.

CHAPTER 20
JOB SAFETY ANALYSIS

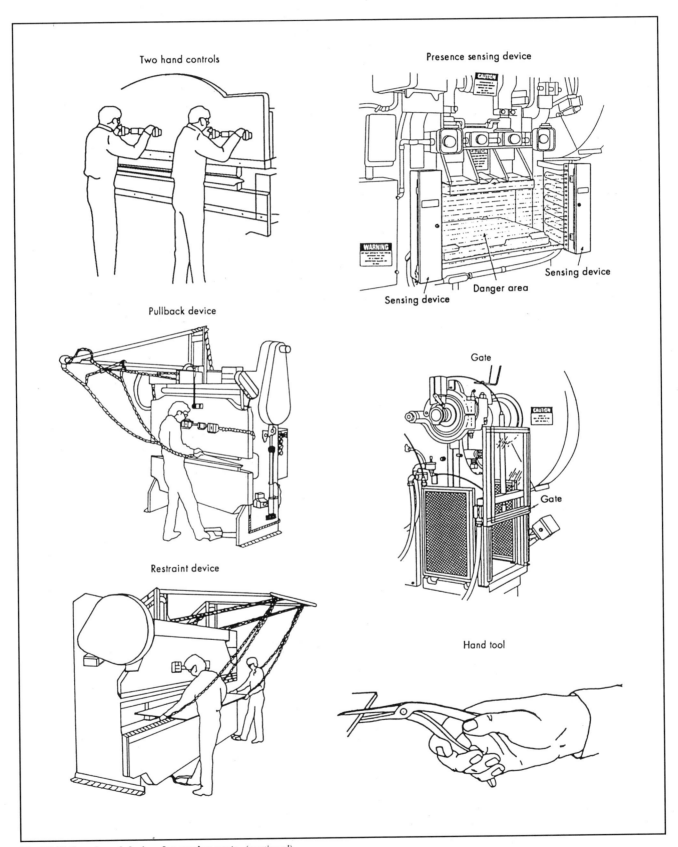

Fig. 20-33 Guards and devices for moving parts. *(continued)*

CHAPTER 20
JOB SAFETY ANALYSIS

JSAs are not difficult to complete but it does take some time and effort to do them properly for each work activity (see Fig. 20-34). However, once they are done, they can start producing immediate pay-backs that over time will far outweigh the effort spent in preparing them.

JSA BENEFITS

There are many benefits that can be gained from JSA that include:

- Identification of hazards and control methods.
- Better understanding and knowledge of each job.
- Improved employee safety attitude with fewer complaints because of their participation in determining safe operations.
- Greater safety knowledge.
- Improved job methods.
- Fewer production interruptions.
- Better quality and productivity.

What is a JSA? A JSA is an observation-based study or analysis of:

1. Selecting and defining jobs to study.
2. Breaking the job down into steps or tasks.
3. Identifying the hazards.
4. Seeking solutions and developing controls for hazards.

DOING THE JSA

Selecting jobs. Identify each job and list in order of priority. Jobs that should be selected include those that have:

- A high number of accidents or serious injuries.
- Near miss incidents.
- High turnover or complaints.
- Considerable material handling or repetitive motion tasks.

Also analyze jobs that are:

- New or changed.
- Seasonal or infrequent, such as repairs or maintenance.

Breaking the jobs down into steps. Observe the job being performed and describe what must be done in the normal order of occurrence. Do not make the steps too small, too detailed, or too broad.

Identifying the hazards. Ask whether the employee can:

- Be struck by or against objects.
- Be caught in moving chains, gears, or machinery points of operation.
- Fall, trip, or slip.
- Be strained or overexerted.
- Be exposed to health hazards.

Also consider if there are other machinery, electrical, mechanical or manual handling, dust, fume, heat, gas, or hand-tool hazards, or other hazards inside or outside the work area.

Developing solutions. Consider possible solutions in order of effectiveness, from the most to the least effective.

1. Can the hazard be eliminated? Is there a better way to do the job that is easier and safer? Are there any new tools, equipment materials, or methods that could be used?
2. Are there any safeguarding methods that could be applied, or layout or other changes made that would control the hazard? This would include guards and other safety devices, ergonomic changes in the work area or job setup, improved lighting or ventilation, or the use of personal protective clothing or equipment.
3. Are additional or continual training and/or procedural changes needed to lessen the hazard?

Other JSA pointers include:

- Involve employees in the JSA. Tell them what is being done. Get their input in group or individual discussion regarding the hazards and possible solutions.
- Review the completed JSA with the employees involved to make certain that no points have been missed.
- Record the information on a JSA form similar to the one shown.
- The JSA can provide documentation for hazard elimination solution recommendations.
- Use JSA as a basis for training all new or transferred employees, and for retraining all other employees at least annually.
- Use the JSA as a guide for accident investigations and update or change it if necessary.
- Review and update each JSA at least once a year and whenever operational changes are made.

LOCKOUT / TAGOUT

A lockout program involves establishing written procedures for isolating machines and equipment from energy sources, and affixing locks to the energy-isolating devices. The purpose of lockout is to prevent employee injuries and fatalities caused by unexpected start-up or release of stored energy during repairs, maintenance, cleaning, and operation of machinery. The major components of an effective lockout program include:

- Shutting down equipment.
- Placing locks on the energy-isolating device(s).
- Releasing stored energy.
- Verifying isolation of equipment.

Lockout will not be complete until the locks are removed in accordance with an established procedure.

Establish a Written Program

Top management must develop and support a written lockout program. In developing the program, review the specific OSHA lockout regulations. The following areas are the most important to include:

- A statement on how the procedure will be used.
- The steps needed to shut down, isolate, block, and secure machines or equipment.
- The procedure for placing, removing, and transferring lockout devices (and who has responsibility for them).
- The specific requirements for testing machines or equipment to verify the effectiveness of locks and other energy control measures.

Locking Out

Survey all machines and equipment that may need locking out. Check to see that there is an energy-isolating device for all power sources leading to each machine and that it, indeed, can be locked out conveniently. An energy-isolating device is a durable and standardized physical device that prevents transmission or release of energy. Examples include:

- Manually operated circuit breakers or switches.
- Disconnect switches.

CHAPTER 20
JOB SAFETY ANALYSIS

Company and location Jones Corporation - Plant 5		Department/Operation Wood products	
Position and/or job titles (or type of specific job/activity) Assembler		Date _____ X Original _____ Revision _____ Recheck	By S. Miller
Steps Sequence of basic job steps	Hazards Potential job hazards and accidents	Controls Recommended safe job procedures and/or operation Improvement/redesign	
1. Get two wood pieces from pallet, carry 6 feet to work bench.	1. Trips and falls, carrying back injury from lifting and bending. Splinters Pinched fingers, hand bruises Dropped wood	1. Keep floor clean Set pallet on platform Gloves Safety shoes	
2. Assemble by hammering in 6 nails.	2. Eye injury Striking fingers Nail punctures	2. Safety glasses Proper nail holding and striking method Gloves	
3. Place finished piece onto adjacent pallet	3. Dropped material Back injury Splinters	3. Safety shoes Set pallet on platform Gloves Improvement recommendations Change layout to set pallet with wood pieces next to workbench Install levelers for both pallets Make jig to hold pieces for nailing Use a nailing machine	

Fig. 20-34 Job safety analysis sheet (JSA).

CHAPTER 20

JOB SAFETY ANALYSIS

- Air, hydraulic, or steam valves.
- Safety blocks.

Energy-isolating devices must be locked out to prevent accidents!

Examples of when lockout procedures are required include machinery being repaired, overhauled, or maintained if that machinery has:

- Electrical, hydraulic, or pneumatic power sources.
- Gas, steam, chemical, and other pipeline systems or pressurized storage tanks.
- Stored-energy sources such as raised machine members, springs, etc.

Lockout should also be used during lubrication, cleaning, and unjamming of equipment if the employee is required to:

- Remove or bypass machine guards or other safety devices.
- Place any part of the body in contact with hazardous machine parts.
- Place any part of the body into a danger zone during a machine cycle.

Make sure that workers can easily reach switches and valves and do not have to climb a ladder to operate them. Label the switches and valves clearly to identify which equipment they control. Symbols and/or bilingual wording can be used to communicate necessary information to the employees.

Issue key-operated padlocks to each person involved. No two locks should have the same pattern. Identify each lock with the owner's name, badge number, and department or trade. Every employee who is working on a machine should place a lock on the switch or valve.

Lockout adapters that hold several locks should be available to each person. Cables or chains may also be needed to secure large valves in the closed position.

Testing

After seeing that no one is in a dangerous position, try the pushbuttons and other normal starting controls to make sure the equipment will not operate. Make sure the operating controls are returned to the neutral or "OFF" position after this test.

Other Sources of Energy

Pressurized air, oil, or other fluids can present hazards unless the pressure is released. Reduce accumulators and air surge tanks to atmospheric pressure, and do not allow pressure to reaccumulate. Secure any loose or freely moveable machine parts. Block rams or slides if they can fall or move. Springs in tension or compression should be considered energy sources and handled accordingly.

Lock Removal

The only employee who should remove the lock is the one who installed it. Authorized personnel should visually inspect the area before power is restored to ensure that everyone is clear of the equipment and the work has been completed properly.

Follow these guidelines during lock removal:

- Do not rely on pulling fuses as a substitute for placing a lock on the energy-isolating device. A pulled fuse does not guarantee that the circuit is dead, nor does it prevent someone from replacing the fuse.
- Do not rely on locking out operating controls instead of power sources.
- Do not assume the job is too small to require lockout.
- If a locked-out job is not completed by the end of a shift, workers going off duty should not remove their locks until those on the next shift have attached theirs.
- Establish and enforce written procedures for safely working on equipment.

These guidelines should apply to both major and minor repairs, even if lockout is not required.

Tags

In the opinion of many safety professionals, using tags without locks is insufficient and could result in serious injuries or fatalities. Although some companies have been able to implement an effective tagout system, there can be problems with using tags alone as a preventive measure against the unexpected start-up of equipment:

- Tags may fall off.
- Someone could remove the wrong tag.
- Workers may forget to use the tag.
- People might ignore the tags.

Tags, like the one shown in Fig. 20-35, can be used along with locks, however, as a supplemental informative aid. Danger tags,

Fig. 20-35 Tags are a good way to supplement lockout, but should not be the only safety measure.

CHAPTER 20

PRODUCTS LOSS CONTROL

containing an explanation of what equipment is locked out, by whom, and for what purpose, can be very helpful.

Training

Provide all employees with initial and ongoing training so that they understand the program's objectives and procedures, and their own responsibilities. Be sure to include new and transferred employees. Retrain workers when there is a change in machines, equipment, processes, or energy control procedures. Document all training in writing to indicate who received the training, who conducted the training, and the date.

On-site employers and outside contractors working together must inform each other of their existing lockout programs. Individuals working for each group must know the requirements of both programs.

Follow-up

Be sure to enforce the written lockout procedures consistently. Audit the program frequently to ensure that all safety rules are followed. Maintain written records of the machines or equipment that were locked out, persons performing the audit, employees who were interviewed, audit findings, recommendations, and the date.

The benefits of establishing an effective lockout program can include fewer worker injuries and fatalities, reduced equipment damage, and improved efficiency.

PRODUCTS LOSS CONTROL

PRODUCT SAFETY PROGRAM

Product safety plays an important part in a continuous improvement program. If a company adopts certain management and operating principles, it can take a long step toward minimizing and even eliminating products liability losses. To do so requires an organized loss control plan that has three significant components:

- A corporate policy on product safety.
- A designated organization with clearly fixed responsibilities.
- Loss control measures.

A Corporate Policy on Product Safety

This is essential to demonstrate top management leadership. It explains objectives such as designing, manufacturing, and selling products that are reliable and meet applicable regulations and standards. It should indicate who in the organization is responsible for product safety, and how the program will be implemented and monitored.

A Designated Organization with Clearly Fixed Responsibilities

In the complex field of products liability, every department plays a vital role, and fixing responsibility is the only way to achieve the desired end result. The size of the company will determine if one individual will be appointed as the Product Coordinator or if a Product Control Committee will be formed. Smaller companies may have one person responsible for product safety along with other duties, and a formal committee may not be needed.

Regardless of what organizational structure is selected, the company must:

- Initiate procedures that support company policy and objectives.
- Develop product safety training programs.
- Learn and communicate changes in codes, regulations, and standards.
- Coordinate product safety hazard analyses.
- Participate in organizations developing improved product standards.
- Communicate with insurance companies.
- Coordinate accident and product failure investigations.
- Maintain loss data including accident reports, warranty claims, etc.
- Alert management to potential product recalls.
- Audit and evaluate the effectiveness of the products program.
- Report progress to management.
- Recommend corrective action.

If a product safety committee is appointed, some of the departments that should be represented are:

- Engineering.
- Quality Control.
- Manufacturing.
- Legal.
- Safety.
- Service.
- Marketing/Advertising.
- Purchasing.

Committee duties may include:

- Product design.
- Quality control.
- Product performance.
- Patent applications.
- Warranty claims.
- Printed matter.
- Accident reports.
- Product modification.
- Program activities.
- Labeling and warnings.

Loss Control Measures

There are four important areas that include:

Identify the exposures. Take a good look at the products to identify possible ways that each one could cause or contribute to accidents or injuries. This evaluation should consider the entire life cycle of the products from design through disposal. It should also include the environment in which the products will be used and, foreseeably, misused. Consider intended product users and also unintended users who may be exposed to potential hazards.

CHAPTER 20

PRODUCTS LOSS CONTROL

Guard against the accident. Understand and apply safety principles concerning product design and manufacturing. Once the hazards are determined, adequate protective devices should be installed. It is recommended that the complete physical protection package (guards, devices, etc.) be included in the product price rather than being listed as optional accessories. The purchase of quality components and establishment of high quality manufacturing standards are essential to meeting performance requirements. As technological safety advances are made, retrofit packages that upgrade older product models should be offered to consumers.

It is also important to identify critical parts of the product and indicate rated capacities clearly, prominently, and permanently. Professionalism in all product sales, engineering, and field services becomes increasingly important. Sometimes, even the most thoroughly designed controls can fail, and a product with serious hazards can be distributed. This situation may require a recall, and the program should include procedures for identifying and tracing specific parts or products. Test the recall program without actually recalling a specific product to check its effectiveness.

When dealing in a retail or wholesale situation, develop a system to assure that recalled products taken off the sales floor are not inadvertently put back out for sale by uninformed sales staff or merchandisers. Inform the lead salespeople and department managers to ask customers if the products being returned for exchange or refund have been involved in accidents or near accidents.

Warn against the hazard. The controls in this area are based on clear, accurate, and complete printed matter and proper packaging and labeling. Advertising and public relations releases should reflect sound accident prevention measures. "Warning" and "Caution" labels and pictorial symbols will help (along with a complete instruction manual) with the products. When packaging and shipping hazardous products, it is essential that the carrier, warehouser, distributor, and consumer know what they are handling and how to use it in a safe manner.

Defend against claims. Since accident prevention may not prevent all accidents and claims, the key to successful defense is advance planning. Work closely on this with the legal counsel and insurance company. Some important areas include:

- Establishing a policy concerning record preservation.
- Preparing a policy of how to handle customer complaints and claims.
- Prompt reporting and thorough investigation of accidents and complaints.
- Establishing a claims philosophy with the insurance company in advance.
- Legal review of product literature, contracts, disclaimers, warranties, etc.

Establishing an effective organization, identifying exposures, guarding against accidents, and warning against hazards are very important steps in preventing and solving products liability problems.

PRODUCT SAFETY COMMUNICATIONS

Written product communications include warning labels, instruction and maintenance manuals, packaging, and advertising. Think of them as part of the product, because of their importance to the manufacturer, distributor, retailer, and user.

Warning labels are not a substitute for safe product design. However, they must be clear, correct, and complete. If the warnings and instructions do not meet these requirements, the product may be considered defective from a legal viewpoint.

Lawsuits involving products quite often include an allegation that the warnings and instructions were insufficient or improper. It is important to understand when and how to warn of hazards. The manufacturer should warn or instruct the user when:

- The product has a significant risk of potential injury or property damage.
- The manufacturer is aware of, or should be aware of, the risk of foreseeable injury or property damage.
- The intended or unintended users are not aware of the hazards and would not normally guard against them.

Communication Requirements

The manufacturer should review all product literature such as warning labels, manuals, packaging, and advertising to make sure they meet all appropriate regulations and standards. This can be accomplished by involving the Product Safety Coordinator (or person responsible for product safety in smaller companies), Engineering, legal counsel, and other departments. Warnings must:

- Identify all hazards that are not obvious to the user.
- Explain what the user should or should not do to avoid injury or property damage, when these methods are not obvious to the user.
- Explain what injuries or property damage may occur if the warnings are not followed.
- Be permanently mounted on the product, in a visible location and designed to last for the life of the product.
- Also be included in instruction manuals and packaging.

Signal Word Selection

Use correct signal words and messages to alert the product user to reasonable and foreseeable hazards. Specific standards discuss the use, size, and color of warnings used on products and in literature; the signal word definitions are:

- DANGER—Indicates an imminently hazardous situation which, if not avoided, will result in death or serious injury.
- WARNING—Indicates a potentially hazardous situation which, if not avoided, could result in death or serious injury.
- CAUTION—Indicates a potentially hazardous situation which, if not avoided, may result in minor or moderate injury. It may also be used to alert against unsafe practices.

Effectiveness

Labels, instruction manuals, etc., need to be effective in communicating the information to the user. Consider who is expected to read the warnings and instructions. The age, language, literacy, technical knowledge level, possible disabilities, etc., of the user should be recognized. International symbols, large print, braille and multi-lingual information can also be used to communicate the messages.

Remember: safe product design is preferable to reliance on warning messages.

Product Safety Signs and Labels

Here are some general ideas and concepts. Refer to ANSI Z535.4 for specific guidelines concerning design, color, size, and wording.

CHAPTER 20

PRODUCTS LOSS CONTROL

Regulations and Guidelines

Writing effective warning labels, instructions, and manuals requires a joint effort between the technical writers, engineers, product designers, and legal counsel who are familiar with the products. The format and content of warning labels are discussed in various regulations and standards. Some of the best ones are:

- ANSI Z535.1 Safety Color Code.
- ANSI Z535.2 Environmental and Facility Safety Signs.
- ANSI Z535.3 Criteria for Safety Symbols.
- ANSI Z535.4 Product Safety Signs and Labels.
- ANSI Z129.1 Hazardous Industrial Chemicals. Precautionary Labeling.

Other important sources of standards include, but are not limited to:

- American National Standards Institute (ANSI), New York, NY.
- American Society of Mechanical Engineers (ASME), New York, NY.
- American Society for Quality Control (ASQC), Milwaukee, WI.
- American Society for Testing and Materials (ASTM), Philadelphia, PA.
- Consumer Product Safety Commission Regulations (CPSC), U.S. Government—Consumer Product Safety Commission, Washington, DC.
- International Organization for Standardization (ISO). Standards available through ANSI.
- U.S. Government Military Specifications (MIL SPECS), Standardization Document Order Desk, Philadelphia, PA.
- National Electrical Manufacturers Association (NEMA), Washington, DC.
- National Fire Protection Association (NFPA), Quincy, MA.
- Society of Automotive Engineers (SAE), Warrendale, PA.
- Underwriters Laboratories (UL), Inc., Northbrook, IL.

PRODUCT RECALLS

A recall program is a systematic procedure for identifying and tracing specific parts or products. It should be developed before a crisis/need requires that product(s) be returned for repair, replacement, or destruction. There are many advantages to having a recall program, including:

- Potentially hazardous products are not sold or used; this helps to prevent personal injuries, property damages, etc.
- Local, state, and federal rules and regulations are met.
- Unnecessary expenses and confusion are minimized.
- The recall process is expedited, reducing the opportunities for injuries and claims.
- Efficient and effective systems help maintain good customer and public relations.

When is a Recall Necessary?

It is good to develop a recall program as soon as possible, before a problem occurs. However, if any of the following situations occur, a recall should be considered:

- A product safety analysis indicates a potential hazard that could cause an accident or injury.
- One or more accidents reveal a hazardous situation that should be corrected.
- Users of the product report unsafe conditions.
- A review of warranty claims, repairs, and service needs indicate potentially hazardous characteristics of specific products. Product liability problems do not cease with the end of the warranty.
- The product does not meet specific local, state, or federal regulations.

Remember to include these elements when putting together a recall program:

1. Prepare a written company policy statement that includes recall procedures and assigns the responsibility for authorizing one. This information should first be reviewed by legal counsel and then provided to appropriate company personnel, distributors, and dealers (if any).
2. Depending on the size of the company, a formal product safety committee may be established. This committee should formulate policies, audit product safety performance, and provide recommendations to top management. Smaller companies may have one individual who is responsible for product safety; larger organizations may split these functions among representatives from different departments that could include:
   - Engineering.
   - Quality Control.
   - Manufacturing.
   - Legal.
   - Safety.
   - Service.
   - Marketing/Advertising.
   - Purchasing.
3. Label components, parts, and completed products to identify the company name, model and serial numbers, date of manufacture, etc.
4. Establish a recordkeeping system that includes:
   - The quantity of each product—by model and serial number—that has been manufactured on specific dates.
   - The materials used and the identification of suppliers and contractors who supply parts, materials, etc. (Batches, processes, and product changes should also be recorded.)
   - The quantity and location of each product held by distributors or consumers. (Although it is difficult to keep track of customers and products that move from place to place, every attempt should be made to keep accurate records.) Traceability methods include user registration cards, repair records, parts sales, and distributor records.
   - Incident reports, accident investigations, and warranty claims that describe how specific products malfunctioned or how the users may not have operated them according to instructions. These reports should also indicate what action has been taken to correct the situations and prevent them from recurring in the future.
5. Develop a method to break down any product into its raw materials and components. It is important to be able to identify which products contain specific raw materials or components. If a problem develops with a specific item, it will be easier to determine if other products are involved.
6. Establish procedures for stopping production, shipment, and sales of recalled products.
7. In a retail situation, develop a system to assure that recalled products taken off the sales floor are not inadvertently put back out for sale by uninformed sales staff or merchandisers. Teach lead salespeople and department managers to ask customers if any products being returned

CHAPTER 20

PRODUCTS LOSS CONTROL

for exchange or refund have been involved in accidents or near accidents.

8. Create a plan for notifying customers, distributors, dealers, and company personnel in writing of a product recall. (This may also include radio, television, newspaper, and magazine advertisements.) Insurance carriers and governmental agencies should be notified when applicable or legally required.

Test the program without actually recalling a specific product to check its effectiveness. Budget expected costs of implementing a recall to minimize financial problems if an actual problem occurs. No one wants to have a recall, but preparing for unexpected future product hazards is an important part of effective product safety management.

PRODUCT SAFETY PROGRAM AUDITS

A program audit is a systematic approach for evaluating the continuing effectiveness of any Product Safety Program. Some goals of conducting these audits include determining if the:

- Company's Product Safety Program is meeting the organization's needs and goals.
- Product accident exposures have been defined and controlled.
- Manufacturing and quality control procedures are adequate.
- Instruction manuals and warnings have been properly prepared.
- Product safety program adequately addresses specific areas such as Recalls, Record Retention, Product Design, Regulations, etc.

Types and Scope of Audits

Product safety audits can vary according to their intended goals, how often they are done, and who completes them. Some companies conduct them on a regular schedule while others perform them on a less frequent, as-needed basis. Certain organizations include them as a part of their normal daily operations.

Depending upon the objectives of the audits, they may focus on one specific operation or department. Other situations may suggest performing an audit that evaluates a wide range of company operations. It is important that the purpose and scope of the audit be clearly determined before it is designed and conducted.

The magnitude and type of audit will help determine how many individuals and departments will take part in the evaluation. The auditors should possess an understanding of the various technical areas that will be studied. Some of the people who could function as auditors include:

- The product safety coordinator.
- Product safety committee member(s).
- Employee teams.
- Outside consultants.

Preparing for the Audit

It is important that the auditors are well prepared to conduct the evaluation. This will help ensure that the audit is successful and that they will receive cooperation from the other individuals involved in the process. Some items that they should be familiar with before conducting the audit include:

- The goals of the product safety program audit.
- The company product safety program.
- Technical information concerning the products being evaluated.
- Actual and potential sources of accidents and injuries.
- Federal, state, and local regulations.

Use of a checklist reduces the possibility of oversights and ensures evaluation of critical items. Each portion of the checklist should be designed and tailored to properly assess the specific operation or department. The checklist is completed through personal interviews and by reviewing records, literature, manuals, policies, procedures, etc.

Conducting the Audit

Auditors should meet with the Operation/Department Managers to review, discuss, and clarify the responses, along with ideas concerning potential corrective actions. A formal report should be prepared and it should include an executive summary and the audit's findings, conclusions, and recommendations with target dates. Items should be prioritized indicating which are the most critical and if some require additional study.

Depending upon the goals and scope of the audit, the existence and effectiveness of a number of different items can be evaluated. Some of these include, but are not limited to:

1. Management.
   - Corporate Product Safety Guidelines exist in writing.
   - A Product Safety Coordinator has been appointed and has defined duties.
   - Management and employees have been trained concerning their product safety responsibilities.
   - A Product Safety Committee exists and has clear guidelines.
   - There is an established Record Retention Policy.
   - A recall program has been developed and tested.
   - Effective lines of communication exist between departments.
   - Accidents and injuries are promptly investigated and reported to the insurance company.
2. Research and Development / Engineering.
   - Product hazard analyses identify and control potential accident exposures.
   - Products conform to applicable safety codes, regulations, and industry standards.
   - New and revised regulations are obtained and implemented.
   - Products are designed to meet ergonomic and human factor principles.
   - Intended product uses and possible misuses are identified and evaluated.
   - Adequate, reliable safety guards and devices have been provided.
   - Quality Control records are used to identify and correct product safety problems.
   - Necessary safety warnings meet applicable codes and standards.
   - Instruction, operating, and maintenance manuals exist and are technically correct.
3. Manufacturing.
   - Products are produced using approved methods and equipment.
   - Parts and materials are transported properly to prevent damage.
   - Quality control records are reviewed to indicate areas that need improvement.

CHAPTER 20

OTHER SAFETY CONCEPTS

- Products have markings to identify the company name, location, manufacturing date, etc.
4. Marketing Department.
   - Advertising and sales literature correctly describe proper use and limitations of the products.
   - Literature and materials are technically correct and are reviewed by legal counsel.
   - Photographs and diagrams illustrate the safe use of the product.
   - Sales individuals are trained concerning the importance of product safety.
   - Guards and devices are included in the base price of the product.
   - Records are kept concerning the names and addresses of product purchasers.
5. Purchasing.
   - Legal counsel reviews wording of purchase orders, hold-harmless agreements, etc.
   - Quality control requirements and engineering specifications are met.
   - Vendors are notified of problems and corrective action is taken.
   - Purchased parts and materials are properly packaged.
6. Quality Control.
   - There is a written Quality Control Program in effect.
   - Critical raw, in-process, and finished materials are tested.
   - Manufacturing flaws are detected at each production and testing stage.
   - Procedures exist for correcting deficiencies.
   - Testing equipment is properly maintained and certified.
   - Packaging protects products during storage and shipping.
   - Rejected items are reworked, destroyed, or handled in a proper manner.
   - Warranty claim information is reviewed for recurring product safety problems.
7. Distributors / Dealers.
   - Customer accidents and complaints are promptly reported to the manufacturer.
   - Information concerning product hazards and recalls is provided to the distributors and dealers.
   - Certificates of insurance are required and obtained from the distributors and dealers.
8. Service / Installation.
   - Field installations are adequately inspected and tested.
   - Service repairs are documented in writing with copies to the customers.
   - Customer refused repairs are documented in writing with copies to the customers.
   - Unsafe operating practices or modifications are recorded in writing with copies to the customer.

Important information concerning the effectiveness of the Product Safety Program can be provided by these types of audits. They can verify if the present program is capable of identifying and controlling product safety exposures. The audit can also specify deficient areas that need improvement or additional study.

OTHER SAFETY CONCEPTS

There are a number of important topics that affect the efficiency, productivity, and safety of the employees, visitors, and facility. This section will review the prevention of falls, housekeeping, and the use of personal protective equipment.

FALL PREVENTION

Falls of all kinds are the second largest cause of worker injuries—and the largest cause of injuries to the public—in society. Two thirds of the falls reported to a large insurance company are same-level (as opposed to between-levels) falls. Same-level falls account for 60% of the total costs of falls claims. Eighty-five percent of these falls occur inside customer premises.

Inside falls of all types are a problem in manufacturing facilities, offices, stores, etc. At these locations, workers, vendors, and customers are routinely exposed to fall hazards. Regardless of who falls, two thirds of all falls can be traced to floor surface slip resistance problems.

Major Causes of Falls

Faulty housekeeping. The most common same-level falls happen when employees or visitors drop or track in objects or liquids that should not be on the floor.

A combination of engineering and behavioral measures is needed to control this problem. Routine sweeping and mopping are important to minimize slip and fall hazards. Aside from making sure that the proper tools (brooms, dust pans, mops, etc.) are available, a sweeping program should be implemented and enforced.

Departmental floor hazards must be periodically sampled, recorded, and information about them fed back to managers and employees on a comparative basis. Using the right cleansing agents on floors can also help prevent falls. The wrong soap or detergent and/or incorrect application can add to the slipperiness of floors.

Throughout the facility there may be many different sources of liquids, chemicals, and spilled materials that pose slip and fall hazards. Cleaning up the hazards is important, but it is even more critical to identify the problems that are causing the spills to occur. Leaking hoses, pipes, clogged drains, and materials being sprayed from equipment and processes should be repaired and kept in good operating condition. It may be necessary to install new drains and splash shields to control the hazards.

When employees are the potential slip and fall victims, footwear is a controllable factor. Some sole and heel materials in contact with certain floor surfaces are inherently less slippery than others. Aesthetics, wear, maintenance, and cost considerations are those most often weighed when selecting flooring. Because safety is many times an afterthought, it is not uncommon to come across flooring materials that seem inappropriate from a safety standpoint.

Inadequate floor mats. To help prevent inside slips and falls, employee and visitors' shoe soles should be cleaned of debris and dried before they come inside. Floor mat systems should perform

CHAPTER 20

OTHER SAFETY CONCEPTS

these functions, but commonly fail to do so because of design and maintenance inadequacies.

Floor mats that perform mechanical cleaning of shoe soles need to be abrasive, while those that dry the soles need to be absorptive. In warm weather climates or seasons, abrasive mats should be placed outside the entrances and the absorptive ones just inside. In cold weather, both types of mats should be placed inside and their order should be reversed—absorptive first, abrasive second.

No floor mat system can perform well indefinitely—Lack of inspection, maintenance, or substitution can make a successful floor mat system into an unsuccessful one. Dirty and wet floor mats must be exchanged for clean, dry ones.

Floor dressings (waxes and polishes). Floor dressings protect the floor surfaces from wear and may make them easier to clean. However, if not selected carefully, they may also make the floors more slippery. Usually, only experience with a floor dressing under actual conditions can fully establish its acceptability. However, in recent years, dressing manufacturers have generally added slip-resistance data to their literature and to the product packaging. Most such data is obtained through Underwriters' Laboratories slip-tests of the dressings under dry conditions.

Imbedded abrasive grits. For improved slip resistance, abrasive grit applications are recommended for normally dry floors and for many types of normally wet or moist floors. One important exception is in some food preparation and service applications, where all but the smallest grit diameters are normally too difficult to clean to Food and Drug Administration or local sanitary standards. In these cases and where mineral floors are involved, retrofit by chemical etching may be advisable.

Adhered abrasive grits should also be avoided where floors are likely to be subjected to frequent high contact, crushing, and turning forces associated with steel-wheeled fork lift or platform trucks. In these slippery floor situations, coarsely trowelled concrete surfacing, abrasive grits imbedded in concrete surfacing, or chemical etching may be more effective. Some situations where grits have worked well to reduce slipperiness are manufacturing plant workstations and ramps for rubber-wheeled vehicular and pedestrian traffic.

Consultation with an abrasive grain manufacturer may be helpful in the selection of the proper grit size. Unusual conditions should be discussed with the supplier. For example, epoxies vary with respect to temperature resistance. Some are damaged by extremely hot water.

There is also a process called chemical etching that uses hydrochloric acid or other acid-based solutions that roughen the floor surface. It can be an expensive process and, because it enlists suction as well as an etching principle, it is more effective on floors that are constantly wet. Depending on the floor's traffic, this process may have to be repeated every few years.

Whatever type of floor is in the facility, it is important to keep it clean and in good condition, and follow the recommendations of the manufacturer.

PERSONAL PROTECTIVE EQUIPMENT

Personal protective equipment (PPE) provides a barrier between the employee and various types of hazards that may cause injuries. Whenever possible, safety hazards should be eliminated from the workplace. PPE should be the control method of last choice. Even though it can be implemented quickly and with minimal initial cost, it is rarely as effective as engineering controls, and ongoing costs can be significant.

Also, the devices themselves are sometimes uncomfortable and may hinder performance to some extent.

In some situations, it may not be possible or feasible to completely eliminate the hazards by using engineering controls. In these cases, it is important that appropriate personal protective equipment be provided, used and maintained. Some examples include, but are not limited to those listed in Table 20-2.

TABLE 20-2
Personal Protective Equipment

Part of body	Hazards	Protection
Eyes / Face	Flying chips	Safety glasses with side shields
	Chemical splashes	Goggles / face shield
	Welding sparks	Helmet
Respiratory	Nuisance dusts	Disposable respirator
	Solvent vapors	Cartridge respirator
	Oxygen deficiencies	Self-contained breathing apparatus
Head / Ears	Falling objects/struck by objects	Hard hat
	High noise levels	Ear plugs and/or muffs
Hands / Arms	Rough / sharp objects	Leather gloves
	Hot / cold objects	Insulated gloves
	Chemicals	Rubber / synthetic gloves
	Electrical shock	Dielectric gloves
Body / Legs	Chemical splashes	Rubber / synthetic aprons
	Cold work areas	Insulated clothing
	Hot work areas	Aluminized body suit
	Falls from heights	Harness / safety belt / lanyard
	Asbestos	Full body suit
Feet	Dropped / falling objects	Safety shoes / boots
	Chemicals / liquids	Rubber boots

CHAPTER 20

OTHER SAFETY CONCEPTS

WRITTEN POLICIES

These policies should include guidelines for the selection, training, use, and maintenance of personal protective equipment. It may also specify equipment limitations and exceptions concerning their use. Requirements concerning visitor's and outside contractors' use of personal protective equipment should also be addressed.

The key elements of an effective PPE program include:

- Establishing written policies and rules concerning the use of PPE.
- Identifying and evaluating workplace hazards.
- Determining if PPE is necessary or if engineering controls can eliminate or adequately control hazards.
- Selecting the proper PPE.
- Purchasing and supplying employees with equipment.
- Training the users of the equipment.
- Enforcing the use of PPE.
- Inspecting, maintaining, cleaning, and storing PPE.

HAZARD ELIMINATION OR CONTROL

It is appropriate to use PPE when engineering controls are not available or feasible, or if there will be a delay in implementing these controls. Even if healthy individuals do not require PPE to perform a given job, appropriate PPE may be useful for providing accommodation for disabled individuals. Certain types of remedial devices such as splints and braces should be fitted by qualified health professionals, even though they are not recognized as PPE.

SELECTING THE PROPER PPE

Actual and potential hazards must be identified before PPE can be selected. If all of the exposures are not properly determined, individuals may use incorrect equipment. They could have a false sense of security, and permanent damage may occur to their hearing, eyesight, respiratory system, skin, etc. Job Safety Analyses and Material Safety Data Sheets (MSDSs) can provide important information concerning the hazards and recommended control measures.

Once the general categories of PPE are determined, specific information and specifications can be obtained from the manufacturers. Many forms of PPE require the degree of hazard or specific type of hazard to be evaluated before proper selection can be made. Examples include foot, eye, and respiratory protection.

Some important considerations in determining the proper PPE include:

- ANSI, NIOSH, etc., certification or approval, if available.
- Selection is based on the type and degree of the hazards.
- The device should not create additional hazards (heat stress, breathing difficulty, etc.); if these hazards cannot be avoided, medical evaluation of the user may be necessary.
- Different sizes must be available to fit the users (PPE that does not fit can cause discomfort and may not properly protect the wearer).
- The device must not prevent the employee from completing the task (heavy gloves may not permit enough manual dexterity).

TRAINING

It is important that the users of PPE are thoroughly trained to understand the importance of using, properly wearing, inspecting, cleaning, and maintaining the equipment. Initial and periodic training should include information such as:

- Explaining the company PPE program, including medical issues.
- Describing the workplace hazards and present controls.
- Discussing the design and limitations of PPE.
- Demonstrating how to wear, use, clean, maintain, and store the equipment.
- Practice using the PPE on the job.
- Policies concerning cost and replacement of the devices.
- What to do in the case of an emergency.

ENFORCING THE USE OF PPE

Individuals are more likely to comply with the PPE program requirements if they understand their importance, the benefits of using the equipment, and the potential health risks of not complying. If the users are included in the process of selecting the PPE, they are more apt to use it voluntarily.

If PPE is simply provided to the individuals without any training or instruction, there may be resistance to using it. Discussing the benefits of using the equipment, and any reluctance of the users, may help the success of the program. Supervisors should set a positive example by wearing the appropriate devices whenever they are required.

Enforcement of the program is necessary to ensure its success. The overall program should include clearly stated disciplinary rules that are explained to each individual. It is important that the rules are fair and uniformly enforced. Remember that even the best PPE is worthless, if the employee will not use it.

INSPECTION AND MAINTENANCE

The manufacturer of the PPE can provide information concerning the proper inspection, maintenance, and cleaning of the devices. The types of hazards and duration of exposure can affect the potential for degradation and permeation of the equipment. Chemical reactions, oxidation, ultraviolet light, etc., can change the original characteristics of the PPE.

Visual inspection, qualitative or quantitative leak testing, etc., may be recommended by the manufacturer or required by OSHA. Any equipment that has damage such as rips, pinholes, etc., should be repaired or replaced immediately.

Attention must be given to properly cleaning and storing the devices. OSHA has developed Bloodborne Pathogen guidelines for safely cleaning, decontaminating, and sterilizing any items that may contain human blood and certain bodily fluids. These devices are treated as if they are known to be infectious for HIV, HBV, and other bloodborne pathogens.

Bibliography

American National Standards Institute. ANSI Guide for Developing User Product Information. New York, NY, 1990.

Apple, James. *Material Handling Systems Design.* New York: John Wiley & Sons, 1977.

Apple, James. *Plant Layout and Material Handling.* New York: John Wiley & Sons, 1977.

Armstrong, Thomas. *Analysis and Design of Jobs for Control of Cumulative Trauma.* Ann Arbor, MI: University of Michigan, 1980.

Ayoub, M.M. "Work Place Design and Posture." *Human Factors,* 1973.

Chaffin, Don B. "Localized Muscle Fatigue-Definition and Measurement." *Journal of Occupational Medicine,* 1973.

Chaffin, Don B. "Ergonomics Guide for the Assessment of Human Static Strength." *American Industrial Hygiene Association Journal,* 1975.

CHAPTER 20

OTHER SAFETY CONCEPTS

Eastman Kodak. *Ergonomic Design for People at Work.* Volumes 1 and 2. New York: Van Nostrand Reinhold, 1983.

Kulwiec, Raymond. *Advanced Material Handling.* Material Handling Institute, 1983.

Kulwiec, Raymond. *Basics of Material Handling.* Material Handling Institute, 1981.

Kulwiec, Raymond. *Material Handling—Loss Control Through Ergonomics.* Schaumburg, IL: Alliance of American Insurers, 1983.

Liberty Mutual Insurance Group. *Loss Prevention Department Research and Customer Literature.* Boston, MA, 02117.

McCormick, Ernest J. *Human Factors in Engineering and Design.* 4th ed. New York: McGraw-Hill Book Co., 1976.

Muther, Richard. *Systematic Layout Planning.* 2nd edition. Boston, MA: Cahners Books, 1973.

National Safety Council. *Accident Facts—1992 edition.* Itasca, IL 1992.

National Safety Council. *Accident Prevention Manual For Industrial Operations.* Volumes 1 and 2, 10th ed. Itasca, IL 1992.

National Safety Council. *Product Safety Management Guidelines.* Itasca, IL, 1989.

OSHA 1910.147—Control of Hazardous Energy (Standard). Washington, DC: Occupational Safety and Health Administration, U.S. Department of Labor.

OSHA 3120—Control of Hazardous Energy (Information Booklet). Washington, DC: Occupational Safety and Health Administration, U.S. Department of Labor.

Putz-Anderson, V. *Cumulative Trauma Disorders: A Manual for Musculoskeletal Diseases of the Upper Limbs.* New York: Taylor and Francis, 1988.

Roebuck, J.A., Kroemer, K.H.E., and Thomson, W.G. *Engineering Anthropometry Methods.* New York: John Wiley and Sons, 1975.

U.S. Government—Consumer Product Safety Commission. *Consumer Product Safety Commission Regulations.* Washington, D.C.

Wiersma, Charles. *Material Handling & Storage Systems—Planning To Implementation.* Mansfield, OH: Vimach Associates, 1984.

INDEX

A

ABC, see: Activity Based Costing
ABM, see: Activity Based Management
Accidents, 20-1; see also Safety
    cost, 20-1, 20-31
    employee indifference, 20-4
    injuries, 20-4, 20-8, 20-30
    prevention, 20-1, 20-2
    problems, 20-3
    safe attitude, 20-3
Activity Based Analysis, 8-36 (Fig. 8-43)
Activity Based Costing, 8-3
    activity mechanics, 8-10 (Fig. 8-14)
    congruency, 8-11
    cost object mechanics, 8-12 (Fig. 8-15)
    cost distribution system, 8-7 (Fig. 8-8)
    data collection, 8-9 (Fig. 8-12)
    definition, 8-3 (Fig. 8-4), 8-9
    design, 8-4
    distortions, 8-11
    framework, 8-4
    principles, 8-3 (Fig. 8-2)
    process measurement, 8-10 (Fig. 8-13)
    report, 8-6 (Fig. 8-7)
    tools, 8-5
    uses, 8-2
    versus traditional costing, 8-11
Activity Based Management, 8-13
    benchmarking, 8-24, 8-26 (Fig. 8-32)
    business process re-engineering, 8-27 (Fig. 8-33)
    cost drivers, 8-16
    cost variability, 8-14 (Fig. 8-17)
    defining activities, 8-14, 8-19 (Fig. 8-23)
    designing for, 8-24
Advanced tooling, 16-22
    cutting tool materials, 16-22 (Fig. 16-14), 16-23 (Figs. 16-15 and 16-16), 16-24 (Figs. 16-17 and 16-18), 16-25 (Figs. 16-19 and 16-20), 16-26 (Fig. 16-21), 16-27 (Figs. 16-22 and 16-23), 16-28 (Fig. 16-24), 16-29 (Fig. 16-25)
    insert technology, 16-28, 16-30 (Figs. 16-26 and 16-27), 16-32 (Fig. 16-28)
Agile manufacturing, 8-33 (Fig. 8-40)
Assembly drawing, 14-2 (Fig. 14-1), 14-7 (Fig. 14-6)
Assembly systems, 19-1
    automatic, 19-19
    cells, 19-19
    critical cycle times, 19-3 (Fig. 19-10)
    ergonomic issues, 19-19
    improvement techniques, 19-6, 19-7 (Fig. 19-4), 19-10, 19-10 (Fig. 19-7), 19-14, 19-17, 19-18 (Fig. 19-12)
    objectives, 19-3
    operations, 19-14, 19-15 (Table 19-3)
    process description, 19-3
    quality pre requisites, 19-5
    semiautomatic, 19-18
    timing, 19-19 (Fig. 19-13)
    variables, 19-5, 19-6 (Table 19-1), 19-14 (Table 19-2)
Attributes, 8-17, 8-20 (Fig. 8-25), 8-20 (Fig. 8-26)
    activity analysis, 8-21 (Fig. 8-28)
    activity drivers, 8-19 (Fig. 8-24)
Autodeposition, 18-43
    benefits, 18-43
    coating chemicals, 18-44
    future developments, 18-43
Automatic Identification, 14-22
    Bar coding, 14-23
    electronic data interchange, 14-23
    radio frequency, 14-23

Automation, 17-4

B

Baldrige, Malcolm, 1-10
Baldrige National Quality Award, 2-6, **13-1**
    assessment tools, 13-1
    criteria, 13-2 (Fig. 13-1), 13-3 (Fig. 13-2)
    examiners, 13-3
    managing expectations, 13-11, 13-12 (Table 13-2)
    scoring, 13-4, 13-11 (Table 13-1)
        approach, 13-4
        examination items, 13-4 (Fig. 13-3), 13-5 (Fig. 13-4)
        evaluation, 13-6 (Fig. 13-5)
        deployment, 13-4
        results, 13-4
Bar coding, 14-23
Behavior modification, 19-14
Benchmarking, 1-13, 6-7, **7-1**, 8-24
    analysis, 7-8
    benefits, 7-2
    best practices, 7-1
    categories, 7-2 (Table 7-2), 7-3 (Fig. 7-2)
    competitive analysis, 7-1, 7-2 (Table 7-1)
    z-chart, 7-8 (Fig. 7-4)
    eight-step process, 7-3
    integration, 7-9
    key lessons, 7-9
    performance indicators, 7-5 (Fig. 7-3)
    pitfalls, 7-2 (Fig. 7-1), 7-10
    planning, 7-4
    prerequisites, 7-2
    selection, 7-11
    visit guide, 7-13 (Fig. 7-9)
Benefits of continuous improvement, 1-2
Best practices, 1-8, 7-1
Budgeting, 8-26
Buffer management, 14-21
Business management processes, 3-3 (Fig. 3-4)
Business process redesign, 8-24
Brazed and soldered joints, 19-27
    automation, 19-29, 19-30, 19-32
    cost comparison, 19-31
    fixture design, 19-33, 19-35 (Fig. 19-34)
    furnace operation, 19-33
    induction, 19-33
    problems, 19-27
    process data, 19-30
    review, 19-27
    rotary indexing, 19-34 (Fig. 19-33, 19-34)

C

Capability studies, 1-12
CARC, see: Chemical agent resistant coating
Case studies–GPT manufacturing, 4-10
Cause and effect diagram, 11-7 (Fig. 11-7)
Checksheets, 1-12
Chemical agent resistant coating, 18-19
Communication, 5-17, 20-40
Competitive analysis, 7-1, 7-2 (Table 7-1)
Constraints, 14-20 (Fig. 14-15)
    Impact, 14-21 (Fig. 14-16)
    Managing, 14-20
    Understanding, 14-21
Control charts, 1-11
Cost of quality, 8-23 (Fig. 8-30), 11-11
    categories, 11-12 (Table 11-1)
    getting started, 11-13
    goal setting, 11-14
    implementation, 11-14
    percent of total spending, 11-14 (Table 11-2)
    reporting, 11-14

    roots, 11-12
    structure, 11-13
Cost variability, 8-26
    activity-based budgeting, 8-28
    cost estimating for budgeting, 8-29
    controlling imprecision, 8-30, 8-30 (Fig. 8-36)
    facility-sustaining costs, 8-28 (Fig. 8-34)
    full absorption costing, 8-29 (Fig. 8-35)
    myths, 8-26
Costing, 8-1, 8-26, 8-29 (Fig. 8-35)
Cultural change, 5-16, 5-17 (Fig. 5-8)
Cumulative trauma disorders, 20-8
    contributing factors, 20-10
    hand tools, 20-11
    incidence, 20-8, 20-9
    postures, 20-10, 20-11 (Figs. 20-4 through 20-6), 20-12 (Fig. 20-7)
    reducing, 20-10
    types, 20-8
    wrist, 20-9 (Fig. 20-3)
Customer demands and profitability, 8-30, 8-31 (Fig. 8-31), 8-32 (Fig. 8-38), 8-32 (Fig. 8-39)
Customer-focused business strategy, 3-2 (Fig. 3-1)
Customer requirements for continuous improvement, 1-9
    automotive industry, 1-10
    ISO 9000, 1-9
Customer satisfaction, 14-26

D

Data acquisition, 14-39, 14-40, 14-42 (Fig. 14-23) 14-43, (Fig. 14-24)
Datums, 14-16 (Fig. 14-12)
    precedence, 14-17 (Fig. 14-13)
Deburring and surface conditioning, 18-41
    upgrading, 18-42
Deming, W.E., 2-14 (Table 2-4), **10-1**
    philosophy, 10-1
        impact on manufacturing engineering, 10-4
        implementation, 10-5
        success and failure, 10-4
        systems integration, 10-3
    14 points, 10-6
    planning for quality, 10-7
    system of profound knowledge, 10-8
    theory of variation, 10-8
    theory of knowledge, 10-9
Design layout, 14-2 (Fig. 14-1), 14-3 (Fig. 14-2)
Design of experiments, 1-12
Detail drawing, 14-3 (Fig. 14-2)
Developing a strategy for innovative machining, 16-1
    the machining processes, 16-1, 16-2 (Fig. 16-1), 16-3 (Fig. 16-2)
    inventory planning and control, 16-2
    pre-production planning and setup reduction programs, 16-2 (Fig. 16-3), 16-4 (Fig. 16-4), 16-5 (Fig. 16-5)
Dimensions, 14-15 (Fig. 14-11)
DOE, see: Design of Experiments
Downtime reduction, 14-39
Drill jig design, 14-8 (Fig. 14-7)

E

EDI, see: Electronic Data Interchange
Electrocoating, 18-14
    anodes, 18-17
    as a primer, 18-18, 18-19, 18-20
    bath, 18-16

INDEX

cathodic tank, 18-16 (Fig. 18-3)
cure, 18-17
dry-film lubricant, 18-19
on tubular products, 18-20
organic, 18-14
specifications, 18-15
testing, 18-20
to replace other finishes, 18-17, 18-19, 18-20
ultrafiltration, 18-17
Electronic Data Interchange, 14-23
Ergonomics, 20-4, 20-7 (Fig. 20-2)
  guidelines, 20-4
  material handling tasks, 20-5
    redesign, 20-6
  workplace design, 20-7, 20-8 (Table 20-1), 20-13, 20-20 (Fig. 20-20)
Employee involvement, 3-8 (Fig. 3-6)

F

Facility layout, 20-13, 20-14 (Figs. 20-8 and 20-9), 20-18 (Fig. 20-16) see also Workplace design
  aisle design, 20-23
  by computer, 20-23
  material flow, 20-14, 20-15 (Fig. 20-10), 20-16 (Fig. 20-11 and 20-12), 20-17 (Fig. 20-13 through 20-15)
    analysis 20-16, 20-18 (Fig. 20-16), 20-19 (Fig. 20-17 and 20-18), 20-20 (Fig. 20-19)
  planning, 20-13
  responsibility, 20-3
  space requirements, 20-23
  types, 20-13
Factors affecting process quality, 16-6
  important machining parameters, 16-6
  machine tool aspects, 16-6
  setup and workholding aspects, 16-6
  process planning aspect, 16-6
  work material condition, 16-6
  value-added concept, 16-6,
Failure Mode and Effects Analysis, 1-12, 14-35, 19-7 (Fig. 19-3)
  forms, 14-38
  guidelines, 14-38
  mechanics, 14-37
  process, 14-39
  purpose, 14-35
  recommendations, 14-38
  vocabulary, 14-36
Fasteners
  hook and loop, 19-51, 19-52 (Fig. 19-54, 19-55)
    improvements, 19-52, 19-53 (Fig. 19-56, 19-57)
    materials, 19-53
Feeders, 19-19
  bowls, 19-20 (Fig. 19-14), 19-21 (Fig. 19-15), 19-22 (Fig. 19-16), 19-24, 19-25 (Fig. 19-27), 19-26 (Fig. 19-29)
    tooling, 19-21
  centrifugal, 19-24
  hoppers, 19-26
  muffling, 19-25
  orientation, 19-22, 19-23 (Fig. 19-17 through 19-26), 19-26 (Fig. 19-28)
  selecting, 19-20
  stands, 19-26
Finishing, **18-1**
  autodeposition, 18-43
  deburring and surface conditioning, 18-41
  electrocoating, 18-14
  honing, 18-35
  induction heating, 18-27
  machinery, 18-3

management, 18-3
materials, 18-2
methods, 18-2
money and manpower, 18-1
paint, 18-4
power brushes, 18-29
pretreatment, 18-4
quality control, 18-22
surface improvement, 18-33
system variables, 18-8 (Table 18-1)
Fishbone diagrams, 1-11
Flow diagram, 10-12 (Fig. 10-1), 10-14 (Fig. 10-5)
Flowcharts, 5-6 (Fig. 5-2), 19-16 (Fig. 19-11), 19-12 (Fig. 19-9)
FMEA, see: Failure Mode and Effects Analysis
Forming, 17-1
  limit diagram, 17-9 (Fig. 17-4)

G

Geometric control, 14-4 (Fig. 14-3), 14-6 (Fig. 14-5), 14-12 (Fig. 14-10)

H

Histograms, 1-12, 10-16 (Fig. 10-7)
History of continuous improvement, 1-3
Honing, 18-35
  abrasive selection, 18-35
  bore errors, 18-36 (Fig. 18-13)
  cross-hatch pattern, 18-36
  cutting pressure versus cost, 18-38 (Table 18-8)
  fluids, 18-38
  honed and bored cylinder, 18-36 (Fig. 18-14 and 18-15)
  machine selection, 18-39, 18-40 (Fig. 18-21), 18-41 (Fig. 18-22, 18-23, 18-24)
  spindle speed, 18-35, 18-37 (Fig. 18-16)
  statistical process control, 18-39
  stock removal rate, 18-37 (Table 18-5); 18-38 (Table 18-7), 18-40 (Table 18-10)
  surface finish, 18-37, 18-39 (Table 18-9), 18-40 (Fig. 18-19)
  time required, 18-38 (Table 18-6)
Hook and loop fasteners, see Fasteners
House of quality, 2-12 (Fig. 2-8)

I

Implementing continuous improvement, **5-1**
  assuring success, 1-13
  communication, 5-17
  cultural change, 5-16, 5-17 (Fig. 5-8)
  flawed approaches, 5-16
  flowcharts, 5-6 (Fig. 5-2)
  issues and traps, 5-15
  management behavior, 5-18
  planning, 5-1
    checklist, 5-7 (Fig. 5-3)
    organizational issues, 5-7
    preparation, 5-4
    purpose, 5-2
    methodologies, 5-5
  principles, 5-14 (Fig. 5-7)
  recognition and rewards, 5-17
  rules, 5-8
  standards and measures, 5-18
  strategies, 5-11 (Fig. 5-5)
  techniques, 5-12 (Fig. 5-6)
  training, 5-17
  user groups, 5-18, 5-21 (Fig. 5-11)
    benefits, 5-19

    small firms, 5-19
    "storyboard," 5-20 (Fig. 5-10)
  vision, mission, and values, 5-8
Importance of continuous improvement, 1-4
Induction heating, 18-24
  coils, 18-27
  equipment, 18-28
  power supplies, 18-26
  process controls, 18-26
  work handling, 18-28
  workstations, 18-26
In-process gaging systems, 16-19
  recognizing process variables, 16-19
  key system components, 16-19
  the grinding cycle, 16-20
  types of in-process gaging systems, 16-20
  gap control systems, 16-20
  in-process considerations, 16-20
  measuring for machining, 16-21
ISO 9000, 1-9, **12-1**
  conceptual foundations, 12-3
  effect of registration, 12-5
  guidelines, 12-10
  improvement, 12-4
    documentation, 12-4
    problem prevention, 12-5, 12-7
    records, 12-4
  registration, 12-9
  standards, 12-2 (Table 12-1)
  supplemental standards, 12-10
  support for continuous improvement, 12-2
  terminology, 12-4 (Table 12-2)
Interfaces machine tool/cutting tool, 16-36
  accurate toolholding, 16-36, 16-37 (Figs. 16-34 and 16-35)
  precision clamping, 16-37, 16-38 (Figs. 16-36 and 16-37)
  in-between adaptation, 16-37

J

Joints, see Brazed and soldered joints
Juran, Joseph M., 1-5, 10-11
  philosophy, 10-11
    identifying customers, 10-13 (Fig. 10-2)
    quality control, 10-12
    quality improvement, 10-12
    system components, 10-14 (Fig. 10-4)
Just-in-Time, **9-1**
  elements of, 9-2
  human/people issues, 9-5
  implementation, 9-7
  organization issues, 9-6
    scheduling, 9-6 (Fig. 9-4)
  technical issues, 9-3
  U-line layout, 9-4 (Fig. 9-2)

K

Kaizen, 1-6
Kinematic control, 14-4 (Fig. 14-4)

L

Laser processing, 16-54
  autofocus, 16-54 (Fig. 16-52), 16-55 (Fig. 16-53)
  part programming, 16-55
  coordination of beam characteristics and positioning systems, 16-55 (Fig. 16-54)
  orbital nozzle assembly, 16-56 (Fig. 16-55)
  use of inert gases, 16-56
  welding-laminar barrier inerting, 16-56 (Fig. 16-56)

INDEX

cutting-high pressure inert assist, 16-56
Laser welding, see Welding

M

Machining coolants, 16-57
   improving quality and productivity through filter media selection, 16-57
   filter media selection criteria, 16-57
   filter media physical properties, 16-57
   filter media manufacturing process characteristics, 16-57, 16-58 (Fig. 16-57), 16-58 (Fig. 16-58), 16-59 (Fig. 16-59)
   filter media composition characteristics, 16-59, 16-61 (Fig. 16-60)
   evaluating filter media, 16-60, 16-62 (Fig. 16-61), 16-63 (Figs. 16-62 and 63)
   comparative study sequence, 16-62
   statistical analysis of evaluation results, 16-62, 16-65 (Fig. 16-64), 16-66 (Figs. 16-65 and 66)
   sludge evacuation to eliminate coolant wastes, 16-64, 16-69 (Fig. 16-67)
   on site clean up of small volumes of oily wastes, 16-68
Machining centers, 14-32 (Fig. 14-17), 14-33 (Fig. 14-18), 14-33 (Fig. 14-19)
Macro design, 8-25 (Fig. 8-31)
Maintenance, 15-1, 20-31
   see also Total Productive Maintenance
Management, 14-18
   constraint-based, 14-19
   global versus local, 14-18
   scheduling, 14-22, 14-26, 14-34 (Fig. 14-20)
Manufacturing Resource Planning II flowchart, 19-12 (Fig. 19-19)
Material flow, 20-14, 20-15 (Fig. 20-10), 20-16 (Figs. 20-11 and 20-12), 20-16, 20-17 (Fig. 20-13 through 20-15)
   analyzing, 20-16, 20-18 (Fig. 20-16), 20-19 (Fig. 20-17 and 20-18), 20-20 (Fig. 20-19)
   evaluation chart, 20-22 (Fig. 20-23)
   From-to chart, 20-21 (Fig. 20-21)
   patterns, 20-26 (Fig. 20-29)
   relationship chart, 20-24 (Fig. 20-25)
   diagrams, 20-25 (Figs. 20-26 and 20-27), 20-26 (Fig. 20-28)
Mechanisms for continuous improvement, 12-5, 12-6 (Fig. 12-3)
   contract review, 12-7
   corrective action, 12-6
   design control, 12-8
   handling, storage, packaging, 12-8
   inspection and test, 12-8
   internal quality system audits, 12-7
   management review, 12-7
   process control, 12-6
   training, 12-8
Metal stamping, 17-1
   automation, 17-4
   blanking operations, 17-1
   equipment selection, 17-1
   lubrication, 17-2
   miscellaneous considerations, 17-8
   press, die, and operator protection, 17-7
   production rate, 17-6 (Fig. 17-3)
   quick die change, 17-3
   rolling bolster operations, 17-5 (Fig. 17-2)
   strip utilization, 17-2 (Fig. 17-1)
MRP II see Manufacturing Resource Planning II

N

Need for further advancement, 16-49
   high speed machining, 16-49
   interfaces machine tool/cutting tool, 16-50 (Fig. 16-50)
   cutting tool material, 16-52, 16-53 (Fig. 16-51)

O

Optimization, 14-19 (Fig. 14-14)
Optimizing processes and parameters, 16-38
   single step-machining, 16-38
   one pass-machining, 16-39, 16-40 (Figs. 16-38 and 16-39)
   proactive finetuning, 16-39, 16-41 (Fig. 16-40)
Overall equipment effectiveness, 19-11 (Fig. 19-8)

P

Paint and Painting, 18-4
   conveyor loading, 18-9 (Table 18-2), 18-13
   efficiency, 18-7
   electrostatic, 18-10 (Fig. 18-1), 18-11 (Fig. 18-2)
   environmental concerns, 18-7
   film testing, 18-20
   hose volume, 18-12 (Table 18-3)
   improvement, 18-6
   performance characteristics, 18-22, 18-24
   pretreatment, 18-4
   quality control 18-22
   quality assurance, 18-22, 18-24 (Table 18-4)
   system analysis, 18-8
   variables, 18-8 (Table 18-1)
   systems, 18-6
   testing, 18-20, 18-24 (Table 18-4)
   waste minimization, 18-7
Parts feeders, see Feeders
Performance indicators, 7-5 (Fig. 7-3)
Performance monitoring, 11-4
   measurement, 11-6
Plan-do-check-act cycle, 1-5, 1-6
Planning, 5-1
Power brushes, 18-29
   abrasive-filled nylon, 18-31
   advantages, 18-29
   applications, 18-30 (Fig. 18-5)
   compared to abrasive wheel, 18-32 (Fig. 18-11)
   design, 18-29
   performance variables, 18-32 (Fig. 18-10)
   recommended surface speeds, 18-32 (Fig. 18-9)
   types, 18-29, 18-30 (Fig. 18-6)
   when to use, 18-31
   wire, 18-30, 18-31 (Fig. 18-7, Fig. 18-8)
Practices of continuous improvement, 1-5
   best practices, 1-8, 7-1
   plan-do-check-act cycle, 1-5, 1-6
   Juran approach, 1-5
   Kaizen, 1-6, 1-8
   Taguchi approach, 1-7
Press forging, 19-36
   components, 19-36
   hydraulic clamping, 19-37, 19-38 (Figs. 19-37, 19-38), 19-40 (Figs. 19-39, 19-40), 19-41 (Fig. 19-41), 19-42 (Fig. 19-42)
   improving precision, 19-41, 19-42 (Figs. 19-43, 19-44), 19-44 (Figs. 19-45, 19-46), 19-45 (19-47), 19-46 (Figs. 19-48, 19-49)
   quick change tool system, 19-36, 19-37 (Figs. 19-35, 19-36), 19-39, 19-43 (Fig. 19-44)
Preventive maintenance, 14-28
Probability, 10-17 (Fig. 10-8)
Problems
   correcting and preventing, 11-5
Problem solving, 19-8 (Fig. 19-5)
Process Appraisal, **11-1**
   cause and effect diagram, 11-7 (Fig. 11-7)
   complexity, 11-8
   correcting problems, 11-5
   factors, 11-4 (Fig. 11-4)
   innovation, 11-8, 11-9 (Fig. 11-8)
   integration, 11-11
   management support, 11-11
   measurement, 11-6
   model, 11-2 (Fig. 11-1)
   monitoring performance, 11-4
   optimization, 11-9
   production, 11-3 (Fig. 11-2)
   understanding, 11-2
   universal factors, 11-4 (Fig. 11-3)
   viewpoint, 11-1
   waste, 11-8
Process development, 19-1, 19-2 (Fig. 19-1)
Process Capability Index, 19-4 (Fig. 19-2)
Process chart, 20-20 (Fig. 20-19)
Process improvement opportunities, 6-7
Process improvement pointers, 16-7
   machining aspects, 16-7
   machine tool aspects, 16-7
   setup and workholding aspects, 16-7
   process planning aspect, 16-8
   work material condition, 16-8
   value-added concept, 16-8, 16-9 (Fig. 16-6), 16-10 (Fig. 16-7), 16-11 (Fig. 16-8), 16-12 (Fig. 16-9), 16-13 (Fig. 16-10), 16-14 (Fig. 16-11), 16-15 (Figs. 16-12 and 16-13)
Process re-engineering, 8-22 (Fig. 8-29), 8-27 (Fig. 8-33)
Product definition, 14-1, 14-7
   Focus, 14-4
   Implementation, 14-9
Product realization cycle, 12-3 (Fig. 12-1)
Productivity, **14-1**
Project management, 1-13

Q

QFD, See: Quality Function Deployment
Quality
   improvement, 10-12
   planning for, 10-7
   program maturity, 15-2 (Fig. 15-1)
Quality assurance programs, 18-22
   benefits, 18-24
   service report, 18-25 (Fig. 18-4)
Quality Assurance Standards, 12-2, (see also ISO-9000)
Quality by design, 14-25
Quality control, 5-10 (Fig. 5-4), 10-12, 10-15 (Fig. 10-6)
   tools, 2-10 (Fig. 2-6), 2-11 (Fig. 2-7), 2-12 (Fig. 2-8)
Quality cost, see: Cost of Quality
Quality Function Deployment, 1-12, 6-7, 19-9 (Fig. 19-6)
Quality system requirement, 12-4 (Fig. 12-2)
Quick die change, 17-3

INDEX

R

Recognition and reward, 5-17
Re-correcting, 14-21
Refined tooling modules, 16-32, 16-33 (Fig. 16-29), 16-34 (Figs. 16-30 and 16-31)
    modular design, 16-32
    toolbody, 16-35 (Fig. 16-32), 16-36 (Fig. 16-33)
Risk, 14-35 (Fig. 14-21), 14-38

S

Safety, 17-7, **20-1**, 20-3, 20-43; see also Accidents
    falls, 20-43
    hazard elimination, 20-45
    housekeeping, 20-33
    Job Safety Analysis, 20-34, 20-37 (Fig. 20-34)
        benefits, 20-36
        procedures, 20-36
    layout planning, 20-13
    lockout/tagout, 20-36, 20-38 (Fig. 20-35)
    loss control, 20-39
    machines, 20-30, 20-31, 20-33
        guards, 20-31, 20-32 (Fig. 20-32), 20-34 (Fig. 20-33)
        tips, 20-33
    planning checklists, 20-25, 20-27 (Fig. 20-30), 20-29 (Fig. 20-31)
    policies, 20-45
    product safety, 20-39
        audits, 20-42
        communications, 20-40
        recalls, 20-41
    protective equipment, 20-33, 20-44, 20-44 (Table 20-2)
        enforcing use, 20-45
        inspection and maintenance, 20-45
        selecting, 20-45
    training, 20-45
Scattergrams, 1-12
Self-assessment, 13-1 (see also Baldrige National Quality Award)
    lessons learned, 13-14
    perspectives, 13-16
    uses of, 13-3
        consensus, 13-13
        develop next steps, 13-14
        develop and define expectations, 13-13
        drafting the Baldrige application, 13-13
        feedback, 13-14
        opportunities for improvement, 13-14
        review, 13-13
        scoring, 13-13
        site visit, 13-14
        training in evaluation processes, 13-13
    versus continuous improvement, 13-12
Simulation, 14-24, 14-24 (Table 14-1)
    applications, 14-33
    extended to suppliers, 14-28
    for scheduling, 14-34 (Fig. 14-20)
    getting started, 14-34
    history, 14-32
    in decision making, 14-31
    in process control, 14-28
    limitations, 14-27
    machining centers, 14-32 (Fig. 14-17), 14-33 (Fig. 14-18), 14-33 (Fig. 14-19)
Shop floor management, 14-18, 14-30, 14-40 (Fig. 14-22)
Soldering, see Brazed and soldered joints
SPC, see: Statistical Process Control
Standards, 14-1, 14-10 (Fig. 14-9), 14-30
    ANSI Y14.5M, 14-1

Standards and measures, 5-18
Statistical Process Control, 18-39
Stringent finish requirements, 16-42
    tight tolerances, smooth surface finishes, 16-42 (Fig. 16-41), 16-43 (Fig. 16-42), 16-44 (Fig. 16-43), 16-45 (Fig. 16-44)
    six sigma and c_{pk}-manufacturing, 16-45 (Fig. 16-45), 16-46 (Fig. 16-46), 16-47 (Fig. 16-47), 16-48 (Fig. 16-48)
    first part/good part, zero defect-production, 16-46, 16-48 (Fig. 16-49)
Suppliers, **6-1**
    certification, 6-1, 6-2 (Fig. 6-1), 6-4 (Fig. 6-3), 6-7
    contract, 6-6
    evaluation criteria, 6-3 (Fig. 6-2)
    performance, 6-3
    reducing the number, 6-6
    support, 6-6
    visits, 6-3
Surface improvement technology, 18-33
    choosing a method, 18-33
    finishing stone, 18-34 (Fig. 18-12)
System of profound knowledge, 10-8

T

Taguchi, Genichi, 1-7, 1-9, 10-14
    quality control processes, 10-15 (Fig. 10-6)
    reduction of variability, 10-14
Teams and teamwork, **3-1**
    formation and growth, 3-4, 3-5 (Fig. 3-5)
    paradigm, 3-1
    performance, 3-6
        cultural integration, 3-12
        diagnosis, 3-7, 3-9 (Fig. 3-8)
        direction, 3-10, 3-10 (Fig. 3-9)
        improvement, 3-7, 3-9 (Fig. 3-7)
        redesign, 3-11, 3-11 (Fig. 3-10)
        review and recycle, 3-12
        support structure, 3-12
    requirements, 3-1
Tools of continuous improvement, 1-11
    benchmarking, 1-13, 7-1, 8-24
    capability studies, 1-12
    checksheets, 1-12
    control chart, 1-11
    design of experiments, 1-12
    failure mode and effects analysis, 1-12
    fishbone diagrams, 1-11
    histograms, 1-12
    project management, 1-13
    Quality Function Deployment, 1-12
    scattergrams, 1-12
Total Productive Maintenance, **15-1**
    equipment effectiveness, 15-4
        calculation, 15-5 (Fig. 15-3), 15-6 (Fig. 15-4)
        measuring, 15-4
        overall, 15-4
    inspection forms, 15-7 (Fig. 15-5), 15-9
    planning, 15-6
    stages, 15-3 (Fig. 15-2)
    training, 15-9
Total Quality Management, **2-1**, 19-12
    history, 2-1
    philosophy, 2-8
    responses, 2-4
        American, 2-4
        crossfunctional management, 2-5
        Japanese, 2-4
    Deming's 14 points, 2-14 (Table 2-4)
    quality control tools, 2-10 (Fig. 2-6), 2-11 (Fig. 2-7), 2-12 (Fig. 2-8)
    techniques, 2-9 (Table 2-2), 2-13 (Table 2-3)
    tools, 2-10 (Fig. 2-6), 2-11 (Fig. 2-7), 2-12 (Fig. 2-8)
    technology, 2-8
Total quality organization, 3-1
Tolerance refinement, 14-10 (Fig. 14-8)
TPM, see: Total Productive Maintenance
TQM, see: Total Quality Management
Training, 4-1, 5-17
    analysis, 4-8
        capturing data, 4-8 (Fig. 4-5)
        key roles, 4-8
        purpose, 4-8
    as a result of CI, 4-2
    best approach, 4-3
    costs, 4-5 (Fig. 4-2)
    levels of learning, 4-4 (Fig. 4-1)
    delivery strategies, 4-6 (Table 4-2)
    design, 4-9
    development, 4-6, 4-9
    levels of improvement, 4-2
    pilot-test, 4-10
    products, 4-1
    project planning, 4-7
        key roles, 4-7
    revision and release, 4-10
    to support CI efforts, 4-5

U

Unification of quality, time, and cost data, 8-31
User groups, 5-18

V

Variability reduction, 10-4
Vision, mission, and values, 5-8

W

Waste, 9-2, 9-3 (Fig. 9-1), 11-8
    minimization, 18-7
Waste elimination, 14-24
    extended to suppliers, 14-28
    opportunities for improvement, 14-25
Welding, 19-45
    laser, 19-45, 19-47 (Fig. 19-50), 19-48, 19-49 (Fig. 19-51), 19-50 (Fig. 19-52)
    material considerations, 19-47
    optimizing, 19-48
    seam welds, 19-51 (Fig. 19-53)
Workholding, 16-12
    workholding technologies for continuous improvement, 16-12
    advantages of preset workholding, toolholding, and part registration to reduce setup time, 16-16
Workplace design, 20-7, 20-8 (Table 20-1), 20-13, 20-20 (Fig. 20-20), 20-21 (Fig. 20-22), 20-23 (Fig. 20-24); see also facility layout